TRIBOLOGICAL DESIGN OF MACHINE ELEMENTS

LEEDS
LYON
LEEDS
LYON

TRIBOLOGY SERIES 14

TRIBOLOGICAL DESIGN OF MACHINE ELEMENTS

edited by

D. DOWSON, C.M. TAYLOR, M. GODET AND D. BERTHE

Proceedings of the 15th Leeds–Lyon Symposium on Tribology held at Bodington Hall, The University of Leeds, UK

6th - 9th September 1988

ELSEVIER Amsterdam – Oxford – New York – Tokyo 1989
For the Institute of Tribology, Leeds University
and
Institut National des Sciences Appliquées de Lyon

ELSEVIER SCIENCE PUBLISHERS B.V.
Sara Burgerhartstraat 25
P.O. Box 211, 1000 AE Amsterdam, The Netherlands

Distributors for the United States and Canada:

ELSEVIER SCIENCE PUBLISHING COMPANY INC.
655 Avenue of the Americas
New York, NY 10010

ISBN 0-444-87435-6 (Vol. 14)
ISBN 0-444-41677-3 (Series)

Printed in the Netherlands.

CONTENTS

Introduction

The fifteenth Leeds-Lyon Symposium on Tribology was held from 6th-9th September 1988 at Bodington Hall, The University of Leeds. On previous occasions each Symposium has focused attention on a current and significant research topic, usually reflecting the interests of the Leeds or Lyon research groups, but this time the vitally important subject of technology transfer was recognized. Delegates appeared to appreciate this rare opportunity to discuss the impact of their studies upon machine design, since some 154 of them from 21 countries attended the Symposium on the "Tribological Design of Machine Elements". We were particularly pleased to welcome a strong group of friends from INSA, Lyon, led by Professor Maurice Godet.

The Symposium was dedicated to the late Professor F T Barwell, who died on 15th January 1988 after a long illness. Freddie Barwell was a gentleman. He did much to promote a sound engineering approach to tribology and his book on "Bearing Systems - Principles and Practice" reflected his interest in applying tribological knowledge to bearing design. We know that he would have approved the topic chosen for the 15th Leeds-Lyon Symposium. He regularly attended and contributed to the Symposia and will be sorely missed.

The Symposium opened in customary style with the Keynote Address on the Tuesday evening. On this occcasion we were pleased to welcome Professor Herb Cheng of Northwestern University, Evanson, USA, to set the scene, with a wide ranging and personal account of his view of "The Tribological Design of Machine Elements". Delegates then travelled to York to enjoy the Symposium Dinner in the ancient Merchant Adventurers Hall. The Guest of Honour at the dinner was Peter Jost, who took the opportunity to comment on the coming of age of tribology. Peter himself chaired the Working Party which produced the now famous report which introduced the word tribology some twenty one years ago.

The somewhat unusual nature of the Symposium provided the organizers with an opportunity to offer a perspective on the current state of tribological design in a number of major industries. This was achieved by inviting selected authors to present Review Papers on Tribological Design in;-

- **The Aerospace Industry** — **(J A Dominy, Rolls Royce Ltd, Derby U K)**
- **The Railways** — **(C Pritchard and T G Pearce, British Rail, Derby, U K)**
- **The Automobile Industry** — **(P A Willermet, Ford Motor Company, Dearborn, U S A)**
- **The Process Industries** — **(J D Summers-Smith, Guisborough, U K)**
- **The Power Generation Industry** — **(P G Morton, GEC Stafford, U K)**
- **The Nuclear Industry** — **(T C Chivers, CEGB, Berkeley, UK)**
- **The Spacecraft Industry** — **(R A Rowntree, E W Roberts and M J Todd, National Centre of Tribology, Risley, U K)**
- **The Electronic Industry** — **(E A Muijderman, A G Tangena, F Bremer, P L Holster, A V Montfoortand, Philips, Eindhoven, The Netherlands)**

. **Information Storage and and Retrieval** **(B Bhushan, IBM, San Jose, USA)**

We are particularly grateful to the authors of these Review Papers for preparing these commentaries on contemporary practice in tribological design.

Some sixteen working sessions were included in the Programme and this necessitated the holding of parallel sessions throughout the morning of Thursday 8th September. The fifty four papers presented nevertheless represented only about fifty percent of those offered, thus facing the organizers with the difficult task of declining many attractive offers. Two sessions were devoted to the Review Papers mentioned earlier, while others dealt with;- Seals; Cams; Belts; Gears; Rolling Element Bearings (2); Plain Bearings (2); Wear; Ceramics; Hydrostatic Bearings; Information Storage and Retrieval/Magnetic Bearings and Knowledge Based Systems. The latter sessions reflect the growing interest in the role of tribology in the computer based information society of the 1980's. We are particularly grateful to the distinguished Chairmen who presided over the Symposium Sessions and whose names are recorded in this volume.

Parallel sessions were also introduced into the Social Programme held on the afternoon of Thursday 8th September. The intricacies of the arrangements were described in a specially arranged Wednesday evening session, which is rapidly becoming an established feature of the Symposia in Leeds, by Mr Brian Jobbins. All delegates were taken to Whitby, where the famous explorer Captain Cook learned his seamanship, but this was achieved by following one of three routes. Most of the delegates travelled by coach to Pickering and then crossed the North Yorkshire Moors on the Railway operating on the line originally built by George Stephenson in 1836. Smaller groups visited Kilburn village to see the hand carving of the famous Thompson (Mouseman) oak furniture, or the Fylingdales Early Warning Radar Station. The journey home was broken for dinner at the Crown Hotel, Boroughbridge after a very full day. As far as we know all delegates returned to Bodington Hall!

Initial discussion of the arrangements for each Leeds-Lyon Symposium usually commences some eighteen months to two years before the Symposium is held. The Proceedings are then published within the following year. This rolling organizational cycle of some three years duration calls for considerable dedication and support. We are particularly grateful for the financial support for the Symposium generously provided on this occasion by:

[1] **British Petroleum Research Centre, Sunbury-on-Thames, UK**
[2] **Fiat Research Centre, Turin, Italy**
[3] **Michell Bearings plc, Newcastle-upon-Tyne, UK**
[4] **SKF Engineering and Research Centre, The Netherlands**
[5] **The US Army Research Development and Standardisation Group, UK**

The Symposium literature was once again presented to delegates in handsome wallets provided by Elsevier, Publishers of the Journal WEAR, and we are pleased to acknowledg this most welcome support.

The smooth running of the Symposium owes much to the enthusiasm and hard work contributed by colleagues in the Institute of Tribology in

The University of Leeds. We would particularly like to express our appreciation to Mrs Sheila Moore, Mrs Catharine Goulborn, Mr Stephen Burridge, Mr Ron Harding, Mr Brian Jobbins, Mr David Jones, Dr John Fisher, our technicians and our current research students and fellows. We are also most grateful to the staff of Elsevier Science Publishers BV, Amsterdam, for their professional and friendly service in producing the volumes of Proceedings of the Leeds-Lyon Symposia.

As we were undertaking the editorial work on the present volume of Proceedings, we heard with great sadness of the death on February 15th 1989 after a long illness of one of our editorial colleagues, Professor Daniel Berthe. Daniel was a great enthusiast for this Anglo-French cooperation and we know that the biennial arrangements in Lyon depended heavily upon him and Professor Godet. His contributions to tribology and his intellect were respected and appreciated by colleagues and friends in Lyon and Leeds. On behalf of all our delegates and his friends in Leeds we would like to send our deepest sympathy to Daniel's family and his colleagues in Lyon. The Leeds-Lyon Symposia on Tribology which Daniel Berthe helped to establish have covered a wide range of topics since their inception in 1974, as illustrated by the following list.

[1]	**1974**	**(Leeds)**	**Cavitation and Related Phenomena in Lubrication**
[2]	**1975**	**(Lyon)**	**Super Laminar Flow in Bearings**
[3]	**1976**	**(Leeds)**	**The Wear of Non-Metallic Materials**
[4]	**1977**	**(Lyon)**	**Surface Roughness Effects in Lubrication**
[5]	**1978**	**(Leeds)**	**Elastohydrodynamics and Related Topics**
[6]	**1979**	**(Lyon)**	**Thermal Effects in Tribology**
[7]	**1980**	**(Leeds)**	**Friction and Traction**
[8]	**1981**	**(Lyon)**	**The Running-In Process in Tribology**
[9]	**1982**	**(Leeds)**	**Tribology of Reciprocating Engines**
[10]	**1983**	**(Lyon)**	**Numerical and Experimental Methods in Tribology**
[11]	**1984**	**(Leeds)**	**Mixed Lubrication and Lubricated Wear**
[12]	**1985**	**(Lyon)**	**Mechanisms and Surface Distress**
[13]	**1986**	**(Leeds)**	**Fluid Film Lubrication - Osborne Reynolds Centenary**
[14]	**1987**	**(Lyon)**	**Interface Dynamics**
[15]	**1988**	**(Leeds)**	**The Tribological Design of Machine Elements**

The 16th Leeds-Lyon Symposium on Tribology will be held in Lyon, France, under the title "Mechanics of Coatings" from Tuesday 5th to Friday 8th September 1989. We greatly look forward to meeting our friends in Lyon once again.

Duncan Dowson
Chris Taylor

SESSION I
KEYNOTE ADDRESS

Chairman: Professor D Dowson

Paper I(i) The Tribological Design of Machine Elements

Paper I(i)

The tribological design of machine elements

H. S. Cheng

History of tribology indicates that many important tribological concepts and theories were stimulated by the needs of new machinery developments, and the new tribological research findings, in turn, have helped to upgrade the design of more efficient and more reliable tribological elements. This mutual dependence is illustrated with some classical examples.

Discussions are then given to the tribological design process of generic elements and its application to determine tribological performance of machine elements, and to the needs in improving this design process to meet the requirements of future machineries in aerospace, automotive, and information processing industries.

1. INTRODUCTION

Good tribological design has often been recognized as the key to the success in developing new machineries. This was true in the days of Reynolds when railroads relied heavily on good journal bearing design of the axles, true in the days of Model T when automotive engines depended strongly on good design practice of main shaft and connecting rod bearings, and is still true today when computers rely critically on good tribological performance of the head/disc interface.

It is now over a century since Reynolds first established the formulation of fluid film lubrication controlling the tribological performance of sliding surfaces. In this past hundred years, a vast amount of design data has been generated in predicting the average lubricant film thickness between the sliding surfaces generated either hydrodynamically or hydrostatically. In some cases, an accurate prediction of the average lubricant film is sufficient to ensure a good tribological design of mechanical components. However, in most other cases, particularly for counterformal contacts, the lubricant film thickness is insufficient to ensure a good design because film thickness alone cannot predict the lubrication breakdown leading to sliding failures by scuffing and wear. It is the lubrication breakdown of the asperity oil film and the surface film which control the failure process. The lack of accurate predictions of the asperity oil film and the surface film breakdown is seen to be the weakest link in the tribological design process.

As new technologies emerge in aerospace, energy, manufacturing, and communication industries, machineries are required to operate under much higher temperature with higher efficiency and reliability. These requirments present new challenges in research and design of tribological elements to discover new lubrication concepts, new materials, new surface modification techniques, and new predictive theories to develop the advanced machineries.

In this paper, some significant historical developments in tribology are used to illustrate the mutual dependence of tribological design and machinery development. Weak areas in current methods of tribological design of machine elements are then indicated and discussed. Some newly emerged tribological concepts for meeting the new challenges in tribological design are also described.

1.1 Historical Developments

In tribological design, one is mostly concerned with two basic elements, the conformal sliding bearings and counterformal rolling and sliding contacts. Historically, evidences of conformal slidings can be traced back thousands years ago when iron journal bearings were used in olive crushing machines in Greece and bronze journal bearings were used as the wheel bearings in the Chinese South Point Chariot[1].

Few would disagree that the most significant historical development in design of sliding bearings is the discovery of a continuous oil film in a journal bearing by Tower[2] and the subsequent derivation of the Reynolds equation[3] which established the foundation for prediction of tribological performance for design of sliding bearings. This most important development was motivated by the problems associated with the journal boxes in railcar axles. Thus, it is a classic example in which a critical need in machinery development triggered a major breakthrough in tribological design which in turn benefits the development of many other future machineries.

Tribological design of rolling and sliding contacts has been evolved mostly from the needs of better rolling element bearings and gears in machinery development. Unlike the sliding bearings for which rational design was established around the turn of the century, there have been no rational criteria for designing the rolling and sliding contacts until almost half a century later when Blok[4] established the critical temperature concept to predict the scuffing threshold and Lundberg and Palmgren[5] established a method to predict the contact fatigue life.

The importance of lubricant film in rolling and sliding contacts was recognized, but satisfactory prediction of lubricant film thickness was not possible until the developments of elastohydrodynamic lubrication[6]. While the detailed influence of lubricant film thickness in contact failures is still being assessed by researchers, its role as an indicater of the severity of asperity contacts is firmly placed in tribological design.

Tribological design of very low speed and heavily loaded sliding contacts depends on the protection of boundary films. Historically, the ties between boundary lubrication and machinery development are less evident than hydrodynamic and elastohydrodynamic lubrication from the first publication on boundary film by Hardy[7]. Nevertheless, these films are the last line of defence against severe wear. Unless rational methods are formed in predicting the conditions of breakdown of these boundary films, the process of tribological design will always be incomplete.

2. TRIBOLOGICAL DESIGN PROCESS

There are two types of elements in tribological design. The first type is simple generic elements involving either conformal sliding surfaces or the counterformal rolling and sliding contacts, as shown in Fig. 1. The second type is tribological machine elements which include all bearings, gears, cams, and other rolling and sliding components. A brief discussion is given to the tribological design process for each of these two types of tribo-elements.

2.1 Generic Tribo-Elements

The processes for designing a generic tribo-element, as illustrated in Fig. 2, begin with the input data, which can be arranged in the following five groups:

1. Geometrical Data, such as the principal radii of the contacts,
2. Roughness Data, such as the average roughness height, average asperity radii, etc.
3. Operating Conditions, such as range of load, speed, temperature, etc.
4. Lubricant Properties, such as viscosity, shear modulus, limiting shear stress, thermal conductivity, etc.
5. Material Properties of Solids, such as elastic and shear modulus, thermal conductivities, specific heat, yielding stress, fracture toughness, hardness, etc.

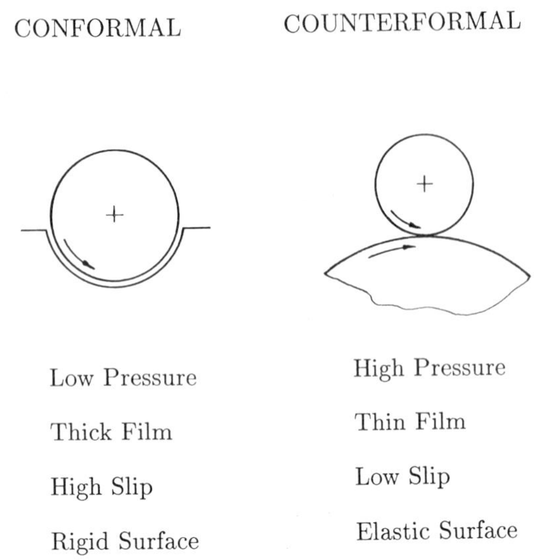

Fig. 1 Generic Tribo-Elements

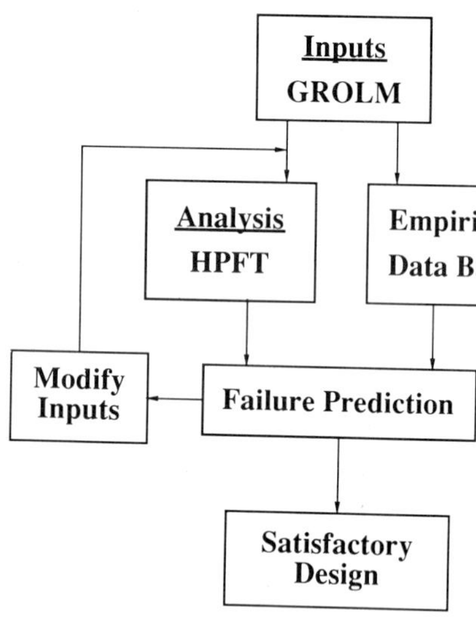

Fig. 2 Design Process for a Generic Tribo-Element

One may use a subroutine named "GROLM" to identify the five major groups of input data needed to initiate the tribological design process.

At the present time, most of these input data is entered manually in the computer programs for tribological design. This inefficient method of handling the input data can be improved if the lubricant and material database is computerized and interfaced directly with the tribo-design program.

The second step in the design process is to determine the tribological performance described by:

1. The film thickness, H
2. The contact pressure, P
3. The friction force, F
4. The contact temperature, T

A subroutine named "HPFT" may be used to identify the calculation of these tribological performance variables which are used to determine whether the element would fail under the given operating conditions.

For contacts operating in thick film lubrication, the distributions of film thickness, pressure, friction, and temperature can be predicted by theories ignoring the surface roughness effects. Indeed, the smooth-surface theories in hydrodynamic, hydrostatic, elastohydrodynamic lubrication have been fully developed in the past century to enable the prediction of these quantities to a high degree of accuracy. Reviews of these developments can be found in the Proceedings of the 1986 Leeds- Lyon Symposium on Tribology[8].

For contacts operating in thin film lubrication, the sliding asperities are no longer separated by a thick oil film. They are protected by a very thin oil film or by a surface film. Tribological performance in the thin film regime, as represented by average quantities of the film thickness, lubricant and asperity pressure, lubricant and asperity shear stress, and surface temperature, $\bar{h}$, $\bar{p}$, $\bar{p}_a$, $\bar{\tau}$, $\bar{\tau}_a$, $\bar{T}_s$, shown in Fig. 3, may not be sufficient to predict tribological failure. There is a need to extend the tribological performance to include the characteristics of these quantities at the asperity level considering each asperity as a micro-contact. These micro-quantities are labelled as asperity film thickness, contact pressure, shear stress, and surface temperature, h^*, p^*, τ^*, T_s^*, as shown in Fig. 3.

Some analyses of tribological performance in thin-film lubrication in terms of the above described quantities are available[9], but the area has not been fully developed. Moreover, the relation of these performance variables with the major tribological failures such as contact fatigue, scuffing. and wear are not fully understood. They are yet to be identified and quantified. Considerable more efforts are needed in this area.

The most important step in tribological design is failure prediction. Ideally, failure predictions should be based on analytical failure models which relate a major failure mode to certain critical tribological variables, such as the relation between contact fatigue life and a critical maximum Hertzian pressure, and the relation between scuffing and a critical total contact temperature. Unfortunately, analytical predictions of sliding failures based on calculated tribological performance variables in the thin-film regime are not always accurate and reliable. One cannot predict analytically the tribological failures as accurately as the structural failures. This most critical block is also the weakest link in the tribological design process.

In the case there are no accurate analytical failure models available, alternative methods, based entirely or partially on failure database obtained experimentally, may be used.

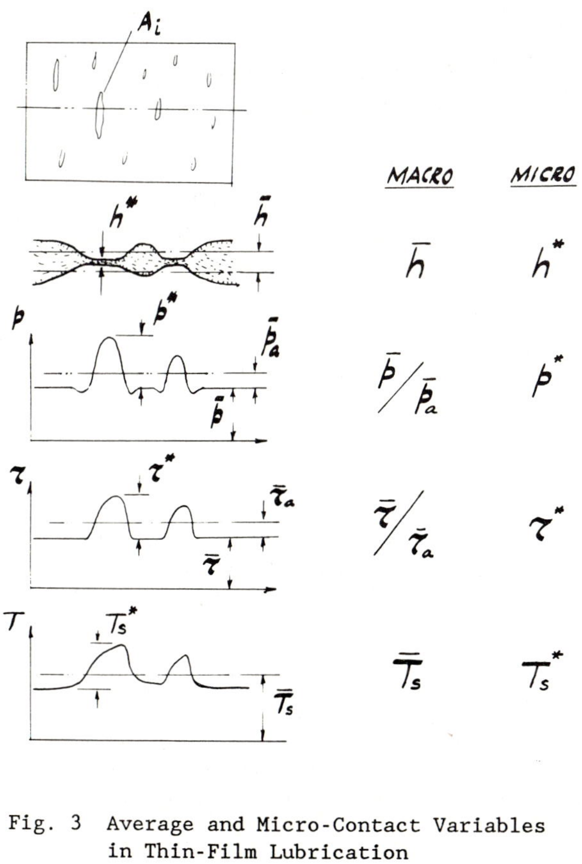

Fig. 3 Average and Micro-Contact Variables in Thin-Film Lubrication

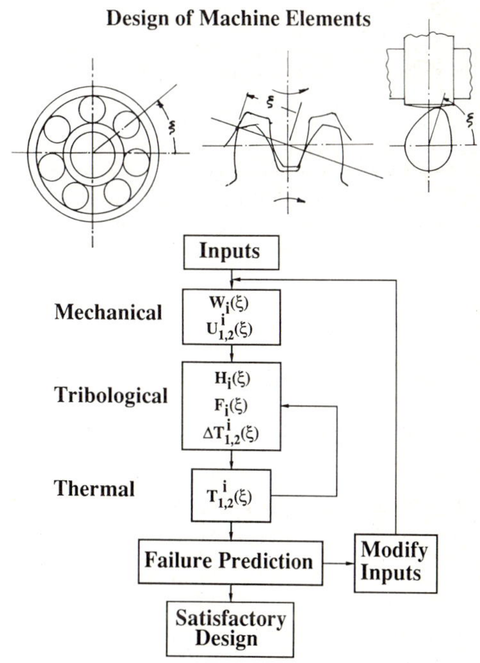

Fig. 4 Block Diagram for Design of Tribological Machine Elements

2.2 Tribological Machine Elements

Fig. 4 shows a block diagram illustrating typical steps for designing a machine element, such as rolling bearings, gears, and cams. As shown in the top figures, each of the elements may contain many generic tribo-elements like EHL contacts which may interact dynamically and thermally.

The design process begins with entering all the geometrical, roughness, operating, lubricant, and material property data, as described earlier for the generic elements. The first step is to determine the cyclic dynamic loads on all the interacting generic elements. In this block, the symbol $W_i(\xi)$ denotes the dynamic load for the ith contact; ξ denotes a space coordinate tracing the contacting path; and $U^i_{1,2}(\xi)$ denotes the surface velocities of the ith contact. The tribological performance together with the bulk surface temperatures $T^i_{1,2}(\xi)$ are determined for all the interacting elements simultaneously by an interactive process. The calculated tribological performance is used to predict failure for each element. A satisfactory design is obtained if no tribological failure is found for all tribo-elements.

The weakest link in this design process, as described earlier, is the lack of reliable analytical failure predictive tools for the tribo-elements. A second weak link in this process is the calculation of the bulk surface temperatures $T^i_{1,2}(\xi)$. The reliability of these predictions is dependent on how close one can predict the surface convective coefficients around all tribo-elements. A slight error in the bulk surface temperature predictions can lead to considerable uncertainties in tribological failure calculations.

3. CURRENT NEEDS

In reviewing the tribological design processes, one can identify a number of weak areas where continuous research is needed to improve the design process. These include:

1. Prediction of Tribological Performance in Thin Film Lubrication
2. Analyticl Models of Contact Fatigue Life
3. Modelling of Scuffing Failure
4. Wear Modelling in Lubricated Contacts
5. Prediction of Bulk Surface Temperature
6. Computerized Tribological Design System

Brief descriptions are given to each of the above listed areas.

3.1 Prediction of Tribological Performance in Thin-Film Lubrication

In thin-film lubrication, the tribological failure is likely controlled by events at the asperity contacts. The distributions of asperity lubricant film thickness, pressure, temperature, and shear stress are pertinent information for establishing failure criteria in this regime. These quantities are, of course, dependent upon the surface roughness topography represented by a set of parameters such as average height, asperity radius, skewness, etc.

At present, there have not been many efforts in studying these distributed quantities at the asperity contacts. For example, the distribution of asperity oil film or micro-EHL film can be calculated for rough EHL contacts by using the classical EHL theories for distributed asperities; however, this has not been fully treated. Computer simulation of asperity pressure and temperature have been achieved by Lai and Cheng[10] for randomly generated rough surfaces and by McCool[11] based on Greenwood and Williamson's model[12] for spherical asperities. However, the use of these methods for determining the characteristics of the asperity pressure and temperature distributions is yet to be achieved.

3.2 Analytical Modelling of Contact Fatigue

In the past three decades, life prediction in lubricated contacts has relied heavily on the Lundberg-Palmgren formula[5] derived from a set of extensive bearing life tests relating the maximum orthogonal shear stress and stress volume to life. Successes in this area include Li, Kauzlarich, and Jamison's application of L-P's relation to asperity contacts to investigate the effects of partial-EHL to contact fatigue life[13], and Harris and Iaannides' recent extension of L-P's model to include the effects of the entire stress field[14]. Fig. 5 shows the predicted trend

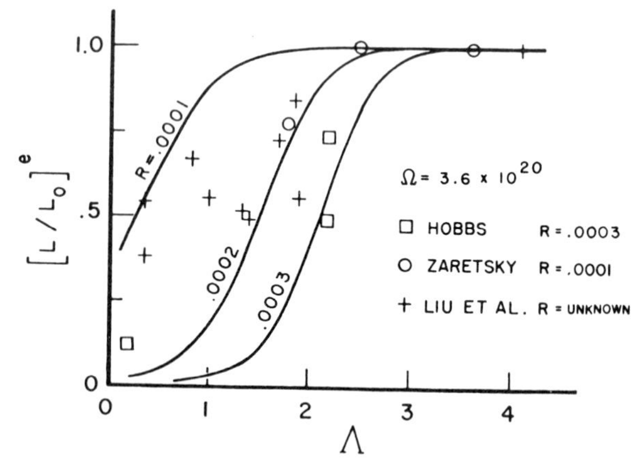

Fig. 5 Predicted Trend of Increasing Fatigue Life with Increasing Λ

of increasing fatigue life with increasing Λ, the ratio of lubricant film thickness to surface roughness. It agrees well with those observed in bearing life tests.

An alternative model to L-P's relation for contact fatigue life was developed by Tallian, Chiu, and Van Amerongen[15]. Their model is considerably more sophisticated than L-P's model, and includes not only the material strength effect and Hertzian stresses but other effects like defect shape, defect density, asperity contact stresses, and traction due to shearing the lubricant and the asperity junctions. Trends similar to that shown in Fig. 6 for the EHL effects can also be predicted from Tallian's theory.

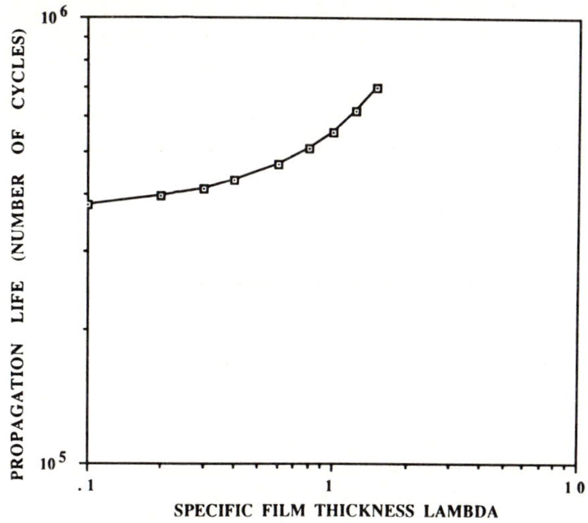

Fig. 6 Analytical Prediction of Life with Specific Film Thickness[17]

While the Lundbery-Palmgren's empirical relation has proven to be a practical tool for life prediction of rolling bearings, it is not an analytical model based on the basic principles governing the fatigue crack growth in contacting solids. Contact fatigue modelling based on basic relations controlling initiation and propagations of surface and subsurface cracks appears to be a more satisfying and rational approach, and should accommodate more easily the effects like lubrication, asperity stress, friction, and material imperfections. Analytical models based on propagation of cracks with a known initial crack length using Paris type propagation laws have been developed by Miller, et al.[16]. The extension of this approach for contacts with a known distribution of initial cracks under both asperity and Hertzian contact stresses has been carried out recently by Blake[17]. A typical life calculation shows a trend similar to earlier results on the effect of EHL film thickness in the low Λ region.

It appears that the status of contact fatigue life prediction is still heavily dependent on empirical methods originated from Lundberg and Palmgren. Analytical models based on understandings of crack propagation yield results encouraging, but are insufficient for accurate life prediction because of the exclusion of initiation life. Continous efforts are needed for developing a complete model for contact fatigue life including both initiation and propagation life.

3.3 Modelling of Scuffing Failure

For lubricated contacts operating at a high slide to roll ratio, the load can be limited by scuffing which is characterized by a sudden transition from low to very high wear rate accompanied by metal transfer. Because of a lack of adequate understanding of the lubrication breakdown mechanisms causing scuffing failure, analytical predictions of the threshold of scuffing have not been entirely satisfactory, as discussed by Dyson[18]. Future success in this area appears to depend on continuous efforts in understanding the breakdown of the following two types of films in protecting the sliding asperities.

The first line of defense is the asperity oil film, also known as the micro-EHL film. The effectiveness of this film is dependent upon the oil viscosity around the sliding asperity. If the oil is pressurized by EHL, asperity oil film would be effective due to a high oil viscosity around the asperity. Dyson[19] used the oil viscosity at the inlet half of the Hertzian conjunction as a measure of scuffing. He employed a convenient parameter, $Q = 1 - (\eta_0/\eta)$ to indicate the inlet viscosity and scuffing: when $Q \to 0$, scuffing would occur. This scuffing criterion based on the parameter Q has found a reasonable correlation with scuffing experiments[20], Fig. 7. Similar results, obtained by Snidle and Rosside[21] and recently by Lee and Cheng[22] also support the "Q" criterion (Fig. 8).

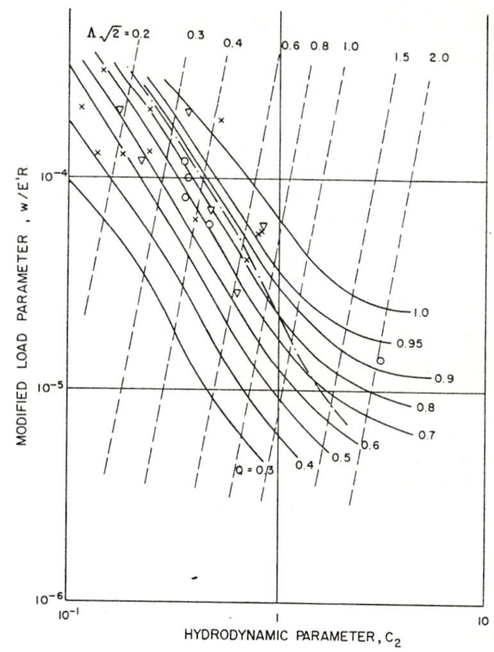

Fig. 7 Correlation of Experimental Scuffing Conditions with Some Hydrodynamic Parameters (o Straight mineral oil; ∇ Mild e.p. oil; × Straight mineral oil; ·——·——· Threshold value of C_2 based on Q = 1.0 with dry contact Geometry; $\sigma/R = 1.8\times10^{-5}$; $\alpha E' = 3333$; $p^*/E' = 7.5\times10^{-5}$; $C_3 = 0$)

The second line of defense is the boundary or surface films which include the monolayer from adsorption of polar species, the chemical film from metal and additive reaction, and the polymeric film from metal/lubricant/oxygen catalytic reaction. Breakdown of these films seems to be all associated with some critical temperature at which these films cease to function.

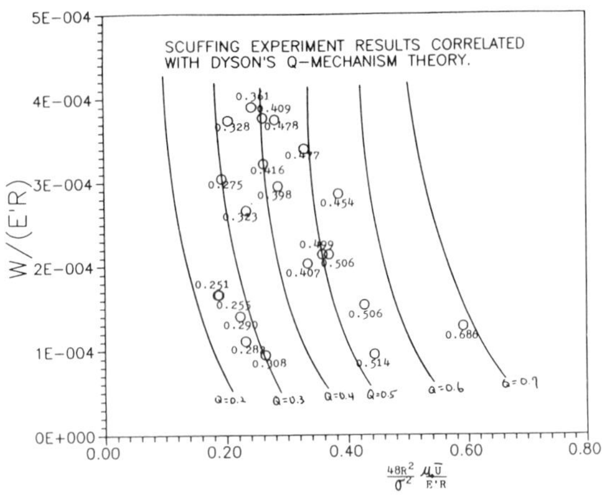

Fig. 8 Further Correlation of Experimental Scuffing Conditions with Dyson's Theory

Among these three types of surface film, only the prediction of critical temperature for the adsorbed film has had some success. Prediction for the breakdown of chemical and polymeric film are difficult because of a lack of surface tools capable of indentifying the exact chemical structure of these films. Since in most cases the threshold conditions at failure are determined by the breakdown of this last line of defense, the surface film, developments of failure criteria for surface films become an extremely important task.

3.4 Wear Modelling in Lubricated Contacts

Reliability prediction of tribo-systems would be much easier if designer can accurately predict and control wear in lubricated contacts. Unfortunately, such wear predictions are not possible at the present time because of a lack of good wear models in the lubricated regime. Tribologists must rise to meet this challenge.

Wear behavior in dry and lubricated contacts, as envisioned by participants in a recent workshop[23], is so complex that a general predictive model appears to be a nearly impossible task. However, a restrictive model, based on a limited number of simulative wear tests, may be successful in predicting wear within a range of operating condictions.

Much of the past modelling efforts in lubricated wear seems to follow this approach. For example, Rowe[24] modified the Archard's wear model for dry wear developed a relation for wear in boundary lubrication constants to be derived from wear experiments. It appears that Rowe's model may be extended to include the effect of EHL in reducing the asperity contacts and effect of micro-EHL in reducing the wear at the asperity contacts. This seemingly simple approach might be a useful first step towards developing an adequate wear model in lubricated contacts.

3.5 Prediction of Bulk Surface Temperature

The significance of bulk temperature in determining the tribological performance has been well recognized by designers. However, there still is a lack of an effective method in handling this problem other than the cumbersome and costly finite element method.

With the current trends towards developing integrated computer softwares for prediction of tribological performance, it appears to be ripe to develop a versatile computation tool to determine the bulk temperatures of a system of interactive tribo-elements using Blok's thermal network concept[25]. Of course, the thermal resistance for each tribo-element within the network can still be determined beforehand by the finite element codes and enter in the network as numeric database.

3.6 Computerized Tribology Information System

In carrying out a tribological design process as shown in Fig. 2, it is often difficult to locate the pertinent numeric database for the lubricant and material properties, the appropriate softwares for calculating the tribological performance, and the pertinent criteria for predicting the failure thresholds. Often designers find themselves spending so much time in searching such information and still end up with not quite the most desirable data or method. This problem can be alleviated if a centralized and computerized tribological information system can be developed. Such need has been pointed out by several countries,and action are now underway to develop such systems for designers. An example of this is ACTIS, known as a computerized tribological information system currently being developed by the National Institute of Standards and Technology with support from the U.S. Department of Energy[26]. As shown in Fig. 9, ACTIS will have six databases with the numeric and design databases aiming directly to tribological design and other databases to assist in production and research.

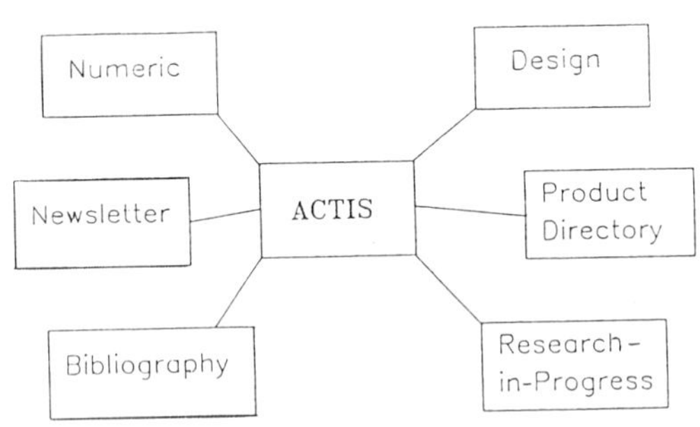

Fig. 9 The Six Database in ACTIS

4. NEW CHALLENGES

In the past, tribological design has centered mainly around metallic pairs lubricated by mineral or synthetic fluids. This situation is changing rapidly due to needs in developing high efficiency and light weight automotive and aircraft engines, precision manufacturing machines, and information processing equipments. Numerous unconventional materials, lubricants, and lubrication concepts have emerged as promising candidates to meet these new needs. Yet, there is very little analytical understandings to guide the design of these tribological applications. There exist many golden opportunities for tribologists to break new grounds in these new areas.

4.1 Mechanics of Soft Films

The use of a low shear strength layer, such as MoS_2 or graphite on a hard substrate to reduce friction is well known. Its friction characteristic can be readily predicted based on a simple linear relation between the shear strength and the normal contacting pressure. Such relation yields a frictional coefficient which decreases asymtotically with pressure to a constant α, as shown in Fig. 10. For some low shear MoS_2 films very small, α have been measured in vacuum[27].

Friction of Doposited MoS_2

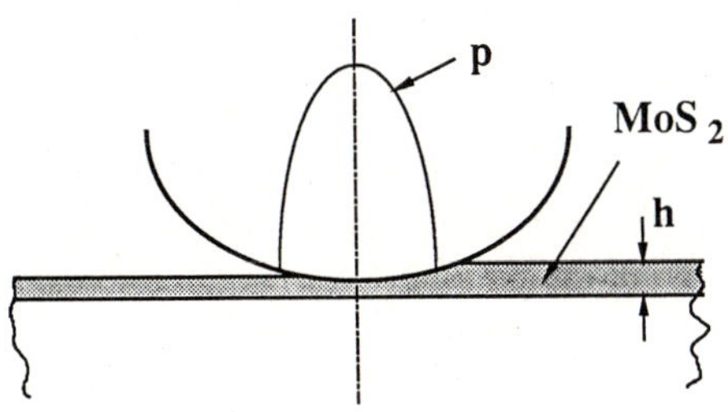

$$\tau = \alpha p + \tau_0$$

$$f = \frac{\int \tau \, dA}{W} = \alpha \int p \, dA + \tau_0 A$$

$$= \alpha + \frac{\tau_0}{\bar{p}}$$

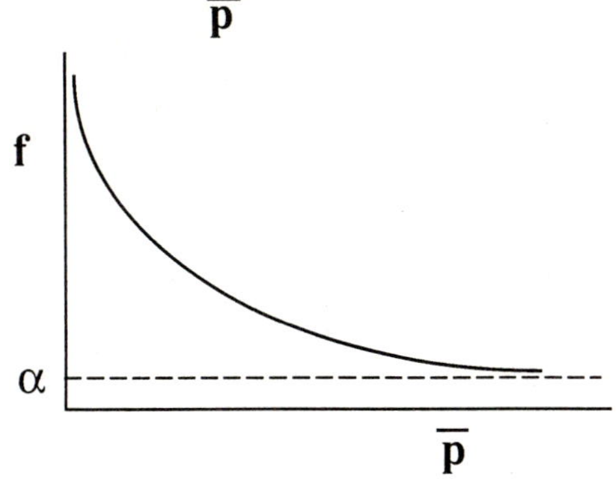

Fig. 10 Friction Characteristics of A Doposited Soft Film

Since life of these soft films is determined by wear, wear prediction becomes essential in design of such films. Wear modelling of a soft film based on analysis of a rigid indentor on a perfectly plastic thin layer may shed some light on the wear process. Coupled with a careful experiment of wear generation between a single conical sliding tip and a soft layer, a semi-empirical wear model may be realizable.

4.2 Mechanics of Hard Coatings

The use of hard coatings to reduce wear is also a well known concept. However, its growing acceptance in tribological applications is quite recent. At present, the selection of coating materials and thickness is achieved largely by trial and error. There is a need of basic understanding of the mechanics of hard coatings to guide the design. Recently, Halling and Arnell[28] introduced a useful concept of effective hardness to guide the selection of coating thickness for both hard and soft coatings. Fig. 11 shows that the effective hardness is depending on the ratio of the coating thickness t to the radius of the indenting slider β. When this ratio approaches 0.03, the coating is fully effective, and the substrate has no effect on the hardness of the layered structure.

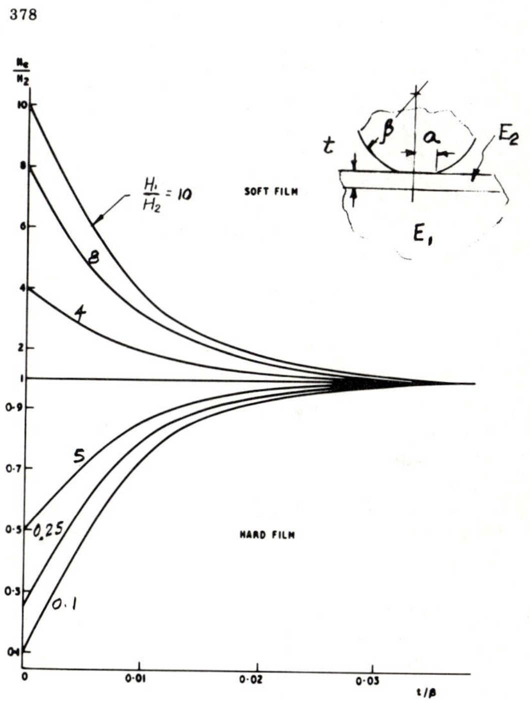

Fig. 11 Effective Hardness of A Soft or Hard Coating

Very little is available to determine quantitatively the beneficial effect of hard coatings on scuffing load, wear rate, and fatigue life. Since hard coatings are anticipated to be used more extensively for wear reduction, it is critical to determine the predominant failure modes, such as interfacial debonding, subsurface spalling, and surface cracking, and to derive some failure prediction methods based on stress field calculated for the layered structure.

4.3 Lubrication and Wear of Ceramics

Ceramics has emerged in the past decades as the most promising candidate for tribo-elements operating at elevated temperatures. In a recent report[29], issued by the U.S.

National Research Council, lubrication and wear of ceramics have been recommended as two key areas for developing methods for design ceramic tribo-elements.

In recent studies of mechanisms in ceramic wear, it is found that there are as many wear modes in ceramics as in metallics[30]. Even for the same type of ceramics, the wear mode can vary depending on its microstructure, state of stress, kinematic conditions, and temperature. For example, a typical silicon nitride which exhibits a grain-pull-out wear mode in pure rolling (Fig. 12) changes into a polishing wear mode when sliding is introduced (Fig. 13)[31]. At present, there are practically no reliable models for predicting the wear rate corresponding to these modes. The developmet of quantitative predictive models for rolling and sliding wear based on fracture of brittle solids under stress field induced by asperities would accelerate greatly the pace in using ceramics as tribo-elements. In lubrication of ceramics at moderate temperatures, it is not always possible to transfer directly the experience in lubrication of metallics with liquids mainly because of the difference in tribochemisty. Prediction of lubrication performance in thin film would depend on further studies of the lubricant and ceramic reactions over a range of temperatures. This area is another challenge to the tribologists.

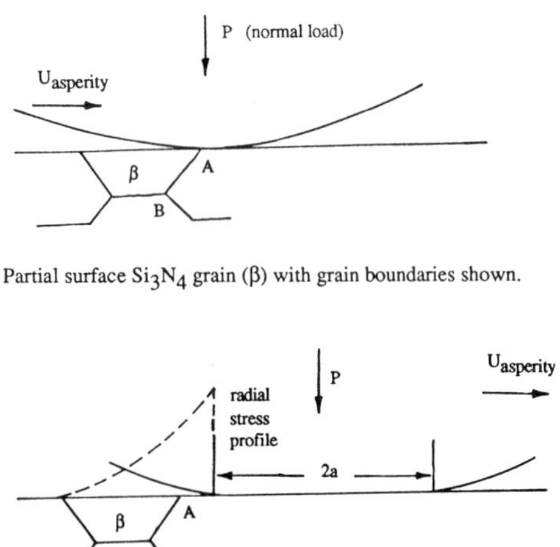

(a.) Partial surface Si_3N_4 grain (β) with grain boundaries shown.

(b.) Crack growth along grain boundary AB after n cycles occuring at trailing edge of asperity contact zone.

(c.) Propagation of crack along the remaining grain boundaries till pullout occurs leaving pit. Dashed line does not represent adhesion to opposing asperity.

Fig. 12 Grain-pull-out in Pure Rolling Silicon Nitride Contact

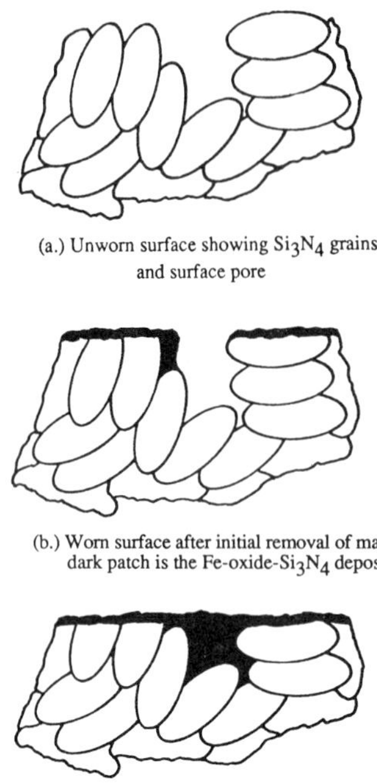

(a.) Unworn surface showing Si_3N_4 grains and surface pore

(b.) Worn surface after initial removal of material, dark patch is the Fe-oxide-Si_3N_4 deposit.

(c.) Filling in the surface pore (dark region) with further polishing of grains

Fig. 13 Pollishing in Sliding Silicon Nitride Contact

At elevated temperatures, liquids cannot be used for lubrication. One must rely on non-circulating solid lubricants or solid films. The mechanics of solid film is seen to be a major challenge in tribological design and has already been discussed earlier.

Recently, a noval concept in lubrication of ceramics at high temperatures was introduced by Graham and Klaus[32]. Both friction and wear were found to be acceptable at temperatures as high as 600°C when hydrocarbon vapor is fed once through the sliding contact. At present, the mechanism of lubrication is not entirely clear, and there is yet no analytical model to characterize the lubrication performance. This appears to be another challenging task.

4.4 Mechanics of Super Thin Oil Film

In computer hard discs, a thin layer of oil film is applied for lubrication. During start, stop, or sudden transient dynamic loading, asperities on the head surface will contact the disc surface. Lubrication of the sliding asperities is provided by a super-thin film, less than 100 A[33]. The formation of such films may be predictable from the cassical Reynolds theory as demonstrated recently by Chan and Horn[34] in studying the super-thin oil film formed by the squeeze film effect.

If the distribution of the thickness of these super-thin films is known, one should be able to derive a new criterion to determine the lubrication breakdown based on the ratio of the film thickness to the micro-roughness. Tribological design in computer head/disc system will no doubt be greatly enhanced.

5. CONCLUDING REMARKS

Early critical needs in railroad, automotive, and gear industries led to the establishment of thick-film lubrication which has served for many years as the foundation of tribological design of conformal and counterformal contacts. Current and future needs in aerospace and information processing industries are presenting tribologists with a set of new challenges due to requirements of high temperature, heavy load, high speed, low friction, and long wear life. These are to:

* predict the characteristics of thin-film lubrication in terms of the average and asperity scale film thickness, pressure, shear stress, and surface temperature;

* develop analytical methods to predict fatigue life in elastohydrodynamic contacts;

* develop tribo-chemical models for predicting wear and scuffing in sliding contacts;

* develop computerized tribological information systems for tribological design in industry;

* understand the mechanics of soft coatings for lubrication and hard coatings for wear reduction;

* develop friction and wear data-base for ceramics, ceramics metal and polymer matrix composites;

* develop new synthetics and noval lubrication methods, such as vapor lubrication, for high temperature applications.

Successes in the above areas are not likely to come from a single discipline, but from multidisciplinary efforts by collaborative researchers from industry and universities.

REFERENCES

[1] Dowson, D., "History of Tribology," 1979, Longman Group, London.

[2] Tower, Beauchamp, "First Report on Friction Experiments," Proc. Inst. Mech. Eng., 1849, 161, pp.59-69; 2nd Report, ibid. 1885, 36, pp.58-70; 3rd Report ibid., 1888, 39, pp.173-205; 4th Report, ibid., 1891, 42, pp.111-140.

[3] Reynolds, Osborne, "On the Theory of Lubrication and its Application to Mr. Beanchamp Tower's Experiments, Including an Experimental Determination of the Viscosity of Olive Oil," Phil. Trans. Roy. Soc. 1886, 177, pp.157-234.

[4] Blok, H., "Theoretical Study of Temperature Rise at Surfaces of Actual Contact under Oiliness Lubricating Conditions," Proceedings of gen. Disc. Lubr., Institute of Mech. Eng. 2, pp.222-235, 1937.

[5] Lundberg, G. and Palmgren, A., "Dynamic Capacity of Rolling Bearings," Acta. Polytech. Mech. Engr. No.1.

[6] Dowson, D. and Higginson, G., "A Numerical Solution of the Elastohydrodynamic Problem," Journal of Mechanical Engineering Science, 1, 6.

[7] Hardy, W.B. and Doubleday, "Boundary Lubrication - The Paraffin Series," Proc. of the Roy. Soc. (London) Ser.A, Vol.100, No.707, 1922, pp.550-574.

[8] Dowson, D., Taylor, C.M., Godet, M., and Berthe, D., "Fluid Film Lubrication - Osborne Reynolds Century," Proc. of the 13th Leeds-Lyon Symposium on Tribology, Sept. 1986.

[9] Cheng, H.S., "Surface Roughness Effects in Lubrication," Proc. of the 11th Leeds-Lyon Symposium, 1984.

[10] Lai, T., and Cheng, H.S., "Temperature Analysis in Lubricated Simple Sliding Rough Contacts," ASLE Trans., Vol.28, No.3, July 1985, pp.303-312.

[11] McCool, J., "Relating Profile Instrument Measurements to the Functional Performance of Rough Surfaces," Journal of Tribology, ASME Trans., Vol.9, No.2, pp.264-270, April 1987.

[12] Greenwood, J.A. and Williamson, J.B.P., "Contact of Nominally Flat Surfaces," Proc. R. Soc. London, Ser. A, 295, pp.300-319, 1966.

[13] Li, D.F., Kauzlarich, J.J., and Jamison, W.E., "Surface Roughness Effects on Fatigue in Partial EHL Lubrication," Trans. of ASME, Journal of Tribology, 98, Oct. 1976, pp.530-537.

[14] Harris, T.A. and Ioannides, E., "A New Life Model for Rolling Bearings," Trans. of ASMe, Journal of Tribology, 107, pp.367-378, 1985.

[15] Tallian, T.E., Chiu, Y.P., and Van Amerongen, E., "Prediction of Traction and Microgeometry Effects in Rolling Contact Fatigue Life," Trans. of ASME, Journal of Lubrication Technology, 100, pp.156-166, 1978.

[16] Miller, G.R., Keer, L.M., and Cheng, H.S., "On the Mechanics of Fatigue Crack Growth due to Contact Loading," Proc. of the Roy. Soc. of London, A. 397, pp.197-209.

[17] Blake, J., "A Surface Pitting Life Model for Spur Gears," Ph.D. Thesis, Northwestern University, June 1989.

[18] Dyson, A., "Scuffing - A Review Part 1 and 2," Tribology International, April 1975, pp.77-87 (Part 1), June 1975, pp.117-122 (Part 2).

[19] Dyson, A., "The Failure of Elastohydrodynamic Lubrication of Circumferentially Ground Discs," Proc. Inst. Mech. Engr., 76,190, No.52.

[20] Cheng, H.S. and Dyson, A., "Elastohydrodynamic Lubrication of Circumferentially-Ground Rough Discs," ASLE Trans., 21, 25-40, 1976.

[21] Rossides, S.D. and Smidle, R.W., "Surface Topography and the Scuffing Failure of Circumferentially Ground Discs," Proc. of Symposium on Surface Roughness Effects in Hydrodynamic and Mixed Lubrication, ASME Publication, 1980, pp.93-144.

[22] Lee, S.C. and Cheng, H.S., "Further Correction between Scuffing Experiments with EHL Analysis of Rough Surfaces," to be submitted for Presentation at the 1989 ASME/STLE Joint Tribology Conference.

[23] Ling, F.F. and Pan, C.H.T., "Approaches to Modelling of Friction and Wear," Proc. of the Workshop on the use of Surface Deformation Models to Predict Tribology Behavior, Springer-Verlag., 1986.

[24] Rowe, C.N.,"Some Aspects of the Heat of Adsorption in the Function of a Boundary Lubricant," Trans. of ASLE, 9, pp.108-111, 1966.

[25] Blok, H. "The Postulate about the Constancy of Scoring Temperature," NASA SP-257, Interdisciplinary Approach to the Lubrication of Concentrated Contacts, pp.153-248, 1970.

[26] Jahanmir, S., Ruff. A.W., and Hsu, S.M., "A Computerized Tribology Information System," presented at the International Conference on Engineering Materials for Advanced Friction and Wear Applications, March 1988.

[27] Singer, I., Private Communication.

[28] Halling, J. and Arnell, R.D., "Ceramic Coating in the War on Wear," Wear, A Celebration Volume, Elsevier Sequoia, 1985, pp.367-380.

[29] Frost, B.R.T., et al., "Tribology of Ceramics," a National Research Council Report, NMAB-435, National Academy Press, 1988.

[30] Braza, J.F., et al., "Mechanical Failure Mechanisms in Ceramics Sliding and Rolling Contacts," Tribology Trans., Vol.32, No.1, Jan. 1989, pp.1-8.

[31] Braza, J.F., "Wear of Ceramics in Lubricated Sliding and Rolling Contacts," Ph.D. Thesis, Northwestern University, June 1988.

[32] Graham, E.E. and Klaus, E.E., "Lubrication from the Vaporphase at High Temperature," ASLE Trans., Vol.29, No.2, April 1986, pp.229-234.

[33] Bhushan, B., "Tribological Design - Information Storage and Retrieval," presented at the 15th Leeds-Lyon Symposium on Tribology, 1988.

[34] Chan, D.Y.C. and Horn, R.G., "The Drainage of Thin Liquid Films between Solid Surfaces," Journal of Chem. Phys., 83, 15, NoV. 1985, pp.5311-5324.

SESSION II
REVIEW PAPERS

Chairman: Professor M Godet

PAPER II(i)	Tribological Design - The Aerospace Industry
PAPER II(ii)	Tribological Design - The Railways
PAPER II(iii)	Tribological Design - The Automotive Industry
PAPER II(iv)	Tribological Design - The Process Industry

Paper II(i)

Tribological design - The aerospace industry

J. A. Dominy

Abstract

The tribology of both aero engines and airframes plays an important role in their operation and reliability in service. Often the tribology of a system influences a components ultimate life rather than its reliability. The primary requirement of a component such as a gear tooth is that it should not suffer immediate and complete failure such as root cracking, whereas a more benign tribological failure of the surface may well be acceptable. The constraints imposed upon the designer by the very limited range of approved lubricant types reduce the range of variables that can be varied to play "tribological tunes" to any significant extent.

Tribology - the definition

In considering tribology in the aerospace industry, the first problem is that of definition. The classical definition of tribology as "the study of friction and wear" with, perhaps, the inclusion of lubrication, is not usually consistent with the structure and individual areas of expertise found in industry. The industrial definition will tend to include component life or load capacity, heat generation and often oil systems. Certainly, as far as engines and helicopters are concerned, the tribologist will probably consider himself to be a transmission and lubrication engineer, possibly with further interests in materials and chemistry.

The problem is compounded when we ask where the tribologist's sphere of influence stops. Consider the simple gearbox and oil system shown in figure 1. The design of the gears and the definition of the teeth are definitely the job of the tribologist, whereas the design of the aerodynamic shape of the cooler duct is not. But what of the design of the gearbox itself, gearbox efficiency and oil flows, where do the responsibilities lie?

Since the author considers himself to be a tribologist yet takes a strong interest in these fringe items, this paper unashamedly takes a liberal view of the definition of tribology.

Areas of Tribology in Aerospace

Tribological design in the aerospace industry can be considered to fall into three main groups of components. The first is the engine which offers a wide variety of

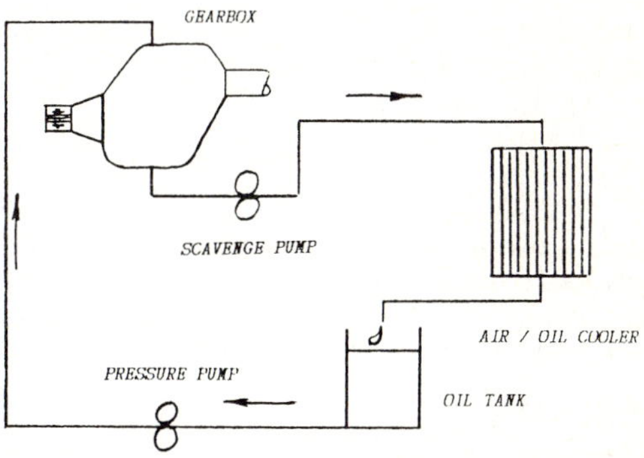

Figure 1 - Diagram of gearbox oil system

tribological puzzles. The second is the gearbox which appears in some engines but is fundamental to the successful operation of the helicopter. The third is the airframe where the problems of helicopters and fixed wing aeroplanes are quite different.

The approach to tribological design will vary from component to component according to how critical is its reliability. The helicopter manufacturer is unlikely to allow the design of the gearbox to go outside the industry since it is critical to the success and safety of his product. However, his bearings, which are often working in established regimes, may well be entrusted to the more reputable suppliers. In contrast, the engine manufacturer will, almost certainly, keep his bearing design in-house since historically these have been a source of anguish and, in terms of temperature and speed, they are often out on a limb. Yet he will consider sub-contracting accessory drive gears. Consequently, each area of the industry has built up distinct areas of expertise.

Engines

Since the introduction of the gas turbine engine in the mid 1940s, the rapid increase in shaft speed and operating temperatures has led to the development of mainshaft bearings capable of operating reliably under adverse conditions. The main advances have been in materials where, in order to achieve high temperature stability, derivitives of tool steels are used as opposed to the carbon chromes more usually found(1). In the U.K. Tungsten tool steel has generally been used (18/4/1) and in the U.S.A. Molybdenum (M.50). More recently commercial pressures from multi-national bearing suppliers and the relative size of the U.K. and U.S. markets has tended to lead to the general adoption of M.50.

In general the design of engine bearings, particularly ball bearings, is dictated by the requirement to achieve a specific fatigue life. There is always a compromise to be made since the controlling of the bearing load will require air flow which will cost the engine performance. There is thus always resistance to reducing bearing loads. If very high loads are imposed upon the bearing it is sometimes necessary to reduce the fatigue life since the internal geometry must be modified to prevent the ball/race contact ellipse extending beyond the race shoulder.

True tribological design is not usually a primary consideration. Film thicknesses are calculated and compared with previous values and experience. The problem is that the bearing designer has relatively little control over the environment and lubrication. Speed, load and temperature are all broadly defined by the engine performance and configuration while the lubricant must be selected from a very limited list of acceptable specifications. The tribologist has very little influence over the performance of the bearing unless a particular problem is either foreseen or encountered in Service.

Heat generation in bearing chambers (which often include gears) is an increasingly important area. Although the heat rejected to the oil is a relatively small proportion of the output power for a big jet engine, the absolute values become large and oil cooling becomes difficult. Traditionally a turbo fan engine has returned the heat in the oil system to the engine cycle by fuel cooling thus pre-heating the fuel going to the combustors. However, as engines become more efficient and fuel flows are reduced this becomes increasingly difficult unless the heat-to-oil is minimised. The resulting penalty is the weight and bulk of external oil coolers. The heat generation can be minimised by the careful shrouding of components such as gears, the design of the bearings themselves and, most importantly, the feed and scavenge systems. It is important that once the oil has lubricated (and cooled) the bearing it be free to return to the scavenge system without further contributing to churning and windage.

An aspect of engine tribological design that does not receive much publicity but is critical to the engine performance - particularly the all important fuel economy - is compressor blade tip seals. Tip clearance is of fundamental importance to compressor efficiency and must be kept as small as possible. The general solution has been to allow the tips to rub against an abraidable material at the worst condition and to set their own clearance. As the pressure ratio increases (to 30 or so) erosion of the liners will occur and the performance of the engine will deteriorate with time. Harder coatings are necessary at the high pressure stages to withstand the temperatures. These will, by their nature, have a less benign wear characteristic. This limits the materials that can be used for the compressor blades and thereby imposes a constraint on shaft speed and engine performance. It would be nice to be able to use titanium blading throughout the compressor but the hard linings force the use of steel for the high pressure stages to prevent the possibility of titanium fires as a result of blade rub.

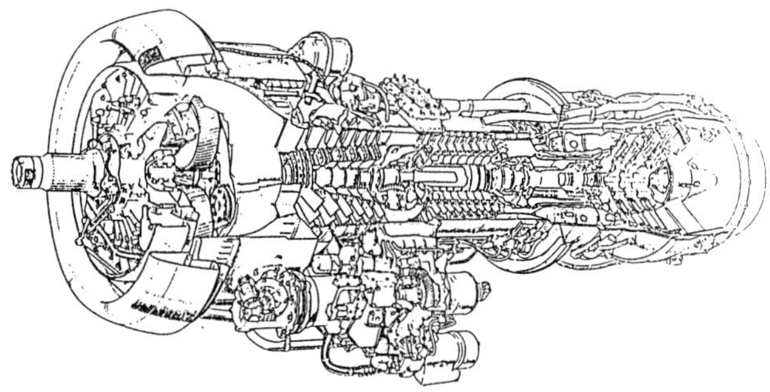

Figure 2 - Rolls-Royce Tyne 4.5 MW engine

Gearboxes

The two applications for high duty gearboxes in aviation are helicopter rotor drives and turbo-prop engine reduction transmissions. Both are unusual insofar as the rotor or propeller loads are fed straight into the gearbox with no intermediate structure to ease the conditions. However, in terms of duty and safety requirements the two are very different. To the helicopter the gearbox is fundamental to the operation and safety in exactly the same way as the wings are to a fixed wing aircraft. In an engine, particularly in a multi-engine application, the gearbox is not so critical as its failure does not directly result in the loss of the aircraft which may be the case in a helicopter. The safety requirements are complex but distinctly different (2,3).

In a turbo-prop or geared-fan engine (figures 2 and 3) the gearbox is necessary to match the speeds of the relatively low speed fan to the high speed turbine. For maximum powerplant efficiency the fan will be as large and slow as possible while the turbine will be small and fast. Figure 4 shows the trend in ratio and transmitted power since 1950. The initial rapid rise in ratio was

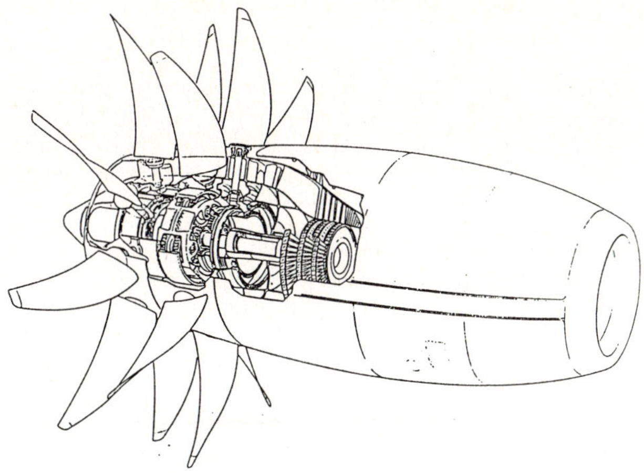

Figure 3 - Design study for a 9.0 MW prop-fan powerplant

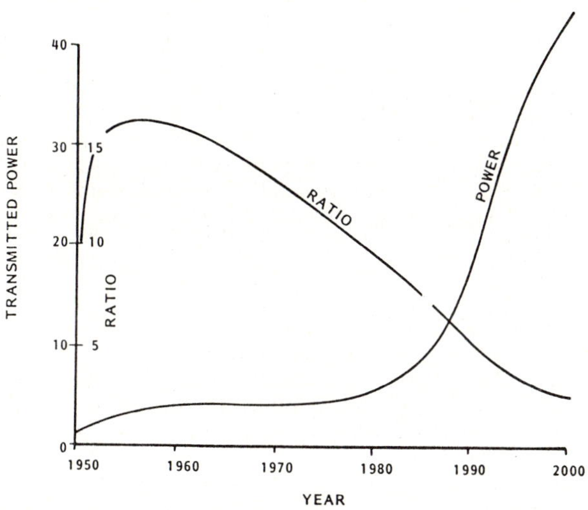

Figure 4 - Trends in power and ratio for existing and proposed engine gearboxes

due to the introduction of the gas turbine with its very high shaft speeds. Since then the increase in engine power (hence size) has tended to contain the shaft speed while the output has generally increased in speed as advances in structures and aerodynamics have allowed higher blade speeds. Consequently ratios have tended to fall over the years to the extent that the very low ratios (less than 2,5:1) envisaged for future generations of engines cannot be achieved with simple epicyclic (and very efficient) gearboxes (4) (figures 5 and 6).

Whereas the engine gearbox will often be of a simple inline configuration the helicopter transmission is an altogether more complex proposition (figure 7). The main drive has to be turned through 90 to the rotor head. It will usually be necessary to combine the drives from two engines and to provide outputs to the tail rotor and aircraft accessories. Very high degrees of reduction are necessary to match the 20,000 r.p.m. of the engines to the 200-300 r.p.m. of the main rotor. Novikov gears (figure 8) are used in

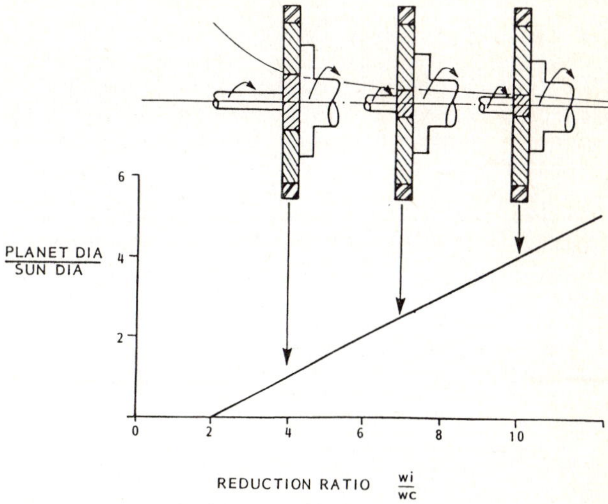

Figure 5 - Range of ratio available from a simple epicyclic gearbox

some instances (5) due to their ability to accommodate more ratio then involutes in a single stage with consequent savings in weight and bulk. Nevertheless, it is still necessary to have some intermediate stages to provide output drives for the accessories.

As in bearings, the tribological design of a gear plays second fiddle to stress considerations. It is more important that a gear be designed so that the teeth do not fail at the roots and fall off than it is that they do not fail in surface fatigue which is a fairly benign failure that can be detected many hours before it becomes critical. Advances in health monitoring (6) are leading to a much improved ability to detect gearbox failures; particularly the cracked gear.

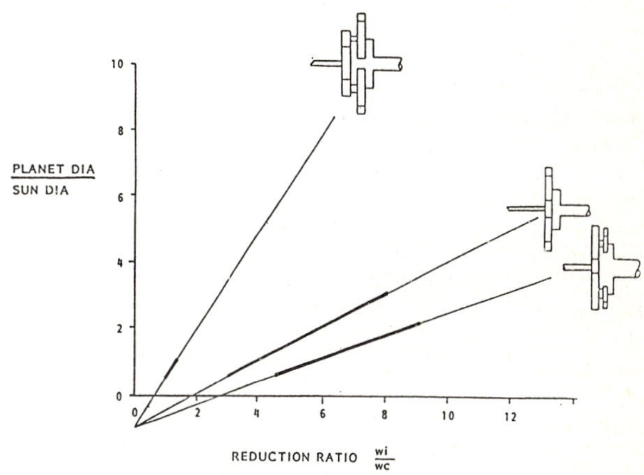

Figure 6 - Range of ratio possible from a compound planet epicyclic gearbox

Calculation of oil film thickness can be and is carried out but the difficulty of estimating thermal effects and possibly starvation at high rolling speeds and slide-roll ratios under very transient conditions render this unreliable in absolute terms. As in the case of the engine bearing, constraints of stressing and lubricant selection leave the designer relatively little leeway to optimise the tribological performance. An area in which the tribology can influence the stress is in traction where the tangential tooth forces can affect the root stress. At the moment, our understanding of traction, particularly for synthetic lubricants and under the conditions imposed at the contact in a high duty gearbox, is not adequate to allow the necessary computation.

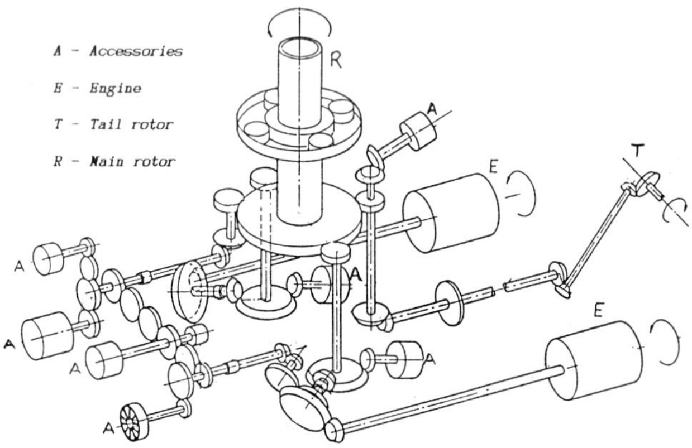

Figure 7 - Diagram of the transmission system in the Agusta A129 helicopter - after (13)

Tribology does, however, play an important part in the estimation of gearbox efficiency which, in turn, is critical in the selection of gearbox configuration. In many cases the viability of an engine project depends upon the gearbox efficiency due to the weight and drag of the associated cooling system (7). A knowledge of traction and the mechanism of the lubrication of a gear contact allows an estimation of the meshing loss within a

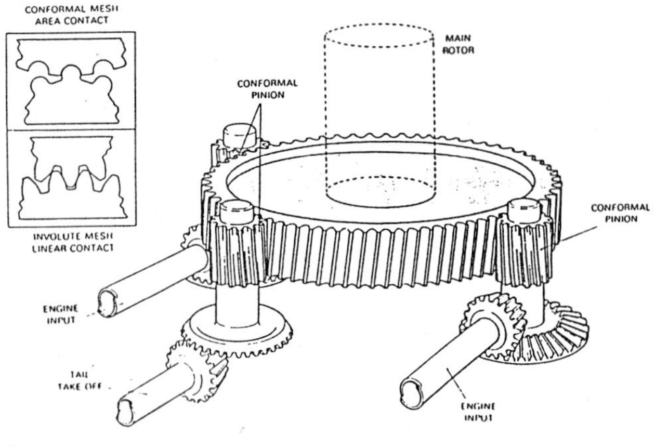

Figure 8 - Novikov gear arrangement - after (14)

gearbox. This accounts for roughly half the total loss - most of the rest being windage. Unfortunately, as mentioned above, the understanding of the traction of synthetic lubrications in hostile environments is not understood sufficiently to allow a "scientific" analysis of meshing loss. This is a priority area for academic/industrial research if large gearboxes are to become a reality. In order to minimise the cooler size the gearbox must run as hot as possible. Once again a fundamental understanding of the tribology of a gear contact is necessary if the effect of temperature on the oil thickness and traction are to be assessed. Again the understanding of the tribology and the thermal conditions at the gear teeth are currently unable to allow us to perform the calculations needed to give the confidence necessary to make a significant change in scavenge temperature (from about 150 C to 250 C). Any such change is further limited by the constraints of available and commercially acceptable (to the operator) lubricants.

In the U.K. both the helicopter and engine industries are keen to extend the understanding of the lubrication of gear teeth; involute and Novikov. A particularly valuable tool in this investigation is the micro transducer capable of measuring the pressure, temperature and film thickness in a real gear contact (8).

Airframes

The tribological requirements of airframes differ between fixed and rotary wing. The control system of a fixed wing aircraft does not have to contend with the continuous cycling of the rotor head while the tyres and brakes of a helicopter do not have to withstand heavily loaded, high speed landings.

A helicopter control system is centred around the rotor heads. The controls must be taken round or through the gearbox (whose design must accommodate this) and then be transferred from the static airframe to the rotating rotor via a swash plate. Thus the controls are continuously moving and the control bearings are required to withstand many cycles albeit at a relatively low load. The fixed wing airframe will subject its bearings to far fewer cycles but often at much higher loads and there is a more arduous requirement for the bearing properties to remain consistent through a range of atmospheric conditions to a very much higher altitude. Control systems generally use plain, self lubricated spherical bearings. As a result of continuous development (9) they have progressed from stainless steel (440C) balls in PTFE liners to Tungsten Carbide balls in composite liners.

Tyres and brakes in fixed wing aircraft - particularly, large, heavily loaded transport aircraft - provide frictional problems of the most fundamental nature. Heat generation in tyres due to slip at touch down and hysteresis in the structure is difficult to contain and hard or high speed landings can cause tyres to deflate (10). Brakes also cause problems although they are assisted by the reverse thrust provided by the engine. The problem of the ever increasing weight and speed of transport aircraft has led to the disc brake and the carbon disc, both of which have since found their way into the automotive industry. The proximity of the brakes and tyres needs careful consideration by the designer if the hub bearings and tyre itself are not to be damaged by heat soak once the aircraft is parked and there is no longer a flow of cooling air. Brakes and tyres generally are the result of continuous product development rather than intensive research.

The design of the helicopter rotor head bearing system has, in some cases, tended to simplify over the last few years due to the introduction of the semi-rigid head (figure 9) which allows some of the motion of the blade to be accommodated elastically rather than through bearings. The rotor head presents a wide variety of problems to the bearing designer from the continous motion of the swash plate to the blade retaining thrust bearings which suffer very high axial loads but little movement with the consequent risk of brinelling. The main rotor bearing is perhaps the most critical single tribological device in the aerospace industry since the whole helicopter literally hangs from it and any major failure will result in the loss of the aircraft.

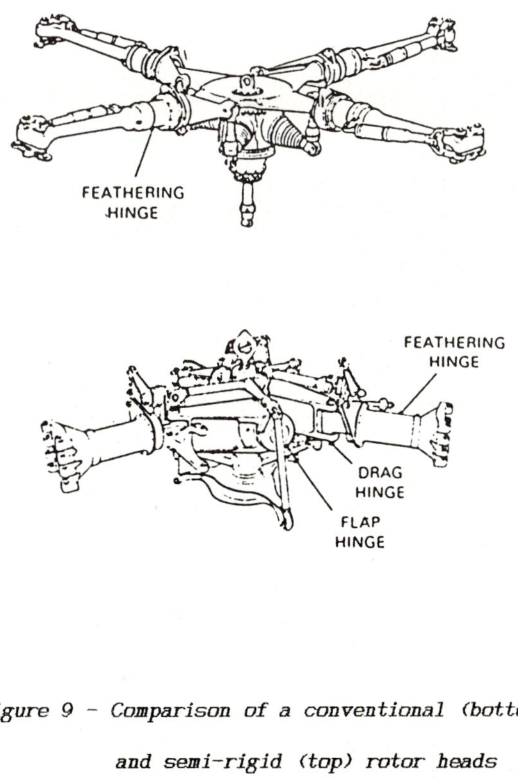

Figure 9 - Comparison of a conventional (bottom) and semi-rigid (top) rotor heads - after (14)

Lubricants

One of the most powerful constraints on the tribological design of engines and gearboxes is the lubricant. Although there may appear to be many brands available these are brewed to a very few, very tight specifications and the real choice is very restricted. To break out of this the aviation authorities must be convinced of the need for a new specification, the oil companies must be able to see a profit at the end and the operator (i.e. the airline) must be convinced that technically there is no alternative and the concept is economically viable. This last point really means that the oil must be readily available at any major airport. The difficulty of overcoming the technical, administrative and commercial problems has inhibited the development of lubricants for some aerospace applications. Since the introduction of synthetic lubricants in the 1940s there has been a progressive increase in the temperature capability (figure 10) but the limit of existing materials is being approached and new base stocks will be needed for significant further advances.

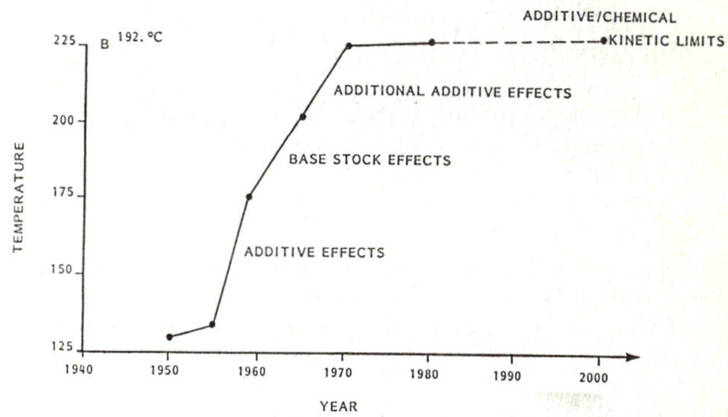

Figure 10 - Temperature limitation of synthetic gas turbine lubricants

The requirements for an engine and gearbox lubricant are fundamentally different. The engine requires a lubricant that is stable to very high temperatures (greater than 200 C), needs resistance to the formation of carbon at these temperatures and must be capable of being pumped during a cold start (-40 C) without causing damage to the system. This imposes a maximum viscosity of 13,000 c.St. at -40 and results in traditionally low viscosities of gas turbine lubricants at their operating temperatures. There is relatively little need for load carrying additives which would often break down at high temperatures, in some cases adding to the problem of carbon formation. The gearbox does not need the high temperature capability and both needs, and can take advantage of, the additive package. Both helicopter and engine gearboxes must maintain capability of a cold start. The requirements of the oil systems also differs since the engine, due to high temperature carbon formation cannot easily take advantage of fine filtration since the filters would block. The gearbox, since it does not run hot enough to form carbon, is able to exploit a high degree of filtration.

Industrial Research in Tribology

So why does industry carry out research into tribology? Surely the subject is already defined? An examination of Tribiological design problems shows this not to be the case. The problem of "real" film thickness in "real gears" is one example, slip in ball bearings is another. The science and/or data does not exist to allow the calculation of all the parameters necessary to carry out the mechanical design.

A good example is the simple cyclindrical roller bearing where co-operation between industry and academia has produced something close to a practical design tool. By the mid 1980s the academic community had provided solutions for:

1. The isothermal, fully flooded film thickness.

2. Thermal correction IF the temperatures were known.

3. Starvation effect IF the inlet meniscus position was known.

4. Traction IF the oil properties were known.

The 'IFs' were not known and the industrial research plugged the gaps that it felt were relevant. Temperature network modelling of the bearing (11), studies of the inlet position at high speed (12) and traction evaluations were all completed in industry. The result of the research means that it is now possible to estimate the film thickness, starvation, slip and to optimise the geometry to achieve a satisfactory compromise at the design stage. This would not have been possible without the fundamental understanding of e.h.d. lubrication in the academic world combining with the real service problems and need to know in industry.

This pattern of co-operation can be seen in many areas. The research will continue but, particularly in the case of engines with their very long lead times, it will often be a stop-go-stop process due to the commercial considerations that inevitably govern industrial research and the funding of the parallel academic work. Considering just the field of e.h.l. there is still much to be done. The prediction of starvation in real machine elements, traction in transient contacts and thermally induced stress, are all areas of importance to the industrial tribologist - areas where our current lack of understanding constrains the mechanical design.

There is still substantial scope for research into the application of our knowledge of tribology.

Acknowledgements

The author wishes to thank Dr. Andy Olver of Westland Helicopters and Mr. Bryan Rayner of Rolls-Royce for their invaluable contributions to this paper. Figures 8 and 9 were reprinted by permission of the Council of the Institution of Mechanical Engineers from "Proc. I.Mech.E., Vol. 199, (1985), No D.I.

References

1. E. N. Bamberger
"Status of understanding for bearing materials"
Tribology in the 80's, NASA CP-2300

2. J. E. Witham
"Safety standards for helicopter and engine transmissions"
I.Mech.E. Seminar "Pushing back the frontiers of failure in aerospace transmissions", 1986

3. G. Tomasson
"Advanced turboprop airworthiness considerations"
I.Mech.E. Seminar "Transmissions technology for propfan and geared fan engines", 1986

4. J. Dominy and B. Rayner
"The technology of advanced prop-fan transmissions"
15th Congress of ICAS , London, 1986, ICAS-86-3.10.1.

5. R. W. Snidle, A. W. Dyson and H. P. Evans.
"Research in support of high conformity gearing technology"
I.Mech.E. Seminar "Pushing back the frontiers of failure in aerospace transmissions", 1986.

6. D. G. Astridge.
"Improving the integrity, life and reliability of helicopter gearboxes"
I.Mech.E. Seminar "Pushing back the frontiers of failure in aerospace transmissions", 1986.

7. J. Dominy.
"Transmission efficiency in advanced aerospace powerplant"
23rd Joint Propulsion Conference, AIAA-87-2043, 1987

8. M. Safa, J. Anderton and J. Leather.
"Transducers for pressure temperatures and oil film thickness measurement in bearings"
Sensors and Actuators, V3, pp 119-128, 1982

9. J. K. Lancaster.
"Composites for increased wear resistance: current achievements and future prospects"
Tribology in the 80's, NASA CP 2300

10. S. Clarke and R. Dodge.
"Heat generation in aircraft tyres under free rolling conditions"
NASA CR-3629

11. R. Nicholson.
"The prediction of operating temperatures in high speed angular contact bearings"
XIII Leeds-Lyon Symposium on Tribology, 1986
"Fluid Film Lubrication - Osborne Reynolds Centenary"
Elsevier Science Publishers, 1987

12. J. Dominy.
"The effect of traction on oil film thickness in high speed roller bearings"
"Friction and traction" Proc VII Leeds-Lyon Symposium on Tribology, IPC Business Press Ltd.

13. A Garavaglia and G. Gattinoni
"Design criteria of the A129 helicopter drive system"
AGARD conference proceedings No 302, paper 12-1.

14. B. Blackwell and V. Rogers.
"Mechanical engineering aspects of helicopter design"
Proc. I.Mech.E., Vol. 199, (1985) No. D1.

Paper II(ii)

Tribological design - The railways

C. Pritchard and T. G. Pearce

The tribological interaction of fundamental importance to railways is that between wheel and rail. A description is given how the friction and wear characteristics of steel wheels and rails, together with relevant forms of rolling contact damage, are taken into account in the design of modern railway suspension systems.

1 INTRODUCTION

Of all industries, railways can be said to have a technology rooted in tribology. F.T. Barwell illustrated this by identifying 47 interacting surfaces in applications ranging from bridge bearers below the rails to the pulleys supporting the overhead catenary on electrified lines (1). Other authors have covered much of the same ground (2,3).

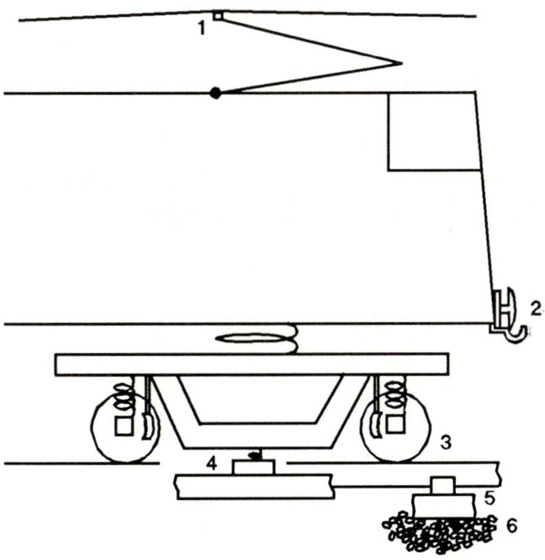

1 & 4. SLIDING ELECTRICAL CONTACTS.
2. BUFFERS AND DRAW GEAR.
3. WHEEL/RAIL CONTACTS.
5 & 6. RAIL/SLEEPER/BALLAST CONTACTS.

Fig. 1 Surface Interactions on Railways

Most of the 47 interactions pose design problems common to industry in general, but some remain unique, or nearly so, to railways (Fig 1). Of these, clearly the most fundamental is the interaction between wheel and rail, for the tribology of the steel-to-steel rolling contact is of prime importance when designing not only for the traction and braking of vehicles but also their guidance. (Rubber tyres have replaced steel wheels on some Metro systems, but these provide traction and braking only, guidance being effected by separate horizontal rollers or flanged steel wheels).

Railway vehicles require a lateral suspension which ensures that the wheels steer smoothly round curves while maintaining a stable path on straight track. Conflicting requirements arise from these two aims. The guidance forces arise largely from friction developed between the wheels and rails, less than ideal performance causing increased wear. Optimisation of the suspension design thus requires an understanding both of the friction and the wear characteristics of the wheel/rail system.

Friction at the wheel/rail interface also enters into the requirements of traction and braking control. Here problems arise when the coefficient of friction is so low that the required traction or braking forces can not be sustained. Performance is restricted, and gross wear of rails or wheel treads occur if wheels are allowed to spin in traction or slide in braking. Recently, however, measurements of the frictional behaviour at high levels of wheel slide/roll ratios are enabling control equipment to be developed to make optimum use of available friction forces.

2 REQUIREMENTS OF THE LATERAL SUSPENSION

2.1 Guidance

Traditionally the two wheels of a railway wheelset rotate together with a common axle. The wheel treads are profiled so that axial displacement results in a different rolling radius at the two wheels and a different contact angle against the two rails (Fig 2). The forces acting to restore the wheelset derive from the difference in the lateral components of the gravitational reaction at the two contact areas and from friction forces developed by "creepage" at the rolling contacts, creepage arising because of small mismatches between the rotational and equivalent translational speed of each wheel (4,5).

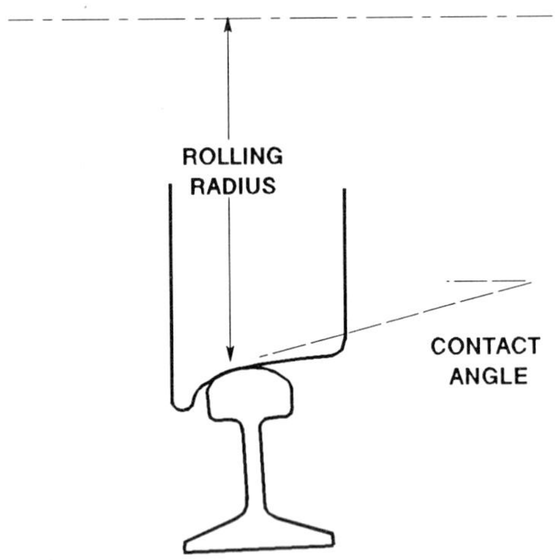

Fig. 2 Lateral Contact

The profile angle is normally designed to vary across the width of the tread, frequently being turned to a "worn profile" which is expected to remain longer within specification. The bearing surface of the rails similarly wears from its initial shape to a more complex flattened profile. The rolling radii and contact angles depend sensitively on these wheel and rail profiles, and computer based solutions are required to determine the contact geometry.

2.2 Position of Contact Area

Apart from the basic dimensions of wheel diameter and rail gauge etc., a description of the wheel and rail profiles is expressed as a series of numerical co-ordinates, obtained from measurements made by suitable stylus based equipment or to a theoretical curve. The computer represents the co-ordinates by a simple polynomial, brings the two surfaces together and determines the point of contact. The rolling radii and contact angles are then calculated as a function of axial displacement. The functions are found to approximate more or less to straight lines, the slope of the rolling radius difference/displacement curve defining an "equivalent conicity" (4).

2.3 Contact Patch Dimensions and Contact Stresses

Where the principal radii are constant the stress distribution and contact patch dimensions can be found by the method developed by Hertz, but with worn wheel and rail profiles computer based numerical integration techniques are employed. The technique used by British Rail is an iterative calculation which begins by proposing a shape generated by the vertical intersection of the two rigid bodies to a small depth Δz (normally taken to be 0.03 mm). Following Hertz the stress distributions along each axis are assumed to be similar semi-elliptical elements, with base lengths varying according to the assumed width of the contact patch. The integrated stresses are equated to the normal load. The vertical deflections are then computed and, usually, a correction must be made to Δz, with further iterations until the calculations converge (6). Fig 3 shows an example of contact patch shapes varying with increasing axial displacement of a wheelset.

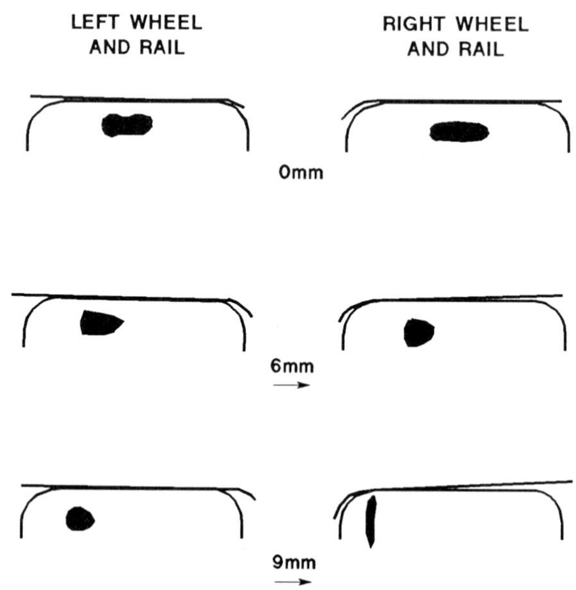

Fig. 3 Contact Patch Shapes for Different Lateral Displacements

A more general formulation of the problem, an inverse iteration method called CONFORM, has been published in which the contact patch and pressure are taken as unknowns and the rigid body approach is specified (7). Comparison of the two methods shows that agreement is close except that CONFORM produces a more accurate prediction when the two surfaces are conformal but, unlike the BR method, fails to predict the division into two areas when flange contact with simple coned wheels occurs (6).

2.4 Creepage Force Calculations

Knowing the contact shapes and interface stresses, frictional forces are calculated (8) as a function of slip defined as the difference between the rotational speed of the wheel tread (taken to be a rigid body) and its translational speed along the rail, divided by the mean of the two. "Creepage" defines a regime where the slip is so small that elastic distortions of the two surfaces are relieved by a small amount of "micro-slip" over only part of the contact area. As slip is increased, the fraction of the area undergoing "micro-slip" and the resulting friction force also increase until a limit is reached where slip spreads over the whole of the contact patch (Fig 4). The friction force is then defined by the limiting coefficient of friction. An initial assumption in the calculations is that the friction coefficient in micro-slip is μ=0.60, the friction coefficient observed in laboratory conditions between clean steel surfaces (9).

Three components are computed: longitudinal creep, caused by the rolling radius difference, lateral creep (lateral velocity component divided by the mean rolling velocity) caused by wheelset yaw, and spin creep (rigid body angular velocity about an axis normal to the contact area, divided by the mean rolling velocity) derived from the angle of contact. The three frictional forces are functions of the contact shape, the normal load, and the elastic constants of the wheel and rail steels. For small creepage and spin the force/creepage relationships are almost linear, the slopes defining "creep coefficients". For large creepages, Kalker's non-linear three-dimensional exact theory includes the effects of all creepages occurring simultaneously (8). A large table of results has been compiled, covering contact patch ellipticity ratios from 0.2 to 10.0 under all possible conditions of longitudinal, lateral and spin creepage. This table can be interpolated quickly and cheaply.

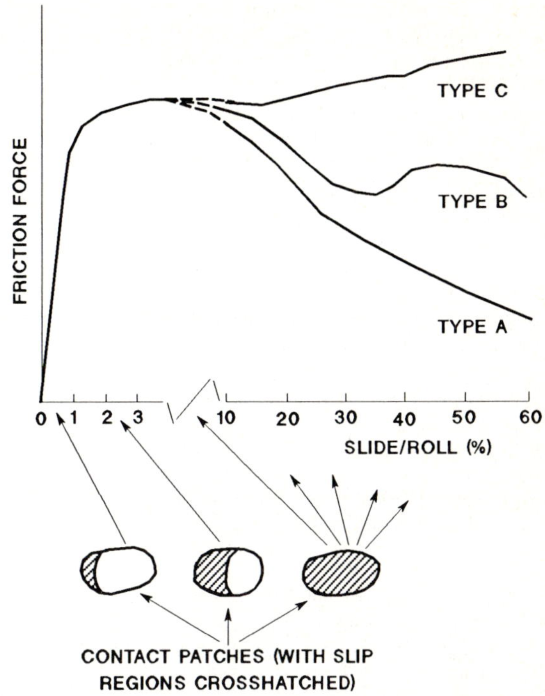

Fig. 4 Friction/Slip

2.5 Creepage Force Measurements

It is a reasonable assumption that creepage forces, depending as they do on friction in micro-slip, depend in turn on the overall limiting friction coefficient (μ), termed the "adhesion". Measurements, however, suggest that, except at uncommonly low adhesion levels, no such dependence exists.

Longitudinal creep coefficients were determined by applying brake forces to a measuring wheelset and by measuring the rotational speed of the wheel to a high accuracy by means of optical encoders on the axle ends. Lateral creep coefficients were measured by the vehicle "Decapod" which is fitted with an independent central wheelset which can be yawed relative to the track. Strain gauged spokes directly measure the resulting creep forces (10).

The longitudinal creep coefficients were found to be constant, exhibiting the theoretically expected values calculated assuming μ=0.60 (with an experimental scatter of $\pm$ 20%), while adhesion values ranged between μ=0.27 and 0.15. Only below

μ=0.15 did creep coefficients fall significantly, to about half the theoretical value as the adhesion approached 0.08. The lateral creep coefficients similarly showed no correlation with adhesion between μ=0.46 and μ=0.12, varying randomly between 66% and 100% of the theoretical Kalker values. Below μ=0.12 there was evidence of the lateral creepage coefficient beginning to be substantially reduced (10).

2.6 The Limiting Friction

Coefficients of limiting friction (the "adhesion") are measured by applying the brakes of a measuring wheelset until slip is just induced. On reasonably well used track, adhesion is found to range between about 0.40 and 0.10 with occasional excursions over short distances down to 0.07 (11). On little used rusty track, or on rails covered with thick leaf contamination in autumn, in moderately wet weather adhesion levels below 0.02 have been measured. Copious rain usually improves considerably on these low values.

These wide variations are accounted for by changes in the amount and the constitution of the surface contamination on the rail head (9,12,13). The solid contaminants are largely made up of rust products, oxides and hydroxides of iron. Mixed with dew or rainwater they form pastes with rheological properties sensitively dependent on the water content (13). When dry they sustain considerable shear, when saturated they are readily squeezed from the wheel/rail contact so that steel to steel contact similarly transmits large shear forces, but in intermediate concentrations they act as viscous lubricants. Oils are not necessary to this low adhesion mechanism but they reinforce and possibly prolong the formation of the viscous rust pastes. Similarly it is probably the cellulose content of leaves (or paper or sawdust) which helps bind rust particles into layers up to 20μm thick on dry rail, providing a ready lubricated surface in the event of dew or light rain.

2.7 Friction in Micro-slip

Section 2.5 reveals that measured creep coefficients approached the "Kalker" values calculated on the assumption that the friction coefficient in micro-slip is 0.60, though the limiting "adhesion" coefficient was commonly less than half this value. The reason is unclear but may reflect the considerable difference in sliding conditions encountered in the two measurements. In creepage the sliding distances are small, less than 0.1 mm in a contact patch of length 15 mm, and in these particular measurements the rolling speed was also low (40 km/h). Shear deformation of any surface feature, and frictionally generated temperatures, are therefore small. Further, in micro-slip the shear of discrete patches of contamination is constrained by the allowable elastic deformations of the two surfaces. Limiting adhesion on the other hand was measured at higher speeds, at slip rates approaching 5%, and with the whole of the contact patch in slip. Both shear deformation and frictional heat would be considerably higher.

Creepage is affected, however, when adhesion falls below 0.15. Contamination on the rail surface, mixed with just sufficient moisture to act as a viscous paste (9,13), has then become so extensive that the contact patch approaches a condition in which it is wholly supported by a quasi-viscous surface layer. Elastic relaxation of the two surfaces is opposed, as is full slip, by the viscous layer, with the result that micro-slip becomes greater than is anticipated under the assumption of frictional contact.

2.8 Suspension Design Package

Over a period of about 20 years computer programs making use of the above ideas have been written by which contact positions and dimensions and the resulting creep forces are computed for wheelsets of any given tread profile traversing track of any given uneveness and curvature and with any rail profile. Vertical as well as lateral suspension calculations are included. The various programs have recently been assembled into one coherent package, named "Vampire", which is available for commercial use (14).

A three-dimensional model of the vehicle is assembled, consisting of general masses, wheelsets and flexible modes connected by springs, dampers and friction forces within the physical constraints provided by the bumpstops. A steady state program calculates the equilibrium position of the vehicle negotiating a constant radius curve. The effects of track cant (banking in curves), twists, wheelset misalignment and external forces can be included. A transient response program, working as a function of time, calculates the response to known vertical, lateral, cross level, curvature and kink alignments of straight track, for a range of vehicle speeds and wheel/rail contact geometries, giving a measure both of ride quality and of the safety of the system.

Using this model the necessary compromises in design are reached, providing curving ability with adequate stability for the required service performance. For modest speeds it is possible to design suspensions which give adequate performance on track of lower quality, with consequent savings in track maintenance costs. For higher speeds the necessary track improvements, and their costs, can be specified. Assessments of safety, for example of derailment risks when curving with a failed pneumatic suspension, can also be made.

3 RAIL AND WHEEL WEAR

An important aspect of the design procedure is to assess the costs associated with wheel and rail wear. Rates of wear are generally low. Rails even on well used main lines can remain in service for tens of years while wheels on some vehicles run hundreds of thousands of miles before reprofiling for tread wear. On sharply curved track, however, rails and wheels can suffer rates of wear so high as to require replacement or reprofiling within weeks.

On a curve the leading wheelset of a bogie moves laterally towards the outer rail and is constrained by the suspension at an angle to the centre line. As the curve becomes more severe, larger creep forces are required to steer the wheelset, and there comes a time when they are insufficient to maintain it in the required near-radial alignment. The outer wheel is then driven into flange contact or, with worn profiles, into contact at the flange root. The resulting creepages then become far higher and the "side wear" of rails and the "flange wear" of wheels can increase by orders of magnitude.

The curving program determines the rapid changes in creepage as the contact approaches the flange root, or as it in some cases divides into separate tread and flange contacts. To determine wear, a great deal of laboratory work, together with service measurements, was required to provide a correlation between a "wear number", involving creepage and friction forces, and the resulting wheel/rail wear.

3.1 Wear Number

Several "wear numbers" have been considered. An early hypothesis was that wear should be related to the energy expended through creepage, which is related in turn to the sum of the products of the individual frictional creep forces (T Newtons) and creepages (γ). The most convenient empirical wear index identified to date is simply $T\gamma$ (15). Wear is expressed as cross-sectional area lost per 1000 cycles (or 1000 axles passed), which is equivalent to material lost per unit distance rolled. (In those laboratory tests where the contact patch was readily calculable, wear has sometimes been expressed as material lost per unit distance rolled per unit contact area (A) and has been plotted as a function of $T\gamma/A$ (16). In practice it is simpler to ignore the contact area and use the factor $T\gamma$ only).

3.2 Laboratory Measurements

Evidence is based on laboratory work using small-scale twin-roller discs (35 mm dia. Amsler specimens) rolling against each other with controlled longitudinal or lateral slip, and from a full-scale rig in which a wheel was rolled repeatedly against a short length of rail under realistic loads. The wheel was given an angle of yaw so that it rolled into flange contact. In other tests a yawed wheel with a cylindrical tread was rolled over a railhead specially profiled to give a well defined contact area. Wear/$T\gamma$ relationships (Fig 5) were found to be of the same form from both small and full scale tests, but the magnitude of the wear was different. In the end, the wear measurements at full scale were adopted as reasonable empirical estimates of that to be expected in service (15).

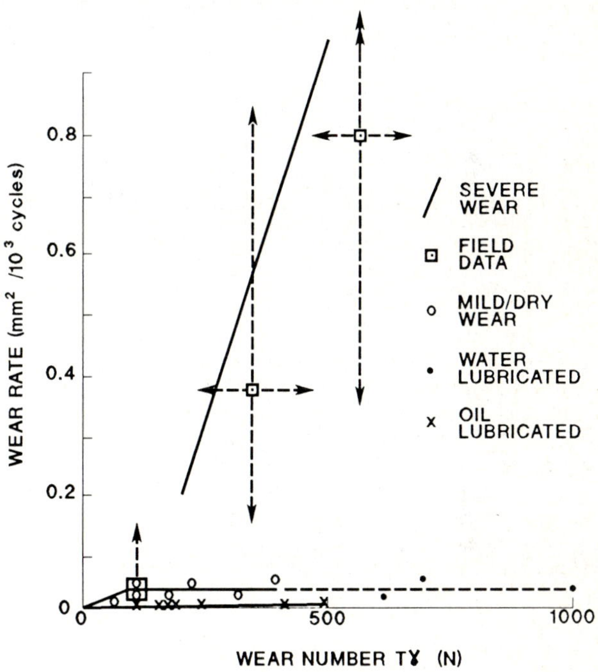

Fig. 5 Wheel/Rail Wear

On clean, dry surfaces the wear associated with low wear numbers ($T\gamma$ below 200N) is of a mild nature. For wear numbers between 200 and 400N the mechanism is either mild or severe, when it increases by a factor of 10 (for design purposes it is taken to be severe). For $T\gamma$ above 400N the wear is invariably metallic and the surfaces rough and abraded. Water, as well as reducing the friction to about $\mu=0.25$, prolongs the mild wear regime to wear numbers as high as 1000N. Oily contaminants reduce the friction coefficient to a value approaching 0.1 and the rate of mild wear by a factor of about 10 (Fig 5); rail side or wheel flange lubrication thus becomes a worthwhile expedient on sharply curved track.

3.3 Wear in Service

Consistent field wear data is difficult to come by but measurements were obtained from rails in several curves of a branch line which in the main carried only one type of coal carrying hopper wagon (17). Four curves from 172 to 893 m radius were analysed, using the two-point contact version of the curving program. Only the outside rails gave $T\gamma$ values indicative of severe wear, and the results are included in Figure 5. The spread in $T\gamma$ caused by the range of wheel/rail profile pairings is indicated, as is the mean wear and scatter. The wear rates fall somewhat below the laboratory data but it is encouraging that the two agree so closely.

Another study became possible when a fleet of electrical multiple units were transferred from a high speed/low curvature service to the more severely curved track of Merseyrail (15). Because of the lower speed the yaw stiffness of the suspension could be reduced without exceeding the stability criterion, and the wear life was then predicted to be doubled. In practice it was found to increase by about a third.

Greasing the side of the rail by traffic actuated lubricators reduces wear on sharp curves by orders of magnitude, and in some cases is essential to maintain an economic service. On Merseyrail the unmodified vehicles mentioned above wore out their flanges in less than 4 weeks when several track lubricators were inadvertently removed.

3.4 Wear Mechanisms

At small scale, three wear mechanisms are identified (16). Mild wear is associated with a predominantly oxidative process involving the generation and destruction of thin oxide films but with some thin (2-3µm) metallic flake debris. Increasing slip and contact pressures give rise to severe wear in which the debris is predominantly metallic and the surface rough. A further order of magnitude increase in wear rate is observed by increasing the slip to about 20%. Only in this third wear regime was the superiority of high manganese steel revealed in small scale testing, exhibiting a wear rate about eight times less than the normal rail steel.

On rails the bright running band on the head is generally smooth, surface platelets being plastically deformed to a depth less than 1 mm. The edges of the platelets provide sites for preferential corrosion. Up to 50% of the running band on fast main lines is transformed to a hard, slow etching "white phase" which resists wear. (Thirty years ago head wear was about three times greater. Possibly the slower speeds of that era resulted in less wear resistant white phase being formed). Towards the inner face of the rail (the "gauge face") there is a well defined zone of subsurface deformation. On the gauge face itself some areas are heavily deformed and some not, and ploughing is evident. These features are similar to the third wear regime found at small scale, but the corresponding increase in the rate of wear is not observed.

4 ROLLING CONTACT DAMAGE

Various forms of railhead damage, where surface cracking can lead to rail fracture, are far more serious than wear. Some damage is associated with discontinuities in the rails, such as at rail joints or switches and crossings, where the unsprung mass of the wheelset is a factor in determining the "running on" impact forces which cause fatigue cracks (18). On plain track, however, cracking of the railhead is usually ascribed to some form of rolling contact fatigue (19).

Thus, "squats", a form of railhead defect virtually unknown up to 1972, initiate either at some small surface defect, typically a depression about 5 mm diameter, or towards the gauge corner where creepage and contact pressures are highest. In either case cracks propagate into the rail at an angle of 10-15° to the horizontal until, at a depth of 3-5 mm, they branch downwards at about 45° and eventually pose a strong risk of fracture. They may be associated with the outer rail in curves but are sometimes correlated with rail welds or "corrugations" of the rail surface on straight track.

Corrugations are undulations formed in the railhead which of themselves cause unacceptably noisy running on "roaring rail". Their origin is a matter for conjecture. The most common form on British Rail has a wavelength between 40 and 80 mm and, when well developed, a peak to trough height of about 0.1 mm. The peaks are often found to exhibit subsurface shear and work hardening to a depth of about 0.25 mm with hard white phase zones commonly evident at the surface. The troughs exhibit no phase transformation and little deformation (19).

4.1 Mechanism of Corrugation Formation

Recent theory postulates a mechanism in which a wheelset running at an angle to the rails (possibly because of a manufacturing or maintenance defect) causes the rails to vibrate laterally in a stick-slip process (20). Wear is high (perhaps "severe" in nature) during the slip compared with the intervening periods of rolling with lateral creepage, and a periodical wear pattern is initiated. The scars take the form of sharply defined gouges, each about 30 mm in length (related to the size of the contact patch).

When subsequent wheels pass, the pattern may be reinforced or reduced, dependent on the magnitude and phase of vertical vibrations of the rails (21). First, it is found that a pattern with a spacing equal to twice the gouge length gives rise to the largest vertical force fluctuations, because the wavelength then has the largest first harmonic content. It is thus hypothesised that a scar spacing of about 60 mm is the one which has the greatest probability of becoming established as a corrugation pattern.

A second factor is the apparent vertical stiffness of the track, "stiff" track inducing the largest force fluctuations. On rails, three vertical modes are important. On open track a natural frequency of about 800 Hz is associated with bending of the rail between its sleeper mountings. With a scar spacing of 60 mm this natural frequency is excited by trains passing at about 170 km/h and it is on track traversed by such trains that corrugations are observed, most pronounced close to sleepers where the vertical stiffness is largest. The other two natural frequencies, about 50 and 250 Hz, give rise to corrugations close to stations where train speeds are about 10 km/h, and on freight lines traversed at about 60 km/h. If this explanation is accepted then the eradication of corrugations can only be achieved by a redesign of track by which some damping of the vertical vibrations is incorporated or some adjustment is made of the phase relationships by suitable choice of rail pads. Support of the rails on concrete "slab track" also eliminates the high frequency "pinned-pinned" mode associated with sleepers.

5 BRAKING AND TRACTION

Braking and traction performance is fundamentally constrained by the value of the limiting "adhesion coefficient". In particular, for safety reasons, braking distances are specified to require adhesion levels of only 0.095 (11). Even then, it is recognised that adhesion below this value is sometimes encountered and allowances are made in the spacing of signals to allow an overrun typically of 15%. Traction demands higher adhesion levels, greater than $\mu=0.30$. In Britain, trains are loaded such that they require up to $\mu=0.24$ on the steepest gradient on any given route.

5.1 Braking and Traction Control

On modern stock low adhesion in braking is countered by "Wheel Slide Protection" (WSP) devices which have the prime purpose of preventing the wheels being braked to a stop while the train is in motion. If wheels do "lock-up" then prominent "flats" are worn in the treads which require the vehicle to be taken out of service and the tread returned.

WSP equipments sense the speeds of individual wheelsets and come into operation whenever a wheelset is decelerating too rapidly or is rolling at a speed substantially slower than train speed. Necessarily, WSP's release the brake for short periods, with the cumulative effect of extending the stopping distances. To minimise this extension, modern WSP's are designed to make the optimum use of the available wheel/rail adhesion. Various philosophies are employed. Some seek to control slip at a value where the adhesion is prejudged to be a maximum, others "peak seek" to maintain slip wherever the optimum may be, while others claim to improve the existing adhesion by "conditioning", by controlling slip at a relatively high slide/roll ratio. Unfortunately, there is as yet no universally agreed picture of the way in which wheel/rail friction varies or is changed by high levels of slip.

In recent years traction control has also been designed to make best use of the available adhesion. With improved traction motor control becoming possible, wheels can be accurately controlled in wheelspin. The same philosophies are employed as with WSP control, and the question remains how the wheelspin should be specified for optimum results. Measurements have therefore been made of adhesion/slip curves in traction as well as in braking. In both cases care has to be exercised that slip is not so excessive as to damage wheels or rails; temperatures can rise so high that brittle martensite is formed which subsequently spalls out of the surfaces.

5.2 Adhesion/Slip Measurements in Braking

In braking, measurements have been made at low speed by British Rail, and at high speeds by French and German railways on behalf of the European Railway Research Organisation ORE. British Rail measures brake forces via a separately suspended brake torque frame pivoted on the axle while French and German railways measure bending moments in the measuring axle (22). In most cases slip is induced to reach momentarily high slip values of as much as 50% (slip to these values is usually calculated as the difference between wheel and train speed divided by the train speed). At least three types of braking behaviour are identified, labelled by ORE as Types A, B and C (23) (see Fig 4).

ORE suggests that curves of Type B are most frequent, exhibiting a secondary maximum friction, Point B. The position of optimum slip is claimed to be a function of the slip energy dissipated between wheel and rail, and tests are planned with WSP equipment microprocessor controlled to compute this energy in real time and then to control the slip to the appropriate optimum. The tests on British Rail have enabled typical adhesion/slip profiles over representative lengths of track to be assembled (Fig 6). A computer based simulation device has been constructed to subject experimental or commercially available WSP's to runs over such typical profiles, in order to compute the resulting stopping distances.

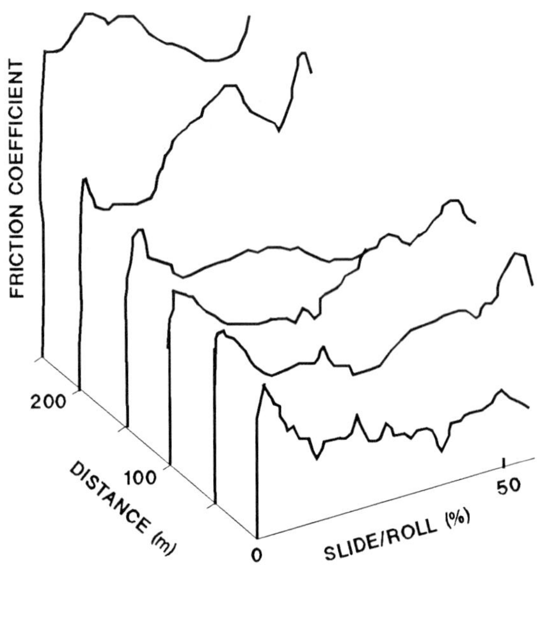

Fig. 6 Friction/Slip Profiles

Experiments with British Rail's tribometer train, in which the adhesion was measured independently ahead of the WSP controlled wheelset, show that the friction available for braking can be improved by controlled high slip. Improvements of about 0.01 have been observed, which is greater than 10% of the adhesion on slippery rail. "Conditioning" of the rail has also been detected when three or more axles are controlled in high slip ahead of the measuring wheelset.

5.3 Adhesion/Slip Measurements in Traction

In traction, slip curves have been obtained by controlling motor speed momentarily to high rates of slip, with measurements of motor torque, suspension forces, and wheel speed allowing friction/slip curves to be drawn. Published results conflict, some exhibiting a gradual increase of friction with slip (24, 25), others a "breakaway" similar to the curves obtained in braking (26, 27). Recent measurements by British Rail support the latter picture, with measurements made in high and naturally low adhesion conditions, in collieries affected by coal slurry, with artificial sprays of water and water/oil mists, and with cellulose based pastes to simulate autumn leaf contamination (28).

As a general summary of the BR work, the friction rises to the breakaway value at a slip of about 1%. If the breakaway friction is low, about 0.10, the friction remains almost constant as slip increases; if greater than 0.10 then the friction falls with slip, to roughly half the breakaway value at a slip of 1.0 (wheel speed twice train speed). Obvious contamination, such as rust or coal slurry, results in the initial slope being reduced so that breakaway is reached at higher slips of about 2%. Tests with the locomotive stationary on the relatively rust contaminated rail of a siding showed that the friction fell rapidly from a breakaway value of μ=0.28 to about 0.10 at a spin speed of 1.4 m/s. On well used rail, in natural or water sprayed conditions, the breakaway friction was reached at a slip speed of about 0.2 m/s. On water/oil sprayed track, however, there was no sharp breakaway and friction reached a maximum at a slip speed of about 2 m/s.

6 CONCLUSIONS

The tribology of the wheel/rail interface exhibits a considerable influence on the way in which railway vehicles are designed, the greater knowledge of recent years enabling optimum levels of performance to be determined. Many aspects of tribology are included: friction forces provide traction, braking and guidance; wear and rolling contact damage provide economic criteria against which performance can be assessed; lubrication influences both factors, while rheology and considerations of hydrodynamic type behaviour are important in understanding "adhesion". Whether apparent or not, a considerable tribological understanding is required, which perhaps further explains something of the fascination of the mechanical working of trains.

7 ACKNOWLEDGEMENTS

The authors are pleased to acknowledge the help of many of their colleagues when summarising the various aspects covered in the paper, and thank the Director, Research, and the British Railways Board for permission to publish.

References

(1) BARWELL, F.T. "The contribution of tribology to the development and operation of railways", Proc. Instn. Mech. Engrs., 1973, 187.

(2) BAIRSTOW, S. "Some examples of the scope of tribology in the railway industry", Industrial Lubrication and Tribology Symposium, London, November 1969.

(3) Symposium on "Tribology in Railways", The Instn. Mech. Engrs, London, 19-20 February 1969.

(4) WICKENS, A.H. and GILCHRIST, A.O. "Railway vehicle dynamics - the emergence of a practical theory", Council of Engineering Institutions MacRobert Award lecture, 1977.

(5) ELKINS, J.A. and GOSTLING, R.J. "A general quasi-static curving theory for railway vehicles", IUTAM Symposium, Vienna, September 1977.

(6) GOSTLING, R.J., PIOTROWSKI, J. and ROSE, K.A. "A comparison of two theoretical methods of calculating the contact area between measured railway wheelsets and track", BR Research Division Technical Memorandum TM DA 31, 1981.

(7) HASHEMI, J. and PAUL, B. "Contact stresses in bodies with arbitrary geometries; Applications to wheels and rails", Federal Railroad Administration Report FRA-ORD-79/23, 1979.

(8) KALKER, J.J. "A survey of wheel/rail rolling contact theory", Vehicle System Dynamics, 1979, 5, 317.

(9) BEAGLEY, T.M. and PRITCHARD, C. "Wheel/rail adhesion - the overriding influence of water", Wear, 1975, 35, 299.

(10) PEARCE, T.G. and ROSE, K.A. "The tangential force - creepage relationships and their use in vehicle response calculations", Proc. 9th Symposium of International Association for Vehicle System Dynamics, Linkoping, Sweden, 1985, 427.

(11) PRITCHARD, C. "Brakes and wheel/rail adhesion", Instn. Mech. Engrs. International Conference on Railway Braking, York, September 1979.

(12) PRITCHARD, C. "Traction between rolling steel surfaces. A survey of railway and laboratory experiments", 7th Leeds-Lyon Symposium, "Friction and Traction", September 1980.

(13) BEAGLEY, T.M. "The rheological properties of solid rail contaminants and their effect on wheel/rail adhesion", Proc. Instn. Mech. Engrs., 1976, 190.

(14) EVANS, J.R. "The vehicle dynamics modelling package", BR Research Division Technical Report TR VDY 001, 1986.

(15) McEWEN, I.J. and HARVEY, R.F. "Wheel/rail wear and lubrication - laboratory studies and their relevance to field situations", 2nd International Symposium on Contact Mechanics and Wear of Rail/Wheel Systems, Kingston, Rhode Island, July 1986.

(16) BOLTON, P.J., CLAYTON, P. and McEWEN, I.J. "The wear of rail and tyre steels under rolling/sliding conditions", ASLE Trans., 1982, 25, 17.

(17) HARVEY, R.F. and McEWEN, I.J. "The relationship between wear number and wheel/rail wear in the laboratory and the field", BR Research Division Technical Memorandum TM VDY 001, 1986.

(18) JENKINS, H.H., STEPHENSON, J.E., CLAYTON, G.A., MORLAND, G.W. and LYON, D. "The effect of track and vehicle parameters on wheel/rail vertical dynamic forces", Railway Engineering Journal, 1974, 3, 2.

(19) CLAYTON, P. and ALLERY, M.B.P. "Metallurgical aspects of surface damage problems in rails", Canadian Metallurgical Quarterly, 1982, 21, 31.

(20) CLARK, R.A. and FOSTER, P. "Mechanical aspects of corrugation formation", Symposium on Rail Corrugation, Berlin, 1983.

(21) SCOTT, G.A., CLARK, R.A. and POOLE, W. "A mechanism for the formation of short wavelength rail corrugations based on slip stick vibrations", BR Research Division Technical Memorandum TM VDY 017, 1988.

(22) "Adhesion during braking and anti-skid devices - Synthesis of current knowledge concerning adhesion", ORE Report B 164/RP1, 1985.

(23) "Adhesion in braking and anti-skid devices - Fundamental laws of adhesion in braking", ORE Report B164/RP2, 1988.

(24) "Synthesis Report : The current state of knowledge about wheel/rail adhesion", ORE Report B44/RP14, 1978.

(25) LOGSTON and ITAMI "Traction adhesion tests" Trans. ASME, 1980, 102, 275.

(26) FIEHN, WEINHARDT and ZEEVENHOOVEN "Three phase drive test vehicle of Netherlands Railways. Adhesion measurements", Brown Boveri publications D.VK.1054.81E.

(27) HARPRECHT, W. "Anfahrverhalten, Leistung und Zuverlassigkeit der Lokomotiven der Baureike der DB", Elektrische Bahnen, 1984, Pt 2, 30.

(28) ARMSTRONG, D.S. "Adhesion/Slip characteristics in traction", BR Research Division Technical Memorandum TM TBC 015, 1988.

Paper II(iii)

Tribological design - The automotive industry

P. A. Willermet

Rapid and continuing change within the automotive industry demands continual improvement in the quality, performance and reliability of vehicles. At the same time, competitive forces demand shorter product cycle times and new organizational approaches to the design process. The introduction of new technology and increased reliance on suppliers demand better methods of evaluating designs and materials. All of these factors lead to increased opportunities for the introduction of improved tribological design methods as well as the introduction of improved designs and materials.

1 INTRODUCTION

The automotive industry is in a period of rapid change. This change is driven by pressures from several sources: increased international competition, higher customer expectations, and social pressures in the form of regulations relating to emissions, fuel economy and safety.

Automotive technology has advanced significantly in the last few years in response to these needs. These advances include items such as electronic engine controls and catalytic converters which are not tribological in nature, but which may impact tribological components. Other advances include the implementation of design changes which directly involve tribology (see appendix).

The challenge faced by automotive manufacturers, however, is not simply one of achieving a certain level of technical sophistication. The challenge is also to develop better ways to implement continual product improvement within the constraints unique to the industry.

This paper will focus on current applications of tribological principles to automotive design, and on ways technology transfer might be improved. Because manufacturers have different approaches and organizational structures, and because much of the technology is proprietary in nature, the discussion will necessarily be somewhat general. However, it is to be hoped that the presentation will be of use to those attempting to facilitate the translation of research results to a practical end use.

2 APPLICATION OF TRIBOLOGY TO DESIGN

2.1 The design process

The general characteristics of a new model and its placement in the product cycle are almost always defined by a management consensus process. This process must take into account many non technical factors, including marketing, financial considerations and the technological vision the corporation aims to imprint on the future product.

Advanced engineering teams propose and evaluate design alternatives. Many factors will enter into the choices made, many of them not immediately obvious. For example, space limitations may exclude certain powertrain options, especially for low drag coefficient front wheel drive vehicles.

In the past, such forward looking teams have tended to carry out the task with only minimum input from outside sources. This approach has been found to lead to inefficiencies in implementing and producing the final design as well as to design shortcomings which are difficult to correct after the fact. The current trend is to bring other organizations into the design process early on. These organizations may include manufacturing, component suppliers, tooling suppliers for manufacturing, and design support groups. In the case of joint ventures, which are becoming increasingly common, even other automotive manufacturers may be included. Ultimately, needs for quality, manufacturing efficiency, and shorter product cycles lead to the adoption of simultaneous engineering, in which product and manufacturing engineering decisions proceed in parallel.

In the final design stage, the selected design is finalized, refined and released for manufacturing. This may be the responsibility of forward model teams, which retain responsibility for monitoring the vehicle in production. These teams do detailed development, calibration for performance and emission objectives and durability testing.

2.2 The role of tribology

At this point one may ask: "Who does the tribology?". Tribology was not explicitly mentioned at any point. Tribology is an organic part of the process, and tribology specialists are brought into the design process as described below.

Component suppliers: Many parts which are subject to friction and wear may be designed and furnished by component suppliers (Table 1). Accordingly, design responsibility may be rather diffuse. When a supplier assumes responsibility

Table 1. Partial list of powertrain parts having tribological significance which are often obtained from suppliers

Engines	Transmissions	Axles
Camshaft	Clutch Plates	CV Joints
Tappets	Transmission fluid	Bearings
Valves	Bearings	Seals
Crankshaft	Seals	Gear Oil
Rods		
Pistons		
Rings	**Brakes**	**Tires**
Seals		
Pumps	Friction materials	Tires
Bearings	Seals	
Engine Oil		

for the design of a part, the automotive manufacturer relies on his knowledge. It is not always easy to know how sophisticated various suppliers are because of a wish to protect proprietary technology on the part of the supplier and lack of a perceived need to know on the part of the automotive manufacturer. Obviously, there will be wide variations. For example, some suppliers have established reputations for depth of understanding in tribology which rests on a substantial body of publications in the open literature. Clearly, visible competence in this area makes a supplier more attractive to the vehicle producer. Other suppliers may not have established reputations in this manner, but the fact that the components supplied function with low failure rates suggests at least some empirical knowledge. Unless the automobile manufacturer has retained expertise in design of the particular component and applies it to evaluate designs and supplier capabilities, it will be difficult to know whether the part functions as well as it could. This will continue to be an issue, as competitive pressures lead to greater reliance on supplier research and expertise.

<u>Design engineers:</u> Few engineers have extensive prior training in tribology. Generally speaking, knowledge is obtained from design practice documents, from colleagues, from vendors and from experience. Forward thinking organizations actively encourage continued education through participation in seminars and courses. In such organizations, design engineers are usually competent practical tribologists in their areas of responsibility. However, the move to ever higher standards of reliability and durability will demand more sophisticated approaches than are commonplace today. In addition, design engineers will need to take the lead in evaluating suppliers for competence in design of tribological components.

<u>Design analysts:</u> Mathematical analysis of tribological interactions is critical for highly stressed contacts such as cam/tappet contacts, journal bearings and gearing. Design analysis computer programs fill these needs and are progressively becoming more powerful and flexible. These will be discussed more fully below.

<u>Tribology specialists:</u> These are understood to include metallurgists, lubrication specials and the like as well as tribologists per se. These resources are typically employed during a specific design exercise if and when a need is perceived by design engineers. If no problems are encountered, these resources often are not employed. If the link between design engineer and specialist is too weak, design issues which would benefit from deeper understanding may be dealt with by an extrapolation of existing methods. The tribology specialist should accordingly make sure that lines of communication with his design colleagues remain open.

<u>Lubricant suppliers:</u> The role of the lubricants supplier is somewhat special, particularly in the case of engine oils. Gear oils, which are rarely replaced, are specified by manufacturers. Automatic transmission performance is critically dependent on the frictional behavior of the fluid/friction material couple. Accordingly, the properties of these fluids are also specified and controlled by manufacturers. Engine oils are periodically replaced by the vehicle owner. They usually are described generically for better customer identification, especially in North America. Suppliers and manufacturers work together to define test methods and sometimes also specifications covering viscosity and performance properties such as antiwear protection, oxidation resistance, bearing corrosion, etc. Because of the quasi-generic nature of automotive lubricants, properties are usually taken as typical for the current generation of fluids during the design process. When possible, unique fluids are avoided because of service availability concerns.

2.3 <u>The design process - summary</u>

Competitive pressures are leading to shorter product cycles, simultaneous engineering practices, and to greater demands for product quality, performance and economy. At the same time, improved productivity on the part of design groups will be required.

The automotive industry relies heavily on the design capabilities of suppliers, and this reliance is growing. Accordingly, ensuring that the supplier community is using the latest design methods and technology should have a high priority.

Knowledge of applied tribology is widespread among design engineers, although few have much formal academic training in the subject. Even so, the drive for continual improvement in the product as well as the importance of critically assessing supplier designs show the need for increased sophistication.

Shortening of the product cycle and implementation of simultaneous engineering make it essential that tribological concepts be applied to the design process at the very beginning. This can be done by:

- Improving the sophistication and capabilities of design tools
- Introducing new tools and methods
- Helping design engineers to upgrade skills
- Evaluating and improving supplier

capabilities
- Improving links between specialists and design engineers

The tribology community can participate by:

- Working with automotive manufacturers and suppliers to develop new and upgraded design tools.
- Organizing courses and seminars aimed at design engineers. This can be done by consultants, suppliers and in-house experts as well as by universities and professional societies.
- Communicating research results and tribology concepts in forms accessible to the general engineering community[1].

2.4 Tribological design methods and design issues

Design tools, generally in the form of computer programs or reference handbooks, are the best means of formalizing the incorporation of tribology concepts and methods into the design process. Considerable advances can be made by application of information already available to the tribology community. Some open issues, however, will require new information for satisfactory resolution.

Design guide: Design guides describe and set limits on parameters such as physical dimensions, materials, stress levels and the like, which are to be used in designing a component. These are intended to be a compilation of the state of the design art as practiced by the manufacturer. Although such documents are not primarily tribological design tools, practices specified therein impact the tribological behavior of the final product. A design guide is, of course, updated periodically and will grow in sophistication as better information becomes available to design engineers. Efforts to encourage the development of expertise through education in its many forms should assist the transfer of new technology into design guides.

Engine/powertrain simulation programs: These allow evaluation of design alternatives early in the design process with a minimum of experimental input. Friction and durability are less important in these global models than combustion efficiency, emissions and power output, but they do play a role. For programs of this sort, reasonable approximations of friction losses or stress levels are more appropriate than precise calculations requiring detailed input. Accordingly, well documented scaling algorithms could be of practical value in this application.

Bearing design programs: Bearing analysis programs calculate the film thickness and temperature rise in dynamically loaded crankshaft journal bearings. This is no small task, since loads and film thicknesses change continually in complex ways (1, 2). These programs are used to optimize a number of design variables, including bearing dimensions. This is particularly important for transversely mounted front wheel drive configurations, which can place severe restrictions on the overall crankshaft length. They have also been important in assessing the impact of reduced engine oil viscosity as part of the North American move toward 5W30 oils. While these programs are very useful, additional work needs to be done to establish quantitative correlation with operating engines. This will be increasingly important as higher loads drive bearing operating conditions further into the mixed lubrication regime, and as higher maximum engine speeds put greater thermal stress on lubricants. Other challenges are to incorporate and verify methods of dealing with non-ideal conditions, such as misalignment (3).

Viscometrics: Improved bearing design programs, lubricant viscometrics and bearing properties are linked issues which are of importance to both automotive manufacturers and suppliers. Lubricant suppliers are particularly interested because engine oil specifications will ultimately reflect the critical parameters determining bearing performance. Improved understanding of the relationships between the performance of multigrade engine oils and viscosity as measured at high temperature and high shear rate (HTHS viscosity) (4) has led to HTHS limits for factory fill oils, and may lead to a redefinition of API viscosity. However, open issues still remain. Is HTHS viscosity an adequate lubricant parameter for defining film thickness, or will further development of viscometric methods be needed (5)? Since EHD film thickness and traction calculations require parameters such as pressure-viscosity coefficient and glass transition pressure/temperature, should oil specifications ultimately include similar parameters? The answers are not obvious to this researcher.

Bearing performance is ultimately a function of oil additive composition, bearing materials and surface finish as well as design and viscometrics. Accordingly, it is necessary to not only determine relationships between film thickness and oil viscosity, but also between film thickness, oil and bearing composition, and bearing performance. This is a difficult experimental task being pursued by manufacturers and lubricant suppliers (1, 2, 6, 7).

Low friction oils: Oils formulated to give lower friction losses have resulted in some improvement in fuel economy (see appendix). There is reason to believe, however, that further improvements in fuel economy may be achievable through improvements in friction reducing additive technology (8). Further improvements in friction reducing oil technology would not only result in improved fuel economy, but would also affect the economics controlling the selection of design options. For example, if the friction losses in sliding cam/tappet contacts were reduced by 50% by low friction oils, then the cost/benefit ratio for roller cam followers would increase by a factor of 2. Accordingly, it is important to know what is the ultimate level of friction reduction attainable with practical oil formulations so that the industry can work toward implementation of improved technology.

[1]. Examples would include Society of Automotive Engineers technical papers and "Lubrication Engineering", published by STLE.

Surface properties: Although considerable progress has been made in surface measurement techniques, experience has shown that the description of engineering surfaces embodied in specifications is often inadequate.

For example, a finishing method may be given, together with a surface roughness value. If the surface is inadequately specified, the finishing method might go out of control without the knowledge of the operator, resulting in inexplicable field problems. Difficulties have been encountered in the industry when implementing new finishing methods because surfaces which appear the same have performed very differently. Potential problems include noisy cams and failed bearings. Because surfaces need to be statistically controlled in production, an ideal result would entail characterization methods sufficiently straightforward to allow use in factories on a routine basis. Failing that, more complex methods could be useful in problem solving or quality control.

Valve gear design programs: A considerable effort has been devoted to developing computational methods for the design and analysis of valve trains (9). These have resulted in proprietary computer programs which are used on a routine basis for analyzing the kinematics, stresses and thermal behavior of cam/tappet contacts. Programs are used on a more limited basis for analyzing more complicated issues related to valve train dynamics. Stress levels and flash temperatures are used as criteria to define design limits, for example, to exclude excessively aggressive cam profiles. These criteria can also be used to select materials appropriate for the demands placed on them by a particular design approach. The utility of such programs could be further enhanced by the application of EHD and mixed lubrication theory to allow estimations of oil film thickness and friction losses to be made as well (10, 11).

Improved computer methods are being developed, spurred by developments in computer technology and by increasingly user friendly software. Since this work does not appear to directly impact material specifications, it is less visible than the efforts referred to above, and is generally proprietary. Finite element analysis and CAD/CAM methods are being applied, as are approaches obtained from studies of elastohydrodynamic lubrication.

Valve gear design options: Valve train friction is a significant fraction of total engine friction. In the United States, government mandated Corporate Average Fuel Economy (CAFE) objectives make low friction design options like roller cam followers attractive in many designs despite their higher cost. The move to 4-valve engines, however, doubles the cost of this particular approach. Accordingly, minimizing sliding cam/tappet friction by attention to design details deserves careful attention.

Piston ring/cylinder bore models: Some manufacturers have in-house models aimed at calculating ring/bore friction and wear (12, 13). These do not appear to be widely used as design tools by the automotive industry, because pistons and rings are generally designed by outside suppliers, who provide needed engineering data to engine designers. Models focussed on evaluating design variables such as side loading, bore distortion and scaling effects, rather than the details of the ring/bore interface, could see wider use.

Other design models: Computer methods are widely used to assist in many design problems. Examples include finite element analysis methods applied to cylinder block distortion and connecting rod stresses. Although these and other issues are not tribological in nature, they may influence tribological behavior. For example, if cylinder bores distort on engine assembly or during operation at temperature, then ring tension must be high to achieve effective sealing. Maintaining round cylinder bores allows the designer to minimize ring tension and thus reduce piston/cylinder bore friction. Similarly, reduced reciprocating mass leads to reduced loads, greater oil film thickness and lower friction in rings and bearings.

Material/surface treatment selection programs: Data bases, especially in combination with improved data retrieval methods, may greatly enhance the ability of design engineers and tribologists to choose better materials and surface treatments. Such programs are not in widespread use at the moment, but current developments suggest that they may be developing rapidly into valuable tools. Some programs are being developed on a proprietary basis, but others, in various stages of development, have been presented in the open literature (14, 15). It seems probable that specialized data bases will see widespread use well before a general tribology data base is far enough advanced to be applied to automotive issues.

For the longer term, the idea of a general tribology data base has considerable appeal, even though the task of constructing such a system appears formidable. An effort sponsored by agencies of the United States government is underway, with input from technical and industrial committees (16). At this point, only preliminary work to define program structure has been completed (17), but even this much progress holds out the hope that the vast body of tribological data may eventually be made generally accessible.

2.5 Design methods and design issues - summary: A substantial body of tribological information, much of it formalized by inclusion in computer programs, is now being applied in the design process. More information must be made available to design engineers to improve design quality. This information must be in readily accessible and easily usable form to ensure its use and to shorten design time. Improved methods for characterizing materials are also needed. These must be sufficiently straightforward to be carried out at a well equipped industrial laboratory and should not require expert interpretation. Additional information in certain areas would also be clearly useful. Some suggestions are as follows:

Well documented engine friction scaling algorithms could be incorporated in engine simulation programs.

- Bearing design programs needs validation for high transient load conditions which may push bearings into the mixed lubrication regime.

  Programs need verified means of dealing with non-ideal conditions such as misalignment. Friction calculation algorithms for mixed lubrication would also be useful.

- What information is required to adequately characterize a lubricant in terms of its tribological behavior? What are the necessary parameters and how should they be measured? A complete answer would include not only viscometrics (perhaps including viscoelastic effects), but also a general method for characterizing friction modifiers.

- What is the lowest "boundary" friction coefficient attainable with the ultimate friction modified oil? How closely can that goal be approached with a practical engine oil?

- What set of measurements will adequately characterize a tribological surface? This issue impacts how materials, coatings and surface treatments are selected, specified and controlled in production.

- Valve train analysis programs could benefit from the incorporation of friction and perhaps durability calculations based on EHD theory and lubricant chemical effects.

- Alternate valve train design approaches need to be explored to reduce friction losses in a cost effective manner.

- Piston ring/cylinder bore models focussed on assessing the effects of design variables and dimensional effects could be of value.

- How can we best optimize material combinations, select materials or coatings for an application and screen new materials? Efforts to introduce new materials or to use old materials in new applications are often wasted because information available to experts is not readily available to designers.

2.6 Technology transfer

New designs and materials: In order to be easily accepted into a vehicle program, a new technology must be low risk and cost effective. If the material or device is to be supplied by a vendor, production capacity must be available. Little time can be devoted to development once a vehicle program begins, and competitive pressures are reducing the available time even more.

A high degree of confidence in the new technology is required because of economic, marketing and safety considerations. This is more true of the automotive industry than it is of many others, and can induce a certain degree of conservatism.

Accordingly, development and introduction of new technology can be a lengthy process. This can be done by working directly with a vehicle manufacturer. However, suppliers are often in a better position to perform development work in their area of specialization. As supplier companies come to work more and more closely with their automotive customer, they become more and more able to initiate development work earlier in the product cycle to anticipate needs.

Automotive companies can promote the introduction of new technology by improving technology management practices. That is, identifying areas of new technology which are promising and adopting management structures and methods which facilitate technology development. Originators of new technology should recognize that multiple input points exist, and that these may be either at the automotive manufacturer or at a supplier. A long term commitment and perhaps a substantial investment in testing and development may be required.

Design methods: New design methods can potentially be adopted as they become available, regardless of the timing of the product cycle. The initial input point may be to any of several areas in the organization: design analysis groups, advanced or forward engineering groups, or research and development groups. The new method does not necessarily need to be a completely finished product to be useful, as long as it expands the ability of the customer to design better products or to design more efficiently. Possibilities exist for cooperative development over extended and imperfectly defined time periods.

Effective internal development and transfer of technology requires good communication channels and a cooperative attitude on the working level as well as good technology management practices on the management level. These are sometimes problem areas in large organizations. Much has been written on the subject, and each organization must find its own ways to improve communication and cooperation. These factors are important not only for the transfer of technology developed in-house, but also for the transfer of technology developed outside from the point of input to the point of application.

Design Engineers need to remain current through participation in professional societies and through the stimulation afforded by cooperative research.

Those outside the automotive industry must communicate with the automotive customer at some input point. While publication in scientific journals is an excellent way to reach researchers, design engineers are often reached more directly by industry publications (eg., the Society of Automotive Engineers).

3 CONCLUSIONS

Significant changes are taking place in the automotive industry and in automotive technology. These changes are driven by intense competition and the resultant need to provide a high quality, economical product in a timely manner. Both the design of tribological components and the design process itself have been affected.

The automotive industry relies heavily on the design capabilities of suppliers and this reliance is growing. Accordingly, ensuring that the supplier community is using the latest design methods and technology should have a high priority.

Increasing competition, the need for continual improvement in product quality and performance as well as the need to critically assess supplier designs show the need for increased knowledge of tribology. Knowledge can be increased by participation in courses and seminars obtainable at universities, through consultants, by use of in-house expertise and through supplier presentations.

Tribological design tools include design guides and computer programs aimed principally at calculating parameters affecting bearing and valve train performance. These are being upgraded and validated on a continuing basis. Data bases for the selection of tribological materials are being developed. Opportunities exist for the tribology community to participate in these developments by introducing established methods and information into design practice and by generating new information in critical areas.

New technology can often be best implemented during the design stage. In order to be easily accepted, the new technology should be low risk, cost effective and available. Accordingly, development work must be close to completion before implementation. Originators of new technology should recognize the existence of multiple input points and should be prepared to make a long term commitment to development of the technology, in cooperation with an automotive company or supplier.

The introduction of new design methods may be less constrained in terms of timing and completeness of product development. Long term development of new or upgraded methods in cooperation with industry may be easier to implement than introduction of new technology. Cooperation in research work is one way for universities to develop contacts in the industry.

REFERENCES

(1) SPIKES, R. H., and ROBINSON, S. M. 'Engine Bearing Design up-to-Date', paper C1/82, from 'Tribology, Key to the Efficient Engine', I. Mech. E. Conference Publications 1982-1, Mechanical Engineering Publications Ltd., London (1982).

(2) GOODWIN, G. and HOLMES, R. 'On Bearing Deformation and Temperature Distribution in Dynamically Loaded Journal Bearings', paper C2/82, ibid.

(3) REASON, B. R., and SIEW, A. H., 'A Numerical Solution for the Design and Performance Evaluation of Journal Bearings with Misalignment', paper C9/82 from 'Tribology, Key to the Efficient Engine', I. Mech. E. Conference Publications 1982-1, Mechanical Engineering Publications Ltd., London (1982).

(4) American Society for Testing Materials Test Methods D4683 and D4741, Coordinating European Council (CEC) Test Method L-36, Institute of Petroleum (IP) Standard IP-370.

(5) HUTTON, J. F., JACKSON, K. P. and WILLIAMSON, B. P. 'Effects of Lubricant Rheology of the Performance of Journal Bearings', ASLE Trans., 29, 1 (1986) 52-60.

(6) BATES, T. W., WILLIAMSON, B. SPEAROT, J. A. and MURPHY, C. K. 'A Correlation Between Oil Rheology and Oil Film Thickness in Engine Journal Bearings', SAE Int. Cong. and Exposition, Detroit, MI, USA, February 24-28 (1986), paper 860376.

(7) SCHILOWITZ, A. M. and WATERS, J. L. 'Oil Film Thickness in a Bearing of a Fired Engine - Part IV: Measurements in a Vehicle on the Road', SAE Int. Fuels and Lubricants Meeting, Philadelphia, PA, USA, October 6-9 (1986), paper 861561.

(8) WILLERMET, P. A., PIEPRZAK, J. M. and DAILEY, D. P., 'Friction Reduction in Valve Trains: the Influence of Friction Reducing Oil Additives', to be submitted to SAE.

(9) FAN Y. CHEN, 'Mechanisms and Design of Cam Mechanisms', 1982 (Pergamon Press, Inc., New York).

(10) STARON, J. T. and WILLERMET, P. A. 'An Analysis of Valve Train Friction in Terms of Lubrication Principles', SAE International Congress and Exposition, Detroit, MI, USA, February 28-March 4 (1983), paper 830165.

(11) DOWSON, D., TAYLOR, C. M., and ZHU, G. 'Mixed Lubrication of a Cam and Flat Faced Follower', paper XX(i), from 'Fluid Film Lubrication - Osborne Reynolds Centenary', D. Dowson, C. M. Taylor, M. Godet and D. Berthe, eds., Elsevier Science Publishers BV., Amsterdam (1987).

(12) PARKER, D. A. and ADAMS, D. R., 'Friction Losses in the Reciprocating Internal Combustion Engine', paper C5/82, from 'Tribology, Key to the Efficient Engine', I. Mech. E. Conference Publications 1982-1, Mechanical Engineering Publications Ltd., London (1982).

(13) TING, L. L., 'Lubricated Piston Rings and Cylinder Bore Wear', from 'Wear Control Handbook', M. B. Peterson and W. O. Winer, eds., The American Society of Mechanical Engineers, New York, pg 609-666, (1980).

(14) SYAN, C. S., MATTHEWS, A. and SWIFT, K. G. 'Knowledge Based Expert Systems in Surface Coating and Treatment Selection for Wear Reduction', Surface and Coatings Tech., 33 (1987) 105-115.

(15) SWINDELLS, N. and SWINDELLS, R. J. 'System for Engineering Materials Selection', Met. Mater. (Inst. Met.) 1, 5, (1985) 301-304.

(16) ANON 'A Computerized Tribology Information System', Lubrication Eng., 1987, 43, 4, 228.

(17) TALLIAN, T. E. 'Tribological Design Decisions Using Computerized Databases', ASME Trans., Journal of Tribology, 1987 109, 381-387.

APPENDIX

A1 RECENT/CURRENT TECHNOLOGY IMPLEMENTATION

Some of the more significant recent applications of tribological principles to automotive design are given below. These show the results of some of the driving forces for change in the industry - the customer driven needs for performance, economy and durability. Also, they illustrate the extent to which the implementation of new technology can depend on cooperation among automotive manufacturers and suppliers.

Roller followers. Roller followers have been widely implemented in North America as a means of reducing valve train friction. Introduction of rollers has required changes in camshaft metallurgy because of the need for improved fatigue resistance. Where fatigue failure has been eliminated by the proper choice of materials, long term durability has been enhanced. Since both roller followers and fatigue resistant camshafts are more expensive, their use entails significant cost penalties; these are considered worthwhile in view of the improved fuel economy obtainable through reduced friction. If use of roller followers became general, this might eventually lead to reduced antiwear protection requirements for engine oils, resulting in less use of antiwear additives such as zinc dithiophosphates.

Low friction ring packs. Low friction rings have been introduced in a number of vehicles. This has been accomplished largely by attention to maintaining bore roundness, thus allowing use of lower tension rings.

5W30 engine oils. SAE 5W30 engine oils have become an ever larger factor world wide. The reduction in viscosity grade not only improves cold start, but results in lower viscous friction losses for most vehicles.

Improved friction reducing oils. Since engine oils are replaced periodically by the customer, who may obtain them from any of a large number of vendors, it was necessary to define categories of energy conserving oils in order to assure availability of these oils to the general public. This has been accomplished through the cooperative work of the automotive, cil and additive industries in SAE and ASTM. The American Petroleum Institute (API) defines an energy conserving oil as one providing a fuel economy improvement of 1.6% over a reference oil and energy conserving II oil as one providing a 2.7% improvement. Both viscometric properties and additive chemistry are important in formulating energy conserving oils.

Continuously variable transmissions. Continuously variable transmissions, long proposed as a means of improving efficiency and performance, are now beginning to enter production in Europe. Current models are based on the Van Doorne metal belt drive, although other types are still under consideration. Friction and wear as well as manufacturing issues have been obstacles to overcome in introducing this technology. In addition to metallurgical, design and manufacturing parameters, transmission fluid composition can also affect the performance of the belt/cone contact by influencing friction and durability.

Lock-up torque converters. Lock-up torque converters have been introduced in a number of automatic transmissions to minimize viscous losses. These designs, together with the desire to maximize fluid life and to achieve the smoothest possible shifts, have led to changes in the composition and performance of automatic transmission fluids. A smooth engagement together with a firm lock-up requires control of the friction versus sliding speed curve. To avoid shudder, the static to dynamic friction ratio should be less than 1. If the ratio is too low, however, slipping can occur. Actual fluid requirements depend on design philosophy. If the weight and size penalties are considered to be acceptable, a lower friction coefficient can be accommodated by increasing the friction material area.

Improved transmission fluids. Ford has recently introduced a new class of automatic transmission fluids. General Motors is expected to in the near future. The Ford fluid is characterized by closely controlled frictional behavior during clutch engagements and by improved stability over time. The General Motors fluid is expected to emphasize smooth shifts by producing an appropriate friction versus sliding speed curve. These demands have required significant changes in transmission fluid additive packages.

Paper II(iv)

Tribological design - The process industries

J. D. Summers-Smith

In general, the process industry does not design its own machines, but has to rely on what is available from machine manufacturers. It is important that the results of operating experience are fed back to machine designers in order that enhanced reliability of future designs is obtained. Examples, based on tribological machine elements, are given to illustrate different ways in which operating problems can be addressed and the experience made available for incorporation in future designs.

1 PHILOSOPHY

I speak from experience in the chemical process industry and, in particular, with machines problems in that industry. In general, apart from a few highly specialised cases, the process industry does not design its own machines, but has to rely on purchasing what is available in the market place, influencing this as far as possible to achieve cost effective reliability. It has to be emphasised that in the present competitive climate in the field of process chemicals, the user is probably more interested in predictability of performance, rather than reliability per se, so that he can select equipment giving the lowest life cycle cost per unit of product, balancing initial capital cost against maintenance costs, downtime costs, particularly unscheduled downtime costs.

In the context of this symposium I want to focus primarily on machine elements and the impact they have on machine performance, not on performance itself nor on other aspects such as strength and corrosion that have to be equally addressed.

The user is in the unique position that he alone can obtain experience of operation under real conditions, where machines are subject to process upsets, operator malpractice, gradual deterioration in service, and other factors, probably unforeseen, outside those specified in the design, rather than the more rigorously controlled conditions of the test bed. In many cases it is not even possible to match the service conditions on the test bed: for example, it may not be practicable to use the actual process fluid or to be able to realise the dynamic foundation conditions that a machine mounted on a marine platform will be subject to.

My starting point is failure, using this in the broadest sense to include not only breakdown, but also the need to restrict output to avoid breakdown. The user has a number of priorities: the over-riding need for a 'quick-fix' in order to get back on line as quickly as possible and minimise production loss; an understanding of the problem in order to prevent its recurrence and, if necessary, to ensure that the manufacturer incorporates the experience in new designs, though in the urgency of the production situation some discipline has to be exercised if adequate time and effort are to be devoted to the last two of these items.

Failure analysis should involve the active cooperation of the machine designer, who in many cases has, first of all, to be persuaded that there is a genuine problem. It is a matter of fact that the normal immediate reaction of a machine manufacturer is that any failure is an operational effect, not a matter of the design or manufacture of the machine. A real understanding of the machine and the functioning of the machine elements by the user is crucial in developing the necessary degree of understanding between the user and the designer.

2 EXAMPLES

I should like to give from my experience some examples that bring out the above points.

2.1 Boundary lubrication in non-oxidising environments

This example concerns two 2-stage, double-acting reciprocating compressors supplied by an internationally known manufacturer. The vast majority of machines are lubricated in an environment of atmospheric air. This machine, however, was to handle very pure nitrogen (< 2ppm O_2, < 2ppm H_2O) so that lubrication of the piston rings had to take place in a non-oxidising atmosphere. The manufacturer's experience was largely in the field of air compressors, but he expected no difficulty with nitrogen, a gas with very similar thermodynamic properties to air. Both machines, however, suffered very severe cylinder scuffing during the initial period of commissioning. The need to ensure adequate boundary lubrication of the piston rings at the ends of the stroke had been overlooked. Boundary lubrication, whether by the natural products found in hydrocarbon oils or by incorporated anti-wear or ep additives requires an oxide film on the contacting surfaces. These machines not only operated in a completely non-oxidising atmosphere, but to ensure purity of the gas the crankcase, from which the lubricant was fed to the cylinders, was also blanketted with nitrogen. In normal circumstances any oxide film that is removed under boundary conditions is rapidly replaced from the atmospheric moisture. In this case, however, there was insufficient oxygen present, boundary lubrication broke down and the cylinders scuffed. The solution was simple once the mechanism of failure had been identified: the cylinder lubricant was supplied from outside the machine where it was in contact with air and the normal dissolved moisture was sufficient to maintain the oxide film on the cylinder wall.

2.2 Load calculation for rolling bearings

My second problem involved failures of roller bearings on the rotors of two plastics extruders (Fig. 1). Examination showed the bearings were failing by fatigue. According to the manufacturer the bearings were operating well within their rated capacity and it was suggested that failure was caused by overload arising from starting without preheating the plactics granules as recommended. In these conditions it was estimated that it would take 15 minutes operation before frictional heating produced the required temperature. A simple calculation, however, showed that operation for 15 minutes per day at the overload trip current of the motor was well within the capacity of the bearings. A Weibull plot of the time to failure allowed a calculation of the bearing L_{10} Life with appropriate confidence limits and from this an estimate was made of the continuous load that would account for the failures. Presentation of these calculations persuaded the manufacturer to reexamine the design and it turned out that an incorrect blending profile between feed and extrusion sections on the rotor could give a high enough axial load to account for the failures. The profile was corrected and the failure problem was cured.

2.3 Sulphur attack of floating-bush seals on centrifugal compressors

My third example concerns the floating-bush seals on centrifugal hydrocarbon process gas compressors. These normally use mineral oil as a barrier fluid and are lined with white metal to prevent damage at starting. Such seals operate in an environment of the barrier fluid and the process gas. Seal failure occurred on a number of machines through reaction between sulphur in the process gas and copper in the white metal forming copper sulphide that took took up the clearance. Chemical reaction is a function of concentration and temperature. The oil film temperature in the seal ring, which is a function of the rate of shear in the oil film and hence the peripheral velocity, could not be measured. It seemed, however, that a plot of sulphur content against peripheral speed might be of use in defining the bounds of the problem.

Fig. 2 gives a summary of experience from a number of machines. The weight of evidence was sufficient to convince the machine manufacturers that there was a genuine problem. Despite the representation of temperature by peripheral speed, and also the lack of definition of the sulphur content of the gas, both in chemical nature and amount (available only to the nearest order of magnitude), it was possible to define a boundary to the problem. The use of a lead-rich, copper-free white metal as a seal ring lining gave some relief, but at the most severe conditions even lead was attacked to form lead sulphide deposits and in these cases a seal redesign was required to reduce the temperature. Fig. 2. shows how three zones were defined that could be used in the purchase specification of new machines.

2.4 Bearing influenced rotor dynamics

Finally, I should like to look at problems arising from the dynamic instabilities in high-speed rotary machines. The simplistic approach to avoid instability was to specify that machines should not have a natural frequency within ± 10% of the rotor operating speed. With increasing sizes and speeds of machines, and particularly with machines capable of operating over a range of speeds, application of such a simple criterion became unrealistic, more particularly as natural frequencies were normally calculated for simply supported rotors with no allowance made for the effect of the dynamic characteristics of the supporting bearing oil films. A change was made to a requirement imposing a limit to the response to specified forcing conditions that were imposed on the test bed. This was clearly an unsatisfactory situation when problems arose on the test bed, or even worse when the machine went into operation, and the problem had to be resolved in an ad hoc way.

It was decided that a more detailed understanding was necessary to aid in the solution of problems and in the development of purchasing specifications that would result in problem-free machines. This came to a head with a steam turbine driver on a process gas stream with a design operating speed of 177 rev/s. The difficulty was compounded in this machine in that the rotor was supported on three journal bearings so that the bearing loads were indeterminate; when instability problems arose, there was no obvious solution and in order to maintain production the speed had to be reduced with a consequent loss in output. The University of Leeds was asked to develop a complete rotor dynamic programme, including the dynamic characteristics of the bearing oil films. This involved a theoretical study that could not be tested economically in the laboratory because of the scale. The advantage of the university-industry co-operation was that the theoretical predictions could be validated with measurements taken from operating plant machines. The final programme was of value not only in guiding remedial actions when vibration problems occurred, but also in allowing a meaningful dialogue between the Company engineers and machine suppliers when new machines were being purchased.

3 CONCLUSIONS

The outcome of failure investigations of the

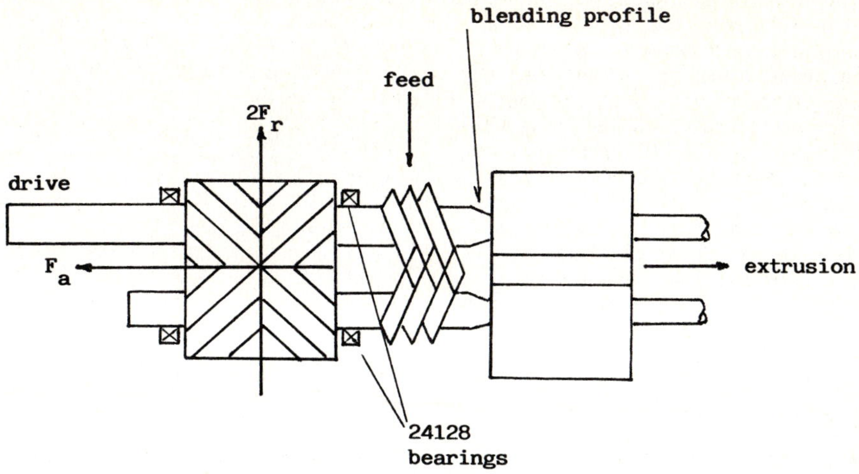

Fig. 1. Schematic arrangement of extruder.

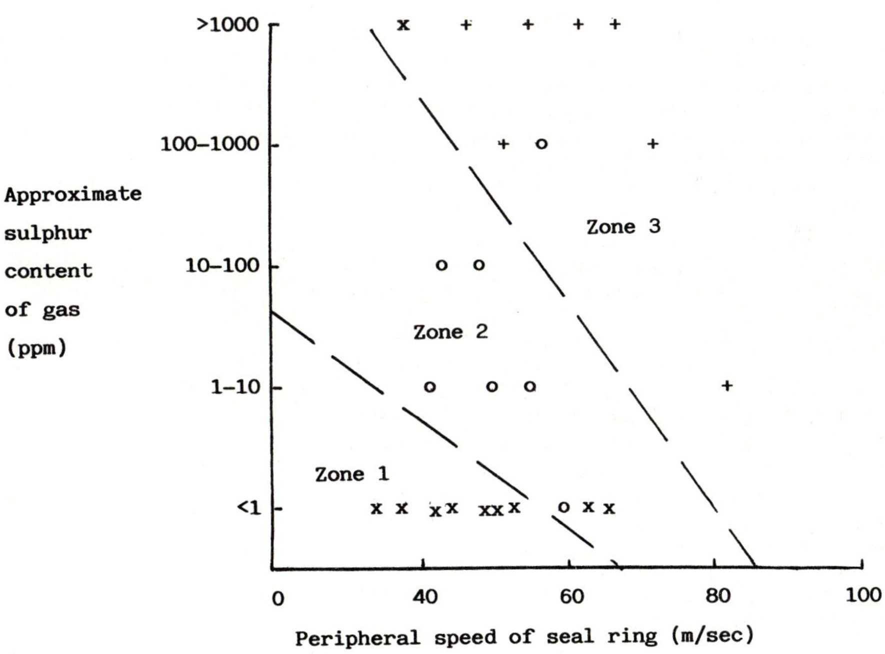

Fig. 2. Attack of white metal lining of floating-bush seal ring by sulphur in process gas.

x no problem
o attack of tin-rich white metal, but no attack of lead-rich white metal
+ attack of lead-rich white metal

Zone 1. No problem
Zone 2. Use copper-free white metal (lead-rich)
Zone 3. Redesign seal to reduce temperature

types described can range from solutions based on an understanding of the mechanisms involved (Examples 2.1 and 2.2), through a crude graphical presentation of the bounds of the problem area using parameters based on a theoretical appreciation of the significant variables involved (Example 2.3) to a research programme aimed at a complete understanding and definition of the problem (Example 2.4).

The final link in the chain is communication. This can be effected in a number of ways. For a one-off problem it may end with rectification in co-operation with the manufacturer. Where the findings are of wider application they can be incorporated in a Company Purchase Specification to prevent future repetition of the same problem, but, in addition, publication of a paper in a journal with a refereeing procedure is of value in making the findings available. In this context encouragement should be given to the acceptance of case history studies, even if these lack some of the theoretical background that tends to be a requirement of many learned societies before papers can be considered. Not only does publication allow the data to be used in design guides, such as the ESDU Data Items (ESDU International Ltd., London), but the fact that publication has been accepted lends weight to a one-to-one discussion with a supplier and allows understandingof the rationale behind the incorporation of additional requirements in purchase specifications.

4 THE FUTURE

These examples have been chosen from a very wide range of past problems that have been satisfactorily solved and are intended to illustrate the approach taken by one user. Inevitably this has been a reactive one and in a forward looking meeting of this type it is probably of more interest to identify possible future problem areas. This necessarily means a certain degree of speculation, otherwise the problems would have already been identified and hopefully solved. Areas that I feel are likely to be of concern in the tribology of machines are sealing, a rather better understanding of the influence of foundation dynamics on the instability of machines, the need to identify the effect of solid particles on the performance of lubricated components in order that more realistic filtration requirements can be specified. The present rather ad hoc treatment of these subjects reflects a lack of satisfactory understanding and the potential for future problems.

SESSION III
SEALS

Chairman: Professor H Blok

PAPER III(i)	Design of Controllable Mechanical Seals
PAPER III(ii)	Micro-Elastohydrodynamic Lubricant Film Formation in Rotary Lip Seal Contacts
PAPER III(iii)	Lubrication of Reciprocating Seals: Experiments on the Influence of Surface Roughness on Friction and Leakage
PAPER III(iv)	Radial Lip Seals, Thermal Aspects

Paper III(i)

Design of controllable mechanical seals

R. F. Salant, O. Giles and W. E. Key

Research into the fluid mechanics of mechanical seals has led to the development of seals in which the thickness of the lubricating film between the seal faces is electronically controlled. In a typical design, a microcomputer-based control system and actuator continuously adjust the film thickness, based on information received from a sensor which monitors conditions in the film. The control system sets the film thickness at an optimum value and responds to changes in operating and environmental conditions so that the film thickness is kept at an optimum value and face contact is prevented. Such a seal will experience less mechanical and thermal damage to the faces, less wear, and a longer life, than conventional mechanical seals.

1. INTRODUCTION

Extensive research over the last thirty-five years has led to substantial understanding of the behavior of mechanical seals. More recently, over the last five years, the wide availability of minicomputers and associated software has allowed the application of the research results to the design and development of practical seals. Furthermore, most recently, this research has led to a new concept in fluid sealing, the controllable mechanical seal.

The earliest mechanical seals were heavily loaded, and operated with mechanical contact between the faces. However, with the development of balanced mechanical seals, the designs evolved such that most modern mechanical seals operate as noncontacting seals at the steady-state design point, with a thin lubricating film of fluid separating the faces.[1] Although there are exceptions in which contacting designs are used (e.g. stern tube marine seals), it is generally preferable to utilize a noncontacting design in order to minimize wear and energy dissipation, and maximize reliability.

Until recently, most seals were designed empirically, utilizing trial-and-error and a great deal of testing. Today, however, a variety of steady state[2,3,4] and dynamic analysis[5,6,7] tools are available to the seal designer, and are used by virtually all major seal manufacturers. Utilizing these tools, the designer seeks a design in which the lubricating film is stable and is of optimum thickness: not too thick, to prevent excessive leakage, and not too thin, to prevent face contact.

The above analyses show that the film thickness is very dependent on operating and environmental conditions. Hence, a seal is generally designed for operation with an optimum film at a single steady-state design point. However, many seals must operate over a range of steady-state conditions, not only at a single point. These therefore frequently operate with nonoptimum films, with periods of face contact and periods of excessive leakage.

Furthermore, all mechanical seals experience transient conditions. At the minimum, this encompasses startup and shutdown. In addition, most seals experience pressure surges, temperature fluctuations, and speed variations. During these transients, the film thickness varies, usually resulting in face contact, and sometimes resulting in excessive leakage.

It is the above off-design operation, both at steady-state and under the influence of transients, which causes the majority of problems with well-designed conventional mechanical seals.

1.1 Notation

h_{ave}	Average film thickness
h_i	Film thickness at ID
h_o	Film thickness at OD
T_c	Chamber temperature
t	Time
V	Voltage applied to actuator
ΔT	Face temperature $-T_c$
ΔT_{HIGH}	Value of ΔT at zero effective film thickness
ΔT_{MAX}	Value of ΔT above which permanent damage to seal occurs
δ	Coning, $h_o - h_i$

2. DESIGN OF CONVENTIONAL MECHANICAL SEAL

As described above, the successful operation of a conventional mechanical seal is closely related to the behavior of the lubricating film. The thickness of that film is determined by the axial location of the floating seal face, which is governed by the forces acting on it. The "closing force," produced by the spring (or bellows) and the sealed pressure acting on the backside of the floating face, forces the floating face towards the fixed face. This force depends on the spring characteristics, the sealed pressure, and the geometry of the floating face cross-section (as measured by the balance ratio). The "opening force," produced by the pressure distribution within the fluid film, opposes the closing force and keeps the film intact (in a successful seal). This force depends on the fluid mechanics of the film. Under steady-state conditions, the film thickness is fixed by a balance between the closing force and the opening force.

Most conventional noncontacting mechanical seals are of the hydrostatic type. These seals have flat, axisymmetric faces. (Hydrodynamic seals have circumferential variations in face geometry, such as a series of slots or pads distributed around the circumference, or circumferential waviness; they are not considered in the present paper.) The pressure distribution within the film of such a hydrostatic seal is associated with the radial flow field induced by the pressure drop across the seal. Thus, the pressure distribution and the associated opening force are dependent on the radial variation in geometry of the gap between the seal faces (or the radial variation in film thickness), as well as on the sealed pressure. The gap may be uniform, converging (going from the high pressure to the low pressure side of the seal), or diverging. However, it is well known that to achieve stable operation, a seal must be designed such that the gap converges[8], as shown in figure 1.

The amount of convergence, or "coning" δ, strongly influences the pressure distribution and opening force. The larger the coning, normalized with respect to the average film thickness, (δ/h_{ave}) the more convex the pressure distribution, and the larger the opening force.[3] Since the closing force is fixed by the spring characteristics, the sealed pressure and the cross-sectional geometry of the floating face (or the balance ratio), under steady-state conditions there is one unique value of the normalized coning for which the opening force equals the closing force and the floating face is in equilibrium. Hence, the steady-state average film thickness must be proportional to the coning: the larger the coning, the larger the film thickness.

Normally, a seal is designed and manufactured such that prior to service the coning is effectively zero. However, once it is in service, coning is produced by mechanical and thermal deformation of the faces and their supporting structures. The coning is therefore determined by the fluid mechanic, heat transfer, energy dissipation, and structural deformation processes. Hence the amount of coning, and therefore the film thickness, is be strongly dependent on the operating and environmental conditions, as well as on the seal design.

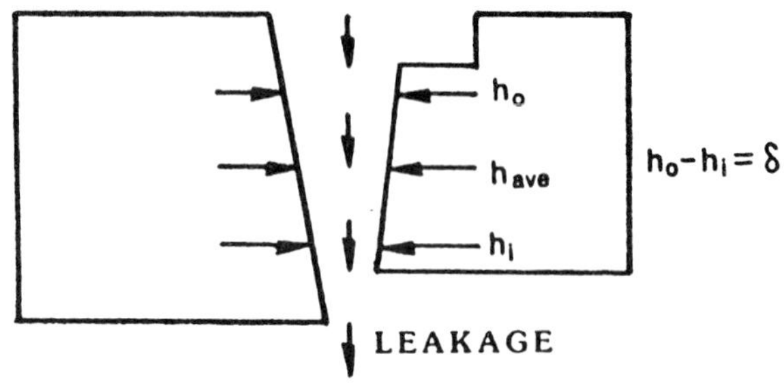

Fig. 1 Converging seal gap

Thus, it is a principal task of the designer to configure the seal in such a way that the faces of the seal deform just the right amount, at the design point, to produce an optimum film thickness. However, even if this is achieved, when operating and/or environmental conditions change (either through a long term or a short term transient) the amount of deformation and the film thickness change to non-optimum values, with the possibility of either face contact or excessive leakage.

3. CONCEPT OF CONTROLLABLE SEAL

To alleviate the problems of conventional mechanical seals, discussed above, the concept of a controllable seal has been developed. A conventional mechanical seal is a passive device. While it can be designed to run at a particular design point in an optimum fashion, it reacts to changes in operating and environmental conditions in an uncontrolled manner, according to the laws of physics. That reaction takes the form of a change in film thickness, often leading to face contact or excessive leakage. Once the seal has been installed, nothing can be done to improve its performance.

In contrast to the conventional seal, the controllable mechanical seal is an active device. It reacts to changes in operating and environmental conditions in a pre-programmed controlled manner, such that optimum operation is achieved at all steady-state operating points (within a specified range) as well as during short term transients. The design of the seal is effectively changed during operation to adapt to local conditions.

Such a controllable seal has a variable film thickness, which can be adjusted explicitly. If conditions are such that the film is too thick or too thin, it is readjusted to its optimum thickness while in service.

The film thickness of a conventional seal is fixed by the location of the floating face, which is governed by the forces acting on the it, as described above. Hence, the film thickness could be controlled by controlling either the closing force or the opening force. However, the authors have found that it is much more effective to control the latter than the former.

Since the opening force and the film thickness are strongly dependent on coning, control is achieved by controlling the coning. This is accomplished by incorporating within the seal an electromechanical actuator to mechanically deform one of the faces and generate coning. The produced coning compensates for the natural mechanical and thermal deformation experienced by the seal faces and supporting structure, and can be adjusted to any desired value by the actuator. Thus, the film thickness can be adjusted to any desired value. It is important to note that the actuator does not directly produce the opening (or closing) force. It is the pressure distribution in the film which produces the opening force, and keeps the two faces separated and the film intact. The actuator regulates the pressure distribution and opening force by controlling the coning. Thus, very precise control of film thickness is achieved.

While the above adjustment of film thickness could be done manually, a much more powerful approach makes use of an adaptive control system. Sensors monitor conditions in the film and in the sealed cavity. Based on those conditions, a microprocessor-based control system causes the actuator to adjust the film thickness, according to a pre-programmed adaptive control strategy. This allows continuous optimization of the film thickness and automatic adaptation to any steady-state operating point. Furthermore, it allows response to short term transients, and especially, the minimization of face contact during such events.

Laboratory-scale seals, based on the above concept, have been built and successfully tested, and are described elsewhere.[9] Most recently, a seal suitable for industrial use has been built and tested, and is described in the following section.

4. SEAL CONFIGURATION

A controllable mechanical seal for the steam generator feed water pump of a nuclear power plant has been built and successfully tested. A controllable seal is especially suitable for this application due to a demand for longer life than that achieved with conventional seals (currently, approximately one year) and due to the economic consequences of a seal failure.

The pump shaft has a diameter of 140 mm (5.5 inches) and runs at a speed of 372 rad/s (3550 rpm). The stuffing box pressure at 100% load is 30.3 bar (440 psig), and temperature is 60°C - 79°C (140°F - 175°F) under running conditions, and 149°C (300°F) under standby conditions.

Figure 2 contains a schematic drawing of the seal. The rotating silicon carbide face is fixed in the axial direction, while the nonrotating carbon graphite face floats axially. The floating assembly, consisting of the nonrotating face, two-piece face holder and actuator, is loaded by ten circumferentially spaced coil springs (not shown on drawing).

To monitor conditions in the lubricating film, a thermocouple is imbedded in the nonrotating face, 1.3 mm below the sealing surface. Its leads are enclosed in stainless steel tubing, which is coiled to accommodate the axial movement of the floating

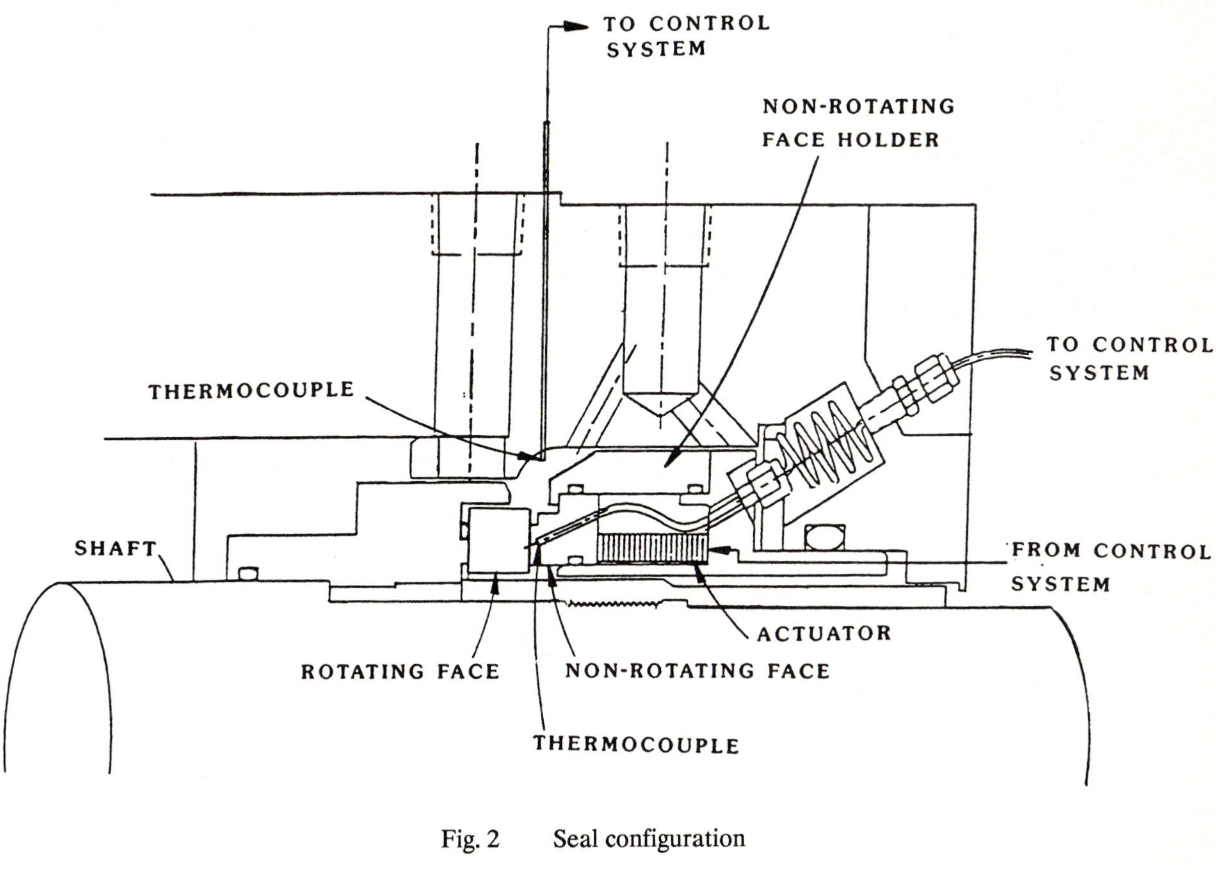

Fig. 2 Seal configuration

assembly. The two leads to the actuator (not shown on drawing) are similarly treated. A second thermocouple monitors the chamber conditions.

Circulation of fluid through a heat exchanger is provided by a pumping ring (not shown on drawing) and cooling passages.

The actuator, contained within the floating assembly, consists of a series of piezoelectric annuli with interspersed brass electrodes. When a voltage is applied to the electrodes, the piezoelectric elements expand and exert an axial force against the backside of the nonrotating face near the ID. Since the OD of the face is restrained from axial movement (relative to the floating assembly) by the face holder, this force produces face deformation contributing to positive coning (converging gap between the faces, going from OD to ID). The larger the applied voltage, the larger the coning and the thicker the fluid film.

The voltage to be applied to the actuator at any instant of time is determined by the control system, shown in figure 3. The outputs of the seal face and chamber thermocouples are processed through a signal conditioner and A/D converter, and then enter a microprocessor. The microprocessor performs all the computational and logic procedures, as specified by the control strategy and control parameters which are stored in memory. The microprocessor also receives information on whether or not the shaft is rotating from a magnetic shaft rotation sensor. The actuator voltage, commanded by the microprocessor, is processed through a D/A converter, amplified by a voltage supply, and applied to the actuator.

5. CONTROL STRATEGY

To manage the many changes in operating and environmental conditions, an adaptive control strategy is utilized. This strategy searches for, and maintains, an operating point with an optimum film thickness to minimize seal face damage/wear and leakage. At this optimum point the film thickness is as small as possible without face contact occurring.

To find the optimum operating point, the strategy first searches for the point where the film thickness is effectively zero, which serves as a base reference point. It then backs away from this reference point a small amount, by slightly increasing the film thickness. This new point is defined as the optimum operating point.

The base reference point, where the film thickness is effectively zero and face contact is imminent, is identified in two ways. One indicator of imminent face contact is the appearance of oscillations (or spikes) in the face temperature, which are produced by a thermoelastic instability[9]. However, depending on the face materials and local conditions, these oscillations may be difficult to detect.

A second indicator of imminent face contact is the magnitude of ΔT, the difference between the face temperature and the chamber temperature. For given face materials ΔT_{HIGH}, the maximum ΔT at which the seal can run without experiencing face contact, can be determined empirically.

Therefore, in searching for the base reference point, if ΔT_{HIGH} is encountered before face temperature oscillations are detected, the point where

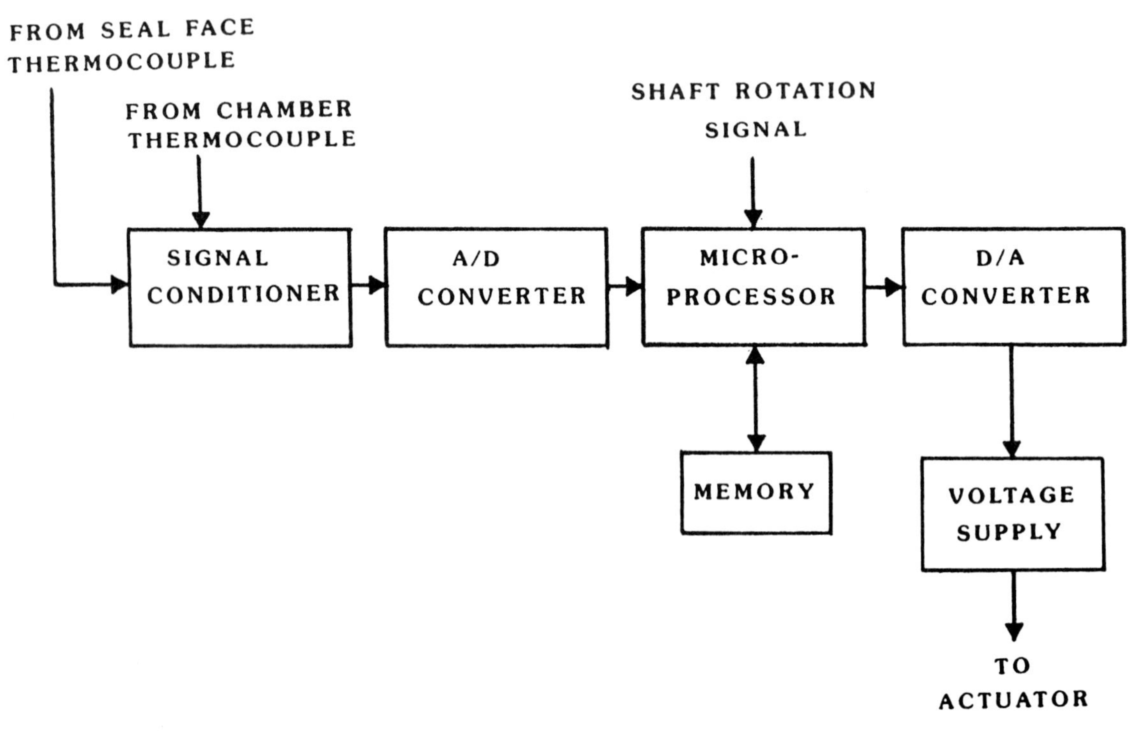

Fig. 3 Control system

$\Delta T = \Delta T_{HIGH}$ is selected as the base reference point.

The control strategy is divided into six modes as shown in figure 4. Each mode specifies an operating state of the system. The "initialization" mode is entered when the system (viz. the pump) is started up. In this mode, a large voltage is applied to the piezoelectric actuator to provide a sufficiently large film thickness so that there is minimum damage/wear to the seal faces during startup.

Following startup, the system enters the "search" mode in which the actuator voltage is systematically reduced, searching for an oscillation in face temperature. If an oscillation is detected, a large step increase in actuator voltage is commanded to increase the film thickness and stop the oscillation, and the voltage level corresponding to initiation of the oscillation is remembered. Control is then transferred to the "optimization" mode. If, as the actuator voltage is decreased during "search", ΔT exceeds ΔT_{HIGH} before oscillations are detected, control is transferred to "optimization".

In the "optimization" mode, the optimum operating point is found. If this mode is accessed following an oscillation, the voltage is decreased gradually (reducing film thickness), until it is slightly above the voltage corresponding to the initiation of the oscillation. If a new oscillation is detected, a large step increase is again applied to the actuator, and the optimization process is repeated. If the "optimization" mode is accessed following an excessive value of ΔT, the voltage is increased until ΔT falls slightly below ΔT_{HIGH}. Once the optimum operating point is found,

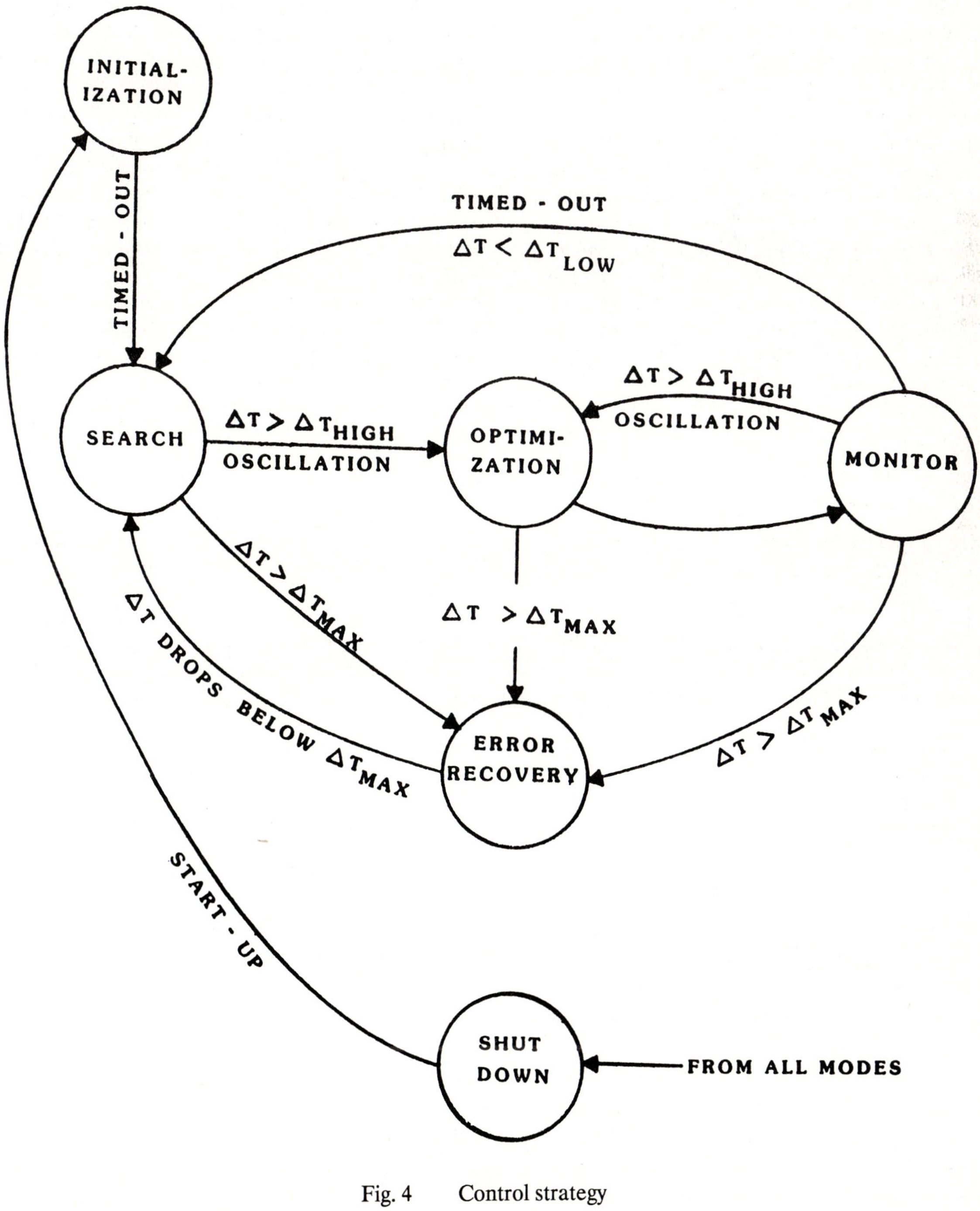

Fig. 4 Control strategy

control is transferred to the "monitor" mode.

In the "monitor" mode, the actuator voltage is held constant. This mode is exited if any of the following conditions occurs:

i. A preselected time interval expires. This allows scheduled optimization of the operating point.

ii. ΔT becomes very small.

iii. ΔT exceeds ΔT_{HIGH}.

iv. An oscillation occurs.

Under conditions i. or ii., control is transferred to "search"; under conditions iii. or iv., it is transferred to "optimization".

The "error recovery" mode is accessed from any mode except "initialization" and "shutdown", when ΔT exceeds ΔT_{MAX}. ΔT_{MAX} corresponds to the temperature above which the seal will be permanently damaged. In this mode, the maximum voltage is applied to the actuator, to minimize face contact. If ΔT is brought below ΔT_{MAX} within a specified period of time, control is transferred to "search". If not, the system stays in this mode until the pump is shut down.

"Shutdown" can be accessed from any mode, and is initiated when the shaft rotation sensor detects that the pump is being shut down. The voltage is increased to a high level to prevent face contact during the shutdown. After a specified time interval, the voltage is set to zero to prevent leakage after completion of the shutdown process.

6. LABORATORY TEST RESULTS

The controllable mechanical seal, described above, has been tested in the laboratory utilizing a conventional microcomputer-controlled double-ended tester. The tester runs through a series of duty cycles, each lasting approximately thirty minutes, with a two minute interval between cycles. Each cycle begins with startup, concludes with shutdown, and contains one step change in pressure, either from 22.1 bar (320 psig) to 27.6 bar (400 psig), or the reverse. Alternate cycles are run at chamber temperatures of 50°C and 75°C.

Figure 5 contains histories of ΔT and the applied actuator voltage in a typical test. Shutdown of one cycle and initialization (startup), search and the beginning of optimization of the next cycle are shown.

Shutdown occurs at t = 2.4 minutes. The actuator voltage is immediately increased to 1200 volt to increase the film thickness and prevent face contact, and ΔT drops rapidly. After 15 seconds, when shutdown is essentially complete, the voltage is reduced to zero, to prevent leakage while the pump is stopped.

In the next cycle, startup occurs at t = 4.5 minutes, and the control system goes into the initialization mode. The voltage is immediately

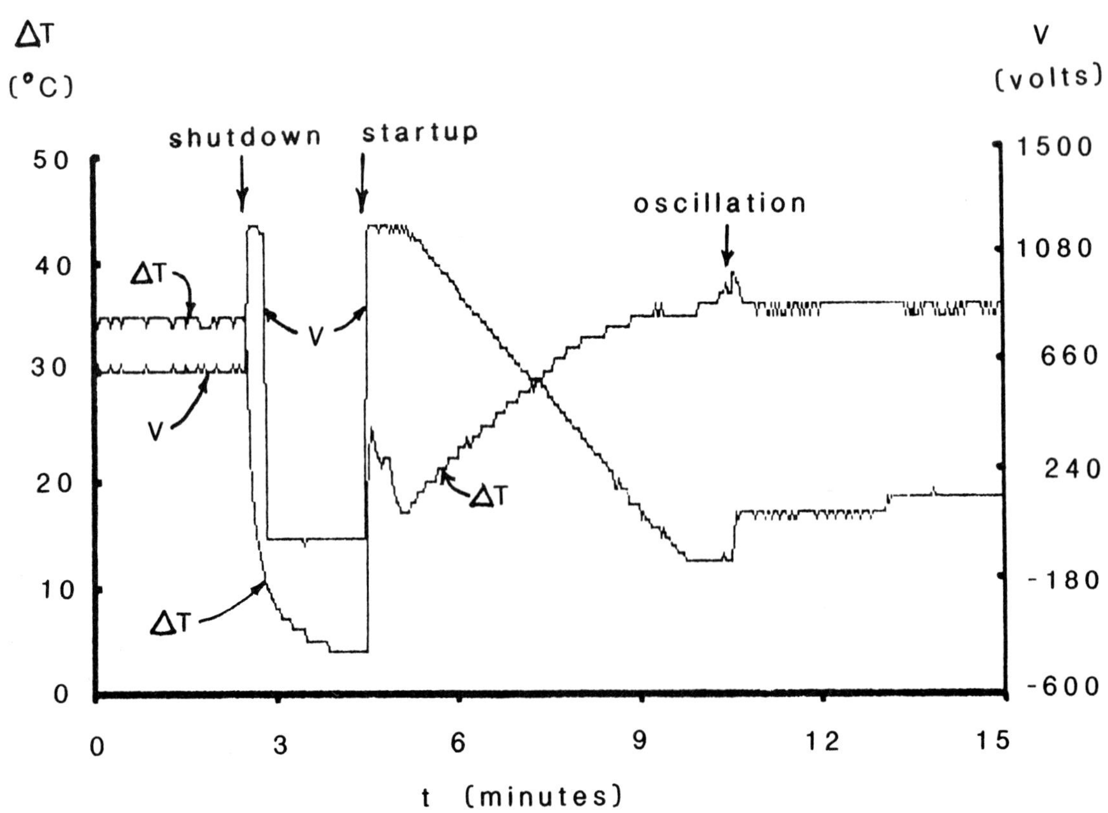

Fig. 5 Laboratory test

increased to 1200 volts to prevent face contact. After startup, the search mode is entered, to find the base reference point (where the film thickness is effectively zero), and the actuator voltage is gradually reduced. During this process, ΔT increases continuously, indicating the film thickness is decreasing. At t = 9.7 minutes ΔT reaches 35°C, which is ΔT_{HIGH}, and the system goes into the optimization mode. The voltage is held such that ΔT is kept slightly below 35°C, and the system is at an optimum operating point. However, at t = 10.4 minutes the system detects an oscillation indicating imminent face contact, and a step increase in voltage of 200 volts is applied, eliminating the oscillation. At t = 13.1 minutes the system detects that ΔT exceeds ΔT_{HIGH}, so the voltage is increased in order to slightly lower ΔT. The system is now at an optimum operating point, and enters the monitor mode (Note that due to the digital nature of the display, temperature resolution is approximately 1°C. Thus, small changes in ΔT are not visible on the plots.)

Figure 6 contains a continuation of the histories displayed in figure 5, and illustrates additional aspects of the monitor and optimization modes. At t = 0 minutes the system is in the monitor mode (as in the previous figure) and stays in that mode until t = 3 minutes, when the system detects the onset of oscillation (indicating imminent face contact), control is immediately transferred to the optimization mode, and a step increase of 200 volts is applied to the actuator. This increases the film thickness, and the oscillation is eliminated. After 2.3 minutes, the voltage is reduced by 150 volts, in order to reduce the film thickness and move closer to the base reference point. However, an oscillation is detected, and another 200 volt increase is applied to eliminate the oscillation. After another 2.3 minute interval, the voltage is again reduced by 150 volts, and again an oscillation is detected and a 200 volt increase is needed to eliminate it. At t = 12 minutes the system detects that ΔT slightly exceeds ΔT_{HIGH}, so the voltage is increased. The system is now at an optimum operating point, and control is transferred to the monitor mode.

The above results illustrate how the film thickness is continually adjusted to avoid face contact, and yet to be as small as possible (to limit leakage). After 200 hours of testing (including steady-state and approximately 100 duty cycles), examination of the faces revealed no observable face damage. Profilometer measurements indicated wear on the silicon carbide face of less than .127 microns (5 microinches) and on the carbon graphite face of less than .254 microns (10 microinches).

7. FIELD TEST RESULTS

Following laboratory testing of the controllable seal, an 800 hour field test has been performed on the main feedwater pump for the steam generator of a nuclear power plant. The pump operates at a speed of 372 rad/s (3550 RPM), and with a chamber pressure of 30.3 bar (440 psig) and temperature range of 76°C - 82°C. Since some of the characteristics of the pump differ from those of the tester (e.g. startup time), the values of some of the control parameters for the field

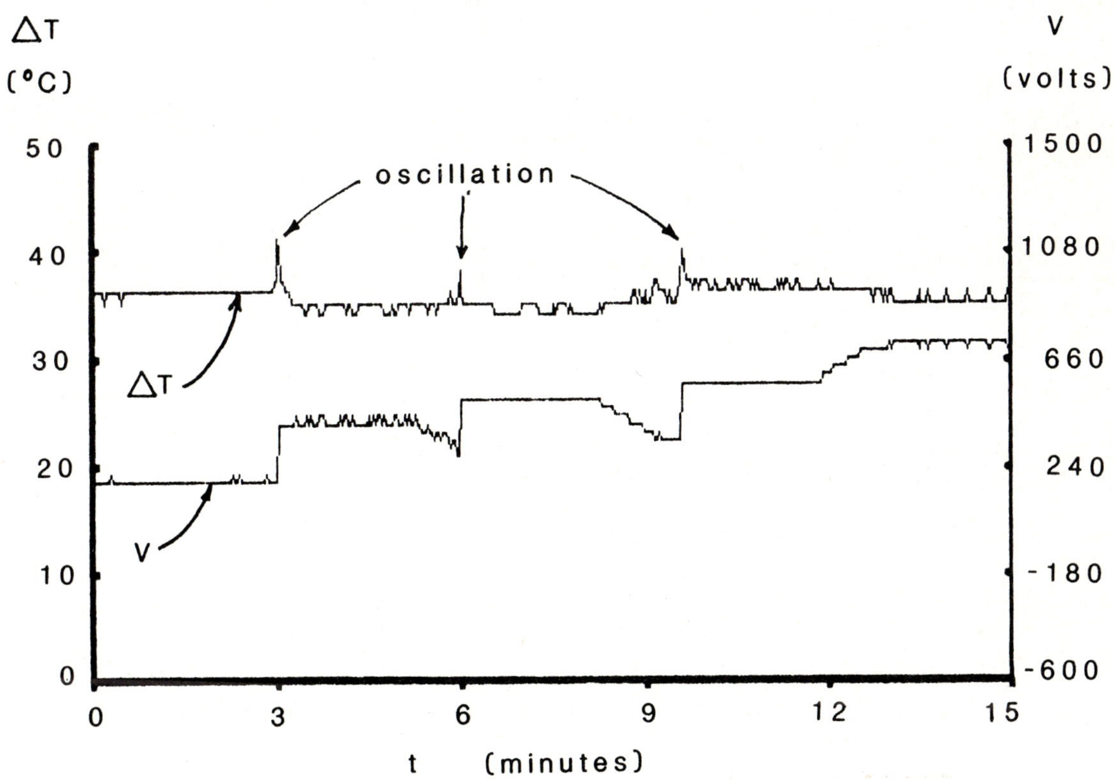

Fig. 6 Laboratory test (continued)

test differ from those of the laboratory test. A microprocessor-based portable data monitoring and acquisition system, especially designed for mechanical seals, is used to collect data.

Figure 7 shows the startup process. At t = .06 hrs., the pump is started and the actuator voltage is immediately raised to 1200 volts to minimize face contact. At t = .34 hrs. control is transferred to the search mode, and the voltage is gradually decreased. This reduces the film thickness, and causes ΔT to

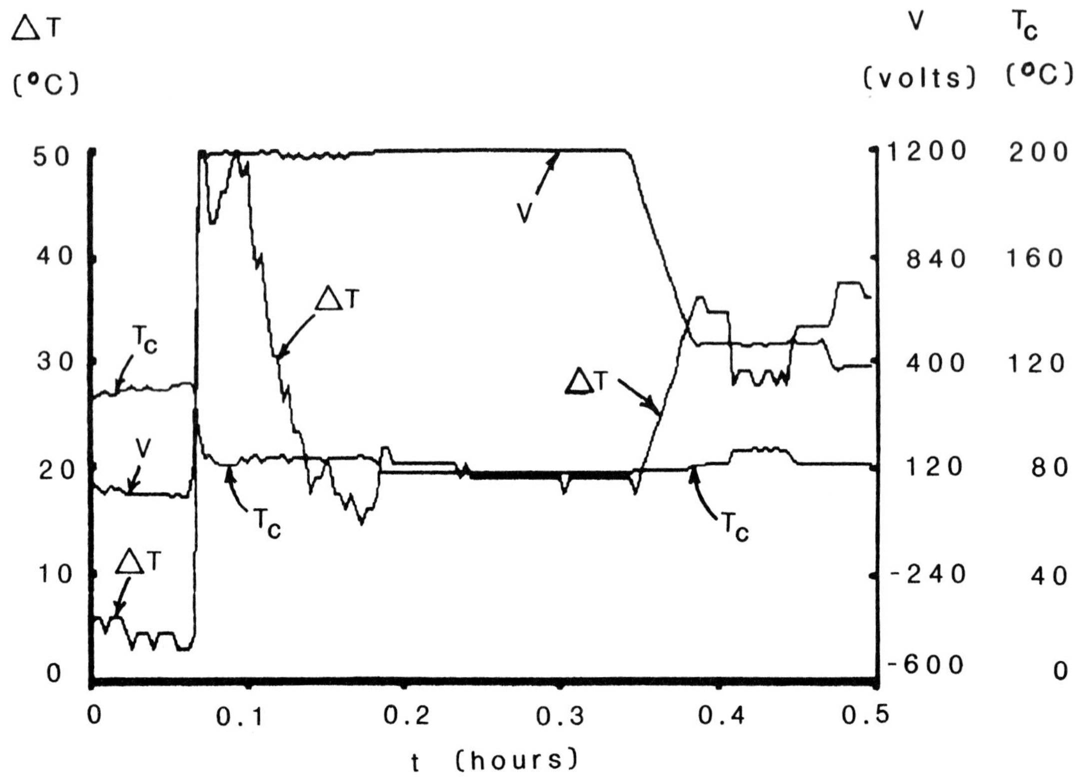

Fig. 7 Field test-startup

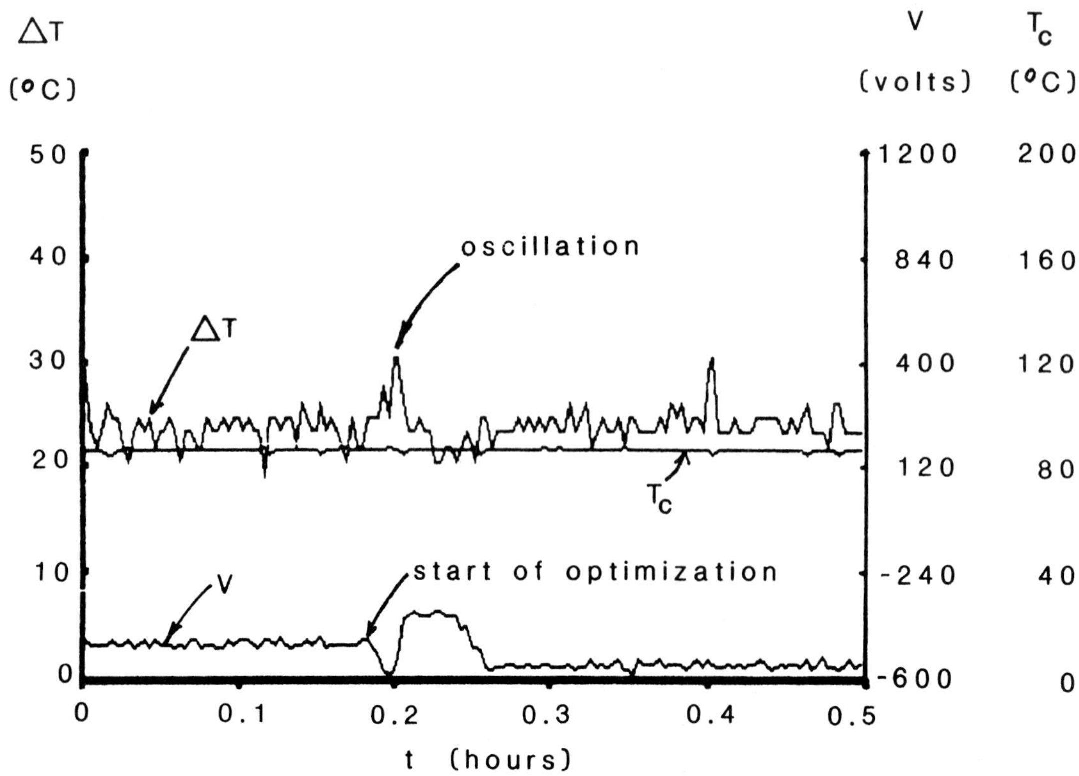

Fig. 8 Field test-optimization

gradually rise. When ΔT reaches 35°C (ΔT_{HIGH}), control is transferred to the optimization and then the monitor mode, and the applied voltage is held constant. The system is now at an optimum operating point. Subsequently, variations in ΔT occur due to fluctuations in the chamber temperature. At t = .46 hrs. a scheduled optimization occurs. The voltage is gradually reduced and ΔT, which had fallen below ΔT_{HIGH}, is increased. When ΔT reaches ΔT_{HIGH}, the voltage is held constant (there is a slight overshoot in ΔT), and the system is again at an optimum operating point.

Figure 8 shows another example of a scheduled optimization. At t = 0 the system is in the monitor mode. At t = .18 hrs. a scheduled optimization begins, and control is transferred to the search mode. The voltage is reduced and an oscillation is detected. The voltage is then increased by 200 volts to eliminate the oscillation, and control is transferred to the optimization mode. To move closer to the base reference point, the voltage is then reduced by 150 volts. Since no oscillation is detected, the system is now at the optimum operating point, and control returns to the monitor mode.

Following 800 hours of testing, examination of the seal faces indicated no observable face damage.

8. CONCLUSIONS

The results presented above demonstrate that the electronically controlled mechanical seal is a feasible approach for applications in which a reduction in seal damage and wear, and an increase in seal life, concurrent with a low leakage rate, is required. It is especially suitable for such difficult applications as those involving frequent transients, critical operations, toxic or hazardous fluids, aseptic requirements, or wide operating ranges.

9. ACKNOWLEDGEMENT

The control system and control strategy described in this paper were developed by J. Kozlowski and A. L. Miller of the Borg-Warner Research Center.

REFERENCES

(1) NAU, B.S., "Hydrodynamic lubrication in face seals," Proceedings 3rd International Conference on Fluid Sealing, 1967, BHRA Fluid Engineering, Cranfield, Paper E5.

(2) METCALFE, R., "The use of finite element deflection analysis in performance prediction for end face seals," Proceedings 3rd Symposium on Engineering Applications of Solid Mechanics, 1976, University of Toronto, Toronto.

(3) SALANT, R. F. and KEY, W.E., "Development of an analytical model for use in mechanical seal design," Proceedings 10th International Conference on Fluid Sealing, 1984, BHRA Fluid Engineering, Cranfield, Paper G3.

(4) SALANT, R.F. and KEY, W.E., "Improved mechanical seal design through mathematical modelling," Proceedings 1st International Pump Symposium, 1984, Texas A&M University, pp. 37-46.

(5) ETSION, I., "Dynamic analysis of noncontacting face seals, Journal of Lubrication Technology, 1982, 104, 460.

(6) DOUST, T.G. and PARMAR, A., "Transient thermoelastic effects in a mechanical face seal," Proceedings 11th International Conference on Fluid Sealing, 1987, BHRA Fluid Engineering, Cranfield, Paper F4.

(7) METCALFE, R., "Dynamic tracking of angular misalignment in liquid lubricated end-face seals," ASLE Transactions, 1981, 24, 509.

(8) METCALFE, R., "Performance analysis of axisymmetric flat face mechanical seals, Proceedings 6th International Conference on Fluid Sealing, 1973, BHRA Fluid Engineering, Cranfield, Paper D1.

(9) SALANT, R.F., MILLER, A.L., KAY, P.L., KOZLOWSKI, J., KEY, W.E. and ALGRAIN, M.L., "Development of an electronically controlled mechanical seal," Proceedings 11th International Conference on Fluid Sealing, 1987, BHRA Fluid Engineering, Cranfield, Paper H3.

Paper III(ii)

Micro-elastohydrodynamic lubricant film formation in rotary lip seal contacts

A. Gabelli

In this paper solution is given to the lubricant fluid film developed under rotary lip seals. The study is carried out considering the less favourable hypothesis of a sealing junction formed by two parallel sliding surfaces. A mathematical model is developed which treats the hydrodynamic pressure required for the generation of a lubricant film as a combination of two separate components. Firstly, the Micro-EHD action originated by the asperities' summits in virtual contact is analysed; secondly, the hydrodynamic pressure generated by the wavy morphology at the sealing interface is studied. The Micro-EHD fluid film developed between colliding asperities is evaluated using up-to-date forms of fluid film calculation of elliptical contacts under isoviscous elastic lubrication conditions. The real area of contact and real pressure originated between the two rough surfaces in match is solved by applying a generalised form of the Greenwood-Tripp stochastic model for contact of two rough surfaces. Results of parametric computations using present algorithms indicate the bounds of validity of the developed theory and its performances when compared with asperity based models for rotary seal lubrication.

1 INTRODUCTION

Rotary lip seals are highly compliant bodies radially loaded in order to adhere perfectly to their sealing counterface, see Figure 1. When second

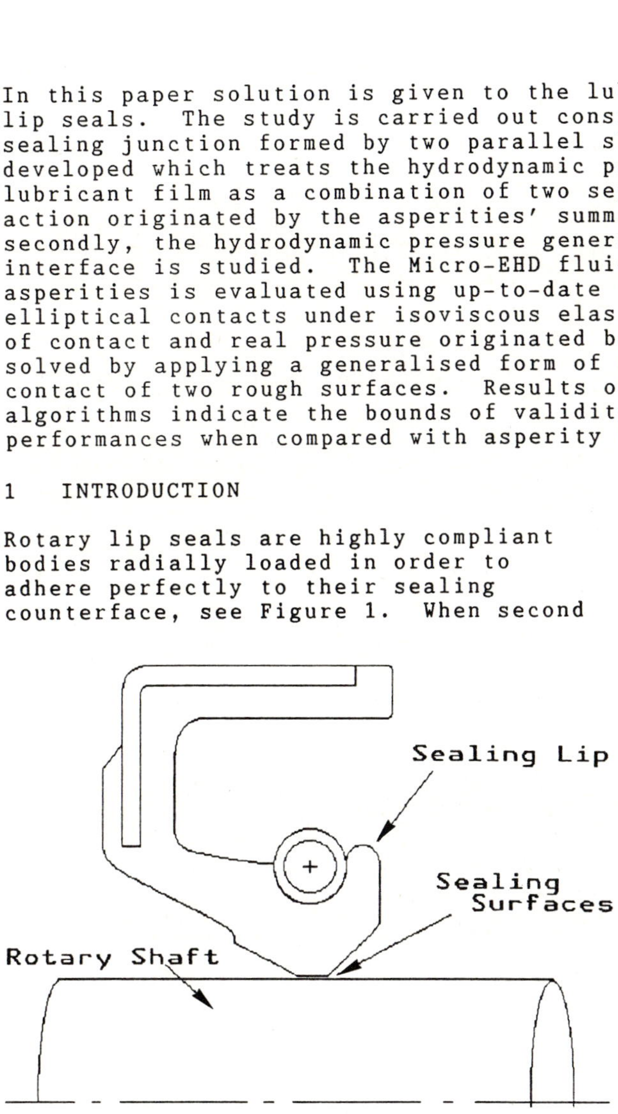

Fig.1 Typical Garter Lip Seal for Rotary Shaft Application.

order effects are neglected, one might associate the geometry of the sealing joint with two surfaces ideally flat and parallel. Within the bounds of this hypothesis the kinematics of the sealing joint is reduced to the parallel sliding of two nominally flat surfaces. This assumption, although reducing the complexity of the problem, also introduces less favourable conditions for the formation and maintenance of a lubricant film. Indeed, under the condition of parallel sliding, the traditional form of the Reynolds' equation fails to predict lubricant film formation. This conclusion, however, is contradicted by the common experience from seal lips and by some new experimental results [1]. Indeed, the latest experiments have shown that these types of seals operate with different regimes of lubrication, varying from simple mixed to full film lubrication.

The most recent works on this subject [1,2] give strong support to the idea of film formation arising from a combined Micro-EHD and Hydrodynamic (HD) action of the surface asperities interacting at the sealing interface. This approach to the problem is by no means new, and indeed many investigators have focused their attention on this basic concept in the past.

One of the formal treatments of the load-carrying capabilities of parallel sliding of rough surfaces is due to Davies (1961), [3]. Davies models the surface asperities as ideal, constant size, one-dimensional triangular shapes. In Davies' approach, the asymmetric deformation of the asperities due to the antisymmetric pressure distribution of the hydrodynamic pressure provides a positive imbalance in the pressure field which leads to the development of a net load support capacity between the two surfaces.

In 1966, Hamilton et al. [4], developed a theory to explain the hydrodynamic pressure found in mechanical face seals working under parallel sliding lubrication conditions. Hamilton considers rigid asperities of simple cylindrical shape. In this model the asperities' dimensions are treated as random statistically distributed quantities. A similar, though simpler, approach was followed by Anno et al. (1968), [5] [6] and Kojabashian and Richardson [7]. In this last work, further simplifications were introduced by considering the minimum film thickness as a constant quantity for all asperities.

In the above models the EHD action of colliding asperities was disregarded. It was not until 1975 that Tsao and Tong [8] introduced the idea of Micro-EHD into the lubrication of parallel surfaces, using a fairly elementary model to allow asperities to get into contact and to deform. Roughness, however, was represented by an oversimplified geometry consisting of a short cylinder with a spherical top.

Lebeck [9,10] recently carried out a comprehensive review of lubrication models for flat surfaces in parallel sliding. Lebeck's study was directed towards the evaluation of both experimental and theoretical works on this subject. His conclusion was that none of the existing theories can fully and satisfactorily explain the behaviour of surfaces in parallel sliding as observed in practice.

In the present study, a fresh approach to the problem is attempted by analysing the combined effect of the Micro-EHD and HD contributions to the load-carrying capacity of the sliding surfaces as a function of their average separation. In both Micro-EHD and HD analysis, the random nature of the surface roughness is introduced using a statistical probabilistic approach, and therefore the analysis is not restricted, as in previous work, to a particular type of morphology of the sealing surfaces.

1.1 Notation

A_o	Apparent nominal unit area (mm^2)
A_c	Area of cavitation (mm^2)
A_h	Area of the hydrodynamic field (mm^2)
A_r	Real area of microcontact (mm^2)
D	Correlation distance (m)
E	Modulus of elasicity (MPa)
E′	Composite plane strain modulus of elasticity for two contacting solids (MPa)
F_r	Real force per microcontact (N)
$\mathbf{F}_n(\underline{h})$	Microcontact Expectancy Function
$f(z)$	Probability Density Function
h	Lubricant film thickness (μm)
h_o, h	Separation between mean planes (μm)
$\underline{h}$	Dimensionless separation
$\mathbf{m}_i$	Roughness Spectral Moments
$G(\omega)$	Power Spectral Density Function
$\underline{p}$	Reference pressure (MPa)
p_c	Cavitated pressure (MPa)
p_h	Hydrodynamic pressure (MPa)
p_{ij}	Nodal pressure component (MPa)
p_r	Apparent contact pressure (MPa)
P_r	Real pressure per contact (MPa)
N_r	Number of microcontacts
$\mathbf{R}_x(\lambda)$	Auto Correlation Function
u_x	Absolute sliding speed (m/sec)
V	Viscosity speed product (N/m)
x,y,z	Coordinate system of the surface
α	Bandwidth parameter
β	Average asperity radius (μm)
η	Density of surface asperities (mm^{-2})
η_o	Dynamic viscosity ($Nsec/m^2$)
λ	Correlation length (m)
λ_x, λ_y	Roughness wave length (μm)
Λ	Dimensionless mean film thickness
μ	Friction coefficient
σ_s, σ	Variance of height distribution (μm)
τ	Shear stress (MPa)
ω_1, ω_2	Roughness spatial frequencies

Indices

h	Hydrodynamic model
r	Roughness contact model
o	Apparent, Nominal
min	Minimum
cen	Central
1,2	Seal, Counterface.
x,y,z	Direction

2 SURFACE ROUGHNESS PARAMETERS

The texture of the sealing surfaces is of complex three-dimensional random geometries, which characterisation requires a statistical rather than a deterministic approach. This gives a greater generality to the solution and limits the number of parameters required to describe a particular case. For our problem, only two functions, namely, the height distribution and the Auto-Covariance Function (ACF), are sufficient for the full statistical characterisation of the surface geometry; see Appendix 1.

The height distribution function has been studied extensively by Williamson [11]. He found that for surfaces finished by an abrasive or a generic cumulative removal process the height distribution is basically Gaussian. Single point and extreme-value processes lead to an only partially Gaussian distribution. This is the case of running-in surfaces where the height distribution is slightly skewed from the simple Gaussian shape. However, peak heights in all cases follow a Gaussian distribution. For the problem of characterisation of sealing surfaces, due to the motion of the sliding surfaces during the wear process, statistical independence is most likely maintained with a resulting texture similar to the one obtained from a fully random abrasive process.

In the case of an anisotropic roughness, nine bispectral moments, defined in Appendix 1, Eq.(4), are required for the full description of the surface. In most practical cases, however, we deal with isotropic or elliptical anisotropic surfaces. This requires only the zeroth, second and fourth spectral moments m_0, m_2, m_4, of the surface profile to get the full statistical characterisation of the roughness texture.

In the definition of the spectral moments of the surface, see Eq.(4) of Appendix 1, ω_1 and ω_2 are the lower and upper spatial frequencies of the roughness profile. These frequencies, which ideally should cover an infinite spectrum, are in reality confined to a finite interval given by the characteristic passband of the measurement system and also by the standard upper and lower cut-off points of the frequencies band admitted to the measurement process. Note that in absence of natural constraints to guide the filtering of the roughness bandwidth, the choice of these frequency limits should be derived from the physical behaviour of the contact following recent development in the concept of functional filtering, [13, 14].

3 MICROCONTACT MODELLING

When two nominally flat rough surfaces come into contact, their separation is characterised by the distance h_o of their reference planes, i.e., the mean plane of the summits' height distribution. It follows that there will be contact at those asperities whose total heights (z_1+z_2) are greater than the characteristic separation h_o, see Figure 2.

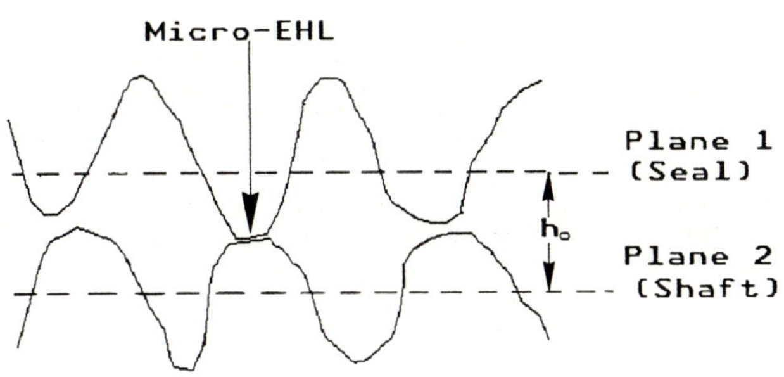

Fig.2 Schematic Representation of the Contact of Two Rough Surfaces.

The problem of prediction of the real area of contact and real pressure developed at the contacting spots of two meshing surfaces has been investigated by many authors, [15-19]. Most of the published work concentrates on the case of a rough surface contacting a perfectly smooth plane. However, no unique theory of general applicability has yet emerged. In the present analysis, the computation of load, pressure and geometry of the contacting spots generated in the sealing interface is carried out Greenwood-Tripp theory [19]; see Appendix 2. The GT model can be regarded as an extension of the well established Greenwood- Williamson analysis of contact of a rough surface, [15,16]. Indeed, the GT microcontact model generalises the GW theory to the more common situation in which both matching surfaces are rough.

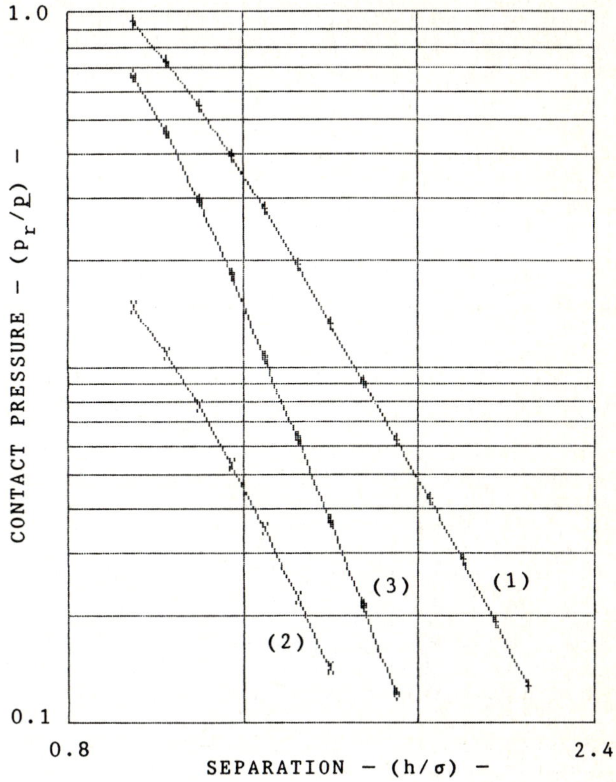

Fig.3 Relationship Between Separation and Apparent Pressure for Various Micro Contact Hypotheses.
(1) Equivalent Surface Against a Plane.
(2) Two Rough Surfaces, (GT Model).
(3) Single Rough Surface Against a Plane

In Figure 3 a comparison between different surface microcontact hypotheses is presented. This test case was performed using roughness parameters of a typical sealing contact and a composite plane strain modulus of elasticity of 18 MPa. The results are similar to those reported in [19]. Curve (1) shows the behaviour of an "equivalent" rough surface against a plane. An "equivalent" roughness is defined as a surface which combines the roughness properties of the two surfaces in match. Figure 3 shows that the load-carrying capacity of such microcontact hypotheses is significantly higher than that obtained by considering both surfaces in contact as rough, i.e., the GT model indicated by curve (2). This is due to the perfect and simultaneous, and therefore "non-genuine", matching of the asperities tips that such hypothesis implies. When two separate rough surfaces are considered according to the GT formulation, curve (2), the natural, real misalignment between the different asperity centres is allowed and thus less load-carrying capacity results. Note that in both cases the stiffness of the coupling remains unchanged, whereas in the case of a single rough surface against a plane, shown by curve (3), the stiffness is obviously noticeably increased due to the change of the geometrical ratio (σ/β) and perfect matching characteristics that such contact hypothesis implies.

All roughness parameters required by the GT model can be obtained from the three characteristic spectral moments m_0, m_2 and m_4 of the combined roughness of the two surfaces, [20], according to:

$$\eta = \frac{m_4}{m_2\ 6\sqrt{3}\ \pi} \qquad \beta = \frac{3}{8}\left(\frac{\pi}{m_4}\right)^{1/2} \tag{3.1}$$

and introducing the bandwidth parameter $\alpha = (m_0 m_4)/m_2^2$, we have:

$$\sigma_s = \left(\frac{m_0(\alpha - 0.8968)}{\alpha}\right)^{1/2} \tag{3.2}$$

Note the difference between the variance of the summit height distribution (σ_s) used in the GT model and the standard deviation of the height z relative to the surface mean plane (σ), i.e., the usual r.m.s. value of the surface height. This last parameter is the most commonly used roughness index and is simply defined as the square root of the zeroth spectral moment of the surface roughness: $\sigma = \sqrt{m_0}$

4 MICRO-ELASTHYDRODYNAMIC FILM

The term Micro-Elastohydrodynamic (μ-EHD) lubrication, introduced by Christensen in 1967 [21], deals with the local pressure and film thickness fluctuations developed at the contacting spots of colliding asperities. The present analysis differs basically in two respects from previous works [22-24]. Firstly, we deal with an Elastic Iso-Viscous (EIV) lubrication regime; secondly, the kinematics of the contact involves pure sliding motion with no rolling. Note that in the case of rolling contacts, the squeeze-film term, due to the normal approach of the asperity tips onto the opposing surface, is the main contributor to the fluid film build-up, [22,23].

In the present analysis the characteristics of the Micro-EHD film developed in the sealing junction are studied by applying new results obtained by Chittenden, Dowson and Taylor [24]. Chittenden studied the case of elliptical conjunctions in the isoviscous elastic regime with entrainment directed along either principal axis. They combined the seventeen results obtained by Hamrock and Dowson [25,26] with the thirty results of entrainment directed along the major axis and six new results for entrainment along the minor axis. Least square regression applied to this extended set of data allowed the derivation of new formulae for the calculation of film thickness in the EIV regime with entrainment directed along either principal axis.

The application of the results [24] to the problem at hand provides the expression of the minimum film thickness formed under the asperities in virtual contact. The mean shearing stress generated over the macrocontact area by the μ-EHD film may also be readily calculated from:

$$\bar{\tau}_r = \frac{A_r}{A_o}\left(\frac{\eta_o\, u_x}{h_{cen}}\right)$$

$$= \pi^2(\eta\ \beta\ \sigma_s)^2 \cdot F_2(\underline{h})\left(\frac{\eta_o\, u_x}{h_{cen}}\right) \tag{4.1}$$

where h_{cen} is the central film thickness under the asperities, as in [24].

5 HYDRODYNAMIC ANALYSIS

According to the hypothesis of our problem, the reference planes of the two sliding surfaces are parallel. However, the geometry of the gap separating these two planes is not constant but is related to the topography of the roughness of the two surfaces in contact, see Figure 2. Herewith we focus our attention on the pressure field and thereby the load-carrying capacity generated between the two parallel sliding surfaces by virtue of the wavy morphology of the surfaces.

5.1 Hypotheses

Let us consider a fully flooded hydrodynamic film having a shape, as far as the global hydrodynamic behaviour is concerned, "equivalent" to the one separating the two matching surfaces of the GT theory. The main geometrical features characterizing such a film are:

- the two wavelengths λ_x and λ_y of the roughness texture in the x and y direction;
- the separation distance $\bar{h}$ between the two mean planes of the two surfaces;
- the combined standard deviation of the surface summit height distribution σ relative to the mean plane.

From this limited set of parameters an HD equivalent shape can be derived for the lubricant film which, while providing a fairly good representation of the actual fluid film found in the sealing contact, will also give an easy to use mathematical formulation. Presently, the chosen form of film thickness is:

$$h(x,y) = \bar{h} + \sigma\cos(2\pi x/\lambda_x)\cdot\cos(2\pi y/\lambda_y) \tag{5.1}$$

In Figure 4 a Scanning Electron Microscope (SEM) immage of the contacting surface of a rolling bearing lip seal is shown. The wear band of the seal is about 150 µm wide and on the surface the wavy structure of the seal roughness is clearly visible. The lubricant film developed under such a surface texture is analytically described in the present model by the undulated shape shown in Figure 5, which is a graphical representation of the function (5.1). The form of this wavy volume is dictated by the roughness parameters of the two surfaces in match. Indeed, in the case of unequal wavelength in the x and y direction, the geometry of the lubricant film would change, as shown by Figure 6.

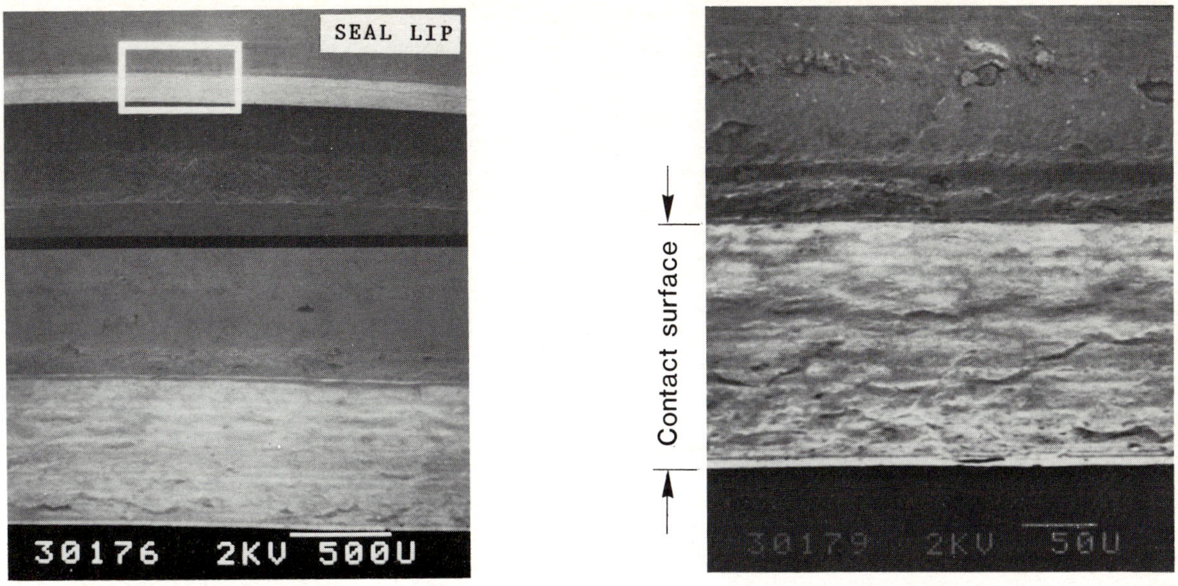

Low magnification

High manification

Fig.4 Scanning Electron Microscope (SEM) Photograph of Typical Contact Surface of Rolling Bearing Lip Seal.

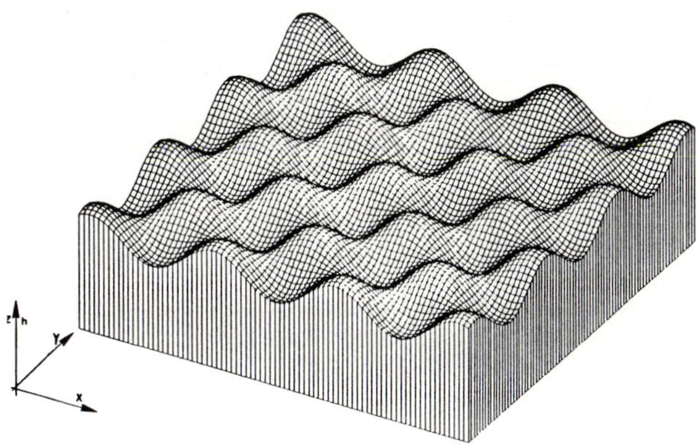

Fig.5 Example of Assumed Shape of the Hydrodynamic Fluid Film of the Seal Lip.

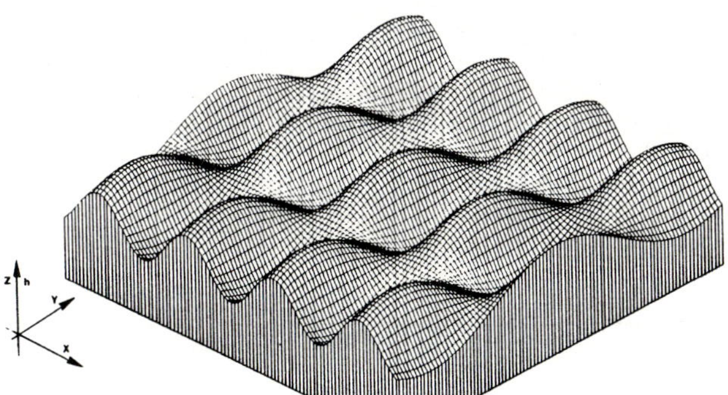

Fig.6 Example of Assumed Shape of the Hydrodynamic Fluid Film in Case of Anisotropic Roughness.

5.2 Numerical

Let us consider a laminar flow of an incompressible Newtonian fluid of constant dynamic viscosity η_o. The Reynolds' equation governing the pressure field in the gap h(x,y) present between the two sliding surfaces is then:

$$\frac{\partial}{\partial x}\left(h^3 \frac{\partial p}{\partial x}\right) + \frac{\partial}{\partial y}\left(h^3 \frac{\partial p}{\partial y}\right) = 6\ \eta_o\ u_x\ \frac{\partial h}{\partial x} \qquad (5.2)$$

Integration of this differential equation was performed numerically using standard finite difference algorithms. The pressure field was discretized over a set K = kn * (2m+1) meshpoints, and since p is given at the boundary $y = -(1/2)\lambda_y$ and $y = +(1/2)\lambda_y$ the problem was reduced to the solution of a set of K linear equations. Moreover, because of the periodic characteristics of the problem, a small number of waves was sufficient for the full analysis. In our computations two wavelengths were used in the numerical integration. In order to accelerate the convergence process a Successive Over-Relaxation (SOR) algorithm was applied in the Gauss-Seidel iterations scheme. Further improvement of the convergence speed was achieved by using an optimised over-relaxation factor.

Cavitation regions were treated by applying the JFO boundary conditions for the separation and restoration of the fluid field. Pressure in the cavitation region was set equal to the vaporisation pressure of oil p_c.

Integration of the nodal pressure over the field provides the amount of load-carrying capacity of the two rough surfaces in relative motion. Its mean value per unit of apparent nominal area is calculated according to:

$$\bar{p}_h = \frac{1}{A_0} \iint_{A_h} p_{ij}(x,y)\ dx\ dy \qquad (5.3)$$

From knowledge of the pressure distribution, the mean shear stress due to the hydrodynamic field acting at z=h is readily obtained as:

$$\bar{\tau}_h = \frac{1}{A_o} \iint_{A_h} \left(\frac{\partial p}{\partial x} \frac{h}{2} + \frac{u_x}{h} \eta_o \right) dx\, dy \tag{5.4}$$

Note that in Eqs. (5.3) and (5.4) A_h is the area where $p > p_c$ or more precisely: $A_h = (A_o - A_c - A_r)$

6 GLOBAL CHARACTERISTICS

The overall lubrication properties of the two parallel sliding surfaces are obtained by simply adding the μ-EHD and HD contributions computed over the relative area fraction of the apparent surface where either μ-EHD or HD lubrication takes place. Furthermore, in order to work with consistent and homogeneous quantities, the coupling of the results has to be carried out by applying proper transformations between the roughness parameters used in the two lubrication models. In other words, relationships should be provided between the dimensionless surface separation relative to the mean summit height plane $\underline{h} = h_o/\sigma_s$, as used in the μ-EHD model, and the dimensionless separation relative to the surface mean plane $\Lambda = h/\sigma$, as used in the HD model. Proper translation of these two expressions of surface separation is obtained immediately when the location of the summits' mean plane $\underline{z}$ [21] is known,

$$\underline{z} = \left(\frac{16 m_0}{\pi \alpha} \right) \tag{6.1}$$

thus:

$$\Lambda = \left(\underline{h} \sqrt{\alpha - 0.8968} + 2.257 \right)^{1/2} \cdot \alpha^{-1/2} \tag{6.2}$$

These expressions show that the relation between $\underline{h}$ and Λ depends only on the bandwidth parameter α. Therefore, in the case of high values of the bandwith parameter, i.e., $\alpha = 10\ \underline{h}$, results in $\underline{h} \simeq \Lambda$. However, for values of roughness usually found in lip seals, α has a mean value of about $5\underline{h}$ which leads to the following transformation between $\underline{h}$ and Λ: $\underline{h} \simeq \Lambda - 1$.

The use of this correspondence between the two definitions of surface separation allows us to carry on the integration of the results derived from the two modes of load generation.

The total specific load-carrying capacity for the coupled surfaces is thus:

$$\bar{p}_{tot} = \bar{p}_r + \bar{p}_h \tag{6.3}$$

Similarly, the total mean shear stress is:

$$\bar{\tau}_{tot} = \bar{\tau}_r + \bar{\tau}_h \tag{6.4}$$

It follows that the global coefficient of friction of the seal lip is the ratio:

$$\bar{\mu}_{tot} = \bar{\tau}_{tot}/\bar{p}_{tot} \tag{6.5}$$

It may be useful to consider also the two partial coefficients of friction which are obtained considering the two separate contributions to the load support given by the μ-EHD and simple hydrodynamics:

$$\bar{\mu}_r = \bar{\tau}_r/\bar{p}_r \quad \text{and} \quad \bar{\mu}_h = \bar{\tau}_h/\bar{p}_h \tag{6.6}$$

7 NUMERICAL COMPUTATIONS

Previously developed algorithms for the lubrication of parallel sliding surfaces were applied on four different types of composed roughess textures. The roughness parameters used in the calculation are given in Table 1. The surface textures were chosen to be representative of typical values found in lip seal contacts. Specifically, the rougnness denoted with (A) is of a very smooth type which might be found in low pressure sealing contacts. (B) and (C) are typical roughnesses encountered in rotary lip seals for standard application, while surface (D) is of a much rougher contact typically found in highly loaded garter seals. In the computations, the effect of changes of the coefficient of elasticity E_1 was evaluated using ranges of values found in rubbers for lip seals. In the hydrodynamic part of the numerical simulation, the study was extended to the effect of the viscosity and speed parameters. The extensive parametric study was carried out in a numerical manner. Therefore only a selection of the most significant results are hereafter presented and discussed in detail.

– **Table 1** –

Roughness Parameters Used for Simulation of Contacting Surfaces.

TYPE	σ [μm]	β [μm]	η[mm^{-2}]	σ/β
A	0.417	120.0	1100	.0034
B	0.667	90.0	916	.0074
C	1.166	60.0	785	.0194
D	2.666	30.0	687	.0888

7.1 Microcontact Results

It is interesting to evaluate the response of the rubber-steel microcontact model in terms of apparent load-carrying capacity against surface separation, i.e., average film thickness of the sealing contact.

The normalised apparent load given by the contacts (A) and (D) is presented in Figures 7 and 8. The curves are related to different values of the composed coefficient of elasticity, varying from 3 MPa to 18 MPa. Note that the dimensionless separation of the two surfaces is given here in terms of dimensionless mean film thickness Λ in order to have homogeneous quantities as discussed in Section 5. The results of

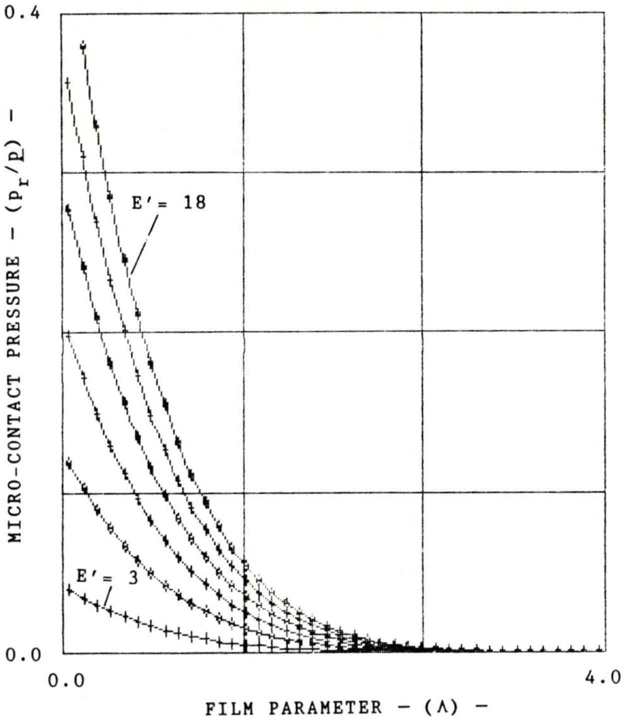

Fig.7 Relationship Between Nominal Micro Contact Pressure (p_r) and Lubricant Film Parameter (Λ). - Surface Type (A) -

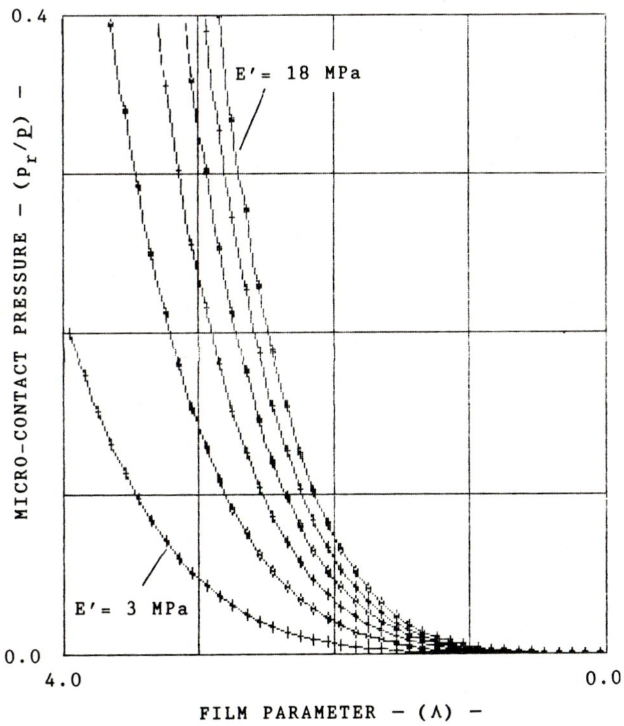

Fig.8 Relationship Between Nominal Micro Contact Pressure (p_r) and Lubricant Film Parameter (Λ). - Surface Type (D) -

both smooth (A) and rough (D) surfaces show, as expected, that a higher value of the modulus of elasticity corresponds to an increased stiffness of the asperities and this is correlated to a consistent increased load-carrying capability for the approach of two surfaces.

A second interesting result of the computations is the increase of load-carrying capability given by the rough contact (D) when compared with the smooth contact (A) for an equal amount of separation (Λ) of the two surfaces. This result is not immediately explainable. Indeed, in the case of the rough contact (D) fewer asperities are in contact compared with the case of the smooth contact (A), see Eq.(5), Appendix 1. However, as shown by Eq.(7) of Appendix 1 and discussed in [13], the stiffness of the asperities increases in the case of contact (D) because also the ratio (σ/β) increases for a rougher surface. The slope parameter (σ/β) thus dominates the stiffness of the contact and the load support achieved for a given separation of the two surfaces.

Additional information obtained by the microcontact model is the prediction of the minimum film thickness developed between colliding asperities in conditions of μ-EHD lubrication. The case presented here is related to surface C which is a good example of roughness found in rolling bearing seals. The curves shown in Figure 9 are a com-

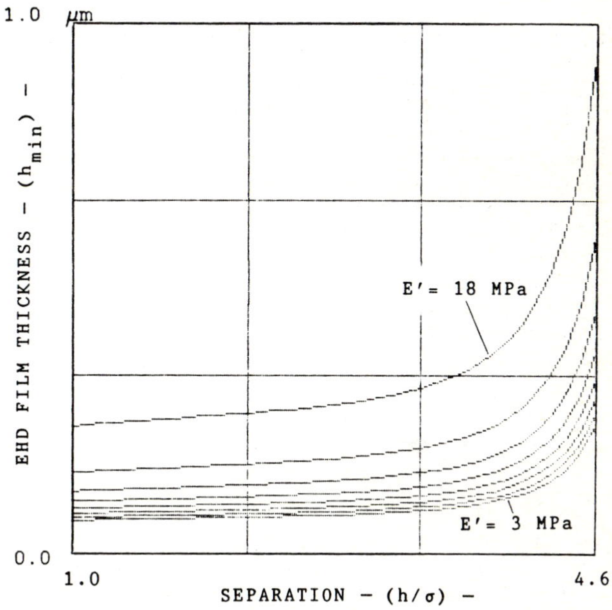

Fig.9 Relationship Between Minimum EHD Film Thickness and Surface Separation. Elastic Coeff. E′= 3+18MPa; U=10mm/s

bination of results obtained for values of the elastic modulus of the rubber materials, varying from 3 to 18 MPa. These results concern a viscosity of 65.38 cPoise and a speed of 10 mm/sec. A speed increase of up to 100 mm/sec leads to a distinct increase in the expected values of the minimum film thickness, as shown in Figure 10. Note also the dominant effects of the elasticity parameter in the film formation behaviour of the contact. In this respect it is interesting to note that localised hardening of the rubber due to temperature effects at the μ-EHD contacts in addition to repeated cycling deformation of the asperities' tips, caused by asperity-asperity collisions, seems to strengthen the understanding and to provide a physical explanation of the fatigue wear process present in elastomeric sliding surfaces under conditions of mixed lubrication.

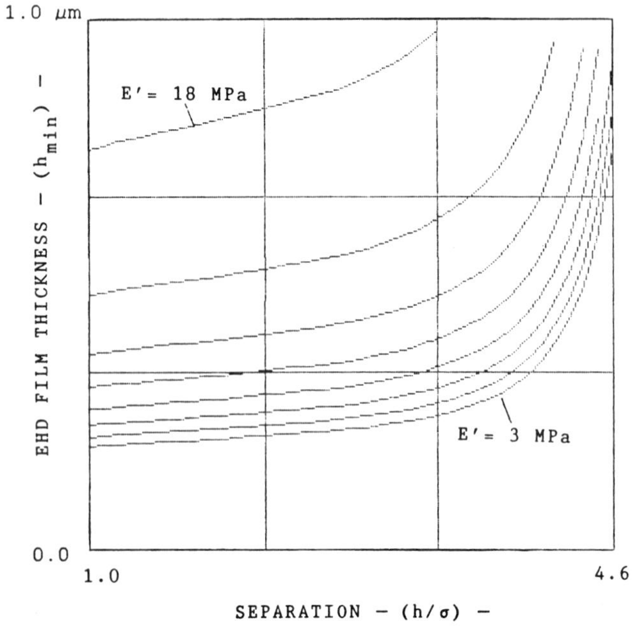

Fig.10 Relationship Between Minimum EHD Film Thickness and Surface Separation. Elastic Coeff. E′= 3+18MPa; U=100mm/s

7.2 Hydrodynamic Results

Parametric evaluation of the hydrodynamic behaviour of the parallel sliding of sealing contact was carried out using the same data set as applied for the microcontact model, Table 1.

The results of the HD unit loading of the sealing contact as a function of the dimensionless film thickness are presented in Figures 11 and 12. These

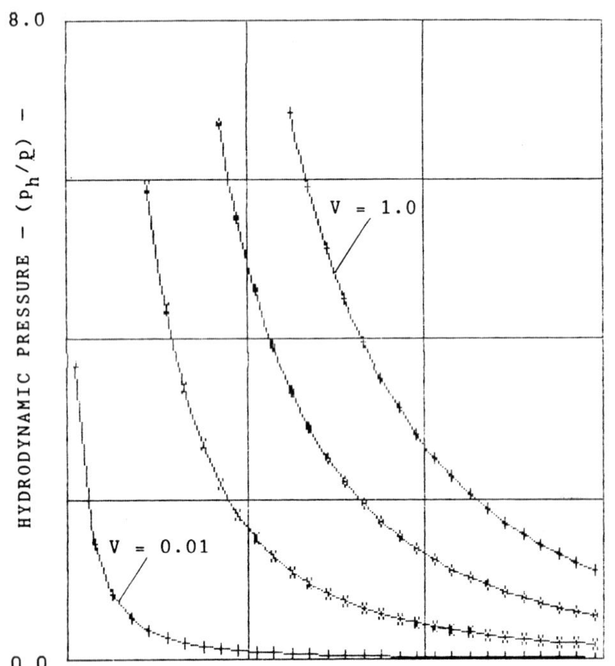

Fig.11 Relationship Between Hydrodynamic Pressure (p_h) and Lubricant Film Thickness (Λ). V=0.01+1.0 Surf.Type (B)

results concern two types of roughness, (B) and (C), and combine four different values of the speed-viscosity parameter

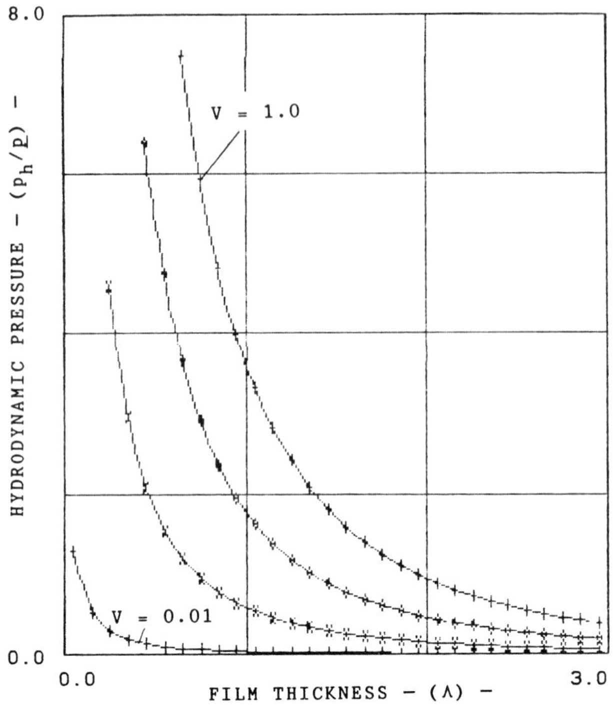

Fig.12 Relationship Between Hydrodynamic Pressure (p_h) and Lubricant Film Thickness (Λ). V=0.01+1.0 Surf.Type (C)

(V = 1.0-0.01). As expected, these results show that the load support is directly proportional to the viscosity-speed parameter while inversely related to the dimensionless film thickness interposed between the two sliding surfaces. The main feature of the load support curves is their hyperbolic asymptotic descending trend which determines the uniqueness of the film thickness for a given nominal load of the seal.

A somewhat unexpected effect highlighted by the HD model is the trend shown by the load support characteristics of the system for an augmented roughness of the contact. Indeed, for the surface of increased roughness, (C), there is an apparently diminished capacity of load compared with the smoother surface, (B). Note that these behaviours are opposite to those obtained by the microcontact load support model.

Explanation for this behaviour may be found in the ratio $\Lambda = h/\sigma$ which diminishes for an increased value of rms rougness of the surfaces in contact. Therefore, a rougher surface requires a thinner lubricant film (Λ) in order to get the same load support as a smoother surface. In other words, the increased roughness causes a shift of the load curve towards the left side of the diagram.

The greatest sensitivity, also in this case as for the μ-EHD model, is to the speed viscosity parameter which for obvious reasons dominates the HD behaviour of the fluid film, as shown by Figures 11 and 12.

7.3 Global Load Support Results

The coupling of the two load generation models as described in Section 6 provides the global amount of load-carrying capacity of the two surfaces in contact. Besides the practical utility of these results, it is of interest to quantify the relative contribution given by the two lubrication modes in order to get a better perception of how the contact load of the sealing lip is shared between microcontact and hydrodynamic generated pressure.

Results combining the computation of surfaces (C), (D) and viscosity-speed parameter (V) of 0.01 and 1.0 are presented in Figures 13 and 14. In the diagram are plotted curves of the μ-EHD load and also of the resulting contribution due to the combination of μ-EHD and HD pressure, i.e., mixed load support. In Figure 13 it is apparent

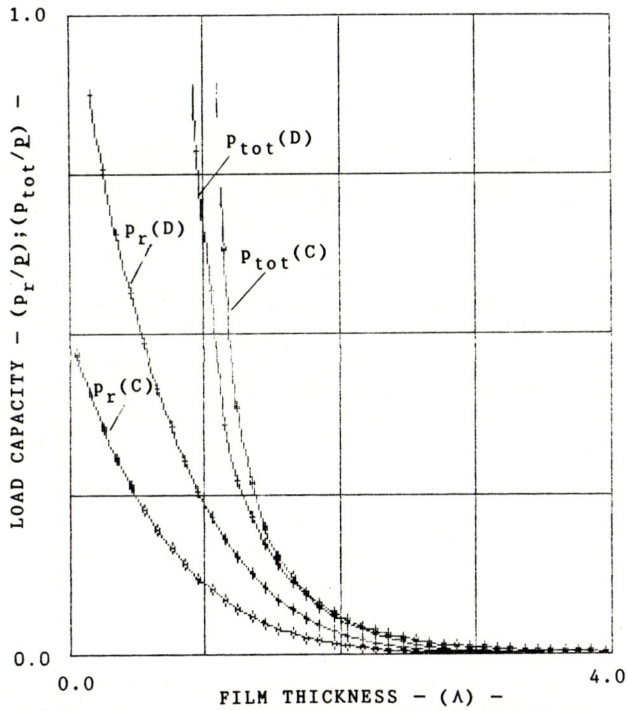

Fig.13 Relationship Between Load Carrying Capacity (p_r - p_{tot}) and Lubricant Film Thickness (Λ). V = 0.01 Surf.Ty.(D),(C)

that for low V values the effect of the asperities is similar to the load support provided by the hydrodynamic generation. This, however, changes dramatically for an increased value of V, Figure 14. In this case the load is almost entirely carried by the hydrodynamic action. In addition there is a noticeable increase in the global load-carrying capacity of the surfaces of about a factor 8.

Further increases in the global load-carrying capabilities of the surfaces are obtained for smoother surfaces (A) and (B), as shown in Figure 15. Here, the effect of the roughess works in the opposite direction and it provokes a further decline for the μ-EHD part of the load, whereas the HD part of the pressure further increases simultaneously.

8 DISCUSSION AND CONCLUSIONS

The modelling of the lubricant film developed under rotary lip seals offers some unique conceptual problems due to

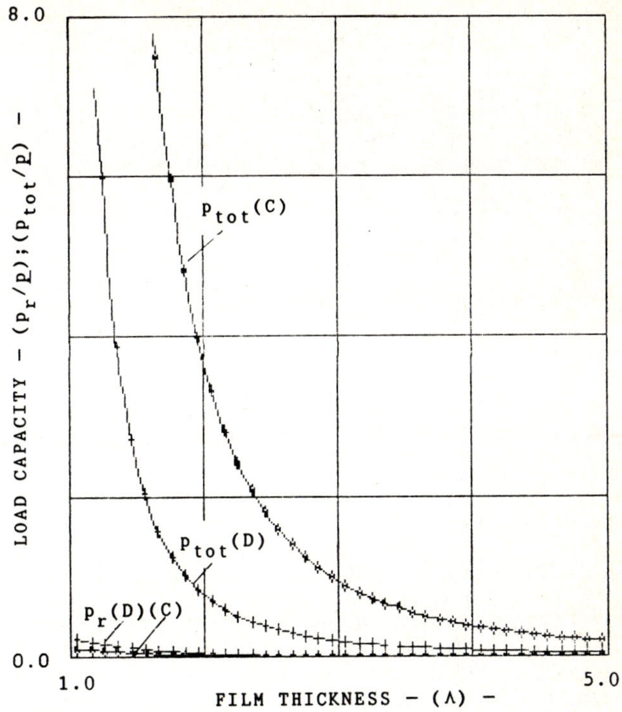

Fig.14 Relationship Between Load Carrying Capacity (p_r - p_{tot}) and Lubricant Film Parameter (Λ). V = 1.0 - Surf.Ty.(D),(C)

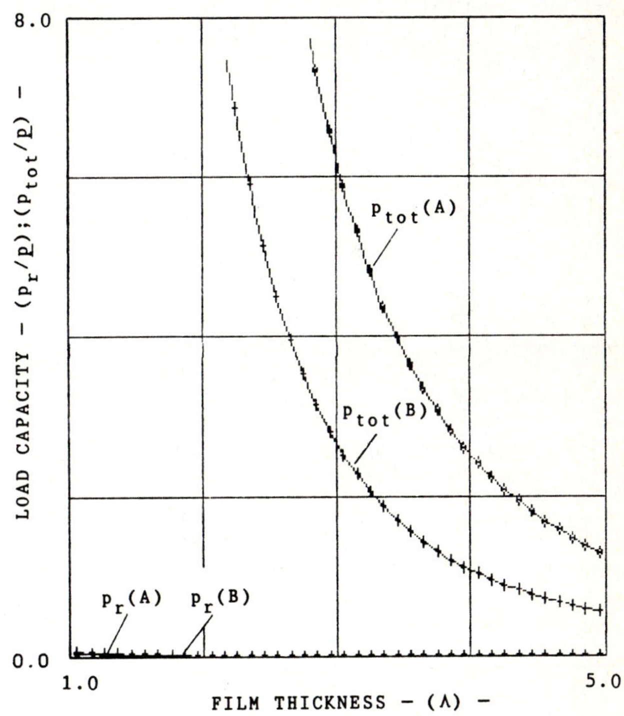

Fig.15 Relationship Between Load Carrying Capacity (p_r - p_{tot}) and Lubricant Film Parameter (Λ). V = 1.0 Surf.Ty. (A),(B)

the peculiar geometry of the contact and the nature of the rubber surface. In the study presented here the problem is reduced to a mixed lubrication case of two nominally parallel sliding surfaces. The parametric computations have demonstrated the capability and versatility of this approach in that it provides information on the effect of the surface roughness, stiffness of the material and nominal contact force of the rubber lip, on the characteristics of the fluid film developed at the sealing contact. This provides a convenient and efficient tool for the tribological evaluation of rotary lip seals, which development, up until now, has been mainly based on elastostatic principles, [27,28]. Additionally, the present solution offers the advantage of being easily adapted to permit its use for different design objectives. In particular, the numerical algorithms can be easily altered to allow the analysis of specific real roughness topographies and their effects on preferential lubricant flows and contaminant transports at the sealing interface.

Finally it should be mentioned that the validity of the model is constrained to a fairly limited rotary speed. Indeed, for increasing speed values the average lubricant film will also increase to the point that the hypothesis of parallel sliding on which this model is developed will gradually lose its validity. However, considering that elastic distorsions of the rubber surface will eventually ease the film formation, the estimation based on the present theory can be regarded as a simple conservative evaluation of the lubricant film formed in sealing contact. The extension to higher rotary speed requires the inclusion in the dynamic calculation of the macro-elastic effects of the rubber surface.

9 ACKNOWLEDGEMENT

The author wishes to thank Dr. I.K. Leadbetter, Managing Director of SKF Engineering & Research Centre, for sponsoring this work and for his kind permission to publish the results of this paper. The author is also indebted to Dr. B. Jacobson and Dr. J. Tripp for the many helpful discussions on the subject of this paper. Finally, the author wishes to thank Dr. E. Ioannides for his encouragement and support.

APPENDICES

APPENDIX 1: Roughness parameter definitions

The Probability Density Function (PDF) of the height distribution of a fully random surface is:

$$f(z) = (2\pi)^{-1/2} \cdot \exp(-z^2/2) \qquad (1)$$

where z is the height relative to the mean and normalized to σ.

Also fundamental in characterising the surface texture is the Auto-Covariance Function (ACF) of the surface profile [12]:

$$R_x(\lambda) = \lim_{L\to\infty} \frac{1}{L} \int_0^L Z(x)\cdot Z(x+\lambda)\cdot dx \qquad (2)$$

where λ is the correlation length, and Z(x) is the height distribution function along x and $Z(x+\lambda)$ is the height at an adjacent coordinate shifted by a length λ. For our case, as for most engineering surfaces, [12], the ACF takes on an exponential form which could be approximated by:

$$R_x(\lambda) = \exp(-\lambda/D) \qquad (3)$$

where D is the correlation distance. The Fourier transform of the Auto-Covariance Function is the Power Spectral Density Function (PSDF) $G(\omega)$ of the rough surface.

The spectral moments of the PSDF are the surface parameters which may be derived, in terms of the PSDF of the surface, using the following expression:

$$m_n = (2\pi)^n \int_{\omega_1}^{\omega_2} \omega^n \cdot G(\omega)\, d\omega \qquad (4)$$

APPENDIX 2: GT Microcontact model

In the GT model, the probability associated with a randomly selected summit of a height in excess of the two surfaces nominal separation h_o is provided by the integration of the Probability Density Function (PDF) of the summit heights. In other words, the expectancy of an asperity-asperity contact is given by:

$$F_n(\underline{h}) = \int_{h}^{\infty} (z-\underline{h})^n \cdot f(z)\, dz \qquad (1)$$

were $f(z)$ is the normalized or standard frequency density function of the asperity heights distribution. Note that $f(z)$ is not necessarily Gaussian. Indeed, Whitehouse and Archard [12] demonstrate that the summits' distribution is not of the fully Gaussian type but can be derived from an assumed Gaussian distribution of the roughness height according to:

$$f(\underline{h}) = \frac{\exp(-\underline{h}^2/2)}{4\sqrt{2\pi}} [1-\mathrm{erf}(\underline{h}/2)]^2 \qquad (2)$$

The apparent pressure acting on the microcontact is therefore:

$$\frac{\bar{p}_r(\underline{h})}{A_o} = \frac{16\sqrt{2}}{15} \pi \eta^2 \beta^{3/2} \sigma_s^{5/2} E' \, F_{2.5}(\underline{h}) \qquad (3)$$

The expected total area of contact per unit of apparent nominal area as a function of the dimensionless separation $\underline{h}$ is:

$$\frac{A_r(\underline{h})}{Ao} = \pi^2 (\eta \beta \sigma_s)^2 \cdot F_2(\underline{h}) \qquad (4)$$

The expected number of contact spots per unit of apparent nominal area A_o is:

$$\frac{N_r(\underline{h})}{A_o} = 4 \pi \eta^2 \beta \sigma_s \cdot F_1(\underline{h}) \qquad (5)$$

The expected mean real pressure per contact spot is:

$$P_r(\underline{h}) = \frac{4\sqrt{2}}{15} \beta^{1/2} \sigma_s^{3/2} E' F_{2.5}(\underline{h}) F_1^{-1}(\underline{h}) A_o^{-1} \tag{6}$$

The expected real force per contact spot is:

$$F_r(\underline{h}) = \frac{16\sqrt{2}}{15} \beta^{-1/2} \sigma_s^{1/2} E' F_{2.5}(\underline{h}) F_2^{-1}(\underline{h}) \tag{7}$$

REFERENCES

[1] OGATA, M., FUJII, T. and SHIMOTSUMA, Y., "Study on Fundamental Characteristics of Rotating Lip-type Oil Seals", Proc. of the 14th Leeds-Lyon Symposium on Tribology, 1987, pp.553-560.

[2] ISHII, T., MIYAGAWA, S. and TAGUMA, I., "Research and Development on the Hydrodynamic Pressure Generation Between Contacting Flat Surfaces", Trans. ASLE, Vol. 29,3, 1985, pp.339-346, .

[3] DAVIES, M.G. "The Generation of Lift by Surface Roughness in Radial Face Seals", International Conf. on Fluid Sealing, B.H.R.A., Cranfield, England, 1961.

[4] HAMILTON, D.B., WALOWIT, J.A. and ALLEN, C.M., "A Theory of Lubrication by Micro irregularities", J. of Basic Eng., Trans. ASME, Series D, Vol. 88, No.1, March 1966, pp.177-185.

[5] ANNO, I.N., WALOWIT, J.A. and ALLEN, C.M., "Micro-asperity Lubrication", J. of Lubrication Technology, ASME Ser.F, Vol.90,2, April 1968, pp.351-355.

[6] ANNO, J.N., WALOWIT, J.A. and ALLEN, C.M., "Load Support and Leakage from Micro-asperity-Lubricated Face Seals", J. of Lubrication Technology, 91, 1969, pp.726-731.

[7] KOJABASHIAN, C. and RICHARDSON, H.H., "A Micropad Model for the Hydrodynamic Performance of Carbon Face Seals", Third International Conf. on Fluid Sealing, B.H.R.A. Cambridge, England, 1967.

[8] TSAO, Y.H. and TONG, K.N. "A Model for Mixed Lubrication", ASLE Trans., Vol.18, No.2, 1975, pp.90-96.

[9] LEBECK, A.O., "Parallel Sliding Load Support in The Mixed Friction Regime. Part 1 - The Experimental Data." ASME J. of Tribology, Vol.109, 1986, pp.189-195.

[10] LEBECK, A.O., "Parallel Sliding Load Support in the Mixed Friction Regime. Part 2 - Evaluation of the Mechanisms." ASME J. of Tribology, Vol.109, 1986, pp.189-195.

[11] WILLIAMSON, J.B.P., "Topography of Solid Surfaces", Interdisciplinary Approach to Friction and Wear, NASA SP-181, 143, 1968.

[12] WHITEHOUSE, D.J. and ARCHARD, J.F. "The Properties of Random Surfaces of Significance in their Contact", Proc. of the Royal Society of London, 97, 1970.

[13] TRIPP, J.H., HOUPERT, L.G., IOANNIDES, E. and LUBRECHT, A.A., "Dry and Lubricated Contact of Rough Surfaces", Proc. Inst. Mech.E. International Conf., 1987, pp.71-80.

[14] THOMAS, T.R. and SAYLES, R.S., "Some Problems in the Tribology of Rough Surfaces", Tribology International 11, 1978, pp.163-168.

[15] GREENWOOD, J.A., "On the Area of Contact Between Rough Surfaces and Flats", ASME J. of Basic Engineering, Pap. 65-LUB-10, 1965.

[16] GREENWOOD, J.A. and WILLIAMSON, J.B.P., "Contact of Nominally Flat Surfaces", Proc. of the Royal Society, A, Vol.295, 1966, pp.300-319.

[17] ONIONS, R.A. and ARCHARD,J.F. "The Contact of Surfaces Having a Random Structure", J. of Physics, D:Appl. Physics,6, 1973, p.289.

[18] GREENWOOD, J.A. and TRIPP, J.H., "The Elastic Contact of Rough Spheres", ASME J. of Applied Mechanics, Vol. 89, 1967, pp.153-159.

[19] GREENWOOD, J.A. and TRIPP, J.H., "The Contact of Two Nominally Flat Rough Surfaces." Proc. Inst. Mech. Engrs., Vol.185, 1970, pp.625-633.

[20] BUSH, A.W., GIBSON, R.D. and KEOGH, G.P., "The Limit of Elastic Deformation in the Contact of Rough Surfaces", Mech. Res. Commun. 3, 1976, pp.169-174.

[21] CHRISTENSEN, H. "Micro-elastohydrodynamics", Chapter 4 of Tech. Rep. MTI 67TR23, May 1967.

[22] CHENG, H.S. "On some Aspects of Microelastohydrodynamic Lubrication", Proc. of the 4th Leeds-Lyon Symposium on Lubrication, April 1977, pp. 71-76.

[23] CHENG, H.S. "Micro-elastohydrodynamic Lubrication", Proc. 9th U.S. Nat. Congress of Applied Mechanics, ASME, June 1982.

[24] CHITTENDEN, R.J., DOWSON, D. and TAYLOR, C.M., "The Lubrication of Elliptical Conjunctions in the Elastic-Isoviscous Regime, with Entrainement Directed Along Either

Principal Axis", 13th Leeds-Lyon Symposium on Tribology, 1986.

[25] HAMROCK, B.J. and DOWSON, D. "Elasohydrodynanamic Lubrication of Elliptical Contacts for Materials of Low Elastic Modulus. Part I - Fully Flooded Conjunction", Trans. ASME, J. Lub. Technology 100/2, 1978, pp.236-245.

[26] HAMROCK, B.J. and DOWSON, D., "Elastohydrodynamic Lubrication of Elliptical Contacts for Materials of Low Elastic Modulus, Part II - Starved Conjunction", Trans. ASME, J. Lub. Technology, 101/1, 1978, pp.92-98.

[27] MEDRI, G.,STROZZI, A., BRAS, J. and GABELLI, A., "Mechanical Behaviour of Two Elastomeric Seals for Rolling Bearing Units", Proc. of the 10th International Conf. on Fluid Sealing, Innsbruck, 1984.

[28] GABELLI, A., "Analytical Modelling of load Deflection Characteristics of Elastomeric Lip Seals", Proc. of the 11th International Conf. on Fluid Sealing, Cannes, 1987.

Paper III(iii)

Lubrication of reciprocating seals: Experiments on the influence of surface roughness on friction and leakage

A. F. C. Kanters and M. Visscher

In the past, most theoretical work on reciprocating seals has been dedicated to the solution of the elastohydrodynamic lubrication (EHL) problem for smooth surfaces. However, measured friction forces and film thicknesses often do not support the hypothesis of full film lubrication. In the present paper, a first attempt is made to investigate the lubrication by experimental determination of friction forces and leakage using three rods with different surface roughness. An accurate method for leakage measurement has been developed because of the lack of accurate film thickness tranducers.

1 INTRODUCTION

The literature on reciprocating elastomeric seals exhibits many experimental data on their frictional behaviour. Regarding the dependency of the friction force at constant fluid pressure on the velocity or on the product of viscosity and velocity, it appears that reciprocating seals exhibit the characteristic behaviour depicted in fig. 1.

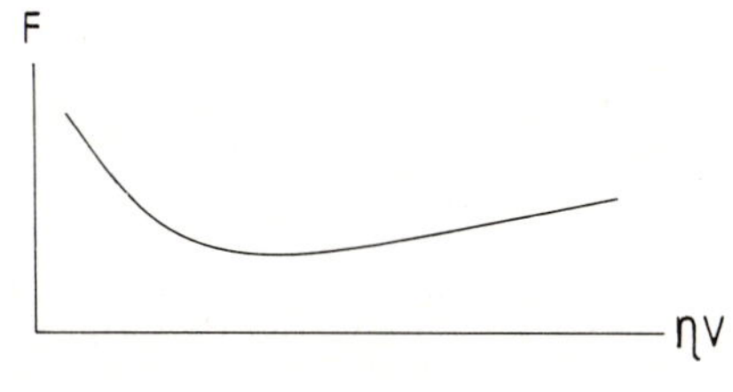

Fig. 1 Characteristic friction curve.

This curve is like a Stribeck curve. For constant viscosity it appears that with increasing speed the lubrication mode of the seal changes from boundary lubrication, via mixed lubrication, to full film lubrication. It has to be noted that it will depend on the specific seal design and experimental conditions whether the entire curve of fig. 1 will be obtained or not during an experimental program.

The tribological process in the seal contact constitues a classic elastohydrodynamic lubrication (EHL) problem if the seal is fully lubricated and if the influence of surface roughness is negligible. This problem is merely complicated by the large seal deformations and material behaviour. Most theoretical work on reciprocating seals has been devoted to the solution of the smooth EHL problem, e.g. (1),(2),(3),(4).

Yet in the early seventies Field and Nau obtained results which makes one reflect on the lubrication of the seal contact. From experiments on annular rubber rings of rectangular cross-section (5) they found that friction was of other than hydrodynamic origin. Nevertheless full lubricant films were observed, which would give rise to friction values by viscous shear at least an order of magnitude smaller. These results were verified in a number of subsequent papers (6),(7). A satisfactory explanation was not given. To our knowledge no further work has been reported on this matter. This is rather amazing since the results of Field and Nau imply the simultaneous occurance of a high friction and a high leakage.

We believe that further investigation into the lubrication of the seal contact is necessary. As a first step friction and leakage measurements on a polyurethan rod seal have been performed. To study the influence of surface roughness three rods have been used, each having a different surface finish.

1.1 Notation

d : rod diameter [m]

$\bar{h}$: average film thickness calculated from leakage data [m]

$\bar{h}_1$: average thickness of leaked film [m]

h : film thickness at the position where the pressure gradient is zero [m]

l : length over which oil is extracted [m]

m : mass of the extracted oil [kg]

p : pressure [Pa]

q : flow per unit of width [m^2/s]

v : velocity [m/s]

x : coordinate [m]

η : dynamic viscosity [Pas]

ρ : specific weight [kg/m^3]

2 TEST RIG

The test rig has been designed for measurements on rod seals. The essential parts of the test rig are represented in fig. 2.

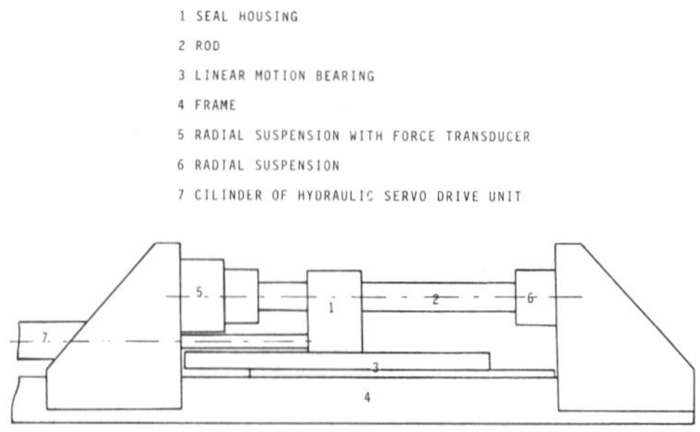

Fig. 2 Essential parts of the test rig.

The seal housing is moved along a stationary rod by a hydraulic servo drive unit and is externally guided by a high precision linear motion bearing, which ensures an accurate positioning between the seal housing and the rod. The seal housing consists of a body and two, easy exchangable, lids. A lid can be equipped with a nut to hold a seal (seal lid) or with a brass clearance bush (leak lid) (fig. 3).

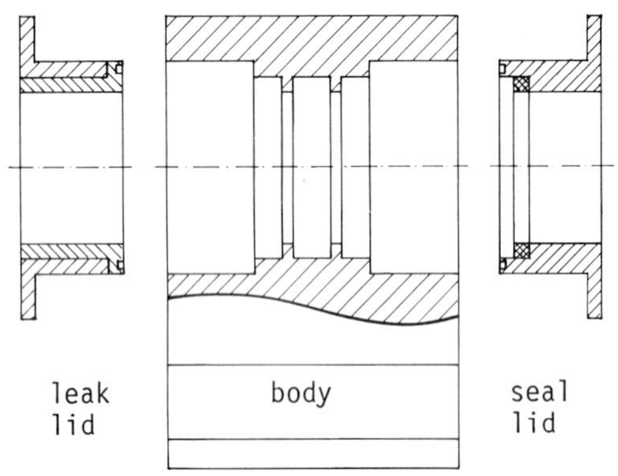

Fig. 3 Housing and lids.

The hydraulic servo drive unit consists of an actuator with very low internal friction, a hydraulic supply unit, an optical position transducer and a control module. It enables the accurate realisation of specific motion patterns over a maximum stroke length of 400 [mm]. The velocity can be varied from 0.5 until 750 [mm/s] with an accuracy of 0.5 %. The control module is linked to a personal computer. An interactive program enables user friendly communication. Automatic operation of an entire test program is possible.

The fluid pressure in the seal housing is applied by a separate hydraulic system, consisting of an electric pump, a buffer, and a cooler. The desired pressure can be obtained by changing a bypass resistance. The pressure is measured by a standard transducer with strain gauges. Pressures up to 40 [MPa] can be realised. Even with a leak lid mounted in the seal housing, the pressure can be kept constant within 0.1 [MPa].

The temperature of the fluid in the seal housing is measured by a Pt thermo resistive transducer. The cooler keeps the temperature constant within 1 [K].

3 FRICTION FORCE MEASUREMENT

The force, exerted by the seal housing on the rod, is measured by a standard force transducer positioned between the rod and the frame at the left hand side (fig. 2). The right hand side of the rod is radial suspended allowing free motion in axial direction. To eliminate disturbances by radial forces on the rod, the left hand side is suspended in the same way just to the right of the force transducer. The friction of the suspensions is negligible.

A relative motion of the rod out of the seal housing will be called an outstroke. A relative motion in the opposite direction will be called an instroke. The friction of a seal will in general depend on the direction of motion. Furthermore, different seals of the same type will not be completely identical and therefore exhibit somewhat different frictional characteristics. Hence, with two seal lids mounted in the housing, only a sum friction of a seal at outstroke and an other seal at instroke can be obtained.

To measure the friction of a single seal one seal lid has to be replaced by the leak lid, the friction of which can be calculated. However at seal outstrokes the clearance bush may become inadequately lubricated. Calculated friction values will then be unreliable. Friction at outstrokes of a certain seal A can alternatively be determined by subtracting the friction of a seal B at instroke from the sum friction of seal A at outstroke and seal B at instroke.

4 LEAKAGE MEASUREMENT

Leakage is, in addition to friction and wear, an important practical parameter to classify the performance of a seal. Leakage along with friction measurements allow some conclusions to be drawn on the lubrication of the sealed contact. However, in this regard measurements of the film thickness throughout the sealed contact have to be preferred. In the past, investigators have tried to measure film thickness in several ways.

Schrader (8) used an iron pin, pressed against the seal surface by a non-ferro spring. The displacement of the pin was measured by an inductive transducer. Disadvantages of this method are among others an additional concentrated mechanical load on the relatively soft seal and a disturbance of the lubrication process at the measuring spot.

Field and Nau (5), (6), (7) and Austin, Flitney and Nau (9) measured the electric capacity over the lubricant film using rubber seals filled with carbon black. The specific resistance of the rubbers used was 2.78 [Ωm] (9), which is low compared to the specific resistance of oil (about 10^7 [Ωm]). The errors in the measurements of Nau vary from about 3 % at low film thicknesses (0.25 [μm]) to about 40 % at film thicknesses of 5 [μm], which is attributed to parasitive capacities in the electric circuit (5). Without explanation they report that the accuracy is pressure dependent. The errors at pressures above 8 [MPa] were estimated to be about 30 % lower then the errors at lower pressures. This might be explained by influence of the pressure on the seal resistance.

Wernecke (10), (11) measured the electric resistance of the lubricant film. The electric conductivity of the rubbers (also filled with carbon black) would be sufficient according to (11). According to (10), however, it would not. This is amazing, for the same rubbers appear to be used in both publications. As in the papers by Nau e.a., no report is made on the influence of the seal resistance on the measurements. Wernecke also does not provide an accuracy of his measurements.

Measuring the film thickness by an electric method is not as straightforward as it might appear and a thorough analysis of the applied method is essential. Often, the accuracy of the performed measurements is not very good. The electric resistance of the seal material must be low, which considerably limits the applicability of the method to commercial seals. Furthermore, due to the dimensions of the sensors (about 0.5 [mm]; (5) and (11)), although being small compared to the width of the sealed contact, only an average signal can be obtained. This may be acceptable under the condition of a full film throughout the entire contact, but may lead to misinterpretation of the results if the contact is only partially lubricated.

At the moment lubricant film thickness measurement is under fundamental investigation (12). However, a less complicated but accurate technique for the measurement of leakage has been developed, and used in practice. The technique is based on the extraction of leaked oil by solution in pure hexane. Next, the hexane and oil are seperated by vacuum evaporation, and the mass of the oil is weighed. Leakage at outstroke (out-leakage) can be obtained by extraction of the leaked fluid film that remains on the rod. The area from which the oil is extracted is defined by the distance between two bands of gummed tape applied to the rod. An average thickness of the leaked film may be calculated (see section 7).

Performed with the necessary carefulness this relatively simple method proves very reliable. To establish the accuracy of the method, several tests, described in sections 4.1 to 4.3, have been performed.

4.1 Rate of oil extraction.

To test wether the oil film is fully extracted or only partly, oil has been applied to a smooth and clean plate of steel of known mass. After weighing the plate with oil, the oil has been extracted with hexane and the plate has been weighed again. The mass of it proved to be the same as the initial mass.

Also, the hexane and oil have been separated by vacuum evaporation and the oil has been weighed, showing a mass equal to the mass of the oil applied to the plate. Taking the dimensions of the plate and the accuracy of the mass measurement (± 0.1 [mg]) into account, the remaining oil film thickness was less than 0.01 [μm].

4.2 Influence of the gummed tape.

The influence of the gummed tape has been determined in the following way. The mass of the clean steel plate and a piece of gummed tape fixed thereon has been weighed. Subsequently the plate has been washed with hexane and weighed again, showing an increase in mass. After removal of the tape and again weighing of the plate and tape, the mass had its initial value.

These results imply that there has been some hexane between tape and plate, which evaporated after removing the tape, and more important that no gum has been solved in the hexane.

4.3 Oil loss during vacuum evaporation.

Some unsolved oil of known mass has been evaporated to determine the rate of oil loss during vacuum evaporation. It appeared to be less then 0.1 % and is thus negligible.

5 EXPERIMENTAL CONDITIONS AND TEST PROGRAM

The seals used in the tests are polyurethan rings having a rectangular cross-section with a single rounded edge (fig.4).

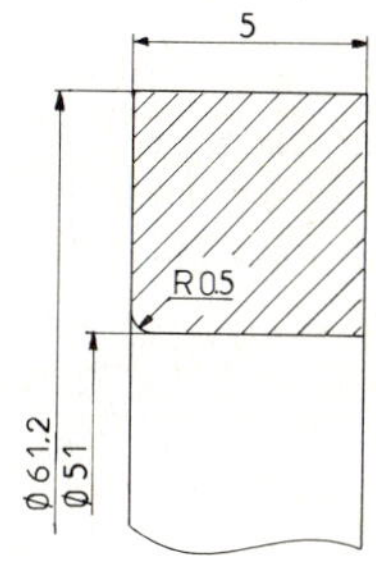

Fig. 4 Test seal.

The rings are mounted in a nut with a nominal diameter of 60 [mm]. During the tests the rounded edge of the seal has always been at the leading edge of the lubricated contact (i.e. at oustroke measurements, the rounded edge was at the oil side and at instroke measurements at the air side). The nominal diameter of the rods is 50 [mm]. The fluid used is a normal hydraulic oil, Shell Tellus T46.

Thesurface roughness of the seal is identical to that of commercial seals made by injection moulding. Roughness profiles have been determined on a test rig enabling computer controlled measurements (13). An optical sensor has been used to prevent deflections of the relatively soft seal material. Three rods, each having a different surface finish, have been used throughout the experimental program. Values of some roughness parameters (according to DIN) obtained from surface profiles measured in the axial direction are given in table 1.

Table 1. Roughness parameters from profiles in axial direction

	rod A	rod B	rod C	seal
R_a [μm]	< 0.01	0.05	0.34	0.54
R_z [μm]	0.06	0.36	2.6	3.8
R_p [μm]	0.03	0.15	2.1	3.4

An identical test program has been performed for the three rods used. The first part of this program was basically performed to establish friction-velocity curves at constant fluid pressure, both at instroke and outstroke. At instroke measurements, extra oil was supplied at the air side of the seal to eliminate the influence of different oil supply at the leading edge of the contact. At pressures of 2.5, 5, and 10 [MPa] motion patterns with constant velocity throughout 80 % of the stroke of 300 [mm] were imposed on the seal housing. The velocity was varied according to the following range: 25, 50, 100, 150, 200, 250, 300, 400, and 500 [mm/s]. The second part of the test program consisted of simultaneous measurements of friction and leakage at outstroke. They were performed at a pressure of 5 [MPa] and primarily within the velocity range, in which the friction-velocity curve displayed a non decreasing tendency. Using rod B additional measurements of out-leakage at a pressure of 2.5 [MPa] have been performed.

6 RESULTS

The results of the friction measurements at instroke and outstroke for the different rods are given in figures 5, 6 and 7.

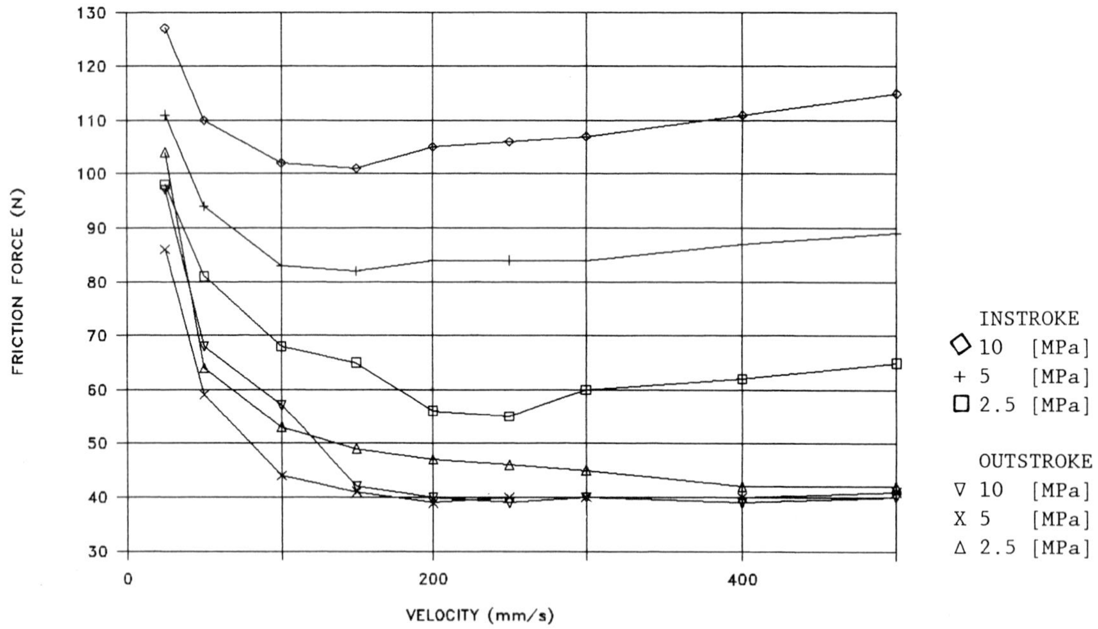

Fig. 5 Friction force (rod A).

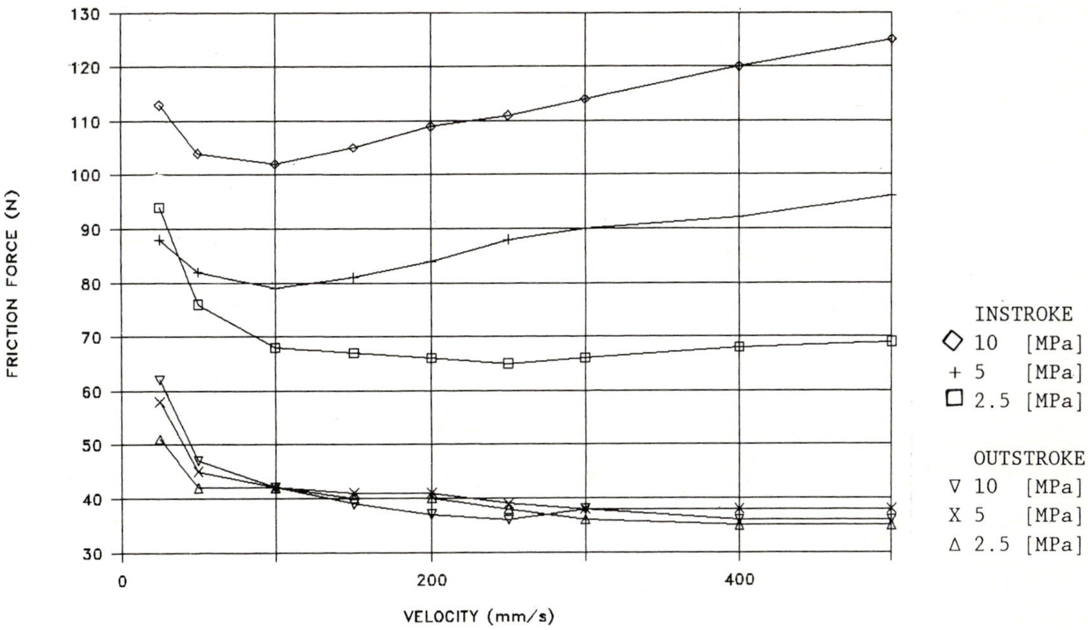

Fig. 6 Friction force (rod B).

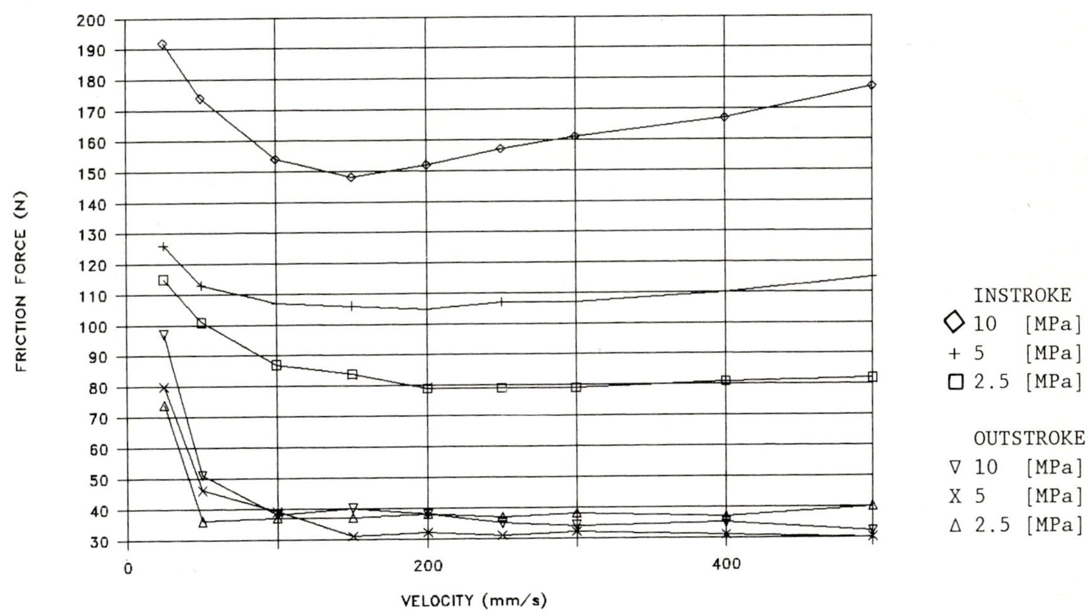

Fig. 7 Friction force (rod C).

The results of the measurements of out-leakage at a pressure of 5 [MPa] for the different rods are presented in figure 8. They are given as average thicknesses of the leaked film, which remains on the rod.

The results of the measurements of out-leakage at two different pressures of 2.5 and 5 [MPa] using rod B are presented in fig. 9.

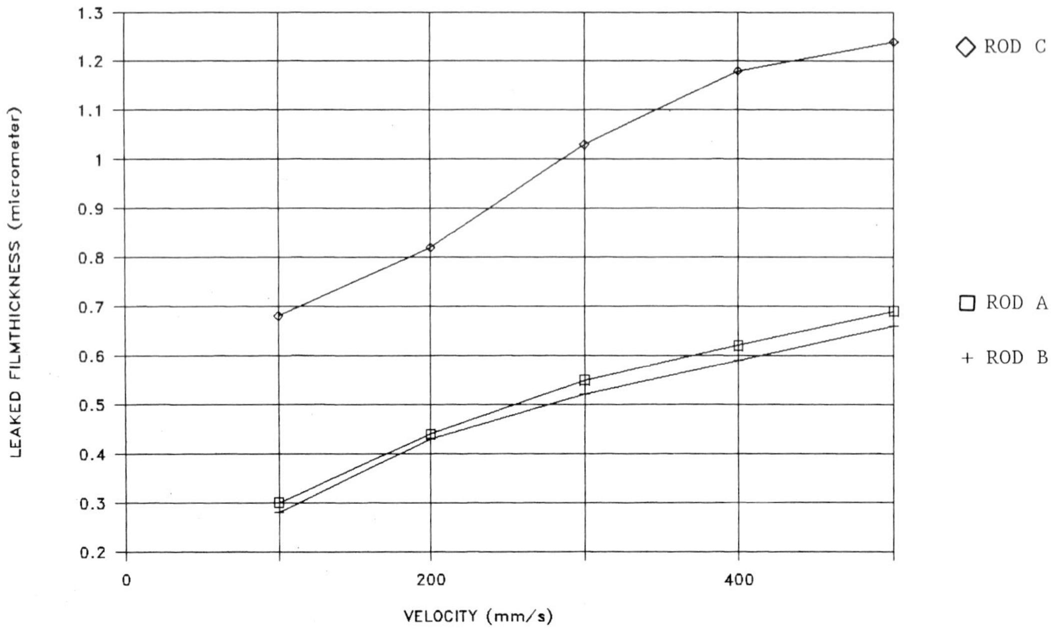

Fig. 8 Out-leakage using different rods at 5 [MPa].

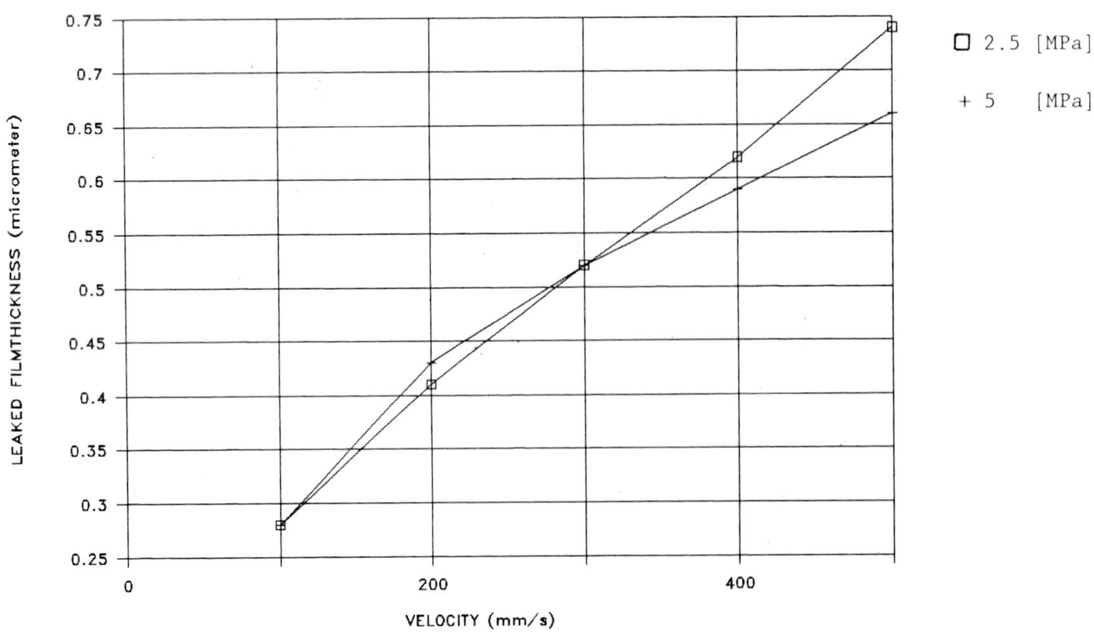

Fig.9 Out-leakage at 2.5 and 5 [MPa] (rod B).

7 DISCUSSION

The experimental friction curves confirm the general frictional behaviour of elastomeric reciprocating seals as depicted in fig. 1. Significant differences are found between instroke and outstroke. The increase of friction at higher velocities is not found at outstroke. Furthermore, friction at instroke is much higher than friction at outstroke. The influence of pressure is remarkable. At instroke friction increases as pressure rises, but at outstroke friction is almost independent of pressure. Similar results were obtained by Field and Nau (6). They found a more pronounced difference in friction levels between instroke and outstroke and a decrease in friction with rising pressure at outstroke. The influence of surface roughness of the rod on friction at higher velocities at outstroke is small, friction decreasing slightly with increasing roughness. At instroke friction clearly increases with roughness. For both directions of motion differences between the results using rod A and B are small.

Out-leakage increases approximately with the square root of the velocity. From fig. 8 it appears that the influence of surface roughness of the rod on leakage becomes important only for the roughest rod C. Using this rod leakage is found to be much higher than leakage using the relatively smooth rods A and B. Using rod B the influence of pressure on out-leakage is negligible (fig. 9). This result is qualitatively in agreement with the observed friction force (fig. 6).

The observed dependence of leakage on velocity is characteristic of smooth elastohydrodynamic lubrication. According to the inverse hydrodynamic lubrication theory (14) the thickness h_0 of the lubricant film at the position, where the pressure gradient is zero and hence the velocity varies linearly over the thickness of the film, can be calculated from:

$$h_0 = \left(\frac{8\eta v}{9 \left(\frac{\partial p}{\partial x}\right)_{max}} \right)^{0.5} \qquad (1)$$

In this formula $\left(\frac{\partial p}{\partial x}\right)_{max}$ stands for the maximum pressure gradient in the direction of velocity at the position where simultaneously the film thickness equals $1.5*h_0$. This position will occur at the leading edge of the lubricated contact. h_0 is directly related to the flow per unit of width through the lubricant film by:

$$q = \frac{v\,h}{2} . \qquad (2)$$

In a highly deformed contact the thickness of the elastohydrodynamic lubricant film may be approximated by h_0 over the major part of its length. Furthermore, in a stationary situation it is evident from conservation of mass that the thickness of the leaked film at outstroke, which remains on the rod and therefore relatively moves with velocity v, must be equal to $0.5*h_0$. Hence, assuming smooth film lubrication, an average film thickness in the seal contact may be calculated from leakage data by:

$$\bar{h}_{c1} = 2\,\bar{h}_1 = \frac{2\,m}{\pi\, d\, l\, \rho} . \qquad (3)$$

Friction at outstroke calculated by viscous shear, using average film thicknesses according to formula (3), is small compared to measured friction. This comparison is presented in fig. 10 for the experimental results obtained at 2.5 and 5 [MPa] using rod B. Similar results may be found for the data obtained using rod A and C.

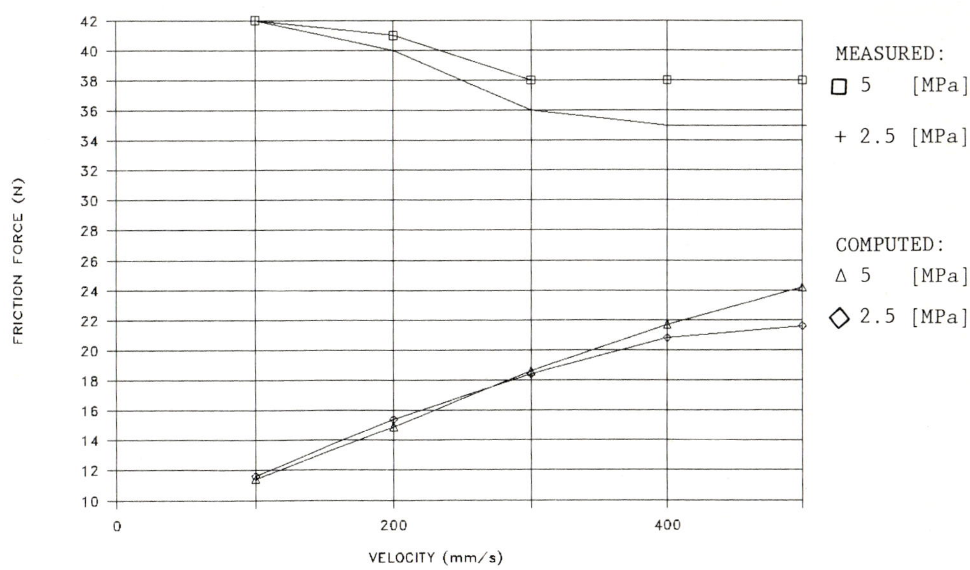

Fig. 10 Friction at outstroke calculated from leakage data assuming full film lubrication versus measured friction (rod B)

Clearly, full film lubrication theory for smooth surfaces is not convenient to describe the lubrication of the sealed conact. This will not come as a surprise regarding the friction-velocity curves, which imply mixed lubrication.

Until now, effects of surface roughness on the lubrication of the sealed contact have not received much attention. In particular the roughness of the seal surface has been disregarded, although commonly being considerably larger than the roughness of the mating surface. Probably it is often assumed that (most of) the roughness of the relatively soft seal disappears under the pressures in the sealed contact. However, for elastomeric materials the deformation of roughness asperities is mainly elastic. If the seal surface is (locally) lifted from its counterface by hydrodynamic effects, previously oppressed asperities may partly reappear. If this is true, then areas of contact may exist even at large nominal separation of the surfaces and consequently large leakage values, because of the dimensions of the (undeformed) roughness asperities (nominal separation is defined here as the distance between reference planes through the roughness height distributions of the mating surfaces). Both viscous shear and adhesion in the contact areas will hence contribute to friction.

At small velocities (< 50 [mm/s]) the contribution from viscous shear will be small. The lowest friction then appears to be found using rod B. If the surface of the rod is smoother (rod A) the total area of contact is larger due to the diminished depth of the roughness valleys. Hence friction will be higher. If the surface of the rod is considerably rougher (rod C) hysteresis effects may become important, causing friction to increase.

As velocity increases, hydrodynamic effects become more important. The total area of contact diminishes, giving rise to the observed friction and leakage characteristics (fig. 5-9). At instroke, the increase in friction at rising velocity may be explained by the merely small variations in total area of contact at increasing separation, which might occur at small separation values. An other explanation may be the occurance of very thin lubricating films in contact areas due to macro-elastohydrodynamics (hydrodynamic-elastic effects on individual roughness asperities). Comparing the roughness of the rods to that of the seal explains the small differences in observed friction and leakage obtained using rod A and B. Using the very rough rod C out-leakage is higher and friction at outstroke is lower due to additional transport of oil in the roughness valleys of the rod. However, at instroke friction is higher, possibly due to increased draping of the elastomeric seal material over roughness asperities of the rod, thereby increasing the contribution of hysteresis.

8 CONCLUSIONS

A test rig has been designed to investigate the tribological behaviour of reciprocating seals. Furthermore an accurate method for the measurement of leakage has been developed.

Friction and leakage measurements have been performed on a polyurethan seal, having a rectangular cross-section with a single rounded edge, using three rods with different surface roughness. The results indicate that it is not convenient to employ EHL theory for smooth surfaces to describe the lubrication of the sealed contact. In the case of rod seals, primarily developed for minimizing leakage, it may be expected that this is often so.

The results may be explained by considering the influence of surface roughness. In particular the relatively large roughness of the seal surface seems important. Measurements using seals with different surface roughness are sceduled for the near future.

For a better understanding of the tribological process in the sealed contact, film thickness measurements appear to be very important. Current film thickness transducers are not very accurate and, because of their dimensions, are probably unable to distinguish between mixed and full film lubrication. Research on film thickness measurements is currently being performed and will be the subject of a future paper.

9 ACKNOWLEDGEMENT

We thank Mr. J. Peels for his indispensable assistance during the experiments. We also thank Parker-Prädifa (Germany) for supplying the test seals.

References

[1] FIELD, C.J. and NAU, B.S., 'A theoretical study of the elastohydrodynamic lubrication of reciprocating rubber seals', ASLE Transactions, Vol 18, No 1, 1974, pp. 48-54

[2] RUSKELL, L.E.C., 'A rapidly converging theoretical solution to the elastohydrodynamic problem for rectangular rubber seals, J. Mech. Eng. Sc., Vol 22, No 1, 1980, pp. 9-16

[3] YANG, Y. and HUGHES, W.F., 'An elastohydrodynamic analysis of preloaded sliding seals, ASLE Transactions, Vol 27, No 3, 1983, pp. 197-202

[4] PRATI, E. and STROZZI, A., 'A study of the elastohydrodynamic problem in rectangular elastomeric seals, Journal of Tribology, Vol 106, October 1984, pp. 505-512

[5] FIELD, C.J. and NAU, B.S., 'An experimental study of reciprocating seals', Proc. Conf. on Elasto Hydr. Lubr., organised by the Inst. Mech. Eng., 1972, paper C5, pp. 29-36

[6] FIELD, C.J. and NAU, B.S., 'Film thickness and friction measurements during reciprocation of a rectangular section rubber seal ring', 6th Int. Conf. on Fluid Sealing, 1973, paper C5, pp. 45-56

[7] FIELD, C.J. and NAU, B.S., 'The effects of design parameters on the lubrication of reciprocating rubber seals', 7th Int. Conf. on Fluid Sealing, 1975, paper C1, pp. 1-13

[8] SCHRADER, K., 'Beitrage zur Klärung des Abdichtvorganges Gummi-elastischer Abdichtungen axial verschiebbarer hydrostatischer Bauteile', Ph. D. Thesis, Dresden Univ. of Technology, 1978

[9] AUSTIN, R.M., FLITNEY, R.K. and NAU B.S., 'Contact stress friction and the lubricant film of hydraulic cylinder seals', 8th Int. Conf. on Fluid Sealing, 1978, paper J2, pp. 11-20

[10] WERNECKE, P.W., 'Untersuchung der physikalischen Vorgänge in Spalten von Hydraulikdichtungen', Ph. D. Thesis, Aachen Univ. of Technology, 1983

[11] WERNECKE, P.W., 'Analysis of the reciprocating sealing process', 11th Int. Conf. on Fluid Sealing, 1987, paper E1, pp. 249-277

[12] VISSCHER, M., 'Filmdiktemeting op de meetopstelling voor translerende afdichtingen' ('Measurement of film thickness on the reciprocating seals test rig'), M. Sc. Thesis, Eindhoven Univ. of Technology, 1987

[13] STRUIK, K.G. and MUCHANG, F., 'Measurement of shape abd roughness by a modified compact disc sensor coupled to a personal computer', 4th Int. Seminar on Precision Engineering, Cranfield Inst. of Technology, UK, 11-14 May, 1987, pp. 81-90

[14] BLOK, H., 'Inverse problems in hydrodynamic lubrication and design directives for lubricated flexible surfaces', Proc. Int. Symp. Lubr. and Wear, by D. Muster and B. Sternlicht (eds.), Houston, 1963, pp. 9-151

Paper III(iv)

Radial lip seals, thermal aspects

M. J. L. Stakenborg and R. A. J. van Ostayen

In this paper the influence of temperature on the seal-shaft contact is studied, using coupled temperature-stress FEM analysis. A thermal network model is used to calculate the seal-shaft contact temperature for steady-state and transient conditions. Contact temperatures were measured under the seal lip, employing different measurement techniques. The experimental and numerical results show good correspondence. This study demonstrates that the thermal network theory provides a useful tool to predict the seal-shaft contact temperature.

Nomenclature

b	Contact width	[m]
c	specific heat	[J/kgK]
h	Film coefficient for convection	[W/m^{2}K]
p_0	Static contact pressure	[Pa]
t	Time	[s]
C_1, C_2	Mooney constants	[Pa]
F_r	Radial contact force	[N]
Q_i	Heat flow i	[W]
R_i	Heat resistor i	[K/W]
T	Torque	[Nm]
T	Temperature	[°C]
α	Thermal expansion coefficient	[1/K]
η	Dynamic viscosity	[Pa.s]
λ	Thermal conductivity	[W/mK]
ρ	Specific mass	[kg.m^{-3}]
ω_1	Shaft angular velocity	[rad.s^{-1}]
ω_2	Seal angular velocity	[rad.s^{-1}]

1 INTRODUCTION

The friction heat generated in the seal-shaft interface results in a raise in seal temperature . Investigating the sealing mechanism of radial lip seals it is important to know the influence of temperature on the contact conditions such as contact force, contact width and contact stress distribution, because they are the boundary conditions for the sealing and lubrication mechanism.
In the first part of this paper the influence of temperature on the seal-shaft contact force is studied, for a standard Nitrile rubber radial lip seal of Nitrile with industrial code BAF 100*10*70, to be used on a shaft diameter of 70mm.
With respect to service life, the seal application engineer needs an indication of the seal-shaft contact temperature which will occur in the seal in practice. In the second part of this paper a thermal network model is employed to calculate the contact temperature. Further, 3 different experimental techniques are utilized to measure the contact temperature.
Finally the numerical results are compared to the experimental results.

2 TEMPERATURE EFFECTS

A distinction can be made between (a) irreversible and (b) reversible temperature effects on the contact conditions. Some examples of irreversible temperature effects are :

1) Influence of temperature on physical ageing, resulting in a permanent change in stiffness of the rubber.
2) Influence of temperature on chemical ageing due to chemical reactions between oil (-additives) and rubber.
3) Influence of temperature on swell of the rubber due to absorbtion of oil (-additives).

The irreversible temperature effects will not be discussed in detail. In this paper we will confine ourselves to the following reversible temperature effects :

1) Influence of temperature on the stiffness of the rubber.
2) Influence of the temperature on the stiffness and initial length of the garter spring.
3) Thermal expansion of the seal material.
4) Thermal expansion of the shaft.

To study the influence of these reversible temperature effects on the contact conditions, information is needed on the temperature distribution in the seal.
In principle the temperature distribution in the seal can be determined in a number of

different ways, e.g. by analytical models [1], by the finite difference method [5], by the boundary element method [10], or by the finite element method [6]. In this paper the finite element method was chosen because this method allows a coupled stress-temperature analysis.

3 TEMPERATURE DISTRIBUTION IN THE SEAL

In this paragraph the temperature distribution in the seal is discussed. In the next paragraph the influence of the temperature distribution on the contact conditions is investigated.
In [9] it is shown that isothermal FEM calculations of the contact force of a Nitrile rubber lip seal using the Mooney-Rivlin material model showed very good correspondence with experimental contact force measurements on a split shaft device. A coupled stress-temperature FEM analysis was performed, assuming that the stress is dependent on the temperature distribution, but that there is no inverse dependence. The thermal stresses only had a very minor influence on the deformed shape of the seal. Therefore it was assumed that the temperature distrubution was not dependent on the stresses. The temperature distribution was calculated using the deformed seal geometry for isothermal conditions at T = 30 °C, see [9]. The calculated temperature field was then used as an input for the coupled stress-temperature analysis. In the coupled analysis the 2 Mooney-Rivlin constants depend on temperature.
The heat problem was assumed to be axisymmetric and the ABAQUS FEM model [12] contained 676 4-node, linear, axisymmetric ring, heat transfer elements. The garter spring and the metallic case were given the thermophysical properties of steel. It was assumed that the thermophysical material properties and the convective boundary conditions are not temperature dependent, thus the heat analysis problem was assumed to be linear. Note however that the stress/strain analysis of the contact problem remains nonlinear.
See table 1 for the (thermo-) physical properties of Nitrile rubber and steel. The following thermal boundary conditions were prescribed on boundaries Γ_1, Γ_2, Γ_3 and Γ_4 (see fig 1) :

1) Γ_1 : Fixed temperature T_m where the seal is in contact with the machine housing.
2) Γ_2 : Fixed temperature T_c in the seal-shaft contact area.
3) Γ_3 : Convection to oil on oilside of the seal, for oil bulk temperature T_{oil} and film coefficient h_{oil}.
4) Γ_4 : Convection to air on airside of the seal, for air temperature T_{air} and film coefficient h_{air}.

The steady-state temperature distribution in the seal was calculated for 5 different sets of boundary conditions as represented in table 2. Figure 1 shows the steady-state temperature distribution in a seal calculated with the FEM model for case 5 of table 2. With case 2 as initial condition the steady-state temperature distribution was reached after a transient time of 25 minutes.
The values of film coefficients h_{air} and h_{oil} are estimates based on formulae given in [2].

4 INFLUENCE OF TEMPERATURE ON THE CONTACT CONDITIONS

The influence of the 4 reversible temperature effects on the contact force for the 5 cases of table 2 is summarized in figure 2. Case 2 was chosen as the starting or reference situation.
Figure 3 shows the E-modulus determined from (quasi-)static iso-thermal uni-axial stress-relaxation tests on Nitrile rubber specimen. Using these experimental data the Mooney constants C_1 and C_2 were determined as function of the temperature. The (reversible) influence of the temperature on the seal material stiffness results in a change in contact force as represented in column 1 of figure 2. The stiffness of the garter spring and its initial length are both a function of temperature. The influence of this temperature effect on the contact force is shown in column 2 . The influence of thermal expansion of Nitrile rubber on the contact force is shown in column 3. Column 4 shows the influence of thermal expansion of the shaft on the contact force. Column 5 shows the influence of the 4 effects together. With 10 nodes (9 elements) in the contact width (b=0.075 mm) no change in contact width was found for all 5 cases, and only very minor changes in contact stress distribution occurred.

5 DISCUSSION OF THE FEM RESULTS

The influence of the temperature on the contact conditions is relatively small for the 5 cases described above. It is worthwhile to consider that :

1) It is the aim of the seal manufacturer to produce a seal that is not very sensitive to changes in temperature. Thus a rubber is chosen which stiffness is not very temperature dependent. In addition the garter spring has received a special temperature treatment to reduce the influence of temperature on the spring stiffness.

2) Only a relatively small zone in the tip of the seal is exposed to higher temperatures, see figure 1.

3) In practice contact temperatures higher than T_c=100 °C of case 5 will often occur. These higher temperatures were not taken into account in the FEM analysis. For temperatures beyond 100 °C the rubber stiffness rapidly changed due to physical ageing (additional vulcanisation), resulting in an irreversible change of rubber stiffness.

4) Although the influence of temperature on the static stiffness is small, temperature can have a strong influence on the dynamic rubber stiffness. This dynamic stiffness may become important in case of periodic excitation of the seal lip due to eccentric movements of the shaft center, or due to out-of-roundness of the shaft, see [9].

It is recommended by seal manufacturers not to use lip seals beyond a certain maximum operating temperature, depending on the composition of the seal material (for Nitrile rubber $T_{max} \approx 90$ °C). Beyond this temperature increased wear will occur, resulting in a considerable decrease in seal service life. The contact temperature which results from a certain dissipated friction heat depends on the heat balance in the entire machine. In the next paragraph a numerical method is presented to calculate an estimate of the steady-state and transient contact temperatures.

6 THERMAL NETWORK METHOD.

In the thermal network method (TNM) a machine is divided into thermal components, forming a thermal network.
The thermal components used in such a network are resistances, capacitors, and positive and negative heat sources. In analogy with electrical circuits, temperature corresponds to voltage and heat rate to current.
Advantages of the TNM method for thermal analysis are : (a) the TNM is easy to conceive and very flexible because of its open structure and (b) the TNM does not demand sophisticated software and hardware. The TNM calculations presented in this paper can be performed on a simple (640 kilobyte) personal computer.
Nevertheless, this method has not yet been used frequently for the analysis of heat problems in mechanical engineering. The main reason is that there are often no adequate analytical models available to describe the different mechanical elements such as gears, bearings and seals in terms of thermal components such as resistances, capacities or heat sources.
The FEM is frequently used for analysing heat problems in mechanical engineering. A disadvantage of the FEM in comparison with the TNM is that the software and hardware are expensive.

In the present paper a combined approach was chosen. The FEM can be used to determine thermal conductivity matrices for relatively complex machine elements, such as seals. These conductivity matrices are then incorporated into the thermal network model.
The size of the conductivity matrix of a machine element depends on the number of temperatures and heat fluxes one wishes to distinguish for that element. The heat flow and the temperature distribution in the seal are assumed to depend on only 4 temperatures: (1) the seal-shaft contact temperature, (2) the temperature of the machine housing, (3) the bulk oil temperature, and (4) the air temperature. Hence, the conductivity matrix of the seal is a 4x4 matrix. This reduces the original 676 element FEM-mesh of the seal to a simple, condensed and more efficient network component.
The thermal network model of the test-rig from figure 4 is shown in figure 5, see also van Ostayen [4]. A part of the test-rig on the lefthand side of the shaft was modelled by one resistor and one capacitor only. The thermal resitance and capacity of these two components have been derived from measurements. This simplification is allowed here, because we are mainly concerned with the relation between the dissipated heat and the temperature in the seal-shaft contact area. The resulting network is smaller and machine elements such as the slip ring unit and the rotating air bearings do not have to be modeled.

The oilfilm between seal and shaft is assumed to have one homogeneous temperature. One node n_0 therefore suffices to describe the oilfilm, see fig. 6. This node is connected to the following components :

1) A heat source Q_0 describing the heat flow dissipated to the oilfilm.
2) A resistor R_1 related to the conductive heat transfer from the oilfilm to the seal. The oil flow in the oilfilm is assumed to be laminar. Then the heat transport from the oilfilm to the seal is conductive and perpendicular to the laminar flow.
3) A resistor R_2 related to the conductive heat transfer from the oilfilm to the shaft.
4) A resistor R_3 related to the heat transfer from the oilfilm to the bulk oil. A part of the heated oil in the oilfilm will be exchanged with the cooler bulk oil. The result is a heat flow from the oilfilm to the oil bulk.

The shaft under the seal is modeled by 82 resistors and 31 capacitors thus enabling a good representation of the large radial and axial temperature gradients. For reasons of simplicity a reduced number of resistors and capacitors are sketched in the shaft of fig. 5.

The other metal elements in the rig such as the oil container and the static air bearing are modeled with less resistors and capacitors because of the very small temperature gradients within these elements. Heat transfer by convection is calculated using the formulae from [2]. Generally these formulae are nonlinear in temperature.

The nonlinear system equations of the thermal network were solved using a standard network program mainly used by electronical engineers: SPICE-2G. It is a general purpose circuit simulation program for steady-state or transient nonlinear analysis. Circuits may contain resistors, capacitors, independent and dependent voltage and current sources and several other electrical components. Nonlinear functions however are limited to polynomials.

7 CONTACT TEMPERATURE MEASUREMENTS

Because of the relatively small dimensions of the contact width b (in practice $0.05 < b < 0.5$ mm) it is difficult to measure the contact temperature. Different experimental methods were employed to measure the contact temperature distribution.

1) NTC thermistors.
2) Thin film thermal transducers.
3) Infra-red temperature measurements.

NTC thermistors are reliable, relatively inexpensive, commercially available temperature sensors (accuracy +/- 1 ^{0}C). A disadvantage of these thermistors is that because of their diameter (d = 2 mm) only an average contact temperature can be measured in a bandwidth which is much larger than the actual contact width.
Therefore, attempts were made to use a Titanium thin film micro-transducer, which was evaporated on the shaft surface, see figure 7. Because of the small dimensions of these sensors (width = 10 μm, length = 600 μm, height = 1 μm) and their accuracy (+/- 0.5 ^{0}C) they seemed very useful for measuring the contact temperature (distribution) under the lip. In order to measure the temperature profile in the contact area the seal was translated in axial direction over the transducer on a rotating shaft. Problem here was that each time the seal was translated, the torque changed, probably because the seal encountered a different local roughness pattern on the shaft, which resulted in a change of dissipated friction heat and in small fluctuations of the contact temperature of +/- 2 ^{0}C. This temperature fluctuation was too large to obtain a valid temperature profile within the contact area. Disadvantages of these micro-transducers sensors are (a) that due to wear they have a limited service life (approximately 2 hours) which is short compared to the transient times before the contact temperature has reached a steady state (approximately 1.5 hours), and (b) that the shaft surface had to be polished (locally) before evaporation, resulting in a local change of shaft roughness, and thus in a change of torque.

In order to obtain an integral temperature profile of the contact temperature of the seal lip contact surface, infra-red temperature measurements were performed via a mirror and a small cylindrical glass window (diameter = 2 mm, length = 1 mm), in a fixed hollow steel shaft. The temperature profile measured at 3 different seal angular velocities is represented in figure 9. The infra-red measurement technique is based on determining temperature differences (resolution 0.5 ^{0}C). However, an accurate determination of the absolute temperature is difficult with this technique without a reference temperature.

8 NUMERICAL RESULTS

The results are valid for the test-rig of figure 4, where the non-rotating shaft was immersed in Shell Tellus 46 oil, over an angle of 0.68 rad of the shaft circumference. The ambient air temperature was held constant at T = 20 ^{0}C.

The torque was measured as function of the shaft angular velocity, see fig 10. Figure 11 shows the steady-state contact temperatures measured with the thermistors, corresponding to the torque of figure 10. In the same figure are sketched the steady-state contact temperatures as calculated with the TNM, for 2 situations : (a) with the seal taken into account and (b) without the seal taken into account in the network model.
In situation (a) the seal was incorporated into the TNM as a 4x4 conductivity matrix as determined by the FEM. This results in a heat flow as represented in figure 12.
In situation (b) the heat transfer through the seal was eliminated by giving the seal a very high resistance (if $R_1 = \infty$ then $Q_1 = 0$). From figure 11 it can be seen that the presence of the seal as a thermal component has only a very minor influence on the calculated contact temperatures. Therefore it can be concluded that the seal does not necessarily have to be taken into account in the network. This simplification eliminates the use of the FEM analysis to determine the seal conductivity matrix, which means a considerable reduction in modeling time.
Figure 13 shows transient contact temperatures at a constant shaft angular velocity for situation (b).
The practical utility of the TNM is illustrated by the good correspondence between measured and calculated results for both steady-state and transient situations.

9 SUMMARY

1) For the Nitrile rubber lip seal under investigation, the influence of the contact temperature T_c on the static contact conditions is neglectably small for $0<T_c<100$ °C.
2) To calculate the seal-shaft contact temperature resulting from a certain dissipated friction heat rate the heat balance of the entire machine has to be taken into account.
3) The thermal network method provides a useful tool to model the heat balance of the machine and to calculate transient and steady-state seal-shaft contact temperatures.
4) The heat transfer through the seal has only a very minor influence on the contact temperature.
5) Accurate experimental determination of the contact temperature is difficult due to the small dimensions of the seal-shaft contact width.

10 ACKNOWLEDGEMENT

The author would like to thank Mr. B. Munneken from the CFT Department of Philips Eindhoven for performing the infra-red temperature measurements.

<u>References</u>

[1] Upper,G. "Dichtlippentemperatur von Radial-Wellen Dichtringen", Thesis University of Karlsruhe, Germany, July 1967.(In German)

[2] Wong, H.Y. "Handbook of essential formulae and data on heat transfer for engineers", Longman, London and New York, 1977.

[3] van Leeuwen, H; Meyer, H; Schouten, M.J.W. "Elastohydrodynamic film thickness and temperature measurements in dynamically loaded concentrated contacts: eccentric cam-flat follower", Proc. 13th Leeds-Lyon Symposium on Tribology, University of Leeds, 8-12 sept. 1986.

[4] van Ostayen, R.A.J. "The use of the thermal network method for modeling the heat balance in mechanical constructions", M.Sc. Thesis, Eindhoven University of technology, August 1988.(in Dutch)

[5] Ozisik, M.N. "Heat Transfer, a Basic Approach", Mc Graw-Hill Book Company, New York 1985.

[6] Bathe, K.J. "Finite Element Procedures in Engineering Analysis", Prentice-Hall, Inc., Englewood Cliffs, New Jersey, 1982.

[7] ten Hagen, E.A.M., "Mechanical Characterization of Synthetic Rubbers" , M.Sc. Thesis, Eindhoven University of Technology, Jan. 1988.(in Dutch)

[8] Treolar, L.R.G. "The physics of rubber elasticity", Clarendon Press, Oxford, 1975.

[9] Stakenborg, M.J.L.,"On the sealing and lubrication mechanism of radial lip seals", Ph.D. Thesis, Eindhoven University of Technology, the Netherlands. Sept. 1988.

[10] Brebbia, C.A. "Applications of the boundary element method for heat transfer problems", Rev. Gen. Therm. Fr. no 280, April 1985.

[11] Luyten, R. "Determination of the contact force, contact width and the contact stress distribution of radial lip seals.", M.Sc. Thesis, Eindhoven Univ. of Techn., Dec 1987.(in Dutch)

[12] Hibbit, Karlson and Sorensen ABAQUS FEM software, user manuals (version 4-5)

		Nitrile	Steel
Specific mass	ρ [kg.m^{-3}]	1460.	7800.
Thermal conductivity	λ [W/mK]	.43	50.
Specific heat	c [J/kgK]	2000.	450.
Thermal expansion coeff.	α [1/K]	9.44 10^{-5}	1.1 10^{-5}

Table 1 : Thermo-physical properties of Nitrile rubber and steel.

case	T_c [°C]	T_{air} [°C]	T_{oil} [°C]	T_m [°C]	h_{air} [W/m^{2}K]	h_{oil} [W/m^{2}K]
1	0.	0.	0.	0.	0.	0.
2	20.	20.	20.	20.	0.	0.
3	50.	20.	30.	20.	10.	200.
4	75.	20.	35.	20.	10.	200.
5	100.	20.	40.	30.	10.	200.

Table 2 : Thermal boundary conditions for 5 cases.

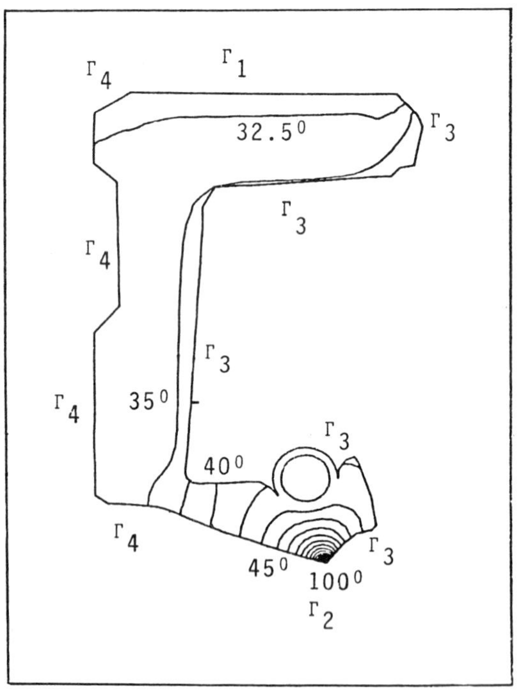

Fig. 1 : Steady-state temperature distribution in the radial lip seal of Nitrile rubber, calculated with the finite element method, for case 5 of table 2.

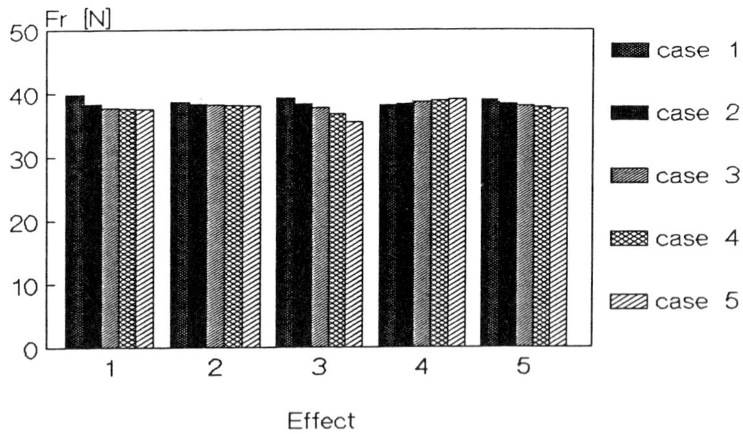

Fig. 2 : The contact force F_r for the 5 different cases of table 2 and for 5 effects, calculated by a coupled stress-temperature FEM analysis. The effects are :

1. **Influence of temperature on material stiffness.**
2. **Influence of temperature on spring stiffness.**
3. **Influence of thermal expansion of the rubber.**
4. **Influence of thermal expansion of the steel shaft.**
5. Combined effect of 1), 2), 3) and 4).

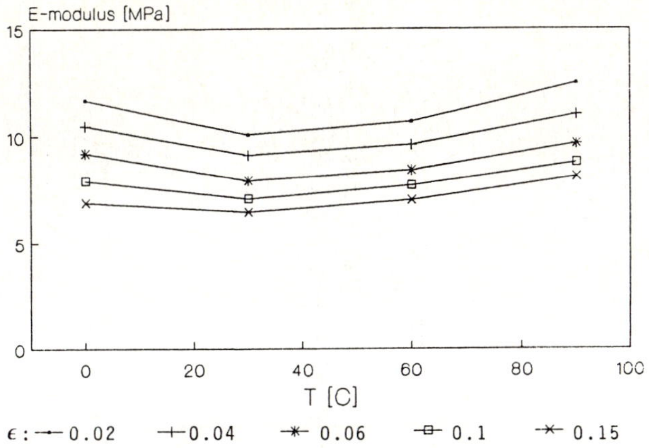

Fig. 3 : E-modulus of Nitrile rubber as function of temperature for different strains $\varepsilon = \Delta l/l$ after a relaxation time t=24 h.

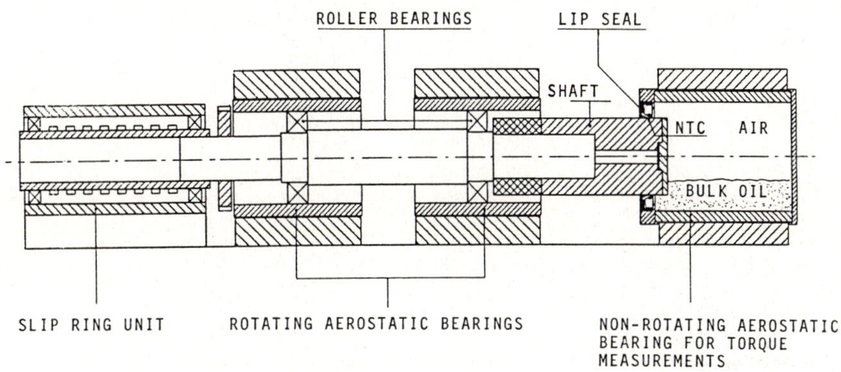

Fig. 4 : Sketch of temperature measurement test-rig .

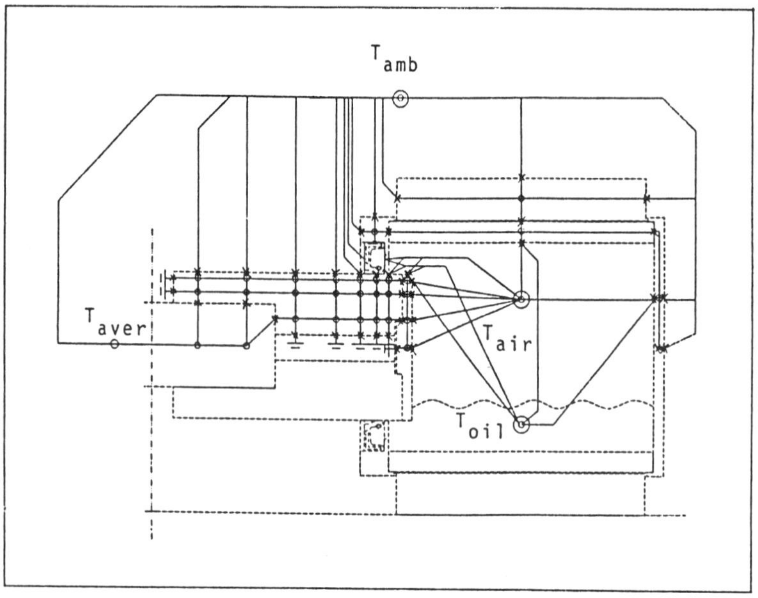

Fig. 5 : Thermal network model of the test-rig from fig. 4

o temperature node with capacity
x temperature node without capacity
___ each line connecting nodes represents a resistor.

T_{amb} this node represents the ambient air surrounding the test-rig, it is assumed to be constant at 20 ^{0}C.
T_{oil} this node represents the average oil bulk temperature.
T_{air} this node represents the average air temperature inside the housing.
T_{aver} this node represents the average temperature of the part of the test-rig that was not modeled: the slip ring unit, the rotating aerostatic bearings and the main shaft.

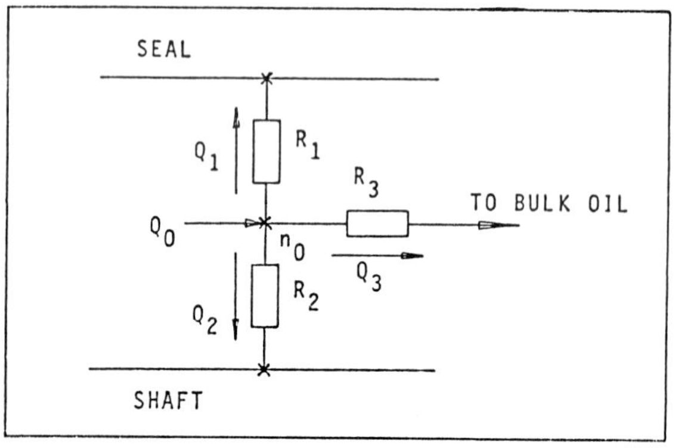

Fig. 6 : Detail of thermal network in the contact area.

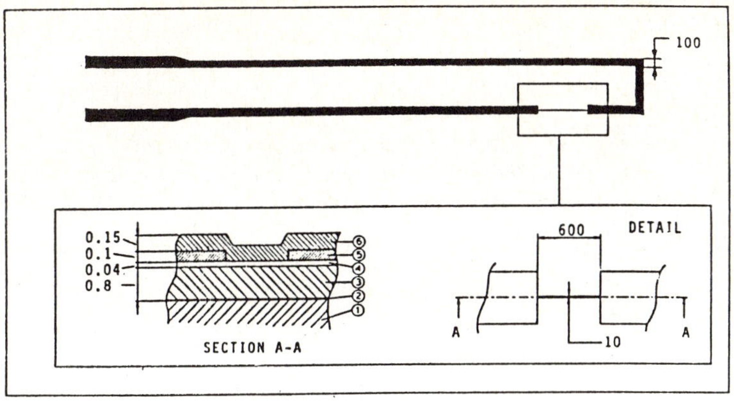

Fig. 7 : Thin film thermal microtransducer (dimensions are in microns). (1) substrate (shaft), (2) adhesive layer of Ti, (3) insulating layer of Al_2O_3, (4) transducer of Ti, (5) conductor pattern of Au, (6) protective layer of Al_2O_3, see [3].

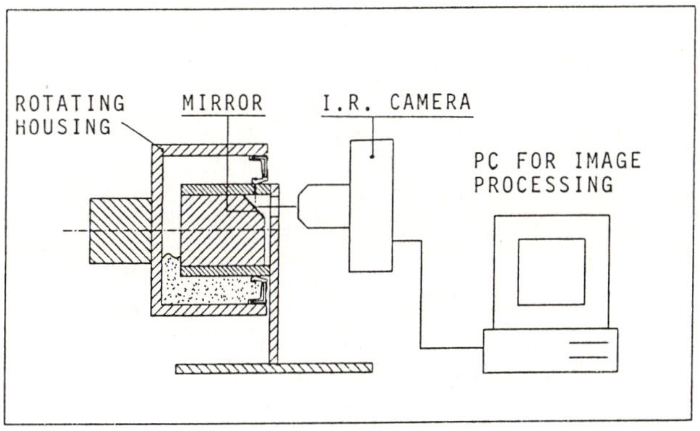

Fig. 8 : Sketch of test-rig for infra-red temperature measurements.

Fig. 9 : Temperature distribution for different seal angular velocities, measured using the infra-red technique.

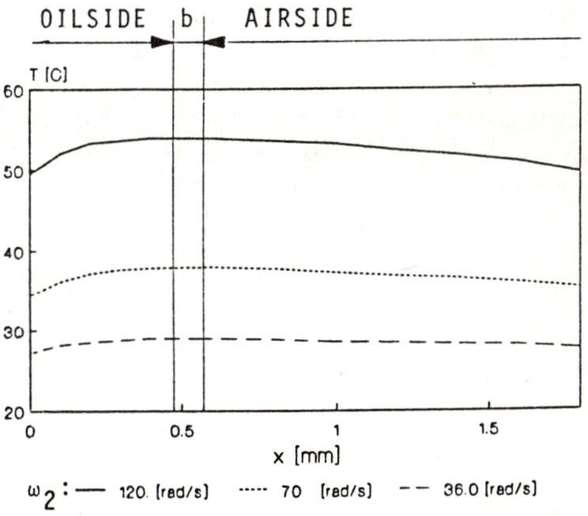

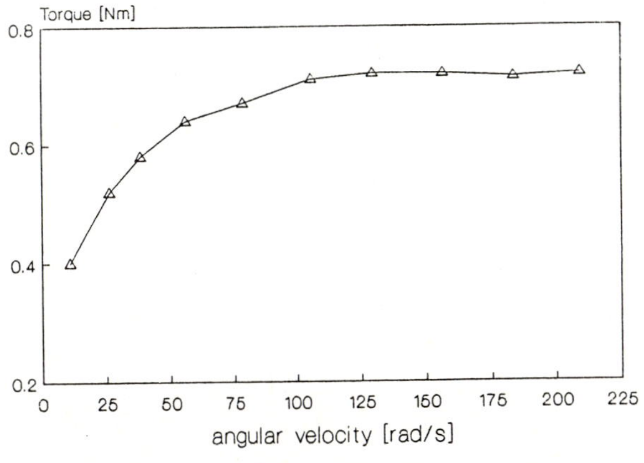

Fig. 10 : Steady-state seal torque as function of the shaft angular velocity as measured on the test-rig.

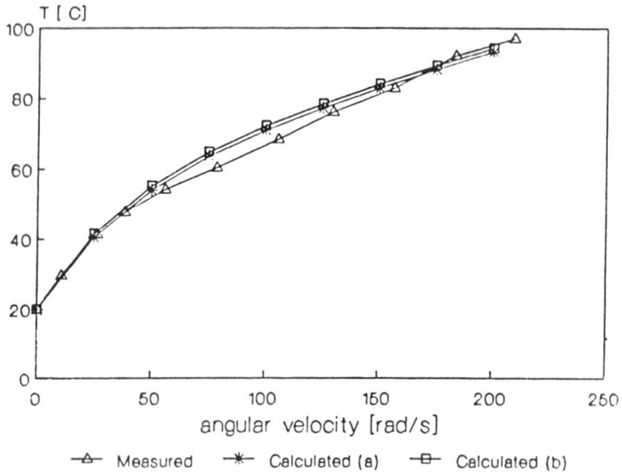

Fig. 11 : Steady-state seal-shaft contact temperature as function of the shaft angular velocity corresponding to seal torque in fig. 10.

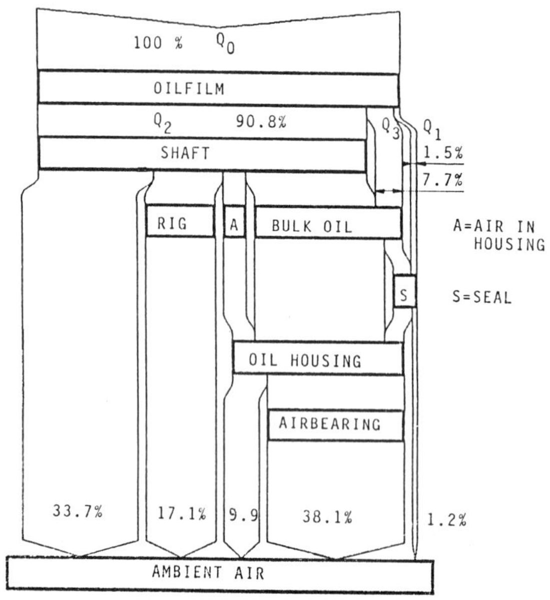

Fig. 12 : Sankey-diagram of the heat flows in the test-rig as calculated with the TNM. ω = 125 rad/s and T = 0.72 Nm, thus $Q_0 = T.\omega = 90$ W.

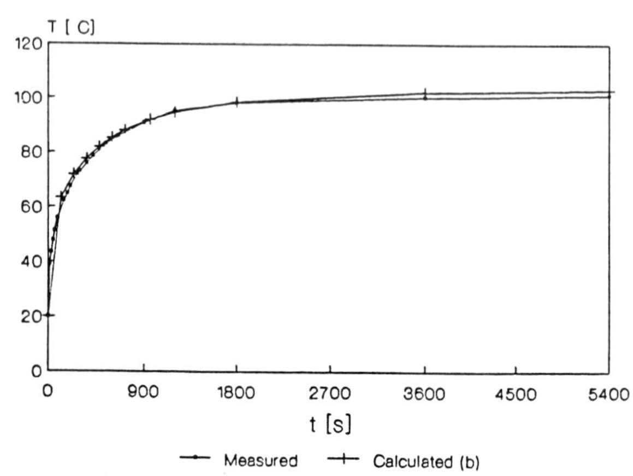

Fig. 13 : Transient contact temperatures for ω = 204.2 rad/s and a steady-state torque T = 0.82 Nm. The steady-state torque was reached after 900 seconds.

SESSION IV
CAMS

Chairman: Professor J F Booker

PAPER IV(i)	Lubrication and Fatigue Analysis of a Cam and Roller Follower
PAPER IV(ii)	Predictions of Cam Wear Profiles
PAPER IV(iii)	Cam and Follower Design

Paper IV(i)

Lubrication and fatigue analysis of a cam and roller follower

B. A. Gecim

The lubrication characteristics of a cam and roller follower interface are investigated. It is established by the use of a lubrication regimes chart that the operating conditions correspond to a heavily loaded elastohydrodynamic contact. The film thickness prediction includes both thermal and squeeze-film effects. Under the conditions investigated, the nominal film thickness increased from 0.05 μm to 0.35 μm as cam shaft speed increased from 250 r/min to 3000 r/min; maximum contact pressures at these speeds were approximately 1.0 GPa.

Required traction forces for no-slip condition are also calculated. The lubrication was close to the boundary mode, which implied a large coefficient of traction. Hence, gross slip at the cam/roller interface was unlikely under the present conditions of practical interest.

Also, the cam torque and the energy losses are predicted. Over a wide speed range, the energy loss was approximately 1.0 Joule per cam rotation.

In addition, a fatigue life equation is derived and incorporated with the kinematics and the dynamics of the roller follower. Under the conditions investigated, the fatigue life at the roller outside surface was on the order of 10^{12} camshaft revolutions. The shaft/needle interface of the needle bearing used in the roller follower had a predicted fatigue life on the order of 10^{8} camshaft revolutions.

Parametric studies are presented which can be used to improve film thickness, contact pressures, and fatigue life.

1 INTRODUCTION

It is relatively recent that the governing kinematic and geometrical relationships of cams and followers are incorporated with elastohydrodynamic lubrication analyses [1-5]. These studies consider bucket and finger type followers where a sliding contact occurs. Muller [4] and Dyson [2,3] demonstrated that the lubrication factor must be considered at the design stage. Muller [4] showed that a cam with smaller nose radius had a higher entrainment velocity, hence, a thicker film around the cam nose. Dyson [2] analyzed the effect of tappet radius on the film thickness and on the contact pressure and showed that a compromise was necessary between these quantities. Dyson [3] presented a similar analysis with a finger follower and showed the effects of changing the follower radius on the entrainment velocity and on the radius of relative curvature. Hamilton [5] measured the lubricant film thickness and the pressure between the cam and the follower. More recently, Dowson and co-workers [6,7] applied the EHD theory to a cam and flat-faced follower. In [6], they used the steady elastohydrodynamic film thickness equation. In [7], the prediction of rigid-surface hydrodynamic lubrication was compared with that of elastohydrodynamic theory including the squeeze film effect. Holland [8] presented the analysis including the squeeze film effect. His predictions matched his measurements. Bair et al. [9] measured the dynamic forces, power loss, and interface temperatures between a cam and flat-faced follower. Some measurements with roller followers indicated significant reduction in friction loss in comparison with the flat followers. Staron and Willermet [10] measured the valve-train torque. They also reported significant reduction in torque and valve-train friction as a result of using roller followers. Sun and Rosenberg [11] measured forces, torques, energy losses, and friction coefficients associated with flat-faced and roller followers.

The present study considers a roller follower. Hence, the prediction of rolling contact fatigue life was one of the purposes of the present study. Low specific film thickness values further emphasized this prediction. In this area, Dawson [12] presented experimental results indicating considerable shortening of fatigue life as the ratio of film thickness to roughness decreased. Tallian [13] examined the causes of specific failure modes from the viewpoint of controlling these through lubrication. Tallian's data on ball bearings [14] and Skurka's data on rollers [15] clearly indicated similar conclusions as to the shortening of fatigue life at small values of specific film thickness. In these studies, the effect of specific film thickness on fatigue life has been presented as an empirical factor that modifies the fatigue life prediction by Lundberg and Palmgren's method [16]. With some modifications, this model is still the standard and widely used tool for rolling element bearings [17]. Further research has been conducted towards finding modifying factors to represent the differences in materials and operating conditions from those used by Lundberg and Palmgren [18]. The sensitivity of fatigue life prediction to materials and processing, for example, is best demonstrated by McCool [19] and Vaessen and DeGee [20].

In addition to the fatigue problem, there are other practical concerns associated with roller followers. These include the possibility of gross-slip and plastic flow at the cam/roller interface and the importance of thermal and squeeze film effects on lubricant film thickness. The present study was motivated by these concerns.

1.1 Nomenclature

- a Hertzian contact half-width $(= 1.596\,(\frac{w r_E}{E})^{1/2})$, m
- D Diameter, m
- E Equivalent elastic modulus $(= 2/[\frac{1-\nu_1^2}{E_1} + \frac{1-\nu_2^2}{E_2}])$, N/m^2
- F_{XR} Bearing reaction force in X direction, N
- F_T Cam/roller traction force, N
- G Materials parameter $(=\alpha^* E)$
- h_o Nominal film thickness, m
- $\bar{h}$ Dimensionless film thickness $(= h/r_E)$
- I Inertia, kg-m^2
- k_L Lubricant thermal conductivity, W/m °C
- L Lift, m
- ℓ Follower width, m
- $\bar{P}$ Dimensionless pressure $(= 1 - e^{-\alpha^* P})$
- Pe Peclet number $(= \frac{(L_H - 1)u}{2\alpha_S^{**}})$
- r_E Equivalent contact radius $(= \frac{r_1 r_2}{r_1 + r_2})$, m
- T_R Reference temperature, °C
- T_B Solid body bulk temperature, °C
- T^* Torque, N•m
- U Speed parameter $(= \mu_o u_R/E/r_E)$
- u_R Entrainment speed $(= 1/2\,(u_1 + u_2))$, m/s
- W Load parameter $(= wE/r_E)$, dimensionless
- w Load per unit contact length, N/m
- z The distance between camshaft center and follower center $(= L + r_F + r_B)$, m
- α^* Pressure-viscosity coefficient, m^2/N
- α^{**} Thermal diffusivity, m^2/s
- β Temperature-viscosity coefficient, °C^{-1}
- Σ Slide-roll ratio $(= 2\,\frac{u_2 - u_1}{u_2 + u_1})$
- σ_m Maximum Hertz stress, $(=0.399\,(\frac{wE}{r_E})^{1/2})$ N/m^2
- σ RMS surface roughness, μm
- ρ Density, kg/m^3
- ω Rotational speed, rad/s
- θ Cam shaft rotation from cam/follower line of centers, degree
- μ Viscosity, N/m^2•s
- μ_o Viscosity at p=0 and $T=T_R$
- τ Shear stress, N/m^2

Subscripts

- B Cam base
- BR Bearing
- C Cam
- F Follower
- PR Push rod
- S Solid

2 CAM/ROLLER FOLLOWER GEOMETRY, KINEMATICS, AND DYNAMICS

This section includes the geometric, kinematic, and dynamic relationships pertaining to the cam and roller follower. The geometric and kinematic relationships are derived by following a similar approach as in [2].

Geometry: The instantaneous radius of curvature of the cam, r_C, is given by

$$r_C = -r_F + \frac{[z^2 + (\frac{dz}{d\theta})^2]^{3/2}}{[z^2 + 2\,(\frac{dz}{d\theta})^2 - z\,\frac{d^2 z}{d\theta^2}]} \tag{1}$$

where

$$z = L + r_F + r_B \tag{2}$$

The equivalent contact radius that enters the EHD calculations is given by:

$$r_E = \frac{r_C\, r_F}{r_C + r_F} \tag{3}$$

Kinematics: The point of contact between the cam and the follower moves in space; hence, the instantaneous relative speed of the cam surface with respect to the point of contact is given by:

$$u_C = (\frac{\omega r_C}{r_C + r_F})\,[z^2 + (\frac{dz}{d\theta})^2]^{1/2} \tag{4}$$

Neglecting resistance from the bearing on which the follower rotates, it is imposed that the follower surface speed is the same as that of the cam.

Dynamics: Figure 1 shows the forces acting on the roller follower and the cam. Balancing the forces on the roller follower yields:

$$\Sigma F_y = 0 \quad , -F_{PR} + F_C \cos\alpha + F_T \sin\alpha = 0 \tag{5}$$

$$\Sigma F_X = 0 \quad , F_T \cos\alpha + F_C \sin\alpha + F_{XR} = 0 \tag{6}$$

The torque balance around the center of the bearing yields:

$$T^*_{BR} - F_T\, r_F = -\,I_F\, \dot{\omega}_F \tag{7}$$

in which

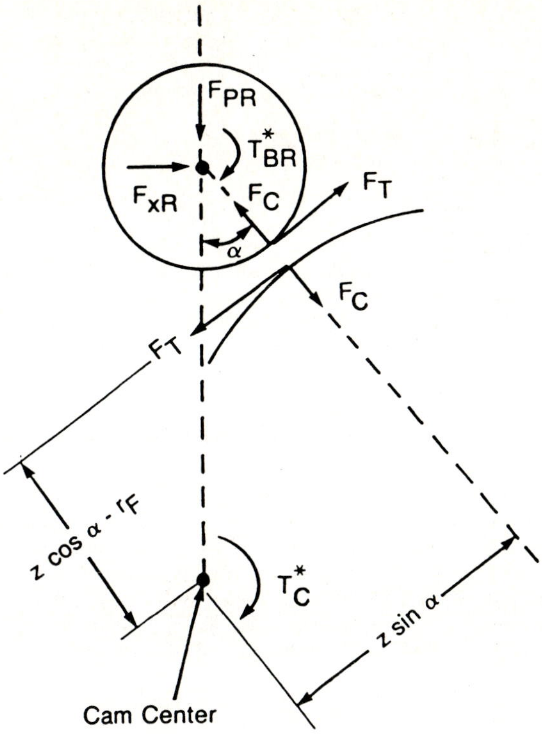

Figure 1. Forces acting on the cam and the roller follower.

$$T^*_{BR} = f_{BR} \frac{D_{BR}}{2} (F_{PR}^{\ 2} + F_{XR}^{\ 2})^{1/2} \qquad (8)$$

where D_{BR} is the bearing bore diameter and $f_{BR} = 0.0045$ [21]. The contribution from the traction force, F_T, to the needle bearing radial load is neglected.

In equation (7), follower inertia, I_F, is given by:

$$I_F = \pi \rho_F \frac{1}{2} [r_F^{\ 4} - (\frac{D_{BR}}{2})^4] \qquad (9)$$

This equation treats the needles and the outer race like a solid body. The angular acceleration, $\dot{\omega}_F$, was calculated by numerically differentiating the follower angular speed, ω_F, where

$$\omega_F = \frac{\omega_C r_C}{(r_C + r_F) r_F} \left[(\frac{dY}{d\theta})^2 + (Y + r_F)^2 \right]^{1/2} +$$

$$\omega_C \frac{(\frac{dY}{d\theta})^2 - Y \frac{d^2Y}{d\theta^2}}{Y^2 + (\frac{dY}{d\theta})^2} \qquad (10)$$

The unknown quantities F_C, F_T, and F_{XR} can be determined from simultaneous solution of equations (5), (6), and (7).

Once the forces F_C and F_T are determined, the cam torque, T^*_C, can be found from the torque balance around the cam center,

$$T^*_C + F_C \ z \sin\alpha - F_T \ (z \cos\alpha - r_F) = 0 \qquad (11)$$

Determination of the pushrod force, F_{PR}, requires the dynamic analysis of the rest of the valve train. This was accomplished by the use of the computer code in [22].

3 LUBRICATION REGIMES CHART

The prediction of isothermal film thickness from available studies requires identification of a certain regime corresponding to the operating conditions. This is done by the use of lubrication regimes charts [23]. It is verified, as shown in figure 2 for one complete camshaft rotation, that the operating conditions at the cam/roller follower interface represent the elastohydrodynamic lubrication of a heavily-loaded contact.

4 SQUEEZE FILM EFFECTS

The cam/follower configuration has three time-dependent EHD parameters; namely, the equivalent contact radius, contact load, and surface speed. To incorporate these transient effects into the present lubrication analysis, the Reynolds equation, including the squeeze film term, was solved, as outlined in Appendix A. The solution is similar to the one presented by Vichard [24].

5 THERMAL EFFECTS

A thermal analysis was performed similar to the study of Murch and Wilson [25]. Their predictions showed a reasonable match with experimental findings by Wilson [26]. Several new features were introduced in the present study:

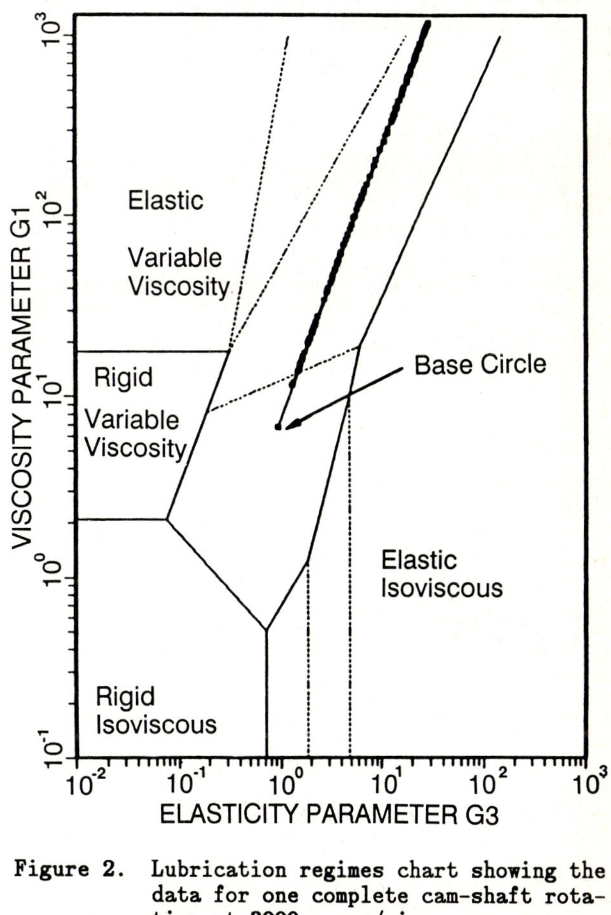

Figure 2. Lubrication regimes chart showing the data for one complete cam-shaft rotation at 3000 cam r/min.

variable surface temperatures, consideration of a bulk solid temperature which can be different than the oil reference temperature, flexibility to consider unequal solid-surface temperatures, and pressure dependent lubricant-thermal-conductivity. The effect of viscous heat dissipation on the film thickness is predicted in the form of a correction factor applied to the isothermal film thickness. The formulation of the thermal-EHD problem is outlined in Appendix B.

6 FATIGUE LIFE PREDICTION

As mentioned in the Introduction, a low specific film thickness shortens the fatigue life [14,15]. The fatigue life depends also on Hertz contact stresses, number of stress cycles, and geometry.

Because the results of Lundberg/Palmgren model are available in forms suitable to roller element bearings only, it was necessary to go back to the fundamental relationships and derive the governing life equation for the roller follower. Lundberg and Palmgren proposed the following expression for the probability Π of surviving N stress cycles:

$$\ell n\ (\frac{1}{\Pi}) = f\ (\tau_o, N, z_o)\ V \qquad (12)$$

They determined the following empirical relationship to match their results,

$$f\ (\tau_o, N, z_o) \sim \tau_o^{\ c}\ N^e\ z_o^{-h} \qquad (13)$$

in which τ_o is the maximum orthogonal shear stress, z_o is the distance from surface where τ_o occurs, V is the stressed volume, and c, e, and h are empirical constants. The stressed volume V is bounded by contact width b, length ℓ, and depth z_o. Hence,

$$\ell n\ (\frac{1}{\Pi}) = K\ \tau_o^{\ c}\ N^e\ z_o^{1-h}\ b\ell \qquad (14)$$

where K is the constant of proportionality. By defining $N = n\Lambda$, where Λ is the life in millions of revolutions, and n is the number of stress cycles per revolution, equation (14) becomes,

$$\Lambda = \frac{[\ell n\ (\frac{1}{\Pi})]^{\frac{1}{e}}\ z_o^{\frac{h-1}{e}}}{K^{\frac{1}{e}}\ n\ \tau_o^{\frac{c}{e}}\ (b\ell)^{\frac{1}{e}}} \qquad (15)$$

Lundberg and Palmgren [16] indicated that $\tau_o = 0.25\ \sigma_m$ and $z_o = 0.5a$. Experiments in [16] with roller bearings yielded: c = 31/3, h = 7/3, and e = 9/8. Hence, for 90 percent probability of survival (Π=0.9), equation (15) becomes, in Newton and millimeter units,

$$\Lambda = 9.673\ 10^{39}\ \frac{(\frac{r_E}{E})^{140/27}}{n\ w^4\ (b\ell)^{8/9}} \qquad (16)$$

where, tedious algebra along the lines presented in [17] yields K to be,

$$K = 1.79\ 10^{-36}\ mm^{20}/N^{31/3}$$

Evaluations indicated that the elastic modulus representative of Lundberg/Palmgren material is $E = 181\ 10^3\ N/mm^2$. Calculation of parameters in equation (16) are outlined in Appendix C.

7 RESULTS AND DISCUSSION

The results presented in this section are based on the input data shown in Table I. The lift data were numerically differentiated to obtain the follower velocity and acceleration which are shown in Figure 3.

Figure 4 shows the lubricant film thickness around the cam circumference. It is clear that the specific film thickness is much smaller than unity at low speeds, even on the base circle. This indicates that boundary lubrication at low speeds is inevitable. At higher speeds, especially on the base circle, conditions are closer to the mixed lubrication regime where the specific film thickness is around unity. The results shown in Figure 4 include thermal and squeeze film effects. The thermal effects were only observed at higher engine speeds (5 percent to 10 percent reduction in film thickness from the isothermal case). Due to the small rates of

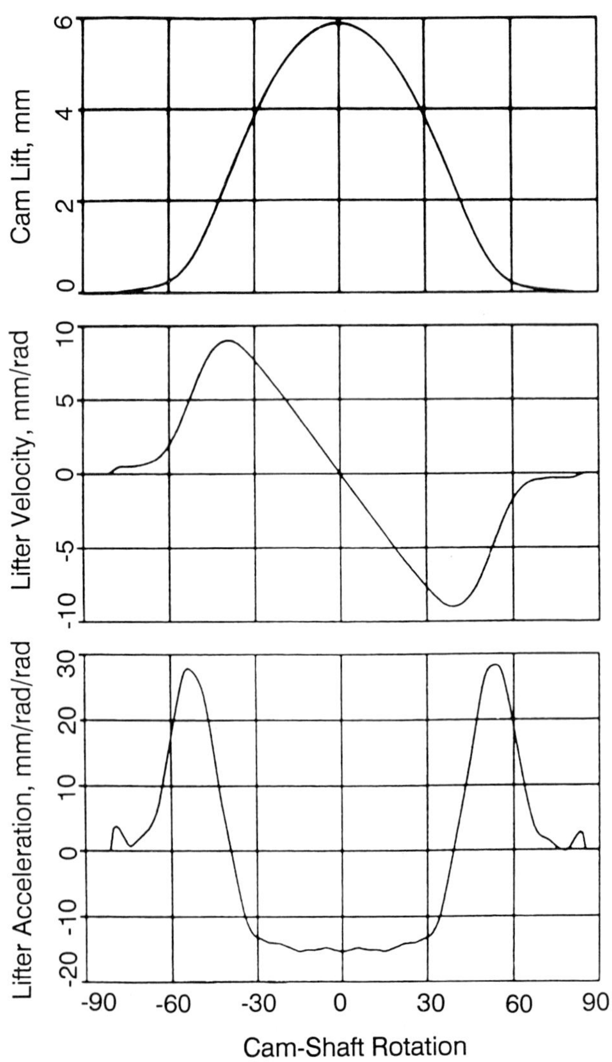

Figure 3. Lift, velocity, and acceleration curves associated with the present cam/roller follower system.

TABLE I

Cam/Roller Follower Operating Conditions

Cam	
Base radius (mm)	17.5
Maximum Lift (mm)	5.8882
Poisson's ratio	0.26
Elastic modulus (GPa)	117.2
Surface roughness, RMS (μm)	0.488
Hardness (HRC)	51.0
Roller Follower	
Outside radius (mm)	8.89
Inside Diameter (mm)	10.7708
Needle bearing shaft diameter (mm)	7.5717
Diametral Clearance (mm)	$24.13\ 10^{-3}$
Total axial length (mm)	12.7
Effective contact length (mm)	8.62
Needle Diameter (mm)	1.5875
Needle effective Contact Length (mm)	8.76
Number of Needles	18
Poisson's ratio	0.3
Elastic modulus (GPa)	203.4
Density (kg/m^3)	7800.0
Needle bearing friction coefficient	0.0045
Surface roughness, RMS (μm)	0.265
Hardness (HRC)	60-64
Valve Train	
Preload (N)	357.9
Spring constant (N/m)	33641.0
Lubricant	
Viscosity at ambient pressure and 100°C (Pa•s)	0.0113
Pressure coefficient of viscosity (m^2/N)	16.0×10^{-9}
Temperature coefficient of viscosity (1/°C)	0.03
Thermal conductivity at ambient pressure (W/m/°C)	0.13

TABLE II

Results of Fatigue life Calculations

	Cam/Roller	Roller/Needle	Shaft/Needle
Average equivalent radius (mm)	6.756	0.93096	0.6562
Equivalent elastic modulus (N/mm^2)	$160.9\ 10^3$	$223.5\ 10^3$	$223.5\ 10^3$
Number of stress cycles per cam revolution	1.049	7.805	11.077
Length of the stressed volume (mm)	55.85	33.837	23.787
Effective contact length (mm)	8.62	8.76	8.76
Average unit load (N/mm)			
250 cam r/min	137.0	30.03	31.78
1700 cam r/min	130.16	28.79	30.48
Fatigue life (number) of cam revolutions)			
250 cam r/min	$2.2\ 10^{12}$	$1.2\ 10^9$	$1.5\ 10^8$
1700 cam r/min	$2.7\ 10^{12}$	$1.4\ 10^9$	$1.8\ 10^8$

change of both load and contact radius, squeeze film effects were negligible over most of the cam cycle except at the transition points between the base circle and the cam flanks. It was verified that the terms in equation (A4) involving the time derivatives are two to three orders of magnitude smaller than the terms representing the steady hydrodynamic behavior except at the transition points where they become comparable.

Figure 5 shows the maximum Hertz pressure around the cam surface. The sudden drop in pressure at the cam nose is due to the change in the direction of the rocker arm friction force. The base circle pressures are generated by the lash adjuster, which is driven by the engine oil pressure. At 250 cam r/min, the contact load is dominated by the spring force; hence, maximum pressure location coincides with the cam nose representing full valve opening. At 3000 cam r/min, the inertia forces associated with the reciprocating valve-train components dominate; hence, the location of the maximum pressure shifts to the flanks where cam acceleration (or deceleration) is greater.

The maximum allowable Hertz pressure for rolling cylinders, from a plastic flow viewpoint, is $\sigma_m=3.1\ k_o$ where k_o is the shear strength, and it is one-half the strength in simple tension and compression. The shear strength in simple tension and compression is one-third the hardness in Vicker scale; hence, it follows that $\sigma_m=3.1*(\text{Vicker Hardness})/6$. Hardness of the cam lobe is approximately 51 HRC (=5.2 GPa); hence, the maximum allowable pressure is 2.7 GPa. The maximum pressure in the speed range shown in Figure 5 is approximately 1.0 GPa. Therefore, the surface deformations stay within the elastic limits.

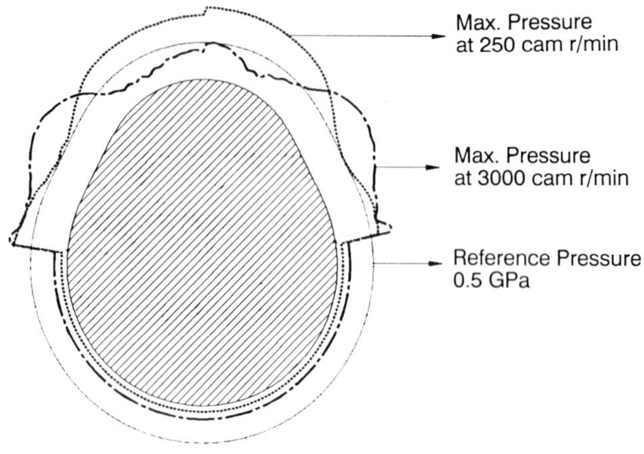

Figure 5. Maximum Hertzian pressure at the cam/roller contact.

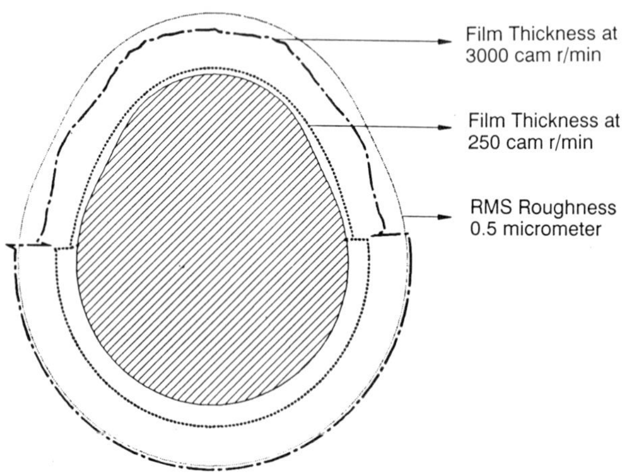

Figure 4. Nominal film thickness at the cam/roller interface, and the representative RMS composite surface roughness.

Figure 6 shows the effects of geometric changes on the film thickness and the pressures at the nose position. Increasing the cam base radius from 17.5 mm to 20.0 mm, and increasing the follower radius from 8.89 mm to 12.0 mm, results in approximately a 20 percent increase in film thickness and a 15 percent decrease in maximum contact pressure.

Figure 7 shows the cam torque as a function of cam angle. The difference between the positive and the negative areas under the torque curve represents the energy loss per camshaft rotation. The overall valve-train energy losses per cycle were found to be 1.04 Joule, 1.20 Joule, and 0.91 Joule at 250, 1000, and 3000 cam r/min, respectively. These are lower than the energy losses with a flat-faced follower. Bair et al. [9] measured the energy losses with flat followers to be 2.5 Joules to 4.0 Joules. They measured the loss with a roller follower to be 18 percent of the loss measured with a flat follower under the same operating conditions. In the present system, the needle bearing contributed approximately 10 percent of the total energy loss in the valve train, and the major source of the energy loss was the rocker arm.

Figure 8 shows the traction force at the cam/follower interface for a no-slip condition. The results are presented in the normalized form as the ratio of the required traction to the normal contact load. At low speeds, inertia is negligible and the only traction required is to overcome the needle bearing friction. It was shown earlier that at low speeds the mode is boundary lubrication; hence, slip at the interface is unlikely. At high speeds the inertia of

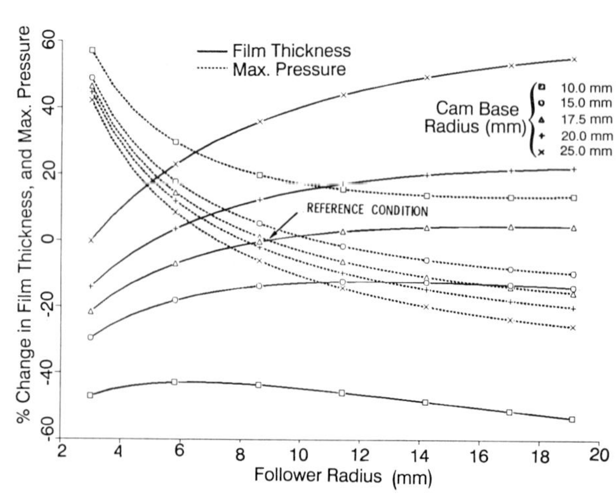

Figure 6. Effects of geometric changes on the film thickness and the Hertzian pressure at the cam nose. Reference condition is: cam base radius = 17.5 mm, follower radius = 8.89 mm.

the follower becomes important. The sign change is due to the reversal of the direction of the traction force during acceleration versus deceleration of the follower. The fluctuations in the required traction are due to the fluctuating normal load at high speed. It was mentioned earlier that at high speed the lubrication mode is, at best, the mixed mode with a specific film thickness around unity; hence, slip still seems unlikely. Lack of inertial forces accompanied by large enough lash adjuster pressures make slip unlikely on the base circle regardless of the cam speed. Winer et al. [27] observed slip only at high speeds and only when the oil temperature was low; even then the amount of slip was negligible. Under their conditions, the present study predicted the specific film thickness to be approximately 3.0 on the flanks. On the base circle, the specific film thickness ranged from 5.0 to 20.0 depending on the lash-adjuster pressure. These values clearly indicate hydrodynamic lubrication which is consistent with their observation of slip.

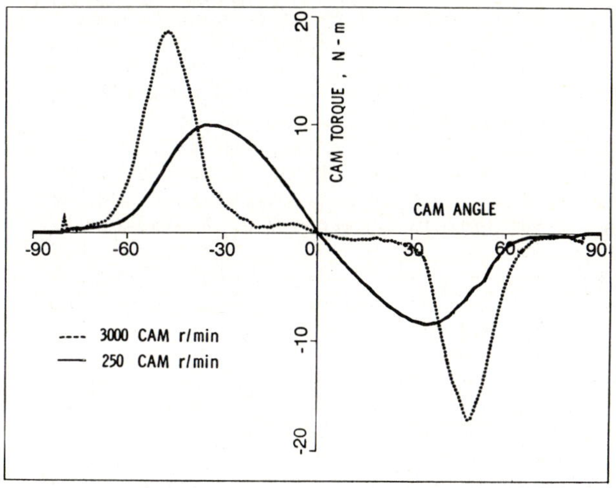

Figure 7. Cam torque versus cam angle.

The fatigue life predictions, in millions of cam revolutions, are shown in Table II. As mentioned earlier, several factors, particularly those related to materials and processing, make fatigue life predictions less than precise. Speculative corrections accounting for these factors are out of the intended scope of this study. Therefore, the fatigue life predictions shown in Table II are reliable for orders of magnitude estimations only. The predicted fatigue life of the roller surface at the cam contact is four orders of magnitude longer than the fatigue life of the shaft. This implies that the shaft is the most likely component to fail by fatigue.

The effect of changing the clearance on the shaft fatigue life is shown in figure 9. Figure 10 shows the effect of number of needles on the shaft fatigue life. The roller outside radius is fixed in all three cases.

8 CONCLUSIONS

The film thickness, especially at low speeds, was smaller than the surface roughness indicating mixed and boundary lubrication modes at the cam and roller follower contact. The coefficient of traction required for no-slip condition was calculated to be 0.002 and 0.04 (max.) at 250 cam r/min and 3000 cam r/min, respectively. Hence, gross slip is unlikely to occur. The squeeze-film effect was noticeable only at the transition points between the base circle and the cam flanks, and the thermal effects became important only at high speeds. The peak contact pressures were approximately 1.0 GPa at both 250 cam r/min and 3000 cam r/min although their locations on the cam surface were different. Plastic flow is not likely to occur because the peak pressures are lower than the yield criterion for the hardened surface. It is demonstrated by a parametric study that design changes can improve the film thickness and the contact pressures.

The maximum cam torques were 10 N•m and 20 N•m for 250 cam r/min and 3000 cam r/min, respectively, and the energy loss per cam rotation was approximately 1.0 Joule over this speed range. This energy loss is approximately 25 percent of what it is with flat-faced followers. The needle bearing in the roller accounted for approximately 10 percent of the energy loss in the valve train.

At 250 cam r/min, roller surface fatigue life was estimated to be $2.2\ 10^{12}$ cam-shaft revolutions. The roller follower shaft's fatigue life was $1.5\ 10^{8}$. At 1700 cam r/min, estimated fatigue life was approximately 20 percent longer. Tighter bearing clearances were shown to prolong roller follower shaft's fatigue life.

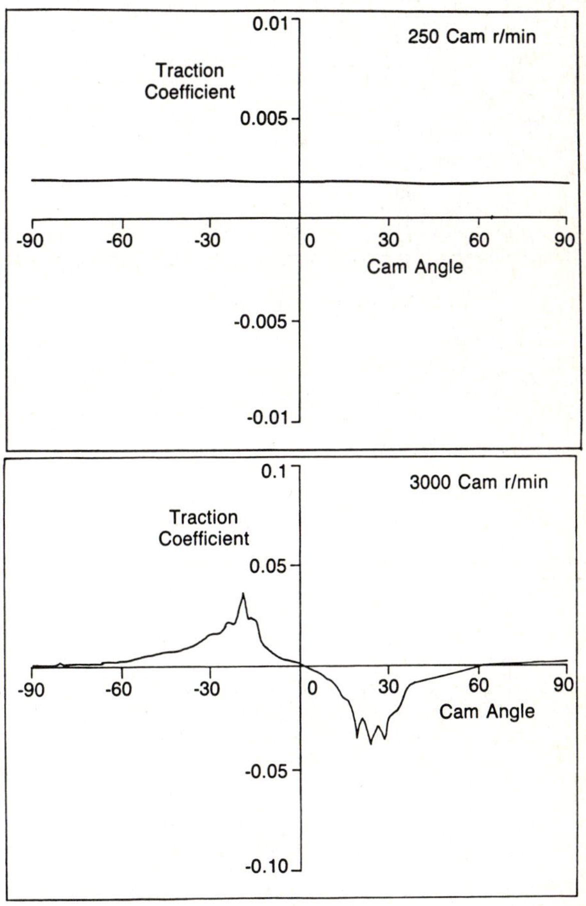

Figure 8. Required traction for no slip.

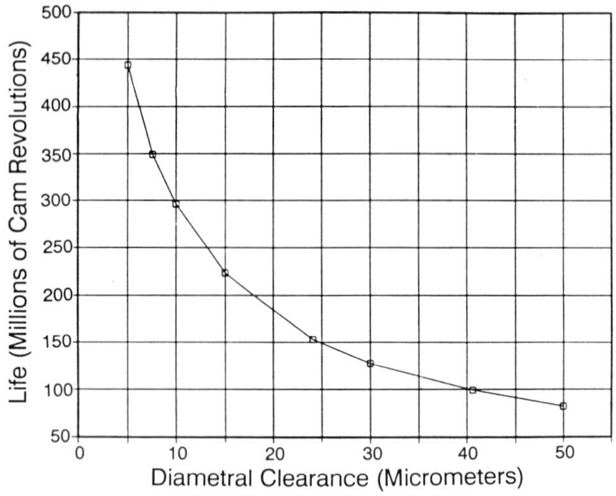

Figure 9. Needle bearing shaft's fatigue life versus needle bearing diametral clearance at 250 cam r/min. Bearing load $F_{c,av}$ = 1181 N.

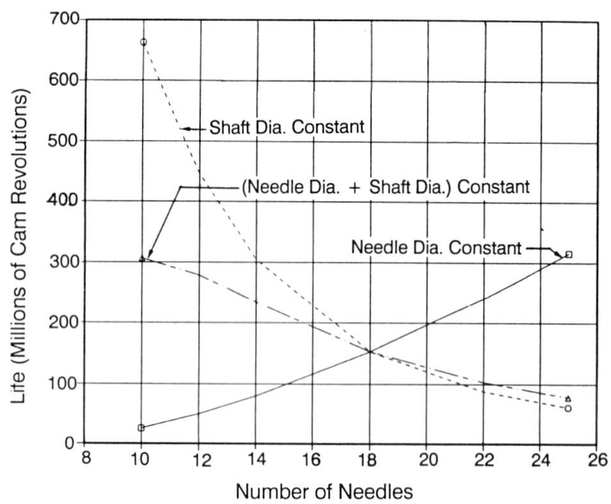

Figure 10. Needle bearing shaft's fatigue life versus number of needles for different design alternatives.

REFERENCES

(1) DYSON, A. and NAYLOR, H., "Application of the Flash Temperature Concept to Cam and Tappet Wear Problems," Proc. Inst. Mech. Engrs., No. 8, 1960-61, 255.

(2) DYSON, A., "EHD Lubrication and Wear of Cams Bearing Against Cylindrical Tappets," SAE Preprint 770018.

(3) DYSON, A., "Kinematics and Wear Patterns of Cam and Finger Follower Automotive Valve Gear," Tribology International, June 1980, 121.

(4) MULLER, R., "Der Einfluss der Schmierverhaltnisse am Nockentrieb. MTZ Motortechnische Zeitschrift, Vol. 27/2, 1966, 58.

(5) HAMILTON, G. M., "The Hydrodynamics of a Cam Follower," Tribology International, June 1980, 113.

(6) DOWSON, D., HARRISON, P., and TAYLOR, C. M., "The Lubrication of Automotive Cams and Followers," Proc. of 12th Leeds-Lyon Symposium on Tribology, 1985.

(7) BEDEWI, A. A., DOWSON, D., and TAYLOR, C. M., "EHD Lubrication of Line Contacts Subjected to Time Dependent Loading with Particular Reference to Roller Bearings, Cams and Followers," Proc. of 12th Leeds-Lyon Symposium on Tribology, 1985.

(8) HOLLAND, J., "Auslegung und Optimierung von Nockentrieben Hinsichtlich des Verschleibverhaltens," MTZ Motortechnische Zeitschrift, Vol. 47/1, 1986, 37.

(9) BAIR, S., GRIFFIOEN, J. A., and WINER, W. O., "The Tribological Behavior of an Automotive Cam and Flat Lifter System," ASME Journal of Tribology, Vol. 108, July 1986, 478.

(10) STARON, J. T. and WILLERMET, P. A., "An Analysis of Valve Train Friction in Terms of Lubrication Principles," SAE Preprint 830165.

(11) SUN, D. C. and ROSENBERG, R. C., "An Experimental Study of Automotive Cam-Lifter Interface Friction," presented at ASLE Annual Meeting at Toronto, Canada, May 12-15, 1986. Paper #86-AM-2B-2, to be published in ASLE Transactions.

(12) DAWSON, P. H., "Effects of Metallic Contact on the Pitting of Lubricated Rolling Surfaces," Journal of Mechanical Engineering Science, Vol. 4, No. 1, 1962, 16.

(13) TALLIAN, T., "Rolling Contact Failure Control Through Lubrication," Proc. Inst. Mechanical Engineers, Vol. 182, Pt. 3A, 1967-68, 205.

(14) TALLIAN, T., "On Competing Failure Modes in Rolling Contact," ASLE Transactions, Vol. 10, 1967, 418.

(15) SKURKA, J., "Elastohydrodynamic Lubrication of Roller Bearings," ASME Transactions, Journal of Lubrication Technology, April 1970, 281.

(16) LUNDBERG, G. and PALMGREN, A., "Dynamic Capacity of Rolling Bearings," Journal of Applied Mechanics, Vol. 16, No. 2, June 1949, 165.

(17) HARRIS, T. A., "Rolling Bearing Analysis, Second Edition," Wiley-Interscience Publication, 1984, 434.

(18) "Life Adjustment Factors for Ball and Roller Bearings," published by The American Society of Mechanical Engineers, 1971

(19) McCOOL, J. I., "Load Ratings and Fatigue Life Prediction for Ball and Roller Bearings," ASME Journal of Lubrication Technology, Vol. 92, No. 1, January 1970, 16.

(20) VAESSEN, G. H. G. and DeGEE, A. W. J., "Rolling Contact Fatigue of Managing Steels of Different Production History: Influence of Film Thickness/Roughness Ratio," Elastohydrodynamic Lubrication, 1972 Symposium. A symposium arranged by the Tribology Group of the Institution of Mechanical Engineers 11-13th April 1972, 40.

(21) PALMGREN, A., Ball and Roller Bearing Engineering, 3rd Edition, S. H. Burbank and Company, Inc., Philadelphia, 1959.

(22) PARANJPE, R. S., "Dynamic Analysis of a Valve Spring with a Coulomb-Friction Damper," GM Research Publication GMR-5965, August 26, 1987.

(23) **GECIM, B.**, "The Review and Enhancement of the Lubrication Regimes Chart," manuscript in preparation.

(24) VICHARD, J. P., "Transient Effects in the Lubrication of Hertzian Contacts," Journal of Mechanical Engineering Science, Vol. 13, No. 3, 1971, 173.

(25) MURCH, L. E. and WILSON, W. R. D., "A Thermal Elastohydrodynamic Inlet Zone Analysis," ASME Transactions, Journal of Lubrication Technology, April 1975, 212.

(26) WILSON, A. R., "An Experimental Thermal Correction for Predicted Oil Film Thickness in Elastohydrodynamic Contacts," Thermal Effects in Tribology, Proc. of 6th Leeds-Lyon Symposium on Tribology, September 1979.

(27) WINER, W. O., personal communication during a consulting visit to GM Research Laboratories, November 1986.

APPENDIX A

Squeeze Film Effects

This Appendix outlines an analytical method to solve the one-dimensional Reynolds equation including the squeeze film term. This equation has the form,

$$\frac{\partial P}{\partial x} = 12\bar{u}\mu \frac{h-h_o}{h^3} + 12\mu \left[\int \frac{\partial h}{\partial t} dx - \int \frac{\partial h}{\partial t} dx \Big|_{x=x_o}\right] \quad (A1)$$

The inlet zone film shape, including the elastic deformation, can be approximated by

$$h = h_o + 2.401 \frac{w}{E} \left(\frac{x}{a} - 1\right)^{3/2} \quad (A2)$$

Applying the condition

$$\frac{\partial P}{\partial x} = 0 \text{ at } x = x_o = a$$

yields

$$e^{-\alpha^* P} \frac{\partial P}{\partial x} = 12 \mu_o \bar{u} \frac{h-h_o}{h^3} + \frac{12 \mu_o}{h^3} \left[a \frac{\partial h_o}{\partial t} \left(\frac{x}{a} - 1\right)\right.$$

$$+ 0.9604 \frac{a}{E} \frac{\partial w}{\partial t} \left(\frac{x}{a} - 1\right)^{5/2}$$

$$\left. - 0.4802 \frac{w}{E} \frac{\partial a}{\partial t} \left(3 \frac{x}{a} + 2\right) \left(\frac{x}{a} - 1\right)^{3/2}\right] \quad (A3)$$

where $\mu = \mu_o e^{\alpha^* P}$

Integrating (A3) in the inlet zone of the EHD contact with the boundary conditions,

$P \to 0$ as $x \to \infty$ and $P = P_I$ at $x = a$

yields, after considerable algebra,

$$\frac{dh_o}{dt} = C_1 h_o^{5/3} + (C_2 - C_3) h_o^{1/3} + C_4 h_o \quad (A4)$$

where

$$C_1 = 0.997 \frac{w^{4/3}}{E^{4/3} \mu_o \alpha^* a^2} \qquad C_2 = 1.7929 \frac{w^{2/3}}{aE^{2/3}} \frac{\partial a}{\partial t}$$

$$C_3 = 1.7929 \frac{w^{2/3}}{E^{2/3}} \frac{\bar{u}}{a} \qquad C_4 = -0.2 \frac{1}{w} \frac{dw}{dt}$$

In the numerical evaluations, the time derivative, dt, is replaced by $dt = d\theta/\omega$, and the derivatives $da/d\theta$ and $dw/d\theta$ are evaluated by numerical differentiation of $a(\theta)$ and $w(\theta)$. The derivative $dh_o/d\theta$ is approximated in the finite difference form by

$$\frac{h_o(\theta=\theta_{i+1}) - h_o(\theta=\theta_i)}{\theta_{i+1} - \theta_i}$$

where the initial condition was taken as the steady operating conditions associated with the base circle.

APPENDIX B

Thermal Effects

This Appendix outlines the formulation of the thermal-EHD problem. The energy equation governing the fluid has the form,

$$\frac{\partial^2 \bar{T}}{\partial \bar{y}^2} = - \frac{A}{1-\bar{P}} e^{-\bar{T}} \left(\frac{\partial \bar{u}}{\partial \bar{y}}\right)^2 \quad (B1)$$

where

$$\bar{T} = (T-T_R)\beta, \quad \bar{u} = \frac{u}{u_R}, \quad \bar{y} = \frac{y}{h},$$

$$\mu = \mu_o e^{\alpha^* P} e^{-\beta(T-T_R)}, \text{ and } A = \left(\frac{\mu_o u_R^2 \beta}{k_L}\right)$$

Equation (B1) is integrated twice with the conditions

$$\bar{T} = \bar{T}_1 \text{ at } \bar{y} = 0 \quad \text{and} \quad \bar{T} = \bar{T}_2 \text{ at } \bar{y} = 1$$

where

$$\bar{T}_{1,2} = \bar{T}_{B_{1,2}} + 1.596 \frac{k_L}{k_S} \left[\frac{(L_H-1)\, W}{2\pi\, Pe_{1,2}}\right]^{1/2} \frac{1}{\bar{h}} \int_o^{\bar{x}} \frac{\bar{q}_{1,2}}{\sqrt{\bar{x}-\bar{\xi}}} d\bar{\xi} \quad (B2)$$

Respective heat fluxes are given by

$$\bar{q}_1 = \frac{\partial \bar{T}}{\partial \bar{y}} \Big|_{\bar{y}=0} \quad \text{and} \quad \bar{q}_2 = - \frac{\partial \bar{T}}{\partial \bar{y}} \Big|_{\bar{y}=1}$$

A partition factor ϵ is defined as $\epsilon = \bar{q}_1/\bar{q}_T$, where

$$\bar{q}_T = \bar{q}_1 + \bar{q}_2 = \frac{A}{1-\bar{P}} \int_o^1 e^{-\bar{T}} \left(\frac{\partial \bar{u}}{\partial \bar{y}}\right)^2 d\bar{y} \quad (B3)$$

Calculation of the total heat flux from equation (B3) requires P, and $(\partial \bar{u}/d\bar{y})$. They are obtained from the solution of the thermal Reynolds equation. It has the form

$$\frac{\partial \bar{P}}{\partial \bar{x}} = \frac{1.596\ UGW^{1/2}}{\bar{h}^2}\left[\frac{(1-\frac{1}{2}\Sigma)(1-\frac{\bar{h}_o}{\bar{h}}) + \Sigma(\frac{\bar{I}_1}{\bar{I}_2}-\frac{\bar{I}_3}{\bar{I}_4})}{\frac{\bar{I}_5\,\bar{I}_1}{\bar{I}_2}-\bar{I}_6}\right] \quad \text{(B4)}$$

where,

$$\bar{I}_1 = \int_0^1\int_0^{\bar{y}} e^{\bar{T}}\,d\bar{y}'\,d\bar{y}\ , \qquad \bar{I}_2 = \int_0^1 e^{\bar{T}}\,d\bar{y}$$

$$\bar{I}_3 = \int_0^{h_o/h}\int_0^{\bar{y}} e^{\bar{T}}\,d\bar{y}'\,d\bar{y}\ , \quad \bar{I}_4 = \int_0^{h_o/h} e^{\bar{T}}\,d\bar{y}$$

$$\bar{I}_5 = \int_0^1 \bar{y}e^{\bar{T}}\,d\bar{y}\ , \qquad \bar{I}_6 = \int_0^1\int_0^{\bar{y}} \bar{y}'\,e^{\bar{T}}\,d\bar{y}'\,d\bar{y}$$

Equations (B1), (B2), (B3), and (B4) are solved by iterations between the temperature and the pressure fields until convergence of film thickness, h_o, is achieved. A least-squares method yields the thermal correction factor ϕ_T,

$$\phi_T = \frac{1.0}{1.0 + 0.2494\ A^{0.7137}}\ ,\quad A = \frac{\mu_o\,u_R^2\,\beta}{k_L} \quad \text{(B5)}$$

APPENDIX C
Calculation of Fatigue Life Parameters

Table II shows the parameters that are used in equation (16) of the text to calculate the fatigue life. The methods to calculate these parameters are outlined below.

Number of Stress Cycles: The number of stress cycles at the shaft/needle and roller/needle contacts are per roller revolution and can be calculated from the information in [17]. The number of stress cycles at the roller outside surface is the number of roller rotations per cam revolution. To find the number of roller rotations, roller angular speed is integrated over the cam event. Letting θ_1 and θ_2 denote the beginning of valve opening and the end of valve closing, respectively, the angle swept by the roller can be written as,

$$\Psi = \int_{\theta_1}^{\theta_2} \left(\frac{\omega_F}{\omega_C}\right) d\theta \quad \text{(C1)}$$

The number of roller rotations during the cam event is $\psi/(2\pi)$. The roller body at the cam/roller interface is stressed 1.049 times per cam revolution. The total number of roller rotations per cam revolution including the base circle is 2.07.

Length of the Stressed Volume: The length of the stressed volume is πD_i and πD_o at the shaft/needle and the roller/needle contacts, respectively. D_i and D_o are the needle bearing shaft's diameter and the roller inside diameter, respectively. The length of the stressed volume at the roller outside surface is $2\pi r_F$.

Average Load: Since the load in the life equation, equation (16) of the text, appears as a fourth power, Harris [17] suggests using a quartic mean for the load as follows:

$$F_{c,av} = \left[\frac{1}{\theta_2-\theta_1}\int_{\theta_1}^{\theta_2} [F_c(\theta)]^4 d\theta\right]^{1/4} \quad \text{(C2)}$$

The average contact force calculated above is taken as the bearing load. The individual needle loads and the average loads at the shaft/needle and roller/needle contacts are computed in the standard way [17].

Average Equivalent Radius: The average equivalent-contact radius at the cam/roller interface is calculated in the same manner as the average load. The fourth power is replaced by the power 140/27 in accordance with equation (16) of the text.

Paper IV(ii)

Predictions of cam wear profiles

R. H. Fries and C. A. Rogers

This paper describes a computational method for predicting the profiles into which cams and followers wear in service. The method is conceptually simple. However, there are several salient features required to model the subtleties of tribological interactions. The first step is to solve the kinematic constraint problem for the cam and follower. This step determines the locations of the contact patch center on the cam and follower for all cam rotation angles. From these results, the follower displacement, velocity, and acceleration are determined. The next step is to solve the cam dynamics problem to obtain the normal forces at the contact patch. The tangential contact forces are determined assuming Coulomb or other friction conditions that may exist in the contact region. A wear model is then employed to estimate the localized wear as a function of the cam angle.

The cam and follower profiles are modified according to the amount of wear estimated, and new cam and follower profiles are obtained. The process is repeated sequentially so that the continuous wear process is approximated by a number of discrete wear steps.

This method has potential for predicting the shapes into which cams and followers will wear in service. The method is applicable to other mechanical components as well, and it has been applied to the wear process of rail vehicle wheels.

1 INTRODUCTION

Cams are used in a wide variety of machines and instruments such as textile machines, printing presses, internal combustion engines, and agricultural equipment. Typically cams are used to provide a specific prescribed motion to an output mechanical component, the follower. The motion is often complex and often requires precise timing. The motion of the cam is transferred to the follower system by direct contact or what is referred to as a higher pair joint (Mabie and Reinholtz, 1987). The contact between a cam and follower system often occurs in harsh environments, i.e., high temperatures, little or no lubrication, and high contact stresses. The contact conditions can affect the performance of the cam system dramatically by introducing wear to the cam and follower system. Generally cam systems are designed for specific applications assuming that the synthesized cam contour and follower optimally satisfy the performance criteria initially, that is with no cam or follower wear. Wear of the cam and follower surfaces will result in deterioration of the initially optimized performance whether that performance be prescribed motion, timing, or dynamic control.

Although wear is a commonplace phenomenon, it is by no means an uncomplicated one; on the contrary, the mechanisms and theory of wear are very complex. Despite voluminous published data reporting individual wear experiments, as yet, no quantitative empirical relations connecting the amount of wear with such operating parameters as load, speed, and material constants have been developed (Chen, 1982). Wear is the result of a complex set of factors that include the specific conditions of service. The properties of the cam material, the mating materials, and the operating conditions can combine to cause different types of wear that may act singly or in combinations. At any particular time, several known wear mechanisms, for example abrasion, corrosion, and adhesion, may be contributing to the overall degradation of performance. However, Chen (1982) claims that "Although all forms of wear may occur at one time, surface fatigue is the dominant cause of failure in cam and follower system."

It is clear, however, that any one theory for wear applied to simple geometries and operating conditions will not be entirely embraced by the tribological community, and agreement for complex mechanical components is not expected to be any more plausible. Therefore, one useful tool for design of cam systems is a computational method for predicting the evolution of worn cams allowing for animation and investigation of degraded performance (i.e., prescribed motion, timing, dynamics), the effect of various wear models (wear assumptions), operating conditions, and design variables. This paper describes such a computational method for predicting the profile of worn cams and followers. The method simulates the physical evolution of the worn profiles, and it contains the versatility to study the effects of assuming various wear modes and models, different operating conditions, and the numerous design parameters.

This work on cams was motivated in part by wear prediction work in rail vehicle wheels (Fries and Davila, 1987). The general approach to wear predictions was developed in the rail vehicle system work. We recognized that cam wear predictions would include a number of similarities. Predictions of worn rail vehicle wheel profiles have shown good qualitative agreement with service worn wheel profiles.

2 METHOD OF SOLUTION

2.1 Kinematics

Joints between two adjacent links are classified as lower pair or higher pair joints depending on the geometry of the contacting surfaces and the nature of constraints for the relative motion of the links. For higher pair joints, such as cams and followers, the geometry of the contacting surfaces is required to be synthesized for a given relative motion between the adjoining links. On the other hand, in lower pair joints, the geometry of the contacting surfaces as well as the degrees of freedom of the relative motion are completely determined by the type of joint specified.

Synthesizing the geometry of one of the contacting surfaces for a given geometry of the other contacting surface and a known relative motion between the adjoining links can be described as the problem of kinematic synthesis in a higher pair mechanism (cam and follower). In the case of a cam mechanism, the problem consists of synthesizing the cam profile for a given geometry of the follower profile and a specified input-output displacement relationship. Numerous graphical, geometrical, and analytical methods have been presented. Notable among these are the geometrical methods of Rothbart (1956) and Kloomok and Muffley (1956). Analytical methods for cam profile synthesis have been developed by Raven (1959) and others while numerical methods, based primarily on finite difference methods, have been presented by Zigo (1969), Johnson (1955), and Chen (1972). However, none of these methods can be extended to the problem of kinematic analysis of higher pair mechanisms (Reinholtz et al., 1978).

The problem of kinematic analysis of higher pair mechanisms, on the other hand, deals with evaluating the relative motion of adjoining links knowing the geometry of their contacting surfaces. In the case of cam mechanisms, this problem consists of finding the output motion of the follower for a given input motion of the cam as well as the specified geometry of the cam and follower surfaces. Although various geometrical and analytical methods have been developed for profile synthesis, only limited work has been reported in the area of kinematic analysis of higher pair mechanisms in general, and of cam mechanisms in particular.

In higher pair mechanisms, solutions to the kinematic analysis problem can be found only for simple geometries of adjoining links and contacting surfaces. In most cases, it is necessary to resort to a numerical technique for the solution of the problem of kinematic analysis (Reinholtz, et al., 1978).

In this work, a somewhat different approach was taken in determining the geometric constraints for the cam mechanism than has been reported in the kinematic literature cited above. Cooperrider et al. (1975) developed an algorithm that provides the means of determining the constraint relationships (contact points, etc.) for railroad wheel and rail geometries. The approach taken to modeling railroad wheel and rails was modified and adapted to determine the location of the contact points between a translating flat-face follower and a planar disk cam. The basic approach begins with describing mathematically the cam and follower profiles by fitting a series of fourth-order polynomials to numerical profile coordinates. Curve fitting is used rather than interpolation between the data points because it provides some numerical smoothing of irregularities in the input data. Once the discreet profile coordinates of the cam and follower are described mathematically in functional form, the contact locations are determined by applying a numerical search procedure that converges to the contact points satisfying the following definition (Cooperrider et al., 1976):

- Position Equation -- a contact point on the cam occupies the same point in space as the corresponding contact point on the follower.
- Rigid Body Constraint -- the cam profile cannot penetrate into the follower profile.
- Slope Constraint -- the profiles are tangent at the contact points.

The numerical procedure to find the contact point locations involves an iterative search. At each iteration step, the orientation and position of the cam and follower are specified and the potential contact points between them are searched to find the point that meets the definition of a contact point.

It is also necessary to know the follower motions. For a given cam position, the cam and follower profile points are fit with splines, and the contact point is found by determining the intersection of the splines. Follower displacements are then obtained for one revolution of the cam. To obtain follower velocity, a least squares fit to the displacement curve is performed and the resulting polynomial differentiated. The results are then smoothed before proceeding further. To obtain the follower accelerations, a least squares fit to the velocity results is obtained and differentiated. The discreet acceleration results are again smoothed.

For a given cam angular velocity, the above method provides follower displacement, velocity, and acceleration, and the location of the contact point on the cam and follower for any cam position.

The kinematic constraint solutions were verified by comparing positions, smoothed velocities, and smoothed accelerations obtained numerically to those obtained analytically for simple cam profiles. The results of our numerical method were quite good and are presented in the results section below.

As the cam and follower wear, their profiles change. In order to simulate the wear process, it is necessary to change the profile definition points and solve the cam kinematic analysis problem for the new profiles.

2.2 Cam Dynamics

In this work a simple cam geometry and mechanical system were selected to illustrate the method of solution. A flat follower with a given mass is spring loaded against the cam as shown in Fig. 1. The cam dynamics are straightforward, and the normal contact force can be computed easily once the cam kinematics

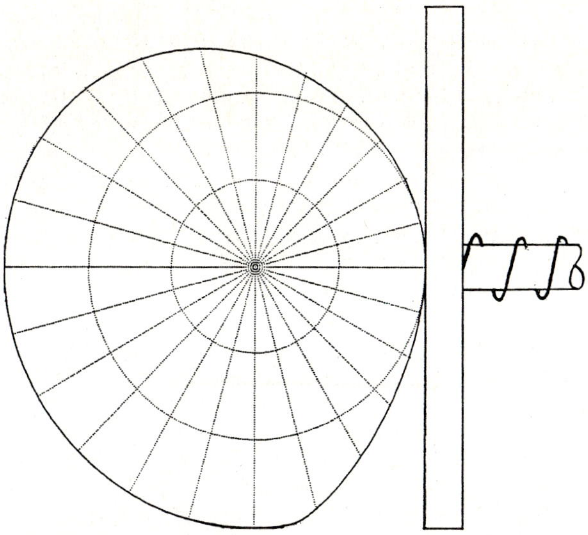

Fig. 1 The cam/follower system

solution is obtained as described above. The contact force is given by

$$F_c = m\ddot{x} + F_{s0} + kx \qquad (1)$$

where

F_c = the contact force

m = the mass of the follower

$\ddot{x}$ = the acceleration of the follower

F_{s0} = the spring preload force

k = the spring rate

and

x = the follower lift

More complex cam dynamics can be accommodated with an increase in computation time.

Coulomb friction conditions were assumed to exist at the contact patch. However, other assumptions could be made and incorporated into the algorithm. The presence of a lubricating film at the contact interface would influence the dynamics. This effect can also be included, but it has been omitted here to focus on the solution method.

2.3 Wear Model

Any one of the numerous wear models available can be incorporated in the algorithm. For the system discussed here, we have assumed the wear process to be described by a simple wear model of the Archard (1953) type wherein wear is postulated to be proportional to normal load and sliding distance. The general method admits the use of essentially any wear model, and the Archard model was used in this work for its simplicity.

For each increment of cam rotation, a wear volume is computed according to the chosen wear model. The points describing the cam and follower profiles are then changed in proportion to the wear volume. The process is repeated sequentially to simulate the continuous wear process and results in an evolution of the worn profile.

The wear model is given simply by:

$$W_V = KF_cD \qquad (2)$$

where

W_V = the wear volume

F_c = the contact force

D = the sliding distance

and

K = a constant of proportionality

The contact force is computed using eq. (1) for each one-degree increment of cam rotation. The sliding distance is obtained as follows. Since the cam rotates about a fixed center, the tangential velocity at the contact point is

$$V_t = R\dot{\theta}_{cam} \qquad (3)$$

where

V_t = the tangential velocity at the contact point

R = distance from center of cam to contact point

$\dot{\theta}_{cam}$ = rotational speed of the cam

See Fig. 2. The velocity of sliding, V_s, is

$$V_s = V_t \cos\gamma \qquad (4)$$

or

$$V_s = R\dot{\theta}_{cam}\cos\gamma \qquad (5)$$

where γ is as shown in Fig. 2. The angle γ is computed as part of the kinematic constraint problem solution. The sliding distance is given by

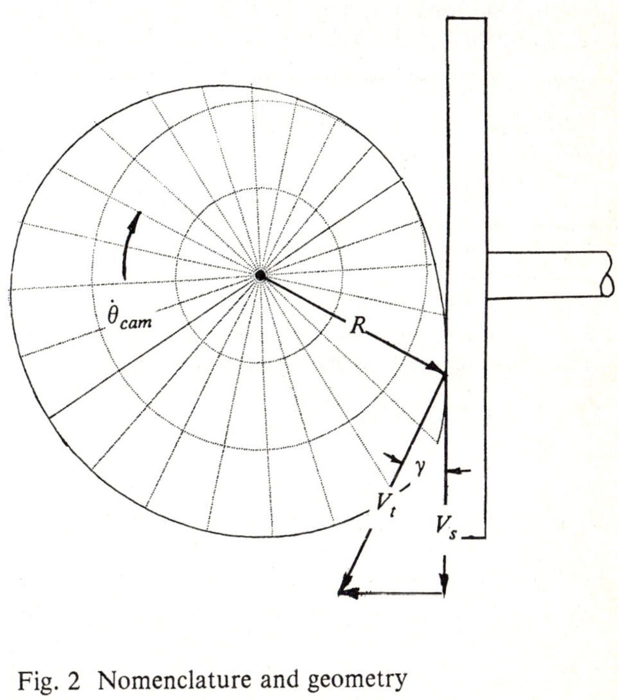

Fig. 2 Nomenclature and geometry

$$D = \int_0^{\Delta t} V_s dt \qquad (6)$$

where

Δt = the time for the cam to rotate by one degree

As an approximation, we have used

$$D = V_s \Delta t \qquad (7)$$

Thus the wear volume is defined. The wear rate is determined by distributing the wear volume over the surface of the cam or follower where contact takes place for each one-degree increment in cam rotation. Since the contact position on the cam and follower are solutions of the kinematic constraint problem, these contact lengths are known.

3 RESULTS

This section of the paper contains wear predictions for one step of the general prediction method described above. This step illustrates the method and brings to light some interesting results. The cam system used in the examples was selected to show how the method proceeds rather than to be representative on an actual cam installed in a known piece of machinery.

3.1 Problem Definition

The cam mechanism selected for demonstration purposes consists of a spring preloaded follower on a planar disk cam as shown in Fig. 1. The cam was synthesized analytically by using the "Theory of Envelopes" presented by Churchill and Hansen (1962). The cam has a base circle of 100 mm and a total lift of 50 mm. The cam imparts cycloidal motion to the follower for a 50 mm lift during cam angles of 0.0° to 180° assuming a constant angular velocity. The angular velocity of the cam is 1200 rpm. The follower dwells from 180° to 270°. The follower returns, or falls 50 mm in the remaining 90°. The mass of the follower is 0.1 kg, and the stiffness of the spring is 4.04 N/mm. A 50-mm spring precompression provides a bias contact force of 202 N.

Cycloidal motion is often used on cams containing dwells because the motion provides zero acceleration at both ends of the action. Therefore, it can be coupled to a dwell at each end without creating a discontinuity in the acceleration profile of the follower. Figure 3 shows typical cycloidal motion characteristics for the rise and return motion and the functional description of the motion that is used in the evaluation of the described computational techniques. Figure 3 was taken from Mabie and Reinholtz (1987).

3.2 Kinematic Analysis

Perhaps the most critical aspect of this computational wear method is the solution to the kinematic constraint problem. The introduction above described the difficulty associated with performing a kinematic analysis of a higher pair. Further review of the literature will show that most numerical methods used to determine the kinematic constraint of cam mechanisms are unable to control a propagation of error that occurs on taking the derivatives for the velocity and acceleration profiles (see Reinholtz et al., 1978). The accuracy of the position of the contact points between the cam and follower is paramount. However, it is equally clear that the accuracy of the dynamic analysis, for example the contact force, is critical to a reliable solution method.

The theoretical lift versus cam rotation angle is superimposed on the numerical solution in Fig. 4. The numerical solution is based upon the discrete cam profile points. It uses the contact point algorithm to determine the lift of the follower. The two curves are indistinguishable and appear as a single curve.

Therefore, Fig. 4 demonstrates the accuracy of not only performing a numerical kinematic analysis of a higher pair mechanism, but it also shows the accuracy of the contact point determination algorithm. Recall that the position of the follower is simply the y-component of the contact point between the cam and follower for an unworn follower.

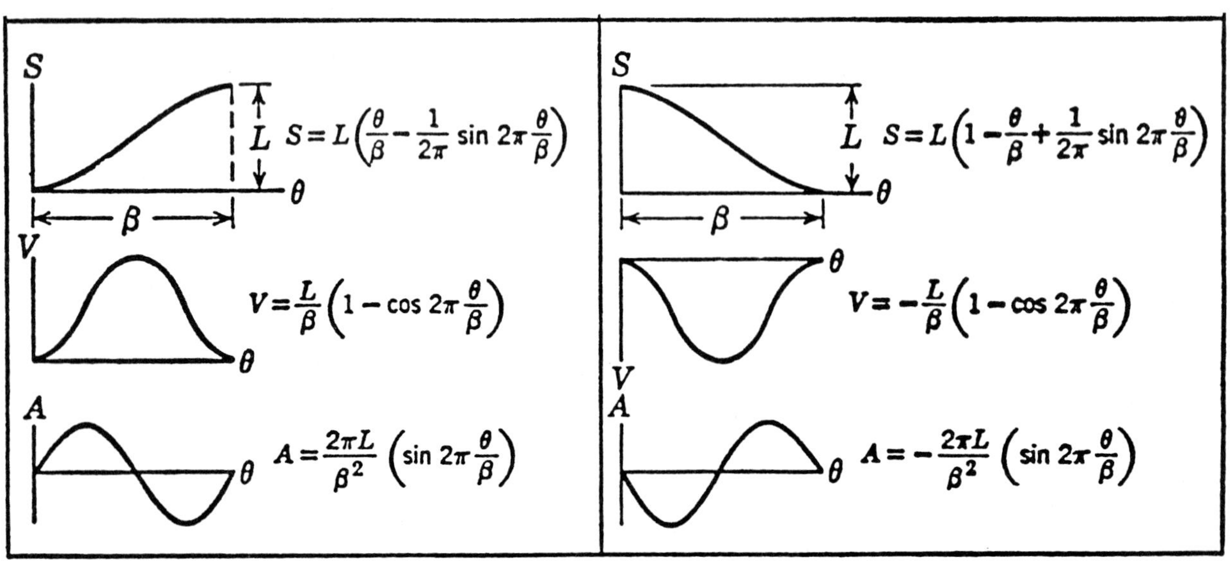

Fig. 3 Cycloidal lift and return definitions

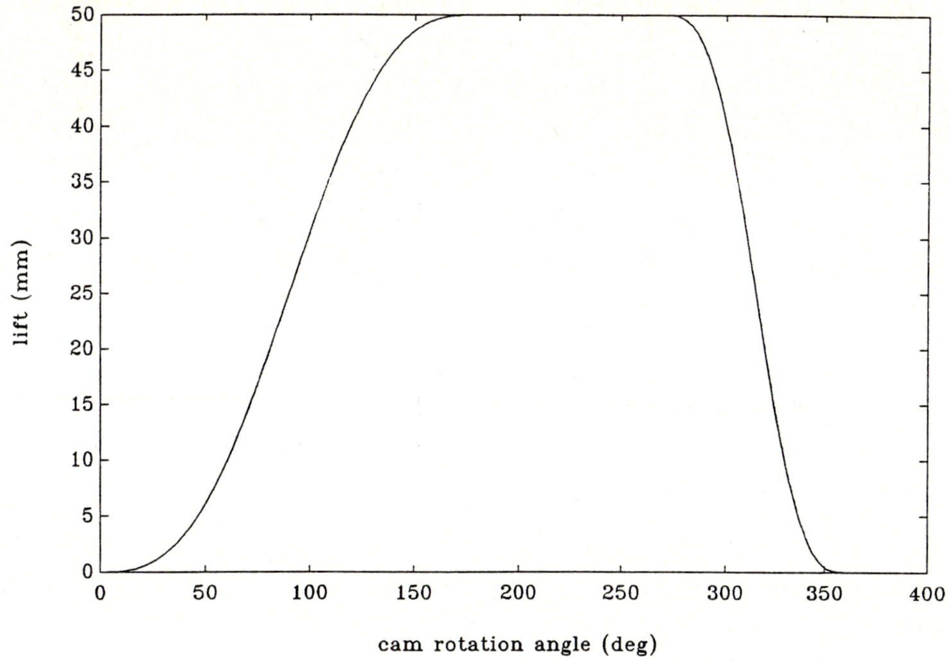

Fig. 4 Comparison of numerical and analytical follower lift

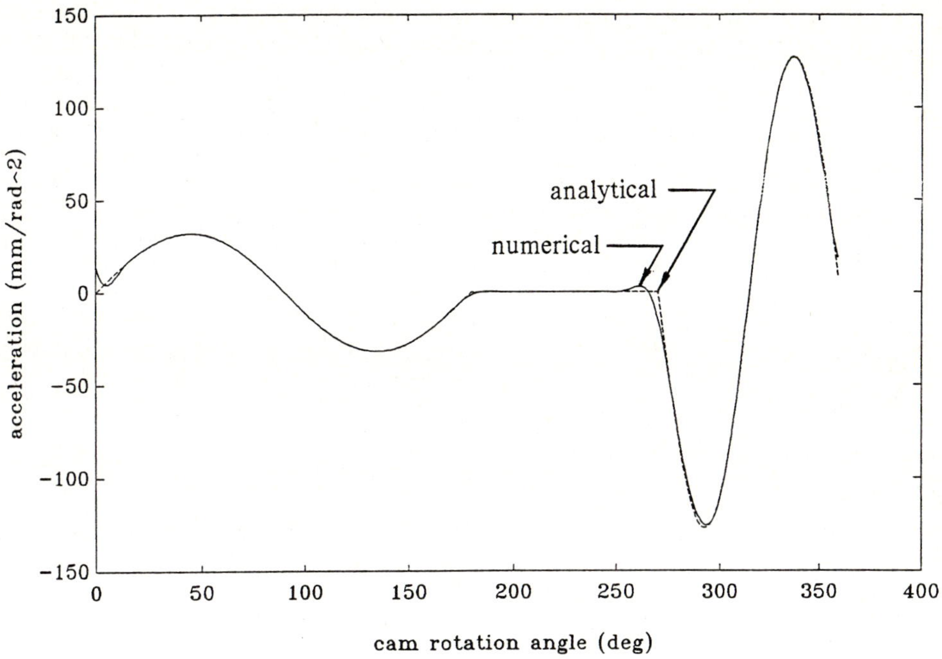

Fig. 5 Comparison of numerical and analytical follower acceleration

Typically, taking the derivatives of the follower motion to determine the velocity and acceleration of the follower generates inaccuracies. The solution method described above to evaluate the acceleration of the follower has proven to be an extremely accurate solution technique for a certain class of follower motion. The theoretical acceleration as shown in Fig. 3 and the numerically determined accelerations are superimposed in Fig. 5 to illustrate both the accuracy of the method and also a current limitation of the technique. Cycloidal motion coupled to a dwell results in a piecewise continuous acceleration profile which in turn would produce a discontinuity in the jerk function. In the chosen motion profile there are three points of slope discontinuity in the accelerations; one each at the coupling point between the cycloidal and dwell motion and one at the coupling point between the return stroke and rise. The method described above for determining the velocity and acceleration profiles involved smoothing the calculated functional values. Therefore, the smoothing operations result in forcing the functions to be continuous. The results shown in Fig. 5 clearly indicate that the solution method devised for the kinematic constraint and subsequent numerical differentiation is a relatively accurate technique, and the results presented below will show that the accelerations and resulting dynamic loads are not nearly as critical to the wear of the mechanical components as the geometric considerations.

3.3 Cam Dynamics

The contact force, F_c, depends upon the spring preload, the spring stiffness, the lift, and the follower acceleration. We plotted the contact force in order to see how it varies as a function of contact angle on the cam. Figure 6 shows this plot. In Fig. 6 the contact force is normalized to 100 percent. The cam profile is shown in mm. Starting at the beginning of the rise, cam force increases until the dwell is reached, after which the force is constant. Immediately upon initiation of the return motion, the contact force decreases sharply because of the inertial forces. Then the force increases as shown, and returns to its initial value.

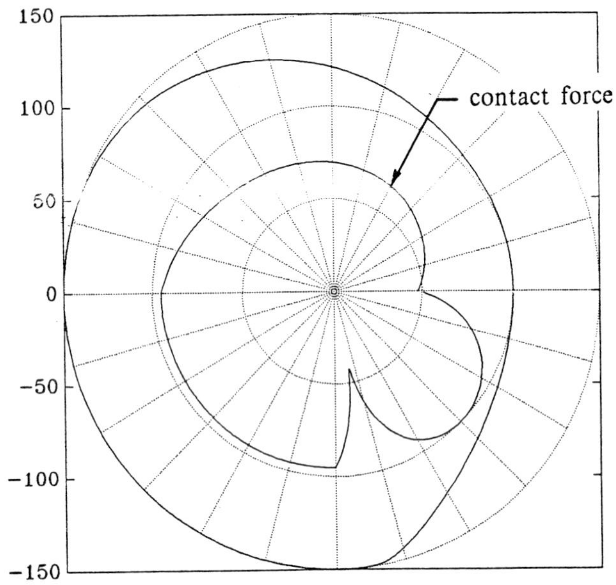

Fig. 6 Contact force

3.4 Cam Wear

At the outset of this work, we thought the wear would be highest at locations on the cam where the contact force is highest. This is in fact the case when the sliding velocities are about the same for two different contact force situations. However, we learned that the contact sliding conditions are far more important than the contact force in the case we have studied.

Figure 7 shows cam relative wear rate plotted as a function of cam contact position. The figure shows that the wear rate is highest near the beginning of the return motion, the same place where the contact force was shown in Fig. 6 to be the lowest. Figure 8 shows the contact force and the wear rate, both normalized to unity, on the same axes. It can easily be seen in Fig. 8 that as follower contact leaves the dwell position at 270°, the contact force decreases sharply and the wear rate increases sharply. The key to understanding this result is knowledge of the contact point location on the cam. Figure 9 shows the contact point location on the cam as a function of cam rotation angle. A substantial dwell occurs at about 280° on the cam. As the cam rotates, the contact point on the cam is nearly stationary. Since wear can occur only where there is contact, the wear rate is very high at the nearly stationary contact point.

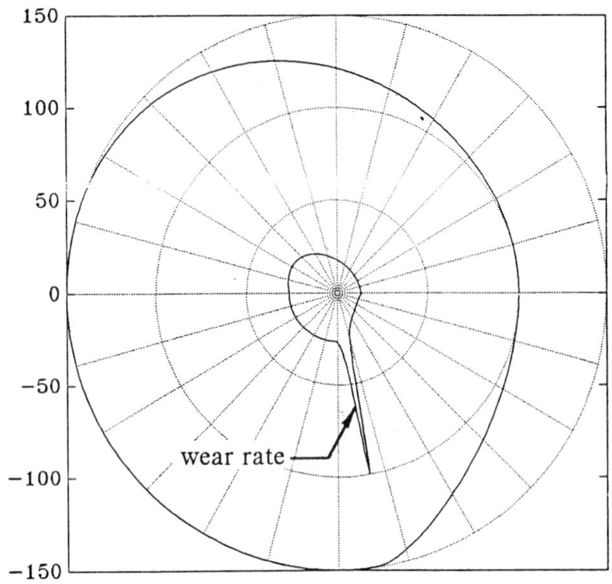

Fig. 7 Cam wear rate

This result has significant design implications. The usual recommendations on cam design focus on the importance of accelerations, and the resulting dynamic contact forces. In designing a cam for long wear life, our results indicate that it may be more important to design cams that avoid nearly stationary contact points.

3.5 Follower Wear

We predicted follower wear using the same general method for predicting cam wear. The cam contacts the follower near its center for most of the cam rotation. Our cam has a fairly long dwell, and the follower contact is at the center for the entire dwell period. Figure 10 shows the relative wear rate distribution on the follower. The wear rate is by far highest at the center where contact occurs most of the time. The follower wear situation is even worse than the cam wear situation because the contact force is high during the 90° dwell at the same time the contact point is stationary on the follower.

4 DISCUSSION

The initial cam and follower profiles would be worn by an amount proportional to the values shown in our relative wear rate plots. To continue the wear prediction we would perform another iteration of the process that we have shown here. The change in profiles should be small between wear steps so that the con-

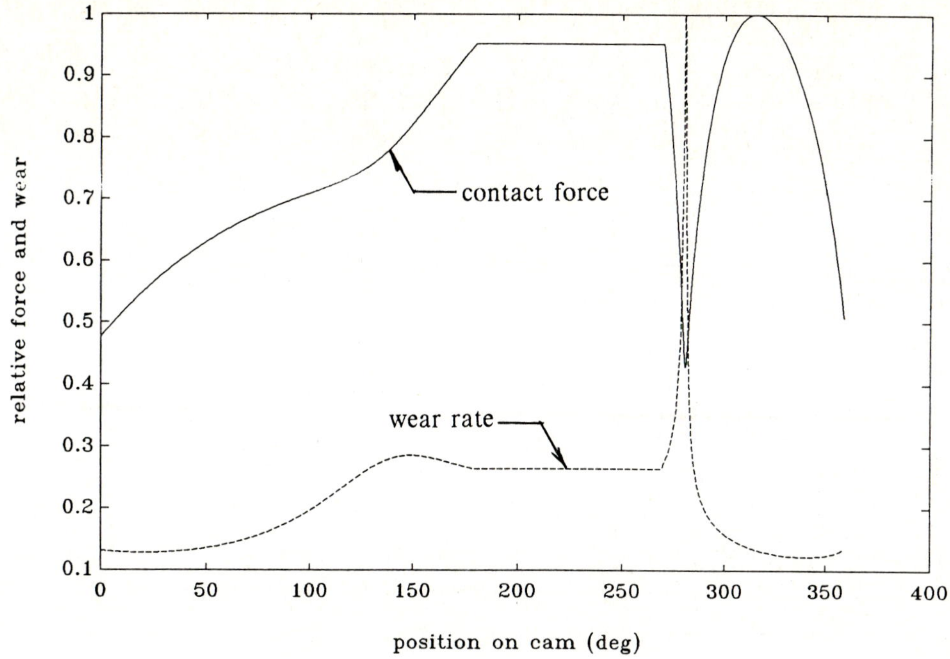

Fig. 8 Relative force and wear rate comparison

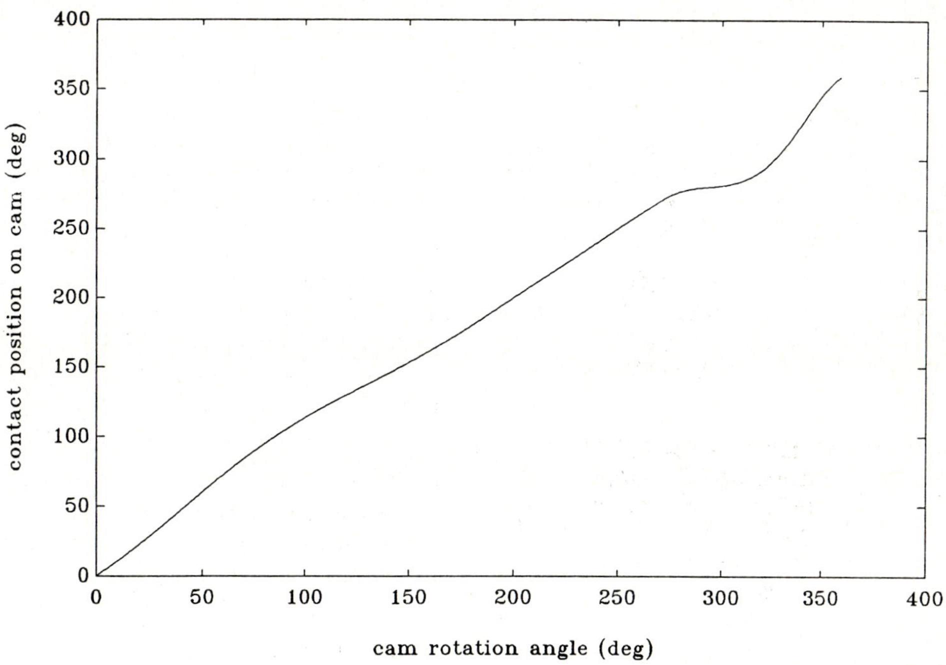

Fig. 9 Contact location on cam

tinuous wear process could be simulated by discrete wear steps. In this work we actually computed worn cam and follower profiles, but the changes in the profiles after one wear step were imperceptibly small. The worn profile predictions are therefore not shown. It is necessary to continue the process for a large number of wear steps to see how the wear would proceed with time.

If this process were repeated, the contact locations on the cam and follower would change. The lift and acceleration characteristics would change also, and worn profiles representative of much longer cam use could be predicted.

This work suggests the possible benefit of cam design based upon wear considerations. The conventional wisdom that keeping the contact forces low will also keep the wear low is correct generally, but a designer wanting a wear-stable cam/follower system would do well to consider the distribution of wear on the cam as well.

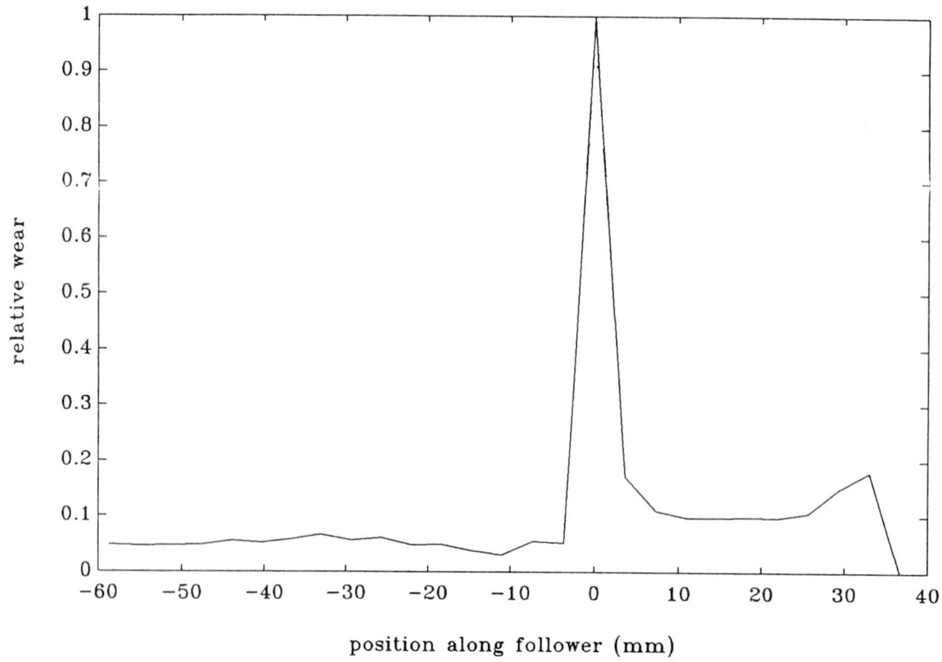

Fig. 10 Follower wear rate

5 CONCLUSIONS

A method of predicting worn cam and follower profiles has been described. The method requires the numerical kinematic analysis of a higher pair cam mechanism, an analysis whose results agree quite well with the correct results which were known for the cam/follower system described here. The wear prediction method contains the flexibility to allow use of nearly any wear model. The method is a refinement of a technique used for railway wheel wear predictions.

Several improvements and refinements are necessary to the solution method before it can be used for multiple wear step simulations. In this work, a single wear step has been demonstrated. This method offers the potential to be a valuable design tool. The technique could lead to optimized design of planar cam and follower systems based upon wear or dynamic performance criteria.

6 ACKNOWLEDGEMENTS

The authors gratefully acknowledge the contributions of Dawn Wales and Paul Fenech in programming and formulation of this computational method.

References

Archard, J. F., 1953, "Contact and Rubbing of Flat Surfaces," *Journal of Applied Physics,* Vol. 24, pp. 981-988.

Chen, F. Y., 1982, *Mechanics and Design of Cam Mechanisms,* Pergamon Press, Inc.

Chen, F. Y., 1972, "A Refined Algorithm for Finite Difference Synthesis of Cam Profiles," *Mechanisms and Machine Theory,* 7, pp. 453-460.

Churchill, F. T., and Hansen, R. S., 1962, "Theory of envelopes provides new cam design equations," *Prod. Eng.,* 35, pp. 45-55.

Cooperrider, N. K., et al., 1976, "Analytical and Experimental Determination of Nonlinear Wheel Rail Geometric Constraints," *Symposium on Railroad Equipment Dynamics,* ASME, pp. 41-70.

Cooperrider, N. K., E. H. Law, R. Hall, P. S. Kadala, and J. M. Tuten, 1975, "Analytical and Experimental Determination of Nonlinear Wheel/Rail Geometric Constraints," FRA-OR&D-76-244, USDOT.

Fries, R. H. and Davila, C. G., 1987, "Wheel Wear Predictions for Tangent Track Running," *J. Dynamic Systems, Measurement and Control,* Vol. 109, pp. 397-404.

Johnson, R. C., 1955, "Method of Finite Differences for Cam Design, '*Machine Design,* 27, pp. 195-204.

Kloomok, M., and R. V. Muffley, 1956, "Determination of Radius of Curvature for Radial and Swinging Follower Cam Systems," *Transactions ASME,* 78, pp. 795-802.

Mabie, H. H., and Reinholtz, C. F., 1987, *Mechanisms and Dynamics of Machinery,* John Wiley & Sons, New York.

Raven, R. H., 1959, "Analytical Design of Disk Cams and Three-Dimensional Cams by Independent Position Equations," *J. Appl. Mech.,* pp. 18-24.

Reinholtz, C. F., et al., 1978, "Kinematic Analysis of Planar Higher-Pair Mechanisms," *Mechanisms and Machine Theory,* Vol. 13, No. 6, pp 619-629.

Rothbart, H. A., 1956, *Cams - design, dynamics, and accuracy,* Wiley, New York.

Zigo, M., 1969, "A Method of Specifying the Dimensions of Cams," *J. Mechansims,* 4, pp. 283-289.

Paper IV(iii)

Cam and follower design

A. D. Ball, D. Dowson and C. M. Taylor

SYNOPSIS A general background to the consideration of tribological factors in cam and follower design is presented as a prelude to the description of the development of computer software for the analysis of the lubrication of any cam and tappet form. Parametric studies of both centrally pivoted and end pivoted follower arrangements are detailed with variations in cam base circle radius, cam width, reciprocating mass, camshaft speed, spring rate and follower radius of curvature considered. The physical interpretation of observed tribological changes are discussed and the paper concludes with a demonstration of how the software can be used to enhance the predicted performance of a given cam and follower geometry through design changes.

1. INTRODUCTION

In the 1970's the western world became increasingly aware of the fact that the earth's reserves of fossil fuels were not limitless. Political pressures from environmental groups and, much more significantly, the pressures brought about by the Middle Eastern countries upon the economies of their consumers during the 'Oil Crisis', caused many governments to campaign for the merits of saving energy. The motor manufacturers were put under increasing pressure from the consumer to design vehicles that not only cost less than their competitors for a similar specification, but also returned better fuel consumption figures.

Coincidentally, around this time, the use of an overhead camshaft (OHC) using pivoting followers to drive two banks of valves had just become very popular within the motor industry. The design was favoured as it allowed the cylinder head to be assembled separately from the block (good from a production point of view), it operated better at high speeds than other single camshaft designs driving two banks of valves (as it was more rigid) and it utilised fewer components and was therefore cheaper. Unfortunately, the design was found to be inherently poor from a tribological viewpoint - many manufacturers suffering from early failures of their (OHC) valve train systems due to excessive wear of the cams and followers.

The problem was therefore defined; the designer was required to design valve trains that had very small frictional losses and that would not wear out within the life of the rest of the vehicle. Regrettably the tools that the designer needed to fulfil the task were not available and similar types of design kept emerging.

Much work has been carried out upon ways in which the efficiency of the internal combustion engine could be increased. Many workers investigated the sources of losses within the internal combustion engine and ways in which they could be reduced (for example, Parker and Adams (1982), Hoshi (1984) and Martin (1985)). Although the majority of the losses were found to be due to thermal inefficiencies, there are still large benefits to be gained from reducing the mechanical losses which account for approximately 15% of the total fuel energy input. Hoshi suggested that the valve train frictional losses account for between 7.5% - 21% of the total engine frictional loss, and so there is still a large scope for improvement.

Whilst there have been major developments in the tribological understanding of the behaviour of many of the engine components, such as the dynamically loaded bearings and the piston assemblies, the amount of work upon the valve train has been relatively small. This may be due to the complexity of the problem as the cam and follower operate in the most arduous tribological conditions within the internal combustion engine. There is still a need for theoretical studies of the cam and follower contact to be undertaken, for the tribology of such contacts is still far from being fully understood. There is also an increasing need for existing knowledge to be put into a form that can be used by designers within industry.

1.1 Notation

A number of symbols are clearly defined in the text and on Figures 1 and 4. Those that are not are included below and full details of derivations reported in the paper may be found in Ball (1988).

b	half width of Hertzian contact
E	Young's modulus
E'	equivalent elastic modulus
F	frictional force
G	dimensionless materials parameter ($\alpha E'$)
h	film thickness
H	power loss
k	spring stiffness
ℓ_v	valve lift
L	cam width
p	pressure
r	radius
R	equivalent radius of curvature of contact
u	fluid velocity in x direction
U	dimensionless speed parameter $\left(\frac{\eta_o V_e}{E'R}\right)$
V_c	velocity of contact point relative to cam

V_e mean entraining velocity $((V_c+V_F)/2)$
V_F velocity of contact point relative to follower
V_s sliding velocity (V_c-V_f)
W load
W' dimensionless load parameter $\left(\frac{W}{E'RL}\right)$
x,y coordinate axes
α pressure-viscosity coefficient
δ initial spring deflection
η dynamic viscosity
η_o dynamic viscosity at atmospheric gauge pressure and reference temperature
μ coefficient of friction
ν' Poisson's ratio
τ shear stress
ψ cam angular rotation from maximum lift
ω camshaft angular velocity

Subscripts
B base circle
cen central
f follower
max maximum
min minimum

2 HISTORY OF VALVE TRAIN DESIGN

Valve train design philosophy has been changing rapidly over the past decade. This has been due mainly to research carried out after many manufacturers experienced serious valve train wear problems with new engines in service. The cost of research therefore paled into insignificance compared with the cost of warranty claims and the loss of sales.

Although throughout this century various different types of engine valving have been tried, the poppet valve has been almost universally adopted by the major automobile manufacturers. Other types of valving such as sleeve or rotary valves have been deemed to have lubrication difficulties, allow excessive engine oil consumption, provide poor sealing and have excessive frictional losses (Buuck (1982)).

Over the years there has been a demand for ever increasing engine speeds in the search for more energy efficient engines. This has caused the rise in popularity of overhead camshaft mechanisms (OHC) at the expense of push-rod systems. The push-rod system had been favoured in the past due to its many virtues; ease of adjustment, the availability of the camshaft to drive accessories such as the oil pump and distributor, and good lubrication and wear characteristics. As the camshaft is located close to the sump it receives a plentiful supply of oil in the form of oil mist and splash from the crankshaft. Also the tappets are free to rotate, thus improving lubricant entrainment and decreasing wear by ensuring that any two points on the cam and tappet surfaces do not continually meet cycle after cycle.

The main disadvantage of push-rod systems is their flexibility brought about by the use of long thin push-rods. This makes them unsuitable for use in very high speed engines. OHC mechanisms are inherently much stiffer. Modern production techniques have also added to the decline in popularity of the push-rod system. OHC mechanisms utilise fewer parts and allow the cylinder head assembly to be built up as a separate unit. Production engineers see these as great advantages (Polak and Letts (1987)). OHC mechanisms appear in two basic forms; either direct acting or via a pivoted follower. One of the major manufacturing and assembly problems with these systems is in the alignment of the camshaft and followers. If the cam lobes and followers are not properly aligned then severe edge loading can occur resulting in damage to the contacting parts. Mercedes patented a novel solution to this problem in 1959. Their design utilised end pivoted followers which located upon spherical ended posts rather than the more conventional rocker shaft. This allowed the followers to self align.

Whilst direct acting OHC mechanisms have proved to be very successful both from a performance and wear point of view, the same cannot be said of pivoted OHC mechanisms. Many manufacturers have experienced major wear problems with these types of valve trains. This has been attributed to many causes. One suggested cause is the higher temperature seen in the cylinder heads of modern engines due to the adoption of thermostatically controlled electric fans and the use of more selective coolant channels through the whole of the engine. These higher temperatures not only increase the bulk temperature of the contacting parts, thus increasing the probability of scuffing (Dyson and Naylor (1960)), but also serve to lower the viscosity of the lubricant. As the followers do not run in bores they have limited means of conducting away heat generated in the contact region - this again leads to high bulk temperatures in the followers. As the camshaft is at the very top of the engine in an OHC arrangement it is also at the end of the lubricant feed path. In many early OHC designs the cam and follower contact had to rely upon oil splashed from the camshaft bearings to provide adequate amounts of lubricant. Other reasons suggested for the untimely demise of such systems have been fuel dilution of the lubricant and oil starvation at engine start up.

Most manufacturers have solved the problem of excessive wear in pivoted OHC systems by the adoption of high specification materials and the use of spray bars and even holes in the cam lobes to supply sufficient lubricant to the contact region. The additional supply not only serves to lubricate the contact but also acts as a coolant. One manufacturer has reverted to the use of manual chokes which perhaps suggests that they blamed fuel dilution for their problems. It is the belief of the authors that if attention were to be directed towards the geometry and kinematics of such systems significant gains in lubricant entrainment could be achieved. This point will be illustrated by the use of parametric studies.

OHC mechanisms, however, do have several drawbacks apart from their frequently poor wear characteristics. The camshaft is often used to drive auxiliaries such as the fuel and oil pump and the distributor. This means that these components must be mounted very high on the engine if they are to be driven from the camshaft. It is therefore often necessary when using an OHC valve train to drive the oil pump from the crankshaft so as to keep the distance from the pump to the sump as small as possible. This unfortunately aggravates overall engine length and may also lead to lubricant aeration problems. The majority of the flexing of the valve train system takes place in the camshaft in an OHC system, which can cause serious ignition timing problems if the distributor is driven from the camshaft.

It can be seen that the choice of a valve train is very complex and that many other engine components and characteristics must be taken into account in making a selection. It is apparent that the valve train designer cannot be isolated from the overall design of the engine. This is becoming more and more apparent as engine specifications improve and engine speeds increase.

3. VALVE TRAIN DESIGN PHILOSOPHY

Several design parameters have usually been decided upon before the valve train of an engine is designed: the engine's displacement, the bore and stroke, the overall engine height, the maximum allowable manufacturing and assembly cost, the desired engine performance, etc. These parameters set restraints upon the final choice and design of valve train.

The performance of the engine is closely related to how efficiently the charge of air and fuel can be drawn into the combustion chamber and how efficiently the exhaust gases can be expelled after combustion. In order to admit and expel the various gases effectively, great care is taken over the design of the combustion chamber and the valve lift curve, and the choice of the number of valves per cylinder and their angle of entry into the combustion chamber. The choice of valve train can often be virtually dictated by these conditions.

If, for example, the performance requirement for an engine dictates that the valve must be opened and closed very rapidly this implies that very high accelerations are required along the cam flanks. It is certain that push-rod systems would be eliminated from the choice of available mechanisms at this point due to their flexibility. OHC pivoted followers may also be eliminated from the choice as the maximum allowable flank acceleration is limited by the permissible concavity of the cam flanks. This may leave a direct acting OHC mechanism as the only choice. The designer must then decide whether the large diameter followers required for this type of system can be fitted into the space available, and whether the luxury of two camshafts can be afforded if the valves are not in line.

The alternative types of valve train mechanisms obviously have differing advantages and disadvantages and the designer must weigh the requirements of his design in order of preference in order to get the right compromise. Below many of the good and bad points of various valve train mechanisms are summarized.

<u>Advantages and Disadvantages of Pivoted Follower OHC Valve Train Systems</u>

- Not as stiff as direct acting OHC mechanisms. This along with cam profile limitations due to cam concavity necessitates the use of modest valve accelerations. These restraints obviously limit the engine breathing. Maximum operating speed is less than that for direct acting OHC.
- Poor wear characteristics.
- Allows the use of only one camshaft even when there are two banks of valves.
- The cam lobes are smaller than those of a direct acting system as a mechanical advantage can be used by having rocker ratios greater than unity. This allows smaller camshaft bearings to be used.
- Adjustment is usually very easy.

<u>Advantages and Disadvantages of Direct Acting OHC Valve Train Systems</u>

- Good engine breathing as valve train is very rigid and therefore high valve accelerations and operating speeds are possible.
- Symmetrical cam profiles give symmetrical lift curves.
- Good lubrication conditions - low wear.
- If the valves are not in line then two camshafts are required unless a direct acting cam and follower is used in conjunction with a cam and pivoted follower.
- The cam lobes for a direct acting system are much larger than any other OHC system. This necessitates the use of large camshaft bearings.

<u>Advantages and Disadvantages of Push Rod Valve Train Systems</u>

- Very flexible. This allows only very low valve accelerations to be used thus limiting engine breathing. The low rigidity also limits the maximum allowable operating speeds.
- Low wear. The cam and followers receive a plentiful supply of lubricant mist and splash from the main engine bearings due to their close proximity to the sump.
- A mechanical advantage may be used by employing a rocker ratio of greater than unity. This leads to small cam lobes.
- Easy adjustment.
- Engine auxiliaries driven by the camshaft suffer less detrimental effects than with any other system.

Once the type of mechanism has been decided upon the geometry of the system must be calculated. If a direct acting OHC system is used then the geometry can be readily decided. The diameter of the follower is dictated by the maximum eccentricity of the cam during the operating cycle. The maximum eccentricity is given directly by the maximum valve velocity (see Dyson and Naylor (1960)). The designer can then calculate whether there is enough space between the valves to accommodate the followers. Once the followers have been sized, an approximate value for the equivalent reciprocating mass of the valve train components can be calculated. This, along with the maximum required engine speed and valve acceleration, gives the minimum allowable spring stiffness to prevent valve bounce at the maximum rated engine speed. The spring stiffness is then chosen by allowing a given amount of spring cover across the nose region of the cam. The actual valve spring dimensions and number of coils are decided by available standard wire diameters, space limitations, an acceptable number of working coils, satisfactory fatigue life and dynamic considerations. Valve spring design is discussed in greater detail by Beard and Hempson (1962).

The cam size is chosen by satisfying the requirements of maximum allowable Hertzian stress, engine height, sliding speed at the contact, cam bearing diameter, and also, hopefully, by studying the lubrication conditions around the cam cycle for the range of operating speeds.

If a pivoted follower system is chosen the decision processes are similar but with additional variables, such as rocker ratios and symmetry to be considered.

Having chosen a particular valve train the designer must then choose the correct materials for the contacting parts. This must be done with a knowledge of the stressing of the components and the sliding speeds and lubrication conditions expected at the contact. The choice of materials for cam and follower pairs is a complex subject worthy of study in its own right, and indeed many learned society papers have been written on the subject. The materials aspects of cam and follower design are considered to be beyond the scope of this paper.

4 DESIGN SOFTWARE

Software has been developed within the Department of Mechanical Engineering at the University of Leeds which allows tribological studies to be carried out upon any of the camand follower designs in common use in todays internal combustion engines. The program is written in such a manner that it can easily be used by any competent professional engineer. It is believed that the program would be highly useful in the design of new valve trains. Full details of the models used in the software may be found in Ball (1988 and 1989), however, a brief desciption is given below.

4.1 Cam and Follower Kinematics

The program uses standard methods of differential geometry to derive the kinematic velocities relevant to the study of lubrication as developed by Dyson and co-workers (1960, 1977 and 1980).

4.2 Contact Loading

The load at the cam/follower interface is determined by summing the force components arising due to the friction at the interface, the spring force and the inertia of the reciprocating parts. To illustrate the method, take the case of a cam acting against a centrally pivoted follower:

Figure (1) shows a cam acting against a centrally pivoted follower. The forces shown are the spring force, (S), the valve inertia force, (F_v), the frictional force at the cam/follower interface, (F_{frict}), the follower inertia force, (F_f), and the load exerted on the follower by the cam, (W). The frictional forces at the valve/follower interface and at the follower pivot point have been neglected as have gravitational and dynamic forces.

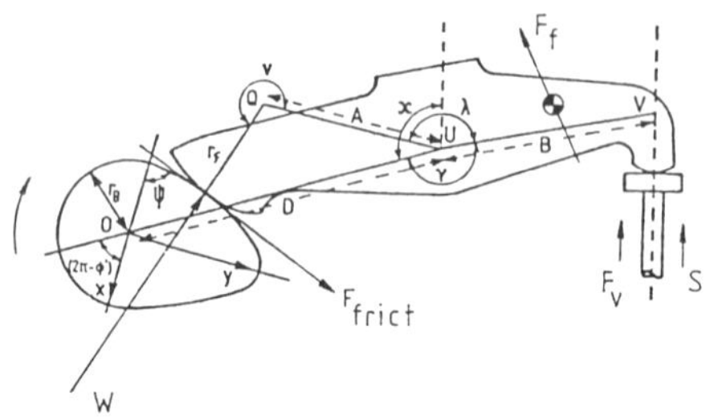

A = 50.06 mm	$\nu'_{follower}$	= 0.3	
B = 54.00 mm	ν'_{cam}	= 0.3	
D = 74.27 mm	$E_{follower}$	= 207 GNm^{-2}	
r_f = 35.56 mm	E_{cam}	= 207 GNm^{-2}	
r_B = 14.30 mm	α	= 15.0 nm^2N^{-1}	
κ = 107.5°	η_0	= 0.013 Nsm^{-2}	
λ = 140.0°	k	= 60.0 kN/m	
L = 17.00 mm	M	= 0.10 kg	
	δ	= 4.00 mm	

FIGURE 1. GEOMETRY OF THE ROVER 2300/2600 VALVE TRAIN

The moments caused by each of the above forces about the follower pivot point, (U), defining a clockwise moment as positive, are:

$$\text{Spring force moment} = -\,S\,B\cos\left(-\frac{3.\pi}{2} + \chi + \gamma\right) \tag{1}$$

$$\text{Valve inertia moment} = -F_v\,B\cos\left(-\frac{3.\pi}{2} + \chi + \gamma\right) \tag{2}$$

$$\text{Cam load moment} = W\,A\cos\left(\frac{3.\pi}{2} - \nu\right) \tag{3}$$

Frictional force moment =

$$-\,F_{frict}\left(r_f - A\cos\left(\frac{3.\pi}{2} - \nu\right)\right) \tag{4}$$

$$= -\,\mu\,W\left(r_f - A\cos\left(\frac{3.\pi}{2} - \nu\right)\right) \tag{5}$$

where $\nu = \frac{\pi}{2} - \phi' - \gamma - \lambda + \psi$

$$\text{Follower inertial moment} = -M_f\,B^2\omega^2\,\frac{d^2\gamma}{d\phi'^2} \tag{6}$$

where (M_f) is the equivalent mass of the

follower at a distance (B) from the follower pivot point (U). Summing the moments and substituting:

$$S = k\,(\ell_v + \delta) \tag{7}$$

and

$$F_v = M\,\omega^2 \frac{d^2\ell_v}{d\phi'^2} \tag{8}$$

gives: W =

$$\frac{\left[M \frac{d^2\ell_v}{d\phi'^2}\,\omega^2 + k(\ell_v + \delta)\right] B\cos\left(\frac{-3.\pi}{2} + \chi + \gamma\right) + M_f \frac{d^2\gamma}{d\phi'^2}\,\omega^2 B^2}{A\cos\left(\frac{3.\pi}{2} - \upsilon\right) - \mu\left(r_f + A\sin\left(\frac{3.\pi}{2} - \upsilon\right)\right)} \tag{9}$$

where (M) is equal to one third of the valve spring mass plus the mass of the valve and any retainers. Similar relationships may be derived for other valve train geometries.

The analysis for the loading at the cam/follower interface for a centrally or end pivoted follower requires that the coefficient of friction at the contact be known. This obviously cannot be calculated without the load at the contact being known. A coefficient of friction of 0.08 is therefore assumed to allow the load to be calculated. This permits the frictional traction at the contact to be determined (see below). The coefficient of friction can then be recalculated and a new load assessed. This procedure is repeated until the loading around the cycle converges. Convergence is very rapid, usually taking only three iterations.

4.3 Evaluation of the Lubricant Film Thickness Between the Cam and Follower

The conditions at the cam and follower contact are very severe. Assuming an adequate supply of lubricant reaches the contact, full separation of the cam and follower around all of the lift cycle is not guaranteed. If the camshaft rotational velocity is high enough then the contact around the cam flanks (where the lubricant entrainment velocity is at its highest) may enjoy elastohydrodynamic lubrication (EHL), otherwise the contact will operate in the mixed lubrication regime. Around the nose of the cam, where the lubricant entrainment is small, some elememt of boundary lubrication can almost always be anticipated.

EHL may occur between lubricated non-conformal contacts. The geometry of the contact is such that very high pressures are generated, leading to elastic deformation of the interacting solids. The pressure generated within the lubricant film may be of the order of hundreds of mega-pascals, which leads to dramatic changes in the lubricant properties. The viscosity of the lubricant increases rapidly with pressure (indeed exponentially according to the Barus relationship, ($\eta = \eta_o e^{\alpha P}$)), and at high pressures exhibits almost solid like characteristics. Dowson and Higginson (1977), have presented a formula for the minimum lubricant film thickness between two cylinders in steady line contact:

$$\frac{h_{min}}{R} = 2.65\, U^{0.70}\, G^{0.54}\, W'^{-0.13} \tag{10}$$

Dowson and Toyoda (1978) presented a similar formula for the film thickness at the centre of the contact:

$$\frac{h_{cen}}{R} = 3.06\, U^{0.69}\, G^{0.56}\, W'^{-0.10} \tag{11}$$

These formulae are obviously not strictly accurate for the situations found in cam/follower contacts, in which the action of squeeze will be very important around the areas where the entrainment of the lubricant into the contact is very small. They are, however, felt to be adequate to allow qualitative judgements of the merits of valve train designs to be made.

Boundary lubrication occurs between two interacting solids when some asperity contact or interaction between surface films takes place as the lubricant film thickness falls to a value less than the composite surface roughness of their surfaces. In such conditions the contact must rely upon the ability of the lubricant and solids to form surface reaction layers in order to prevent severe wear taking place. In boundary lubrication the laws of dry friction may be taken to apply since the coefficient of friction is independent of load, speed, and apparent area of contact. This regime of lubrication can almost certainly be expected around the nose of the cam. The transition from boundary lubrication to full EHL does not take place instantaneously. As the load and entrainment velocity at the cam/follower interface became more favourable a larger proportion of the load is carried by the pressure of the lubricant within the contact, with less and less of the load being borne by surface asperities.

4.4 Evalution of the Hertzian Stress at the Contact

Once the load at the cam/follower interface has been evaluated then the maximum Hertzian stress at the contact can be found. This also allows the dimensions of the contact zone to be calculated according to the theory of Hertz (1882).

4.5 Frictional Traction

In fully lubricated contacts the frictional forces acting upon the interacting components are a function of the velocity gradients across the contact zone. The frictional force consists of contributions due to rolling and sliding of the components. In the contact between cams and followers much shearing of the lubricant film takes place due to the sliding action of the cam and follower. This would, in reality, lead to a lowering of the lubricant viscosity due to the associated temperature rise. The model used in the present study, however, assumes that isothermal conditions exist within the contact region.

As the elastic deformation of the cam and follower is large in comparison to the lubricant film thickness, the contact can be approximated by a lubricated Hertzian contact, Ball (1988), without considering the minimum film thickness nip known to occur in practice. At the inlet to the contact the frictional force arises almost entirely due to the rolling of the components. In the long parallel zone the contribution of rolling is negligible, especially when the speeds encountered with cams and followers are taken into account. Thus, the frictional traction calculations are restricted to the sliding contribution from the 'Hertzian' region taking the Hertz pressure distribution.

The shear stress acting upon the solid boundaries is given by:

$$\tau = \eta \frac{du}{dy} \qquad (12)$$

where (du/dy) is the velocity gradient across the lubricant film thickness. Substituting the Barus relationship for the viscosity term and V_s/h_{cen} for the velocity gradient, we obtain,

$$F = \int \frac{\eta_o V_s e^{\alpha p}}{h_{cen}} dx \qquad (13)$$

upon integrating the shear stress along the length of the contact.

The Hertzian pressure distribution across the contact is given by:

$$p = p_{max} \left(1 - \frac{x^2}{b^2}\right)^{\frac{1}{2}} \qquad (14)$$

Therefore,

$$F = \frac{\eta_o V_s e^{\alpha p_{max}}}{h_{cen}} \int e^{(1-x^2/b^2)^{\frac{1}{2}}} .dx \qquad (15)$$

Equation (15) can be solved numerically using Simpson's rule to give the instantaneous frictional force at a given instant during the cam cycle.

As the Barus viscosity relationship is exponential, the viscosity term in Equation (15) can become very large at high pressures. This leads to very high frictional forces being predicted. The model, therefore, calculates a boundary friction value taking a limiting friction coefficient:

$$F = \mu W$$

The value of the coefficient of friction (μ) was taken to be 0.08. Experimental evidence (Zhu (1988)) and work on piston rings at the University of Leeds show this value to be reasonable. If the frictional force predicted by the boundary lubrication model was less than that predicted by the lubricant shearing model then the former was taken as the frictional force arising at the interface.

4.6 Power Loss

The instantaneous power loss due to friction is given by,

$$H = Fr\omega \qquad (16)$$

where (r) is the perpendicular distance from the cam centre of rotation to the frictional load vector. This can be integrated around the cam lift cycle (using Simpson's rule) to give an average power loss:

$$H = \frac{1}{2\pi} \int^{2\pi} Fr\omega \, d\phi \qquad (17)$$

5 PARAMETRIC STUDIES

Muller (1966) was the first to show that, theoretically, the shape of the cam profile could affect not only the Hertzian Stress but also the lubricant film thickness at the cam/follower interface. It therefore became apparent that if a cam mechanism was to work satisfactorily attention should be paid to limiting the Hertzian stress whilst sustaining an adequate lubricant film. Dyson (1977) showed how both the equivalent radius of curvature and the lubricant film thickness at the contact varied as several important design parameters were changed for a cam acting against a domed follower. He then showed (1980) how this theoretical work could be extended to a cam and end pivoted follower system. Harrison (1985) carried out a comprehensive parametric study of a cam and flat faced follower system and showed how the frictional power loss, lubricant film thickness at the cam nose and Hertzian stress at the cam nose varied as several design parameters were altered. Dowson, Harrison and Taylor (1985) developed a table explaining and summarising the important findings of this study - a useful aid for anyone wishing to design a cam and flat faced follower mechanism.

In this paper parametric studies of two valve trains are presented. These are a centrally pivoted follower design (Rover 2300/2600) and an end pivoted follower design (Ford Pinto). In order to demonstrate how useful tribological parametric studies would be at the design stage, a suggested improved Pinto design is also presented.

5.1 A Parametric Study of a Cam and Centrally Pivoted Follower System.

The cam and follower adopted for this study are in use in a current engine (Rover 2300/2600) and are the same as those studied by Lim et al (1983), the valve lift data being given as a fitted exponential function:

$$\ell_v = \exp\,(a + b\phi + c\phi^2 + d\phi^3 + e\phi^4 + f\phi^5 + g\phi^6)$$

where

a = 2.317111	b = -0.8104×10^{-3}
c = −1.526837	d = −0.1552985
e = −1.789897	f = 0.1171634
g = −0.2422081	

The geometry of the system is shown in Figure (1). This is taken to be the reference condition. Each design parameter was changed from its datum whilst the others remained constant, and the effect upon the frictional power loss, lubricant film thickness and Hertzian stress at the cam nose was studied. The parameters changed were:

(a) Cam base circle radius,
(b) Cam width,
(c) Equivalent mass at the valve,
(d) Camshaft speed,
(e) Spring rate, and
(f) Follower radius of curvature.

As the cam base circle radius is changed, the valve train geometry must be changed to take up the clearance between the cam and follower. This was done by moving the follower contact radius along the line between its centre of curvature and the cam centre of rotation. i.e.

$$\gamma_B + \chi = \text{constant},$$

and,

$$\psi_B + \phi_B + \chi = \text{constant}$$

Similarly, as the follower radius of curvature is changed, the position of its centre of curvature moves in the direction of the line from the cam centre of rotation to the follower centre of curvature, i.e. B, D, r_B, χ and (ψ_B + ϕ_B) are constant.

Figure (2) - in three parts - shows the cam operating characteristics at the datum condition for a camshaft speed of 41.67 Hz (2500 rpm). A parametric study was carried out at 41.67 Hz (2500 rpm). Figure (3) shows the results of the parametric study. Each design parameter was changed from its datum value by -70% through to 300% (0% change being the datum value, -50% change being half the datum value, +50% being one and a half times the datum value, etc). In Ball (1988) other detailed studies are presented for more restricted changes and a different speed of rotation. It is recognised that the magnitude of changes reported here may be unrealistic in a practical sense.

The effect of changing each of the parameters is discussed in turn below and is summarized in Table (1).

5.1.1 Changes in Cam Base Circle Radius

Figure (2h) shows how the surface velocities at the point of contact vary throughout the cam cycle. The curve describing the lubricant entrainment velocity (V_c + V_f - twice the mean entraining velocity) is of a typical shape for a pivoted follower. It can be seen that the pivoting follower introduces a degree of asymmetry to the curve. This makes the interpretation of the results from the parametric study a little difficult. In the parametric studies the value of the film thickness at the position of maximum lift is taken to be indicative of the film thickness across the whole of the cam nose. Due to the asymmetry the film thickness is not constant across the nose, however, it can be seen from Figure (2f) that the lubricant film thickness at the maximum lift position gives a reasonable approximation to the mean value of the film thickness across the whole of the cam nose. It should be noted that the value for the Hertzian stress at the cam nose is not the maximum value across the nose (see Figure (2j)).

As the base circle radius of the cam is increased from its datum value the cam radius of curvature increases and both the velocities of the point of contact relative to the cam surface (V_c) and relative to the follower (V_f) increase which leads to an increase in the entrainment velocity, see Figure (2h). This means that the magnitude of the entrainment velocity around the cam nose falls at first until the entrainment velocity is positive (i.e. in the same direction) for the whole of the cam cycle. As the entrainment velocity changes we see a corresponding change in the lubricant film thickness at the cam nose. As the base circle radius is increased from its datum value the film thickness at the cam nose falls at first down to zero (neglecting squeeze effects), corresponding to the value of (r_B) that gives zero entrainment velocity at the nose, and then rises as the entrainment velocity increases.

The increases in the radius of curvature of the cam at the cam nose also contributes to the enhancement of the lubricant film at this point. This also causes the decrease in the Hertzian stress at the nose.

The frictional power loss increases almost linearly with increasing (r_B). The limiting coefficient of friction is applied around the majority of the cam cycle at this camshaft rotational speed with the datum value of (r_B). The loading at the contact does not change as (r_B) changes but the radius at which the frictional traction force is applied increases as (r_B) increases, thus the frictional power loss increases. At very high values of (r_B) the lubricant entrainment velocity is enhanced to such an extent that the limiting coefficient of friction is no longer applied and so the rate at which the frictional power loss increases as (r_B) increases levels off.

5.1.2 Changes in Cam Width

As would be expected the Hertzian stress at the cam nose will decrease as the cam lobe width is increased because the load per unit width decreases. The film thickness at the cam nose and the frictional power loss decrease as a direct result of this.

5.1.3 Changes in Reciprocating Mass

Increasing the reciprocating mass at the valve causes higher loads on the cam flanks but smaller loads at the cam nose. As the load at the cam nose decreases with increasing reciprocating mass the lubricant film thickness will increase and the Hertzian stress will decrease. The frictional power loss decreases due to the limiting coefficient of friction being applied around the cam nose and so the power loss is proportional to the normal load around this portion of the cam which represents the majority of the cam cycle (time wise).

5.1.4 Changes in Camshaft Speed

The acceleration of the valve around the cam nose is negative. Therefore as the camshaft speed increases the valve acceleration becomes more negative and the contact load at the cam decreases. A point is reached as the speed is increased at which the inertia of the valve becomes so great that the cam and follower part and valve bounce occurs. (It should be noted that the loading on the cam flanks increases as the camshaft speed increases). This decrease in load at the cam nose leads to an increase in lubricant film thickness and a decrease in Hertzian stress. At lower camshaft speeds the frictional power loss increases with speed as would be expected. As the speed increases the loading on the cam nose becomes less and less until a point is reached at which the limiting coefficient of friction is no longer applied around the nose and so the power loss starts to fall with increasing speed. It should be noted that this relationship between the camshaft speed and frictional power loss is highly dependent upon the valve lift curve. A

CAM OPERATING CHARACTERISTICS
ROVER 2300 - 2500rpm

Cam Base Radius (mm)	-	14.30
Maximum Valve Lift (mm)	-	10.15
Cam Width (mm)	-	17.00
Rotational Speed (rpm)	-	2500.0
Spring Stiffness (kN/m)	-	60.000
Initial Spring Disp. (mm)	-	4.0
Equiv. Mass At Valve (kg)	-	.100
Lubricant Viscosity (Ns/m²)	-	.013
Press. Visc. Coeff. (/Pa)	-	15.0E-9
Youngs Mod. (Cam) (GPa)	-	207.0
Youngs Mod. (Foll.) (GPa)	-	207.0
Poissons Ratio (Cam)	-	.30
Poissons Ratio (Foll.)	-	.30
Frictional Power Loss (W)	-	58.78

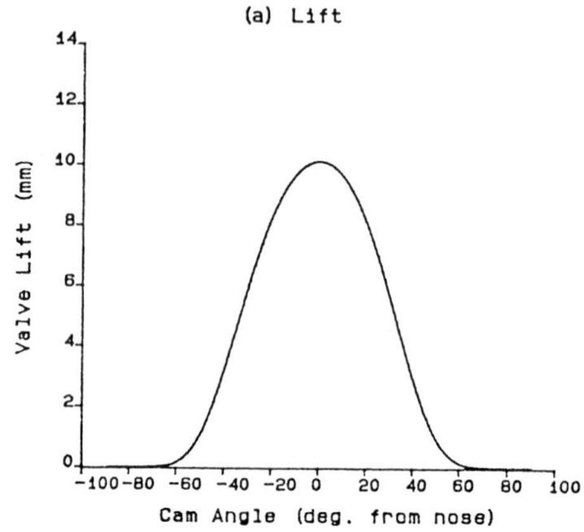

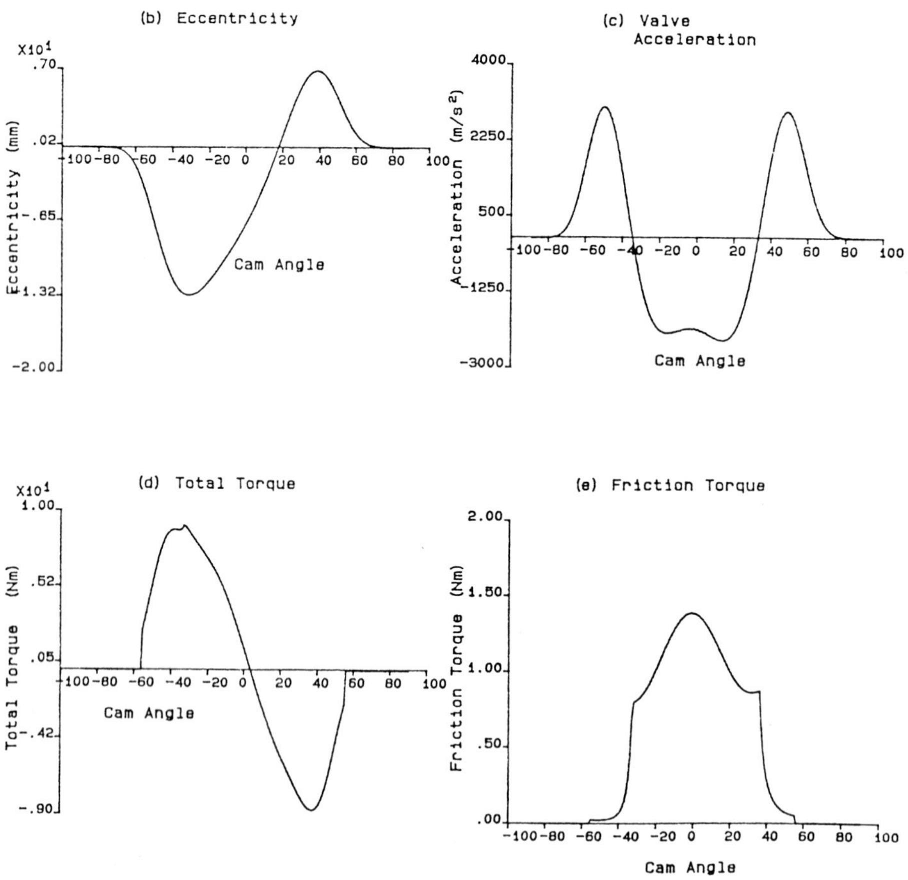

FIGURE 2(a) ROVER 2300/2600 CAM OPERATING CHARACTERISTICS AT THE DATUM CONDITIONS (41.67 Hz (2500 r.p.m.))

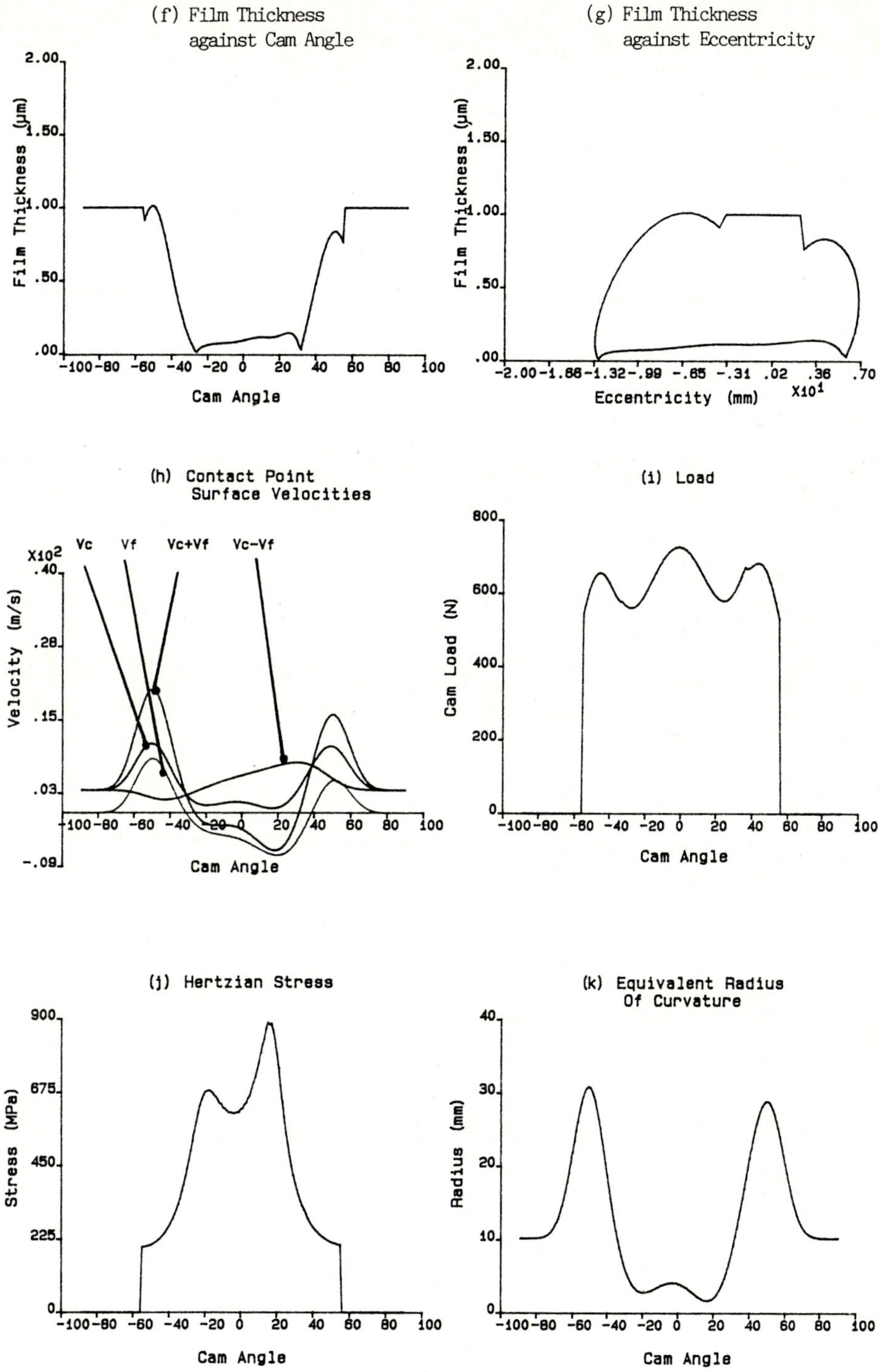

FIGURE 2(b) ROVER 2300/2600 CAM OPERATING CHARACTERISTICS AT THE DATUM CONDITIONS (41.67 Hz (2500 r.p.m.))

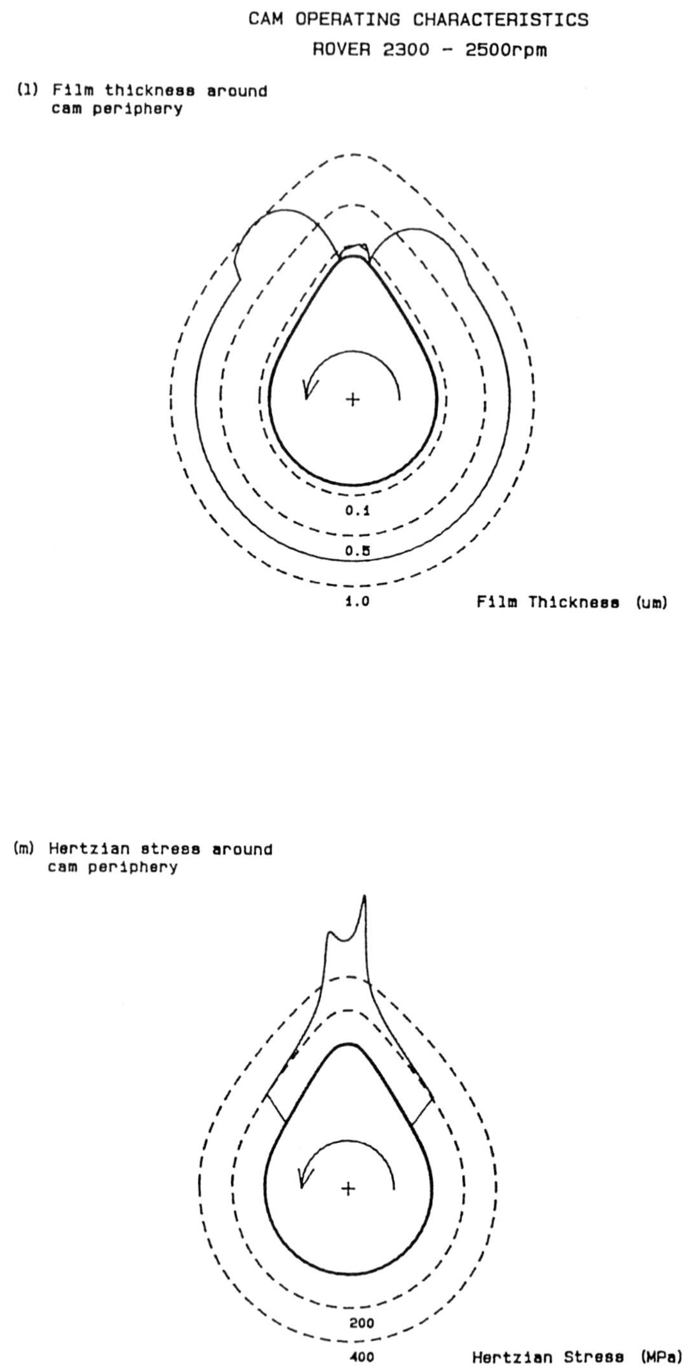

FIGURE 2(c) ROVER 2300/2600 CAM OPERATING CHARACTERISTICS AT THE DATUM CONDITIONS (41.67 Hz (2500 r.p.m.))

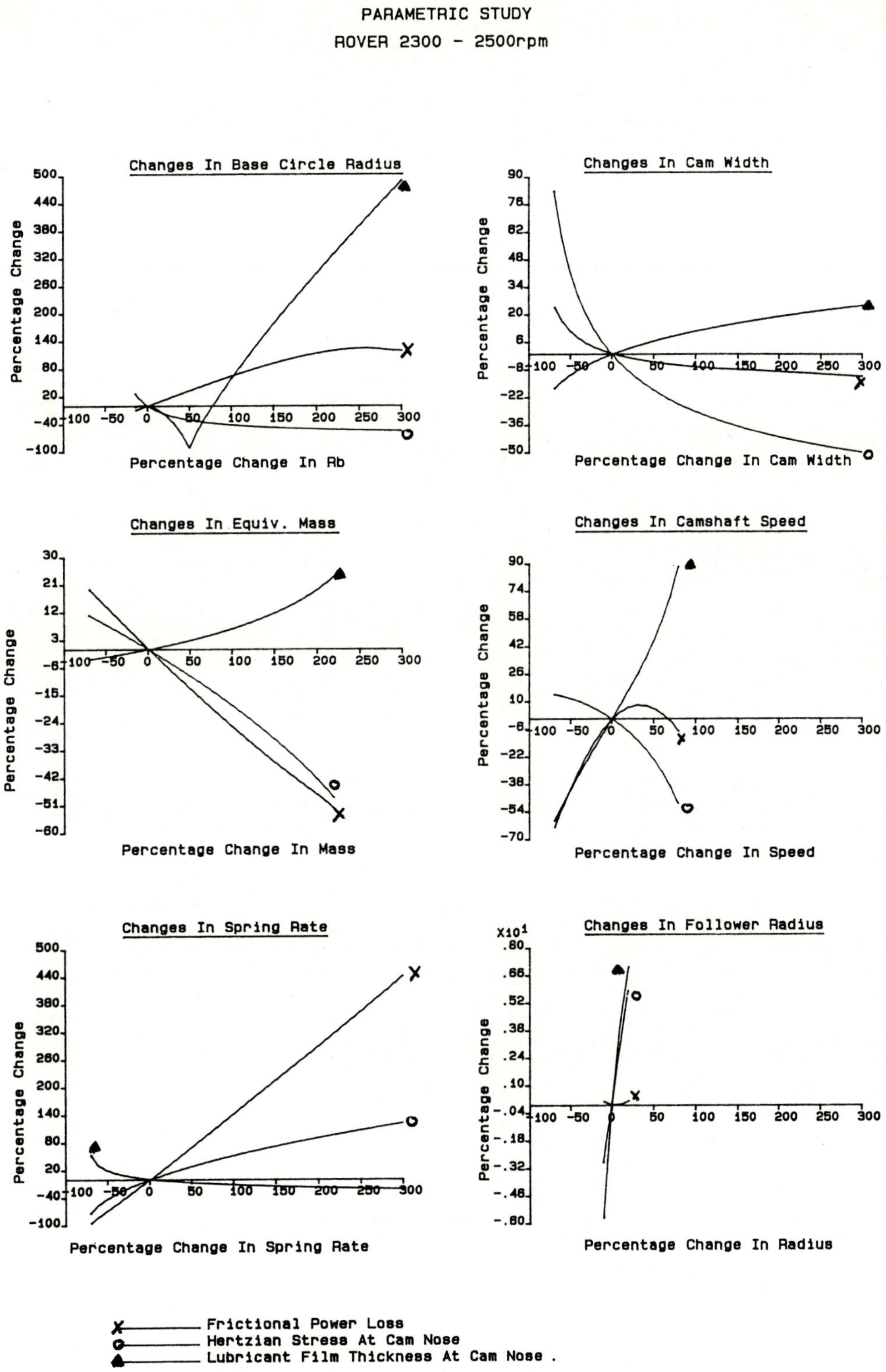

FIGURE 3 PARAMETRIC STUDY OF ROVER 2300/2600 CAM AND FOLLOWER AT 41.67 Hz (2500 r.p.m.) CAMSHAFT ROTATIONAL FREQUENCY

	MINIMUM FILM THICKNESS AT MAXIMUM LIFT POSITION	MAXIMUM HERTZIAN STRESS AT MAXIMUM LIFT POSITION	POWER LOSS	OTHER COMMENTS
BASE CIRCLE RADIUS	A decrease from the design value causes a fall in film thickness as the equivalent radius of curvature falls. An increase causes the film thickness to fall dramatically, before rising, due to the mean entraining velocity across the nose passing through a zero value.	The Hertzian stress is inversely proportional to the square root of the equivalent radius of curvature. Increasing the base circle radius therefore reduces the Hertzian stress at the cam nose.	The limiting coefficient of friction is applied over most of the cycle, hence the friction is a function of load rather than film thickness. The increase in power loss with base circle radius for a given speed is due to the increase in instantaneous radius of curvature (the distance at which the frictional force acts). If the base circle radius is increased sufficiently the entrainment velocity is enhanced adequately such that the limiting coefficient of friction is no longer applied over any of the cycle.	Increasing the base circle radius will increase the overall engine height. If on assembly, the camshaft needs to be fed through the camshaft bearings, then larger diameter bearings must also be used. A larger cam base circle radius allows a larger diameter camshaft to be used which will increase the rigidity of the valve train system; however at the expense of extra weight. Minimum value of base circle radius is dominated by the lift curve characteristics and the necessity for the cam to become concave along the flanks.
CAM WIDTH	As the cam width increases the load per unit width at the contact decreases. This leads to a small increase in the elastohydrodynamic film thickness.	As the cam width increases the load per unit width at the contact decreases, and so, therefore, does the Hertzian stress.	At low rotational frequency the limiting coefficient of friction is applied over most of the cam cycle. Hence, an increase in cam width only marginally reduces power loss. At higher rotational frequencies the proportion of the cycle governed by boundary lubrication falls with increasing cam width and there is a corresponding fall in power loss.	The cam lobe width is limited by the number of cam lobes to be fitted into a given space. i.e. the cam lobe width is governed by the valve spacing. A larger cam width necessitates a larger diameter follower which leads to a larger reciprocating mass.
RECIPRO-CATING MASS	An increase in reciprocating mass causes an increase in the film thickness due to the inertia force reducing the cam load at the nose.	Increasing the reciprocating mass reduces the cam load and hence lowers the Hertzian stress at the cam nose.	Increasing the reciprocating mass generally causes a fall in the power loss as the loading at the cam nose is reduced.	Increasing reciprocating mass increases the load on the cam flanks. Decreasing mass may cause strength problems, increasing mass inertia problems.
VALVE SPRING STIFFNESS	Increasing valve spring stiffness reduces the film thickness as the cam load is increased.	A reduction in spring stiffness reduces the cam load and hence reduces the Hertzian stress.	An increase in spring stiffness causes the power loss to increase as the cam load increases.	The valve spring must provide sufficient load to restrain valve bounce and to control the valve along the cam flanks.
FOLLOWER RADIUS OF CURVATURE	An increase from the design value causes an increase in film thick-ness as the entraining velocity increases.	An increase from the design value causes an increase in Hertzian stress as the equivalent radius of curvature falls as the cam nose narrows.	Very little influence.	The size of the contact patch on the follower increases with increasing follower radius of curv-ature, therefore, larger followers may be needed.
CAM ROTATIONAL SPEED	Increasing speed increases entraining velocity and reduces cam load at nose. Film thickness therefore increases with increasing speed. At high speeds the cam load falls rapidly and therefore the increase in film thick-ness is rapid.	As the cam rotational speed increases the load at the cam nose decreases and therefore the Hertzian stress decreases.	Power loss increases with cam speed. At high speeds the rate of increase falls as valve bounce is approached.	Camshaft speed is limited by the point at which valve bounce occurs and the cam no longer controls the valve. This is due to the inertia of the reciprocating parts being greater than the valve spring load.

TABLE 1 ROVER 2300/2600 - SUMMARIZED RESULTS OF PARAMETRIC STUDY

lift curve producing very high negative accelerations around the cam nose, as seen here, causes this portion of the lift curve to dominate the overall power loss around the cam cycle. If, however, a cam lift curve which produces relatively low negative accelerations around the cam nose is used then the contribution of the power loss at the cam flanks becomes more important as will be seen in the parametric studies in the next section.

5.1.5 Changes in Spring Rate

As the spring rate is increased the loading at the cam/follower interface increases and so the Hertzian stress at the cam nose increases. This increase in load also causes the lubricant film thickness to decrease. The changes in lubricant film thickness are not large ((h_{min}) only being proportional to load to the power −0.13 (Dowson and Higginson (1977)). The frictional power loss increases as both (h_{cen}) decreases and (p_{max}) increases. As the limiting coefficient of friction is applied throughout the majority of the cam cycle the change in frictional power loss with changing spring stiffness is almost linear.

5.1.6 Changes in Follower Radius of Curvature

Changing the follower radius of curvature necessitates a change in the valve train geometry as explained earlier. This causes the cam profile to be changed, and so the radius of curvature of the cam also changes. As the follower radius of curvature is increased the cam nose becomes broader and the cam flanks become flatter and flatter until they eventually become concave and finally the maximum cam concavity is reached (this is equal to the minimum allowable radius of the grinding wheel used to machine the cam). As the follower radius of curvature is increased the nose of the cam becomes narrower until it becomes a sharp edge with zero radius of curvature.

The Hertzian stress at the cam nose increases as the follower radius of curvature increases because the radius of curvature at the cam nose becomes smaller. The film thickness at the cam nose increases with increasing follower radius of curvature as the entrainment velocity is enhanced. The frictional power loss changes very little with changing follower radius of curvature.

It should be noted that the extent of the cam's travel across the follower face increases proportionally to the follower radius of curvature. This will require a larger area of the follower surface to be ground and may necessitate a larger follower.

5.2 A Parametric Study of a Cam and End Pivoted Follower System

The cam and follower mechanism used in this study is in use in a current production car engine (the Ford 2.0l Pinto engine as fitted certain of the Sierra and Granada range of cars). The engine is a four cylinder in-line engine using two valves per cylinder operated by a single overhead camshaft via finger followers. During its early years on the market it suffered badly from camshaft failures characterised by severe scuffing of the cam and followers. This situation was overcome to a satisfactory level by using better cam and follower materials and by increasing oil flow to the cam/follower contact by fitting a spray-bar. A redesign of the lubricant galleys also helped by feeding oil to the contact immediately at engine start up, the previous design having suffered from starvation at start up. The geometry of the system is shown in Figure (4).

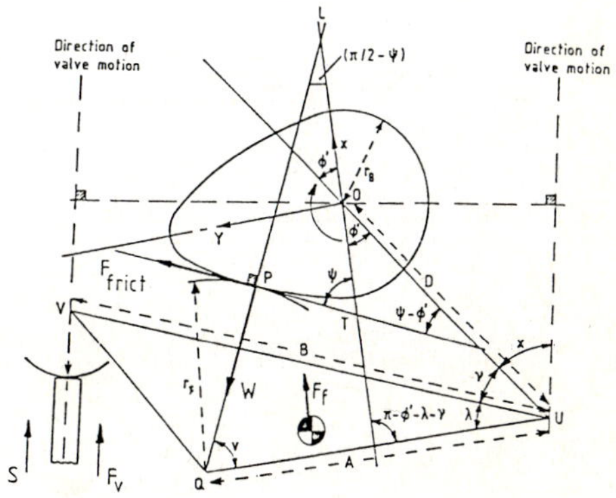

A = 51.80 mm	$\nu'_{follower}$ = 0.3
B = 61.80 mm	ν'_{cam} = 0.3
D = 46.20 mm	$E_{follower}$ = 207 GNm^{-2}
r_f = 44.00 mm	E_{cam} = 207 GNm^{-2}
r_B = 15.00 mm	α = 22.0 nm^2N^{-1}
κ = 39.7°	η_0 = 0.013 Nsm^{-2}
λ = 31.9°	k = 40.0 kN/m
L = 21.00 mm	M = 0.15 kg
	δ = 11.0 mm

FIGURE 4. GEOMETRY OF THE FORD 2.0 LITRE PINTO VALVE TRAIN

The information given in Figure 4 is taken as the reference condition. Again the effect of changing cam base circle radius, cam width, equivalent reciprocating mass, camshaft speed, spring rate and follower radius of curvature has been investigated. Each of the parameters was changed in turn whilst keeping the rest constant. The valve lift curve remained the same throughout, the cam profile being altered to effect this. Each parameter was changed from its datum value by decreasing its magnitude by 70% and observing the effects upon lubricant film thickness and Hertzian stress at the cam nose, and frictional power loss, at intervals of 10% through to a 300% increase in the parameter magnitude from its datum value. Figure (5), again in three parts, shows the cam operating characteristics at the datum condition for a camshaft speed of 41.67 Hz (2500 rpm), whilst Figure (6) shows the results of a parametric study carried out with the same camshaft speed. As the value of base circle radius is changed the geometry must be changed to take up the clearance between the cam and follower. This is done by moving the follower contact radius along the line between its centre of curvature and the cam centre of rotation i.e. as described previously.

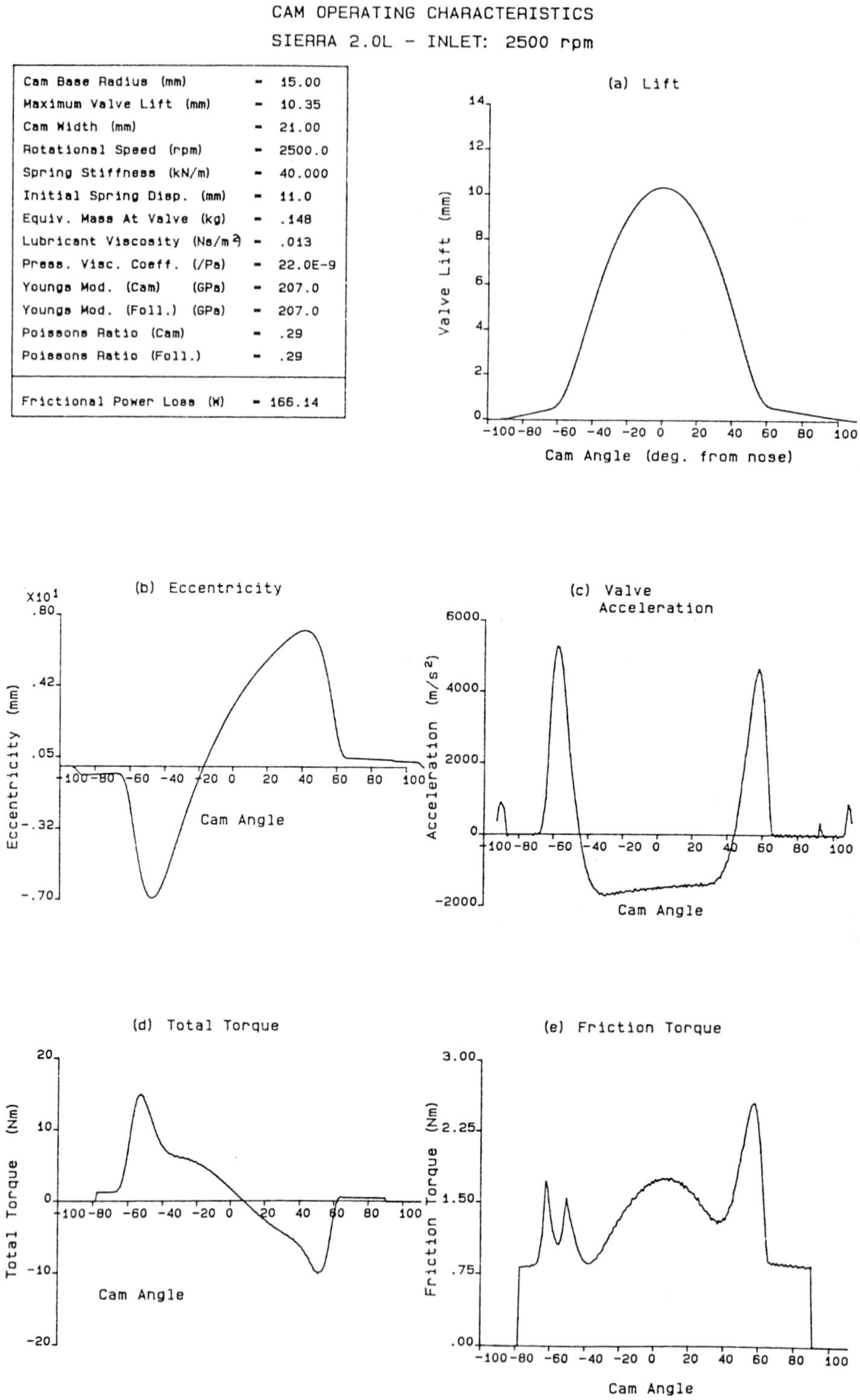

FIGURE 5(a) FORD 2.0 LITRE PINTO CAM OPERATING CHARACTERISTICS AT THE DATUM CONDITIONS (41.67 Hz (2500 r.p.m.))

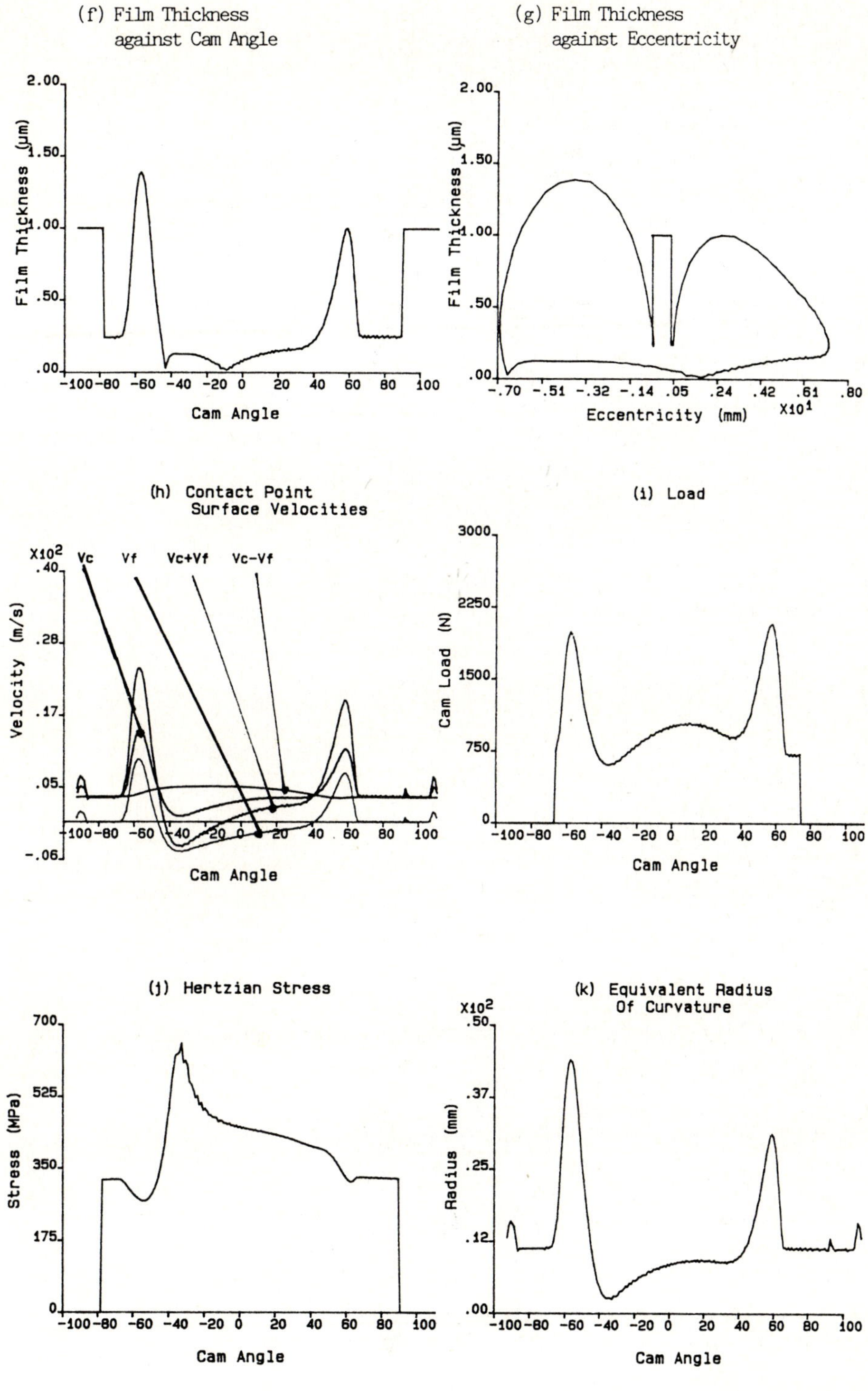

FIGURE 5(b) FORD 2.0 LITRE PINTO CAM OPERATING CHARACTERISTICS AT THE DATUM CONDITIONS (41.67 Hz (2500 r.p.m.))

CAM OPERATING CHARACTERISTICS
SIERRA 2.0L - INLET: 2500 rpm

(l) Film thickness around cam periphery

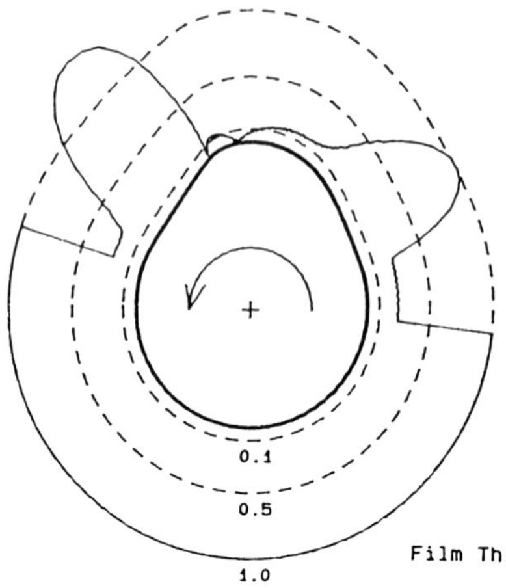

(m) Hertzian stress around cam periphery

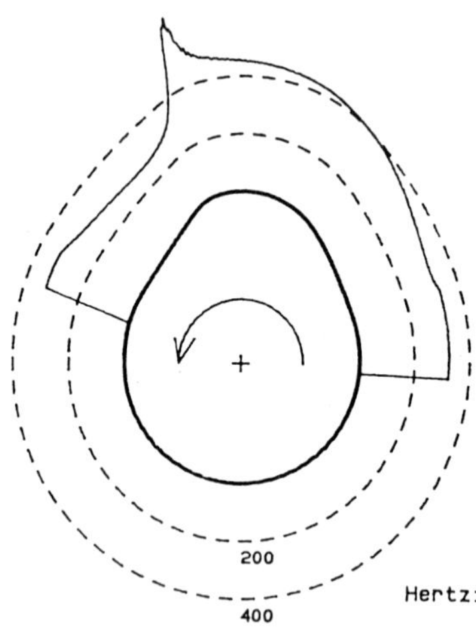

FIGURE 5(c) FORD 2.0 LITRE PINTO CAM OPERATING CHARACTERISTICS AT THE DATUM CONDITIONS (41.67 Hz (2500 r.p.m.))

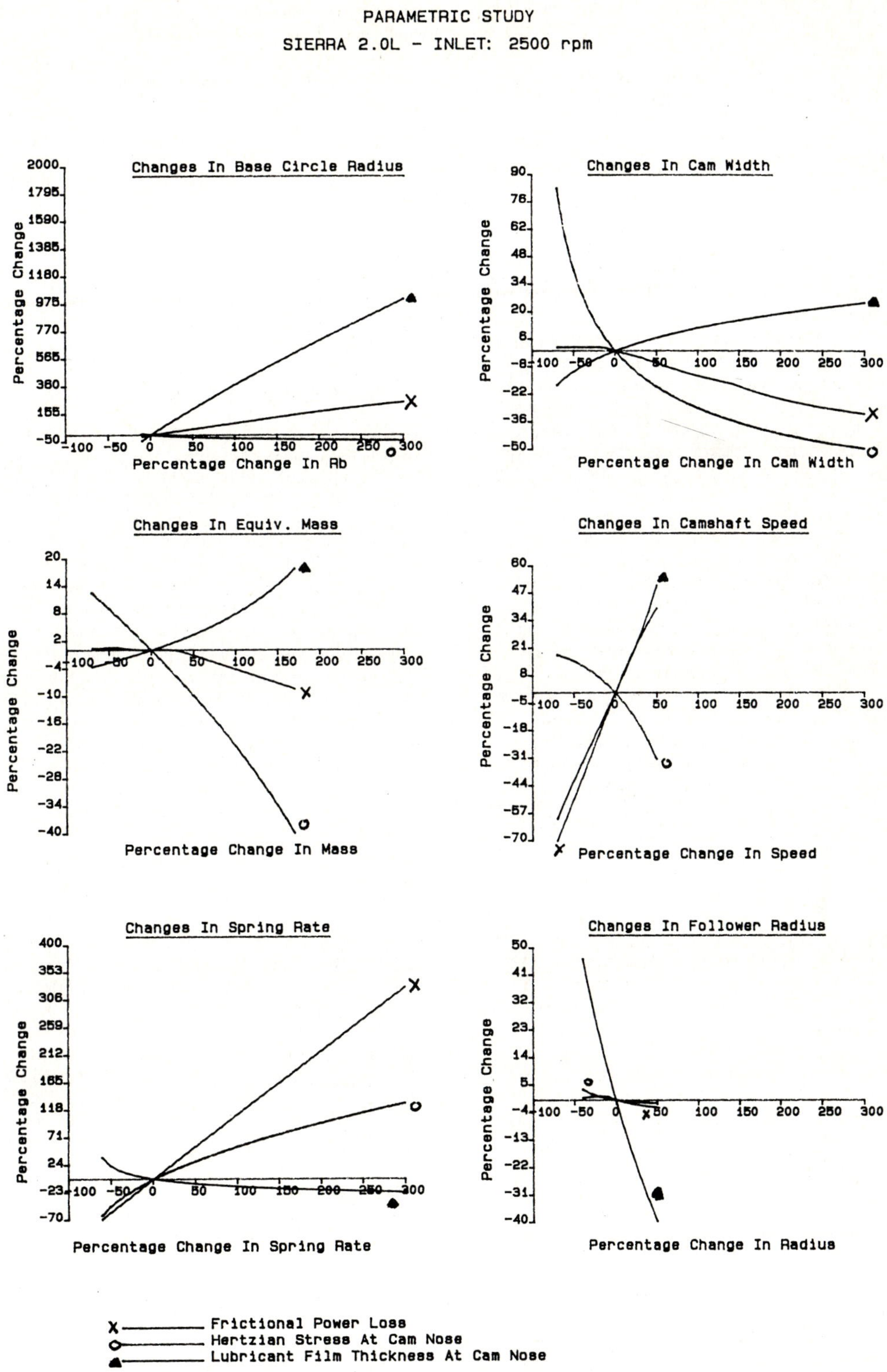

FIGURE 6 PARAMETRIC STUDY OF FORD 2.0 LITRE PINTO CAM AND FOLLOWER AT 41.67 Hz (2500 r.p.m.) CAMSHAFT ROTATIONAL FREQUENCY.

Examining the effects of changes to parameters upon lubricant thickness and Hertzian stress at the cam nose for a finger follower system again is not easy. This is due to the asymmetry introduced by the pivoted follower.

The effect of changing each of the parameters is discussed in turn below and is summarised in tabular form in Ball (1988).

5.2.1 Changes in Cam Base Circle Radius

The change in lubricant film thickness at the maximum lift as the cam base circle radius (r_B) changes can give a misleading view of the lubrication conditions across the whole of the cam nose. This is due to the shape of the lubricant entrainment velocity curve (Figure (5h)). As (r_B) becomes larger the cam radius and the larger value of equivalent radius of curvature at the contact (R) cause the whole of the entrainment velocity curve to be raised around the cam nose, causing the average entrainment velocity around the nose to fall. It should be noted that as the value of (r_B) increases the positions of the two zeros of entrainment velocity get closer together which will not enhance the lubrication and life expectancy of the cam/follower pair. The maximum allowable decrease in the cam base circle radius is limited by the permissible concavity of the cam flanks. The frictional power loss increases almost linearly with increase in (r_B). The limiting coefficient of friction is applied around the majority of the cam cycle at 41.67 Hz (2500 rpm) (this is similar to the centrally pivoted follower). The loading at the contact does not change as (r_B) changes but the radius at which the frictional traction force is applied increases as (r_B) increases, thus the frictional power loss increases.

5.2.2 Changes in Cam Width

As would be expected the Hertzian stress at the cam nose will decrease as the cam lobe width is increased. The film thickness increases and the frictional power loss decreases consequent upon the decrease in load/unit width.

5.2.3 Changes in Reciprocating Mass

Increasing the reciprocating mass at the valve causes higher loads on the flanks but smaller loads at the cam nose. As the load at the cam nose decreases with increasing reciprocating mass the lubricant film thickness will increase and the Hertzian stress will decrease. The frictional power loss decreases due to the load around the majority of the cycle decreasing. It can be seen in Figure (6) that when the reciprocating mass is increased, a point is reached where due to the increasing inertia of the system, the frictional power loss decreases more rapidly as the coefficient of friction at the cam nose becomes less than the limiting value.

5.2.4 Changes in Camshaft Speed

As the camshaft speed is increased the loading on the cam nose becomes less due to the negative lift acceleration. It can be seen that valve bounce occurs at approximately 67.5 Hz (3750 rpm (7500 rpm crank)). (Again it should be noted that the loading on the flanks increases). This reduced loading leads to a larger film thickness and reduced Hertzian stress at the cam nose. The power loss increases as the cam speed increases as would be expected. At speeds approaching valve bounce the rate at which the power loss increases starts to fall as the loading becomes very small on the cam nose.

5.2.5 Changes in Spring Rate

As the spring rate increases the Hertzian stress (p_{max}) at the cam nose increases due to the load increasing. This increase in load also causes the film thickness to decrease but the changes are not large, ((h_{min}) only being proportional to load raised to the power -0.13). The frictional power loss increases. As the limiting coefficient of friction is applied throughout the majority of the cycle then the change in frictional power loss is, as would be expected, almost linear.

5.2.6 Changes in Follower Radius

As the radius of curvature of the follower is increased the cam nose becomes broader hence the radius of curvature of the cam becomes larger in this area. This leads to a fall in the maximum Hertzian stress at the cam nose. It can be seen from Figure (5k) that there is a point on the cam, at the transition from the rising flank to the nose, where the radius of curvature is small. This radius of curvature decreases as the follower radius of curvature increases and limits the amount by which the follower radius can be increased. Decreasing the radius of curvature of the follower causes the cam flanks eventually to become concave, therefore the maximum allowable decrease in follower radius of curvature is limited by the acceptable concavity of the cam flanks.

The change in the lubricant film thickness at the maximum lift position is very misleading. Although the lubricant film thickness is enhanced at the maximum lift position as the follower radius of curvature (r_f) is decreased, the lubrication of the cam nose taken as a whole is much poorer. This is due to the shape of the entrainment velocity curve. As (r_f) is decreased the entrainment velocity at the maximum lift position is greater but over the whole of the cam nose the entrainment velocity is very close to zero for the entire nose period. Also the points where the entrainment velocity passes through zero become closer, which is not advisable as it brings together two points of distress on the cam (and follower), Ball (1988). It is therefore better to increase the follower radius of curvature to gain better lubrication conditions across the whole of the nose region.

The frictional power loss decreases as (r_f) increases due the better lubrication conditions at the cam flanks brought about through increased entrainment velocities. The increase in radius of curvature at the cam nose causes the frictional power loss across this region to increase as the frictional traction force acts at a greater radius and so the frictional torque is increased. This effect is not as great as the decrease in power loss on the cam flanks as the changes in cam radius of curvature are small.

5.3 Enhancement of an End Pivoted Follower Design

By using the information gathered by parametric studies it is possible to comment on valve train designs and to suggest possible improvements. By way of an example the end pivoted follower design used in the parametric study presented above was taken and the effect of changing several parameters at once was investigated.

It was found that because of the necessary geometry changes involved with changing the cam base circle radius and the follower radius of curvature the results of changing these parameters could not be investigated separately and then presumed to be additive. It was therefore necessary to choose these parameters together. Changes to other parameters such as the spring stiffness could obviously be investigated separately. It was also necessary to use a good deal of compromise between the Hertzian stress at the cam nose and the lubricant film thickness across the nose.

It was decided to increase the follower radius of curvature in order to enhance the film thickness across the cam nose. This also had the added advantage of decreasing the Hertzian stress across the cam nose. Unfortunately it caused a dramatic increase in the Hertzian stress at the transition between the cam rising flank and nose where the radius of curvature was already very small. To counter this change it was necessary to increase the base circle radius of the cam. This change decreased the lubricant film thickness across the cam nose and brought the two zeros of entrainment velocity closer together requiring a further increase follower radius of curvature. This process was repeated until a satisfactory compromise was found.

An obvious change that brings improvements to both the Hertzian stress at the cam nose and the lubricant film thickness is a decrease in the valve spring stiffness.

The improved design shown was arrived at by decreasing the valve spring stiffness by 10%, increasing the follower radius of curvature by 20%, and increasing the cam base circle radius by 12%. The resulting geometry becomes,

$A = 57.98$ mm
$B = 61.80$ mm
$D = 47.19$ mm
$r_f = 52.80$ mm
$r_B = 16.80$ mm
$\chi = 37.86°$
$\lambda = 38.45°$

A summary of important changes in predicted performance is recorded in Table (2) below.

Changes to Reference Design	Base Circle Radius	Follower Radius of Curvature	Valve Spring Stiffness
	+ 20%	+ 12%	- 10%
Predicted Performance Changes	Hertzian Stress at Cam Nose	Frictional Power Loss	Lubricant Film Thickness at Cam Nose
	- 19%	- 3%	+ 17%

Table (2). Suggested Redesign of Ford 2.0 Litre Pinto Engine.

It can be seen that the suggested changes give improved predicted performance characteristics of all the major factors studied. Ball (1988) gives a detailed interpretation of the influences observed. These changes significantly improve the distribution of Hertzian stress around the cam whilst also improving the lubrication of the cam. Also the two points where the lubricant film thickness falls to zero (neglecting squeeze effects) are spread further apart. There is also a small decrease in the frictional power loss.

The disadvantages of the new design are that the overall size of the valve train is increased as the cam base circle radius is increased and the maximum camshaft speed is reduced, although the maximum engine speed is still above 110Hz (6600 rpm). The increase in cam base circle radius will necessitate the use of larger camshaft bearings to enable the assembly of the system, although this may be an advantage as the valve train relies upon only three bearings.

Although the operating conditions of the cam are significantly improved by the above changes it must be realised that they are still extremely severe. If the designer of the original valve train mechanism had the above information at his disposal it is quite possible that the whole cylinder head would have been redesigned to accommodate a different valve train geometry.

A similar exercise to suggest an improved design for the Rover 2300/2600 centrally pivoted follower system has also been carried out (Ball (1988)).

6 Conclusions

Details of a computer program and its graphical output designed to enable the assessment with ease of the influence of the variation of a range of design parameters have been given. The basis of the computer program has been described. This comprises only the use of established design formulae, with no numerical analysis, and provides a fast and effective design tool for automotive engineers.

Parametric studies have been presented for two valve train types; a cam acting against a centrally pivoted follower and a cam acting against an end pivoted follower. The results of these parametric studies have been summarised. It has been shown how the use of these parametric studies can bring about improvements to valve train design by the use of an example in which the predicted performance of an end follower mechanism was enhanced.

It is felt that great benefits could accrue if designers of valve trains were given access to design tools, such as the one presented, that allow variations in the design parameters to be studied. Although there are obviously other considerations to be made in the design process, for example combustion requirements, it would allow the designer to design for the best possible tribological conditions at the cam/follower contact within the constraints imposed, thus increasing the life of the components and decreasing the risk of failure. This design procedure would probably have saved several motor manufacturers substantial warranty claims, lost sales and redesign costs, during the early years of production of new engines. Many manufacturers have learnt through costly mistakes of the consequences of not undertaking proper consideration of the tribological conditions at the cam/follower interface during the design of valve-trains.

7 Acknowledgements

The study described in this paper was jointly funded by the Science and Engineering Research Council and the Ford Motor Co. Ltd. It is part of a continuing research programme in association with the Ford Research and Engineering Centre at Dunton and we would like to thank Mr N Weaver, Mr R Windett, Mr J-P Pirault (formerly of Fords) and other staff at Dunton who have given generously and enthusiastically of their time and skills to benefit the work.

References

(1) BALL, A D (1988), "A Tribological Study of the Design and Performance of Automotive Cams", Ph.D Thesis, Department of Mechanical Engineering, University of Leeds.

(2) BALL, A D (1989), "A Computer Program for Valve Train Analysis" Internal Report, Department of Mechanical Engineering, University of Leeds.

(3) BEARD, C A, and HEMPSON, J G G, (1962), "Problems in Valve Gear Design and Instrumentation", S.A.E. National Power Plant Meeting, Philadelphia.

(4) BUUCK, B A, (1982), "Elementary Design Considerations for Valve Gears", S.A.E. 821574.

(5) DOWSON, D, HARRISON, P, and TAYLOR, C M, (1985), "The Lubrication of Automotive Cams and Followers", Mechanisms and Surface Distress, 12th Leeds-Lyon Symposium on Tribology, pp 305-322, Butterworths.

(6) DOWSON, D, and HIGGINSON, G R, (1977), "Elastohydrodynamic Lubrication", SI Edition, Pergamon Press.

(7) DOWSON, D, and TOYODA, S, (1978), "A Central Film Thickness Formula for Elasto-hydrodynamic Line Contacts", Elasto-hydrodynamics and Related Topics, 5th Leeds-Lyon Symposium on Tribology, pp 60-65, MEP (I.Mech.E.).

(8) DYSON, A and NAYLOR, H, (1960), "Application of the Flash Temperature Concept to Cam and Tappet Wear Problems", I.Mech.E., Proc. Auto. Div. No 8, pp 225-280.

(9) DYSON, A, (1977), "Elastohydrodynamic Lubrication and Wear of Cams Bearing Against Cylindrical Tappets", S.A.E. 770018.

(10) DYSON, A, (1980), "Kinematics and Wear Patterns of Cam and Follower Automotive Valve Gear", Tribology International, June 1980, pp 121-132.

(11) HARRISON, P, (1985), "A Study of the Lubrication of Automotive Cams", Ph.D. Thesis, Department of Mechanical Engineering, University of Leeds.

(12) HERTZ, M, (1982), "Über die Berührung Fester Elastischer Körper", ("On the Contact of Elastic Solids"), J. reine und angewandte Mathematik, Vol 92, pp 156-171.

(13) HOSHI, M, (1984), "Reducing Friction Losses in Automobile Engineers", Tribology International, Vol. 17, No. 4, pp 185-189.

(14) LIM, C E, EVANS, H P and SNIDLE, R W, (91983), "Kinematics and Lubrication Conditions at Cam Contact in a Centrally Pivoted Cam-Finger Follower", S.A.E. 830309.

(15) MARTIN, F A, (1985), "Friction in Internal Combustion Engine Bearings", in Combustion Engines - Reduction of Friction and Wear, Paper C67/85, I.Mech.E. Conference Publications, pp 1-7.

(16) MULLER, R, (1966), "The Effect of Lubrication on Cam and Tappet Performance", Motor Tech. Z, Vol. 27, Pt 2, pp 58-61, MIRA Translation, No 27/66.

(17) PARKER, D A, and ADAMS, D R, (1982), "Friction Losses in the Reciprocating Internal Combustion Engine", in Tribology - Key to the Efficient Engine, Paper C5/82, I.Mech.E. Conference Publication, pp 31-39.

(18) POLAK, T A, and LETTS, A, (1987), "When Did You Last Change Your Camshaft?", Automobile Engineer. Vol. 12, No. 3, June/July 1987, pp 6-8.

(19) ZHU, G, (1988), "A Theoretical and Experimental Study of the Tribology of a Cam and Follower", Ph.D. Thesis, Department of Mechanical Engineering, University of Leeds.

SESSION V
BELTS

Chairman: Professor J F Booker

Paper V(i)

Power transmission by flat, V and timing belts

T. H. C. Childs and I. K. Parker

A review of aspects of the mechanics of contact of belts on pulleys is presented. Power loss in flat and V belt drives and tooth loading in timing belt drives is related to belt tension and torque transmission, to belt elastic and friction properties, and to pulley radius and arc of contact. Particular attention is given to the effects of belt carcass distortion, in addition to the generally recognised effects of belt tension member extension. Problems in the mechanics of flat and V belt power transmissions are identified the solution of which would aid the rational design of efficient drive layouts.

INTRODUCTION

The mechanical performance of belt drives has not been as extensively studied as that of other types of power transmissions such as gears. Manufacturers' catalogues, although they give clear guidance on the selection of suitable belts for a given power transmission, do not indicate the basis of the guidance; what failure mode is being designed against or what life might be expected. Most textbooks on machines limit their coverage of belts to the capstan formula derived by Euler (1) for the tension ratio in ropes skidding over cylindrical pulleys

$$F_t/F_r = \exp(\mu\theta_T) \qquad (1)$$

or to its modification for ropes skidding in pulley grooves (2)

$$F_t/F_r = \exp(\mu\theta_T/\sin\beta), \qquad (2)$$

and in the absence of skidding, to the speed loss due to belt extension first explained by Reynolds (3) which for transmission between two equal pulleys is

$$\Delta\omega/\omega = (F_t-F_r)/c \qquad (3)$$

and to the consequent division of the belt-pulley contact into slip and no slip arcs developed by Swift (4).

The annual manufacture of belts is increasing, mainly led by timing belts (5). The capacity and durability of belts continue to improve with advances in both tension member and elastomer carcass materials engineering. It is feasible that there could be greater uses of belts if better design information were available. This review however is not directly concerned with design but with aspects of belt/pulley contact mechanics within the authors' experience: with power losses in flat and V belt drives and tooth loading in timing belt drives. It is presented in the belief that it is useful to bring together the width of studies in this area and with the hope that it might stimulate further study of the mechanics of belt drives and lead to a better level of design information.

1.1 Notation

c	Belt extension modulus; $F = ce$
d	Flat or timing belt width
e	Belt extension strain
g	Belt radial compliance, $g = xR/F$
p	Contact pressure
t	Flat belt carcass thickness
x	Belt radial displacement
A	$(G/c)^{\frac{1}{2}}(d/t)^{\frac{1}{2}}$
B	V belt width
E	Youngs modulus
EI	Bending stiffness
F	Belt tension, with tight and slack side values F_t and F_r
G	Shear modulus
H	V belt depth
L_b, L_p	Timing belt and pulley dimensions, fig. 11
$P, \Delta P$	Power and power loss
P_b, P_p	Timing belt and pulley tooth pitch
Q_i	Timing belt tooth load
R	Pulley pitch radius, ie measured to belt tension member
$T, \Delta T$	Torque and torque loss
V	Belt tension member speed, with tight and slack side values V_t and V_r
α	Timing belt pressure angle
β	V pulley groove semi-angle
β'	Defined in figure 10
γ	Shear strain
λ	Traction coefficient = $(F_t-F_r)/(F_t+F_r)$
λ_i	Timing belt tooth deflection
θ_T	Angular arc of contact
θ'	Defined in figure 10
θ_S	Angular arc of slip
$\omega, \Delta w$	Angular speed and angular speed loss
τ	Traction stress
μ	Sliding friction coefficient

2 BELT CONSTRUCTION AND ELASTICITY

The great majority of modern belts are composite constructions. Load is transmitted through a stiff tension member, traction through an elastomeric facing sometimes protected by a textile cover. This section starts with an introductory description of belt types and finishes with some guidelines of typical belt elastic properties.

2.1 Belt construction

Flat belts are of two main types of construction. Either, as shown in figure 1a, a flat strip or ribbon tension member, usually nylon, is faced with elastomer, the bond between the two being achieved by an intermediate layer of nylon woven textile; or, figure 1b, an elastomer carcass is moulded round a wound cord (polyester, kevlar, glass fibre or steel) tension member, the cords in many cases being held together by textile backing. For the strip tension member belt, the elastomer is roll-coated to the textile either before or after the textile is bonded to the strip and may be either smooth faced or embossed; the belt is manufactured in sheets, cut to width and made into a loop of any required length by a welded lap joint. Wound cord belts are truly endless and are made to specific lengths, for example by winding the cord and laying the elastomer round a mandrel before curing the elastomer. The cord itself is a twisted rope construction.

Most V belts are of wound cord tension member construction. Belts for industrial machinery power transmission are usually completely wrapped in a textile cover, figure 1c, to increase the wear resistance of the driving faces. Pressure from the automotive industry to run belts on ever smaller diameter pulleys, and problems of distortion of wrapped belts on small radii pulleys, has led to the introduction for automotive auxiliary power transmissions of raw edged belts (belts with no textile cover on their driving faces) and, for greater flexibility, cogged raw edged belts: the latter type is sketched in figure 1d. It is shown later that radial motion of a V belt in its pulley groove is a source of power loss: in practice this is reduced by axial reinforcement of raw edged belt carcasses, either by cellulose fibres or by textile laminations which also may serve to increase wear resistance.

Timing belts are exclusively of wound cord tension member construction. The meshing faces of belts used for power transmission are textile covered but for motion transmission, for example for office machinery, are not. A main classification of timing belts is according to tooth form. The initially developed trapezoidal tooth shape, figure 1e, is now giving way to a near semi-circular form, figure 1f. The latter not only meshes more easily with its pulley tooth, enabling greater torques to be transmitted without the tooth failing to mesh (tooth jumping), but also distributes load more uniformly from pulley tooth to belt tension member, giving greater resistance to tooth fatigue failure.

2.2 Belt elasticity and friction

A belt's extension modulus, c, is important to its mechanical behaviour on a pulley. c for a strip tension member is the product of the tension member's Youngs modulus and cross-section area. The room temperature Youngs modulus deduced from manufacturers' data and from our own experience of nylon strip tension members is in the range 0.6 to 0.9 GPa. Thus for a typical light duty belt, 20mm wide and with a tension member 0.5mm thick, c = 6 to 9 kN.

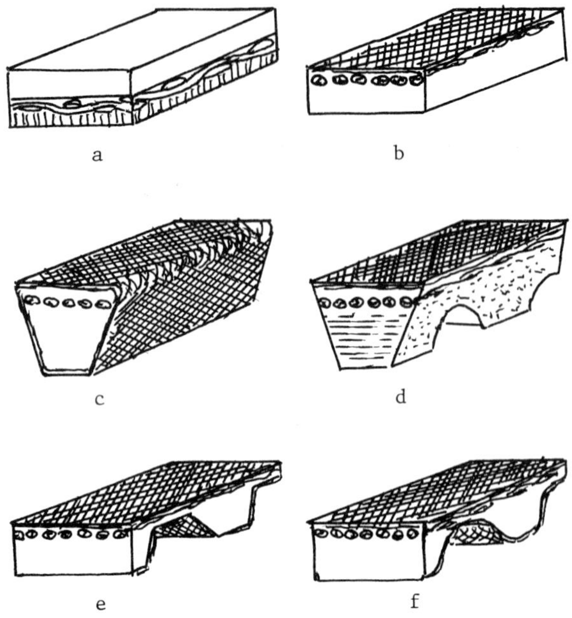

Fig. 1 Examples of flat, V and timing belt types

The extension modulus of wound cord tension member belts depends not only on cord material and cross-section area but on cord twist. Our own measurements on a range of polyester, glass and kevlar cord belts give Youngs moduli in the range 6 to 16 GPa, and similar figures may be deduced from published work (6-8). Thus automotive V Belts, 10mm wide, with polyester cords of 1mm diameter have c values from 60 to 90 kN, while cam shaft drive timing belts, 20mm wide and with glass fibre cords 1.4mm diameter, have c values typically from 180 to 250 kN.

A range of values for the room temperature static Youngs modulus of elastomer traction members may be quoted: values from 3 to 12 MPa have been indirectly deduced for roll coated layers on strip tension member flat belts (9); a value of 6 MPa has been obtained for the unreinforced carcass of an automotive V belt (9) while values of 13 and 38 MPa have been measured for two industrial V belts (10). A reinforced automotive V belt carcass has been found to have a Youngs modulus of 25 MPa perpendicular to and 65 MPa parallel to the fibres (9). A timing belt tooth rubber has been quoted as 10 MPa (7).

The bending stiffness of a V belt is as expected from bending a carcass with the neutral axis located in the tension member: thus a raw edged V belt, 10mm wide and 8mm deep has EI = 15 kNmm^2 (although EI is reduced to 6 kNmm^2 by cogging the carcass) (11); and a wrapped belt 14mm wide and 10.5mm deep has EI = 50 kNmm^2 (6). Other derived carcass elastic properties will be introduced later in the appropriate places.

Belt friction coefficients may vary with temperature, speed and load but a major difference lies between textile covered and direct elastomer contacts with pulleys. Textile covered belts generally have high running-in friction coefficients, while excess rubber is worn away, but after that show values in the range of 0.2 to 0.35 (6,7,11); typical friction coefficients of raw edged V belts are 0.4 to 0.6 (11) while values up to 0.9 may be found with flat belts (9 and manufacturers' catalogues).

3. FLAT AND V BELT DRIVES

The introduction records that tension member extension is recognised as a source of power loss in flat and V belt drives which shows itself as a speed loss between driving and driven pulleys. Distortion of the carcass gives rise to extra speed losses and also to torque loss. In the next section relations between power loss, speed loss and torque loss are developed in such a way as to distinguish between losses related to tension member and to carcass elasticity. These act a a focus for separate considerations in subsequent sections of losses in flat and V belt drives.

3.1 Power loss preliminaries

Figure 2 show a belt tension member and carcass circulating round the driving and a driven pulley of what could be a many pulley system. The variable notation is the same on each pulley but the actual values of the variables can of course differ. In classical theory (3,4), in the absence of skidding, belt and pulley speeds are the same where the belt runs on to the pulley: V_t or $V_r = \omega R$ respectively on a driving or driven pulley. Further, in an ideal, no power loss system, $V_t = V_r$ and $T = (F_t - F_r)R$; a power balance gives at every pulley

$$\omega T = V_t F_t - V_r F_r \qquad (3a)$$

In practice equation 3a is not obeyed. There is a power loss at each pulley

$$\Delta P = V_r F_r + \omega T - V_t F_t, \text{ driving pulley} \qquad (4)$$
$$\Delta P = V_t F_t - \omega T - V_r F_r, \text{ driven pulley}$$

These power losses arise from both tension member and carcass elasticity. Extension of the tension member under load coupled to conservation of mass flow gives

$$V_t = V_r\,(1 + (F_t - F_r)/c) \qquad (5)$$

while distortion of the carcass allows differences ΔV to develop between belt tension member and pulley speed where the belt runs on to the pulley and can also results in torque losses ΔT:

$$\left.\begin{aligned} \omega R &= V_t + \Delta V_t\ , \text{ driving pulley} \\ \omega R &= V_r + \Delta\, V_r, \text{ driven pulley} \\ T &= (F_t - F_r)R + \Delta T, \text{ driving pulley} \\ T &= (F_t - F_r)R - \Delta T, \text{ driven pulley} \end{aligned}\right] \qquad (6)$$

Substituting equations 5 and 6 into 4 leads to

$$\left.\begin{aligned} \Delta P &= \omega T(-F_r/c + \Delta V_t/V_t + \Delta T/T), \text{ driving pulley} \\ \Delta P &= \omega T(F_t/c + \Delta V_r/V_r + \Delta T/T), \text{ driven pulley} \end{aligned}\right\} \qquad (7)$$

For a general system of N pulleys, 1 driving and (N - 1) driven, summation of the power losses on each pulley leads, after ignoring products of small quantities and noting that on this basis ωT may be written $V_o(F_t - F_r)$ where V_o is the nominal belt speed, to

$$\sum_{i=1}^{N} \Delta P_i = V_o \sum_{i=1}^{N} (F_t - F_r)_i \left[\left[\frac{F_t - F_r}{2c} \right]_i + \left[\frac{\Delta V}{V} \right]_i + \left[\frac{\Delta T}{T} \right]_i \right] \qquad (8)$$

where $\Delta V/V$ is always taken to be the speed loss ratio of the belt running on to a pulley.

In the special case when $\Delta V = \Delta T = 0$ and $N = 2$, there being one driving and one driven pulley, equation 8 reduces to equation 3. The first term on the right of equation 8 is clearly the extension creep term. The ΔV and ΔT terms arise from carcass distortion and their form is considered in the next sections.

It has been assumed in this section that speed losses ΔV are small and thus any consideraton of skidding conditions has been excluded; this section applies equally to flat and V belts. Considerations of skidding are best considered separately for flat and V belts and are thus also held back to the next sections.

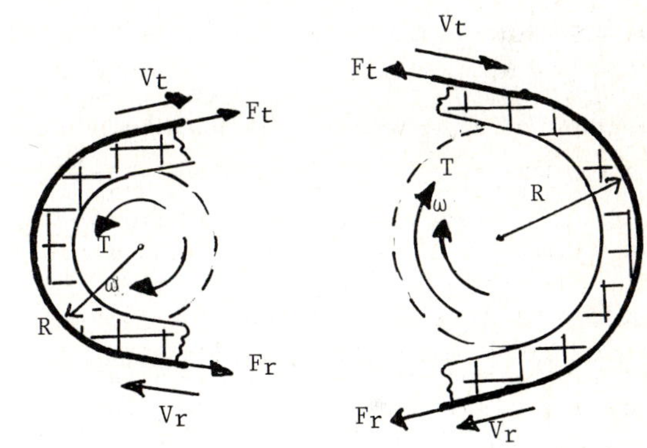

Fig. 2 Power circulation in driven and driving pulleys

3.2 Flat belt drives

Circumferential shear of the carcass is believed to be the main source of carcass - related speed loss in flat belt drives. This speed loss is therefore called shear creep. Figure 3 shows a flat belt running, for definiteness, on a driving pulley. There is a slip arc of angular extent Θ_s and a no slip arc $(\Theta_T - \Theta_s)$; the belt tension at the boundary between the two is F_s and the traction stress is τ_s. If a speed loss ΔV_t is allowed to occur between the belt tension member and pulley, a section oo' will develop a shear strain as it rotates with the pulley in the no slip arc. The shear strain gives rise to a shear stress which in turn alters the belt tension force. A relation between fractional speed loss $\Delta V_t/V_t$, $(F_t - F_r)$ and carcass shear properties is obtained by requiring F_s and τ_s to have the same values whether calculated from relations in the slip or in the no slip arcs.

In both arcs, circumferential equilibrium gives

$$dF/d\theta = \tau R d \qquad (9)$$

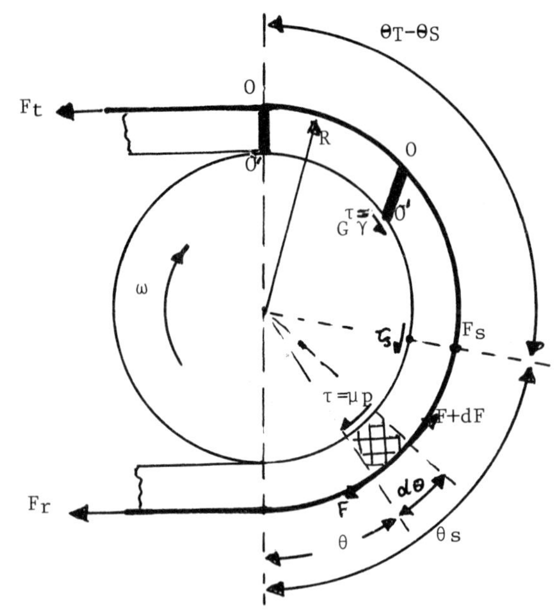

Fig. 3 Flat belt power transmission with shear creep

In the slip arc,

$$\tau = \mu p \text{ and } p = F/(Rd) \quad (10)$$

Substitution of equation 10 into 9 gives

$$dF/F = \mu d\theta \quad (11)$$

which after integration over the whole of the slip arc gives

$$F_s/F_r = \exp(\mu\theta_s)^{*} \quad (12)$$

Substitution of equation 12 into 10 gives

$$\tau_s = \mu F_r \exp(\mu\theta_s)(Rd) \quad (13)$$

In the no slip arc, as has already been written, a shear strain develops when a speed difference ΔV_t occurs at entry to the belt/pulley contact. The consequent variation of belt tension round the no slip arc causes variation of belt tension member extension and hence variation of the speed difference between tension member and pulley. Firbank (12), assuming γ=0 at entry to the contact, has shown that the combined effects of a speed differential and belt extension leads to

$$\tau = G\gamma = [\Delta V_t/V_t][G/(At)]\sinh[AR(\theta_T-\theta)] \quad (15)$$

Substituting equation 15 into 9 and integrating over the no slip arc gives

$$F_s = F_t - 2c(\Delta V_t/V_t)\sinh^2(AR(\theta_T-\theta_s)/2) \quad (16)$$

Equating equations 16 and 12 for F_s and 15 and 13 for τ_s leads to two equations:

*When θ_s grows to θ_T, $F_s = F_t$, skidding follows and the well known capstan formula, equation 1, is recovered. It is common to introduce a traction coefficient $\lambda = (F_t - F_r)/(F_t + F_r)$ with a maximum value at skidding:

$$\lambda_{max} = \frac{\exp(\mu\theta_T) - 1}{\exp(\mu\theta_T) + 1} \quad (14)$$

The expectation of extension creep theory, equation 3, is that $c(\Delta\omega/\omega)/(F_t + F_r)$ when plotted against λ should for $\lambda < \lambda_{max}$ yield a line of gradient unity and at $\lambda = \lambda_{max}$ should jump to infinity.

$$\frac{F_t}{F_r} = \left[1+\left(\frac{2\mu}{AR}\right)\frac{\sinh^2(AR(\theta_T-\theta_s)/2)}{\sinh(AR(\theta_T-\theta_s))}\right]e^{\mu\theta_s} \quad (17a)$$

$$\frac{c(\Delta V_t/V_t)}{(F_t + F_r)} = 0.5\,\frac{(F_t/F_r - e^{\mu\theta_s})}{(F_t/F_r+1)\sinh^2(AR(\theta_T-\theta_s)/2)} \quad (18a)$$

For given belt properties A and μ and belt/pulley geometry R and θ_T these are two equations in the three variables F_t/F_r, $c(\Delta V_t/V_t)/(F_t+t_r)$ and θ_s. The relationship between $c(\Delta V_t/V_t)/(F_t+F_r)$ and F_t/F_r is found parametrically by substituting values of θ_s from 0 to θ_T into equations 17 and 18. Similar equations are obtained for driven pulleys:

$$\frac{F_t}{F_r} = \left[1-\left[\frac{2\mu}{AR}\right]\frac{\sinh^2(AR(\theta_T-\theta_s)/2)}{\sinh(AR(\theta_T-\theta_s))}\right]^{-1}e^{\mu\theta_s} \quad (17b)$$

$$\frac{c(\Delta V_r/V_r)}{(F_t+F_r)} = 0.5\,\frac{((F_t/F_r)e^{-\mu\theta_s} - 1)}{(F_t/F_r+1)\sinh^2(AR(\theta_T-\theta_s)/2)} \quad (18b)$$

As an example of the effects of shear creep calculations have been carried out for a typical light duty textile machinery drive belt, assuming a 20 mm wide, 0.5 mm thick nylon strip tension member to be bonded to a 0.5 mm thick rubber traction face: c = 10kN, G = 2 MPa, d = 20 mm, t = 0.5 mm. μ has been taken as 0.6. Figure 4 shows calculated contributions of shear creep for a (a) driving and (b) driven pulleys, plotted in such a way that the magnitude of speed loss due to shear can be compared with that expected from extension creep: the dotted lines in the figures are those from equations 2; the dashed

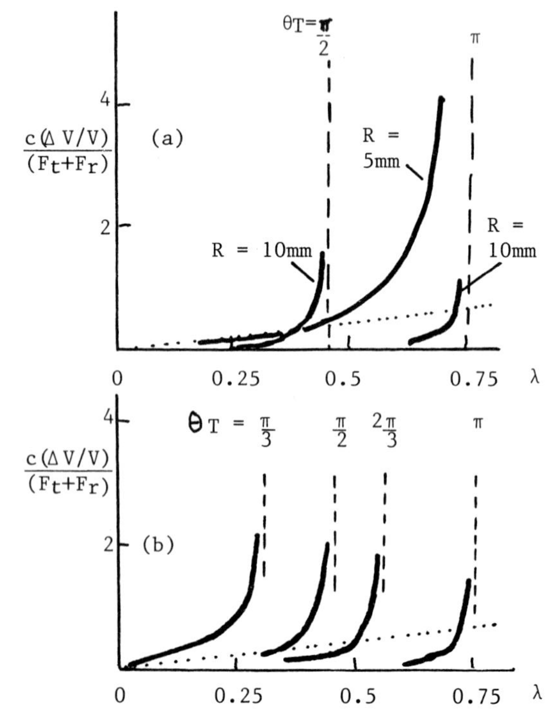

Fig. 4 Shear creep losses for a thin flat belt described in text on (a) 5 and 10 mm radius driving pulleys and (b) 5 mm radius driven pulleys, arcs of contact as marked.

lines are skidding limits from equation 14. In every case the shear creep contribution increases rapidly as skidding is approached; shear creep is the means by which a steady transition from extension creep to skidding occurs.

Further for the driving pulley in this case shear creep is negligible for most of the useful torque transmission range when the arc of contact between belt and pulley is Π and the pulley radius is 10 mm but it dominates the losses when the radius is reduced to 5 mm. It begins to be of importance for a 10 mm radius pulley when the arc of contact is reduced to Π/2. Shear creep losses are less significant on driven pulleys; thus in figure 4b shear creep becomes important for 5 mm radius driven pulley only when the arc of contact is reduced to Π/2. The theory thus predicts minimum radii and arcs of contact for efficient drive layouts.

As a second example, figure 5 shows experimental measurements, published previously in another context (11), for the fractional speed loss between two 50 mm radius pulleys connected by a thick, wound cord tension member, flat belt: c = 85 kN, d = 10 mm, t = 5.8 mm, μ = 0.59. The carcass of this belt was a fibre reinforced rubber used in raw edged V belts. Parallel to the fibre direction E = 65MPa, perpendicular to the fibre direction E = 25MPa. E for the unreinforced matrix was 6MPa. Data from tests at different total tensions fall, as expected, on a universal curve when plotted in the manner of figure 5. The dashed lines are the expectation of extension creep theory; the solid lines are ths sum of extension and shear creep terms (i) taking G = 8 MPa and (ii) taking G = 2 MPa. (ii) agrees with experiment but G = 8 MPa is the value expected on the basis that G = E/3. 2MPa is closer to the value of G expected for the unreinforced rubber. There thus remains some uncertainty about either the completeness of the creep theory presented here or of the appropriate material properties to use in the theory.

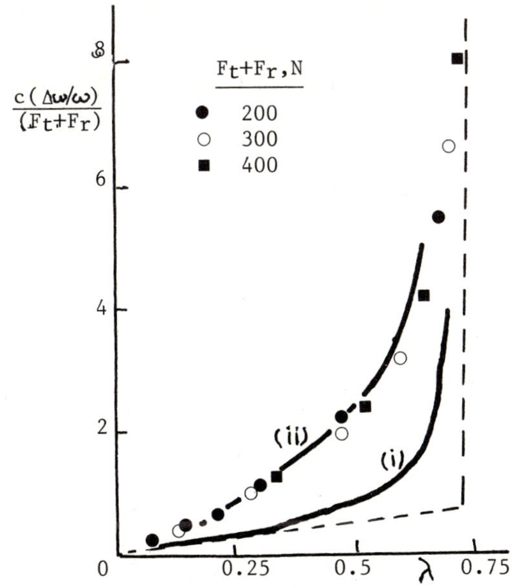

Fig. 5 Angular speed loss between 50 mm radius pulleys connected by a thick flat belt. Symbols are experimental data, solid lines are theory for (i) G = 8MPa and (ii) G = 2MPa.

It is usual to ignore torque losses in flat belt power transmissions. However torque loss was also measured in the tests which generated figure 5. Torque loss between the two pulleys did not depend on torque transmission but increased, as shown in figure 6, with total belt tension. Extrapolation of the results to zero tension leaves a finite torque loss ΔT_H which is attributed to bending hysteresis losses in the belt. Subsidiary tests were carried out in which the carcass was cut from the belt and the cord-only part, with a thin rubber face, was run on the pulleys. The results are also shown in figure 6: torque loss was almost independent of belt tension and can be attributed to bending hysteresis in the cord part of the belt. Comparison of the two sets of results show that for the complete belt a main part of the hysteresis loss occurred in the cord section but that the tension dependent part of the loss must be related to the carcass.

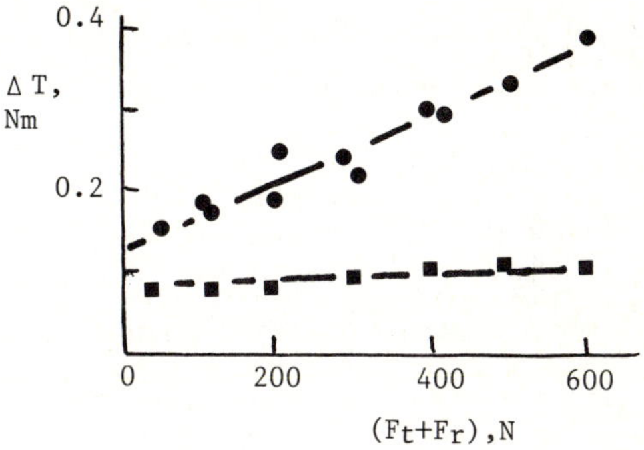

Fig. 6 Torque loss between 50 mm radius pulleys connected by (●) a thick flat belt and (■) the cord-only part of the belt.

A model experiment at low rotation speed and with no torque transmission has been performed to study in part the carcass deformation on the 50 mm radius pulley. A dial gauge rotating with the pulley showed that the belt was radially compliant. As illustrated in figure 7 the displacement x of the belt tension member from its nominal no load position increased linearly with belt tension. Further there was an assymetry between the displacements at belt entry to and exit from the pulley. An additional displacement δ occurred at exit, within an arc length of one to two belt carcass thicknesses, which did not occur at entry; it was independent of load over the range studied.

The torque loss $F\delta$ due to the assymetry of motion is an order of magnitude too low to explain the losses of figure 6. It is believed that as with V belts (6) torque losses are dominated by assymetries of belt curvature at entry and exit caused by the effects of EI, the belt bending stiffness. Belt radial movement has been mentioned here to introduce the radial complaince g of a belt and for later comparison with the behaviour of V belts. An empirical equation has been developed for the torque loss between two equal radius pulleys for both flat and V belts:

$$\Delta T = \Delta T_H + 0.75(F_t+F_r)R \quad (gEI/R^4)^{\frac{1}{2}} \qquad (19)$$

Thus for flat belt drives equations 8, 17 and 18, and 19 exist for the prediction of speed

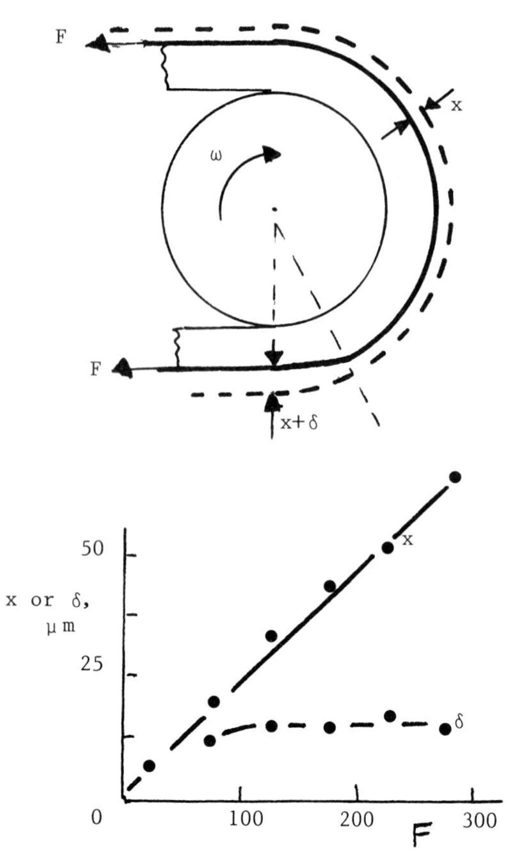

Fig. 7 Measured radial displacements of a thick flat belt on a 50 mm radius pulley, rotating in zero torque transmission conditions.

and torque losses as a function of pulley torques, radii and lay out, and of belt tension, geometry, elastic and friction properties. They could be used as a rational basis for design to avoid inefficient operating conditions. Away from such conditions belt life would be limited either by bending fatigue of the tension member or of the bond between tension and traction member, or by wear due to microslip associated with extension creep. The development of life laws and their incorporation in design guides remains for the future.

3.3 V belt drives

Figure 8 shows the speed loss measured between two 50 mm radius V grooved pulleys connected by an automotive raw edged AV10 V belt (B = 10 mm, H = 8 mm, $\beta = 18^{o}$) with ($F_t + F_r$) = 400 N (11). The speed loss is compared with that for the similar section and construction flat belt shown in figure 5 and with the capstan and extension creep expectations of equations 2 and 3. The wedging action of the V belt in the pulley groove has increased the value of λ for skidding almost to 1.0 but otherwise the V and flat belt behaviours are similar, both showing greater creep than expected from tension member extension, and ranges of linear and non-linear creep with increasing λ. However the traditional explanations of speed loss in excess of extension creep have been completely different for V and flat belts.

For V belts shear creep possibilities have been ignored; instead radial motion of the belt in the pulley groove has been considered as the cause of speed loss (13); and of torque loss too (6): radial motion occurs as belt tension changes because the carcass cross section area changes with belt tension. In this section these and related empirical developments are reviewed under three headings: (i) the onset and development of non-linear speed losses, (ii) speed losses in the linear range, and (iii) torque losses. The purpose is twofold: to record as far as is currently possible equations for calculating power losses, that may be used for determining the limiting conditions for efficient power transmission; and to point the way for what remains to be done.

The development of non-linear speed loss may be followed in figure 8 by considering conditions of slip at the three points labelled a, b and c. At point a in the linear range both a no slip and a slip arc are supposed to exist. If relative movement between the pulley and belt in the slip arc were circumferential, circumferential equilibrium would give

$$dF/F = (\mu/\sin\beta)d\theta \qquad (20)$$

However radial movement of the belt with changing tension results in a non-circumferential slip, so that equation 20 overestimates dF/F. At point b, at the end of the linear creep range, the no slip arc of contact is reduced to zero but

$$F_t/F_r < \exp(\mu\theta_T/\sin\beta) \qquad (21)$$

because of the non-circumferential slip just mentioned. As increasing torque is applied to the pulley, circumferential slip increases relative to radial slip and the effective circumferential friction coefficient increases until at point c, where gross skidding occurs, equation 2 is recovered. Thus for V belts the

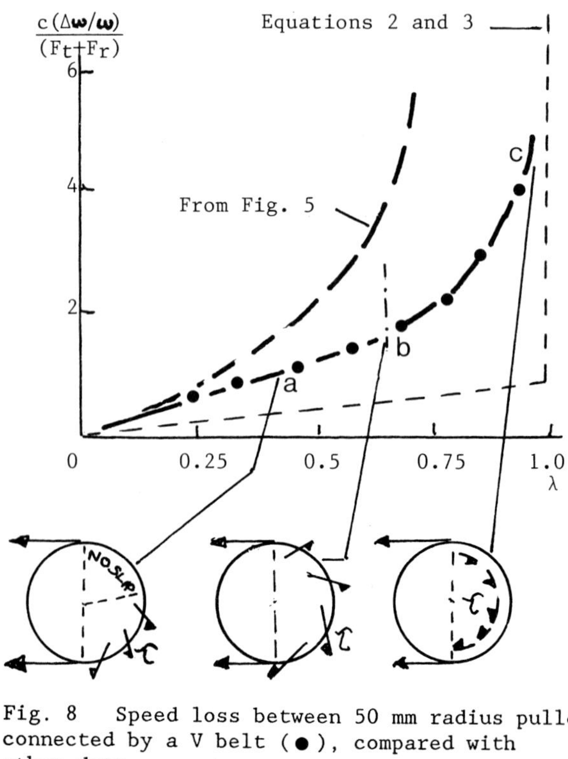

Fig. 8 Speed loss between 50 mm radius pulleys connected by a V belt (●), compared with other data.

non-linear creep region is ascribed to a varying effective friction coefficient whereas with flat belts it is ascribed to carcass shear.

The maximum useful traction coefficient is an important quantity for the design of belt systems. If it is defined as the traction coefficient for most efficient power transmission, it is clearly less than the skidding limit. Experiments (11) show that it is somewhat greater than the linear limit, point b in figure 8; and in any case the linear limit requires an involved numerical calculation for its determination (13). Gerbert (14) has suggested a weak non-linear limit obtained from

$$F_t/F_r = \exp(\mu_{eff}\theta_T) \tag{22}$$

where μ_{eff} depends on μ, belt elastic properties and pulley radius. For the AV10 belt which is the subject of figure 8, μ_{eff} may be calculated by Gerbert's method to be

$$\mu_{eff} = 0.61\ R^{0.174}\ \mu^{\frac{1}{2}} \tag{23}$$

where R is in mm. However experiments (11) show that a traction coefficient derived from equations 22 and 23 is associated with excessive slip on small radii pulleys. A reliable relationship between maximum useful traction coefficient, belt properties, pulley radius and arc of contact remains a goal of V belt contact mechanics.

Within the linear range, radial movement is expected to lead to larger speed losses for V than for flat belts. Experiments and theoretical modelling (11,13) both show that radial displacement in the no slip arc obeys

$$x = g\ (F/R) \tag{24}$$

qualitatively also as found for the thick flat belt behaviour shown in figure 7. Consequently, as an example, in a drive between two nominally equal radius pulleys, a V belt, because it runs on to the driving pulley at a greater tension than on to the driven pulley, seats at a smaller radius on the driving pulley than on the driven one. It is readily shown that this, with extension creep, results in a speed loss

$$\Delta\omega/\omega = (1/c + g/R^2)(F_t - F_r) \tag{25}$$

Equation 25 is found to be satisfactory for belts with large arcs of contact round pulleys and running at high levels of tension. However experiments, to be described in more detail elsewhere, on a drive between two equal pulleys, in which the arc of contact has been varied by the positioning of an idler pulley, have shown larger speed losses than equation (25) at small tensions and with small arcs of contact. Equation 26 has been empirically derived from tests on an AV10 raw edged belt.

$$\frac{\Delta\omega}{\omega} = \left[\frac{1}{c} + \frac{g}{R^2} + (2.5 \pm 1)\frac{(gEI/R^4)^{\frac{1}{2}}}{(F_t+F_r)\theta_T}\right](F_t - F_r) \tag{26}$$

A second goal of V belt contact mechanics is to explain the form of equation 26. A particular need is to determine what part of the speed loss additional to that of equation 25 is attributable to losses on the driven pulley and what part is attributable to the driving pulley. Then the empirical observations could be extrapolated to the determination of losses in drives between more than two pulleys and between pulleys of unequal radius. It is possible that the explanation of the additional losses will involve carcass shear, as with flat belts.

Torque losses are a significant component of the losses in V belt drives. They are commonly attributed to the wedging in and snatching out of a belt at entry to and exit from a pulley groove. Calculations of torque loss (13) which neglect belt bending stifness greatly underestimate the loss. Calculations including bending stiffness (6,11) have not been universally accurate. Experiments on the other hand show, perhaps unexpectedly, that the torque loss of a V belt is close to that of a similar section flat belt running on a cylindrical pulley, for which wedging can play no part. Figure 9 shows the torque loss between two 50 mm radius V grooved pulleys connected by an AV10 raw edged belt; the dashed line is reproduced from figure 6. Equation 19 represents the V belt loss as well as that of the flat belt. A third goal of contact mechanics is to analyse the nature of the slip in the entry and exit regions of a V belt and pulley contact, which is responsible for torque loss. Useful insight would be gained by first solving the geometrically more simple problem of a thick flat belt running on to and off a cylindrical pulley: the patterns of slip in the entry and exit regions are neither understood nor known.

A design aid for assessing the efficiency of proposed V belt drives would have two parts. Firstly equations like 22 and 23 would be used to assess if there were sufficient belt tension for the avoidance of excessive slip. Then equations related to 19 and 26 would be applied to every pulley in a drive to determine the total power loss. However equation 23 is not safe to use in all circumstances and at the present time it is not known how the torque and speed losses of equations 19 and 26, obtained from measurements on a drive between two equal pulleys, are apportioned between the two pulleys.

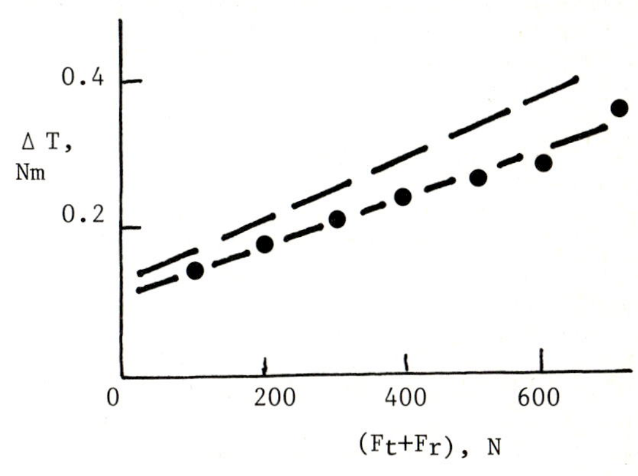

Fig. 9. Torque loss for a V-belt (●) connecting 50 mm radius pulleys compared with data from Fig. 6 (- - -)

4 TIMING BELT DRIVES

The meshing between belt and pulley teeth in a timing or synchronous belt drive enables traction coefficients to be achieved greater than the limiting values found from the capstan formula, without any speed loss at all, limited only by tooth jumping. Tooth jumping follows from the failure of a belt tooth to mesh in a pulley groove, usually on entry to a driven pulley. Instead it rides over the pulley tooth and synchronous transmission is lost. There has been no mechanical analysis directed to predicting the onset of jumping conditions but an experimental study (8) has shown that maximum traction coefficient depends on initial belt tensioning and on tooth form. Neither has there been any analysis directed to the prediction of torque losses that must occur as the result of sliding between belt and pulley teeth at meshing and unmeshing. In contrast to the studies of flat and V belts, but more analagously to the study of gears, mechanical studies (15-19) have been directed to the prediction of contact loads between belt and pulley teeth, on the assumption that these determine belt life either through tooth failure or through sliding wear of the textile covered tooth surface.

The object of the analyses is, with respect to figure 10, to determine how tooth load Q_i varies with tooth number i round the arc of contact. It is important to realise firstly that belt extension under load significantly increases its tooth pitch relative to that of the pulley and it is therefore common to specify that the unloaded belt tooth pitch be less than that of the pulley tooth pitch; and secondly that the patterns of slip between the pulley teeth top lands and the belt are quite different from those in a flat or V belt drive.

Figure 11 shows a developed view of a belt tooth i. The top part shows the belt unloaded and out of contact with a pulley tooth: it shows the belt pitch P_b less than the pulley pitch P_p and divides both pitches into parts S and L. The bottom part shows the belt stretched under a tension F and forced into mesh with the pulley by tooth deflections λ_i and λ_{i+1} relative to the tension member. Sliding is assumed to occur over S as considered further below, in a direction to generate a friction stress τ in the direction shown so that by the capstan formula

$$F_i' = F_i \exp(\mu\beta') \tag{27}$$

In current models Q_i is assumed to act on the tension member as a point force, causing a discontinuity in F as shown. Circumferential equilibrium gives

$$Q_i \cos\alpha = (F'_{i-1} - F_i) \cos\theta' \tag{28}$$

Modelling both the tension member and the belt tooth as linear springs gives firstly the extension of the belt pitch ΔP_b under the force distribution shown in Figure 11 as

$$\Delta P_{bi} = [L_p + \frac{R}{\mu} (\exp(\mu\beta') - 1)] \frac{F_i}{K_b P_b} \tag{29}$$

where K_b, the tension member spring constant, is c/P_b; and secondly

$$Q_i \cos\alpha = K_t \lambda_i \tag{30}$$

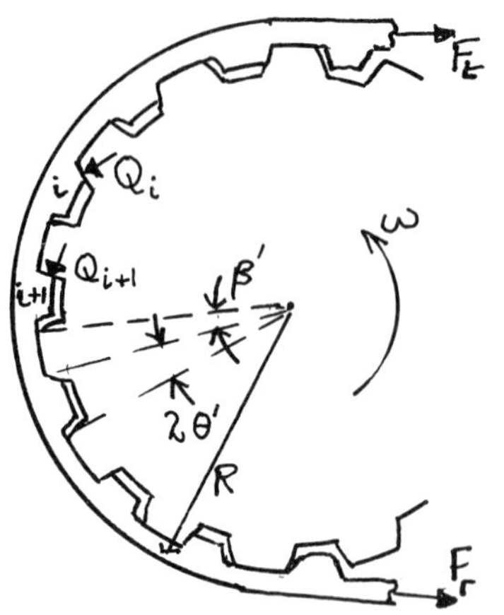

Fig 10. A driving timing belt pulley.

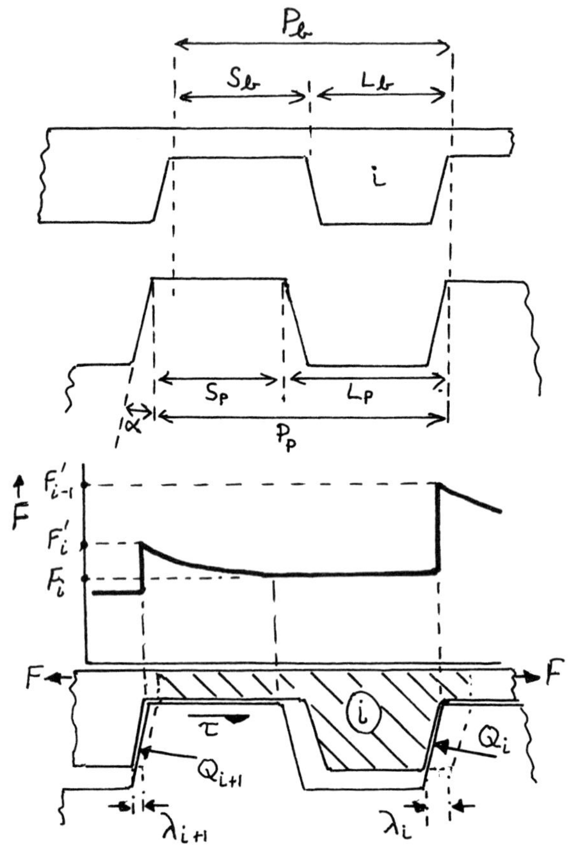

Fig. 11 Meshing and tension behaviour between a timing belt and pulley.

where K_t defines the tooth spring constant. Compatibility of displacements in meshing gives

$$P_p - P_b = \Delta P_{bi} + \lambda_{i+1} - \lambda_i \tag{31}$$

Equation 31 may be converted to a recurrence relation between F_{i-1}, F_i and F_{i+1} by substituting in it equations 27 to 30 and solved subject to initial and final values of F either analytically if K_b and K_t are assumed independent of load (18) or numerically (16). Once all the tensions F_i and F_i' are calculated, the tooth loads are obtained from equation 28.

The outline above has glossed over two problems. Firstly it has assumed a particular direction of slip on the land S. A rule may however be derived for this, in the case of steady state power transmission. Consider, in figure 10, that a tooth i rotates with the pulley to take up the position an instant later of the tooth initially at i+1. If $\lambda_i > \lambda_{i+1}$, the belt element S_{bi} must have advanced over the pulley top land to create a friction stress on the belt in the direction drawn in figure 11. Inspection of equation 31 shows that $\lambda_i > \lambda_{i+1}$ while $\Delta P_{bi} > (P_p - P_b)$. As a special case there might occur $\lambda_i = \lambda_{i+1}$. Then there would be no sliding on the pulley top land and equation 31 with 29 would give

$$F_i \simeq K_b(P_p - P_b) \qquad (32)$$

It is concluded that on the driving pulley slip of the belt over the pulley top land is in the direction of belt movement while F_i is greater than the value of equation 32 and opposite to the direction of belt movement for smaller values of F_i. The same is true on a driven pulley. In those cases when the F value of equation 32 lies between F_t and F_r there will be a point in the arc of contact where the direction of slip reverses. In that part of the contact where slip is opposite to that shown in figure 11, equation 29 must be modified and the recurrence relation between F_{i-1}, F_i and F_{i+1} has different coefficients. The recurrence relations must be solved separately in the two slip regions and matched at their interface.

The second problem concerns the condition of partial meshing that usually exists at the start and end of the arc of contact. Equations 27 to 31 are all altered for teeth in partial mesh (17), leading either to greater complexity or to fuzziness in the solution of the recurrence relations.

However when slip reversals and partial meshing are taken into account timing belt life observations may be rationalised in terms of tooth loading. Figure 12 is derived from an experimental study of the life of a classical L section polychloroprene belt with a glass fibre tension member, of 9.5 mm tooth pitch, transmitting power steadily between two 36 tooth pulleys (20). The figure shows how the number of tooth loading cycles for belt failure depends on torque transmitted per unit width of belt and the unloaded pitch difference between the pulley and belt. Pitch difference is seen to have a major effect on belt life. When, however, cycles to failure are based on the calculated maximum tooth loading, a single life curve is obtained.

To be able to rationalise life observations for a single belt type is not the same thing as to be able to analyse differences in performance between different belt types. For this, better models of belt tooth deflection and of transfer of load from belt tooth to tension member and of partial meshing are required. Newer designs of almost semi-circular toothed belts fill the pulley grooves more completely than do the teeth of L section belts: the modelling outlined here has been less tested for such belts (8). An extension of the modelling to fluctuating torque transmissions is needed for application to automotive cam shaft drive belts. Research in these areas is currently active.

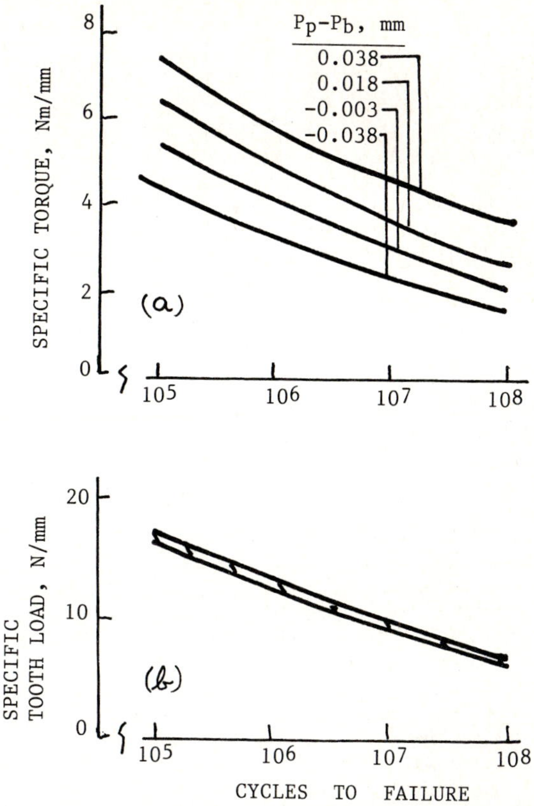

Fig. 12 (a) Dependence of failure of an L section timing belt on pulley and belt pitch difference and a torque per unit belt width and (b) the universal life curve obtained in terms of maximum tooth load per unit belt width.

5 CONCLUSION

General knowledge of power transmission by belts recognises only the capstan formula for the skidding condition of flat and V belt drives and belt tension member extension as a cause of inefficiency. However modern flat, V and timing belts transmit load from tension member to pulley via a compliant rubber carcass. Distortion of this gives rise to further losses which manifest themselves through enhanced wear or fatigue of the belt. At entry to and exit from the contact between a pulley and a flat or V belt, the interaction between belt bending, carcass distortion and slip, which determines torque loss, has not been successfully analysed. Within the contact, carcass shear is understood as a cause of speed loss in flat belts but it is not certain that current analysis is definitive. Carcass radial compliance is responsible for V-belt speed loss at large belt tensions and with large contact arcs but additional distortions, as yet unanalysed, cause additional losses at small tensions and with small contact arcs. The continuing development of tooth loading contact mechanics for timing belts is still resulting in the evolution of new tooth shapes. A better understanding of belt contact mechanics would lead to a better understanding of belt performance, possibly to better designs of belting and to the development of better design guides for the use of belts.

References

1. EULER, L. 'Remarques sur l'effect du frottement dans l'equilibre', Mém. Acad. Sci. Berl. 1762, No 18, 265-278, cited in DOWSON, D. "History of Tribology", Longmans, London, 1979.

2. EYTELWEIN, J.A. 'Handbuch der Statik fester Körper', Bd. 1, 311, Reimer, Berlin 1832, cited in reference 13.

3. REYNOLDS, O. "On the efficiency of belts or straps as communicators of work", The Engineer 1874, 38 Nov 27, 396.

4. SWIFT, H.W., "Power transmission of belts: an investigation of fundamentals", Proc. I.Mech.E. 1928, Vol 2, 659-699.

5. KOYAMA, T. and MARSHEK, K.M. "Toothed belt drives - past, present and future" Mech. Mach. Theory 1988, 23, 227-241.

6. GERBERT, B.G. "Power loss and optimum tensioning of V-belt drives", Trans ASME Jnl. Eng. Ind. 1974, 96, 877-885.

7. GERBERT, G. et al. "Load distribution in timing belts", Trans ASME Jnl. Mech. Des. 1978, 100, 208-215.

8. KOYAMA, T. et al. "Life, transmission error and noise of toothed belt drives for automobile engines", ASME publication 84-DET-217.

9. CHILDS, T.H.C. and COWBURN, D. "Contact observations on and friction of rubber drive belting", Wear 1984, 100, 59-76.

10. GERBERT, B.G. "Pressure distribution and belt deformation in V-belt drives", Trans ASME Jnl. Eng. Ind. 1975, 97, 976-982.

11. CHILDS, T.H.C. and COWBURN, D. "Power transmission losses in V-belt drives" Proc. I. Mech. E. 1987, 201 PtD, 41-53.

12. FIRBANK, T.C. "Mechanics of the belt drive", Int. J. Mech. Sci. 1970, 12, 1053-1064.

13. GERBERT, B.G. "Force and slip behaviour in V-belt drives", Act Polytechnica Scandinavica, Mech. Eng. Series No 67, Helsinki 1972.

14. GERBERT, B.G. "A complementary large slip solution in V-belt mechanics", ASME Paper No. 77-DET-162, 1977.

15. GERBERT, G. et al "Load distribution in timing belts", Trans ASME Jnl. Mech. Des. 1978, 100, 208-215.

16. KOYAMA, T. et al. "A study on strength of toothed belts, 2nd report", Bull. JSME 1979, 22, 982-987.

17. KOYAMA, T. et al. "A study on strength of toothed belts, 4th report", Bull. JSME 1980, 1235-1239.

18. NAJI, M.R. and MARSHEK, K.M. "Toothed belt load distribution" Trans ASME Jnl. of Mechanisms 1983, 105, 339-347.

19. KAGOTANI, M. et al "Load distribtion on toothed belt drives under a state of initial tension", Bull. JSME 1984, 27, 1780-1787.

20. KOYAMA, T. et al, "A study on strength of toothed belts, 5th report", Bull. JSME 1980, 1240-1244.

Paper V(ii)

Power rating of flat belt drives - A wear approach

B. G. Gerbert

Modern flat belts consists of a kernel of fiber reinforced elastomeric material which takes up belt tension. One or both of the flat surfaces are covered with a thin layer of high friction material.

The power rating procedure of flat belt drives described in manufacturers catalogues is based on testing and experience. It appears that the procedure might be adopted to a linear wear model of the surface layer.

1 INTRODUCTION

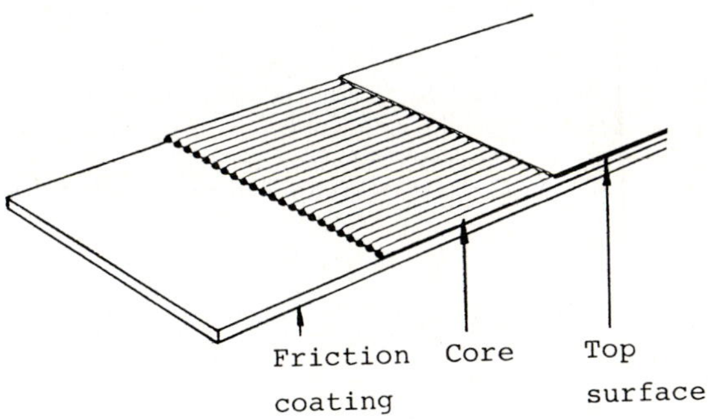

Figure 1. Cross section of flat belt

A modern flat belt is built up of several functional layers according to fig. 1. The tension layer (core) consists of a high strength fiber reinforced elastomer. The friction layer is specially prepared to give high friction against the pully (usually the coefficient of friction $\mu > 0.6$). The top layer protects the tension layer from mechanical damage.

The power rating procedure of flat belt drives is outlined in catalogues from different manufacturers. One method is described in Appendix. No standard method is adopted. The procedures seem to be based on extensive testing, experience and general engineering feeling. There is obviously no thorough theoretical analysis behind them.

In this paper a linear wear concept is applied on the flat belt drive. Wear out of the friction layer leads to decrease in friction and gradual improper operation (excessive slip). This should be possible to detect by inspection during stop and possibly also during operation.

Fatigue of the tensile layer is another failure mechanism. This leads to belt rupture and a sudden malfunction of the belt which might cause hazardous situations.

Thus, belt ruptures should be avoided and friction layer wear preferred as a failure mechanism. In the paper the two failure modes are compared to a practical power rating procedure. The wear concept is in reasonable accordance with the practical one.

2 ELEMENTARY FLAT BELT MECHANICS

Elementary analysis of variation of belt force and contact pressure along the contact between the belt and the pulley are found in many textbooks e.g. ref [1]. More comprehensive analyses of flat belts and V-belts are given in refs. [2, 3]. The elementary analysis is adequate for the purpose of this paper and it will be briefly reproduced.

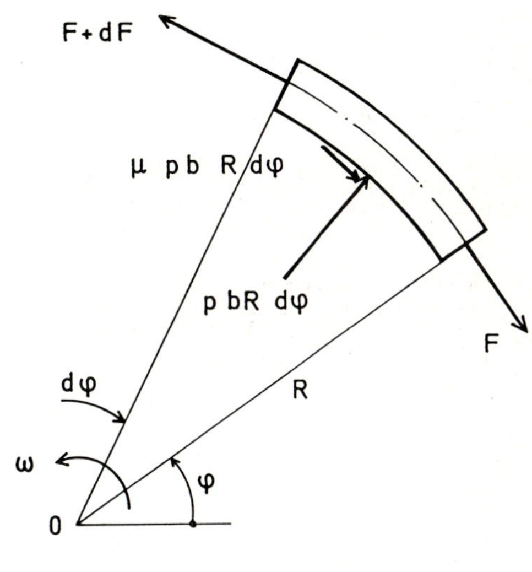

Figure 2. Forces acting on a small flat belt element.

When power is transmitted in a belt drive there must be a difference in belt tension of the free strands between the pulleys, F_2 = tight side tension and F_1 = slack side tension. Power P, belt tension $F_{1,2}$ and belt velocity V are related according to

$$P = (F_2 - F_1)\, V \tag{1}$$

The difference in belt force

$$F_u = F_2 - F_1 \tag{2}$$

is called effective pull and it is caused by friction between the belt and the pulley.

Consider equilibrium of the small belt element in fig. 2 (mass forces are neglected).

$$\text{Tangent: } dF = \mu\, p\, b\, R\, d\varphi \qquad (3)$$
$$\text{Radius: } F\, d\varphi = p\, b\, R\, d\varphi \qquad (4)$$

The second equation gives the contact pressure variation

$$p = F / (b\, R) \qquad (5)$$

between the belt and the pulley.
Divide the two equations. Then

$$dF / F = \mu\, d\varphi \qquad (6)$$

Integration gives

$$F / F_1 = \exp(\mu\, \varphi) \qquad (7)$$

Here the boundary condition $F = F_1$ at $\varphi = 0$ has been applied. Friction is present during the active arc ϕ. Thus

$$F_2 / F_1 = \exp(\mu\, \phi) \qquad (8)$$

which is the well known Eytelweins equation. It is required that $\phi < \alpha$ = contact angle (wrap angle). Within the angle $(\alpha - \phi)$ there is no friction and the belt force is constant.

Tension variation in the belt causes variable elongation ε along the belt. Assume a linear force-elongation relationship

$$F = c\, \varepsilon \qquad (9)$$

where c = strain stiffness. The pulley is rigid so the belt slides on the pulley (which causes friction as shown in fig. 2). The sliding velocity is (creep)

$$v = V(\varepsilon - \varepsilon_1) = V(F - F_1) / c \qquad (10)$$

3 WEAR OF FLAT BELTS

Archard [4] introduced the linear wear rate law which states that wear is directly proportional to normal load and sliding distance. Here we apply the relationship

$$w = p\, v / W \qquad (11)$$

where

w = wear rate
p = contact pressure
v = sliding velocity
W = wear strenght

In flat belt application the contact pressure is given in eq. (5). Knowledge of the sliding velocity means that the wear can be predicted. In flat belt drives there are different sliding mechanisms

- belt elongation
- pulley crowning
- belt transverse contraction

The three mechanisms all act simultaneously. At present there is no unified theory available which considers all of them so we treat them separately in the following.

Belt elongation

The sliding velocity is given in eq. (10) yielding the wear rate

$$w = \frac{1}{W} \cdot \frac{F}{b\, R} \cdot \frac{V(F - F_1)}{c}$$

By passing a pully the belt slides within the active arc ϕ. Thus the wear during passage of one pulley will be

$$s_S = \int_0^{\phi} w\, dt = \int_0^{\phi} w\, r\, d\varphi / V =$$

$$= \frac{1}{W\, b\, c} \int_0^{\phi} F(F - F_1)\, d\varphi$$

According to eq. (6) $dF = \mu\, F\, d\varphi$ so

$$s_S = \frac{1}{\mu\, W\, b\, c} \int_{F_1}^{F_2} (F - F_1)\, dF$$

$$s_S = \frac{(F_2 - F_1)^2}{2\, \mu\, W\, b\, c}$$

The strain stiffness c is proportional to belt width. Introduce

$c' = c / b$ = specific strain stiffness

$F_u' = F_u / b$ = specific effective pull

Then the wear due to the belt elongation will be

$$s_S = \frac{(F_u')^2}{2\, \mu\, W\, c'} \qquad (12)$$

Pulley crowning

The steering of the belt on the pulley is accomplished by crowning the pulley. This means that the surface has a slight curvature in the axial direction. Put Δ to the hight of the warping over the width of the pulley as shown in fig. 3.

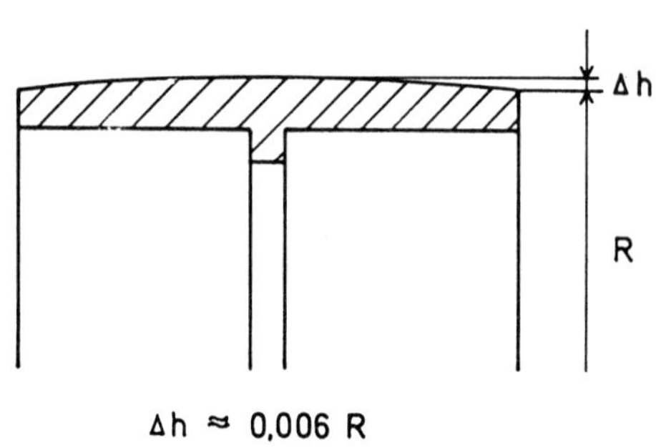

Figure 3. Crowning of flat pulleys

The steering properties of crowned pulleys has been known for years but there is no theory available which consider the phenomenon. An initial attempt is made in ref. [5]. What is the best value of Δ? In practise it is recommended that

$$\Delta / R \approx 0.006 \qquad (13)$$

We here make some simple considerations. The surface velocity of the pulley varies over the width of the pulley but the velocity of the belt is constant. On top of the pulley we put the maximum sliding velocity

$$v = V \Delta / (2 R) \tag{14}$$

We still regard the pressure $p = F / (bR)$ as constant over the width. Thus the wear rate becomes (eq. 5)

$$w = \frac{F}{W b r} \cdot \frac{V \Delta}{2 R}$$

Integrate over the contact arc. During the active arc ϕ we have $dF = \mu F d\varphi$ (eq. 6). Thus the wear due to crowning will be

$$s_C = \int_0^{\phi} \frac{F}{W b R} \cdot \frac{V \Delta}{2 R} \cdot \frac{R d\varphi}{V} =$$

$$= \frac{\Delta}{2 \mu W b R} \int_{F_1}^{F_2} dF = \frac{\Delta (F_2 - F_1)}{2 \mu W b R}$$

Introduce the specific effective pull $F'_u = (F_2 - F_1) / b$. Then

$$s_C = \frac{\Delta F'_u}{2 \mu W R} \tag{15}$$

Neglect the contribution from the rest of the contact i.e. the arc $(\alpha - \phi)$.

Belt transverse contraction

Fazekas [6] observed that a flat belt must slide in the transverse direction. The contribution to the wear can be considerable since on one hand the Poisson ratio is high for polymers and on the other hand the belt might be very wide. We do not carry out any analysis in this paper, but we can expect that the contribution to the wear is of the same character as the one from belt elongation.

Total wear

By summing the two contributions s_S and s_C from eqs. (12) and (15) we get the wear

$$s = \frac{1}{2 \mu W} \left[\frac{(F'_u)^2}{c'} + \frac{\Delta}{R} F'_u \right] \tag{16}$$

when passing one pulley.

Let s_{al} be the total wear allowed before improper function of the belt is expected. Thus the number of passages (cycles) is

$$N = s_{al} / s \tag{17}$$

if we want to express the service life N_h in hours then

$$N_h = \frac{N L}{3600 V}$$

The bending frequency f_B is obtained from the relationship (z = number of pulleys)

$$f_B = z V / L$$

Thus we alternatively have

$$N_h = \frac{N z}{3600 f_B} \tag{18}$$

Introduce the notation

$$A = \frac{s_{al} z}{1800 f_B N_h} \cdot \mu W \tag{19}$$

The combination of eqs. (16) to (19) leads to

$$\frac{(F'_u)^2}{c'} + \frac{\Delta}{R} F'_u = A$$

From this equation we obtain the specific effective pull

$$F'_u = [\sqrt{4c'A + (c'\Delta/R)^2} - c'\Delta/R] / 2 \tag{20}$$

Since $\Delta/R \approx 0.006$ according to eq. (13) we notice that F'_u in this equation is independent of pulley size.

4 POWER RATING CONSIDERATIONS

Wear

The practical procedure for rating flat belts according to ref. [7] is briefly reproduced in Appendix. It appears that for a certain belt type (size) a maximum specific effective pull F'_u is prescribed. The value of F'_u is

- directly proportional to pulley diameter D
- independent of belt velocity V
- independent of belt length L
- valid for a restricted bending frequency f_B
- valid for a two pulley drive (z=2)

Then obviously a minimum prescribed service life N_h is expected.

Compare this procedure with the specific effective pull calculated from eq. (20) which is based on a wear approach. For a given belt and the considerations above we might regard the quantity A in eq. (19) as a constant. Furthermore $\Delta/R \approx 0.006$ according to eq. (13). Thus for a given belt the calculated value of F'_u is constant. So both practical experience and analysis based on limited wear lead to the same result regarding the allowed specific effective pull. Thus the rating procedure of flat belts seems to coincide with a wear approach except what concerns the influence of pulley diameter. This must be considered from other points of view.

Tensile fatigue

The tight side tension F_2 causes a tensile stress

$$\sigma_2 = F_2 / (b h) \tag{21}$$

where h = thickness of the load carring part of the belt. (We consider the load distribution as constant over the cross section.) Maximum contact pressure is (eq. 5)

$$p_{max} = F_2 / (b R) \quad (22)$$

Since h << R then $p_{max} << \sigma_2$ and the contact pressure has almost no influence on the stress situation in the belt.

On top of the belt the bending stress is

$$\sigma_b = E h / D \quad (23)$$

where E = modulus of elasticity. Thus the total tensile stress will be

$$\sigma = \sigma_2 + \sigma_b \quad (24)$$

In V-belt applications ISO has applided the fatigue relationship

$$N_0 / N = (\sigma / \sigma_0)^p \quad (25)$$

where

N_0 and σ_0 = material data
N = fatigue life in cycles
p = exponent (p = 11)

Adopt this relationship for the flat belt giving (eqs. 21 to 25)

$$F_2 / (bh) + Eh / D = \sigma_0 (N_0 / N)^{1/p} \quad (26)$$

Introduce the coefficient of traction (cf eq. A6 in Appendix)

$$\lambda = (F_2 - F_1) / (F_2 + F_1) = F_u / (F_2 + F_1)$$

giving

$$F_2 = F_u (1 + \lambda) \quad (27)$$

Furthermore apply the service life N_h in eq. (18) and the specific effective pull $F_u' = F_u/b$. Then eq.(26) is transferred to

$$F_u' = (B - E h^2 / D) / (1 + \lambda) \quad (28)$$

where

$$B = h \sigma_0 \left[\frac{N_0 z}{3600 f_B N_h} \right]^{1/p} \quad (29)$$

According to the practical considerations we can regard B as a constant. Thus F_u' in eq. (28) is influenced by the pulley diameter D.

Wear - fatigue comparison

Figure 4. shows the main idea of rating flat belts according to Appendix (cf fig. A1). It is there shown that the straight line

$$F_u' = k D$$

is equivalent of prescribing a maximum contact pressure of

$$p_{max} = 2 (1 + \lambda) k$$

The two failure mechanisms discussed, wear eq. (20) and fatigue eq. (28), are also shown in figure 4.

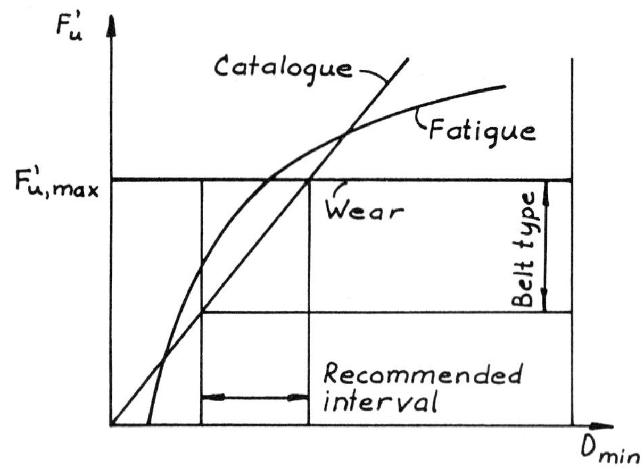

Figure 4. Competing failure mechanisms

The fatigue failure mode should be avoided since this can lead to hazardous belt rupture. Wear of the friction layer leads to a gradual malfunction of the belt drive and is to be preferred. Thus the allowable specific effective pull F_u' should be below the fatigue limit as indicated in figure 4. The maximum value $F'_{u,max}$ for a certain belt type could be obtained from the wear consideration in eq. (20).

Pull a straight line $F_u' = kD$ through the maximum value $F'_{u,max}$. This line represents constant contact pressure as indicated before. It also, in a reasonable way, seems to avoid the fatigue limit within a certain interval of minimum pulley diameters.

There is no failure mechanism easily related to the contact pressure alone. The pressure is much smaller than other stresses. Wear is a combination of pressure and sliding. Thus the wear approach seems to be a reasonable concept as a basis for power rating of flat belts.

5. CONCLUSION

An example of power rating of flat belt drives from a manufacturers catalogue reveals that the procedure is based on a prescribed specific effective pull F_u' for different belt sizes. Moreover, it appears that F_u' is proportional to the minimum pulley diameter which is equivalent of a specifying a maximum contact pressure. However, the contact pressure is much smaller than other stresses in the belt. So a failure mechanism must be based on another quantity.

A common way to rate V-belts is to relate the service life to tensile fatigue. In modern flat belts the tension is carried by a strong tension layer. Fatigue of that layer leads to belt rupture. This happens suddenly and may cause unwanted damage.

One or both surfaces of the belt is covered with a high friction layer. Wear out of that layer reduces the friction and gradual decrease in proper function might be observed.

An equation for specific effective pull is developed in the paper. The equation is based on the linear wear rate law and different sliding mechanisms between the belt and the pulley. For a given belt type and known operating conditions a value of F_u' can be calculated. This value might be the one which should be prescribed in the catalogue. Thus a wear based power rating method might be in accordance with the procedure outlined by the manufacturers.

REFERENCES

[1] JUVINALL, R.C. 'Fundamentals of machine component design', 1983 (John Wiley & Sons Inc., New York).

[2] FIRBANK, T.C. 'Mechanics of the belt drive'. International Journal of Mechanical Science, 1970, 12, No 12, 1053-1063.

[3] GERBERT, B.G. 'Force and slip behaviour in V-belt drives'. Acta Polytechnica Scandinavica, Mech. Eng. series No 67, Helsinki 1972.

[4] ARCHARD, J.F. 'Contact and rubbing of flat surfaces', Journal of Applied Physics, 1953, 24, 981-988.

[5] GERBERT, B.G. 'Axial motion of an idling flat belt running on two conical pulleys', Report No 88-06-01, Division of Machine Elements, Chalmers University of Technology, Göteborg, Sweden, 1988.

[6] FAZEKAS, G.A.G. 'On the lateral creep of flat belts'. Trans ASME, Journal of Engineering for Industry, 1963, 85, 307-313.

[7] EXTREMULTUS, Catalogue order No 201-Issue 01.85/2, Siegling, Hannover, 1985.

APPENDIX

PRACTICAL POWER RATING OF FLAT BELTS

The practical procedure of rating flat belts is outlined in different manufacturers catalogues The following description is taken from the company Sieglings traded belts EXTREMULTUS. We are only interested in the main principles so minor details are omitted.

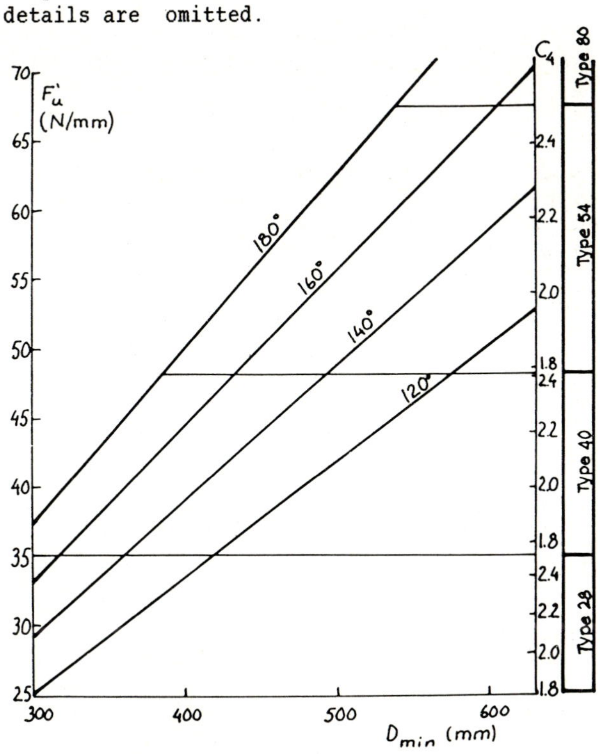

Figure A1. Specific effective pull F'_u versus minimum pulley diameter D_{min} for different belt types

The effective pull is defined as (eq. 2)

$$F_u = P / V \tag{A1}$$

The main design quantity is the specific effective pull

$$F'_u = F_u / b$$

Different belt types (sizes) are capable of transmitting a specific effective pull which is directly proportional to the minimum pulley diameter (see fig. A1) i.e.

$$F'_u = k D \tag{A2}$$

Thus different belt types are recommended within certain intervals of the diameter.

The specific numbers in fig. A1 reveal that the factor of proportionality is

$$k = 0.125 \text{ N/mm}^2$$

It appears that k for some reason is proportional to the arc of contact α of the small pulley i.e.

$$k = 0.125 \alpha / \pi$$

but k is independent of belt length L and belt velocity V.

The bending frequency (z = number of pulleys)

$$f_B = z V / L \tag{A3}$$

should not exceed a preset value $f_B < 30$ Hz.

The belt width is calculated from

$$b = F_u / F'_u \tag{A4}$$

The relationship between shaft load F_W and effective pull F_u is (neglect inertia forces)

$$F_W = 2 F_u \tag{A5}$$

The number 2 is basically the factor C_4 in fig. A1.

Introduce the coefficient of traction

$$\lambda = F_u / F_W \tag{A6}$$

Thus the flat belt drive is designed for $\lambda = 0.5$. The tight side tension F_2 and the slack side tension F_1 thus become

$$F_2 = (1 + \lambda) F_u = 1.5 F_u$$
$$F_1 = (1 - \lambda) F_u = 0.5 F_u \tag{A7}$$

How can we interpreat the factor k? According to eq. (5) the contact pressure is

$$p = F / (b R) \tag{A8}$$

Maximum pressure is

$$p_{max} = \frac{F_2}{b R} = \frac{2 (1 + \lambda) F_u}{b D} = \frac{3 F'_u}{D} = 3 k \tag{A9}$$

We identify k related to the maximum pressure. Thus the design procedure is based on an allowable maximum pressure $p_{max} = 0.375$ N/mm^2.

SESSION VI
GEARS

Chairman: Professor B Jacobson

PAPER VI(i)	The Relationship Between Uneven Tooth Contact Loading and Surface Durability in Flexible Gear Designs
PAPER VI(ii)	Temperature and Pressure Measurements in Gear Contacts with Thin Film Transducers
PAPER VI(iii)	A Static and Dynamic Analysis of Misaligned Gears with Partial Contact Areas

Paper VI(i)

The relationship between uneven tooth contact loading and surface durability in flexible gear designs

J. F. Harrop and A. Tam

A 3D finite element analytical procedure developed by PWC was used to predict uneven tooth contact loading in a flexible high contact ratio helical gear mesh design. Tooth scoring predictions were supported by evidence of tooth damage experienced on a reduction gearbox during early engine development testing.

By stiffening the gear web, rim, shaft and changing the tooth design from high to low contact ratio, a significant improvement in tooth contact load distribution and reduction in tooth scoring probability was shown by analysis. The need for high quality tooth surface finish and adequate oil film thickness was clearly demonstrated.

1 INTRODUCTION

Aero engine gearboxes must be compact, lightweight, durable, easy to maintain and low cost in order to meet the demands of a highly competitive gas turbine market. At Pratt and Whitney Canada, helical gearing is popular because of its capacity to transmit high torque at high speed, balance engine thrust loads and minimize the generation of noise and vibration.

To minimize weight, material is removed from the rim, web and barrel of the gear shaft and pinion. Unfortunately, this contributes to non-parallel flexing and distortion between the loaded pinion and gear which then leads to uneven load distribution in the tooth mesh.

Traditional methods of analysis do not accurately predict tooth contact loading in helical gearing in particular when the gear teeth have a flexible foundation (rim, web, shaft).

An analytical procedure(5) has been developed at PWC using the finite element method with 3D representation of the complete gear stage, to better assess the gear tooth contact load distribution.

This paper discusses the analytical process for predicting tooth contact load distribution in the second stage helical gear mesh of an aero engine reduction gearbox.

The paper also includes a study of tooth surface durability using pitting, scoring and spalling criteria. These criteria relate to local contact loading, friction, sliding velocity, oil film thickness, contact temperature, surface finish and case hardness. The effects on tooth contact loading due to misalignment, lead correction and bearing support stiffness are also evaluated.

1.1 Notation

T_f	=	Critical flash temperature (°F)
T_b	=	Gear blank temperature (°F)
w	=	local tooth contact load intensity (lb/in)
W_{te}	=	effective tangential load (lb)
F_e	=	effective face width (in)
S, S_p, S_g	=	Surface roughness for pinion and gear (μ-ins)
S_c	=	composite surface roughness (μ-ins)
n_p, n_g	=	pinion and gear speeds (rpm)
ρ_p, ρ_g	=	radius of curvature for pinion, gear (ins)
ρ	=	equivalent radius of curvature (in)
N_p, N_g	=	number of pinion and gear teeth
ϕ	=	transverse pressure angle (degrees)
V_s	=	sliding velocity (ins/sec)
V_t	=	sum velocity (ins/sec)
λ	=	surface distress parameter
h_{min}	=	min. oil film thickness (μ-ins)
α_o	=	pressure viscosity coefficient (ins^2/lb)
μ_o	=	absolute viscosity of oil (centipoise)
E_p, E_g	=	Young's Modulus of Elasticity for pinion and gear (lb/ins^2)
E_r	=	equivalent Young's Modulus (lb/in^2)

ν_p, ν_g	=	Poisson's ratio for pinion and gear
τ	=	sub-surface shear stress (lb/ins^2)
d	=	depth to maximum shear stress (in)
k	=	material/geometry parameter
CAEDS	=	Computer-Aided Engineering Design System
CATIA	=	Computer-Aided Three-Dimensional Interative Application
HCR	=	High Contact Ratio
LCR	=	Low Contact Ratio
PWC	=	Pratt and Whitney Canada
L.T.	=	Leading Tooth
M.T.	=	Middle Tooth
T.T.	=	Trailing Tooth
K.S.I.	=	Kilopounds per ins^2
D.M.F.	=	Deflection Magnification Factor

2 DESIGN PHILOSOPHY

The aero-engine gearbox referred to in this paper was designed to transmit power through a 2 stage 6.6:1 offset gear train at a maximum output torque of 555 lb-ft. Both stages of gearing have helical gear teeth, which help balance out the engine thrust load and also minimize noise transmission and vibration.

Thrust loads transmitted through the second stage gearing (Figs 1 & 3) are in fact quite small and the required helix angle for the meshing teeth is only 9.6 degrees. In order to achieve a minimum contact ratio greater than one, a 17 degree pressure angle, high contact ratio involute tooth form was selected. The main advantage of this extra long gear tooth is that an increased number of teeth share the load and as a result, Hertzian contact stresses and tooth bending stresses are reduced. On the other hand, since the teeth are longer and tooth sliding velocities increase towards the tips and roots of the mating teeth, this leads to higher tooth scoring probability. Preliminary analysis based on American Gear Manufacturers Association (AGMA) criteria indicated that the tooth design was acceptable.

However, it should be noted that the afore-mentioned methodology does not take into account the effect of gear tooth foundation flexibility. In order to save weight, an analytical parametric study was carried out to optimize the gear rim and web for weight and strength. 3D F.E. analysis was used to calculate the cyclic stresses which were then related to high cycle and low cycle fatigue design criteria.

The gears were manufactured from the conventional carburising material AMS 6265 with a gear tooth surface hardness of 81 to 85 HRA. The gear teeth are finish ground and honed to a surface finish of 5 μ-ins (after running in), however it should be noted that gears tested on early development engines did not meet this standard.

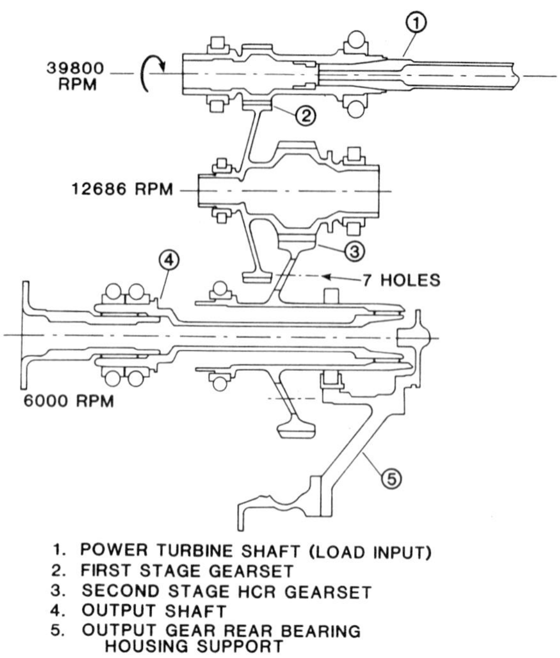

FIG 1: SPEED REDUCTION GEARTRAIN

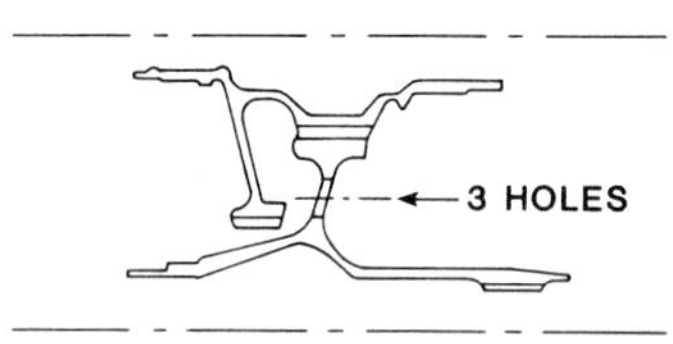

FIG 2: SECOND STAGE LCR REDESIGN

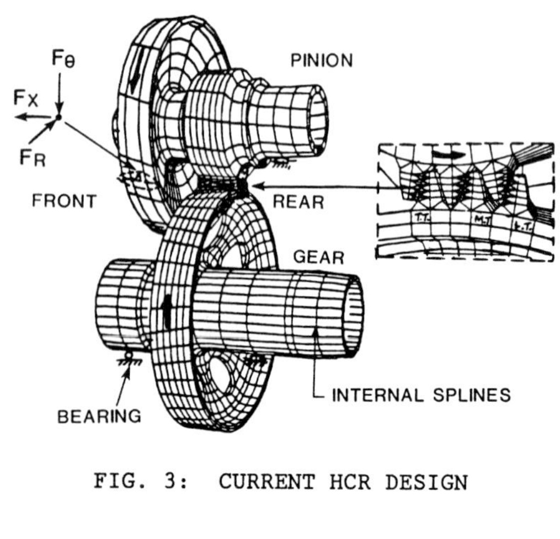

FIG. 3: CURRENT HCR DESIGN

3 SECOND STAGE MESH RE-DESIGN

The second stage HCR gear and pinion suffered severe tooth scoring during early engine endurance tests. The gear set was re-designed using low contact ratio helical teeth with 25 degree pressure angle (details are shown in table 1, see also Figures 2 and 4). The gear rim was thickened and the gear web/shaft stiffened. The number of holes in the gear web was reduced from 7 to 3 and the size of these holes was reduced from 1.00 inch to 0.625 inch in diameter. The change to LCR tooth design reduces sliding velocities and hence lowers tooth scoring probability. Tooth contact stresses and fillet bending stresses still meet the required design criteria. Unfortunately, since the helix angle is still low (only 9.6 degree), minimum tooth contact ratio is 1.0 and the advantage of the helical tooth mesh with regards to minimization of noise and vibration no longer exists. The helix angle is still required to balance engine thrust loads.

The shaft, web and rim have been significantly stiffened to minimize excessive deflections during torque transmission.

GEAR PARAMETER	HCR MESH		LCR MESH	
	Pinion	Gear	Pinion	Gear
Speed (rpm)	12686	6000	12686	6000
No. of teeth	35	74	35	74
Helix angle (degrees)	9.59	9.59	9.59	9.59
Pressure angle (degrees)	17.22	17.22	25.31	25.31
Diametral Pitch (per inch)	11.83	11.83	11.83	11.83
Chord tooth thickness (in)	.1354	.1144	.1387	.1089
Axial Face Width (in)	1.117	1.117	1.117	1.117

TABLE 1. GEAR DATA

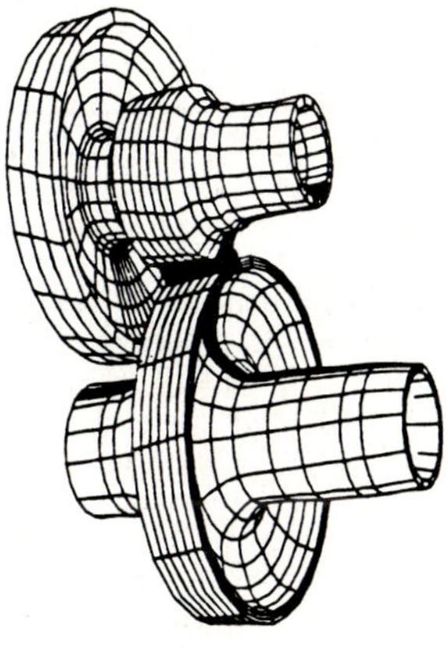

FIG. 4: LCR REDESIGN

4 GEAR LUBRICATION

There are two standard oils used widely throughout the aerospace industry; type I (MIL-L-7808) and type II (MIL-L-23699). Both are synthetic oils and type II oil is extensively used in PWC gearboxes mainly because of its higher viscosity and greater resistance to tooth scoring. However, it should be noted that an increase in viscosity is unfortunately accompanied by an increase in frictional resistance, hence reduced gearbox efficiency.

Due to limitations in oil pump capacity for cost and weight reasons the delivery rate restricts the oil jets velocity to 66 ft. per second. There are two jets supplying inter-mesh lubrication (Fig. 5) and each jet has a flow rate of 2.85 lb/min. The jets are sized to avoid brooming and carefully positioned relative to the gear teeth to provide optimum lubrication and cooling at the working surface. The tangential velocity of the gear teeth is approximately 165 ft. per second and this is significantly greater than the oil jet velocity available. Careful analysis was required to find the optimum angle at which the oil jet should be directed to obtain 100% wetting of tooth surfaces.

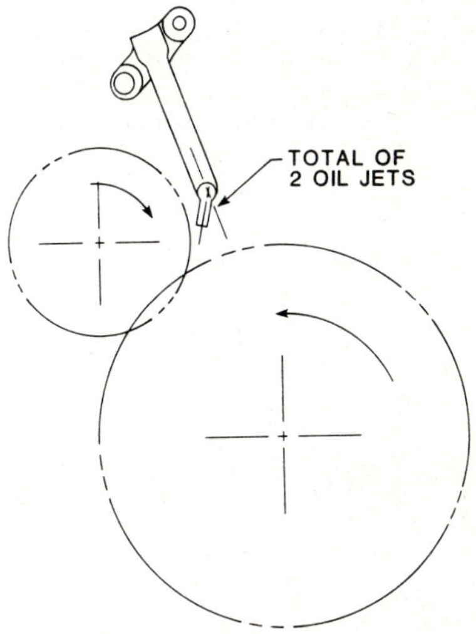

FIG. 5: GEAR LUBRICATION

5 DEVELOPMENT EXPERIENCE

Pits, deep radial scratches, welds and tear marks developed in the front ends of the second stage helical gear and pinion teeth during early engine endurance tests (Figures 6 and 7). Further examination of the hardware revealed that only a quarter to a half of the tooth face width had been in contact but there was no evidence of inadequate lubrication. The gear teeth were clearly experiencing misalignment during engine operation and steps were taken to re-inspect the hardware to ensure that tooth dimensions and shaft alignment were in keeping with normal design requirements. All dimensions were within the tolerances specified on the design drawing with the exception of the tooth surfaces which measured approximately 32 AA μ-ins or .813 AA μm roughness. This (ground) surface finish is acceptable for AGMA Q12 rating for highly loaded high speed gears in the aerospace industry. However, in aero engine main reduction gearboxes, a tooth surface finish of 16 AA μ-ins (.406 μm) or better (AGMA Q 13) is normally required to avoid surface distress, under marginal oil lubrication conditions.

It was then suspected that casing deflection may have been the cause of tooth misalignment and it was noted that the rear bearing housing support for the output gear seemed rather flexible in design. A static test was carried out in the Structures Rig and it was determined that casing deflection would create 0.002 radian misalignment in the gear teeth during maximum torque transmission. Steps were immediately taken to re-design the casing to eliminate this problem. Meanwhile on an interim basis, lead corrections of 0.002 radian and 0.004 radian respectively were applied to the gear teeth and further development tests were carried out. Improvements in tooth contact distribution were observed, but even with 0.004 radian lead correction wear patterns still indicated heavier tooth contact at the front end. More recently, gears have been produced to a much improved surface finish (approximately 5 AA μ-ins or .127 AA μm) and subsequent development tests on gears with 0.004 radian lead correction exhibit uniform tooth contact and minimal surface distress after endurance running. These gears have also been treated with a 'flash' silver coating on the contacting teeth.

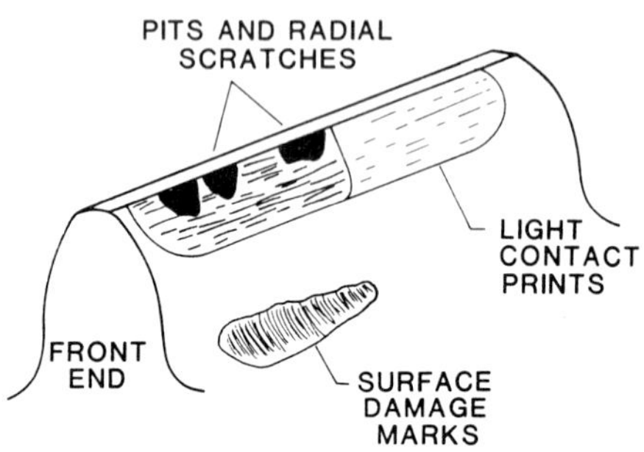

FIG. 6: TYPICAL DAMAGE PATTERN OF HCR PINION

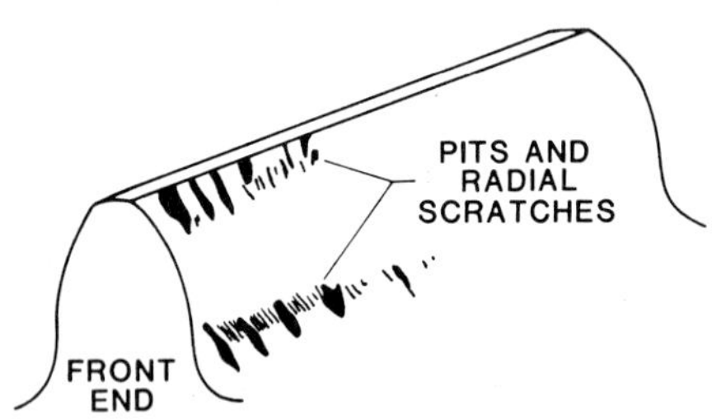

FIG. 7: TYPICAL DAMAGE PATTERN OF HCR GEAR

6 ANALYTICAL MODEL

The high and low contact ratio reduction gear set designs were both modelled using 3D solid finite elements as shown in figures 3 & 4. The computer program uses isoparametric solid elements with curved quadrilateral or triangular cross sections. Sub-structuring techniques are used to permit modifications to be made without having to regenerate the entire grid.

The tooth geometries as defined in table 1 and the 2D cross sections of the gear/pinion foundation assemblies for both the HCR and LCR designs provided the basic input for grid network generation. The 3D grids for the teeth, the HCR pinion foundation as well as the LCR pinion/gear foundation assembly were generated automatically using the computer program. To generate the 3D grids for the HCR gear foundation with the holes in the web (fig.3), the CATIA/CAEDS system was used. The 3 access holes in the LCR gear web were not modelled in the F.E. grid (Figure 4), due to time restraints. However, the stiffness of the elements in the hole locations was modified to simulate the presence of the holes.

Lines of contact are established in the "just loaded" state using involute geometry relations. However, in the loaded condition, tooth contact patterns and load distribution is a function of the combined stiffness of the mating gear and pinion. Contacting nodes are assumed lying along the lines of contact in the "just loaded" state and contacting compressive forces are calculated in the loaded condition by a process of iteration, i.e. contacting nodes in tension are released during this iteration process. The reactive forces at the nodes are converted to distributed line load intensities by satisfying force, energy and moment equivalences. Contact stresses are then computed using standard Hertzian formulae, for spur and helical gears, built into the program. Using a special restart feature within the program, the effect of misalignment and lead correction are readily evaluated by modifying the contact node boundary conditions, i.e. introducing specified gaps between some of the nodes.

7 F.E. ANALYSIS/RESULTS

Referring to figure 3, loading was applied at the first stage mesh location and the relative torque of 555 lb-ft was reacted at the output gear rear spline location by imposing appropriate boundary conditions. Five different cases were analysed to study the effects of casing flexibility, lead correction and geometric misalignment (table 2).

On the original HCR mesh design three teeth are shown in contact at a specific instant in time and fig. 8 provides the tooth contact lines at this critical mesh position. Figs. 9 and 10 show the calculated tooth contact loading for each different case. Peak load intensity occurs in the leading tooth in all cases. The loading is concentrated in the front end of the teeth. Case 4 (stiffened casing and .002 radian lead correction) offers the best solution for reducing peak load intensity (table 3). At the other end of the scale, case 5 demonstrates the serious effects of geometric misalignment. However, .002 radian geometric misalignment is an extremely severe condition and geometric tolerances are normally restricted to less than .001 radian. Increased casing stiffness clearly reduces misalignment and consequently lowers peak load intensity particularly in the leading tooth.

CASE	DESCRIPTION	SYMBOLS
1	Original (flexible) casing, no lead correction.	— x — x —
2	Stiffened casing, no lead correction.	————
3	Original (flexible) casing, .002 rad. lead correction.	— — —
4	Stiffened casing, .002 rad. lead correction.	—– —–
5	Stiffened casing, .002 rad. misalignment.	- - - - - -
6	50% input load, stiffened casing, no lead correction	—••—••—

TABLE 2. LOAD CASES

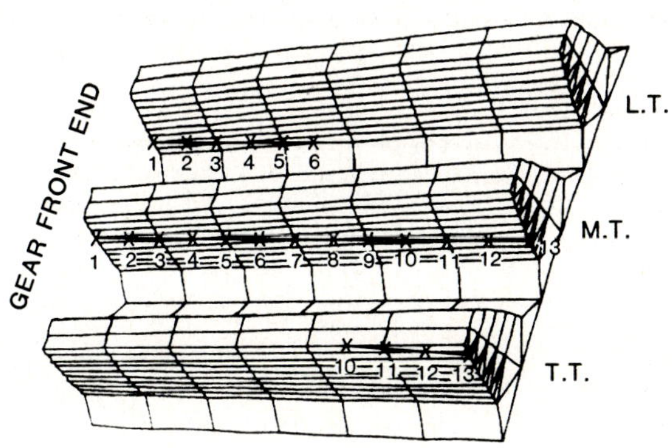

FIG. 8: HCR GEAR TOOTH CONTACT LINES

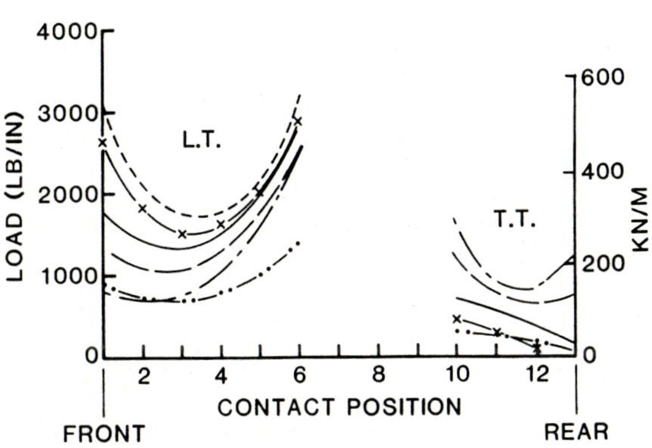

FIG. 9: HCR GEAR
TOOTH CONTACT LOAD INTENSITIES

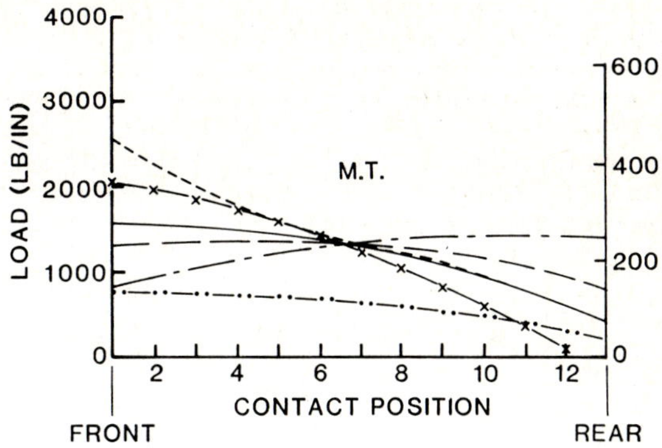

FIG. 10: HCR GEAR
TOOTH CONTACT LOAD INTENSITIES

Figs. 11 and 12 show corresponding Hertzian contact stress distributions for the HCR mesh. Assuming a dynamic factor of 1.0 which is normal for Aerospace quality helical gearing, peak stress levels just meet a recommended 225,000 lb/in^2 (1551 MPa) pitting criteria for high cycle fatigue.

Cases 2 and 6 in figs. 9 and 10 demonstrate an expected linear relationship between torque and tooth contact loading.

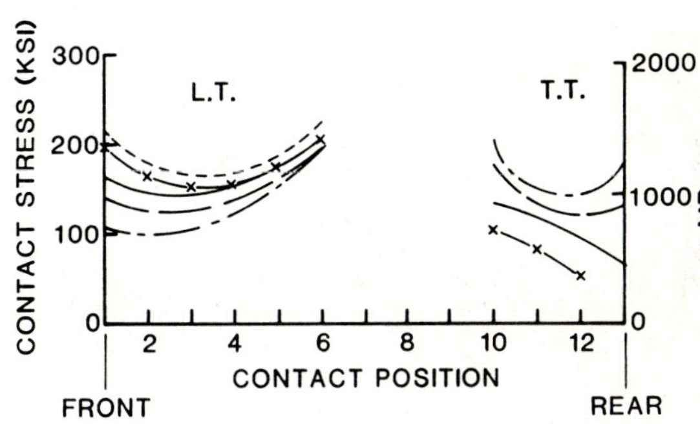

FIG. 11: HCR GEAR TOOTH CONTACT STRESSES

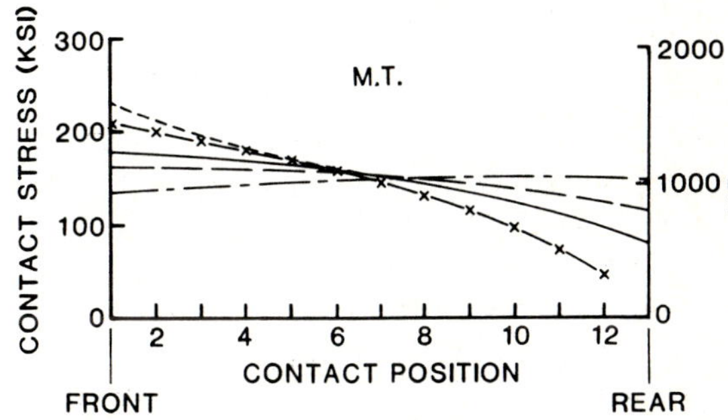

FIG. 12: HCR GEAR CONTACT STRESSES

Fig. 13 shows that tooth contact on the LCR gear mesh is concentrated on one tooth at worst conditions. The stiffened web and rim contribute significantly towards an improvement in tooth contact distribution. As stated previously, the .002 radian misalignment condition is very severe and well outside normal tolerance requirements (fig. 14).

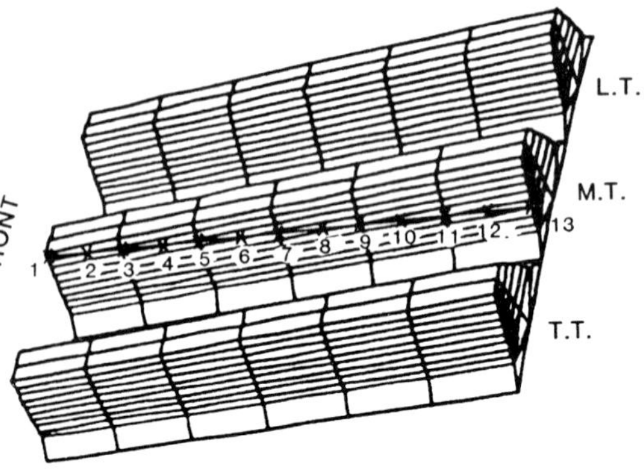

FIG. 13: LCR GEAR TOOTH CONTACT LINE

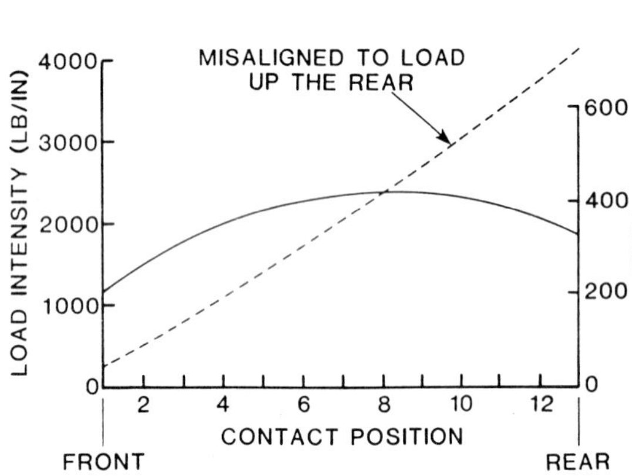

FIG. 14: LCR GEAR
MIDDLE TOOTH CONTACT LOAD INTENSITIES

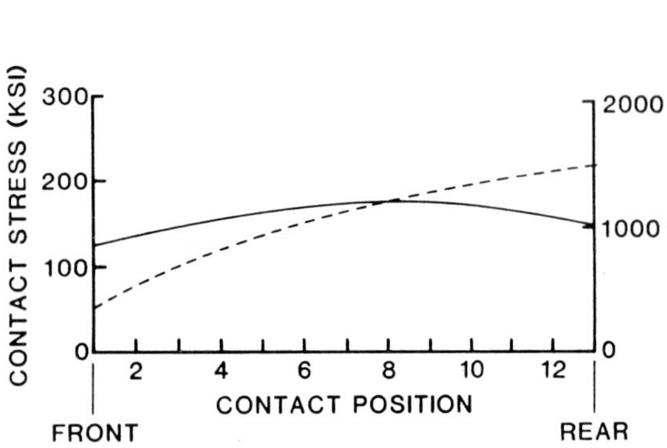

FIG. 15: LCR GEAR
MIDDLE TOOTH CONTACT STRESSES

During load transmission for the HCR helical gearset, the axial component load forces the flexible gear rim to deflect forward at the mesh and twist the gear ligaments (figs. 16, 17 & 18). The radial component load further aggravates the amount of twisting in the ligaments, causing the helix angle of the gear to increase relative to that of the stiffer pinion. The overall deflection tends to separate the tooth mesh at the rear end and load up the front end of the teeth.

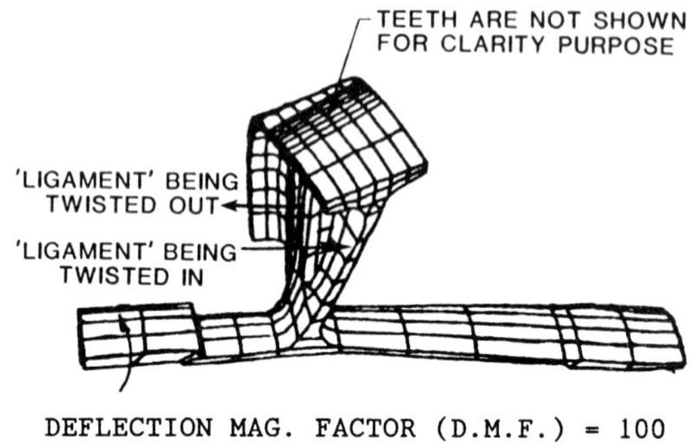

FIG. 16: HCR GEAR RIM AND WEB DEFLECTIONS

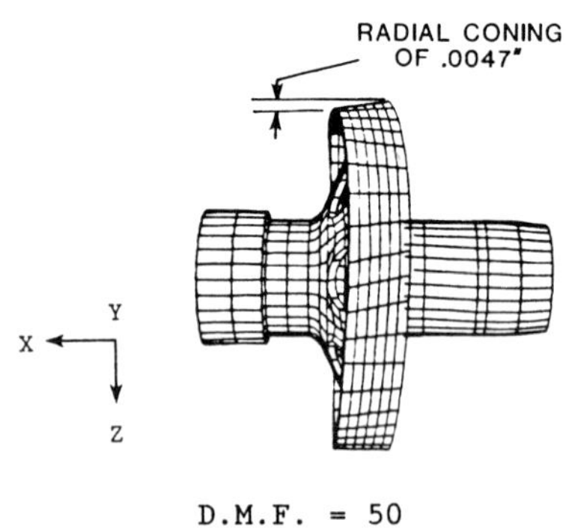

FIG. 17: HCR GEAR DEFLECTION

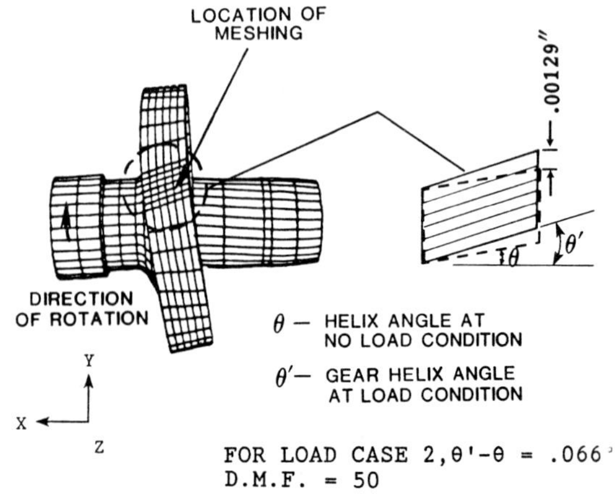

FIG. 18: HCR GEAR DEFLECTION

It is noted that gear deflections are much less severe on the stiffened LCR gear design (figs. 19 and 20) and it can be seen clearly why tooth contact distribution has been so significantly improved.

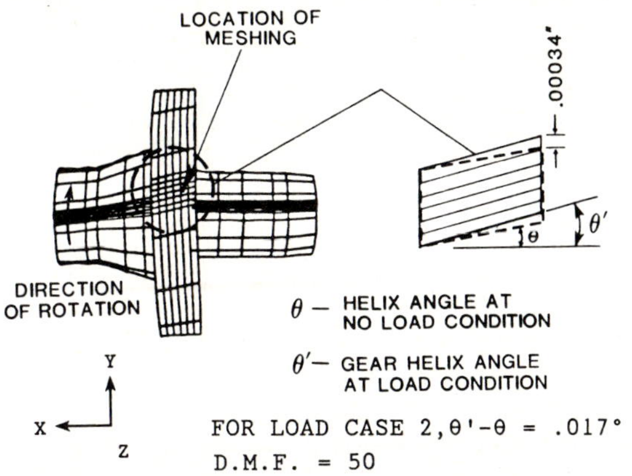

FIG. 19: LCR GEAR DEFLECTION

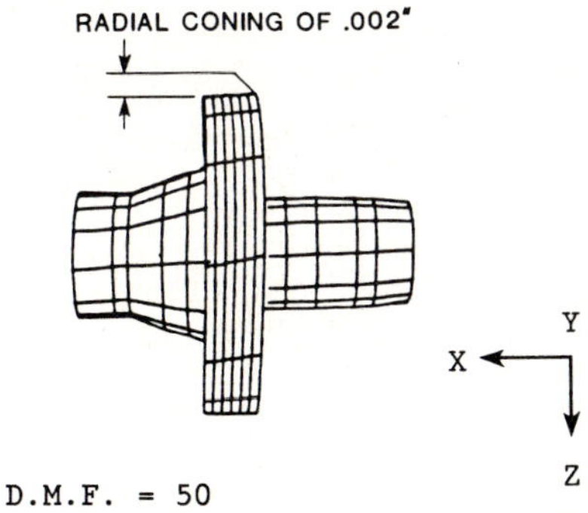

FIG.20: LCR GEAR DEFLECTION

8 TOOTH SCORING PROBABILITY

The following equations were extracted from Reference 1.

1) <u>For Friction Coefficient of .06</u>

$$T_f = T_b + \left[\frac{W_{te}}{F_e}\right]^{.75} \left[\frac{50}{50 - S}\right] Z_t \left[n_p\right]^{.5}$$

$$Zt = \left| \frac{.0175 \left[\sqrt{\rho_p} - \sqrt{\frac{N_p}{N_g} * \rho_g}\right]}{(\cos\phi)^{.75} \left[\frac{\rho_p * \rho_g}{\rho_p + \rho_g}\right]^{.25}} \right|$$

2) <u>For Local Friction Coefficients</u>

$$T_f = T_b + f \left[\frac{W_{te}}{F_e}\right]^{.75} \left[\frac{50}{50 - S}\right] \left[Z\right] \left[n_p\right]^{.5}$$

$$Z = \left| \frac{.2917 \left[\sqrt{\rho_p} - \sqrt{\frac{N_p}{N_g} * \rho_g}\right]}{(\cos\phi)^{.75} \left[\frac{\rho_p * \rho_g}{\rho_p + \rho_g}\right]^{.25}} \right|$$

Where local friction coefficient

$$f = .0127 \log_{10} \left[\frac{3.17 * 10^8 * w}{\mu_o V_s (V_t)^2}\right] \quad \ldots.(6)$$

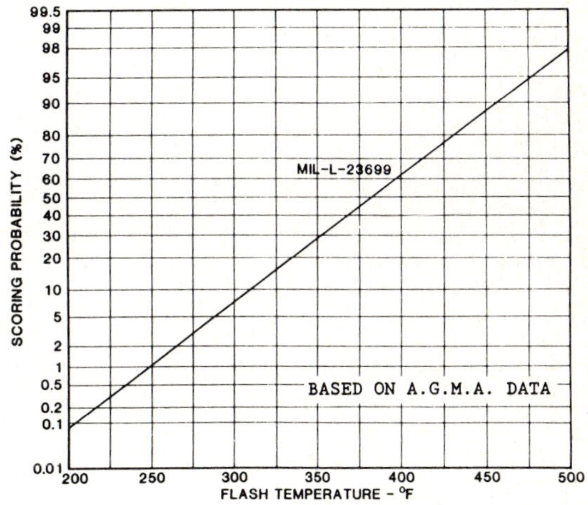

FIG 21: SCORING PROBABILITY VS. FLASH TEMPERATURE

In the above formulae, W_{te}/F_e was replaced with the local tooth contact load intensities (w) to calculate the flash temperatures. The curve in figure 21 was used to estimate tooth scoring probabilities. As stated previously early development hardware had a tooth flank surface finish of 32 µ-ins whilst current parts are produced at a much higher quality standard (5 µ-ins).

Results of analysis in tables 3 and 4 emphasize the importance of high quality surface finish. Despite the fact that the LCR gear is less prone to scoring because of the lower sliding velocities, predicted scoring probabilities are still high for a 32 µ-ins surface finish. There is some question whether the formulation by Benedict & Kelley (6) for local frictional coefficients should be used in preference to the traditional constant value of .06; both cases have therefore been considered. Tooth scoring probability is very high for all cases shown for the HCR gear set. However, case 4 records 24% scoring probability assuming varying friction coefficient and a 5 µ-ins roughness. This hardware has .002 radian lead correction with a stiffened casing. Recent development hardware had high quality surface finish, .004 radian lead correction, "flash" silver coating and a stiffened casing. It is perhaps not too surprising that this hardware did not experience tooth scoring problems during endurance tests.

Scoring probability for the redesigned LCR gearset with 5 μ-ins surface finish is low.

9 MINIMUM OIL FILM THICKNESS AND SURFACE DISTRESS

Scoring is very sensitive to surface finish particularly for gears in gas turbine engines where relatively light oil with low viscosity is used for lubrication and cooling purposes. When loaded, the thickness of the low viscous oil film will not be sufficient to cover the asperities of the contacting surfaces if the surface finish is poor. This allows metal to metal contact and promotes scoring, pitting and even spalling of the gear teeth. The calculated oil film thickness for the HCR and LCR gearsets varies from 7.1 micro inches to 12 micro inches as summarized in tables 3 and 4. The surface distress parameter(λ) varies from 0.15 to 1.69. Full hydrodynamic lubrication is established when λ is 4 or higher and asperity contact is negligible.

For λ of 1 or less, there is a high risk of surface distress. The analysis indicates that the surface distress parameter (λ) at the rear end of the HCR gearmesh is also low (severe), but this is not reflected in the damage patterns. Nevertheless the results do suggest an overall thin film thickness. Gear tooth surface roughness of 5 microinches can be achieved after honing and initial run-in. However surface roughness of worse than 32 microinches was found on early test gears and it is believed that the poor surface finish is one of the major contributing factors causing the premature tooth failure of these gearsets.

The following equations (based on original formulation by Dowson) extracted from reference 2 were used to calculate λ.

$$\lambda = \frac{h_{min}}{Sc}$$

$$Sc = (Sp^2 + Sg^2)^{.5}$$

$$hm = \frac{26.5\, \alpha_o^{.54} (\mu_o V_t)^{.7} \rho^{.43}}{W^{.13} E_r^{.03}}$$

where: $$\frac{1}{E_r} = \frac{1}{2}\left[\frac{1-\nu_p^2}{E_p} + \frac{1-\nu_g^2}{E_g}\right]$$

and $$\frac{1}{\rho} = \frac{1}{\rho_p} + \frac{1}{\rho_g}$$

10 TOOTH SPALLING

A criteria was developed in reference 4 which relates sub-surface shear stresses to shear strength, this being derived from the gear tooth hardness gradient. Sub-surface shear failure could result in relatively large pieces of metal being detached at the surface of the tooth and this type of tooth surface distress is sometimes referred to as spalling.

Sub-Surface Shear Stress Criteria

$$\tau_{max} = .3\sqrt{\frac{WK}{\pi}}$$ Equation extracted from (3).

depth where maximum subsurface shear stress occur.

$$d = 1.6\sqrt{\frac{W}{\pi k}}$$ Equation extracted from (3).

$$\text{Brinell hardness} = \frac{\tau}{145}$$ Equation extracted from (4).

$$k = \frac{\left[\frac{1}{\rho_p} + \frac{1}{\rho_g}\right]}{\frac{1-\nu_p^2}{E_p} + \frac{1-\nu_g^2}{E_g}}$$

Tooth contact load intensities were fed into the formulae (from reference 3) in the computer program to calculate the sub-surface shear stress gradients. The minimum margins of safety for tooth spalling are summarized in tables 3 and 4. In all cases for both HCR and LCR designs there is sufficient margin of safety to conclude that gear failure would not initiate due to tooth spalling.

LOAD CASE	CONTACT TOOTH	PEAK LOAD INTENSITY (lb/in)	MAX. HERTZIAN CONTACT STRESS (KSI)	*MAX. SCORING PROBABILITY (%) SURFACE ROUGHNESS of 5μ-in. FRICTION COEFFICIENT f = .06	*MAX. SCORING PROBABILITY (%) SURFACE ROUGHNESS of 5μ-in. LOCAL FRICTION COEFFICIENT	*MAX. SCORING PROBABILITY (%) SURFACE ROUGHNESS OF 32 μ-in. FRICTION COEFFICIENT f = .06	*MAX. SCORING PROBABILITY (%) SURFACE ROUGHNESS OF 32 μ-in. LOCAL FRICTION COEFFICIENT	MIN. FILM THICKNESS CRITERIA OIL FILM THICKNESS (μ-in.)	MIN. FILM THICKNESS CRITERIA SURFACE DISTRESS PARAMETER λ 5μ-in. ROUGHNESS	MIN. FILM THICKNESS CRITERIA SURFACE DISTRESS PARAMETER λ 32μ-in. ROUGHNESS	SUB-SURFACE SHEAR STRESS MAX. MAGNITUDE (KSI)	SUB-SURFACE SHEAR STRESS DEPTH (in.)	SUB-SURFACE SHEAR STRESS ** MARGIN OF SAFETY (MIN.)	RADIAL CONING (in.) (Fig.17)
1	L.T.	2943	217.5	99	77	>99	>99	9	1.27	.20	65	.007	0.50	0.0058
	M.T.	2067	204.5	20	7.5	97	56	7.5	1.06	.17	61	.005	0.61	
	T.T	480	109.0	10	2.0	77	12	8.1	1.14	.18	32	.002	2.05	
2	L.T.	2749	210.0	99	68	>99	98	9	1.27	.20	63	.007	0.55	0.0047
	M.T.	1549	177.0	13	3.5	85	31	7.8	1.10	.17	53	.005	0.86	
	T.T.	708	132.0	18	3.5	96.5	29.5	7.7	1.08	.17	39	.003	1.49	
3	L.T.	2503	200.0	99	59	>99	>99	9.2	1.30	.20	60	.006	0.63	0.0045
	M.T.	1348	164.0	9	3.5	72	21	7.9	1.11	.17	49	.004	1.02	
	T.T.	1291	178.0	57	12	>99	83	7.1	1.0	.16	53	.004	0.85	
4	L.T.	1661	162.0	94	24	99	98	9.7	1.37	.21	48.2	.005	1.04	0.0034
	M.T.	1425	153.0	8	3	66	18	8.3	1.17	.18	45.2	.005	1.20	
	T.T.	1429	188.0	67	17	99	91	7.0	1.0	.15	56	.004	0.75	
5	L.T.	3191	226.0	99	82	>99	>99	8.9	1.25	.20	67	.007	0.45	0.0059
	M.T.	2517	226.0	30	10	>99	76	7.3	1.03	.16	67	.005	0.47	
	T.T.	"	"	"	"	"	"	"	"	"	"		"	

TABLE 3 - HCR GEAR TOOTH LOADS AT 555 LB.-FT. TORQUE

" NO CONTACT AT THE TRAILING PAIR

* BASED ON OIL VISCOSITY AT 235°F AND GEAR BLANK TEMP AT 235°F

** MARGIN OF SAFETY = (ALLOWABLE STRESS/APPLIED STRESS) -1

NOTE: 5μ-ins = .127 μm
32 μ-ins = .813 μm

LOAD CASE	CONTACT TOOTH	PEAK LOAD INTENSITY (lb/in)	MAX. HERTZIAN CONTACT STRESS (KSI)	MAX. SCORING PROBABILITY (%) SURFACE ROUGHNESS of 5μ-in. FRICTION COEFFICIENT f = .06	MAX. SCORING PROBABILITY (%) SURFACE ROUGHNESS of 5μ-in. LOCAL FRICTION COEFFICIENT	MAX. SCORING PROBABILITY (%) SURFACE ROUGHNESS OF 32 μ-in. FRICTION COEFFICIENT f = .06	MAX. SCORING PROBABILITY (%) SURFACE ROUGHNESS OF 32 μ-in. LOCAL FRICTION COEFFICIENT	MIN. FILM THICKNESS CRITERIA OIL FILM THICKNESS (μ-in.)	MIN. FILM THICKNESS CRITERIA SURFACE DISTRESS PARAMETER λ 5μ-in. ROUGHNESS	MIN. FILM THICKNESS CRITERIA SURFACE DISTRESS PARAMETER λ 32μ-in. ROUGHNESS	SUB-SURFACE SHEAR STRESS MAX. MAGNITUDE (KSI)	SUB-SURFACE SHEAR STRESS DEPTH (in.)	SUB-SURFACE SHEAR STRESS MARGIN OF SAFETY (MIN.)	RADIAL CONING (in.) (Fig.20)
2	L.T.													
	M.T.	2421	171	10	3	77	18	12	1.69	.27	51	.007	0.85	0.002
	T.T.													
5	L.T.													
	M.T.	4147	217	40	10	98	80	12	1.69	.27	65	.010	0.46	0.001
	T.T.													

TABLE 4 - LCR GEAR TOOTH LOADS AT 555 LB.-FT. TORQUE

11 CONCLUSION

The tooth load distribution obtained for the HCR gear mesh indicates heavily concentrated loading at the front end of the gear teeth. The results match well with damage patterns found on the test gears. The load distribution can be improved by incorporating a stiff bearing housing support to reduce gear mesh misalignment and by applying appropriate lead modifications to the gear teeth. However, even with these improvements, scoring risk is still high because of excessive flexing and rim coning. The tooth mesh tends to separate tangentially at the rear end and force the contact loading towards the front.

The redesigned LCR gear mesh features a thick gear rim, less access holes, a shorter web and consequently a much stiffer gear. The 3D F.E. results indicate a more uniform load distribution in the teeth. The LCR teeth are shorter which reduces tooth sliding velocities and hence improves scoring resistance. Tooth scoring resistance is also significantly improved by better surface finish. Surface distress due to scoring, pitting or spalling is of low probability on the redesigned LCR gear set.

12 REFERENCES

1. LYNWANDER, P. 'Gear Tooth Scoring Design Considerations', American Gear Manufactures Association (AGMA), P219.10, October, 1981.

2. KU, P.M. 'Tribology of Gears', Gear Manufacture and Performance, P. 73.

3. ELKHOLY, A. 'Case Depth Requirements in Carburized Gears', American Gear Manufacturer Association (AGMA), Aerospace Gearing Committee Meeting, San Diego, California, 1982.

4. SHARMA, V.K., WALTER, G.H., BREEN, D.H. 'An Analytical Approach for Establishing Case Depth Requirements in Carburized Gears', ASME Publication, 77-DET-152, 1977.

5. Sundararajan, S. YOUNG, B. 'Finite Element Analysis of Large Spur and Helical Gear Systems', AIAA/SAE/ASME/ASEE, 23rd Joint Propulsion Conference, 1987, San Diego, California. AIAA-87-2047.

6. Benedict & Kelley, 'Instantaneous Coefficients of Gear Tooth Friction', ASLE Transaction 4, 59-70, 1961.

Paper VI(ii)

Temperature and pressure measurements in gear contacts with thin-film-transducers

H. Peeken and P. Ayanoglu

The project has the aim of investigating the temperature and pressure distributions in the elastohydrodynamic lubricating film in gear contacts. A specially designed gear wheel allows a thin film coating of the gear tooth. The transducers can thus be sputtered directly onto the tooth surface. The temperature and pressure coefficients of every transducer are determined before the experiments, to enable an exact evaluation of the results. The tests are being carried out on a gear rig with a continuously variable belt drive to load the gears. The experiments with manganin transducers showed typical EH pressure distributions, with much higher peaks than the calculated Hertzian values, a result of the unequal load distribution over the tooth width. Quite high temperature rises were registered, the temperature increasing with growing load, speed and lubricant viscosity.

1 INTRODUCTION

Precise knowledge of the lubricating conditions is an important design criterion for gears. Other than tooth breakage, most forms of gear failure such as wear, scuffing or pittings are closely related to the condition of the elastohydrodynamic lubricating film seperating the teeth in mesh.

The experimental investigation of the tooth contact in respect to the elastohydrodynamic lubrication was previously been confined mainly to two-disc test rigs, where it is simulated.

The project intends to measure temperature and pressure distributions in the real tooth-contact, using sputtered thin film transducers, which can be applied in various positions onto the tooth surface. Temperature and pressure distributions can thus be obtained for different points on the line of action. This is necessary because of the instationary elastohydrodynamics in gear contacts with changing radii, rolling and sliding speeds and slip along the line of action.

2 TEST STAND AND TEST GEAR

The experiments are carried out on a purpose-built gear-rig (Fig. 1), driven by a direct-current electric motor. The gears are loaded by means of a belt-drive with a continuously variable ratio. During acceleration from stand-still, the tooth flank with the transducer is not under load, as the transducer can otherwise be destructed in the mixed friction zone.

The shot of the opened test rig in Fig. 2 shows also the specially designed test gear wheel with offset teeth. This was necessary to enable the coating of the tooth surface in the sputtering apparatus. The teeth are limited to one half of the gear width, with four teeth off-set to the other half (Fig. 3). The driving gear is also split in the middle; when the test tooth (with the transducer) is in mesh, it replaces the lacking tooth on the other side, without influencing the rigidity or elasticity of the mechanism.

3 TRANSDUCER FABRICATION AND CALIBRATION

Thin-film-transducers have been in use for over 20 years for measurements in hydrodynamic and elastohydrodynamic contacts. They are Ohm-resistances, which change under pressure- and temperature influence. They are applied onto the test-object surface using the PVD-coating processes and enable measurements in the tightest gaps, without affecting the geometry or strength of the parts involved. While vapour deposition was the most commonly used process previously, sputtering is getting more and more acceptance nowadays. Sputtering is the erosion of a solid surface under bombardment of highly energized particles. The energy of thus sputtered molecules ore atoms is many times greater than that of thermally evaporated. Consequently, sputtered layers have much better wear resistance and adhesion than the latter.

The tooth flank is first covered with an isolating alumina-film. Transducer material is manganin for pressure- and titanium for temperature-transducers. In order to produce the defined contours of the transducer, either a mask is to be placed on the substrate surface or the tooth surface is covered with etch-resistant photopaint. The transducer contours are to be coated with the transducer material. The lithographic method renders transducers with very fine dimensions possible, which can also be positioned anywhere on the tooth-flank. Finally a thin, protecting alumina-layer is sputtered over the transducer.

Fig. 4 shows a thin film transducer on the tooth flank. It consists of the contact areas on both sides and a very fine bridge connecting them, which is the active part of the transducer, having a much higher resistance (100 - 1000 Ohm) than the contact areas. (In this case, a wire strain gauge is also attached to the root of the tooth, to measure dedendum strain).

This very fine connecting wire, the active part of the transducer is about 20 µm wide, 1 mm long and has a height of ca. 0,1 µm (Fig. 5). It reacts, according to material, primarily to temperature or pressure changes, but it has to be taken into consideration that a pressure transducer has a certain temperature sensitivity and vice versa. Using two transducers with different sensitivities the pressure and temperature distributions in the EH-contact can be determined. To compute the temperature and pressure values from the recorded resistance changes, the temperature and pressure coefficients of both transducers must be known. In other words, each transducer must be calibrated for temperature and pressure.

A special apparatus has been built for the pressure calibration (Fig. 6): A defined oil pressure can be applied to the transducer through a small pressure chamber pressed upon its active part. The test-pressure of this apparatus is limited to about 1400 bar (0,14 GPA), but it enables the calibration of the actual transducers on the tooth flanks.

As the transducers have to bear pressures of over 10 000 bar (1,0 GPa) in the EH-contact, possible non-linearities in their behaviour at this high pressure range should also be determined.

The high-pressure calibration of the thin film transducers up to 1,0 GPa has been undertaken in an autoclave with a chamber diameter of 20 mm. For this purpose the transducers were sputtered onto gage blocks which fitted into the autoclave chamber.

The pressure calibration curves for titanium and manganin transducers are seen on Fig. 7. Manganin has a linear relationship between its electrical resistance and pressure. Titanium shows a non-linear behaviour. But by linearizing in the important zone between 0,15 GPa and 0,65 GPa titanium transducers can also be used in the experiments.

The temperature calibration is done in an oil-bath.

4 RESULTS

One titanium and 3 manganin transducers were used in experiments so far. They were all positioned above the pitch point, their positions being given on the diagrammes of the recorded results. The pitch-radius is 80 mm.

The measurements proved to be very good reproducable as long as they were carried out with the same tooth of the mating gear. Between temperature or pressure measurements taken with different mating teeth, quite big differences were possible under otherwise similar conditions.

The reasons for these variations are to be found in the various errors of the gears used. The first measurements of the gear geometry showed directional errors in the profiles of the mating gear teeth (up to 15 µm over a tooth width of 20 mm), in other words teeth with a certain inclination, which results in unequal load distribution over the tooth width. This effect wouldn't be so noticable under higher loads. The tooth with the transducer had a crowned profile.

Another error is variations in circular pitch, this having a great influence if the transducer is in the double-mesh-region.

Fig. 8 shows pressure distributions recorded with the same mating tooth at a constant speed under three different loads. The curves are typical of elastohydrodynamical pressure distribution. In the parallel gap the curve has an elliptical shape similar to the Hertzian pressure distribution, but with a slowly growing gradient in the inlet zone and an abrupt fall in the outlet region. The peak pressure is in this case 0,4 GPa.

Similar pressure distributions were registered for another opposing tooth under the same parameters, but the peak pressure is in this case 0,7 GPa (Fig. 9).

In Fig. 8 and 9, two sets of values are given for the specific normal tooth force: The nominal force, calculated from the measured torque and the force, which is calculated by integrating the pressure curve. The latter exceeds in some cases the nominal force by quite a high margin. Such a high specific force over the whole width is of course not possible. The contact width in this case must be much narrower than the tooth, as a result of the geometrical errors already mentioned. It can also be seen, that the ratio of the integrated pressure to the nominal force decreases with increasing load. The contact area becoms larger with increasing load.

In Fig. 10 the same curves as in Fig. 9 are normed after the corresponding Hertzian pressure and contact width. It is noticable that they all have similar shapes, the peak pressures being about 1,2 times the Hertzian peak.

Temperature measurements were carried out with a titanium transducer. The variations between values taken with different teeth were here also noticable. Much higher temperatures were measured in the tooth contact than, for example in roller bearings. While in the roller-ring contact the highest measured temperature rise was 5 K, up to 35 K have been registered in the gear contact (at a specific force of 46 N/mm and a curcumferential speed of 8,4 m/s). The load-influence on the contact temperature can be seen on Fig. 11.

In Fig. 12 speed is the parameter, while the load is constant in all cases. The curves have different widths because they are not time-corrected. The slip is in this position about 40 %. The temperature increases with the slide-speed.

The experiments with the temperature-transducer were at first carried out an oil-inlet-temperature of 20°C. At this temperature the oil viscosity is 310 mPa s.

The last figure shows temperature distributions at 3 different oil inlet temperatures and viscosities. The temperature increase in the contact becomes less with decreasing lubricant viscosity (Fig. 13).

5 CONCLUSIONS

The experiments have so far shown the typical elastohydrodynamic pressure distributions in gear contacts, as well as the relatively high temperature rises as a result of the high sliding speeds.

The sometimes unexpectedly high pressure peaks show the effects that geometrical errors can have on the load distribution over the tooth flank.

Emphasis in the future will be put on the investigation of pressure and temperature distributions over the whole line of action and in lateral direction, so appropriately positioned transducers will be built. It is planned to measure dedendum strain with strain gages to get a better indication of the actual force on the tooth than the previous torque measurement on the shaft.

References

DOWSON, D., HIGGINSON, G.R. 'Elasto-hydrodynamic Lubrication', SI-Edition, Pergamon Press Ltd., 1977

RODERMUND, H., 'Beitrag zur elastohydrodynamischen Schmierung von Evolventenzahnrädern', Dissertation TU Clausthal, 1975

SIMON, M., 'Messung von elastohydrodynamischen Parametern und ihre Auswirkungen auf die Grübchentragfähigkeit vergüteter Scheiben und Zahnräder', Dissertation TU München, 1984

KANNEL, J.W., ZUGARO, F.F., DOW, T.A., 'A Method for Measuring Surface Temperature Between Rolling/Sliding Steel Cylinders', Transaction of the ASME 100 (1978) pp. 110-114

BAUER, P., 'Theoretische und experimentelle Untersuchungen zu tribologisch relevanten Betriebsgrößen an verkanteten Zylinderrollenlagern', Dissertation RWTH Aachen, 1987

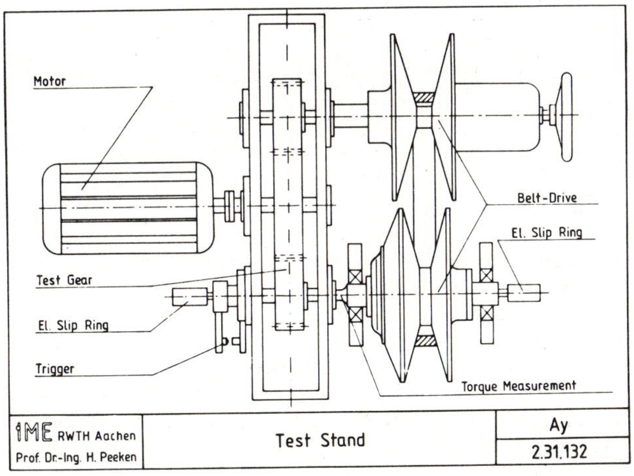

Fig. 1

Fig. 2

Fig. 3

Fig. 4

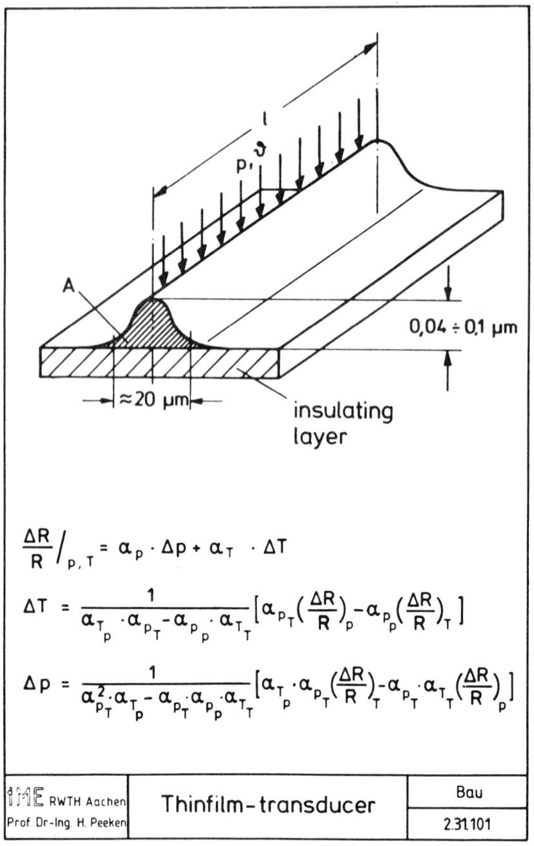

Fig. 5

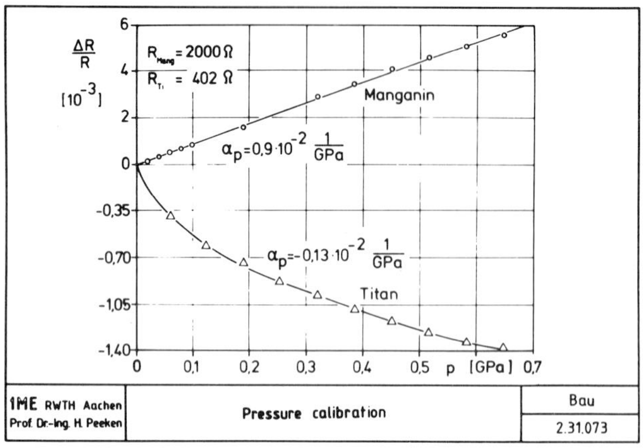

Fig. 7

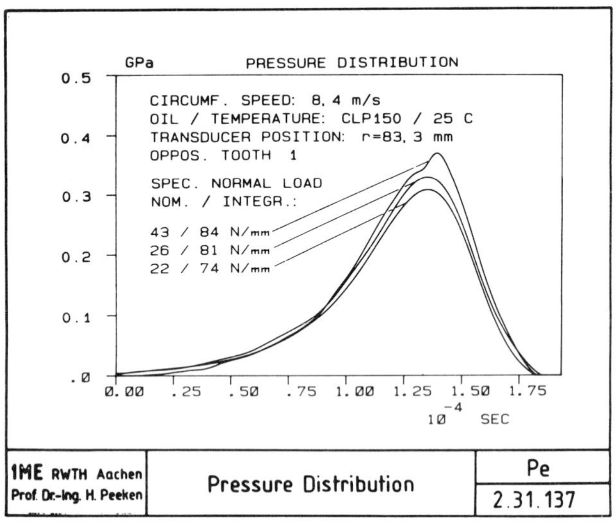

Fig. 8

Fig. 6

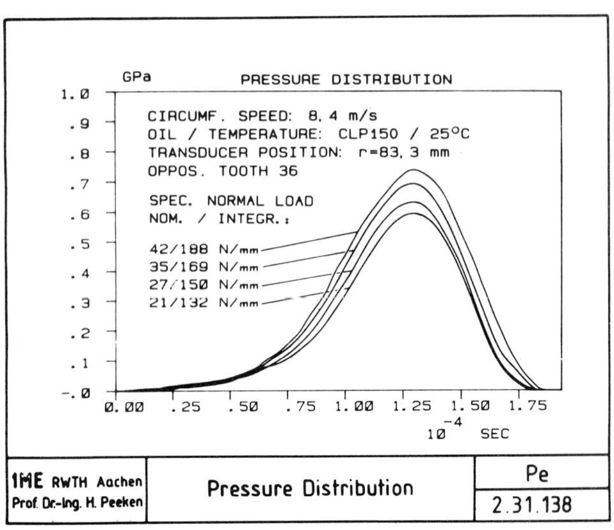

Fig. 9

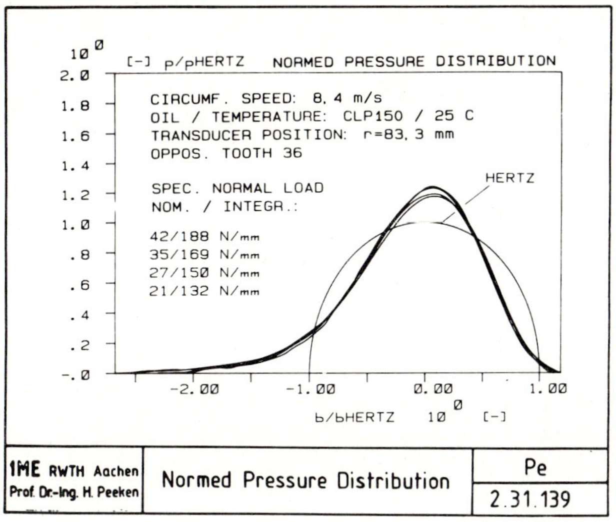

Fig. 10

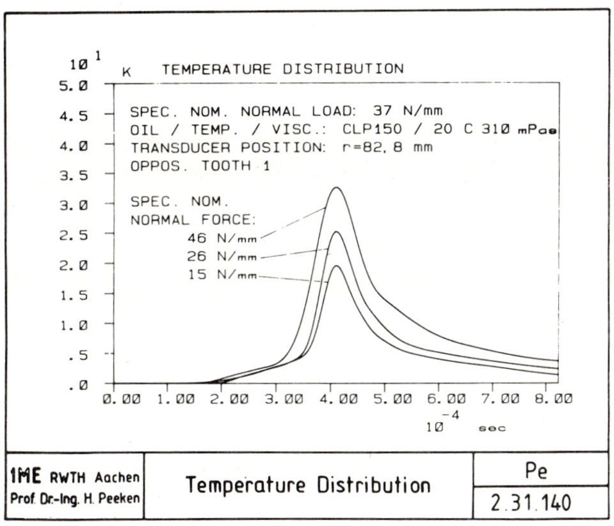

Fig. 11

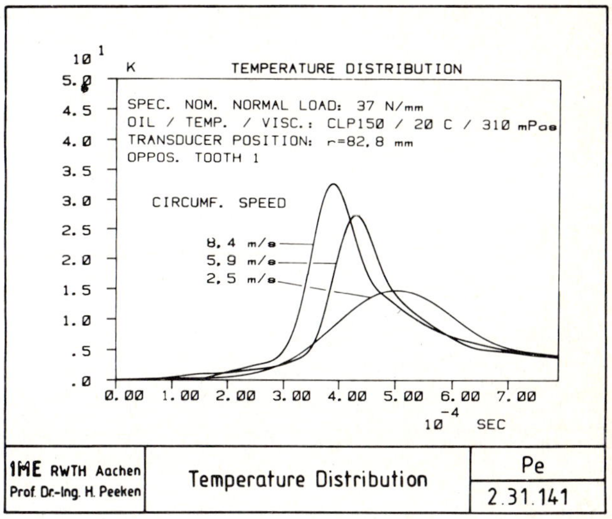

Fig. 12

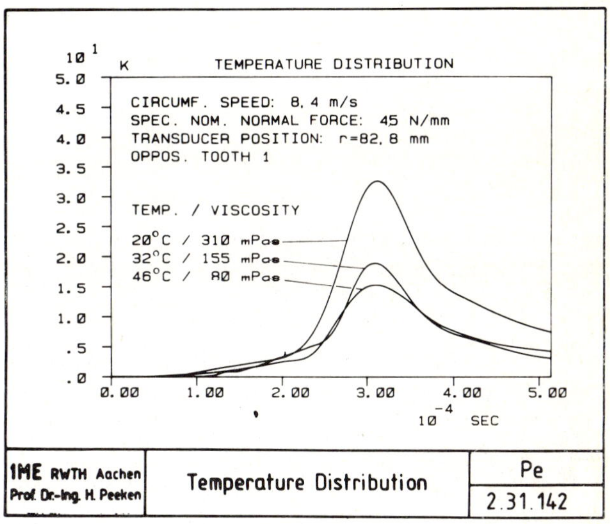

Fig. 13

Paper VI(iii)

A static and dynamic analysis of misaligned gears with partial contact areas

Ph. Sainsot, Ph. Velex and D. Berthe

This paper presents an evaluation of the influence of misalignments on tooth load distribution in gear systems.

Static and dynamic analyses are conducted using a normal contact algorithm including local and global effects.

This algorithm is shown to be well adapted to determine the pressure field between surfaces with any manufacturing errors, inversely, profile modifications can be calculated to have a given load distribution.

The dynamic problem is treated on a simplified reduction unit model, some examples of instantaneous tooth loading, variations of contact stiffness, and critical rotation speeds are given.

Results show that contact conditions may strongly depend on the overall mechanical environment.

1 INTRODUCTION

The load distribution in gear trains is rarely known. It depends on the manufacturing errors, rotation speeds, and on the masses, inertia, and stiffnesses of the reduction unit components.
To calculate that load distribution, a normal contact algorithm is extended and inserted in both static and dynamic analyses. The local displacements near the contacts are found using Boussinesq relations and structural compliances are calculated by F. E. M.
In the static approach, the effect of misalignment on the pressure field between contacting teeth of spur involute gear is shown. Inversely, the longitudinal profile modifications which yield a given pressure distribution are computed.
The dynamic problem is treated with a simplified model in which small misalignments are introduced in the constraint equation that guaranties contact during motion. The corresponding instantaneous tooth loading and critical speeds of rotation are derived and compared with the response given by aligned gears.

1.1 Notations

- c : longitudinal tooth modification
- E_i : contribution of small misalignments to the constraint equation ensuring contact at the i th point on the plane of action
- k_i : meshing stiffness at the i th point of contact
- L : tooth face width
- M : module
- P_i : pressure at the i th point of contact
- R : total maximum instantaneous load versus total static load on the plane of action
- $r(t)$: total instantaneous load versus total static load on the plane of action
- R_{b1}, R_{b2} : base cylinders radii for pinion and wheel
- v_k, u_k : small translations of gear k
- x : addendum modification coefficient
- Z_1, Z_2 : number of teeth for pinion and wheel
- F^* : additionnal second member induced by misalignments
- α : pressure angle
- β : helix base angle, helix angle
- ϕ_k, ψ_k, θ_k : small rotations of gear k
- ϕ^*_k, ψ^*_k : small constant angular error (misalignment)
- Σ_i, Σ_i^* : relative approach and modified relative approach (misaligned cases) at the ith point of contact
- $\varepsilon_{\phi i}$: pseudo-modal damping factor
- ω_m : meshing pulsation
- Ω_1 : pinion rotation speed.

2 GEAR TEETH DEFLECTIONS, MESHING STIFFNESS

In the study of loaded gear teeth, one can define two different scales of analysis, namely : hertzian deformations in the vicinity of the contacts and structural phenomena involving teeth displacements, hub deformations, bending and torsion of the shafts,...
For full E. H. D lubrication, local analysis is conducted by assuming that the tangential loading on teeth may be ignored when compared to the normal one. Semi-infinite bodies are considered so that Boussinesq solutions can be used [1]. The potential areas of contact are discretized into rectangular cells on which normal pressure is assumed constant (fig. 1).
On the other hand, structural deflections are computed by finite elements analysis (fig. 2). Their contributions are introduced in a compliance matrix filled up with the normal displacements due to unit concentrated load which is successively applied on each node of the potential contact areas.
For dynamic analyses, teeth stiffnesses are used and their values are defined at various points on the active flanks by the load to total normal displacement ratio (fig. 3).

3 STATIC AND DYNAMIC SOLUTION OF NORMAL CONTACT

Contact problems between gear teeth are solved by imposing geometrical compatibilities (non-interpenetration of the deformed surfaces), static or dynamic equilibrium of the structure and by imposing positive pressures on the contact areas.
On each element i of the contact surfaces, the following conditions must be achieved :

$$yi = hi + ui - S0 = 0 \qquad (1)$$

$$Pi > 0$$

if the element belongs to the real contact areas,

$$yi > 0 \qquad (2)$$

$$Pi = 0$$

if the element is outside the contact areas.

The static or dynamic equilibrium must be satisfied for the whole mechanism. In particular, we have :

$$\Sigma \; Pi . \Delta Si = F \qquad (3)$$

where :

- F is the static or dynamic load transmitted by the meshing teeth,
- yi is the distance between the deformed surfaces,
- hi is the distance between the undeformed surfaces,
- ui is the relative elastic displacement of the surfaces, it is the sum of the local and structural contributions,
- S0 is the rigid body displacement which ensures contact between the deformed teeth,
- Pi is the normal pressure field.
- ΔSi is the area of the ith cell.

The numerical solution is derived by using an iterative algorithm as decribed by KALKER [2].

4 STATIC ANALYSIS

The reliability of gear trains is highly dependent on the state of stresses : contact and tooth fillet stresses. The load distribution between mating teeth and along teeth width is solved first.
Then, the main purposes of the static analysis developed in this paper are :

- for given geometries of the contacting bodies, the resolution of the load distribution.
- conversely, the calculation of profile modifications for a desired pressure field.

Some examples of longitudinal load distributions for misaligned involute spur gears, and longitudinal profile corrections are given. Gear characteristics are listed in table I.

	pinion	wheel
Z	21	31
M (mm)	4	
α (°)	20	
x	0.0599	0.0880
b (mm)	45	

Table I - Gear characteristics

In this paragraph, the contact line is at the top of a single tooth contact of pinion, and the tooth width is supposed short enough so that the torsion between the two hub faces may be ignored.
The potential contact area is divided in 20 cells on the tooth width and 8 cells on the perpendicular direction. In the formulation of the contact problem, misalignments are introduced as additional terms modifying the hi vector (i.e. initial undeformed shapes).

Figure 4 and 5 show the pressure distributions between aligned teeth submitted to a load of 607.5 daN, and results obtained with a 15 µm alignment error. Comparisons with previous results given by TOBE and INOUE [3], the AGMA [4] and ISO [5] strength rating formulas, and NIEMANN experimental works [6] show quite good agreements.
The contact algorithm also serves to calculate the longitudinal profile modification, modelled by the vector hi, needed to have a desired pressure distribution along the face width when the teeth are pressed together.
Figure 6 shows the pressure field on the contact area, and the corresponding longitudinal tooth geometry for a transmitted load of 1800 daN. With the hypothesis used, corrections are symmetric with respect to the tooth center, while wider flanks will induce larger corrections at the side of the power entrance due to torsional effects.

5 DYNAMIC BEHAVIOUR OF MISALIGNED GEARS

5.1 Constraint equation

A meshing gear with rigid, uniformly spaced involute teeth, mounted on rigid supports transmits uniform angular motions, however, real tooth profiles deviate from involute surfaces, rotating bodies may be misaligned and every part of the reduction unit deforms when power is transmitted.
All these phenomena produce excitations and unsteady motions. Assuming small displacements and rotations near stationary motions, the condition for maintaining contact at the i th point on the plane of action may be expressed in terms of the generalized displacements of the two rotating wheels by [9], [10], [11] :

$$\left[v_1 - v_2 - Rb_1\,\theta_1 - Rb_2\,\theta_2\right]\cos\beta_b + \left[u_2 - u_1 - Rb_1\phi_1 - Rb_2\phi_2 - \ell_{1i}\psi_1 - \ell_{2i}\psi_2\right]\sin\beta_b - \eta_i\,(\psi_1 - \psi_2) - \Sigma_i = 0 \qquad (4)$$

Σ_i represents the value of the relative approach of meshing gear teeth, equation (4) shows that flexural ($v_1, v_2, \phi_1, \phi_2, \psi_1, \psi_2$), torsional (θ_1, θ_2) and axial (u_1, u_2) vibrations are coupled.

5.2 Mechanical model

A 12 degree of freedom simplified dynamic model (S.D.M) is used (fig. 7). Elastic properties of the shafts are approached using Rayleigh's Method [12]. Contacts are assumed to occur only on the action plane where contact lines are discretized by elements of constant stiffness ki. Their numerical values are found by the superposition of contact and structural stiffnesses of the mating teeth as defined in § 2.
When a gear rotates, the number of teeth in contact does not remain constant while the stiffness of an individual tooth pair varies as the location of their mutual line of contact changes during the motion. Therefore, the global meshing stiffness $k = \Sigma ki$ is time dependent and periodic (fig. 8).
Neglecting tangential loads, the potential energy of elastic deformation of the mating teeth is :

$$U = 1/2 \sum_i ki(t) \,.\, \Sigma i^2 \qquad (5)$$
$$\Sigma i > = 0$$

and the normal tooth load distribution is :

$$F = \sum_i ki \,.\, \Sigma_i \qquad (6)$$
$$\Sigma \; i > = 0$$

The equations of motion are :

$$[M]\, x^{oo} + [C]\, x^{o} + [K(t)]\, x = F(t) \qquad (7)$$

with

- x : vector of the generalized displacements
- [M] : mass matrix
- [C] : damping matrix introduced from pseudo-modal properties [9]
- [K(t)] : periodic stiffness matrix which is the mathematical consequence of the internal excitation induced by teeth meshing.
- F(t) : second member vector made by applied torques and external excitations.

5.3 Introduction of misalignments

By assuming small angular errors so that the rigid body kinematics remain unchanged, misalignments can be introduced in the expression of the relative approach between two points in contact as :

$$\Sigma^* i = \Sigma i - Ei \qquad (8)$$

with :
- Σi^*, Σi = relative approach at the i th point of contact with and without alignment error.
- Ei = forcing terms due to small misalignments.

$$Ei = (Rb1\, \phi_1^* + Rb2\, \phi_2^*) \sin\beta b + (l1i \sin\beta b + \eta\, i)\psi_1^* + (l2i \sin\beta b - \eta i)\, \psi_2^*$$

Lagrange's equations give the additional excitation vector F*.

$$F_k^* = \sum_i k_i . E_i \,.\, \frac{\partial}{\partial q_k} \Sigma_i \qquad (9)$$
$$k = 1, 12$$

with
$q = \langle u1, v1, w1, \phi1, \psi1, \theta1, u2, v2, w2, \phi2, \psi2, \theta2 \rangle$

5.4 Resolution and numerical results

Static load distribution solutions show that depending on misalignment amplitude the entire tooth width may not always be in contact (fig. 9). For rotating gears, this may change the meshing stiffness instantaneously, then modify both the locations of critical speeds and excitation spectra. This non-linear problem is solved by coupling a time step by step integration (Newmark's trapezoidal rule) with the contact algorithm presented in § 3.
Although that the dynamic model uses simplified descriptions of geared systems, the static loading distributions found with the S.D.M are generally close to those obtained by other authors (fig. 5).
Meshing stiffness evolutions obtained for various values of alignment errors are plotted in figures 10 and 11 , one notes that the corresponding internal excitation form and amplitude vary with rotation speeds and misalignment amplitude. The resulting dynamic tooth load diagrams, fig. 12 and 13, exhibit a threshold which separates the behaviours of gear trains with partial contact areas and those with full contacts especially in the high frequency domain.
The corresponding response curves show that a static full contact may become partial at critical speeds (fig. 14), these instantaneous evolutions are then dependent on the meshing history.

6 CONCLUSIONS

A normal contact algorithm including hertzian and structural compliances of gear trains, is used to solve static and dynamic problems of load distributions assuming negligible tangential effects. For the vibrational analysis of misaligned spur and helical gears, a simplified model of a one-stage reduction unit is proposed. The solution is conducted step by step in time.
Results may be summarized as follows :
- the influence of alignment error on the static pressure distribution is shown, numerical results compare favourably with previous theoretical and experimental works.
- the procedure developed can calculate the longitudinal profile corrections needed to produce a given pressure field on the active flanks.
- the variations in dynamic loads on the action plane for different misalignment amplitudes can be computed.
- the critical rotation speeds are derived with particular attention to the influence of partial contact areas for high meshing frequencies.

7 ACKNOWLEDGEMENT

The authors wish to thank the D.R.E.T. (Mr. Durouchoux) who financed this research.

References

[1] **CARNEIRO-ESTEVES (A.).** Resolution du contact elastique entre deux corps rugueux, Thèse de Doctorat, **I.N.S.A Lyon**, Nov. 1987, 157 p.

[2] **KALKER (J.J).** -Two algorithms for the contact problem in elastostatics. **Reports of the Department of Mathematics and Informatics**, 1982 Delft, n° 82-26.

[3] **TOBE (T.), INOUE (K.).** - Longitudinal load distribution factors of spur gear teeth, **Congrès Mondial des Engrenages**, Paris, June 1977 Vol.1, p 211-225.

[4] **AGMA** 215-01, 1966

[5] ISO / TC 60 / WG 6 / N° 169 E, 1974.

[6] **NIEMANN (G.)**, **REISTER (D.)** Konstruktion, 18 (1966), 95.

[7] **GREGORY (R.W.)**, **HARRIS (S.L.)**, **MUNRO (R.G.).**- Dynamic behavior of spur gears. **Proc. Inst. Mech. Engrs.**, 1963-64, Vol. 178, n° 8 pp 261-266.

[8] **WOOD (B.)**, **HUNT (T.M.).**- Excitation of resonant vibrations in spur and helical gear. Vol.178, Part 3J, pp 189-201.

[9] **VELEX (P.).**- Contribution à l'analyse du comportement dynamique de réducteurs à engrenages à axes parallèles, Thèse de Doctorat, I.N.S.A Lyon, Juillet 1988, 185 p.

[10] **LUND (J.W.).**-Critical speeds, stability and response of a geared train of rotors. **A.S.M.E, J. Mech. Design,** Vol. 100, July 1978, pp 535,539.

[11] **KIYONO (S.)**, **AIDA (T.)**, **FUJII (Y.).**- Vibration of helical gears. Part 1 : Theoritical analysis and Part 2 : Experimental investigations, **Bull. J.S.M.E**, 1978, Vol. 21, n° 155 pp 915-930.

[12] **LALANNE (M.)**, **BERTHIER (P.)**, **DER HAGOPIAN (J.).**- Mechanical vibrations for engineers, New-York, John Wiley, 1983, 266 p.

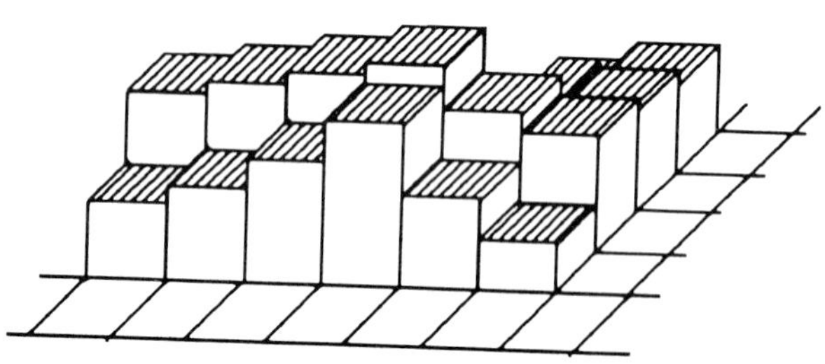

Fig. 1 : Discretization of the contact area in rectangular cells of constant pressure

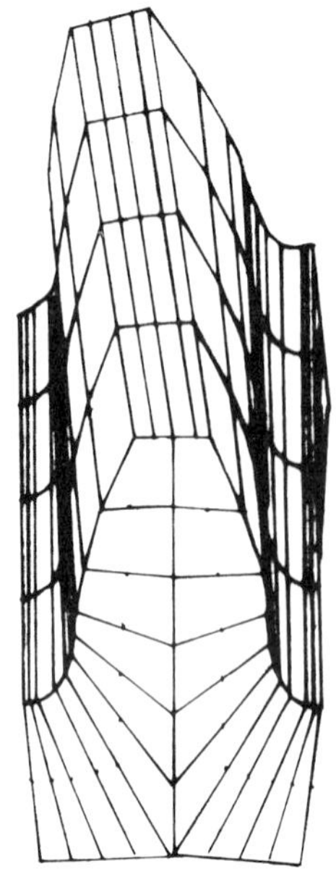

Fig. 2 : Finite element model of an helicoidal gear in 20 nodes bricks (Z = 25, α = 20 °, ß = 15 °)

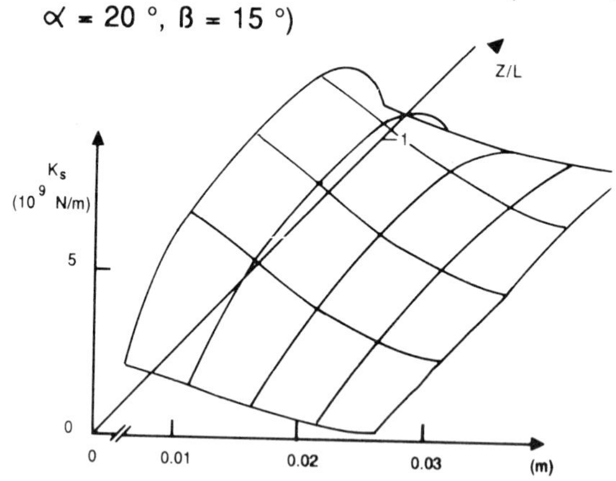

Fig. 3 : Bending stiffness of an helicoidal gear versus the radius of curvature and the position along the face width of the considerated point (Z = 25, α = 20 °, ß = 15 °)

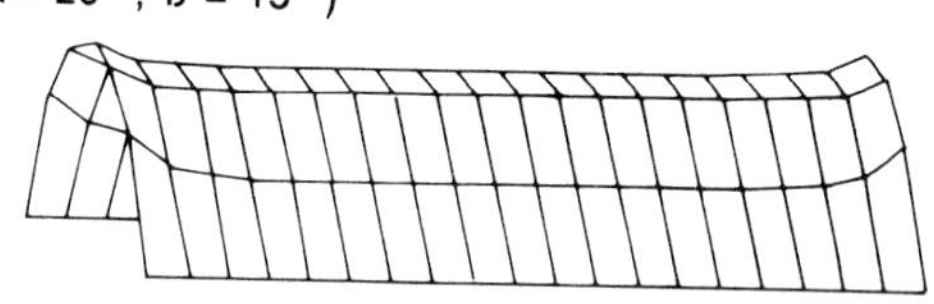

a : aligned surface

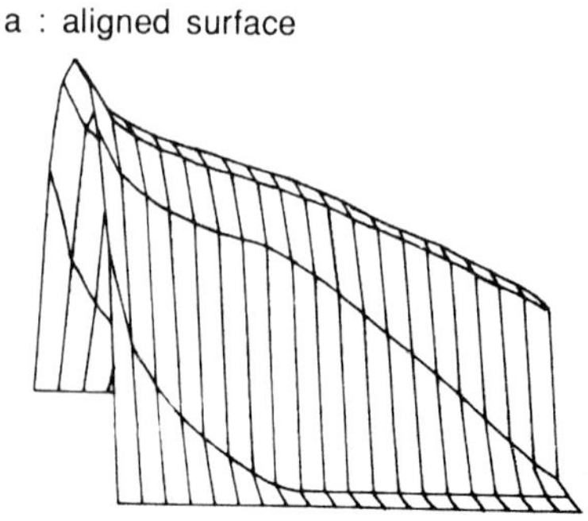

b) : misaligned error of 15 μm

Fig. 4 : Examples of pressure field between meshing teeth (F/L = 13.75 daN/mm)

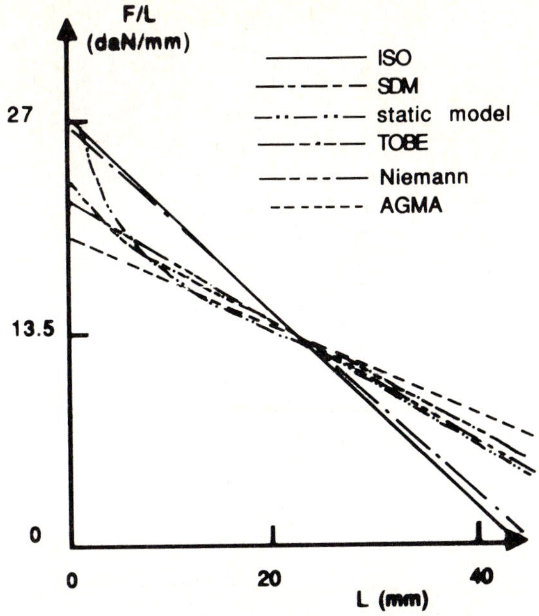

Fig. 5 : Load distribution between the meshing teeth (F/L = 13.75 daN/mm). Comparison with TOBE, NIEMANN, ISO, AGMA and Simplified Dynamic Model (SDM)

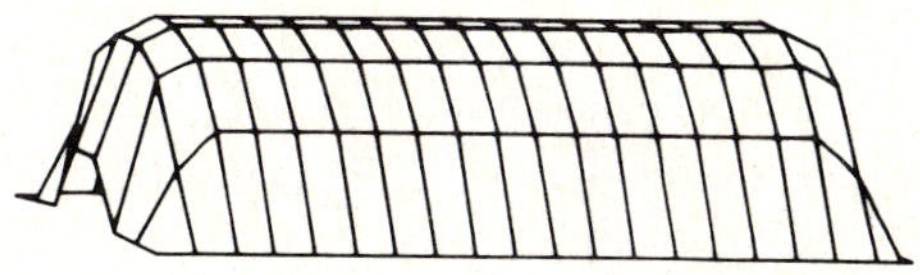

a) : desired pressure field

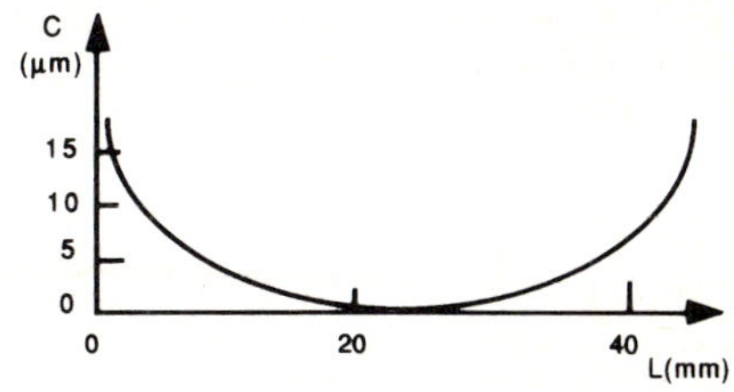

b) : Longitudinal modification of the meshing teeth

Fig. 6 : Tooth modifications for a given pressure field

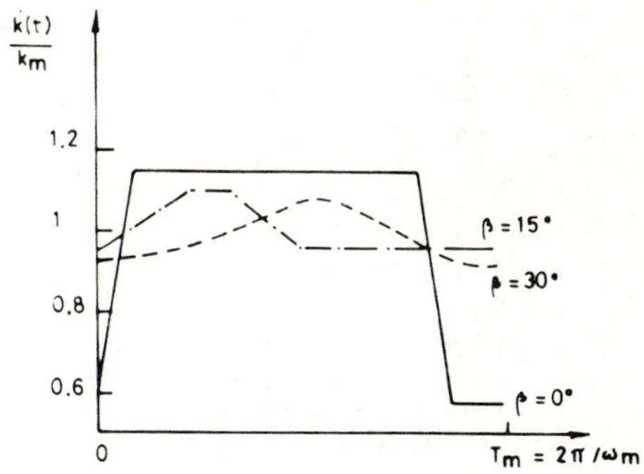

Fig. 8 : Meshing stiffness evolution for various helix angle

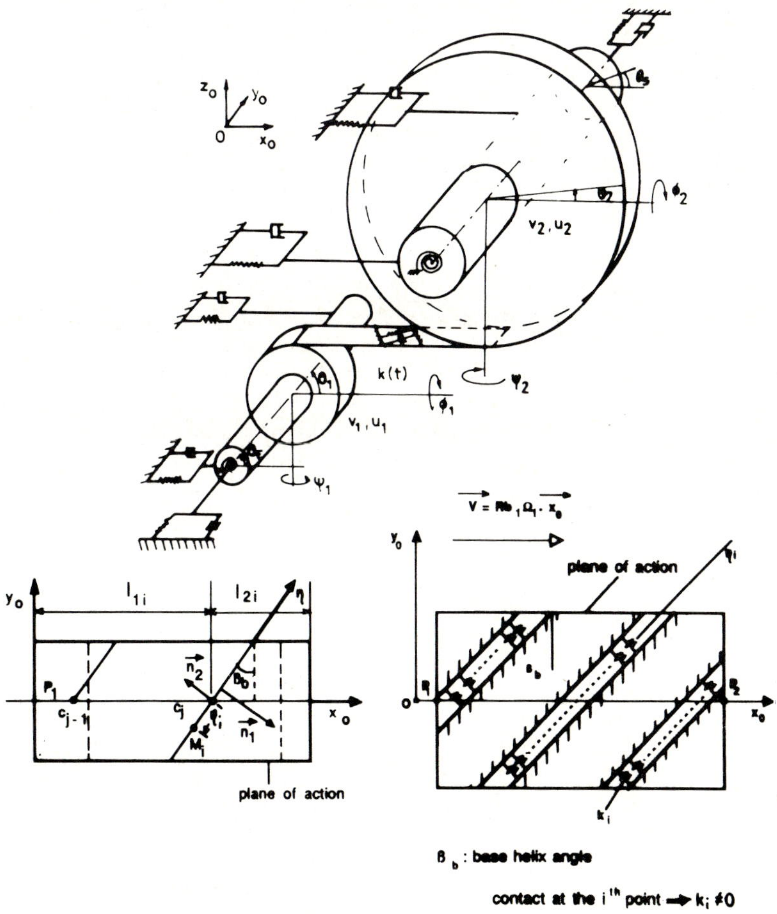

Fig. 7 : Simplified dynamic model of single stage gearing train

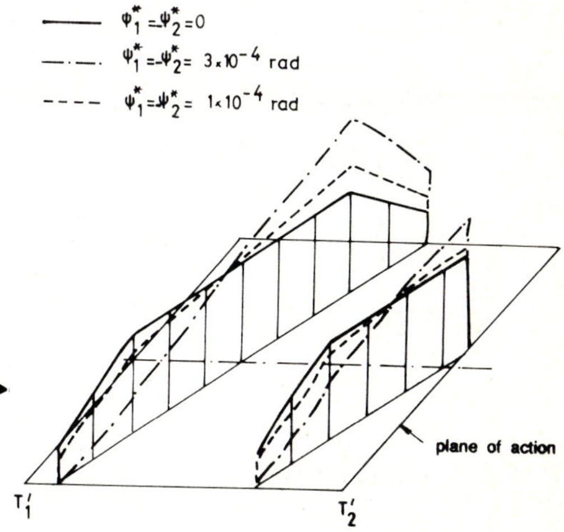

Fig. 9 : Static load distribution for measaligned helicoidal gears (Z_1 = 25, Z_2 = 150, = 20°, β = 15°

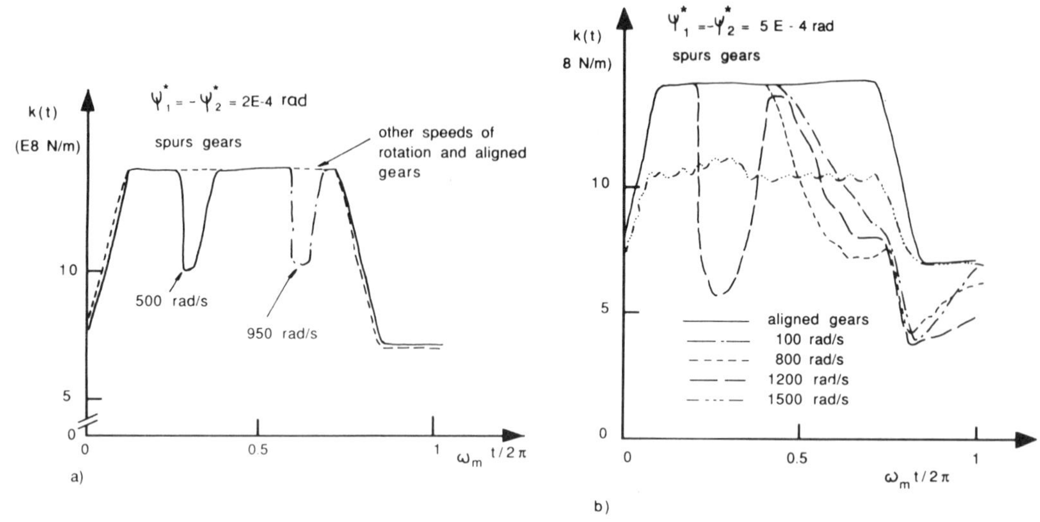

Fig. 10 : Global meshing stiffness (Z = 25, Z = 150, = 20°) for various rotating speeds and misalignments

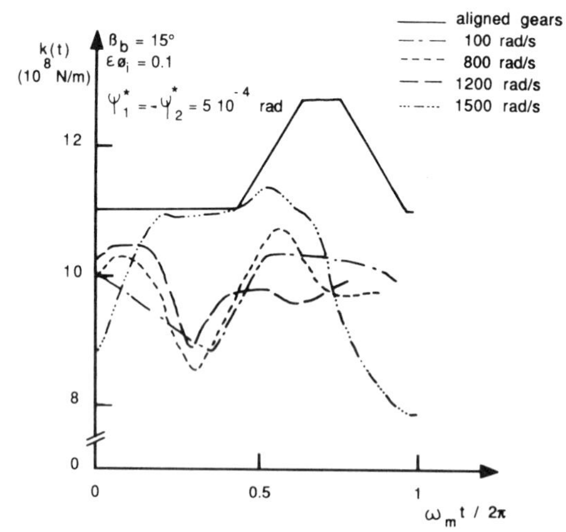

Fig. 11 : Global stiffness ($Z_1 = 25$, $Z_2 = 150$, $\alpha = 20$ °, $\beta = 15°$)

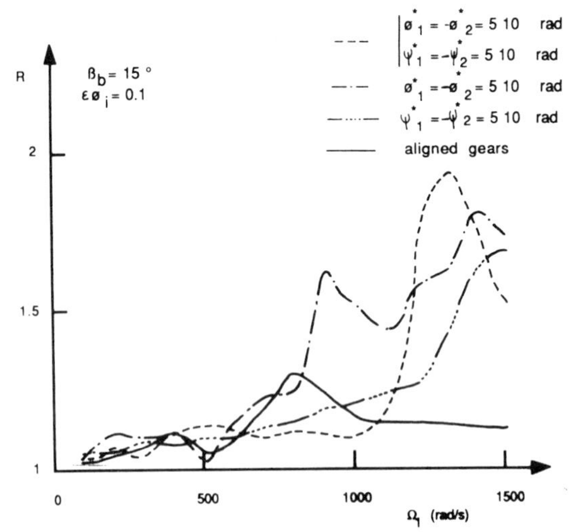

Fig. 12 : Dynamic response for misaligned helicoidal gears ($Z_1 = 25$, $Z_2 = 150$, $\alpha = 20°$)

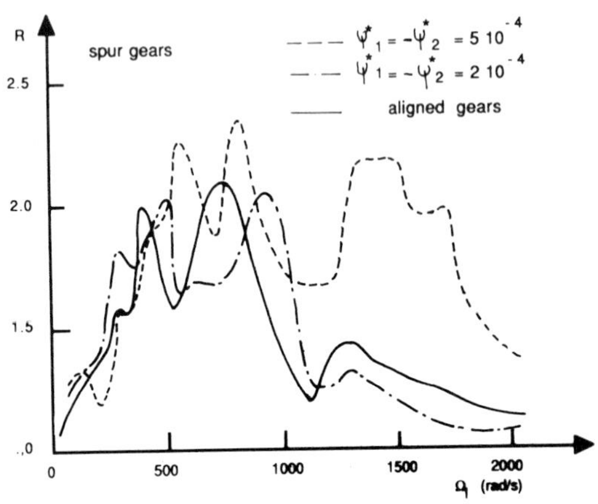

Fig. 13 : Dynamic response for misaligned spurs gears ($Z_1 = 25$, $Z_2 = 150$, $\alpha = 20°$)

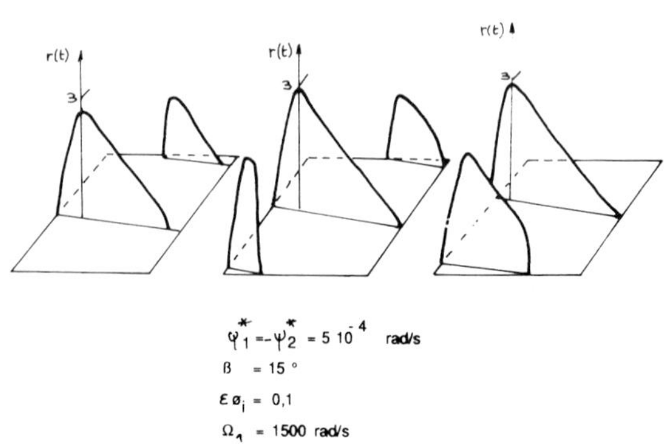

Fig. 14 : Instantaneous load distributions for misaligned helicoidal gears ($Z_1 = 25$, $Z_2 = 150$, $\alpha = 20$ °)

SESSION VII
ROLLING ELEMENT BEARINGS (1)

Chairman: Dr J D Summers-Smith

PAPER VII(i)	The Palmgren-Miner Rule Derived
PAPER VII(ii)	Prediction of Rolling Bearing Life Under Practical Operating Conditions
PAPER VII(iii)	Surface Damage on Rolling Elements and its Subsequent Effects on Performance and Life
PAPER VII(iv)	Debris Denting - The Associated Residual Stresses and Their Effect on the Fatigue Life of Rolling Bearings: An FEM Analysis

Paper VII(i)

The Palmgren-Miner rule derived

J. J. Kauzlarich

The Palmgren-Miner linear damage rule predicts fatigue failure of the component when the summation of the cycles of reversed stress amplitude, N'_i, to the cycles of stress causing failure at each stress amplitude, N_i, equals unity, i.e., $\Sigma_i N'_i/N_i=1$. It is shown that the failure strength curve for a rolling element bearing represents an "energy of failure." As such it is possible to sum the total "energy of failure" for a variable loading duty cycle and derive the P-M Rule. It is shown that the Rule assumes the rated fatigue life curve exponent incorrectly, but that it is possible to revise the Rule to take into account the correct rated life exponent. The revised P-M Rule predicts a somewhat longer fatigue life under varying loading conditions than the standard Rule. The method of derivation of the revised P-M Rule allows one to fit experimental data.

1 INTRODUCTION

A concept of cumulative fatigue damage was proposed by Palmgren in 1924 (1, p.89) for ball bearings, and by Miner in 1945 (2) for beams. The hypothesis is called the Palmgren-Miner cycle ratio summation theory, or more popularly, the Palmgren-Miner linear damage rule, and it is still widely used. The Palmgren-Miner Rule (P-M Rule) predicts failure of the component when

$$\Sigma_i N'_i/N_i = 1 \qquad [1]$$

where, for a given cyclic stress or load amplitude, N'_i is the number of cycles applied and N_i is the number of cycles causing failure at this stress or load amplitude.

The P-M Rule does not take into account prior stress history or sequence of loadings so that when applying the rule to gross cycles with few load changes it can be highly inaccurate, with the summation varying all the way from 0.18 to 23. However, if the various load amplitude cycles are mixed in a quasi-random manner the summation tends to approach unity at the time of failure; falling in the range of 0.6 to 1.6 (3, p.242). In ball and roller bearing applications, there are many situations where the bearing will be subjected to a duty cycle that repeats every few minutes, and the P-M Rule would be expected to give reasonable results.

The P-M Rule assumes that the bearing is properly lubricated, aligned, and loaded as well as kept free of abrasives, moisture, and any corrosive agents. It then reduces the problem of the life of the bearing to fatigue damage due to cyclic loading of the balls and raceways during rotation. Although many of the assumptions are often violated in a typical bearing application, it is also convenient to assume these conditions so as to concentrate only on fatigue failure of the bearing. In the next sections, a semi-qualitative development of the nature of the stresses and loads applied to a rolling element bearing is undertaken in order to show that the P-M Rule has a basic foundation in fatigue theory which allows a derivation as opposed to simply hypothesizing the Rule.

2 ROLLING BEARING LIFE

The rolling contact fatigue strength of 7205B inner rings taken from data by Lorosch (3) is shown in Figure 1. In these tests oil filters with a nominal mesh of 25 microns were used. This figure corresponds to the typical stress/cycle curve for fatigue with the cubic power law line for the Anti-Friction Bearing Manufacturers Association (AFBMA) rated life as shown. Above a maximum Hertz contact stress of 3.0 GPa, plastic deformation of the raceway was reported.

The AFBMA standard load rating for bearings is a probability equation in which the rated load at which 90% of a group of apparently identical bearings will give a life of at least 10^6 revolutions before the first evidence of failure by fatigue is detected, and is

$$L_{10} = 10^6\ (C/P)^3 \qquad [2]$$

where C is the rated dynamic load listed in tables of standards and P is the actual load. In Figure 1 the experimental data exceeds the rated life, but not greatly. It was reported that for this series of tests inspection of the rings showed that fatigue failure was initiated by indentation due to rolling over a particle of solid contaminant. For failure initiated by indentation of the race by contaminant, the characteristic crack initiation time associated with fatigue failure is essentially eliminated. When contaminant particles larger than 5 microns were filtered out of the oil during the testing the fatigue damage was found to originate below the surface of the raceway, and L_{10} life increased up to 4 times the rated life. The rated life equation shown in Eq. 2 is remarkably simple, and the adequacy of the exponent of 3 has been shown to apply to ball bearings as well as roller bearings (4, p.452).

In order to gain a better understanding of Eq. 2 let us next examine the Hertz stress field

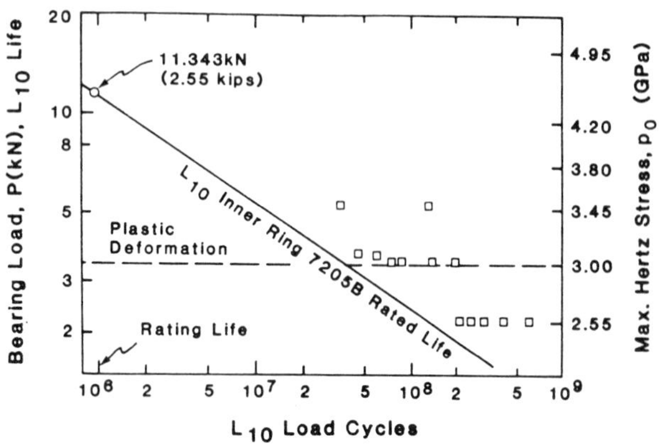

Fig. 1 Rolling contact fatigue of 7205B inner rings, after Lorosch (3).

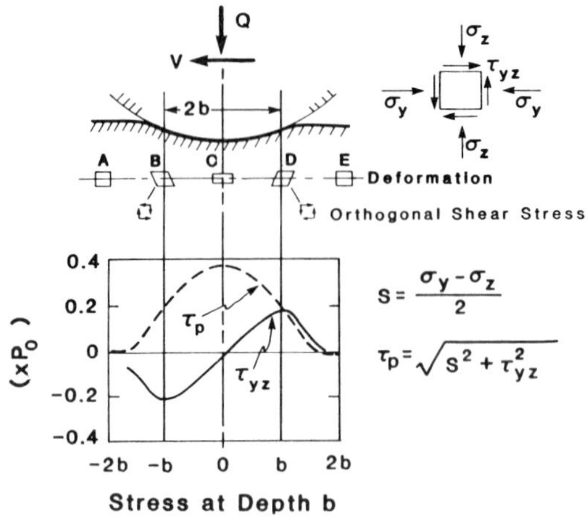

Fig. 2 Hertz orthogonal shear stress due to a loaded rigid cylinder.

below the ball or roller and raceway contact.

3 HERTZ STRESSES

For a rigid cylinder loaded against an elastic solid the stresses at the surface of contact, which were first derived by H. Hertz, and the resulting subsurface stress field have been calculated, e.g., see Johnson (5). There is a significant shear stress reversal of the orthogonal shear stress below the surface, τ_{yz}, and this is shown in Figure 2. The maximum amplitude of the orthogonal shear stress is found to be $0.256p_o$ at y=0.85b at a depth z=0.5b (11, p.379). The term p_o is the maximum stress at the surface of contact. The principal in plane shear stress, τ_p, is shown for comparison with τ_{yz}.

During normal operation of a rolling element bearing the rolling elements must pass in and out of the line of applied load and tractive forces are also developed. Including tractive forces has a significant effect on the subsurface stress field. For example, Castleberry (6) finds that the Smith-Liu equations which calculate contact stresses due to load and friction show that with a coefficient of friction of 0.08 the amplitude of the orthogonal shearing stress increases by a factor of 1.67 for a particular example.

It will be of interest later to consider the amplitude of the distortion energy density, W_S, associated with the orthogonal shear stress, where

$$W_S = \frac{1}{2G}\tau_{yz}^2 \qquad [3]$$

G is the shear modulus of elasticity.

Since the stress field for a Hertz contact is tridimensional, a calculation of the maximum strain energy density amplitude associated with the octahedral shearing stress gives another approach to fatigue failure, where

$$W_S = \frac{3}{4G}\tau_o^2 \qquad [4]$$

and

$$\tau_o=(1/3)\,[(\sigma_x - \sigma_y)^2+(\sigma_y - \sigma_z)^2+(\sigma_z - \sigma_x)^2+6(\tau_{xy}^2 + \tau_{xz}^2 + \tau_{yz}^2)]^{1/2} \qquad [5]$$

But, for our purposes the less complicated equation of the maximum strain energy density amplitude due to the orthogonal shear stress given by Equation 3 will suffice.

For static Hertz stresses due to a roller on a raceway the maximum surface stress at the contact, p_o, is

$$p_o = \sigma_{max} = 0.564[Q' \,\Gamma\, E^*]^{1/2} \qquad [6]$$

where Q' = Q/L, the maximum roller load per unit length of the roller, and E* is the reduced modulus of elasticity given by

$$1/E^* = (1 - \mu_1^2)/E_1 + (1 - \mu_2^2)/E_2 \qquad [7]$$

where 1 and 2 refer to properties of the roller and raceway respectively. Poisson's ratio is μ. The geometry factor Γ defined by the radius R for the roller and inner raceway configuration is

$$\Gamma = 1/R_1 + 1/R_2 \qquad [8]$$

A relation between radial bearing load and maximum ball or roller load is given by Harris (4, p.165 & 169) as

$$Q = 5\,P/Z \qquad [9]$$

where Z is the number of rollers.

Using the fact that the maximum orthogonal shear stress amplitude is $\tau_{yz}=0.256p_o$, then using Eqs. 6 and 9, and letting P'=P/L, results in

$$\tau_{yz} = 0.3229[P' \,\Gamma\, E^*/Z]^{1/2} \qquad [10]$$

From Eq. 3, the maximum orthogonal shearing stress strain energy density amplitude is given by

$$W_S = 0.0521\frac{P'\Gamma E^*}{GZ} \qquad [11]$$

where G applies to the race or roller depending upon the fatigue site.

The most important aspect of the analysis to this point is that the external load on the bearing, P ($P' = P/L$), is directly proportional to the subsurface strain energy density amplitude, W_S (Eq. 11), which is occurring in the raceway or rollers, so that

$$N\,P \equiv \text{"Energy of Failure"} \qquad [12]$$

Thus, the product of N times P on the fatigue strength plot of Figure 1 is proportional to energy, and has all of the characteristics of a scalar function.

A relation between strain energy density and fatigue failure will be investigated next.

4 LINEAR ELASTIC FRACTURE MECHANICS

A fatigue crack that has been initiated by contaminant indentation of the raceway or other sources will grow under sustained cyclic loading until it reaches catastrophic size. Collins (7, p.290) finds that the rate of crack growth can be described by

$$dc/dN = C_o K^{p'} \qquad [13]$$

where c is the crack length, N is number of load cycles, C_o and p' are empirical constants that depend upon material properties and secondary variables, and K is the stress intensity factor amplitude. For the roller bearing problem K can be related to bearing load as follows. The stress intensity factor amplitude due to the orthogonal shearing stress (7, p.62) is

$$K = C_K[\tau_{yz}^2\ \pi\ c]^{1/2} \qquad [14]$$

where C_K is a function of geometry and crack displacement mode. Substituting from Eq. 3, the stress intensity factor may be rewritten in terms of strain energy as

$$K = C_K[2\ G\ W_S\ \pi\ c]^{1/2} \qquad [15]$$

Introducing Eq. 15 into Eq. 13, and collecting the constants in C_o' the integrated form of the crack growth rate equation becomes (let $(p'/2)=p$)

$$N' = \frac{C_o'}{G^p\ W_S^p}\left[\frac{1}{c_1^{p-1}} - \frac{1}{c^{p-1}}\right] \qquad [16]$$

This approach is similar to one taken by Kauzlarich and Thacker (8) and Freakley and Payne (9, p.106) for the dynamic crack growth in rubbers.

The initial crack length c_1 is very much smaller than the final crack length at failure so that the fatigue life in terms of cycles to failure N is

$$N = \frac{C_o'}{G^p\ W_S^p\ c_1^{p-1}} \qquad [17]$$

Substituting for W_S from Eq. 11, and absorbing all constants in C_o'', get

$$N = \frac{C_o''}{[P'\ \Gamma\ E^*/Z]^p\ c_1^{p-1}} \qquad [18]$$

Comparing Eq. 18 with the empirical rated life in Eq. 2, we see that the exponent should be p=3.

The results of this section give an insight into the fatigue crack mechanism applicable to roller bearings and ball bearings with highly conforming raceways.

The analysis shows that crack growth is an inverse function of both load and crack length. Referring to Eq. 16, each load cycle will produce a corresponding crack extension which will affect the subsequent crack extension. If the load duty cycle is very short the sequence of loading has a small effect. In this paper it will be assumed that the load duty cycle is short so that the crack length effect can be neglected.

Turning back to the results given by Eq. 11, we can now investigate a derivation for the Palmgren-Miner Rule.

5 PALMGREN-MINER RULE

Consider a simple duty cycle applied to a roller bearing where the load continually varies between P_1 for N'_1 revs and P_2 for N'_2 revs, where the number of revs to failure at the respective loads is N_1 and N_2, and after some unknown number of revs, N, a detectable fatigue failure occurs. From the previous analysis the product of N'_1P_1 represents an amount of energy associated with progressive crack propagation, so that

$$P_1N'_1 + P_2N'_2 = P_{avg}N \qquad [19]$$

Equation 19 is shown schematically on Fig. 3.

Defining

$$n_i = N'_i/N_i \qquad [20]$$

and making use of Eq. 2 where $L_{10} = N$, Eq. 19 becomes

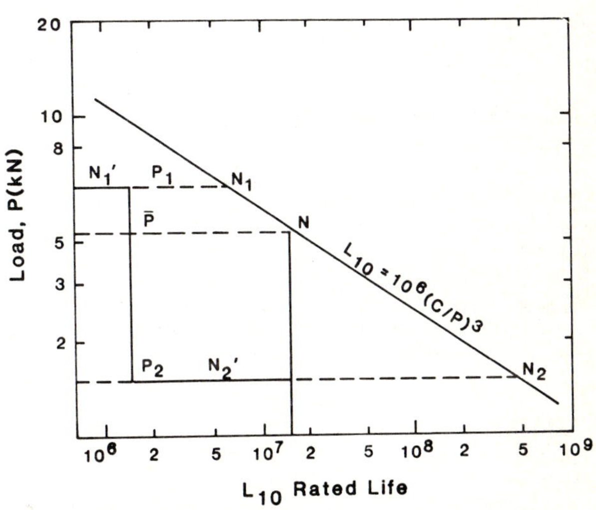

Fig. 3 Effect of duty cycle on fatigue life.

$$n_1(P_1/P_{avg})^{1-p} + n_2(P_2/P_{avg})^{1-p} = 1 \quad [21]$$

If is assumed that the value of the exponent p which comes from the equation of rated life, Eq. 2, was p=1, then Eq. 21 would become

$$n_1 + n_2 = 1 \quad [22]$$

Or, in general

$$\Sigma_i n_i = 1 \quad [23]$$

Equation 23 is the same result as expressed by the Palmgren-Miner Rule in Eq. 1, and it is seen that the Rule makes a drastic assumption about the slope of the rated life equation, as well as neglecting the effect of crack growth length during progressive fatigue crack extension.

In order to use the P-M Rule for rolling element bearing selection, we define

$$\alpha_i = N'_i/N \quad [24]$$

Then, from Eqs. 2, 20, 23, and 24, get

$$10^6\ C^3/N_{P-M} = \Sigma_i\ \alpha_i P_i^3 \quad [25]$$

Applications of Eq. 25 are shown in modern machine design texts (10).

Since it is not necessary to neglect the slope of the rated life curve in the derivation we will consider the possibility of revising the P-M Rule to take this into account in the next section.

6 P-M RULE REVISED

Now reconsider Eq. 21, with the possibility that p=3, where the generalized form of the equation is

$$\Sigma_i\ n_i\ (P_i/P_{avg})^{1-p} = 1 \quad [26]$$

The unknown average load is

$$P_{avg} = \frac{1}{N}\ \Sigma_i\ P_i\ \alpha_i\ N = \Sigma_i\ \alpha_i P_i \quad [27]$$

Substituting Eq. 27 into 26 with p=3, and using Eqs. 2, 20, and 24, results in a new equation for rolling element bearing selection with a known duty cycle, as

$$10^6\ C^3/N_{rev} = [\Sigma_i\ \alpha_i P_i]^3 \quad [28]$$

The effect of the revised P-M Rule on bearing selection as compared to the usual form of the P-M Rule is discussed below.

7 DISCUSSION

Taking the ratio of the revised and standard Palmgren-Miner Rule from Eqs. 28 and 25 gives an equation useful for comparison, and is

$$N_{rev}/N_{P-M} = \Sigma_i\ \alpha_i P_i^3/(\Sigma_i\ \alpha_i P_i)^3 \quad [29]$$

Equation 29 is plotted for a two load duty cycle in Fig. 4. The result shows that the revised P-M Rule will predict a higher value for fatigue life depending upon the loads and cycle ratio encountered; for the example, 1.39 times larger. A search of the literature did not result in any test data for variable loaded bearings, but all of the bearing manufacturers recommend that the P-M Rule be used when designing for variable load. Tests of materials in rotating beam fatigue machines in which the duty cycle is short in comparison to the cycle life tend to give results in fair agreement with the P-M Rule (7, p.242) and (11, p.221).

The plot of the rated life curve on Fig. 1, makes the typical assumption that rolling element bearings do not exhibit an endurance limit. In fatigue tests of corrosion combined with stress concentrations, Collins (7, p.206) shows that steel specimens exhibit no endurance limit even under conditions where the corrosive agent was simply tap water. In the author's experience, a new ball bearing submerged in oil developed pits in a few days after water was inadvertently allowed to drip into the oil from an overhead cold water pipe. Since there is always some water in bearing oil, and an initial crack represents a severe stress concentration, it appears that a bearing life curve without an endurance limit is not an unreasonable assumption. See Beercheck (12) for a discussion of the effect of dirt and water on bearing life.

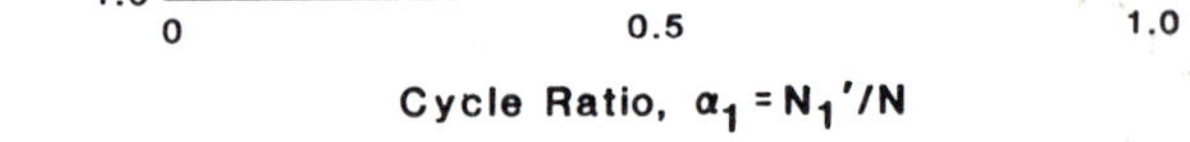

Fig. 4 Cycle failure comparison predicted by revised and standard Palmgren-Miner Rule.

The revised P-M Rule, Eq. 29, can be applied to empirical curves with other than an exponent of p=3. At high loads, an empirical fatigue strength curve for rolling element bearings with an exponent of p=4 or higher will fit the data better (3), and the revised P-M Rule can handle this case.

The approach used in this paper can also be applied to rotating beam fatigue analysis, by substituting the square of the alternating stress amplitude for the bearing load.

The revised P-M Rule gives a method for predicting crack extension under variable cyclic loads, but does not predict crack initiation. In the case where crack initiation occurs early due to oil born contaminants causing indentation of the bearing raceway, or due to other causes, the revised P-M Rule is expected to give reasonable results. If early crack initiation does not occur the revised P-M Rule is expected to be erroneous but give conservative results.

References

(1) Palmgren, A., Ball and Roller Bearing Engineering, 3rd Ed., SKF Industries, Inc., Philadelphia, PA, 1959.

(2) Miner, M. A., "Cumulative Damage in Fatigue," J. of Applied Mechanics, ASME, V 67, 1945, pp. A159-A164.

(3) Lorosch, H. K., "The Life of the Rolling Bearing under Varying Loads and Environmental Conditions," Ball and Roller Bearing Engineering, V 20, # 1, 1981, pp. 17-23.

(4) Harris, T. A., Rolling Bearing Analysis, 2nd Ed., Wiley, NY, 1984.

(5) Johnson, K. L., "One Hundred Years of Hertz Contact," Inst. of Mech. Eng., Proceedings, V 196, # 39, 1982, pp. 363-378.

(6) Castleberry, G. A., "Analyzing Contact Stresses More Accurately," Machine Design, V 56, # 7, April 2, 1984, pp. 92-97.

(7) Collins, J. A., Failure of Materials in Mechanical Design, Wiley, NY, 1981.

(8) Kauzlarich, J. J. and J. G. Thacker, "Wheelchair Tire Rolling Resistance and Fatigue," J. of Rehabilitation R & D, V 22, # 3, July, 1985, pp. 25-41.

(9) Freakley, P. K. and A. R. Payne, Theory and Practice of Engineering with Rubber, Applied Science, London, 1978.

(10) Spotts, M. F., Design of Machine Elements, 6th Ed., Prentice-Hall, NY, 1967.

(11) Juvinall, R. C., Stress, Strain, and Strength, McGraw-Hill, NY, 1967.

(12) Beercheck, R. C., "How Dirt and Water Slash Bearing Life," Machine Design, V 49, # 13, July 6, 1978, pp 2-7.

Paper VII(ii)

Prediction of rolling bearing life under practical operating conditions

E. Ioannides, B. Jacobson and J. H. Tripp

The rolling contact fatigue life model of Ioannides and Harris [1] evaluates the expected life of any stressed volume from the survival probability of individual independent volume elements determined by the local stress level and fatigue limit. Since the model is not restricted to Hertzian stress fields arising from contact between dry, smooth, semi-infinite elastic half-spaces, it has been extended in the present work to studying effects of lubricant films, roughness, contaminants and internal plastic stresses on life, using various fatigue stress criteria. Predictions of bearing life are discussed in relation to those obtained using current ISO bearing life methods, as well as in comparison to available experimental results.

1 INTRODUCTION

The Lundberg-Palmgren (L-P) method [2,3] for assessment of the expected life of rolling element bearings grew out of the need for life estimates, taking into account the then prevailing influences of materials, manufacturing processes and operating conditions. In the intervening 40 years, the L-P approach has been successful in making relative life predictions for bearings with different sizes and loads, the variables appearing in the original life formulation. Current ISO life rating standards are indeed based on this method. Improvements in material properties, in particular, cleaner steels, have resulted in a higher proportion of bearings suffering surface-initiated fatigue compared with the sub-surface-initiated damage in earlier materials. Other influences, such as bearing surface finish, lubricant distribution and bearing kinematics, as well as improvements in the operating environment of the rolling elements themselves through better filtration and sealing, have as a result acquired a greater importance. Their effects combine to extend bearing life very considerably; in some cases, practically and functionally infinite life is observed.

To describe such improved life performance and to compare life under, for example, different raceway finishing operations, it has become necessary to go beyond the so-called 'life adjustment factors', which were introduced in an empirical way to describe the observed longer life. An understanding of the physical influence itself, in this example surface finish, on bearing performance is required. As far as possible, the effects compressed into the life adjustment factors should be extracted in a quantitative way - the number of unknowns remaining hidden should be reduced to a minimum.

The present work links several effects to life through the stress fields they produce in the bearing material. Thus, in one case, the surface stress distribution from the dry contact of two rough surfaces is calculated from the measured profiles. In another, rheological models going beyond the ideal Newtonian viscous fluid are introduced into the calculation of EHL films, from which surface stress distributions may be extracted. Such surface stresses are then used to generate the volume stresses determining the expected life of the bearing material. When contaminant particles are present in the film, plastic deformation of both the particle and the raceway will generally occur. The residual stresses arising from the residual strains (deformed particle and dent shapes) again are used to find the volume stress distributions in subsequent over-rolling. From these stress fields comes a clearer understanding of the various influences on life.

1.1 Notation

a	Hydrostatic stress criterion factor
a,b	Semi-axes of Hertz contact ellipse, [m]
a_i	Life adjustment factors, i = 1-3
A	Probability normalization factor
c,e,h	Lundberg-Palmgren exponents
C	Basic dynamic load rating, [N]
$G(N)$	Elementary survival probability
$H(x)$	Heaviside step function
L_n	Life (n% Failure), [Mrev]

N	Number of cycles
p	Load-life exponent
p_H	Hydrostatic stress, [Pa]
p_o	Maximum Hertz pressure, [Pa]
P	Bearing equivalent load, [N]
r	Position vector, [m]
R_q	Root-mean-square roughness height, [m]
R_n	normalized n th moment of height distribution
$S(N)$	Survival probablity
V	Volume [m^3]
x	Transverse coordinate, [m]
z	Depth coordinate, [m]
z_o	Depth of maximum Hertz shear stress, [m]
z'	Stress weighted average depth, [m]
α	Viscosity pressure coefficient, [Pa^{-1}]
γ	Shear strength pressure (traction) coefficient
Δ_q	Root-mean-square roughness slope
Λ	Film parameter
σ	von Mises stress, [Pa]
σ_{yt}	Maximum surface traction, [Pa]
τ	Shear stress (orthogonal, fatigue limit, elastic limit, etc.), [Pa]

2 FORMULATION

The L-P theory of bearing life begins with the assumption that the probability for a given volume element ΔV to survive N stress cycles and to fail in the next dN is proportional to its size, and is a function not only of its location r but also of N itself. The probability actually increases slowly with N, this memory effect being an essential part of fatigue behaviour. Integrating over all volume elements to obtain the survival probability of the volume V thus yields:

$$\ln \frac{1}{S(N)} = \int_V G(N,r)\, dv \qquad (1)$$

where $G(N,r)\Delta V$ is the accumulated survival probability over N cycles for element ΔV. In the absence of knowledge of actual subsurface stresses, L-P chose simply to give a functional form to the right side of Eqn.(1) as follows:

$$\ln \frac{1}{S(N)} \propto \frac{\tau^c N^e}{z_o^{\,h}} V \qquad (2)$$

where τ is some characteristic stress putting an elementary volume at risk (the stress criterion) and z_o is a measure of the depth below the raceway to which this stress extends. V is the effective volume within which τ differs sensibly from zero. The exponents c, e and h (assumed to be history-independent) were determined to give a best fit to the then available test data.

From the form of Eqn.(2) it is clear that τ can only represent an average stress throughout the volume so that the life N at any given survival level S cannot reflect sensitivity to localized domains of high stress. Moreover, since τ and z_o were calculated from Hertz theory, no account could be given of surface shear stresses arising from frictional sliding or lubricant viscosity. Finally, according to Eqn.(2), no bearing life can be infinite.

The new life theory extends the L-P model in two essential respects. First, Eqn.(2) is interpreted locally so that τ becomes a local variable and the right side of Eqn.(2) becomes a 'risk function'. Second, a fatigue limit τ_u is introduced, patterned on structural fatigue initiation, such that any volume element for which $\tau < \tau_u$ makes no contribution to the risk function. With these new features, the survival probability becomes:

$$\ln \frac{1}{\Delta S} = A \frac{(\tau-\tau_u)^c N^e}{z_o^{\,h}} H(\tau-\tau_u)\Delta V \qquad (3)$$

where H(x) is the Heaviside step function. As a first approximation, this may be written in integral form as follows:

$$\ln \frac{1}{S(N)} = \bar{A}\, N^e \int_{V_R} \frac{(\tau-\tau_u)^c}{z'^{\,h}}\, dV \qquad (4)$$

where $\bar{A}$ is the volume-averaged value of A and the integration runs only over the region where the stress criterion exceeds threshold. The fatigue limit itself may also assume local values. The factor z_o, originally introduced to include the propagation interval between internal damage initiation and its appearance in the surface, is retained but should now be regarded as a stress-weighted average depth, z'. The L-P exponents remain the same.

The L-P formulation leads to a relation between N and S depending on bearing design and load, which is readily reduced to the simple form,

$$L_{100(1-S)} = \left(\frac{C}{P}\right)^p \qquad (5)$$

Here, L is the expected life (N in MRev) under load P, while C, the basic dynamic load rating is the load for which (100 S)% of a population of bearings will survive for at least 1 MRev. The ISO standard is based on this expression with S = 0.9 (10% failure rate defines the L_{10} life) and, using Hertz analysis, τ is the maximum orthogonal shear stress and p, the load-life exponent, is 3 or

10/3 respectively for ball or roller bearings. In full, the ISO definition is written,

$$L_{na} = a_1 a_2 a_3 \left(\frac{C}{P}\right)^p \qquad (6)$$

which includes the familiar three life adjustment factors, a_i, i = 1,2,3. On a logarithmic plot, the basic (non-adjusted) life rating thus decreases linearly with load. The a_i factors serve to shift this line to longer life, but with no change in slope.

2.1 Fatigue criteria

In order to carry out actual life calculations, it is necessary to determine the proper fatigue criterion to use in Eqn.(4). The choice becomes even more critical when the model is applied to the localized type of stress fields encountered in roughness and contaminant studies. Examples of criteria proposed in the literature are:- p_o, maximum Hertzian pressure; τ_o, orthogonal shear stress (ISO); τ_{max}, maximum shear stress; σ_{yt}, maximum surface traction; σ, von Mises stress; τ_a, maximum shear amplitude regardless of plane and direction. In the general multiaxial stress case there have been many other suggestions for criteria, involving both instantaneous and cycle-average values of the stress components [4]. Most direct material tests, however, have been performed under conditions where the deviatoric stress components are proportional to, and thus in phase with, each other. Such conditions do not pertain to rolling contact fatigue, so that the influence of the principal stresses on endurance strength must be explicitly included.

An illustration of the differences between the old and new life models is given in Figure 1, which shows both the basic and adjusted life ratings of a 6309 DGBB as a function of contact pressure according to the standard catalogue procedures, Eqn.(6), together with the new model load-life behaviour. For the model calculations, τ_a was selected as the fatigue criterion, while local modifications of the fatigue limit were made according to the absolute maximum value, τ_{max}, of this shear stress in the cycle. Thus, τ_u = 0.35 GPa for τ_{max} below the elastic limit τ_e, decreasing linearly to zero at the fracture strength τ_f. (This form for τ_u was based on independent data fitting [1]). The curve predicted by the new model is normalized to a single experimental point obtained at 2.9 GPa (2900 N/mm^2). Also displayed are some experimental data of Zwirlein and Schlicht [5] taken from a 7205 B ACBB, with a basic life rating virtually indistinguishable from the DGBB. This shows good agreement with the predicted behaviour in the region of overlap, particularly with respect to slope. At the lower stress levels, where test results become progressively more difficult to obtain, the predicted curve asymptotically approaches the fatigue limit, corresponding to normal pressure $4 \times \tau_u$ =1.4 GPa. The effect of the life adjustment factors in this case is clearly inadequate, both quantitatively and qualitatively. Calculations based on the von Mises stress criterion for this example led to rather similar conclusions.

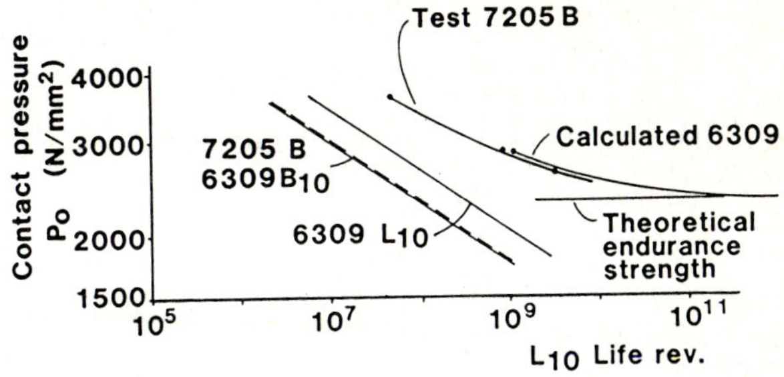

Figure 1 Variation with normal contact pressure of rating and theoretical lives of a 6309 DGBB. The rating and experimental values for a 7205 B ACBB are shown for comparison.

In the following example, a simple multiaxial criterion is introduced which modifies the damaging effect of the shear stress with a term linear in the local hydrostatic pressure:

$$\tau'_a = \tau_a + a p_H \qquad (7)$$

The value a = 0.3 is chosen to agree with many experiments carried out with shear stress in combination with compressive or tensile stress [6,7].

The effect of the 'a' term in Eqn.(7) is illustrated in Figure 2. showing the L_{10} life v. Load relation for a 6309 DGBB, computed from Eqn.(4) using Eqn.(7) with a = either 0 or 0.3. Values have been normalized to the Catalogue value for C/P = 2.8. The first case (upper curve) is thus essentially the same as given in Figure 1. The second case shows life reduction factors of no more than about 4 at low loads, while at higher loads the differences become negligible. This is in general agreement with the lack of sensitivity to criterion noted in the previous example. Both are based on the nominal ideal raceway geometry.

2.2 Internal and residual Stresses

The criterion of Eqn.(7) has been used, also with a = 0.3, to simulate experimental results obtained from modified NU 209 rolling bearing inner races, subjected to different hoop stresses and run in a five roller fatigue test (at 5000 rpm, Hertz pressure = 0.7 GPa, Λ = 3) as described in [8]. Table I compares relative fatigue life measured in these experiments with predictions using either no fatigue limit or the rather low value τ'_u = 70 MPa. This value was determined as giving the best overall agreement between experiment and theory. The important conclusion is

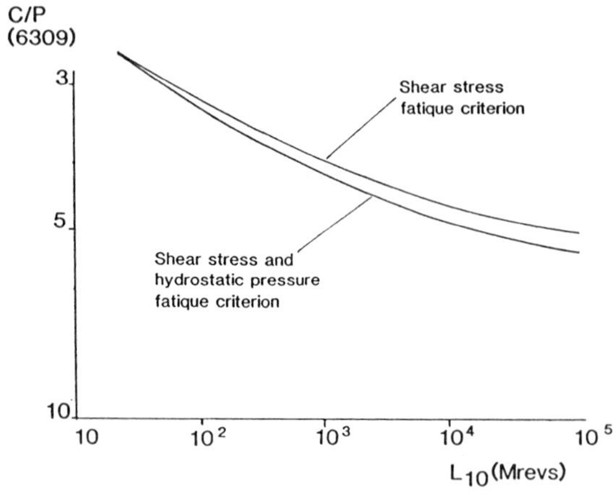

Figure 2 The effect on 6309 L_{10} fatigue life of modifying the criterion by including a hydrostatic pressure term.

thus that the hoop stress alone cannot account for the observed life reduction. Only the combination of internal stress with the fatigue limit shows the correct stress-life behaviour at the three measured levels.

TABLE 1: Experimental and predicted relative L_{10} life of an NU 209 inner ring under different tensile stresses.

CLAMPING PRESSURE (MPa)	HOOP STRESS (MPa)	EXPERIMENTAL RELATIVE L_{10}	PREDICTED RELATIVE L_{10} ($\tau_u' = 0$ MPa)	PREDICTED RELATIVE L_{10} ($\tau_u' = 70$ MPa)
5	21.4	1.0 *	1.0	1.0
40	163.0	0.07	0.329	0.084
80	325.0	0.0125	0.102	0.0096

* L_{10} = 3.2 MRev

3 EHD LUBRICANT FILMS

Modification of the stress fields in bearing elements by an EHL film can be considerable. Parameters of lubricants most influential on the stresses at the film boundaries include:- viscosity (and its coefficients of pressure, α, and temperature variation), elasticity, shear strength (and its coefficient of pressure increase, γ) and solidification pressure. The original L-P calculations assumed that the film pressure distribution was the same as the dry, frictionless Hertz contact solution. However, the dynamics of a viscous fluid film imposes a relationship between the pressure and the shape of its boundaries which is different from that of dry elastic contact. Lubricant films in fact require a local film thickness decrease near the outlet, resulting in very steep pressure gradients and often a local "spike" [9]. Such rapid pressure variations produce high shear as well as normal stress fields in the bearing material close to the surface. Lubricants with high α-values building up large pressure spikes also tend to show large γ-values, which permits high shear stresses in the oil and correspondingly on the boundaries. The result is a particularly damaging combination. The solidification pressure of the lubricant can also be important. As long as the oil in the contact remains liquid, rather high slip velocities can be accommodated without causing excessive shear stresses.

Taking again as example a 6309 DGBB, some of the effects of EHL films with different rheology may be demonstrated. First, for a given bearing load, the pressure distribution in the most heavily loaded ball contact is computed. The fatigue life, relative to the measurement at C/P = 2.8, is then calculated from the subsurface stresses based on this distribution, using the criterion of Eqn.(6), with a = 0.3. Next, the surface stresses are modified by adding a shear stress proportional in magnitude to the normal stress and in the direction opposite to the local microslip velocity. This is a good approximation at high pressure and high slip, when the lubricant operates at its shear strength limit. A typical value of traction coefficient for poly-alpha olefinic oils is γ = 0.05, while for oils with stiffer molecules, such as a naphthenic raffinate, values as high as 0.20 are possible. Stress fields for these two values of traction are then used to obtain relative load-life curves. Results are plotted in Figure 3, normalized to the experimental point for which γ = 0.05 is assumed. The Catalogue Rating Life is also included.

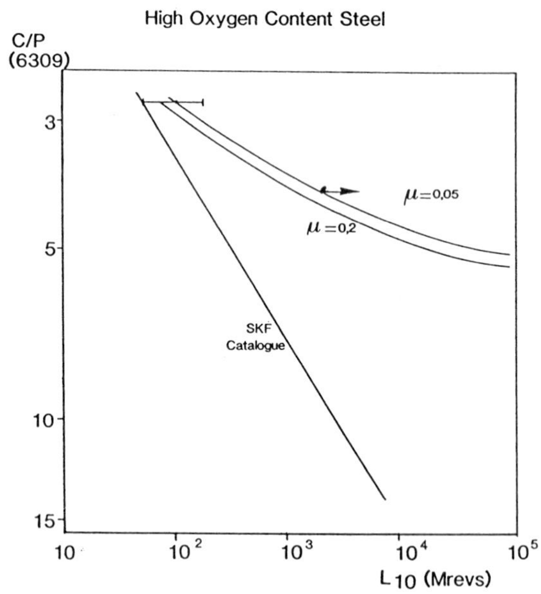

Figure 3 The effect of surface traction on 6309 L_{10} life.

Comparison with Figure 2 shows that, at low loads, the effects of increasing friction from 0.05 to 0.20 are comparable to the effects of modifying the fatigue criterion according to Eqn.(7). Life reduction by a factor of about 4 is predicted. The effect does not, however, disappear entirely at higher loads, where a reduction of about 30% is still seen.

4 SURFACE ROUGHNESS

As for EHL films, surface roughness also has a large influence on the local stresses close to the bearing surfaces, producing significant deviations from the Hertzian stress distribution of smooth body contact. The actual topography imposed by the manufacturing finishing processes - grinding, honing, polishing, etc. - as obtained directly from surface profilometry provides a sampling of heights representing the true surface within the usual limitations of random signal analysis. Such a height sample is regarded as containing all the geometrical information obtainable on the surface, so that interpolation between sampled points, such as that implied by introducing an asperity model, is at best uninformative and at worst dangerous. The gap between two unloaded rough surfaces is known from their step height representations, so that specification of a nominal geometry and load suffices to determine the normal contact stress, computed also as a step distribution [9]. From such stresses, accurate subsurface stress fields may be calculated in the raceway. Discretization effects are confined to depths less than the sampling interval.

An example of these calculations is given in Figure 4, showing the surface height and contact pressure profiles along a transverse section of a 6309 DGBB inner raceway (sampling interval 20 µm). The honing process which follows grinding was in this case stopped before some deep marks left by grinding had been completely removed. Three such circumferential grooves, each about 1 µm deep, may be seen in the loaded zone. Corresponding to the steep local slopes associated with these features, large local pressure deviations, of the order 2 GPa relative to the smooth Hertzian distribution, are produced. The smooth Hertz maximum pressure here has the realistic value of 2.7 GPa, so that contact at the bottom of the groove towards the outside of the track is almost lost. The effect of these strong local features on the subsurface shear stress field is still evident at a depth of order b, here about 200 µm.

Table 2 shows the effect on the fatigue life of such topography-generated stresses. The L_{10} life given is relative to the ideal smooth contact geometry, a reduction in life by a factor of nearly 4. The rms height, R_q, and slope, Δ_q, values also appear. For comparison, the same calculation was performed for an unhoned ring with no deep marks. Corresponding to rms slope some 2½ times that of the first surface, the pressure profile across the track of this ring fluctuates with an amplitude exceeding 1 GPa relative to the Hertz distribution, some 2½ times the amplitude seen in Figure 4. Despite these large stress variations, the calculated life of this raceway is only slightly reduced.

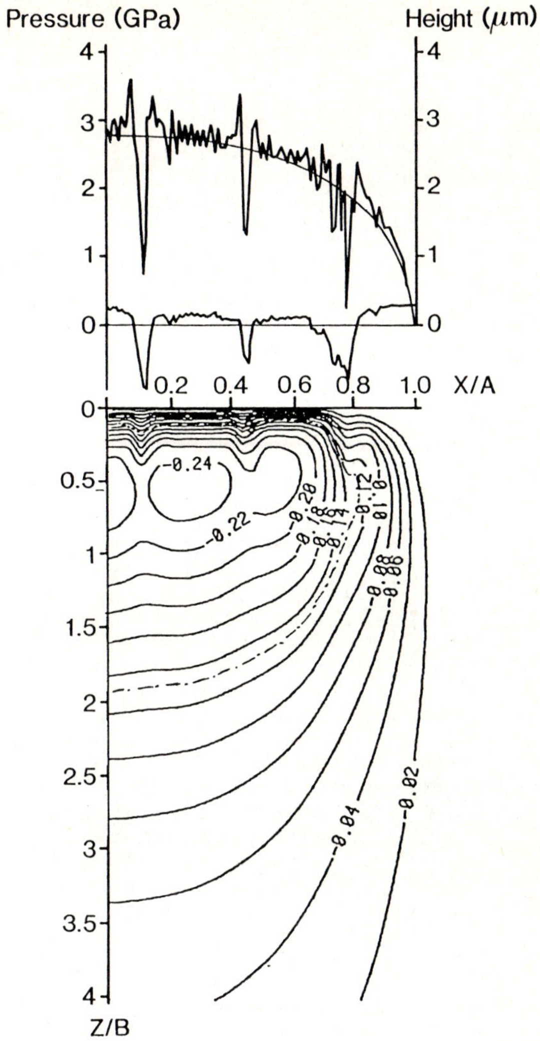

Figure 4 Dry contact pressure distribution across the raceway of a honed 6309 ring, showing a profile with residual grinding marks. The subsurface shear stress field, normalized to p_o, is also given.

TABLE 2: Effect of surface topography on bearing life, $L_{10}(\tau'_a)$, relative to smooth surface (R_q = 0).

6309 DGBB INNER RING	R_q (µm)	Δ_q (deg)	R_{10}	$L_{10}(\tau'_a)$
Honing, with residual grinding grooves	0.271	0.391	5330	0.288
Normal grinding	0.523	0.913	360	0.734

Contrary to expectations, the reduction of life in these two cases is smaller for the surface with larger R_q and Δ_q, roughness parameters containing both vertical and horizontal surface information. Because of the large value, 31/3, of exponent 'c' in Eqn.(4), volume elements where the fatigue criterion is large receive very large weighting in the risk integral. A few vulnerable points in the risk volume, corresponding to a few extreme surface events, may thus completely dominate the life behaviour, reflecting the physical nature of the fatigue process itself. In support of this, some high order moments R_n of the raceway height distributions were obtained. Normalized values of the 10th moment have been included in Table 2 as an example, showing the expected inverse relationship with L_{10}. [The normalized 2n th moment of a Gaussian height distribution has the value (2n-1)!!] This indicates that damage due to surface finish may result more from isolated defects than from average surface properties, an indication borne out in actual experience. The statistical description of such events is clearly a task much harder even than conventional surface characterization.

5 CONTAMINANTS

Lubricant film thicknesses lie typically in the range 0.1-3.0 μm, while contaminant particles carried through the contact may well be an order of magnitude larger, in the range 1-30 μm. Studies, both experimental and theoretical, of raceway deformations caused by entrained particles have shown the occurrence of quite severe denting, even when the raceway has greater hardness than the squashed particle [11,12]. Indeed, particles as soft even as common plastics can permanently deform the bearing surfaces if their size lies toward the top of this range. For hard particles, such as steel debris or ceramic fragments from a grinding wheel, it is necessary that their size be only slightly larger than the oil film thickness before plastic deformation occurs. Such damage features are always surrounded by a local residual stress distribution which does not disappear within the life of the bearing.

Calculation of the residual stress distribution around such dents has in fact revealed the existence of regions of tensile stress at the surface near the dent edge [13]. As already mentioned, reduction of compressive principal stresses by superposition of tensile components decreases the endurance strength of bearing steels. Thus, a tensile residual stress not only raises the local stress but also decreases the local fatigue limit of the material. Such dents thus have a particularly deleterious effect, as may be demonstrated by including the residual stress in the fatigue criterion of Eqn.(3). This has been used to evaluate the risk function at each point beneath a roller as a function of time, i.e., as a function of the distance of the contact patch from the dent as the roller overolls it [14]. For certain positions of the roller, the risk of damage near the edges of the dent is dramatically enhanced.

6 CONCLUSIONS

With the computation facilities currently available, it is possible to obtain very detailed stress fields associated with surface features as well as to combine them with internal stresses or residual stresses arising from plastic deformations. Moreover, these stresses can be combined with surface stresses coming from the pressure and shear stress in the lubricant. The present fatigue life model can also handle a detailed description of the local material properties. However, at present such exact knowledge of material behaviour is not available but must be deduced, chiefly from bearing tests.

The individual effects of surface topography and lubricant parameters on the estimated life of a bearing actually amount to less than one order of magnitude each. These effects may, however, combine with those of residual or other internal stresses to produce more drastic reductions on the endurance life of rolling bearings.

Thus, considerable progress has been made towards the goal of extracting the hidden physical content of the life adjustment factors, which hitherto have been the empirical means of making quantitative estimates of a number of unknown factors. In addition, this approach is applicable to life prediction for lubricated machine elements in general.

7 ACKNOWLEDGEMENTS

The authors would like to thank Dr. I.K. Leadbetter, Managing Director of the SKF Engineering and Research Centre, for permission to publish these results. Thanks are also due to Mr. E. van Amerongen for valuable discussions on experimental aspects of this work.

REFERENCES

[1] IOANNIDES, E. and HARRIS, T.A., 'A new fatigue life model for rolling bearings', J. Tribology 1985, 107, 367-378.

[2] LUNDBERG, G. and PALMGREN, A., 'Dynamic capacity of rolling bearings', Acta Poly. (Mech. Eng. Ser. 1), Roy. Swedish Acad. Eng. Sci. 1947, 7, 5-32.

[3] LUNDBERG, G. and PALMGREN, A., 'Dynamic capacity of rolling bearings', Acta Poly. (Mech. Eng. Ser. 2), Roy. Swedish Acad. Eng. Sci. 1952, 96, 5-32.

[4] KAKUNO, H. and KAWADA, Y., 'A new criterion of fatigue strength of a round bar subjected to combined static and repeated bending and torsion', Fatigue of Engng. Mat. and Structures 1979, 2, 229.

[5] ZWIRLEIN, O. and SCHLICHT, H., 'Material stressing during rolling loading - influences of friction and residual stresses', Z. Werkstofftech. 1980, 11, 1-14.

[6] BURNS, D.J. and PARRY, J.S.C., J. Mech. Engng. Sci. 1964, 6, 293.

[7] DANG VAN, K., GRIVEAU, B. and MESSAGE, O., 'Bi-axial and multiaxial fatigue', Proc. 2nd Intl. Conf. on Multiaxial Fatigue (Dec. 1985, Sheffield), Mech. Eng. Publ. 1988.

[8] CZYZEWSKI, T., 'Influence of a tension stress field introduced in the elastohydrodynamic contact zone on rolling contact fatigue', Wear 1975, 34, 201-214.

[9] HOUPERT. L., IOANNIDES, E., KUYPERS, J.C. and TRIPP, J.H., 'The effect of the EHD pressure spike on rolling bearing fatigue', J. Tribology 1987, 109, 444-451.

[10] TRIPP. J.H., HOUPERT, L.G., IOANNIDES, E. and LUBRECHT, A.A., 'Dry and lubricated contact of rough surfaces', Proc. Inst. Mech. Engrs. Intl. Conf. 1987-5, 'Tribology - 50 years on', 1987, 171-79.

[11] HAMER, J.C., SAYLES, R.S. and IOANNIDES, E. 'Deformation mechanism and stresses created by 3rd body debris contact and their effect on rolling bearing fatigue', Proc. 14th Leeds-Lyon Symp. on Trib. (Lyon 1987, to appear).

[12] HAMER, J.C., SAYLES, R.S. and IOANNIDES, E. 'Particle deformation and counterface damage when relatively soft particles are squashed between hard anvils', J. Tribology 1988, 110, (to appear).

[13] HAMER, J.C., LUBRECHT, A.A., IOANNIDES, E. and SAYLES, R.S. 'Surface damage on rolling elements and its subsequent effects on performance and life', Proc. 15th Leeds-Lyon Symp. on Trib. (Leeds 1988, this conference).

[14] KO, C.N. and IOANNIDES, E., 'Debris denting - the associated residual stresses and their effect on the fatigue life of rolling bearings: an FEM analysis', Proc. 15th Leeds-Lyon Symp. on Trib. (Leeds 1988, this conference).

Paper VII(iii)

Surface damage on rolling elements and its subsequent effects on performance and life

J. C. Hamer, A. A. Lubrecht, E. Ioannides and R. S. Sayles

The influence of lubricant contamination and subsequent surface damage on rolling bearing fatigue has formed the basis of several studies over recent years. Webster et al (1) calculated the elastic subsurface stress fields from real dent profiles and used these as input to a fatigue life model to determine the reduction in lives. In this paper a slip line field analysis has been used to calculate the subsurface residual stresses for different idealised dent/roller combinations. These stress fields were superimposed upon those calculated from a dry contact analysis of the overrolling of the dent which used novel multi-level techniques to accelerate convergence, and the resultant stress fields provided input to the fatigue life model.

The influence of the dent itself on life appears to be small and only becomes significant on inclusion of the residual stresses. These have a particularly marked effect as the roller radius and load are reduced, suggesting expected lives may not increase as rapidly with decreasing load as would be expected from conventional models.

1 INTRODUCTION

Lubricant contamination and subsequent surface damage is increasingly recognised as having a significant effect on bearing fatigue lives which would not normally be predicted by the traditional Lundberg and Palmgren model (2). In (3) Ioannides and Harris presented an important generalisation of the classical model in which the stress perturbations resulting from non-perfectly smooth contacting surfaces can be included. Essentially the new model is an elemental form of the earlier analysis where the probability of failure is calculated from the cumulative contribution of stresses above a threshold value in small volume elements of the material. Therefore to apply the model a complete history of the stress field underneath the dent as it passes through the contact is required. This field will be effectively made up of two components the residual stress field from the indentation and the EHD stress field from the subsequent overrolling of the dent. Unfortunately these are not necessarily independent as further plastic deformation of the dent edges and some shakedown will probably occur, but a useful first approximation can be made by treating them as such. This problem was first tackled by Webster, Ioannides and Sayles (1), where they measured the profiles of dents from a real bearing surface. This information was the input to a numerical contact model coupled to a finite element analysis to determine the state of stress in an indented bearing raceway. The stress information was used in conjunction with the modified fatigue model to determine the reduction in fatigue life.

In (4) approximate solutions were presented for the interfacial pressures and deflections resulting from debris particles being squashed in the inlet to an EHD contact. A plane strain approach was used to allow the application of 2 dimensional extrusion theory. The problem is similar to strip rolling, but in this case the rollers cannot be expressed as circular arcs as the elastic deflections due to the EHD and extrusion pressures are significant. The solution was found by numerically iterating between the pressure and deflection equations until convergence.

As the rolling surfaces continue to approach one another in the inlet, the interfacial pressures may increase sufficiently such that plastic deformation of the raceways can occur. The critical value of pressure may be estimated from indentation experiments or theoretical considerations such as (4). By applying these criteria to the model, the approximate dent shapes and pressure distributions may be found and compared with those measured experimentally (5). The measured profiles of these dents closely approximate a circular arc similar to the impression left by a rigid cylindrical indenter. Dumas and Baronet (6) produced a finite element solution to the circular indenter problem and found that at significant depths of depression, where the subsurface zone was entirely plastic the calculated pressure profile was fairly constant over most of the interface. At smaller depths of depression the pressure profile is not flat, but reaches a maximum at the centre where the pressure is approximately 50% greater than it is near the periphery of the dent. This suggests that for relatively small indentations the interfacial pressure is still increasing towards the centre of the dent, thus generating a radial extrusion force on the debris particle resulting in a surface tractive force on the bearing raceway. At greater depths of indentation the pressure profile flattens and the resultant surface traction force diminishes. Outside the indented zone where the deflections are entirely elastic, debris deformation will continue to occur and the pressure profile will be defined by the extrusion and elasticity equations as discussed in (4).

The residual stresses resulting from the indentation especially in the presence of a surface traction force may encourage fatigue crack initiation as postulated by Olver (7). In this paper a slipline field model based on that developed by Olver was used to estimate the plastic stresses in the subsurface and the resultant residual stresses on unloading.

After indentation the squashed debris particle is lost and the damaged surface repeatedly passes through the EHD contact. Modelling this process is not practicable at present, however as the film thickness is small compared to the elastic deformation of the surface and will probably be further reduced around the edges of the dent, a dry contact approximation may not be too unrealistic. Clearly, although the residual stress distribution is fixed relative to the dent the superimposed contact stress field will be changing during overrolling and the full history will be required for the life calculations.

1.1 Notation

A	constant
b	half-width of Hertzian contact, m
dB	incremental area of the cross-section of the ring, mm^2
e	life exponent
H	dimensionless film thickness
H_{00}	constant in dimensionless film thickness equation
k	yield shear stress, N/m^2
N	fatigue life
P	dimensionless pressure
R	reduced radius of curvature, m
S	probability of survival
x,x'	coordinates in rolling direction, m
X,X'	dimensionless coordinates
y	coordinate in axial direction, m
z	coordinate normal to the surface, m
z'	stress weighted average depth, mm
Ω	domain
Ω'	extended domain

2 SLIPLINE FIELD

In the slipline field analysis the debris indentation of the bearing surface is modelled by a rigid die impressing a rigid-plastic halfspace coupled with a traction force extending from the centreline. If the traction force is zero this reduces to the classical field for a rigid indenter as presented by Hill (9) in the deforming zone. In (7) Olver presented a solution for a single uni-directional traction force applied to the interface, figure 1. As the sliplines will not meet the interface at 45^0 an extra fan must be drawn in (region BCF). In the region DCFE the lines were calculated numerically by a finite difference technique proposed by Hill (9) and the elastic boundary (BE) was derived analytically by Johnson (8). If the traction force is made equal to zero the sliplines meet the surface at 45^0 and the field reduces to that proposed for a purely normal indentation.

If the traction force extends radially from the centre then the zone below the centreline will be plastic and the sliplines must meet at 90^0, figure 2. From symmetry a square section BEFG can be drawn and as before a fan is constructed between BC and BE. The sliplines in regions CDHE and EHIF can be calculated numerically by finite differences. Although the slip line field could be extended indefinitely the actual zone of plasticity will only exist directly below the indenter. Defining the elastic/plastic boundary exactly is not possible, but an approximation can be made by applying the interface pressures to an elastic halfspace and calculating where the yield shear stress is exceeded, as suggested by Olver (7). Significant plastic deformation will probably only occur in the region OADCB, whereas in the remainder of the field the material may be plastic but will not deform.

Having constructed the field the stresses may be calculated in the normal manner see (7). During unloading, a tensile normal force equal in magnitude to the plastic pressures is effectively applied to the interface. Consequently the residual stresses can be found by summing the stress distribution resulting from this elastic tensile force to the plastic stress distribution. The elastic stresses were calculated analytically as described by Ford (10). A map of the principal residual stress distributions for traction coefficients of 0 and 0.2 are shown in figures 3(a) and (b) and a contour map of the orthogonal residual shear stresses in figure (4). The effect of the surface shear stress is to reduce the tensile residual stresses immediately below the surface but to increase those below the dent shoulder.

3 DRY CONTACT EQUATIONS

To solve the dry contact problem it is necessary to find a pressure distribution P(X), that satisfies the film thickness equation (1), ie solve the integral equation (2) for the pressure P.

$$H(X) = 0 \qquad X\in\Omega \tag{1}$$

in the one dimensional case, this equation reads:

$$H_{00} + \frac{X^2}{2} - \frac{1}{\pi}\int_{\Omega} P(X')\,K(X,X')\,dX' = 0 \qquad X\in\Omega \tag{2}$$

where $K(X,X')=\ln|X-X'|$

Unfortunately the domain on which equation (1) holds is generally not known in advance. Therefore (1) is extended to (3) such that (1) holds in the original domain and $P(X)=0$ $X\in\Omega'$

$$\begin{array}{lll} P(X) > 0 & H(X) = 0 & X\in\Omega \\ P(X) = 0 & H(X) > 0 & X\in\Omega' \end{array} \tag{3}$$

A second condition that has to be satisfied is the force balance equation:

$$\int_{\Omega} P(X)\, dX = \frac{\pi}{2} \tag{4}$$

The interaction of these equations is identical to the lubricated contact case, see for instance (11).

In this dry contact analysis the film thickness is set to zero over the domain (Ω) and then only the pressure distribution has to be calculated. The problem has traditionally been solved by direct methods (Newton Raphson), resulting in long computing times for large problems. As a large number of solutions is required in applying the life model the computing time can become excessively long. To alleviate this problem an alternative iterative type solution was used which has been described in (12). Convergence was further accelerated by the application of novel multi-level techniques which have been applied previously to the solution of differential equations.

The computing time for the usual iterative solution tends to be dominated by the integral computation, so the solution time is proportional to order n^2 where n is the number of points, as can be seen in Table 1. However it is possible to reduce the computing time to order nlogn by the application of multilevel techniques, specifically Multilevel Multi-Integration (MLMI). The basic idea was given in (13) and worked out in detail in (12). As can be seen from Table 1, column 3 significant time savings can be obtained from level 6 onwards, so the approach is most useful for problems with many grid points. The calculation of the subsurface stresses is a task similar to the film thickness solution, when one considers only one particular value of the depth z, at a time. Plots of the dry contact surface pressure and associated subsurface orthogonal shear stress distribution are shown in figure 5.

4 LIFE PREDICTIONS

In the traditional Lundberg and Palmgren bearing fatigue life model, the probability of failure can be expressed in terms of the stressed volume and the magnitude and depth below the surface of the maximum orthogonal shear stress. In (2) Ioannides and Harris proposed a generalised model in which the stressed volume is divided into discrete volume elements in which the maximum stress is calculated according to some stress related fatigue criterion. In common with structural fatigue life predictions for steels in reversed bending or torsion, a threshold stress value is defined below which failure will not occur. Each element is weighted according to its depth below the surface and the probability of failure is expressed in terms of the integral of the elemental stresses over the entire volume.

A modified life criterion has been used to compute the L_{10} bearing lives in the present case. The maximum shear stress amplitude τ_a is calculated during the overrolling of the dent. In common with structural fatigue, the fatigue stress threshold τ_u is modified according to the absolute value of the shear stress. τ_u is assumed to remain unchanged if τ_{max} does not exceed the yield stress and to diminish linearly to zero for τ_{max} varying between τ_e and the fracture strength τ_f. As the crack might be expected to be created more easily in the presence of a tensile rather than compressive stress field, an additional hydrostatic weighting was included in the model. In this the critical stress τ_a was modified to $\tau_a + a.Hp$, where Hp is equal to the hydrostatic pressure and a is taken as a=0.3. Using these values, the probability of survival of the inner ring can be expressed as:

$$\ln\frac{1}{S} = AN^e \int_{B_r}\int_{l} \frac{(\max(\tau_a - \tau_u))^c}{z'^h}\, dl\, dB$$

The effective perturbation on the global pressure distribution by the dent will depend very much on the ratio of the dent width to the Hertz contact size. To assess this effect four dent/roller combinations were chosen. An artificial circular dent of 200 micron width and 3 micron depth was overrolled by rollers of 2, 4, 8 and 16mm radius. The overrolling of the dent is simulated using 9 different positions of the rolling element with respect to the defect, in order to pick up the maximum stresses. The position of the centre xc of the rolling element is given by:

$$xc = b\,(n-5)/2 \quad \text{for} \quad n=1,2,\ldots,9.$$

The stress history in each point is analysed with respect to these nine positions and then the life integral is calculated. As the number of positions in time and space are relatively small, 9 and 49x17 respectively, the values of the life integrals are rather jumpy and consequently the numerical results should be interpreted with care.

5 RESULTS

The life integrals were calculated for four different values of the reduced radius of curvature and for six different loads (corresponding to Hertzian pressures ranging from 2.0 to 3.3 GPa). Three cases were examined, a smooth raceway, a raceway with one dent and a raceway with one dent and associated residual stress field. The effect of these stress fields on predicted lives can be graphically expressed in terms of risk maps. In these a section of the x, z plane is drawn on a grid with the 'fatigue criterion' stress expressed as the y co-ordinate. Each map is normalised to the smooth case by a scaling factor. The smooth case risk map is shown in figure 6(a) where as expected the highest risk occurs at the position of the maximum orthogonal shear stress, 0.8b below the bearing surface. As the dent (figure 6(b)) and the dent plus residual stresses (figure 6(c)) are included the map is modified, particularly around the dent shoulders. As a result the scale factor, effectively a measure of the increased risk, increases dramatically, by a factor of almost 50 on the inclusion of the residual stresses.

The predicted lives for each dent/roller combination are plotted in figure 7. From this data an approximate map of relative lives can be constructed in terms of the dent size, contact size and roller radius of curvature (figure 8). As can be seen the life of the smooth raceway increases rapidly with decreasing load and the life generally increases with increasing radius. The influence of the dent without the associated residual stresses on life is minimal. A significant life reduction only occurs for the smallest radius under the three lightest loads.

This changes dramatically when the (tensile) residual stress field below the dent is taken into account. The residual stresses were obtained assuming no radial traction force at the interface, ie a flat pressure distribution. Under high loads the influence of the residual stress fields are relatively small, but as the load is reduced the lives decrease markedly relative to the smooth cases. As the radius of the contact is increased the influence of the residual stresses and of the dent geometry is diminished.

Inspection of the orthogonal shear stress contours (figure 9) helps explain why this might be so. At higher loads although the orthogonal shear stresses of the modified Hertzian field are of a higher magnitude, the maxima are situated well below the surface. The stress concentrations from the shoulder of the dent and particularly the tensile stresses of the residual stress field lie much more closely to the surface and cannot therefore combine with them to generate high values of the fatigue criterion. As the load is reduced the stress contours lie more closely to the surface and can combine with the tensile residual stresses to cause more damage.

The results should be interpreted with some caution because of the simplifications made and the coarse grid numerics and strict quantitative conclusions should not be made. However, the qualitative results are interesting enough to continue the research in this direction, incorporating more realistic dent shapes, residual stress fields and more accurate calculations on finer grids.

6 CONCLUSIONS

The effect of both dent size and subsurface residual stresses have been added to the original life reduction work. The slipline field analysis can only be realistically applied to deep transverse indentations where the assumptions of plane plastic strain can be justified and the dry contact analysis is probably only reasonable for relatively thin film conditions. However the trends in the life reduction factors are distinctive and the explanation for them would seem reasonable and applicable to any dent profile/roller combination. The most striking outcome is that where failure is initiated through surface indentation and associated residual stresses (and considerable evidence exists to suggest this is so) the expected lives may not increase with decreasing load as rapidly as would be predicted by conventional models. The reduction in expected lives is very sensitive to the size of dent in relation to the roller radius and this may well have important consequences in terms of critical particle size and safe and unsafe levels of filtration.

To be able to draw quantitative conclusions further research is needed that uses more realistic residual stress fields and finer grids in the life calculations.

7 ACKNOWLEDGEMENTS

We would like to thank Dr Andrew Olver for his helpful advice in this work and to register our gratitude to SKF-ERC, The Netherlands, who have sponsored this work, and to Dr Ian Leadbetter, Managing Director of SKF-ERC for permission to publish.

APPENDIX

References

[1] Webster, M. N., Ioannides, E. and Sayles, R. S., (1985), "The Effect of Topographical Defects on the Contact Stress and Fatigue Life in Rolling Element Bearings", Proceedings of the 12th Leeds-Lyon Symposium on Tribology, Lyon, Butterworths, Vol. 12, pp. 121-131.

[2] Lundberg, G. and Palmgren, A., (1947), "Dynamic Capacity of Rolling Bearings ", Acta Polytechnica, Mechanical Engineering series , Royal Academy of Engineering Sciences, Vol. 1, No 3, 7.

[3] Ioannides, E. and Harris, T. A., (1985), "A New Fatigue Life Model for Rolling Bearings ", ASME Journal of Lubrication Technology, , Vol. 107, pp. 367-378.

[4] Hamer, J. C., Sayles, R. S. and Ioannides, E., (1985), "Deformation Mechanisms and Stresses Created by 3rd Body Debris Contacts and Their Effects on Rolling Bearing Fatigue ", Proceedings of the 14th Leeds-Lyon Symposium on Tribology, Lyon, Butterworths, Vol. 14.

[5] Hamer, J. C., Sayles, R. S. and Ioannides, E., "Particle Deformation and Counterface Damage When Relatively Soft Particles are Squashed Between Hard Anvils", Trans ASME/STLE to be published .

[6] Dumas, G. and Baronet, C. N., (1971), "Elasto-plastic indentation of a half-space by a long rigid cylinder", International Journal of Mechanical Sciences, Vol. 13, 519.

[7] Olver, A. V., (1986), "Wear of Hard Steel in Lubricated, Rolling Contact", Phd Thesis, Imperial College.

[8] Olver, A. V., Spikes, H. A., Bower, A. and Johnson, K. L., (1986), "The Residual Stress Distribution in a Plastically Deformed Model Asperity", Wear, Vol. 107, pp. 151-174.

[9] Hill, R., (1950), "The Mathematical Theory of Plasticity ", Oxford University Press.

[10] Ford, H., (1963), "Advance Mechanics of Materials", Longmans .

[11] Lubrecht, A. A., "The Numerical Solution of the Elastohydrodynamically Lubricated Line and Point Contact Problem, Using Multigrid Techniques", Phd Thesis, Twente University,1987, The Netherlands.

[12] Brandt, A. and Lubrecht, A. A., "Multilevel Multi-Integration and Fast Solution of Integral Equations", to be published in the Journal of Computational Physics.
[13] Brandt, A., "Multigrid Techniques: 1984 Guide with Applications to Fluid Dynamics", monograph available as GMD studien 85 from GMD postfach 1240 , Schloss Birlinghofen D-5205 St. Augustin 1 BRD.

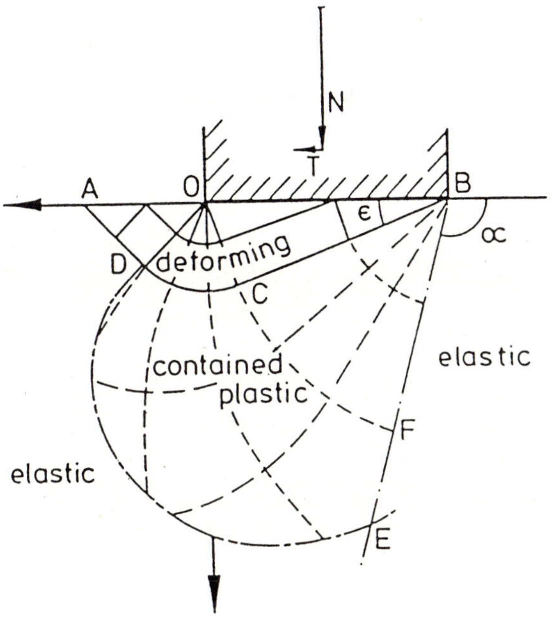

Figure 1 Olver slipline field for sliding asperity.

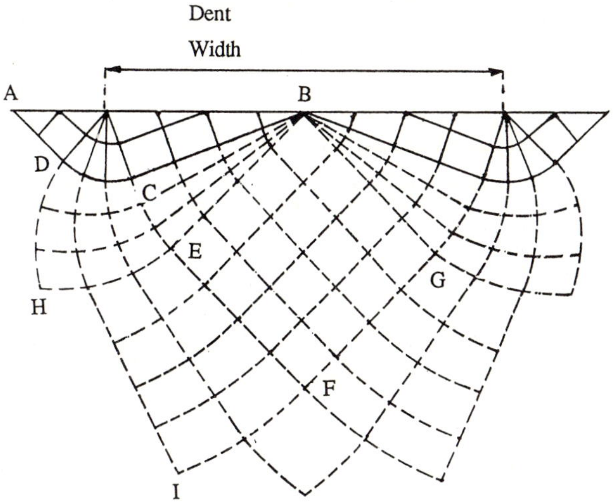

Figure 2 Slipline field for normal indenter plus radial traction force.

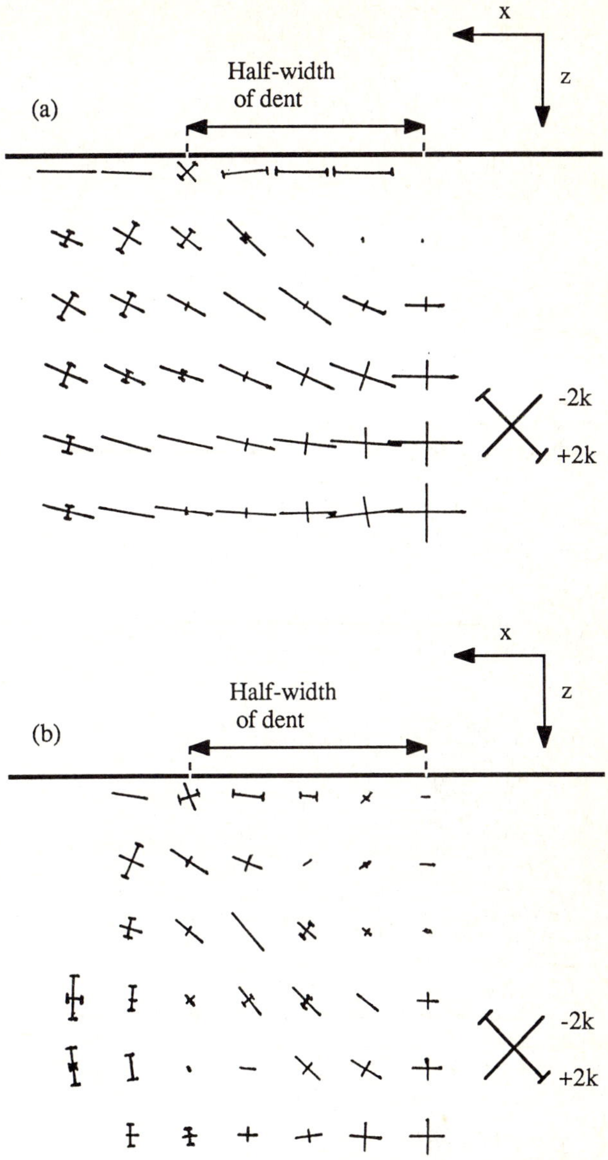

Figure 3 Residual stress distribution for radial traction coefficient (μ) of: (a) $\mu = 0$ and (b) $\mu = 0.2$.

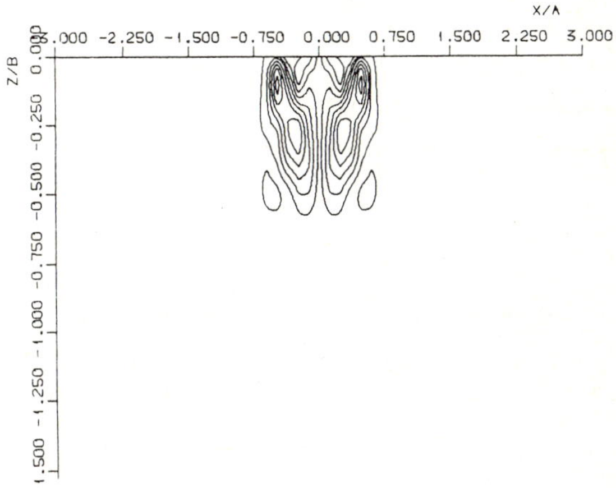

Figure 4 Residual orthogonal shear stress (τ_{xz}) contour map.

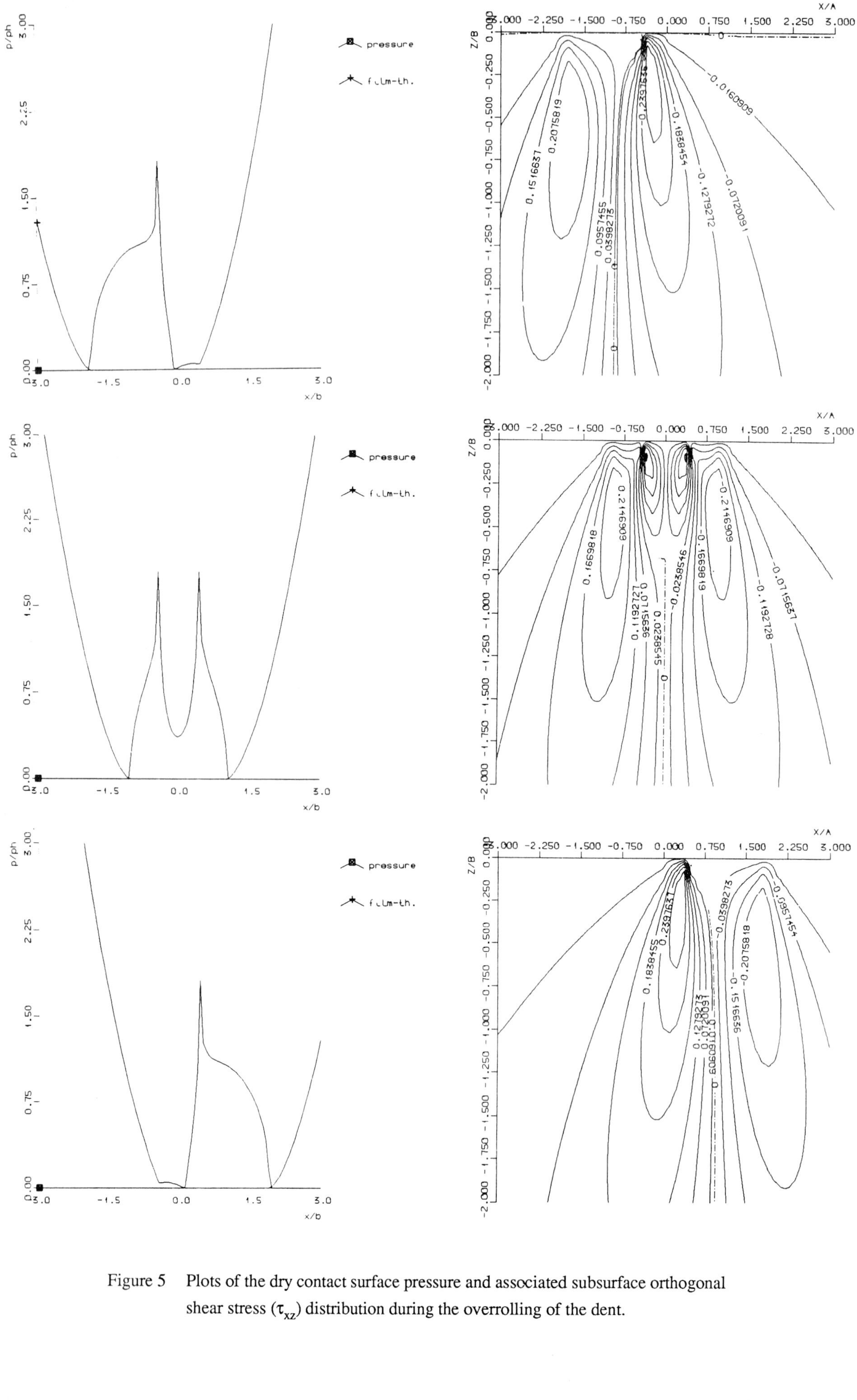

Figure 5 Plots of the dry contact surface pressure and associated subsurface orthogonal shear stress (τ_{xz}) distribution during the overrolling of the dent.

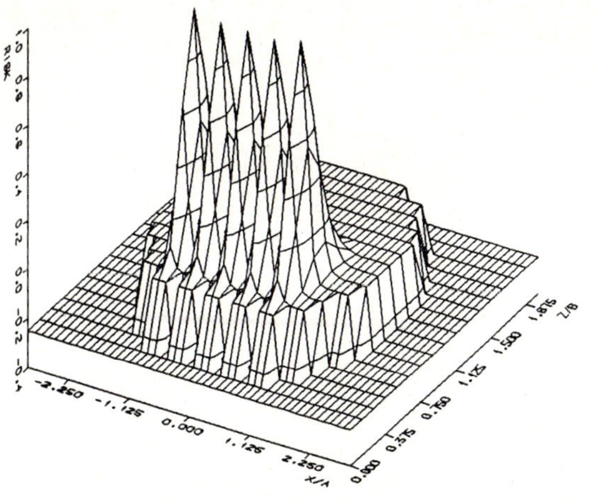

Figure 6(a) Risk map of smooth surface.
Scale factor=1

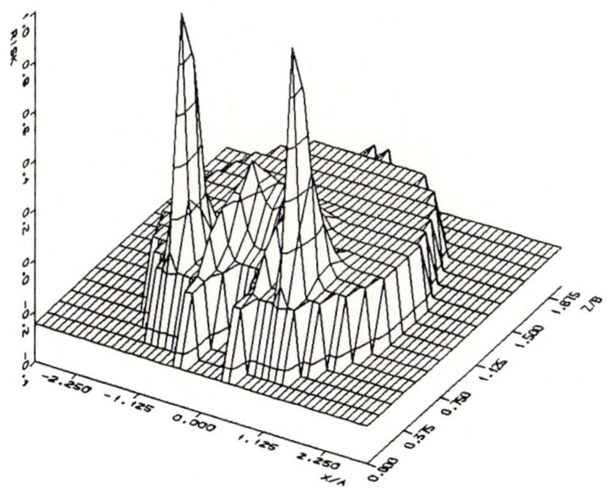

Figure 6(b) Risk map of indented surface.
Scale factor=5.7

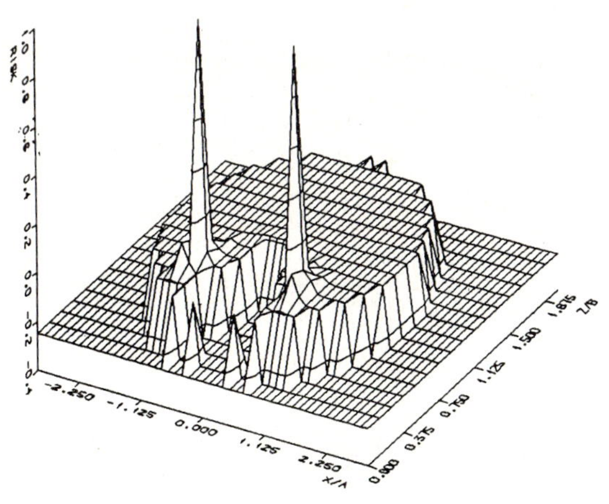

Figure 6(c) Risk map of indented surface plus residual stresses.
Scale factor=49.7

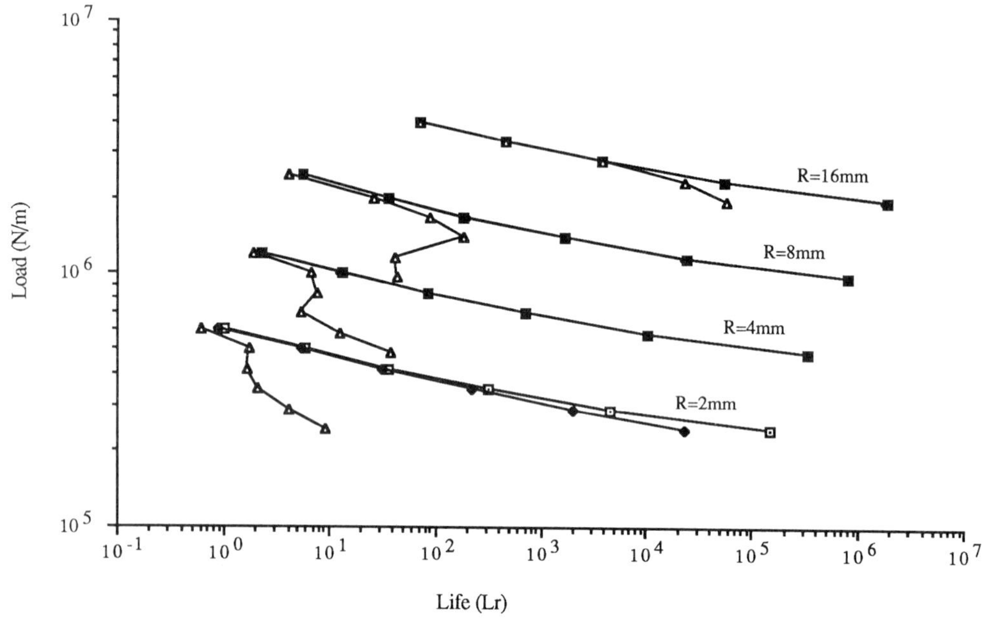

Figure 7 Normalised lives (L_r) versus load at differing radii of curvature (R) for; ▫ smooth surface, ♦ indented surface, ▲ indented surface plus residual stresses.

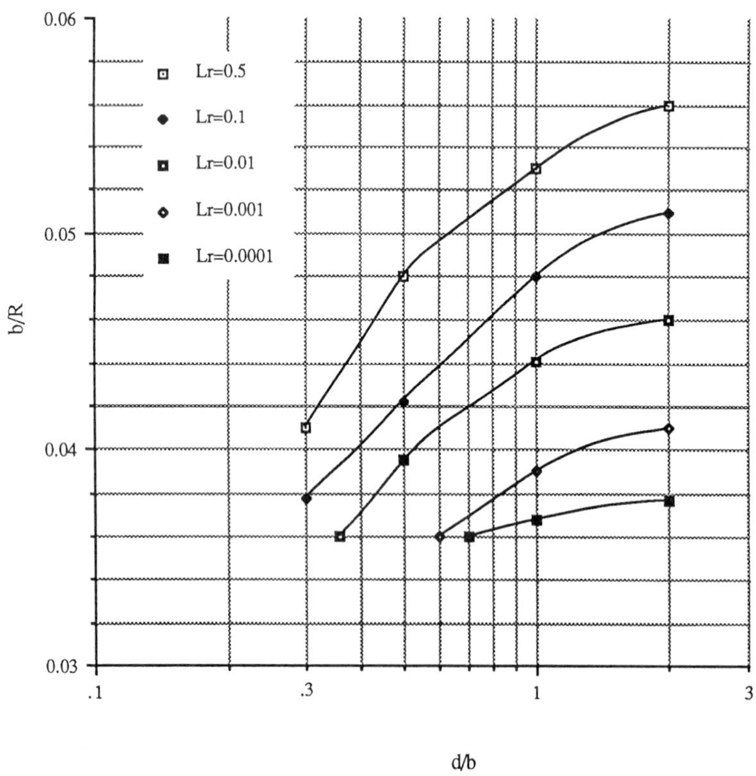

Figure 8 Map of relative lives (L_r) of indented surfaces to smooth surfaces as a function of b (Hertz contact), R (reduced radius) and d (dent size).

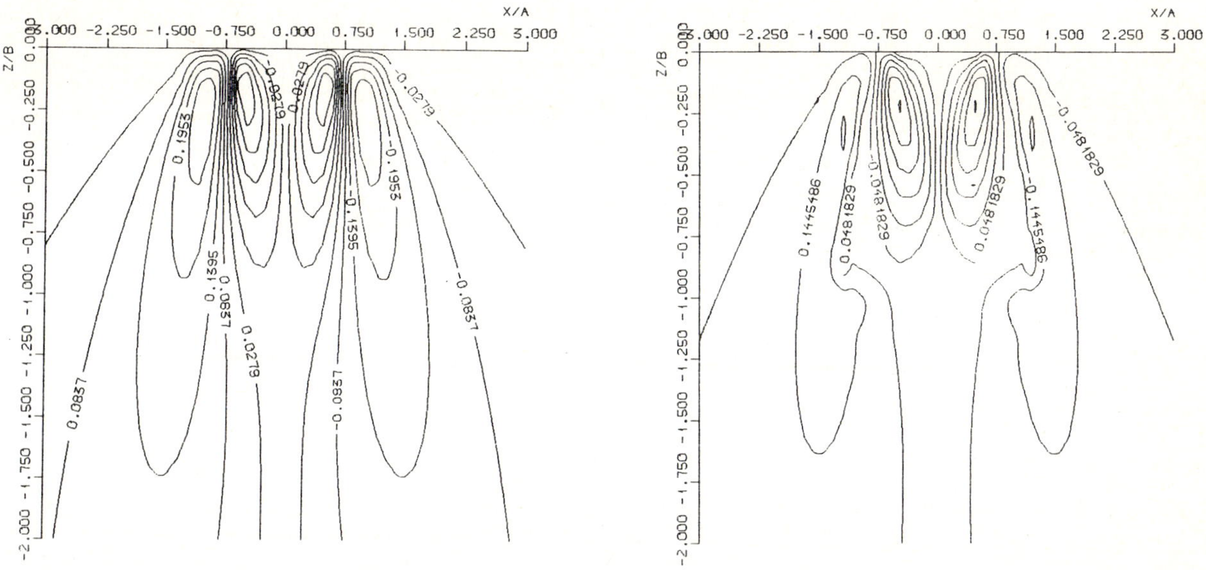

No residual stresses   Plus residual stresses

Figure 9   Plots of maximum orthogonal shear stress (τ_{xz}).

L	n	time	time *
2	9	0.8	-
3	17	1.5	-
4	33	2.4	-
5	65	3.9	-
6	129	8.4	6.2
7	257	23.0	12.3
8	513	79.0	24.3
9	1025	306.0	48.4

Table 1:   Computing time as a function of the level (L), the number of points n, in seconds on a VAX 785, for the smooth dry line contact problem, with (*) and without MLMI.

Paper VII(iv)

Debris denting - The associated residual stresses and their effect on the fatigue life of rolling bearing: An FEM analysis

C. N. Ko and E. Ioannides

The squashing of contaminant ductile particles in rolling bearings was simulated using FEM (Finite Element Methods). Calculated dent shapes from this elastic-plastic analysis are in good agreement with experimentally measured dent shapes. The calculated residual stress fields associated with the dents are discussed and used to estimate rolling bearing life reduction.

1 INTRODUCTION

Evaluation of rolling contact fatigue life has historically been based on elastic stress analysis. The elastic assumption is no longer valid when the stress level exceeds the yield limit. High stresses can be caused by high contact loads or by contacting local geometrical features such as dents, scratches or other defects which may be present in the rolling contacts. When the stresses become high enough, plastic flow will occur and cause subsequent permanent deformations. This in turn will change the contact conformity, produce residual stress fields, alter the material properties by strain hardening. All these effects will naturally contribute to the endurance ability of the machine element and should be considered in the rolling contact fatigue analysis.

The present investigation is focused on denting caused by debris and thus associated with contamination in rolling bearings. Contamination is highly detrimental to bearing operations and it is strongly connected to local plastic deformation. Overrolled contaminant particles form dents on the raceways which cause surface-initiated spalling and thus shorten bearing lives. It has been demonstrated by Hamer et al [1] that even soft particles can cause dents if they are large enough. A zone of plastically deformed material is generated under the dent and the localized plastic strains in the raceway will subsequently produce residual stresses.

Two major effects on fatigue initiation caused by the denting process can be identified: (i) the dents generated will become local stress raisers during overrolling, and (ii) local self-equilibrating residual stress fields will develop in the material. There has been a suggestion that the existence of tensile residual stresses thus generated will have a significant influence on crack initiation [2], and ultimately on fatigue life. The primary objective of this investigation is to model with the use of FEM the deformation process leading to the formation of a dent. This allows the calculation of residual stresses which can be subsequently used in the estimates of the fatigue lives of raceways with dents.

In previous work, researchers have also calculated the distribution of residual stress fields resulting from rigid indentors pressed onto a finite or infinite half space. However, the indentor solution due to Hill [3] and also Johnson's cavity solution [4] are quite limited as they only apply to 2-dimensional problems. Moreover, such solutions depend crucially on the positioning of the elastic/plastic boundary. Such limitations are not necessary in FEM analyses. The present analysis was carried out using the multi-purpose computer program ABAQUS [5-7]. In this respect the present work resembles other FEM analyses of contact problems [8,9], but it differs from them in that the contact pressure profiles used in the generation of the dents have been extracted from the work of Hamer et al [1] and are applicable to particles undergoing heavy plastic deformation.

1.1 Notation

a	Contact width of the axisymmetric soft indentor [mm]
a'	Contact width of the axisymmetric rigid indentor [mm]
b	Contact width for ideal line contact [mm]
E	Young's Modulus [N/mm^2]
p_H	Maximum Hertzian pressure [N/mm^2]
P_y	Load to initiate yield [N]
r,z,θ	Cylindrical coordinate system for the axisymmetric models
R	Rigid indentor radius [mm]
x,y,z	Rectangular coordinate system for the plane strain model

Y	Yield stress (Plastic limit) [N/mm^2]
σ_{rr}	Radial stress [N/mm^2]
σ_{zz}	Axial stress [N/mm^2]
$\sigma_{\theta\theta}$	Circumferential stress [N/mm^2]
τ_{rz}	Shear stress in the rz plane [N/mm^2]
ν	Poisson's Ratio

2 METHOD OF ANALYSIS

Most of the debris deformation takes place in the inlet of an EHD contact where both the convergence angle and the surface separation between the rolling element and the raceway of a bearing are very small. The simplification of treating the two surfaces as parallel can therefore be made. Under these conditions a simple extrusion solution to the problem of a small sphere denting two flat platens has been obtained in the past [1]. Dent shapes calculated from this extrusion method agree favourably with dent shapes observed in experiments conducted by the same authors [10]. Similar simplifications were exploited in the present work where FEM were used in axisymmetric and plane strain cases, as shown schematically in Figures 1 and 2 respectively.

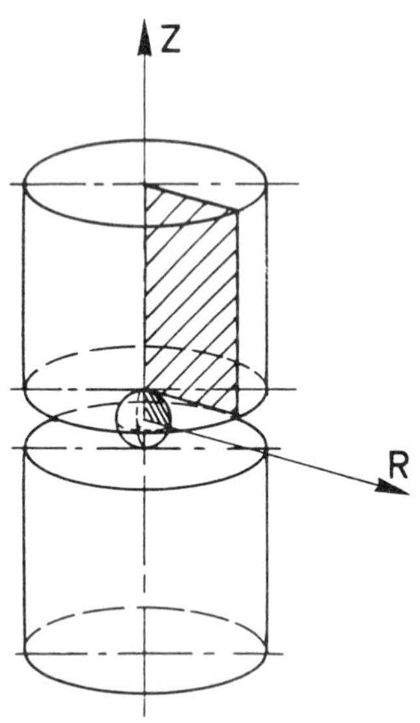

Figure 1 Axisymmetric model.

In the present work, however, instead of solving the complete extrusion problem of three bodies, the contact pressure as derived in [1] was applied directly to one of the platens (both in the plane strain and the axisymmetric cases). Furthermore, symmetries were exploited as indicated in Figures 1 and 2 so that the computational effort was kept to a minimum.

The low carbon steel platens of hardness 340 Vickers are modelled as isotropic, elastic-perfectly plastic

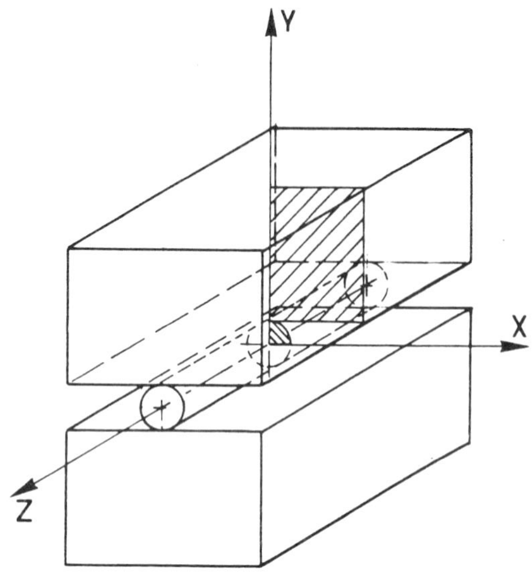

Figure 2 Plane strain model.

bodies with E = 207000 MPa, Y = 1000 MPa, and ν = 0.3, whilst the material of the particle is assumed to be perfectly plastic. As indicated earlier on, only the interfacial pressure between the particle and the platen was used to simulate the denting, which was obtained from the work of Hamer et al [10]. An incremental loading procedure was followed during which this pressure was increased in steps until its full magnitude was reached. No attempt was made to follow the real loading history experienced in the actual squashing process, and the frictional stresses were ignored.

Finally, in the fatigue life calculation the combined stress fields of the roller-induced stresses and the residual stresses were used. It should be noted that in this work no secondary yielding is allowed for as this will involve additional time-consuming elastic-plastic calculations. Therefore, when the combined stresses exceed the plastic limit the life estimates should be interpreted with caution and used only as guidelines.

3 FINITE ELEMENT ANALYSIS

As indiciated previously the computational effort can be substantially reduced by appropriate use of symmetry. In the case of spherical particles, axisymmetric models are used, Figure 1, whilst in the case of cylindrical particles plane strain models are used instead, involving only half of one of the platens, Figure 2.

Three different FEM models were used in the present analysis. Two of them are axisymmetric with different mesh densities, AXFINE (fine) and AXCOARSE (coarse), and a plane strain model DENT for life calculations. The axisymmetric models were intended for comparison with the measurements obtained in [10], while the plane strain models were used in fatigue life predictions. The details of these models are given in Table 1.

TABLE 1: Details of the Finite Element Models

MODEL	DIMENSIONS (widthxheight)	ELEMENT TYPE	NO. OF ELEMENTS	NO. OF NODES	DOF*
AXFINE Fig.(3a)	10mm x 10mm	8 noded parabolic axisymmetric with reduced integration	2240	7021	14042
AXCOARSE Fig.(3b)	10mm x 10mm	4 noded bilinear axisymmetric with full integration	1521	1600	3200
DENT Fig.(3c)	100mm x 230mm	4 noded bilinear plane strain with full integration	1645	1728	3456

* DOF = Total number of degrees of freedom

For the elastic-plastic calculations the von Mises yield criterion was used, together with the Prandtl-Reuss flow rule [5]. The boundaries of the platens were set sufficiently far away to prevent spurious stresses from developing at the boundaries which will influence the distribution of the residual stress fields. The dimensions of the models have therefore been carefully chosen, as shown in Table 1 above. Axis or planes of symmetry, as well as supporting boundaries, were restricted only in the perpendicular directions, Figure 3.

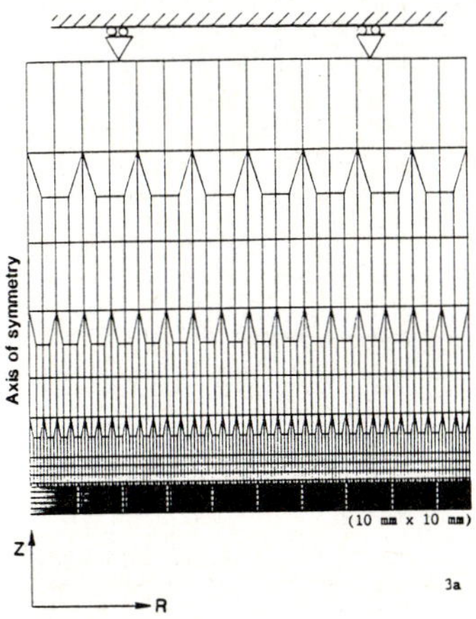

Figure 3
FEM mesh for the three models:
(a) AXFINE,
(b) AXCOARSE,
(c) DENT.

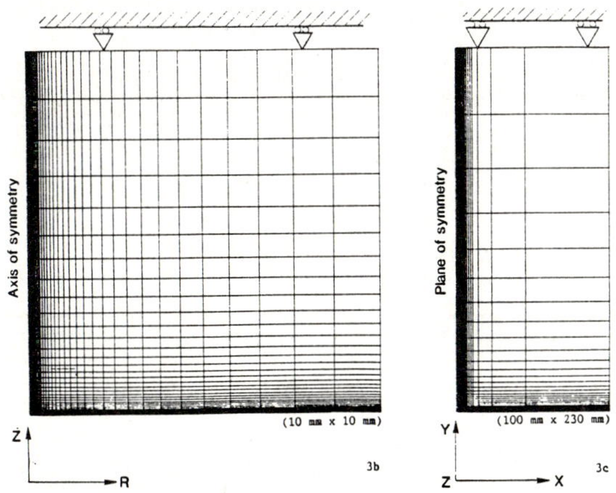

The platen was first loaded by progressive increase of pressure until its full value and then unloaded by complete removal of the pressure. The pressure loading for the axisymmetric models has a peak pressure of 2790 MPa and its distribution is shown in Figure 4a. For the plane strain model, where life evaluations were to be carried out, there are 4 different pressure loadings (see Figure 4b) to simulate 4 different denting processes.

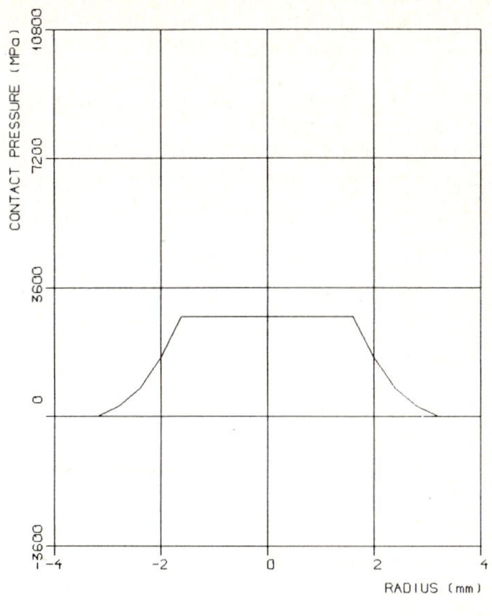

4a

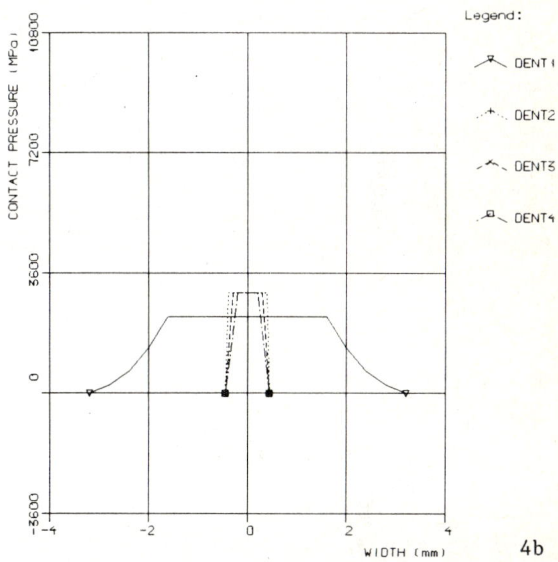

4b

Figure 4 Contact pressure distribution for the:
(a) Axisymmetric models;
(b) Plane strain dent life models.

4 RESULTS AND COMPARISONS

Prior to the elastic-perfectly plastic calculations, the suitability of the FEM mesh was checked by reproducing the Hertzian solutions for both the axisymmetric and plane strain models. In both cases a particle of 2 mm in diameter was used to simulate the contact. The maximum contact pressure difference between the FEM and the Hertzian solutions was 7%.

After these preliminary numerical tests the axisymmetric models were used to simulate the experiments of Hamer et al [10] where actual dents were produced and measured by squashing particles. Before comparing with the experimental results, the two axisymmetric models (AXFINE and AXCOARSE) were first compared and checked against each other. The stresses and the deformations predicted by AXFINE and AXCOARSE are almost identical, except from differences (approximately 13%) in high stress gradient regions where AXFINE seems to be more capable of picking up the peak stresses. This is expected because AXFINE has a finer mesh as well as higher order elements. In addition, the elastic deformations obtained by AXFINE and AXCOARSE are practically the same, being 47.2 μm and 47.1 μm respectively. The dent depth predicted by AXFINE is slightly higher than that of AXCOARSE; the difference is about 3.4%. This is because AXFINE repesents a more flexible structure than AXCOARSE as it has more degrees of freedom. On the whole, it can be concluded that these 2 axisymmetric models are converging towards the same results and are thus reliable.

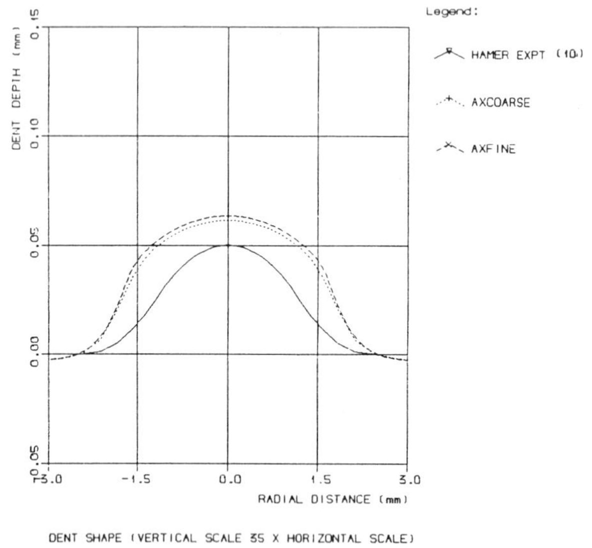

Figure 5 Comparisons of dent depths: experimental [10], AXFINE, AXCOARSE.

From Figure 5 it can be seen that the calculated dent depths are very close to the value measured experimentally. The depths calculated using AXFINE and AXCOARSE were 63.5 μm and 61.4 μm respectively. The measured value was estimated from Figure 8(c) in [10] to be 50 μm. Moreover, from the same figure it can be seen that the measured diameter of the dent is approximately 2.3 mm, which agrees well with the calculated values of 2.49 mm.

The difference of 20% in dent depth between the calculation and the measurements can be attributed mainly to the lack of strain hardening in the model (elastic-perfectly plastic). In reality, strain hardening will take place, inhibiting the plastic flow and thus reducing the dent depth.

Next, the predictions from the FEM models were compared with the analytical indentor solution [3], cavity solutions [4] and other FEM indentor solutions [8,9]. The main difference between the present work and the indentor solutions is that the shape of the pressure distribution caused by the plastically deforming particle, Figure 4, is different from the one caused by the rigid indentor.

In order to compare the subsurface stress and strain fields with other solutions [4,8,9], the indentor radius must be known. The shape of the dent obtained by FEM analysis appears substantially spherical, and consequently the dent can be compared to one caused by a rigid sphere with a certain radius. From the deformed shape of the platen under full load, Figure 6, the radius of

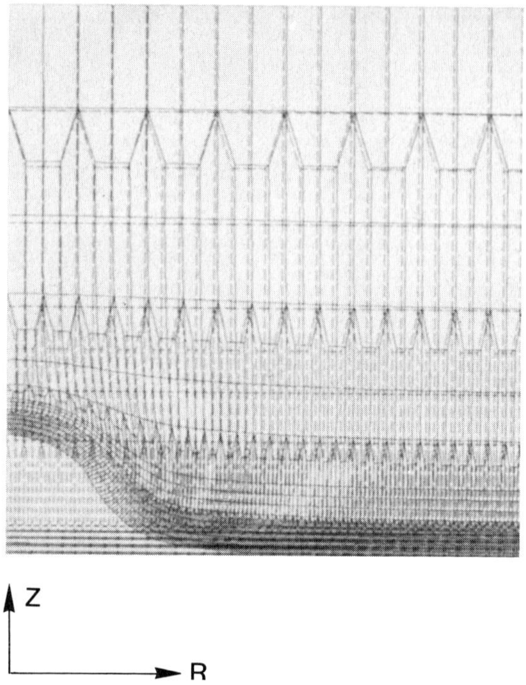

Figure 6 Deformed and undeformed shapes of AXFINE under full load (magnification factor = 20).

a rigid indentor can be estimated by curve fitting of the dent shape with a sphere. The radius of this sphere was approximately 33 mm (with a correlation coefficient of 0.98).

For axisymmetric contact of solids of revolution, the load to initiate yield (P_y) according to the von Mises criterion is:

$$P_y = \frac{\pi^3 R^2 (1.6Y)^3}{6 E^{*2}}$$

where R = indentor radius
Y = yield limit of the contacting bodies
E* = $E/(2(1-\nu^2))$ where E = Young's Modulus of the contacting bodies; ν = Poisson's ratio of the contacting bodies

In our axisymmetric analyses, P_y is 1782 N. The applied load is 44000 N, which corresponds to a modest 24.7 P_y. According to Figure 6.17 in [16], we fall well within the early part of the elastic-plastic regime, which implies that the effects of elastic strains cannot be ignored. This is because the elastic strains are felt over a much bigger domain than the plastic ones. Moreover, the ratio between the elastic and the plastic strains is between 10 to 30%, suggesting that the rigidly-perfectly plastic material model may not be appropriate for the current load level. This further confirms the limitations of the rigid-plastic assumption of the indentor solution of Hill [3] and the cavity solution of Johnson [4].

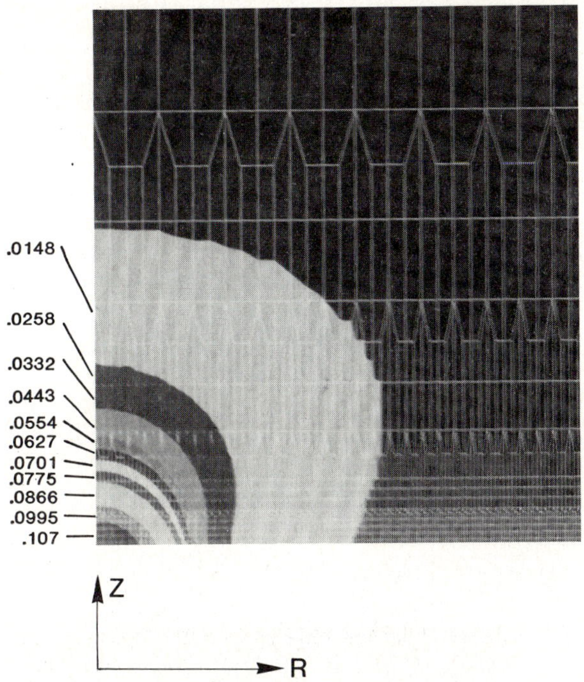

Figure 7 Total displacement (mm) distribution of AXFINE under full load (magnification factor = 1).

From Figure 7 it can be seen that the displacements produced under load are approximately radial from the bottom of the dent, with roughly hemi-spherical contours of equal strain. The results match the observations of Samuels and Mulhearn [11]. These observations were the basis of the cavity solution which was later developed by Marsh [12] and Johnson [4]. One of the major difficulties in the cavity problem, as mentioned before, is to find the elastic-plastic boundary in the body. In contrast, this boundary is obtained as part of the FEM solution, as can be seen in Figure 8. Also, from Figure 9 it can be seen that there is a spherical layer of elastic material with rather high elastic hoop strains surrounding the plastic core (region A in Figure 9). The magnitude of such circumferential elastic strain is found to be between 0.00156 and 0.00257 and the corresponding stress is between 122 and 303 MPa.

The elastic deformation obtained by the current FEM models is 47 μm. The total deformation due to a rigid indentor with a radius of 33 mm on an elastic half space under the same load (44000 N) is found to be 86 μm, and the corresponding maximum contact pressure 4600 MPa. In reality, the maximum contact pressure should be less since the platen is elastic-plastic. According to Hardy et al [8], the maximum pressure corresponding to a load of 24.7 P_y should amount to 2310 MPa which is 20% less than our maximum contact pressure (2790 MPa). However, the contact pressure distribution of a rigid indentor is much more uniform and looks rather like a rectangular distribution. Thus, in order to keep the same contact load of 44000 N, the contact width (a′)

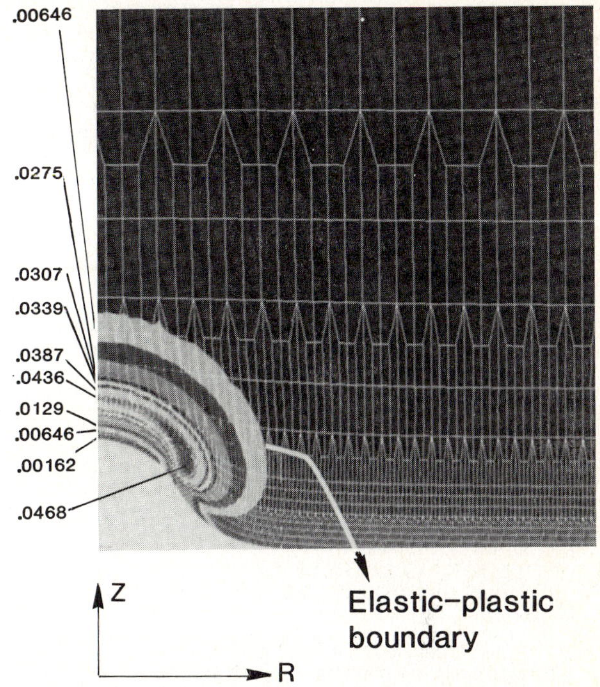

Figure 8 Plastic octahedral shear strain of AXFINE under full load (magnification factor = 20).

for the rigid indentor has to be smaller and should be equivalent to 0.77a, where a is the contact width of our soft indentor. Again according to Hardy et al [8], the depth of the plastic core is estimated to be 6a′ or 4.7a, whereas the present model's prediction is 1.3a. The

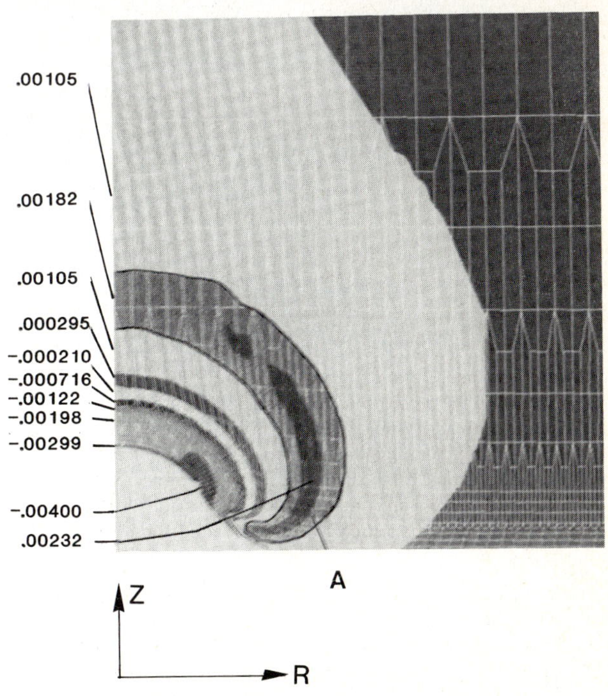

Figure 9 Elastic circumferential strain of AXFINE under full load (magnification factor = 20).

difference is attributed to the rectangular shape of the pressure distribution imposed by the rigid indentor. Consequently, the amount of plastic flow developed in the core under the rigid indentor will be larger and hence the plastic core itself will be larger. Additional supporting evidence has been obtained from the denting processes simulated in the fatigue life calculations, the detailed results of which will be discussed in the next section. It was there observed that the dent depth as an indicator of the size of the plastic core depends mainly on the shape of the pressure distribution and not directly on the total contact load.

When the plastic limit is first exceeded, the plastic zone is small and fully contained by the elastic material surrounding it. The elastic strains and the plastic strains are therefore expected to be of the same order of magnitude. When the load is further increased, the pressure beneath the indentor will increase to accommodate the necessary expansion. Eventually the pressure inside the plastic zone will become so intense that plastically flowing material will break through the free surfaces near the edge of the indentor. From Figure 10 it can be seen that a substantial amount of plastic strain is accumulated at the edge of the contact, even for the present moderate load.

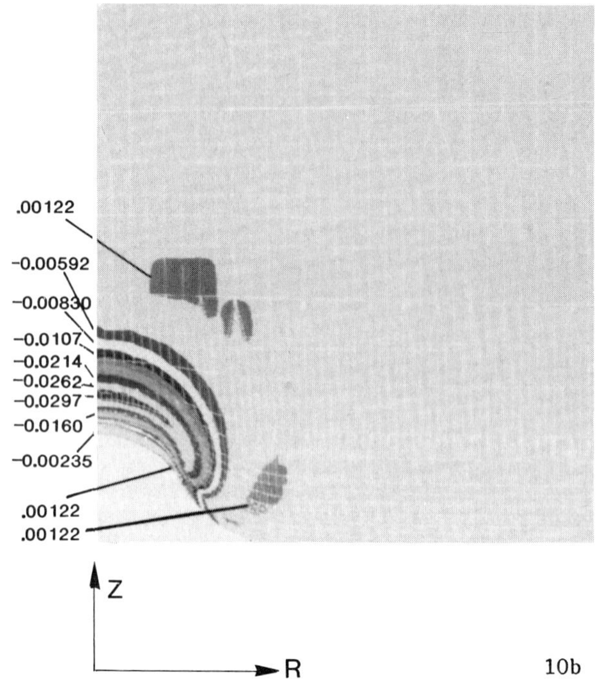

10b

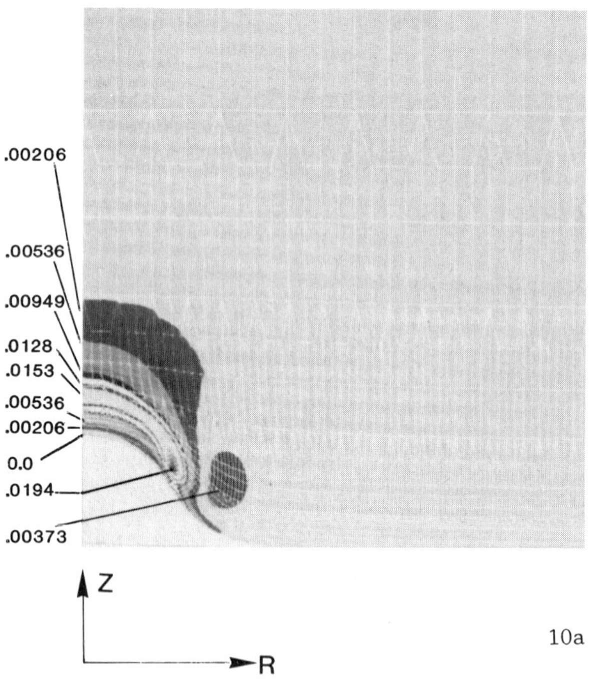

10a

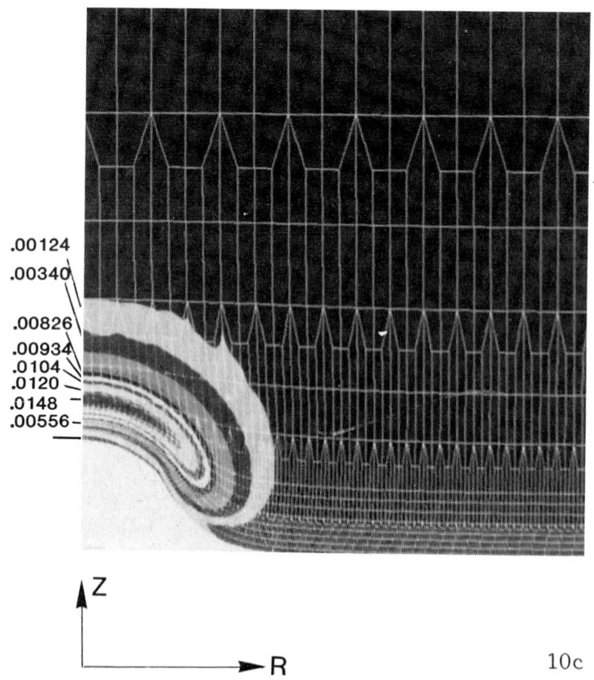

10c

Figure 10 Plastic strains of AXFINE under full load: (a) radial, (b) axial, (c) circumferential (magnification factor = 20).

Along the axis of symmetry, σ_{rr} is found to be equal to $\sigma_{\theta\theta}$, and τ_{rz} is found to be zero. These results are not surprising because of symmetry. Similar results have been obtained by Hardy [8] and Sinclair [9] in their FEM analyses, as well as by Johnson in his cavity solutions [4]. In addition, σ_{zz} is found to be zero at the bottom of the dent in the unloaded state. This is expected as equilibrium has to be maintained at the centre of the dent.

The distributions of σ_{rr} and σ_{zz} along the axis are shown in Figure 11. Within a thin layer of 0.05a below the surface, the radial stress is found to change sharply from -2500 MPa to -1800 MPa when the platen is under full load. This stress gradient is only slightly reduced when the platen becomes unloaded and the residual radial stress varies from -446 MPa to 40 MPa. This suggests that the indentation creates

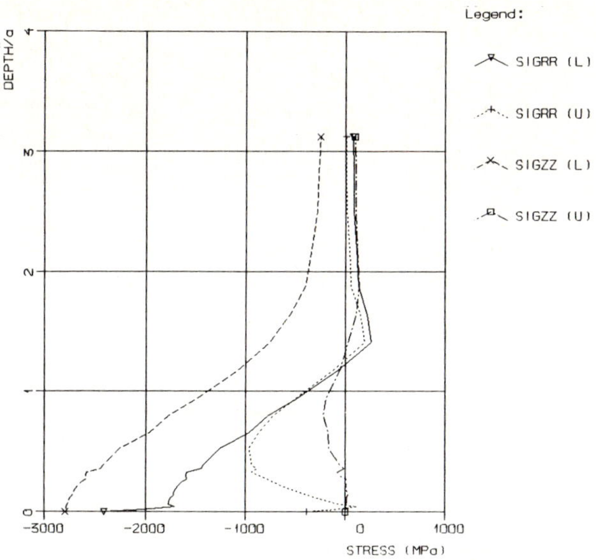

Figure 11 Distribution of σ_{zz} and σ_{rr} against the axis of symmetry of AXFINE for both loaded (L) and unloaded (U) states.

rather high compressive residual radial stresses near the dent bottom, in a layer very close to the surface. Moreover, in the residual state, small but tensile radial residual stresses are found at two locations along the axis of symmetry. In these locations, sharp stress gradients are present which are similar to the ones obtained in Hardy [8] but not in Sinclair [9]. Such residual stress patterns closely resemble the axial stress distribution of an elastic-plastic beam after severe bending.

5 EFFECTS OF RESIDUAL STRESSES ON ROLLING CONTACT FATIGUE

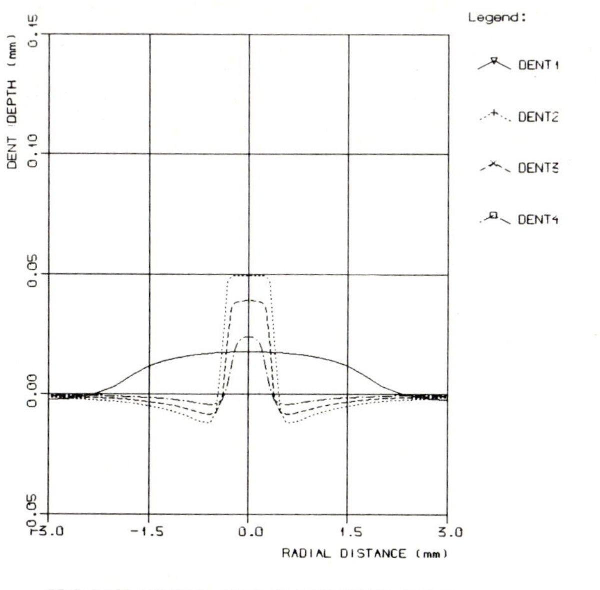

Figure 12 Dent shapes for fatigue life predictions.

Only a plane strain model with 4 noded elements is used in the investigation of the effects of residual stresses on rolling contact fatigue because of CPU time constraints. Thus various denting pressure distributions (see Figure 4b) have been used to simulate the formation of different dents and their effects on fatigue life. The corresponding dents are shown in Figure 12.

A previously published fatigue life model [13] which was recently extended [14] was used to evaluate bearing lives. For this evaluation the residual stress field was added to the stress field induced by the load as a roller passes over the dent. This is the same procedure as that used in [15] but now the residual stresses are calculated from the FEM analysis rather than from the simpler slipline method of [15]. As in [15], no secondary yielding is allowed and therefore the life calculations are only indicative and should be treated with care when large combined stresses are encountered. For the fatigue evaluation, a roller with a reduced radius of 8.0 mm (p_H = 3000 MPa, b = 0.425 mm) was selected. The results of the present fatigue life calculations are summarized in Table 2.

TABLE 2: Fatigue life ratios of raceway with dents to raceway without dents.

	NO RESIDUAL STRESS	WITH RESIDUAL STRESS
DENT1	1.0	1.0
DENT2	0.0013	0.0013
DENT3	0.0044	0.0052
DENT4	0.024	0.12

In this table the fatigue lives of the four dents are shown under two conditions: (i) when only the geometry of the dent is taken into account (no residual stresses), and (ii) when the residual stresses are considered together with the geometry of the dent. It can be seen in this table that DENT1 with its very smooth transitional curvature at the edges does not affect the life predictions as compared to the smooth case. In contrast, DENT2 with very pronounced bulging at the edges reduces life by three orders of magnitude and this reduction remains unaffected by the introduction of the residual stresses.

The geometry of the dent is therefore the cause of the life reduction in this calculation. (The very high stress which develops during overrolling will produce secondary yielding and therefore the present life estimates should be treated with care.) From the same table it can also be seen that for DENT3 the residual stresses improve the life slightly whilst for DENT4 this improvement is substantial.

This result contrasts with the findings of [15] where the reduction in life was attributed to the residual stresses rather than to the roller stresses. It can be generally stated that the life reduction of dented raceways can be attributed to two effects: the residual stresses generated during the denting process and the higher stresses generated when a roller passes over the dent. The contribution of each of these two effects depends on the particular case and cannot be generalised. In [15] the tensile residual stresses calculated near the surface are higher than the ones calculated here, and furthermore the shallowness of the assumed dent shape (a circular arc) makes the residual stresses the dominant feature in the life calculations. In contrast, here we have calculated smaller tensile residual stresses and more realistic dent shapes with bulging edges, which may account for the differences in life reduction.

Finally, as the roller-induced stresses (which depend on the size and the load of the bearing) are superposed on the residual stresses, it is clear that the importance for the life predictions shifts from these individual stress fields to their combination. For example, hard particles (which may more closely resemble rigid indentors) may produce dents and residual stress distributions which are different from the ones produced by softer particles, and consequently can be more detrimental to life. The present solution methods can provide answers to these important questions relating to bearing performance in practical operating conditions.

6 CONCLUSIONS

The current elastic-perfectly plastic FEM model has been found to agree well with experiments in terms of both dent depths and diameter. The small differences between calculations and measurements could be improved if strain hardening were introduced into the calculations. The calculated residual stresses are in broad agreement with previous solutions which indicated the presence of tensile residual stresses surrounding the plastic core, in a central region underneath the dent, as well as in small regions of high intensity under the edges of the dents.

A dependence of fatigue life reduction on both the geometry of the dent and the residual stress is indicated by the fatigue life calculations of the dents studied. The relative importance of these two causes appears to vary from case to case. It is therefore of interest to establish in general terms the effect of the hardness of the debris particle, its size, and of the applied load and the size of the rolling elements. This can point the way to the filtration requirements of particular applications and to the bearing material requirements when contamination cannot be prevented.

These aspects of debris denting, the modification of dent geometry and the associated residual stresses during overrolling, as well as the extension of life calculations to 3-dimensional dents are subjects of continuing research into the effects of contamination.

7 ACKNOWLEDGEMENTS

The authors would like to thank Dr. I.K. Leadbetter, Managing Director of the SKF Engineering and Research Centre, for permission to publish these results. The authors are also grateful to Hibbit, Karlsson & Sorensen, Inc. for making available the newest version of their multi-purpose finite element program ABAQUS. Thanks are also due to Dr. A. Lubrecht for his help in performing the life calculations.

REFERENCES

[1] HAMER, J.C., SAYLES, R.S. and IOANNIDES, E., "Deformation mechanisms and stresses created by 3rd body debris contacts and their effects on rolling bearing fatigue", Proc. of the 14th Leeds-Lyon Symposium on Tribology, Lyon, 1987. Butterworths, Vol.14.

[2] OLVER, A.V., SPIKES, H.A., BOWER, A. and JOHNSON, K.L., "The residual stress distribution in a plastically deformed model asperity", Wear, Vol. 107, pp.151-174 (1986).

[3] HILL, R., "The mathematical theory of plasticity", Oxford University Press.

[4] JOHNSON, K.L., "The correlation of indentation experiments", J. of Mechanics and Physics of Solids.

[5] HIBBIT, H.D., KARLSSON, B. and SORENSEN, P., "ABAQUS Theory Manual", Version 4.6, Providence, Rhode Island, USA (1987).

[6] HIBBIT, H.D., KARLSSON, B. and SORENSEN, P., '"ABAQUS User's Manual", Version 4.6, Providence, Rhode Island, USA (1987).

[7] HIBBIT, H.D., KARLSSON, B. and SORENSEN, P., '"ABAQUS Example Manual", Version 4.6, Providence, Rhode Island, USA (1987).

[8] HARDY, C., BARONET, C.N. and TORDION, G.V., "Elastoplastic indentation of a half-space by a rigid sphere", J. of Numerical Methods in Engineering, 3,451 (1971).

[9] SINCLAIR, G.B., FOLLANSBEE, P.S. and JOHNSON, K.L., "Quasi-static normal indentation of an elasto-plastic half space by a rigid sphere - II. Results", Int. J. of Solids and Structure, Vol.21, No.8, (1985).

[10] HAMER, J.C., SAYLES, R.S. and IOANNIDES, E., "Particle deformation and counterface damage when relatively soft particles are squashed between hard anvils", J. of Tribology, 110, 1988 (to appear).

[11] SAMUELS, L.E. and MULHEARN, T.O., "The deformed zone associated with indentation hardness impressions", J. of Mechanics and Physics of Solids, 5,125 (1956).

[12] MARSH, D.M., "Plastic flow in glass", Proc. of Royal Society, A279, 420 (1964).

[13] IOANNIDES, E. and HARRIS, T., "A new fatigue life model for rolling bearings", ASME J. of Lubrication Technology, Vol. 107, pp.367-378 (1985).

[14] IOANNIDES, E., JACOBSON, B. and TRIPP, J.H., "Prediction of rolling bearing life under practical operating conditions", Proc. 15th Leeds-Lyon Symposium on Tribology, (Leeds 1988, this conference).

[15] HAMER, J.C., LUBRECHT, A.A., IOANNIDES E. and SAYLES, R.S., "Surface damage on rolling elements and it subsequent effects on performance and life", Proc. 15th Leeds-Lyon Symposium on Tribology, (Leeds 1988, this conference).

[16] JOHNSON, K.L., "Contact Mechanics", CUP (1985).

SESSION VIII
PLAIN BEARINGS (1)

Chairman: Mr F A Martin

Paper VIII(i)

Axially profiled circular bearings and their potential application in high speed lubrication

S. Basri and D. T. Gethin

Summary

The isothermal and thermal characteristics of an axially profiled circular bore bearing are determined using the finite element technique. These are compared with the predicted performance characteristic of a cylindrical bearing operating over the same range of conditions. The results obtained lead to the conclusion that the present analysis may be used to obtain general design data for an axially profiled bore bearing operating at high sliding speed. Axial profiling reduces the bulk operating film temperature by a small amount and it decreases the load carrying ability of the film and frictional losses also.

Principal Notation

C	radial clearance
C_f	lubricant specific heat capacity
D	journal diameter
H, $\bar{H}$	Power loss, non-dimensional power loss
K_f	Lubricant thermal conductivity
L	bearing length
N	Journal rotational speed
P, $\bar{P}$	Load, non-dimensional load
Q, $\bar{Q}$	side flow, non-dimensional side flow
Q_{co}	lubricant carried through the cavitating film
Q_{in}	lubricant inflow
R	radius of journal
T_{co}	oil temperature upstream of the groove
T_f	Lubricant feed temperature, °C
U	Journal sliding speed m/s
a,b	constant (Walther equation)
ℓ	cylindrical land width
p	film pressure
ϵ	eccentricity ratio
ρ	lubricant density
μ	lubricant dynamic viscosity
ψ	attitude angle
ν	lubricant kinematic viscosity

Introduction

The application of numerical methods to synthesise journal bearing behaviour has been established for a number of years. The early models were usually isothermal in nature and were directed mainly towards cylindrical bore bearings operating at slow or moderate sliding speeds. More recently models have been developed to analyse bearing stability and thermal effects. Thermal effects become important when the shear stresses in the film are large which usually occurs when the sliding speed is high, for example, in turbo alternators, gearboxes and pumps. Under this circumstance, oil film stability can be a problem, unstable behaviour resulting in the journal precessing around the clearance space in the bush causing machinery vibration which can lead to premature bearing failure. Prediction of this along with methods of establishing film stiffness and damping characteristics has also been described in the literature. The latter are important when the design of rotodynamic systems is addressed.

Under conditions of high shear stress (or sliding speed) it is important to be able to estimate the maximum film temperature accurately since this can establish the operating limit for the bearing [1]. Additionally, the power loss associated with shearing of the lubricant film will be reflected in the size of the cooling system for the bearing. The parameter which may be readily controlled in the design and which can be chosen to limit temperature rise is the film clearance ratio; a large value gives a reduced temperature rise since it permits a larger oil through flow. However, this is also reflected in a reduction of load carrying ability and, as mentioned above, the need for an oil supply of large capacity.

With the development of modern manufacturing technologies, it is now practical to produce circumferentially profiled bore bearings accurately [2]. The same technology may be used to produce bearings which are profiled over their axial length also, indeed this is a much less demanding manufacturing process. The advantage of these bearing types is that they may be designed to have a thick film centre section to maintain cool operation and a thinner film close to the edge to retain load carrying ability and to reduce lubricant sideflow. The thrust of the present investigation is to determine theoretically the effect of axial profiling on the performance of a circular bore bearing. Before this is described, a review of numerical models for synthesising profiled bore bearing behaviour will be considered.

Over the last 30 years, a series of investigations have been directed to predict the characteristics of profile bore bearings. In 1956 Pinkus [3] reported an analysis of elliptic bearings and later in 1959 [4] he reported an investigation of a three lobe bearing. In his papers it was demonstrated as a method of solving Reynolds' equation using a numerical technique for a fluid film of finite length. Since then, a number of investigations have used isothermal models to analyse various profile bore configuration. A few (for example references [5-8]), have studied selected cases of profile bores for dynamic performance and associated stability. The latter is summarised by Garner et al [9] for a variety of bore profiles and loading conditions. Apart from some very limited experimental studies and a few theoretical investigations of the types described in references [3-9], so far, little thermal modelling has been carried out for hydrodynamic journal bearings having circumferential profile geometries. Even fewer have considered the effect of axial profiling.

For thermal analysis, Mc Callion et al [10] have presented a solution method with uncoupling of the energy equations from the Reynolds equation. This approach is applicable for bearings of modest aspect (L/R) ratios operating at moderate eccentricity. Conversely, Medwell and Gethin [11] have presented a more comprehensive thermal solution for a cylindrical bore bearing by using the finite element method featuring the rigorous solution of both the equations of fluid mechanics and heat transfer.

These studies utilise the fact that in practice, the viscosity at any point depends on local film temperature which is coupled hydrodynamically to the film pressure and in some cases shear rate [12]. The temperature in film lubricated bearings increases rapidly with speed due to the increased rate of frictional heat production, which on constant viscosity theory is proportional to speed squared. At higher speeds, severe temperature gradients are set up, both across the film because of heat removal by conduction and in the plane of relative motion because of increased shear stress and reduced convective heat transfer.This arises from the reduced oil flow associated with the thin lubricant film.

The finite element method has been used in a wide range of scientific and engineering problems. The technique is suited to solve complex coupled engineering problems [13]. It has simplicity in concept, elegance in development and potency in application. A number of investigators over the years have reported the results of their studies on the application of FEM in fluid film lubrication as applied to Newtonian or Non-Newtonian lubricants [see references 14-17]. In this paper, the isothermal and thermal analysis of axially profiled hydrodynamic journal bearing using the finite element technique is presented . This has received little attention previously and which has motivated the present study.

2. Analysis

The Navier Stokes equations [18] are fundamental to the analysis of problems in fluid mechanics. For hydrodynamic lubrication, these may be simplified as explained in [19] to give the classical Reynold equation which may be written with respect to a cartesian coordinate system (see Figure 1) for laminar flow as

$$\frac{\partial}{\partial x}\left[\frac{h^3}{\mu}\frac{\partial p}{\partial x}\right] + \frac{\partial}{\partial z}\left[\frac{h^3}{\mu}\frac{\partial p}{\partial z}\right] = 6U\frac{dh}{dx} \qquad 2.1$$

As explained in [10], when the thermal effects are included a number of models have been proposed to synthesise it. Where the film thickness is comparatively large (C/R = 0.004) an equation which equates convective heat transfer to viscous generation is usually adequate since heat removal is mainly by convection. As derived in [20], this is expressed by the equation

$$\rho C\left[Q_x\frac{\partial T}{\partial x} + Q_z\frac{\partial T}{\partial z}\right] = \frac{\mu U^2}{h} + \frac{h^3}{12\mu}\left[\left[\frac{\partial p}{\partial x}\right]^2 + \left[\frac{\partial p}{\partial z}\right]^2\right] \qquad 2.2$$

In equation 2.2, the volumetric flowrates are given by:

$$Q_x = \frac{Uh}{2} - \frac{h^3}{12\mu}\frac{\partial p}{\partial x} \qquad 2.3$$

$$Q_z = -\frac{h^3}{12\mu}\frac{\partial p}{\partial z} \qquad 2.4$$

Equations 2.1 and 2.2 are coupled hydrodynamically by the pressure terms. Additionally, they are linked by the lubricant viscosity and its dependence on temperature can be derived from the Walther equation which relates kinematic viscosity with temperature

$$\text{i.e. } \log_{10}\left[\log_{10}(\nu + 0.6)\right] = a \log_{10} T + b \qquad 2.5$$

The constants a and b can be determined from the measured lubricant viscosity.

When equations 2.1 and 2.2 are solved using the finite element method, they are usually discretised using the Galerkin weighted residual method which is well documented in the literature [21]. For each equation this yields a set of matrix equations which may be solved to yield either pressures or temperatures appropriately.

Closure of the equation set is completed by a relationship which expresses the variation of film thickness. At some position relative to the maximum film thickness this is given by an equation of the form

$$h = C(1 + \epsilon \cos \theta) + f(z) \qquad 2.6$$

The latter term (f(z)) is included to incorporate the axial variation of film thickness due to profiling in this direction. Admissible profiles will be discussed later in section 3.

Before the equations can be solved it is necessary to prescribe boundary conditions. For a journal running aligned with the bush, advantage can be taken of centreline symmetry and consequently the boundary prescription can be written with reference to Figure 1 as

For equation 2.1

$$\frac{\partial p}{\partial z}(x,0) = 0 \qquad 2.7a$$

$$p\left(x, \frac{L}{2}\right) = 0 \qquad 2.7b$$

$$p(0,z) = 0 \qquad 2.7c$$

$$p(x,Z) = \frac{\partial p}{\partial x}(x,Z) = 0 \qquad 2.7d$$

Similarly for equation 2.2, they may be written

$$\frac{\partial T}{\partial z}(x,0) = 0 \qquad 2.8a$$

$$T(0,z) = T_{in} \qquad 2.8b$$

Equations 2.7a and 2.8a reflect centreline symmetry while 2.7b expresses the ambient pressure condition. At the downstream end of the film, equation 2.7d expresses the Reynold boundary condition while at the film inlet, equation 2.7c embodies the assumption that the feed pressure is small in comparison with expected excursions in the film under realistic loading conditions. Also at the film inlet, equation 2.8b embodies the assumption that there are no significant axial temperature excursions. The method of calculating T_{in} is based on an enthalpy balance at the feed groove which may be expressed mathematically for a constant lubricant density as

$$T_{in} = \frac{[Q_{co}\,T_{co} + Q_f\,T_f]}{Q_{in}} \qquad 2.9$$

The solution of the equation set was encoded into a Fortran program and the strategic steps in the solution procedure are:

1. assume an initial viscosity field

2. solve Reynolds equation (2.1)

3. update the temperature boundary conditions (2.8b)

4. solve the energy equation

5. update the viscosity field and film thickness profile

This procedure was repeated until solution convergence which was assumed when the changes in the pressure field between successive iterations was less than 5%.

On completion of the solution, the usual bearing design parameters of load capacity, power loss, and flowrate were determined. For the case of isothermal operation, these parameters were cast in dimensionless form according to

$$\bar{P} = \frac{P}{LD}\left[\frac{C}{R}\right]^2 \frac{1}{\mu N} \qquad 2.10a$$

$$\bar{H} = \frac{H}{\mu N^2 LD^2}\left[\frac{C}{R}\right] \qquad 2.10b$$

$$\bar{Q} = \frac{Q}{2CLDN} \qquad 2.10c$$

These equations require the specification of a radial clearance C which is not straightforward to define for an axially profiled bore. This will be considered further in section 3.

3. Results and Discussion

The performance of three types of axially profiled circular bore bearings was considered and the shapes and associated nomenclature used are illustrated in Figure 1. Profile 1 incorporates axial film thickness variation over the full bearing width while profiles 2 and 3 are shaped over a central band only. The reason for this is that under stationary or start up conditions, for profile 1, the load will be supported on a very small part of the bearing edge which may result in local damage and consequently less effective bearing operation. For profiles 2 and 3 the cylindrical lands at either side of the bearing are intended to be used for supporting the stationary shaft or for start-up conditions. These can be of different width and in the present study two were considered l/L = 0.25 and l/L = 0.5 (i.e. profile 2 and 3 respectively). The axial profile was assumed to be made up of a circular arc since this is convenient to manufacture. The radius of the arc was derived by specifying a deviation Δ (see Figure 1). In both cases lubricant supply was assumed to be by a twin axial groove arrangement as shown.

In the following sections both isothermal and thermal studies are addressed and these will be considered separately.

3.1 Isothermal Studies

A series of calculations were carried out for the condition of an isotheral film and an aspect ratio (L/R) of 1.0 was assumed throughout. The load-carrying part of the film was mapped by 100 eight node isoparametric elements (5 axially by 20 - circumferentially). Figure 2 illustrates the pressure contours for the full width bearing in the load carrying part of the film for a cylindrical bore, and the 3 axial profiles. Detailed operating conditions are tabulated on the figure where the clearance, C, is for convenience that at the edge of the bearing and the profiling extent is the same for each i.e. $\Delta/R = 0.002$. From this it can be seen that the contours are quite different for the various film shapes. The cylindrical bore shows the most gentle fall-off at the bearing edge whereas the introduction of profiling results in a more uniform distribution over the load carrying area of the bearing but with steep gradients at the side of the film. These results are similar to those presented for a bearing having a deformable shell [22] where the pressure field gives a film thickness profile which shows similar axial variation. Also, the axial pressure profiles under the load line are shown in Figure 3. These are presented for a nominal dimensionless load ($\overline{P}$) of 5.07 and these show clearly the axial form of the pressure profile for the different bearing geometries. Clearly to support an equivalent load, the different bearing types will operate at different eccentricity ratios and these are listed in Figure 3 also. For the nominal load case presented ($\overline{P}$ = 5.07) the choice of axial profile 1 results in the need for the most eccentric operation and therefore the thinnest film occurs under this circumstance. However, for profile 1, the small increase in eccentricity over that for the other profiles gives a very significant reduction in peak pressure in the film while the load carrying ability is not changed so dramatically. This accrues since for profile 1, there is little axial variation in pressure over the film width. This has clear advantage where there is a need to minimise bearing bush deflection due to hydrodynamic action. This may occur in polymeric bearing liners, for example, or it may happen naturally during operation [22]. Additionally, the highest pressure under the load line is associated with profile 3, however, from the form of the contours (Figure 2) this higher pressure does not extend into the region close to the minimum film thickness and therefore the centreline pressures associated with this profile are lower than those for a cylindrical bore.

Figure 4 illustrates the variation in bearing parameters for the various profiles when operating at different eccentricity ratios. For load carrying ability, the most noticeable changes occur at extreme eccentricity ratios ($\epsilon > 0.85$) where the effect of profiling is to reduce load carrying ability at these extreme conditions. The parameter which is affected most significantly is the power loss associated with the bearing. The profiled bores give reduced power loss and this occurs since the shear stresses on the bearing centreplane are reduced with the thicker film. This is a significant factor since it may be reflected in lower operating temperatures in the film. This will be considered in further detail in the following section. Also, it is evident that with axial profiling, hydrodynamic leakage from the film exceeds that for a cylindrical geometry. This accrues since a larger gap section is presented at the upstream edge and the steeper pressure gradients at the bearing edge result in increased sideflow from the film. This can be considered to be a disadvantage for the design schemes featuring axial profiles, fortuitously it is not too significant.

Additionally, Figure 5 illustrates the effect of using different profiling parameters (Δ/R) for axial profiles 1 and 3. Clearly, its influence is not too significant for either geometry.

3.2 <u>Thermal Considerations</u>

Calculation was carried out using the thermal model described in section 2. A supply temperature of 30°C was assumed for the complete set of calculations. Also, the coefficients in the Walther Equation were assigned values of
$a = -3.79$ and $b = 9.64$ which are pertinent for an oil of ISO viscosity grade 32. In the series of calculations, speed, eccentricity ratio and aspect ratio were considered to be design variables.

Figure 6 illustrates isotherm contours for the four profiles considered. As expected, the contours for the cylindrical bearing show no axial variation whereas those for all the profiles considered do. Additionally, the effect of profiling is to reduce centreline temperature with the greatest effect being obtained by profile 1. The reason for this has been suggested in 3.1 and accrues from the reduced shear stress on the bearing centreline. When the temperature excursions on the bearing edge are considered, examination of the contours suggests that the maximum excursion takes place on the bearing edge for profiles 1 to 3 and it becomes similar to that for a cylindrical geometry. However, it is clear from this calculation that due to the axial temperature excursions the usual assumption of zero axial temperature gradient is not applicable to simplify the more rigorous analyses which will be reflected in the need to develop more sophisticated thermo-hydrodynamic models when thinner film bearings are considered.

For the cases shown, the associated performance parameters are listed below in Table 1.These show similar trends to those predicted from isothermal analysis.

Load (kN)	Shaft Torque (N.m)	Sideflow (l/min)	Attitude Angle (°)	Comment
5.27	1.15	2.02	40.80	Cylindrical
4.33	0.98	2.13	40.65	Profile 1
4.83	1.02	2.09	40.70	Profile 2
5.21	1.13	2.04	40.80	Profile 3

Table 1 Bearing Performance Parameters

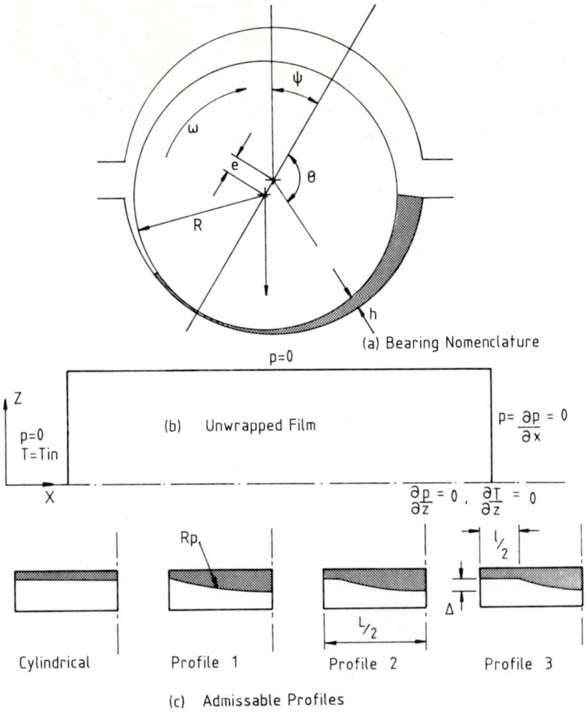

FIGURE 1. BEARING NOMENCLATURE AND AXIAL PROFILES CONSIDERED.

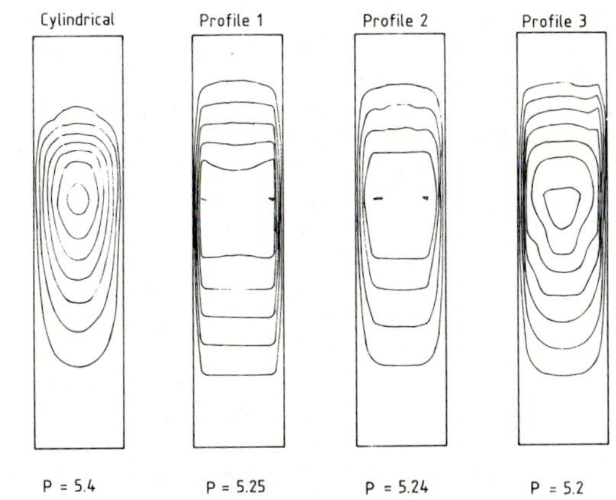

FIGURE 2. FILM PRESSURE CONTOURS FOR THE AXIAL PROFILES CONSIDERED ;
$L/R = 1.0$; $\varepsilon = 0.7$; $C/R = 0.004$; $\Delta/R = 0.002$

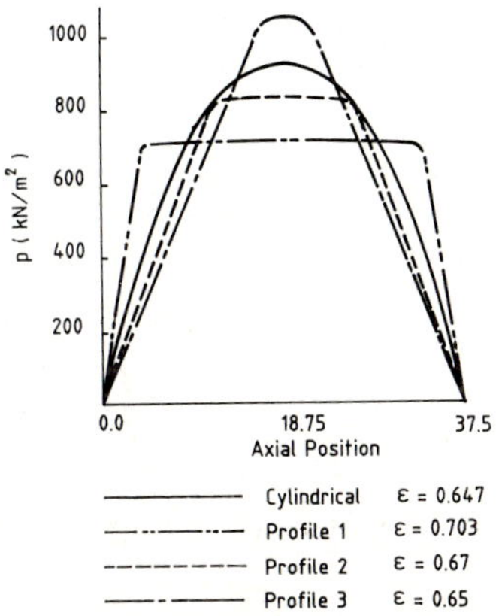

FIGURE 3. AXIAL PRESSURE PROFILES UNDER THE LOAD LINE FOR NORMALLY EQUIVALENT LOAD
$L/R = 1.0$; $\bar{P} = 5.07$; $\Delta/R = 0.002$

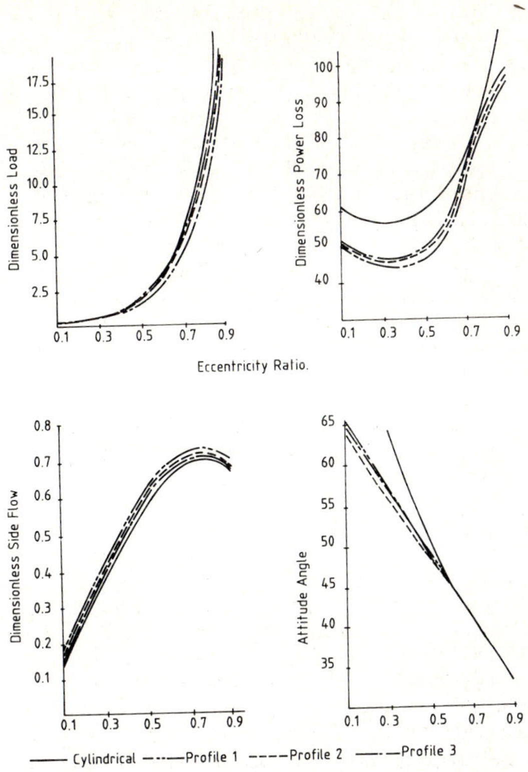

FIGURE 4. BEARING PERFORMANCE FOR DIFFERENT PROFILE SHAPES
$L/R = 1.0$ $\Delta/R = 0.002$.

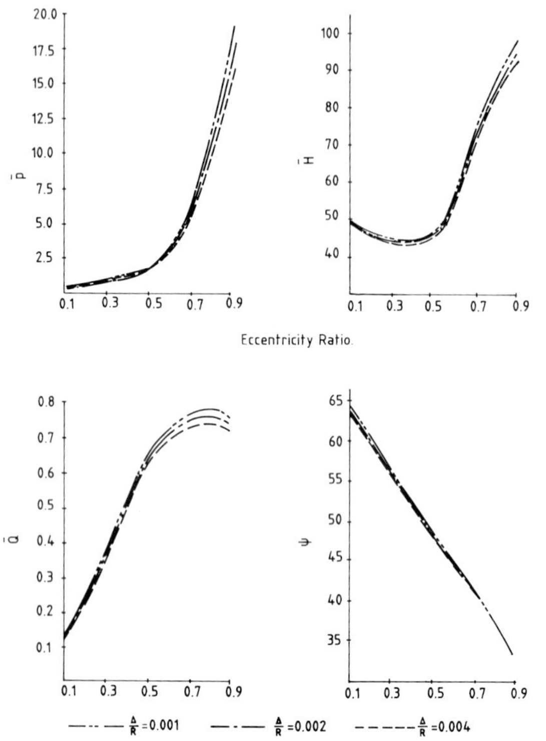

FIGURE 5a. BEARING PERFORMANCE FOR VARIOUS PROFILE PARAMETERS
L/R= 1.0 ; PROFILE 1.

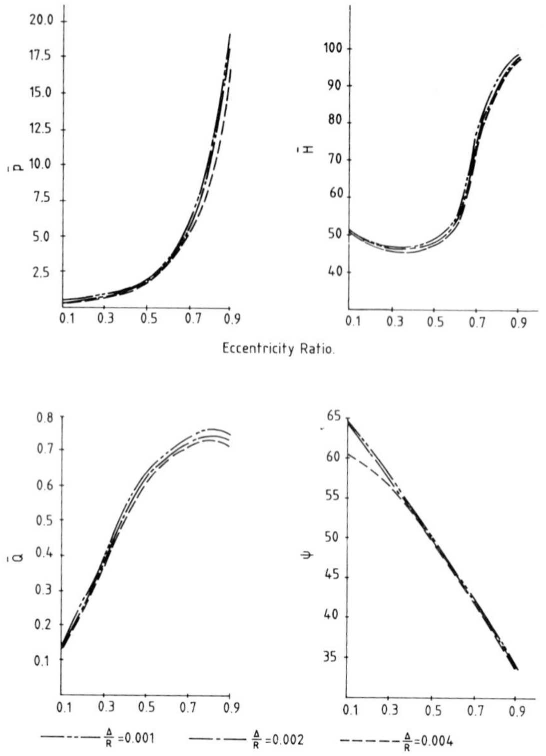

FIGURE 5b. BEARING PERFORMANCE FOR VARIOUS PROFILE PARAMETERS
L/R= 1.0 ; PROFILE 3.

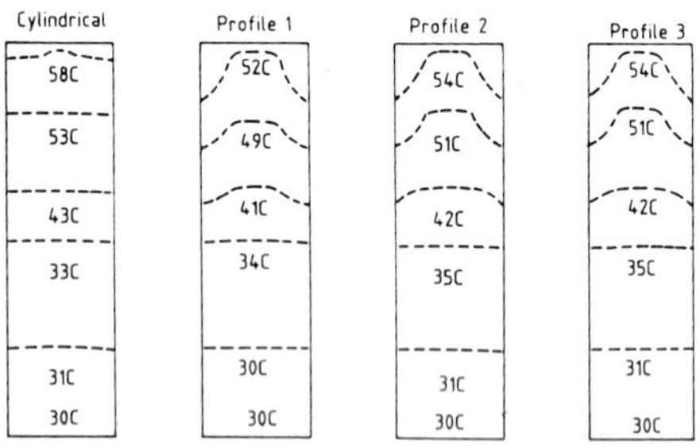

FIGURE 6 ISOTHERM MAP FOR THE AXIAL PROFILES
CONSIDERED (L/R=1.0 , ε =0.7)

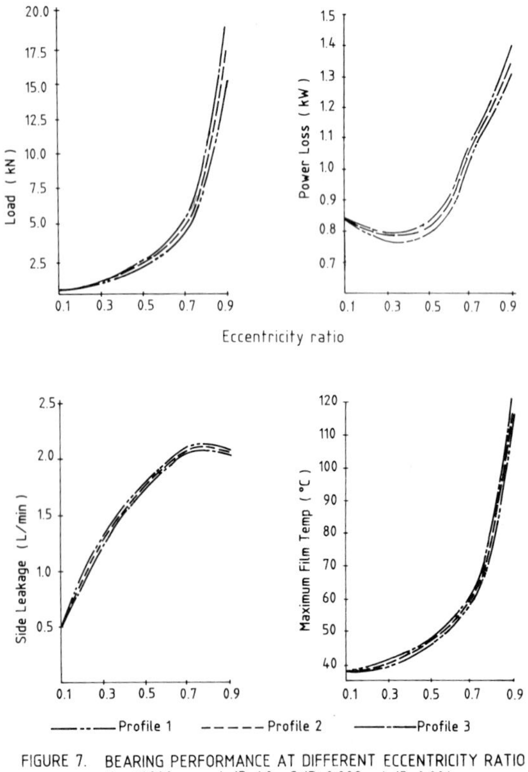

FIGURE 7. BEARING PERFORMANCE AT DIFFERENT ECCENTRICITY RATIO
N = 10000rpm , L/R=1.0 , C/R=0.002 ; Δ/R=0.001 .

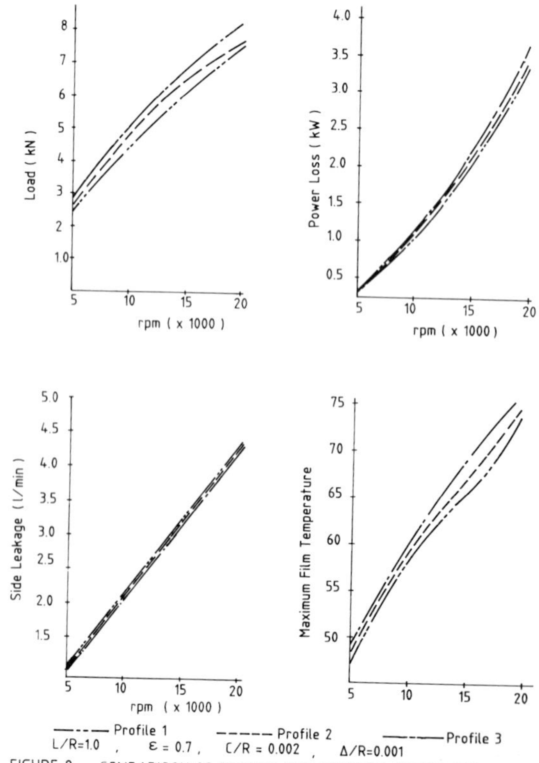

FIGURE 8. COMPARISON OF BEARING PERFORMANCE TRENDS FOR
DIFFERENT ROTATIONAL SPEEDS.

Figure 7 illustrates performance trends with eccentricity ratio changes for profiles 1, 2 and 3. In comparing the load carrying ability, it can be seen clearly that profile 3 has the best load carrying ability with profile 1 having the lowest. This occurs since profile 3 is closest to a cylindrical form. The difference in load carrying ability is most noticeable at the extreme eccentricity ratio where for an eccentricity ratio of 0.8, for the case considered, the load carrying capacity for profile 1 is about 80% that for profile 3.

Similar results are presented for power loss in that profile 3 has the highest loss associated with it while profile 1 has the lowest. Despite the introduction of cylindrical lands, the reduction in sideflow is only marginal, however, the trends shown exclude any pressure induced components.

Additionally, Figure 7 shows the variation in maximum film temperatures on the bearing centreline for the three profiles and again there are only small differences. Profile 3 gives the highest temperatures while the lowest is associated with profile 1. For the axial profiling shown, the difference in thermal behaviour is too small to affect bearing performance trends significantly and therefore, the reduction in load capacity arises mainly from the profiling effect.

Figure 8 illustrates the performance of the bearing types over a range of rotational speeds and again performance trends are bounded by profiles 1 and 3. However, at the more extreme operating conditions more significant differences between models is evident particularly with regard to load carrying ability and maximum centreline temperature. The highest film temperatures and load carrying ability are associated with profile 3.

The trends in Figures 7 and 8 have been computed for a film having only a small amount of axial profiling ($\Delta/R = 0.001$) and a comparatively thin film ($C/R = 0.002$) and therefore the bearing performance trends have shown little difference being close to that for a cylindrical geometry. Figure 9 illustrates bearing performance trends for two clearance ratios ($C/R = 0.002$ and $C/R = 0.004$) and for different axial profiles ($\Delta/R = 0.001$ to $\Delta/R = 0.004$).

Clearly the effect of using a bearing having a large clearance ratio is to reduce both the maximum centreline temperature and the load carrying ability. This is most noticeable for profile 1. As expected, maximum centreline temperature falls off with an increase in axial profiling since under this circumstance the shear stress is reduced and a larger volume of oil is available to remove the heat from this part of the film.

Conclusions

In this paper, a study of the performance characteristic of an axially profiled circular bore bearing has been presented. Where appropriate the results in this paper are presented as a comparison between the cylindrical bore bearing and an axially profiled counterpart. Since information on cylindrical bearing performance is readily available [1], this comparison shows a direct indication of the behaviour of an axial profile bore bearing.

The main conclusions of the study are as follows:

- Axial profiling affects bearing performance most noticeably by reducing load carrying capacity and power loss.
- Axial profiling gives lower bearing centreline temperatures particularly where the deviation from a cylindrical geometry is significant.
- Axial profiling may be introduced with benefit to reduce bearing temperatures where extreme operating conditions are encountered.

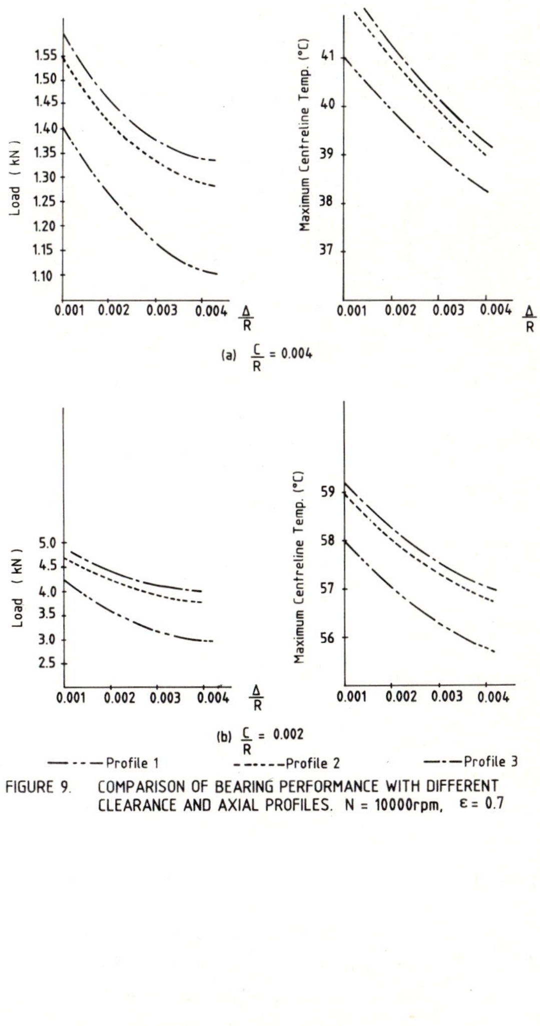

FIGURE 9. COMPARISON OF BEARING PERFORMANCE WITH DIFFERENT CLEARANCE AND AXIAL PROFILES. N = 10000rpm, $\varepsilon = 0.7$

References

1. Martin, F A and Garner, D R. Plain Journal Bearings Under Steady Loads: Design Guidance for Safe Operation, 1st European Tribology Congress. I Mech E paper C 313/73, 1973.

2. Albin, F T, Champbell, J and Garner, D R. The Technical Development and Market Exploitation of Novel Manufacturing Techniques for High Speed Plain Bearings; Proc.I.Mech E. 200, 1986, p77-78.

3. Pinkus, O. Analysis of Elliptical Bearings; Trans. ASME, 78, 1956, p965-973.

4. Pinkus, O. Analysis and Characteristic of the Three Lobe Bearing; Trans ASME, Journal of Basic Engineering, 81, 1959, p49-55.

5. Malik, M; Sinhasan, R and Chandra, M. Design Data for Three Lobe Bearing. Trans. ASLE, 24, 1981 p.345-353.

6. Malik, M; Chandra, N and Sinhasan R. Performance Characteristics of Tilted Three Lobe Journal Bearings. Tribology International, 14, 1981, p.345-349.

7. Malik, M. The Analysis of Symmetric and Tilted Four-Lobed Journal Bearing Configurations ASLE Transaction, 26, 1983, p.264-269.

8. Mehta, N P and Singh, A. Stability Analysis of Finite Offset Halves Pressure Dam Bearing. Trans. ASME, (JOLT), 108, April 1986, p.270-274.

9. Garner, D R; Lee, C S and Martin, F A. Stability of Profile Bore Bearings. Influence of Bearing Type. Selection: Tribology International 13, 1980, p 204-210.

10. McCallion, H; Yousif, E and Lloyd, T. The Analysis of Thermal Effects in A Full Journal Bearing Trans. ASME (JOLT), 92, 1970, p578.

11. Medwell, J O and Gethin D T. A Finite Element Analysis of Journal Bearing Lubrication Proceeding of the Leeds-Lyon Symposium Fluid Film Lubrication 1983, (Eds. Dawson et al).

12. Ng, C W and Pan, C H T. A Linearised Turbulent Lubrication Theory. Trans ASME Journal of Basic Engineering, 87, 1965, p.675-688.

13. Huebner, K H. Finite Element Analysis Fluid Film Lubrication - A Survey, Finite Element in Fluids-Volume 2 p.225-254 (Ed. R H Gallagher, J T Oden, C Taylor and O C Zienkiewicz).

14. Allan, T. The Application of Finite Element Analysis to Hydrodynamic and Externally Pressurized Pocket Bearing. Wear, 19, 1972, p.169

15. Tayal, S P; Sinhasan, R and Singh, D V. Analysis of Hydrodynamic Journal Bearing having Non-Newtonian Lubricant (Prandt Model) by a Finite Element Method, Journal Mech. Engineering Science, 1981 p63-68.

16. Booker, J F and Huebner, K H. Application of Finite Element Methods to Lubrication: an Engineering Approach. Trans. ASME (JOLT) 94, 1972, p.313.

17. Gethin, D T. An Application of Finite Element Method to the Thermo-hydrodynamic Analysis of a Thin Film Cylindrical Bore Bearing running at High Sliding Speed. Trans. ASME (JOT), 1987 p 283-289.

18. Schilichting, H. Boundary Layer Theory, published by Pergamon Press, 1955.

19. Dowson, D. A Generalised Reynolds Equation for Fluid Film Lubrication. Int. J Mech. Eng. Sci., 4, 1962, p.159-170.

20. Constaninescu, V N. Basic Relationship in Turbulent Lubrication and Their Extension to include Thermal Effects. Trans. ASME (JOLT), 95, 1973, p147-154.

21. Zienkiewic, O C. The Finite Element Method, 3rd Edition, 1977 (McGraw Hill).

22. Jain; S C, Sinhasan, R and Singh, D V. A Study of EHD Lubrication in a Journal Bearing with Piezovisions Lubricants; Trans ASLE, 27, 1984 p.168.

Paper VIII(ii)

Analysis of partial arc journal bearings

E. W. Cowking

Two computer programs are described which carry out thermohydrodynamic analyses of multi-arc journal bearings. A validation procedure for the programs is also described. Some thermohydrodynamic results are presented as possible benchmarks.

1 INTRODUCTION

Before software can be used in an engineering organisation it is usually necessary that quality assurance requirements be met which include a full validation of the results. The purpose of this paper is to describe a possible approach to the validation of bearing analysis software, as applied to two partial arc journal bearing programs.

The two computer programs were developed to analyse steadily-loaded journal bearings with fixed partial arcs. They both provide four thermal options ranging from isothermal up to fully thermohydrodynamic. They also produce linear dynamic coefficients for use in rotordynamic calculations. The first program carries out a two-dimensional solution and has been described previously in (1). The second program, which has been completed more recently, carries out a similar solution in three dimensions. The methods of analysis are described more fully in section 2, but the main emphasis of this paper is on the validation process described in section 3.

In validating the programs, the aim has been to rely on published results and on checks which can be carried out by hand. Comparisons are made with the isothermal results of Lund and Thomsen (2) and the simple adiabatic solutions of Hakansson (3) which predicted the average temperature across the film. Each program uses the same subroutines for calculating the pressure, oil forces, power loss and oil flow for all temperature options. Thus the checks on simple results are generally useful in validating the program. Further work which has been done to validate the full thermohydrodynamic options is also described.

Ideally, it should be possible to benchmark thermohydrodynamic solutions using published results. However, analyses of partial arc bearings involve empirical data for the inlet region which will vary between different implementations. Some sample results are provided for a single partial arc under adiabatic conditions. These are intended to provide some checks on the thermohydrodynamic analysis of the oil film alone.

1.1 Notation

Cp	Specific heat of lubricant
C	Radial clearance of bearing
D	Journal diameter
F	Shear flow factor in generalised Reynolds equation
h	Film thickness
k	Thermal conductivity of lubricant
L	Axial length of bearing
p	Pressure in fluid film
Q_x, Q_z	Flows per unit length in fluid film
R	Journal radius
T	Temperature in fluid film
(u,v,w)	Velocity component relative to (x,y,z)
v*	Transformed velocity
(x,y,z)	Local cartesian co-ordinates
(X,Y)	Fixed co-ordinates with Y vertically upwards
α, β	Pressure and temperature coefficients of viscosity
ΔR	Pad radius difference
ζ	Transformed y- co-ordinate
η	Fluid viscosity
ρ	Fluid density
χ	Non-dimensional viscosity-temperature parameter
ω	Angular velocity of journal

2 DESCRIPTION OF PROGRAMS

Both programs analyse a journal bearing with up to five fixed partial arcs. These arcs are usually separated by inlet gutterways which are produced by taking a cylindrical cut out of the bearing surface as shown in Fig. 1. The end-lands which seal the inlet gutterways can generate small hydrodynamic forces but in the present analyses this is neglected and the arcs are treated as separate rectangular regions of fluid film.

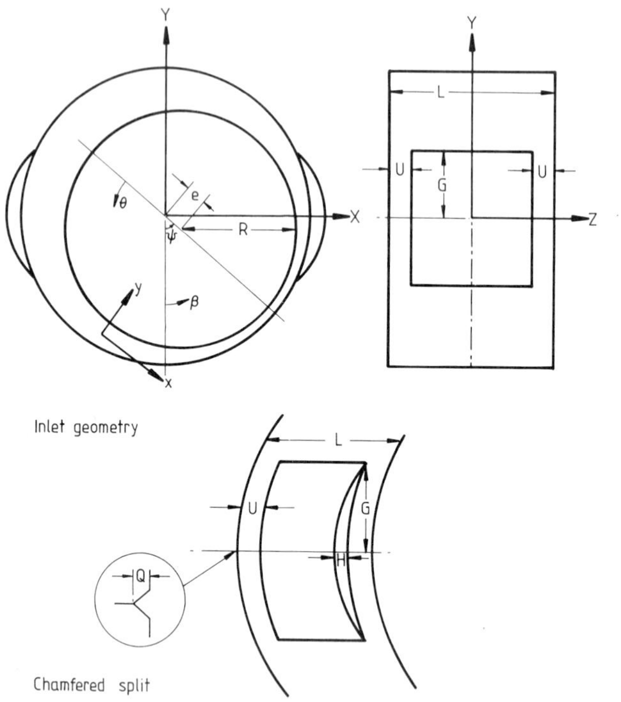

Fig. 1 Example of Bearing Geometry

A separate pressure-temperature iteration is carried out for each of these rectangular regions of film. The programs support three alternative types of thermal analysis depending on the input parameter ITEMP:

<u>ITEMP</u>

0 Isothermal solution

1 Temperature averaged across the film thickness

2 THD analysis of film

3 THD analysis of film with heat conduction in a cylindrical bearing housing

The programs treat each inlet gutterway as a lumped system at a single temperature which is determined from a heat balance. The programs also carry out an overall heat balance iteration to ensure that the temperatures in the inlets are cyclically consistent.

The outermost iteration in the programs determines the journal position for a given load using a root-finding algorithm in two dimensions. Alternatively, if this iteration is not requested, the programs produce the fluid film forces for a given journal position.

After the steady-state solution has been obtained the programs can calculate linearised dynamic coefficients by numerically differentiating the force components.

A block flow diagram which shows the various iterations involved is given in Fig. 2. All the iterations are optional, dependent on the data so that the programs can be used for a wide variety of journal bearing problems.

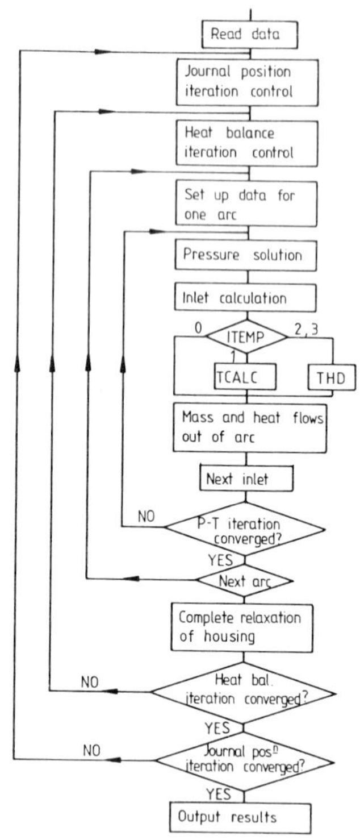

Fig. 2 Block Flow Diagram

2.1 Fluid Film Equations

The equations are formulated relative to a locally rectangular co-ordinate system and the effect of film curvature are neglected. The fluid film is transformed into a rectangular region using the change of variable:

$$\varsigma = \frac{y - y_0}{h} \qquad \ldots\ldots\ldots\ldots(1)$$

where y is radially inwards, y_0 is the position of the bearing surface and h is the film thickness. It is shown in (1) that this leads to a transformed continuity equation:

$$\frac{\partial(\rho h)}{\partial t} + \frac{\partial(\rho h u)}{\partial x} + \frac{\partial(\rho v^*)}{\partial \varsigma} + \frac{\partial(\rho h w)}{\partial z} = 0 \qquad \ldots(2)$$

where the transformed transverse velocity v^* is:

$$v^* = v - v_0 - \varsigma\left(\frac{\partial h}{\partial t} + u\frac{\partial h}{\partial x} + w\frac{\partial h}{\partial z}\right) \qquad \ldots(3)$$

This transformed velocity component has the useful property of being zero at the two surfaces $\varsigma = 0$ and $\varsigma = 1$.

The thermohydrodynamic analysis is carried out using values for the density, specific heat and thermal conductivity which are averages across the film thickness. This simplifies the formulae, with only a minor effect on accuracy.

For laminar flow, the velocity components in the fluid film are:

$$\left.\begin{aligned} u &= \omega R \frac{I_o(\varsigma)}{J_o} - \left[\frac{J_1}{J_o} I_o(\varsigma) - I_1(\varsigma)\right] h^2 \frac{\partial p}{\partial x} \\ w &= \qquad\qquad - \left[\frac{J_1}{J_o} I_o(\varsigma) - I_1(\varsigma)\right] h^2 \frac{\partial p}{\partial z} \end{aligned}\right\} \quad ..(4)$$

where

$$I_n(\varsigma) = \int_0^{\varsigma} \frac{\varsigma^n d\varsigma}{\eta} \qquad J_n = \int_0^1 \frac{\varsigma^n d\varsigma}{\eta} \quad ..(5)$$

The generalised Reynolds equation for laminar flow can be written in the form

$$\frac{\partial}{\partial x}\left(\frac{\bar{\rho} h^3}{12\eta_e}\frac{\partial p}{\partial x}\right) + \frac{\partial}{\partial z}\left(\frac{\bar{\rho} h^3}{12\eta_e}\frac{\partial p}{\partial z}\right) = \frac{\partial}{\partial t}(\bar{\rho} h) + \omega R \frac{\partial}{\partial x}(F\bar{\rho} h) \quad ..(6)$$

where $\eta_e = \frac{J_o}{12(J_2 J_o - J_1^2)}$ is an effective viscosity and

$F = 1 - \frac{J_1}{J_o}$ is a flow factor.

The program can also handle turbulent flow using the linearised turbulent lubrication theory of Ng and Pan (4). In this case effective viscosities η_x and η_z are used instead of η_e

The energy equation for laminar flow in the film is used in the form:

$$\bar{\rho}\bar{C}_p \frac{dT}{dt} = \eta\left[\left(\frac{\partial u}{\partial y}\right)^2 + \left(\frac{\partial w}{\partial y}\right)^2\right] + \bar{k}\frac{\partial^2 T}{\partial y^2} \quad ..(7)$$

$\bar{\rho}$, $\bar{C}_p$ and $\bar{k}$ are average values across the film.

The convection term retains its original form after the change of variable (1):

$$\frac{dT}{dt} = \frac{\partial T}{\partial t} + u\frac{\partial T}{\partial x} + \frac{v^*}{h}\frac{\partial T}{\partial \varsigma} + w\frac{\partial T}{\partial z} \quad ..(8)$$

However, the partial derivatives with respect to x and z have different values when they are evaluated with ς fixed.

In the case of turbulent flow the programs calculate the average temperature across the film from the equation:

$$\bar{\rho}\bar{C}_p\left(q_x \frac{\partial T}{\partial x} + q_z \frac{\partial T}{\partial z}\right) = \eta_\tau \frac{(\omega R)^2}{h} + \frac{h^3}{12}\left[\frac{1}{\eta_x}\left(\frac{\partial p}{\partial x}\right)^2 + \frac{1}{\eta_z}\left(\frac{\partial p}{\partial z}\right)^2\right] - H \quad ..(9)$$

H is the heat flux to the metal surfaces which is calculated from empirical heat transfer coefficients. The simple temperature calculation uses this formula with $\eta_x = \eta_z = \eta_\tau = \eta$. The validation process will concentrate on laminar flow and details of the turbulent flow option have been kept correspondingly brief.

2.2 Solution Methods for Fluid Film Equations

The two dimensional program implements a variational solution of Reynolds equation which is described in (1). This leads to a second order ordinary differential equation which is solved by standards numerical methods. The three dimensional program solves Reynolds equation by relaxation methods. Cavitation is simulated by setting pressures equal to p_{cav} when they drop below that value during the relaxation process.

The THD energy equation (7) is solved by a simple implicit method. Backflow regions are handled by setting negative values of the tangential velocity to zero whenever they occur. This is found to be reasonably accurate since backflow regions are not usually very significant in their effect in partial arcs of relatively small length/diameter radius. In the THD analysis the journal surface temperature is assumed not to vary circumferentially. Its value can optionally be determined from a simplified analysis of heat conduction in the journal. The boundary condition at the bearing surface is based on a specified heat transfer coefficient when ITEMP = 2. When heat conduction in the housing is included (ITEMP = 3) a standard heat conduction boundary condition is used.

The simple energy equation (9) is solved by marching in the direction of rotation.

2.3 Inlet Heat Balance

Each inlet region is treated as a lumped thermal system at a mean temperature T_p. The mass flow and inlet flow balances are as follows:

$$\left.\begin{aligned} M_2 + M_s &= M_p + M_1 \\ C_p(M_2+M_s)(T_p-T_s) &= H_2 + H_p - H_J - H_B \end{aligned}\right\} \quad ..(10)$$

where M_2, H_2 = Mass and heat flow at trailing edge of previous arc

M_1 = Mass flow at leading edge of next arc

M_s = Mass flow of fresh fluid

M_p = Side-leakage from inlet

H_p = Power loss due to shear flow

H_J, H_B = Heat conducted to journal and bearing

The inlet is assumed to thoroughly mix the fluid so that the leading edge temperature of the next arc is uniform. This leads to step discontinuities in temperature at the journal and bearing surfaces. This idealization mainly affects the heat flux to the journal which is predicted to peak sharply at the leading edge of the arc.

The flow in the inlet pocket is assumed to normally be turbulent and the power loss and heat transfer coefficients are calculated from empirical data for turbulent shear flows. The power loss and flow over the end lands are calculated from lubrication theory.

2.4 Film Shape

The two dimensional program analyses a well-aligned bearing with circular arcs. The film shape for the nth arc is calculated from the standard formula:

$$h = \Delta R_n - e_n \cos(\beta - \psi_n) \qquad ..(11)$$

where R_n is the radius difference of the arc and ( ) is the position of the journal centre relative to the centre of the arc.

The three dimensional program analyses an arc of a general film shape. The film thicknesses is calculated from the formula:

$$h = h_b - (-e_y + \phi_x z)\cos\beta - (e_x + \phi_y z)\sin\beta \qquad ..(12)$$

where h_b is a basic film shape which can be calculated from a formula or read as data. ϕ_x and ϕ_y are the angles of tilt of the journal about the X and Y axes. e_x and e_y are the rectangular co-ordinates of the journal centre relative to the bearing centre.

2.5 Lubricant Properties

The viscosity of the fluid is calculated by the programs from Roelands' formula (5), with a linearised pressure term:

$$\log_{10}\eta + 1.2 = \frac{G_o(1 + Zp/1961)}{\left(1 + \frac{T}{135}\right)^{S_o}} \qquad ..(13)$$

where η is the viscosity in centipoise, T in °C and p in bar. G_0, S_0 and Z are constant parameters.

The density and specific heat are assumed to vary linearly with temperature. These lubricant properties can all be determined from two initially specified values at two different temperatures, except for Z which is specified separately.

3 VALIDATION

The programs contain a wide range of options so it is necessary to adopt a step-by-step approach to validation. Comparisons were first made with the results for a single circular arc published by Hakansson (3). Then it was shown that the programs build up solutions correctly for multi-arc bearings. The heat balances for the inlet regions were checked manually. Dynamic coefficients for an isothermal solution were compared with Lund and Thomsen (2). Finally, the thermohydrodynamic analysis was checked as far as possible by comparison with the simple temperature calculation and by checking mass flow and heat balances.

3.1 Single Partial Arc Results

Hakansson (3) provides tables of non-dimensional results for a single circular arc with laminar flow and variable viscosity. A cavitation condition is applied and the average temperature across the film was evaluated assuming no heat loss by conduction. He uses the exponential formula for viscosity:

$$\eta = \eta_i \exp[\alpha p - \beta(T - T_i)] \qquad ..(14)$$

where T_i is the temperature at the leading edge of the arc.

He tabulates results as functions of the non-dimensional viscosity-temperature parameters:

$$\chi = \beta \frac{\eta_i \omega}{\rho C_p}\left(\frac{R}{\Delta R}\right)^2 \qquad ..(15)$$

The present programs require data in SI units and use a different formula (13) for the viscosity. The following test case was therefore considered:

Diameter, D = 200 mm
Length, L = 100 mm
Radius difference, ΔR = 0.2 mm
Arrangement: centrally loaded 120° arc
Leading edge temperature, Ti = 60°C

Lubricant data:

T(°C)	η(cP)	ρ(kg/litre)	Cp(kJ/kg°C)	k(W/m°C)
40	27.4	0.83	2.1	0.13
100	4.37	0.83	2.1	0.13

A value of = 0.030°C^{-1} was chosen to fit the Roelands curve closely over the range 60-90°C as shown in Fig. 3.

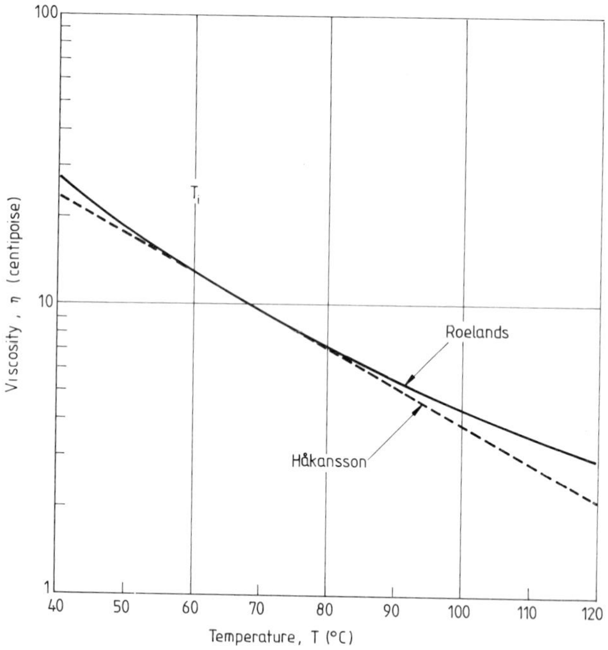

Fig. 3 Lubricant Viscosity

Hakansson's results are such that $\rho C_p \alpha/\beta$ = 1.2. This corresponds to a value of Z = 0.80 in the present case. Results were produced for χ = 0 (isoviscous solutions) and χ = 0.05 (at 8558.5 rpm with ITEMP = 1). The isothermal solutions were also obtained at 8558.5 rpm. Hakansson quotes results at a series of eccentricity ratios whereas the present program calculates the journal position for a given load vector. The programs were run at a series of loads:

W = 2,5,10,20,50,100,200 kN

Hakansson's results are on page 160 of (3).

They are non-dimensional, at specified eccenticity ratios, and are best compared graphically with the present results. This is done in Figs. 5-10 in terms of Hakansson's non-dimensional parameters. The results from the three dimensional THD program alone are plotted since differences between the two programs would not show up on graphs. The values of the relevant parameters are as follows:

Quantity	Divisor	
Load	$\eta_i \omega \left(\frac{R}{\Delta R}\right)^2 L R$	= 29050 N
Flow	$\omega R L \Delta R$	= 107.9 litres/min
Power loss and Heat flow	$\eta_i \frac{(\omega R)^2 L R}{\Delta R}$	= 5.226 kW
Temperature	$\frac{\eta_i \omega}{\rho C_p}\left(\frac{R}{\Delta R}\right)^2$	= 1.667°C

(The value of η_i at 60°C is 12.920cP)

The agreement in Figs. 5-10 is clearly good. Slight differences can be discerned at high loads, probably due to the different viscosity formulae.

3.2 Two Arc Bearing Example

A bearing with two 120° circular arcs was used as an example for the checking of the inlet analysis and the combined results. This was chosen with both horizontal and vertical arc displacements in order to provide a more general check on the programs. A sketch of the bearing is shown in Fig. 4a. The data were as follows:

Diameter, D = 200 mm
Length, L = 100 mm
Arc radius difference, ΔR = 0.2 mm
Configuration: two 120° arcs displaced by 0.2 mm horizontally and by 0.1 mm vertically.
Inlet geometry: G = 50 mm
(Fig 1) H = 3 mm
U = 12.5 mm

Lubricant properties as in 3.1 above
Supply temperature, T_s = 50°C
Supply pressure, p_s = 1 bar
Shaft speed, N = 3000 rpm
Load, W = 10 kN

Adiabatic conditions were assumed.

The results from the two programs, referred to as 2Dthd and 3Dthd, are shown in Table 1. It is possible to check flow and heat balances directly from this table. Spot checks were also carried out on the results for the inlets by hand. There are some differences between the two programs which were found to be mainly due to detailed differences in the way the inlet geometry is specified to the two programs.

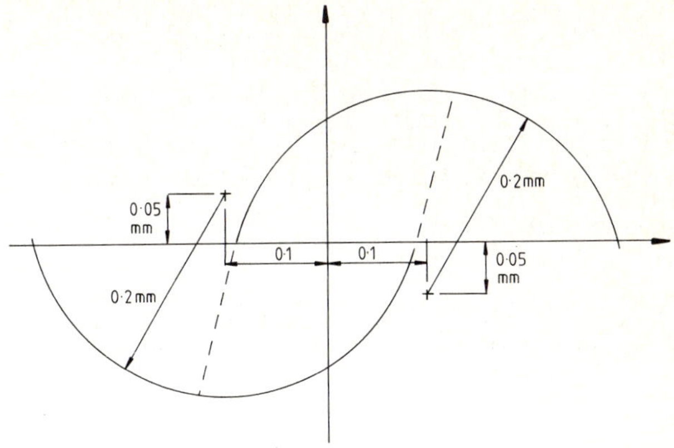

a) Example of two-arc bearing

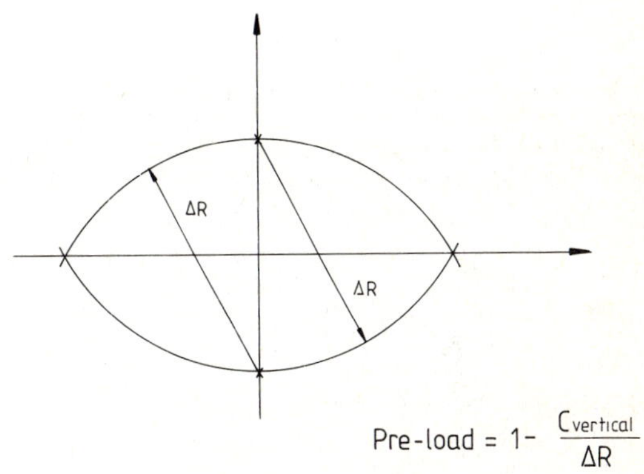

b) Lemon-bore bearing with pre-load of 0·5

Fig. 4 Arc Geometries

3.3 Dynamic Coefficients

The most accurate published coefficients appear to be those produced by Lund and Thomsen (2). These were obtained by direct differentiation of Reynolds equation with respect to displacement and velocity components. This avoids errors due to differentiation of the force components. A comparison was made for a lemon bore bearing with L/D = 0.5, 160° arcs and a preload factor of 0.5, as shown in Fig. 4b . The results are plotted non-dimensionally in Figs 13 and 14. Note that Lund and Thomsen's results are non-dimensional with respect to the arc radius difference. The good agreement confirms that the force derivatives are evaluated correctly and accurately by the program.

In thermohydrodynamic analyses some assumption has to be made concerning temperature variations in the oil film under dynamic conditions. The resulting coefficients cannot be checked externally.

3.4 Thermohydrodynamic Analysis of the Oil Film

The two arc example discussed in 3.2 above was also used to check the thd analysis of the oil film (ITEMP = 2). The journal was kept fixed at the same position so that the results could be compared more directly with those of 3.2. The journal surface was assumed to be at a single, fixed temperature which was determined so that there is no net heat flow to the journal. The results are given in Table 2.

The calculated forces are reduced by thermohydrodynamic effects and the maximum pressure, power loss and oil flows are also lower. The mean temperatures across the film calculated in the thd cases agree closely with those determined directly by the simpler analysis of 3.2.

The thd analysis determines the heat conducted to the journal from each arc (HCFJ) and from each inlet (HCIJ). The heat flows can be seen to balance when these are included.

These types of checks enhance confidence in the validity of the results, but ideally a full external comparison would be preferable. A possible approach to such external benchmarking is suggested in the next sub-section.

4 THERMOHYDRODYNAMIC BENCHMARKS

Single arc results are most suitable as benchmarks since they do not involve empirical assumptions. Because of the large number of parameters involved, only dimensional results will be presented. These are for the same arc geometry and lubricant as in 3.1 above. Adiabatic conditions were assumed with no net heat conducted to the journal. The operating conditions were taken as follows:

Speed, N = 3000 rpm
Load, W = 1, 2, 5, 10, 20 kN
Inlet temperature, T_i = 60°C
Inlet pressure, p_i = 0

The finite difference mesh had 30 circumferential intervals, 10 intervals over half the bearing length and 16 intervals across the film. The results from the two programs are shown in Tables 3 and 4.

The results do not cover the analysis of inlets or of heat conduction in the bearing housing or journal. However, these can be checked separately with relative ease.

5 CONCLUSIONS

Two computer programs are described which carry out thermohydrodynamic analyses of partial arc journal bearings. A validation procedure for the programs is discussed which relies mainly on comparisons with published results and on checks which can be carried out by hand. Adiabatic solutions for a single partial arc have been found to provide useful checks on the solution methods. Some examples are provided of thermohydrodynamic results of this type.

REFERENCES

(1) COWKING, E.W. 'Thermohydrodynamic analysis of multi-arc journal bearings', Tribology international, August 1981.

(2) LUND, J.W. and THOMSEN, K.K. 'A calculation method for the dynamic coefficients of oil lubricated journal bearings', Topics in Fluid Film Bearing and Rotor Bearing System Design and Optimus action, ASME, 1978, pp 1-28.

(3) HAKANSSON, B. 'The Journal Bearing considering Variable Viscosity', Trans. of Chalmers University of Technology, Nr 298, 1965.

(4) NG, C.W. and PAN, C.H.T. 'A linearized Turbulent Lubrication Theory', ASME J. of Basic Engineering, September 1965, p 48.

(5) ROELANDS, C.J.A. 'Correlational aspects of the viscosity - temperature - pressure relation of lubricating oils', PhD Thesis, Technische Hogeschool te Delft, April 1965.

Table 1. Two Arc Bearing. ITEMP = 1

	2Dthd Program			3Dthd Program		
	Arc 1	Arc 2	Bearing	Arc 1	Arc 2	Bearing
Eccenticity, e(mm)	0.1339	0.1073	0.0471	-	-	0.0472
Attitude angle, (deg)	43.84	-91.84	-8.88	-	-	-8.68
Oil FX(N)	1467	-1469	-1.4	1452	1454	-1.5
Force FY(N)	-15244	5247	-9997	-15273	5274	-9999
Power Loss, E (kW)	2.91	1.83	5.86	2.90	1.81	5.65
Inlet , E (kW)	0.54	0.57		0.47	0.47	
Oil Q1	18.31	26.57		18.68	27.33	
Flows Q2	6.62	11.15		6.59	11.13	
(l/min) Qs	11.69	15.42	39.10	12.06	16.27	36.73
Qs(inlet)	5.02	6.96		3.84	4.57	
Heat H1	2.45	2.92		2.51	3.07	
Flows H2	3.12	2.58		3.11	2.55	
kW Hs	2.24	2.18	5.87	2.32	2.34	5.69
Hs (inlet)	0.67	0.77		0.52	0.51	
Pmax(bar)	23.04	7.92	23.04	23.03	7.82	23.03
Inlet Temp T (°C)	54.61	53.79		54.63	53.86	
Mean Film Tmax (°C)	66.26	57.88	66.26	66.58	58.06	66.58

Table 2. Two Arc Bearing. ITEMP = 2

	2Dthd Program			3Dthd Program		
	Arc 1	Arc 2	Bearing	Arc 1	Arc 2	Bearing
Eccenticity, e(mm)	0.1339	0.1073	0.0471	-	-	0.0472
Attitude angle, (deg)	43.84	-91.84	-8.88	-	-	-8.68
Oil FX(N)	963	-1345		881	-1329	-448
Force FY(N)	-13743	4942	-8801	-13687	4990	-8698
Power Loss, E (kW)	2.84	1.80	5.75	2.83	1.78	5.56
Inlet , E (kW)	0.54	0.57		0.47	0.47	
Oil Q1	18.42	26.65		18.80	27.45	
Flows Q2	7.19	11.53		7.12	11.51	
(l/min) Qs	11.23	15.11	38.28	11.63	16.01	35.99
Qs(inlet)	5.02	6.90		3.84	4.52	
H1	2.48	2.72		2.52	2.83	
Heat H2	2.77	2.57		2.74	2.51	
Flows Hs	2.20	2.14		2.25	2.27	
(kW) HCFJ	0.33	-0.19	5.73	0.31	-0.19	5.50
Hs (inlet)	0.68	0.71		0.51	0.47	
HCIJ	-0.05	-0.09		-0.05	-0.08	
Pmax(bar)	20.89	7.37	20.89	20.66	7.29	20.66
Inlet T (°C)	54.64	53.52		54.61	53.54	
Mean Film Tmax (°C)	66.41	58.74		67.12	59.12	
White metal Tmax (°C)	74.75	64.05	74.75	75.89	65.31	75.89

Table 3. THD Benchmarks. 2Dthd Program

Case Number	1	2	3	4	5
Load, W (kN)	1	2	5	10	20
Eccenticity, e(mm)	0.0370	0.0614	0.1012	0.1301	0.1543
Attitude angle, (deg)	67.97	57.36	45.70	38.91	32.82
Power Loss, E (kW)	1.350	1.496	1.891	2.373	2.976
Oil { Q1	20.22	19.98	18.54	16.76	14.54
Flows { Q2	16.28	13.89	10.12	7.36	4.97
(l/min) { Qs	3.94	6.09	8.42	9.39	9.56
Heat { H1	0	0	0	0	0
Flows { H2	1.188	1.244	1.487	1.838	2.219
(kW) { Hs	0.156	0.244	0.391	0.546	0.804
$\underline{P}max$ (bar)	1.18	2.44	6.81	15.36	35.34
$\bar{T}max$ (°C)	63.63	64.35	66.82	70.88	78.30
Tmax (°C)	67.74	68.85	72.26	77.24	85.65
T_J (°C)	60.92	61.16	62.02	63.64	67.46

Table 4. THD Benchmarks. 3Dthd Program

Case Number	1	2	3	4	5
Load, W (kN)	1	2	5	10	20
Eccenticity, e(mm)	0.0370	0.0615	0.1014	0.1302	0.1542
Attitude angle, (deg)	68.29	57.47	45.92	39.17	33.15
Power Loss, E (kW)	1.350	1.496	1.893	2.382	3.008
Oil { Q1	20.24	20.00	18.59	16.85	14.72
Flows { Q2	16.28	13.88	10.09	7.33	4.94
(l/min) { Qs	3.92	6.05	8.40	9.37	9.56
Heat { H1	0	0	0	0	0
Flows { H2	1.189	1.241	1.473	1.786	2.113
(kW) { Hs	0.162	0.249	0.391	0.538	0.761
$\underline{P}max$ (bar)	1.18	2.45	6.81	15.36	35.34
$\bar{T}max$ (°C)	63.83	64.65	67.19	71.09	77.93
Bearing, Tmax (°C)	68.12	69.44	73.07	77.90	85.64
Journal, T_J (°C)	60.98	61.20	61.94	63.33	66.58

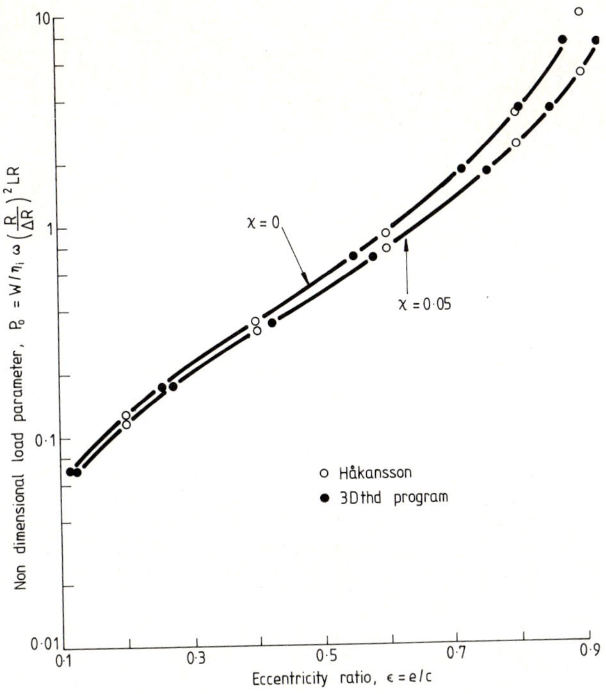

Fig. 5 Load Capacity

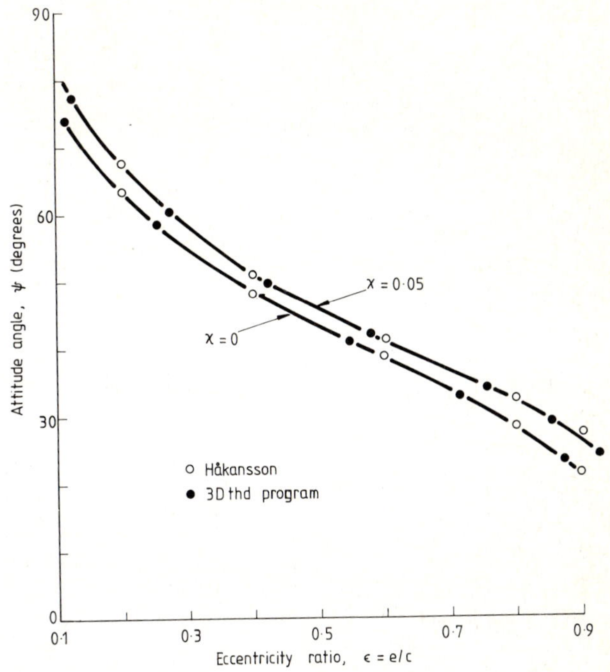

Fig. 6 Attitude Angle

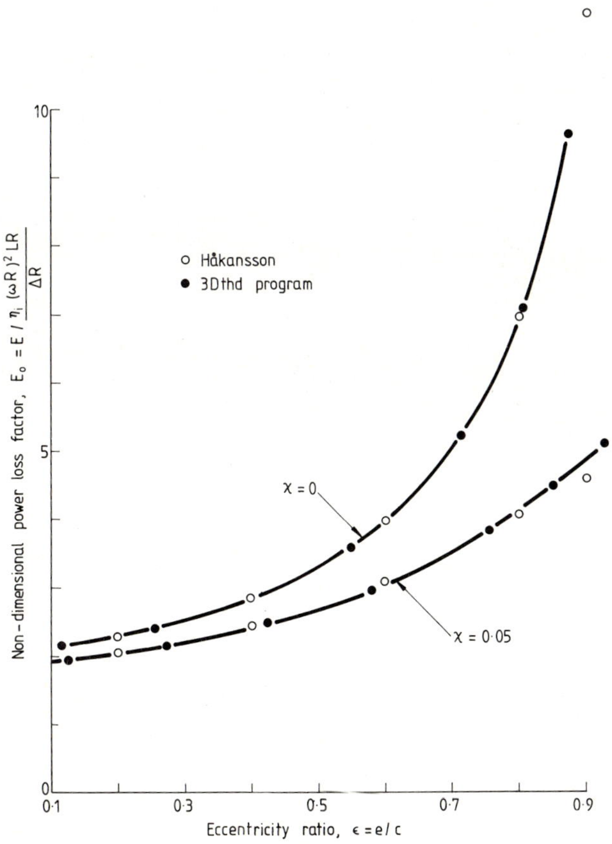

Fig. 7 Power Loss

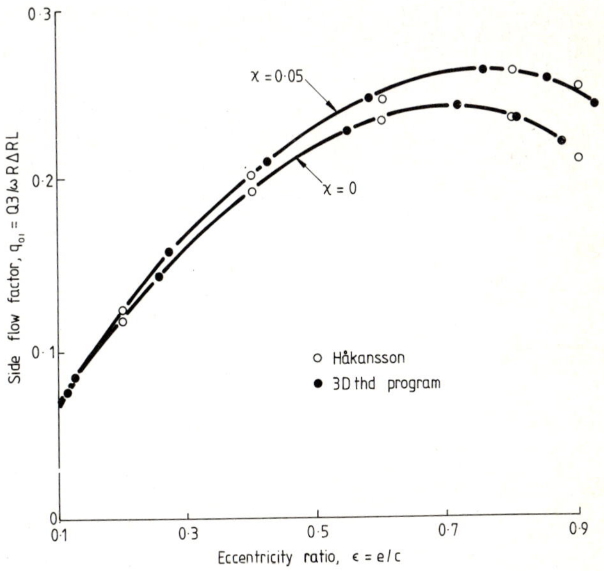

Fig. 8 Oil Flow

Figs. 5 - 8 Comparison with Hakansson

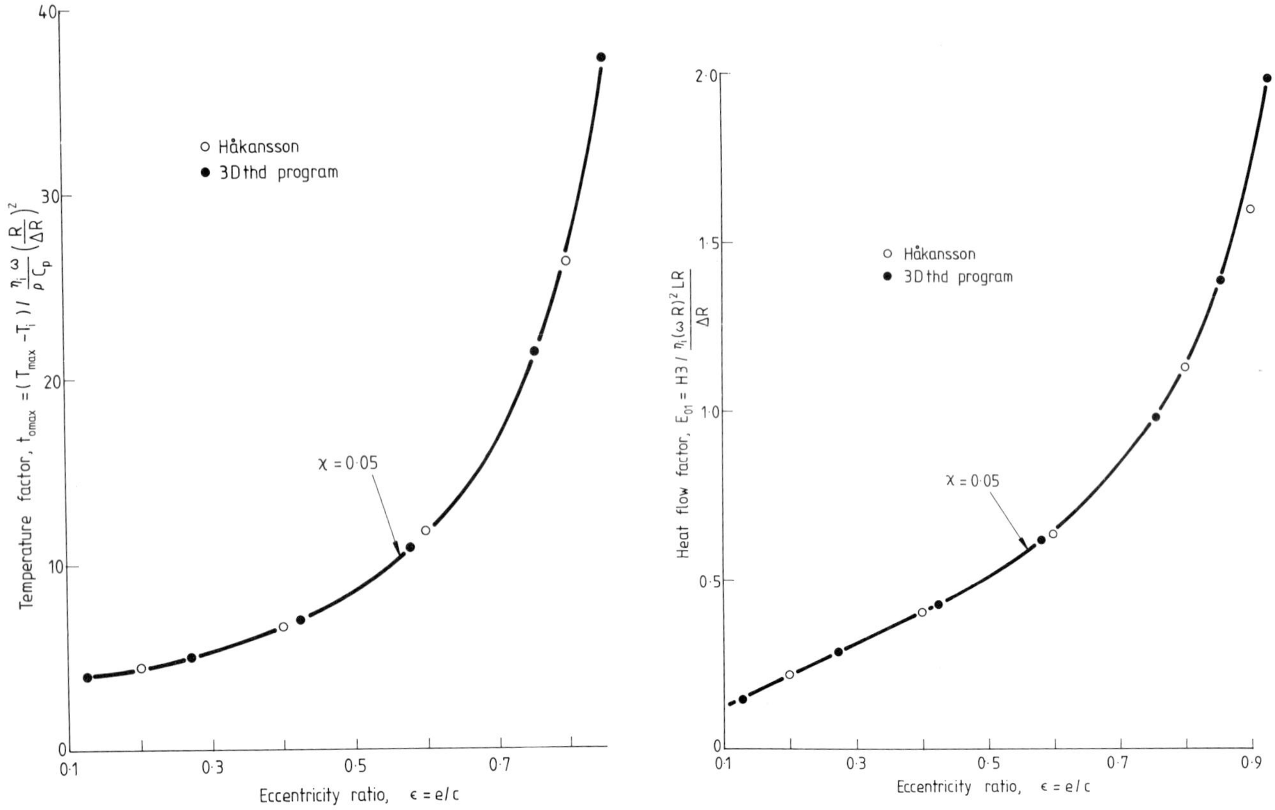

Figs. 9 and 10   Comparison with Hakansson

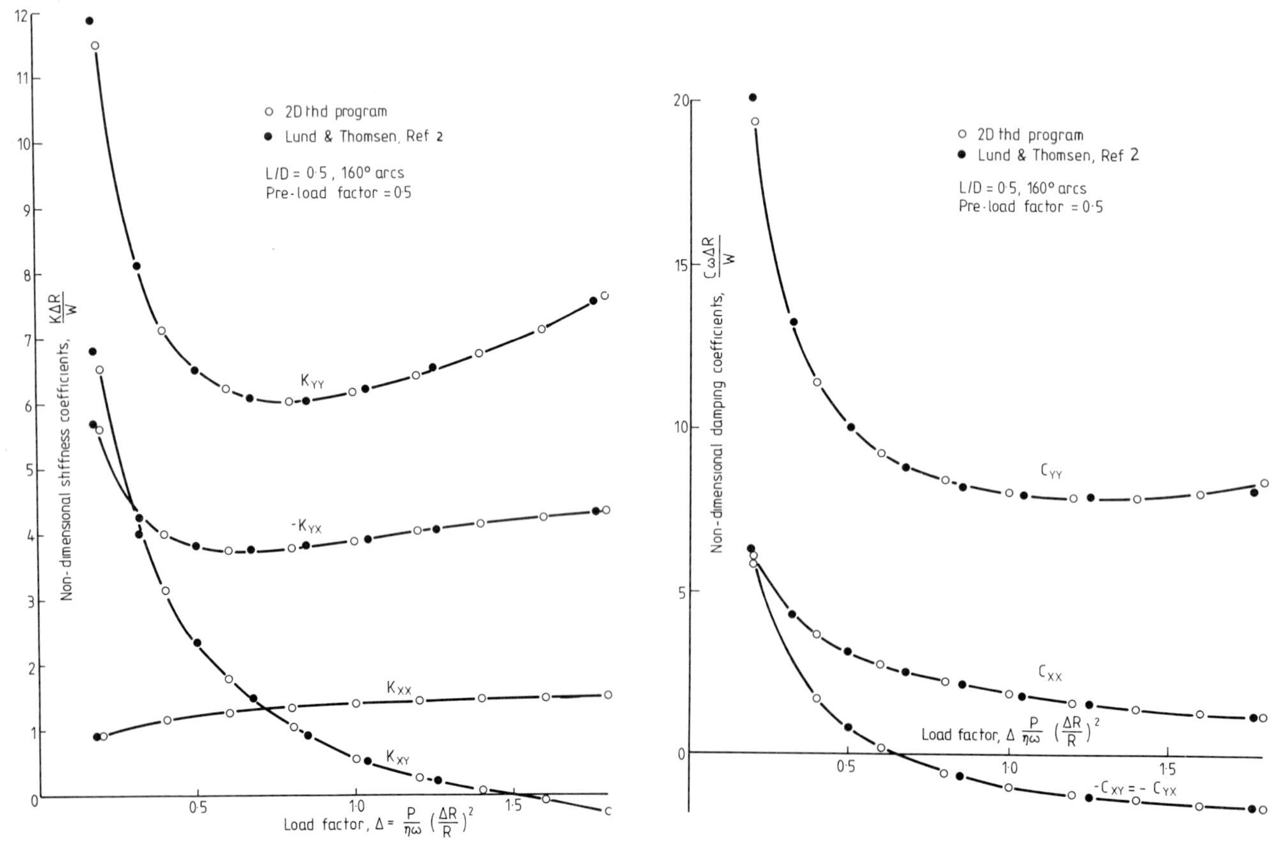

Figs. 11 and 12 Comparison with Lund and Thomsen

Paper VIII(iii)

Elapsed time for the decay of thermal transients in fluid film bearing assemblies

C. M. M. Ettles, H. Heshmat and K. R. Brockwell

The decay time of thermal transients in a bearing assembly is of practical interest, since premature measurements could be misleading. In addition, the thermal deformation that occurs during a transient does not vary smoothly or monotonically. In severe cases this can lead to an instability and failure of the bearing.

The principal factors governing thermal transients are inferred from classical solutions. More accurate predictions may be made from a coupled solution of the film and bounding components. Experimental data from six thrust bearings and two journal bearings are examined. It is found that, following a step change of operating condition, the heating or cooling rate varies approximately exponentially.

1. INTRODUCTION

When the operating conditions of a bearing are changed, a finite time must elapse before thermal equilibrium is reestablished. In the transient period that follows, say, a step change of load or speed, the bearing film thickness is affected by varying viscosity and varying thermal deformation. Thermoelastic effects will always lag on observed changes of surface temperature, and may not blend smoothly from one condition to the next. Temporary reductions in clearance can occur, which in extreme cases lead to seizure of the bore from a thermal ratchetting process.

Two aspects of transient thermal behavior are studied in this paper:

- The time delay for some proportion (50%, 70%) of the film temperature change to occur, following a step change in conditions.
- Transient thermoelastic deformations and their effect on the time delay.

In the first section some simple, classical solutions of thermal transients in solids are considered. These give guidelines concerning the important parameters and boundary conditions. The results are used in setting up the second section of the paper, which is a more detailed, numerical analysis that couples the separate components of pads, film and rotor of a thrust bearing in the time domain. The numerical solution is considered to be more accurate than the classical solutions, but it is, of course, applicable only to particular cases and generality is lost.

In the final section of the paper a collection of experimental results for thrust and journal bearings is considered. An attempt is made to correlate all the results with Newton's law of heating (or cooling). This law states that the current rate of temperature change of a body is proportional to the current temperature difference between the body and ambient. This results in an exponential change of temperature with time which, to an approximation, was the observed behavior of most of the experimental results.

1.1 Notation

A	Bush outer radius/bore
c	Specific heat of shaft, pad or bush material
C	Journal bearing radial clearance
k	Thermal conductivity of shaft, pad or bush material
k_f	Thermal conductivity of lubricant
F	Fourier number, Eq.(3)
G	Surface heat transfer coefficient
H	Pad thickness
Nu	Nusselt number, GR/k or GH/k
r	Radial coordinate
R	Radius of shaft
t	Time
T	Temperature
u	Radial displacement
α	Coefficient of expansion
ΔT_F	Overall temperature change
ϵ_T^*	Thermal strain, Eqs.(5), (6)
ρ	Density of shaft, pad or bush material.

2. CLASSICAL SOLUTIONS FOR THERMAL TRANSIENTS

The equation of transient heat conduction with axisymmetry is

$$\frac{\partial T}{\partial t} = \frac{k}{\rho c}\left(\frac{\partial^2 T}{\partial r^2} + \frac{1}{r}\frac{\partial T}{\partial r}\right) \tag{1}$$

When this is set in nondimensional form, the result is

$$\frac{\partial T^*}{\partial t^*} = \frac{\partial^2 T^*}{\partial r^{*2}} + \frac{1}{r^*}\frac{\partial T^*}{\partial r^*} \tag{2}$$

where nondimensional time t^* is defined as

$$t^* = F = \frac{k}{\rho c} \cdot \frac{t}{R^2} \tag{3}$$

where t is time and R is a reference radius. This parameter is also known as the Fourier number F, and for given boundary conditions is the most important parameter in any transient heat solution.

2.1 Boundary Conditions

To simulate a step change in a journal bearing the most appropriate boundary condition is (apparently) to allow the surface of the shaft or bush to be set in contact with fluid at a temperature ΔT_F, with a convection coefficient G effective at the surface. If G is infinite the surface temperature rises to ΔT_F instantaneously. When set in nondimensional form, the convection coefficient G appears as GR/k = Nu, the Nusselt number.

2.2 Results

Figures 1a, 2a show the surface temperature change of a journal and bush respectively for various Nu. The radius of the journal is taken as R and the inner and outer radii of the bush are taken as R and 2R, with the boundary condition $\partial T/\partial r = 0$ at the outer radius.

The rate of surface temperature rise is dominated by the value of Nu, so the correct choice of value for Nu is important for these results to be meaningful. Some classical solutions exist for Nusselt number Nu_f in the fully developed laminar flow of fluid between parallel walls a distance h apart.

For uniform wall temperature:

$$Nu_f = \frac{G \cdot 2h}{k_f} = 7.54$$

For uniform heat flux:

$$Nu_f = \frac{G \cdot 2h}{k_f} = 8.24$$

Here, k_f is the thermal conductivity of the fluid, 2h is the hydraulic depth and heat transfer takes place to or from the fluid. To set Nu_f in terms of Nu, these simplifications could be made:

$$\text{Clearance ratio, } 2h/R = 2C/R = 2\delta \times 10^{-3} \quad (4a)$$

$$\text{Lubricant conductivity, } k_f = 0.15 \text{ W/m}\,^\circ\text{C} \quad (4b)$$

$$\text{Journal \& bush conductivity, } k = 50 \text{ W/m}\,^\circ\text{C} \quad (4c)$$

$$Nu_f = 0.5\,(7.54 + 8.24) = 7.89 \quad (4d)$$

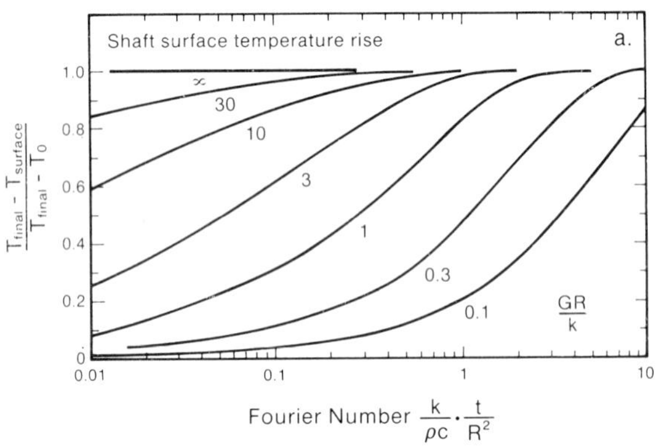

Fig.1a The Surface Temperature of a Shaft Following Exposure to an Ambient Temperature ΔT_F. The numerical values are Nusselt number effective at the surface, GR/k.

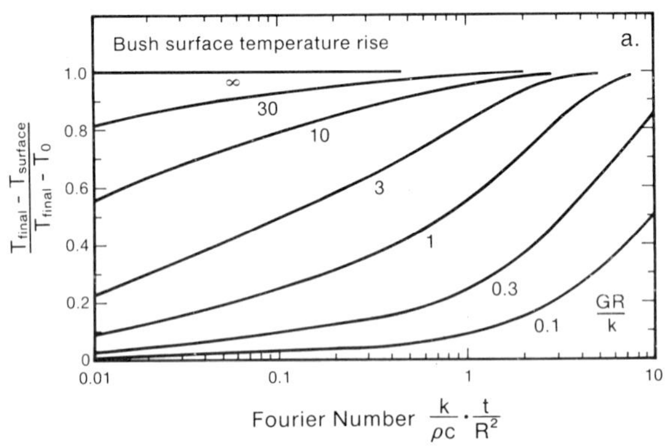

Fig.2a The Inner Surface Temperature of a Bush Following Exposure of the Bore to an Ambient Temperature ΔT_F. (Outer radius/inner radius = A = 2.) The numerical values are Nusselt number effective at the bore, GR/k.

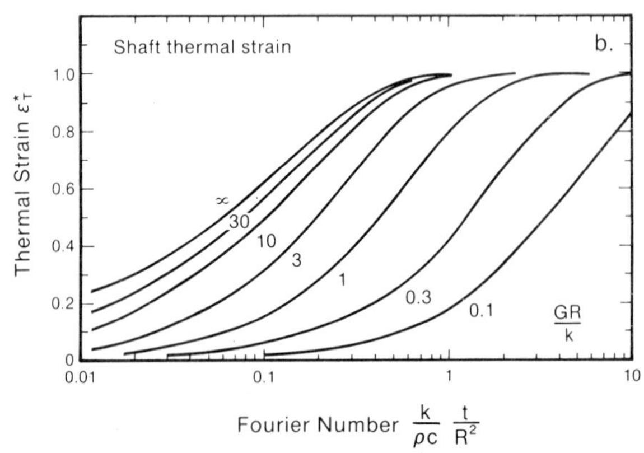

Fig.1b The Proportional Thermal Strain

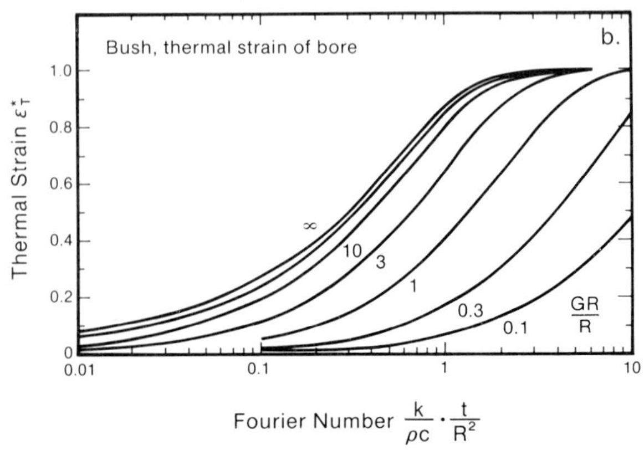

Fig.2b The Proportional Thermal Strain of the Bore

This gives the value of Nu to be used in Figs.1 and 2 as $Nu = GR/k = 11.85/\delta$, where δ is the clearance ratio defined by (4a).

Unfortunately this can give only an upper bound of Nu, since the fluid in which the journal or bush is suddenly "immersed" is taken to be at the final temperature, ΔT_F. In practice, the temperature difference between the midpoint (say) of the film and the surfaces is much less than this (as will emerge from the numerical solution). In summary, the value of $Nu = GR/k \doteq 11.85/\delta$ is probably reasonable, but the temperature difference "across" G is always less than the assumed value.

The experimental data to be presented suggest an effective value of Nu in the range 1 - 5. For Nu = 1, the 50% response time for the surface of the shaft is F = 0.29 and for the surface of the bush is F = 0.79. Since the particular bush considered here has a greater thermal capacity than the shaft, the 50% response time for this assembly will be closer to 0.79 than 0.29.

2.3 Thermal Strain

The thermal strain of the shaft is given by

$$\epsilon_T = u_{r=R}/R = \alpha\Delta T_F \cdot \epsilon_T^* = \alpha\Delta T_F \cdot 2 \int_0^1 T^* r^* \cdot dr^* \qquad (5)$$

When the expansion or contraction is complete, the thermal strain is $\alpha \cdot \Delta T_F$ and $\epsilon_T^* = 1$.

The thermal strain of the bush is (for zero radial stress at the inner and outer boundaries):

$$\epsilon_T = u_{bore}/R = \alpha\Delta T_F \cdot \epsilon_T^* = \alpha\Delta T_F \cdot (2/(A^2 - 1)) \cdot \int_1^A T^* \cdot r^* \cdot dr^* \qquad (6)$$

where $A = R_{outer}/R = 2$ in Figure 2. When the deformation is complete, the thermal strain is $\alpha\Delta T_F$ and $\epsilon_T^* = 1$.

For the particular conditions analysed (no restraint of the bush at the outer radius), when the transient process is complete the radial clearance of a journal bearing will be unchanged, since both components deform by the same amount, $\alpha R\Delta T_F$. However the clearance will temporarily expand following (say) a decrease of speed or will temporarily contract following an increase of speed. The timewise behavior of an expansion or contraction may be found from Figures 1 and 2 by subtracting ϵ_T^* for the two components. Under severe conditions a contraction may be unstable and lead to seizure. The closure of the clearance from a thermal ratchetting process is a potential danger with tilting pad journal bearings [1]. With this type of bearing, the only conduction path from the film to the housing is through the pivots, consequently the housing temperature and bore size is only slightly increased following a rise in speed. However the shaft will expand outwards, while the pads will expand "inwards," which under normal circumstances will increase the bearing preload. Examples of unstable increase are given by Conway-Jones and Leopard [1].

Figures 1b and 2b show an upper bound to the rate at which a shaft or bush will expand. This corresponds to $Nu = \infty$, or a stepwise increase of the surface temperature to the final value.

2.4 Transient Thermal Deformation of a Beam

To explore the transient deformation of thrust pads it is convenient to treat them as beams. Figure 3 shows the thermal deflection of a beam subjected to a step increase in temperature on one face. The deformation reaches a peak quite early in the process at F = 0.086 and subsequently decreases as equilibrium is established.

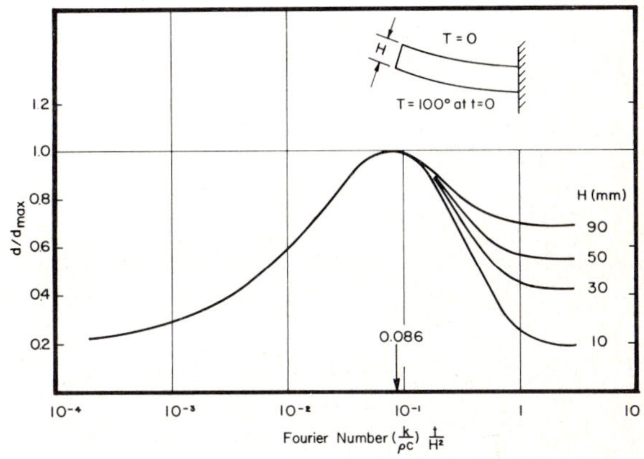

Fig.3 The Transient Thermal Deflection of a Steel Beam Subject to a Sudden Temperature Increase on one Face

This perhaps unexpected behavior has been verified experimentally by Zerbe [2], using beams of various materials, heated suddenly on one face by a stream of hot water under laminar flow conditions. The convection coefficient G for flow over a plate has been widely researched. The closed form result for G was used in subsequent calculation of the Nusselt number, GH/k where H is the thickness of the beam.

Figure 4 shows Zerbe's experimental results of peak deflection and time to peak deflection against the theoretical curves. The agreement is quite good, although not many data points are shown. The results of Figures 3 and 4 indicate that transient thermal deflection in a thrust bearing might precipitate failure if the pads are already highly deformed from thermal causes before some change in the operating conditions is made.

3. NUMERICAL MODEL OF TRANSIENT CONDITIONS

The simple solutions in the previous section are useful for giving guidance as to the important parameters and boundary conditions, but in practice the film and the two components are coupled, which is best treated numerically. A numerical solution can give more realistic predictions although, as is usual in any thermohydrodynamic analysis, generality is lost. Previous numerical models of transient effects have been described by Ezzat and Rhode [3] for thrust bearings and by Czeguhn [4] for journal bearings.

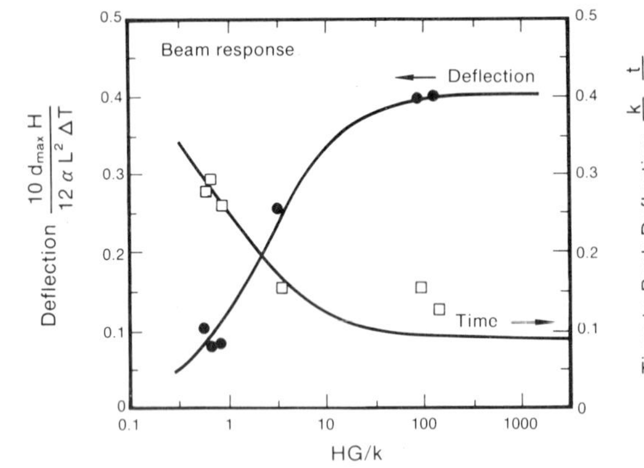

Fig.4 The Peak Thermal Deflection of a Beam and the Time Required for Peak Deflection to Occur. The points are experimental data and the curves are results of a numerical model.

The results shown in this present paper are taken from Ettles [5], and concern a thrust bearing viewed in elevation. This allows the appropriate flux continuity conditions to be applied at the fluid-solid interfaces and at the exterior surfaces. The analysis is two-dimensional, with coordinates normal to the film (and through the pad and rotor) and in the direction of sliding. The equation of transient heat conduction is solved in the pad and rotor at each time step.

The analysis has these features:

a) The thermal inertia of the film is neglected. At the completion of a time step, a new film shape and film temperature distribution are found to satisfy the current load requirement. During this iteration the surface temperature of the pad and rotor are maintained constant and the temperature within the film is allowed to vary. At the completion of this iteration the local heat flux at the boundaries of the solid components is known, which allows the next time step to be made.

b) The time step is made by calculating the new temperature distributions in the shoe and rotor, using the updated flux values at the boundaries.

c) The temperature in the film is then reevaluated as in a), using the updated surface temperatures. The operations in a), b) form those for a time step.

d) Thermoelastic deflection of the shoe is considered, together with hot oil carryover. Reverse flow is allowable and is accommodated by up-winding. More complete details are given in [5].

Figures 5a,b show the change in bearing temperature, deflection and minimum film thickness consequent to a sudden doubling of the load at constant speed. The results are plotted to a base of Fourier number, where the characteristic dimension is H = 30 mm and is the shoe or rotor thickness (set equal for all cases to be shown). A time scale in seconds is also given, and the expected equilibrium temperatures at the end of the transient period are indicated.

The deflection reaches a slight peak at $F \approx 0.2$, which corresponds approximately to the behavior of a beam, as discussed in the previous section. Figure 5c shows a corresponding case where the load is suddenly halved.

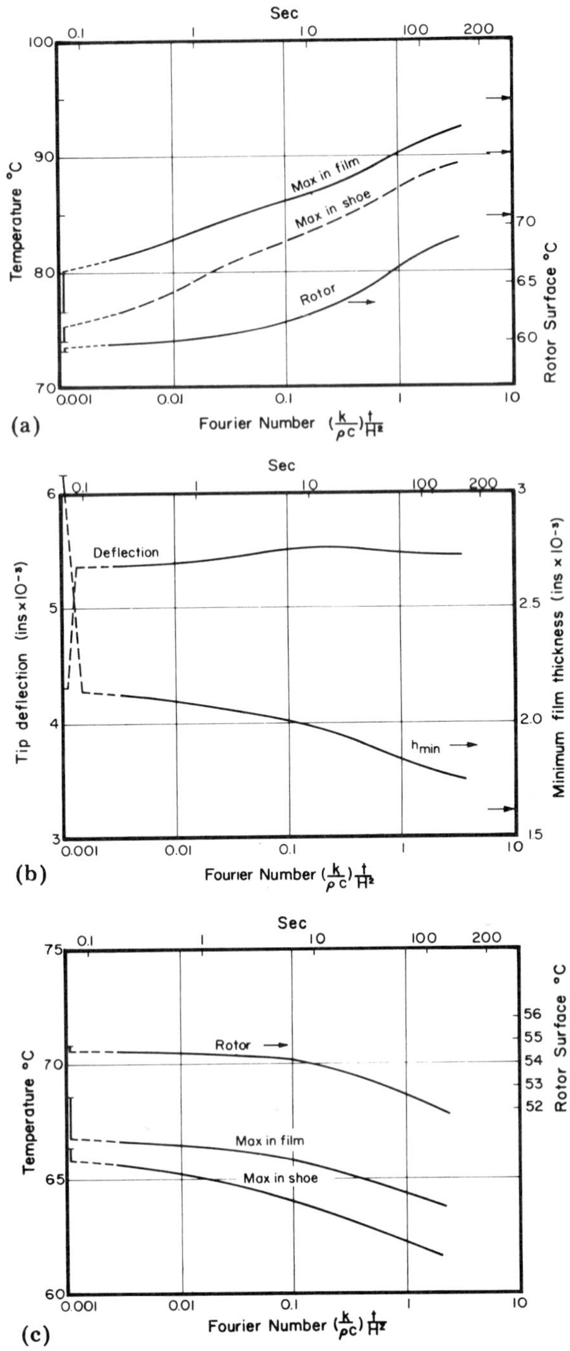

Fig.5 Model Result Showing Changes in (a) Temperature and (b) Deflection and Minimum Film Thickness in a Thrust Bearing, Following a Step Increase in Load from 2.8 to 5.6 MPa. The pad is 120 mm long and 30 mm thick, speed is 30 m/s. (c)
(c) Changes in temperature of same bearing when load is halved from 1.4 MPa.

Figure 6 shows a similar result for a much larger pad of 250 mm thickness. The similarity with Figure 5 is quite striking. In both cases the proportional changes of temperature $\Delta T/\Delta T_F$ appear to be correlated against the Fourier number. This suggested a possible treatment of experimental results, in which the proportional temperature rise is plotted against Fourier number, using the thickness of the pads or rotor (whichever being the greater) as the characteristic dimension.

4. EXPERIMENTAL RESULTS

4.1 Thrust Bearings

The authors were able to collect transient results for six bearing assemblies that varied in size by about an order of magnitude. Two sets of results appear in the open literature [6,7]. The remainder are proprietary.

Figure 7 shows temperature data from Chambers and Mikula [6], taken during the start-up of a hydroelectric turbine. The thickness of the pad H is approximately 80 mm. The results shown are for a thermocouple on the mid radius within 9.7 mm of the face. The measurement point is sufficiently far from the film to "damp" the sharp initial rise shown in the model solutions. The response, in the form $\Delta T/\Delta T_F$, resembles the model solutions. The dashed line is the characteristic:

$$\frac{\Delta T}{\Delta T_F} = 1 - e^{-1.89\ F} \qquad (7)$$

which is set to pass through the 60% response point. The matching of this curve with the experimental data is quite good.

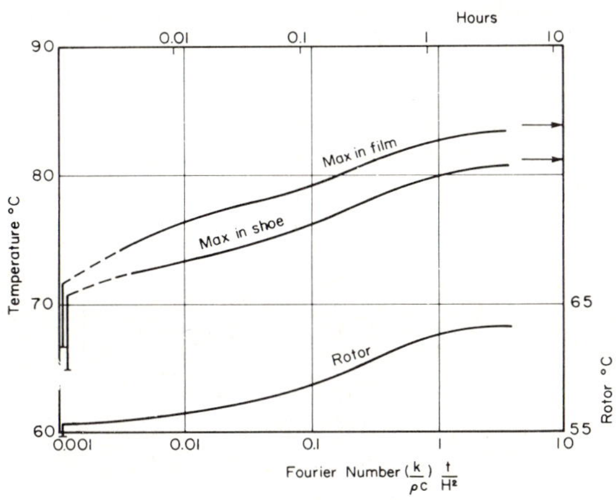

Fig.6 Model Result Showing Changes in Temperature for a Large Shoe (Length 1 m, Thickness 250 mm) Subsequent to a Sudden Increase in Load from 1.8 MPa to 3.6 MPa. (The arrows show known asymptotic values.)

To further check the use of the Fourier number as a common time scale, the 50% response time of all six bearings was plotted against the pad thickness H, as shown in Figure 8. The line drawn through the points has the characteristic $t \propto H^2$, and for the data shown:

$$\frac{\Delta T}{\Delta T_F} = 1 - e^{-1.6\ F} \qquad (8)$$

This gives the following nondimensional times F for 50%, 70% and 90% response:

% Response	F
50	0.433
70	0.752
90	1.44

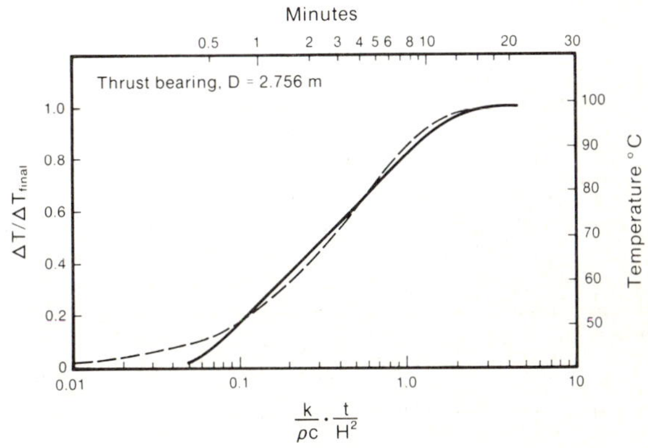

Fig.7 The Increse of Thrust Bearing Temperature Following the Start-up of a Hydrogenerator. The temperature is measured on the mid radius at approximately the 70% position. The dashed line is the exponential law, Eq.(7). From Chambers and Mikula [6].

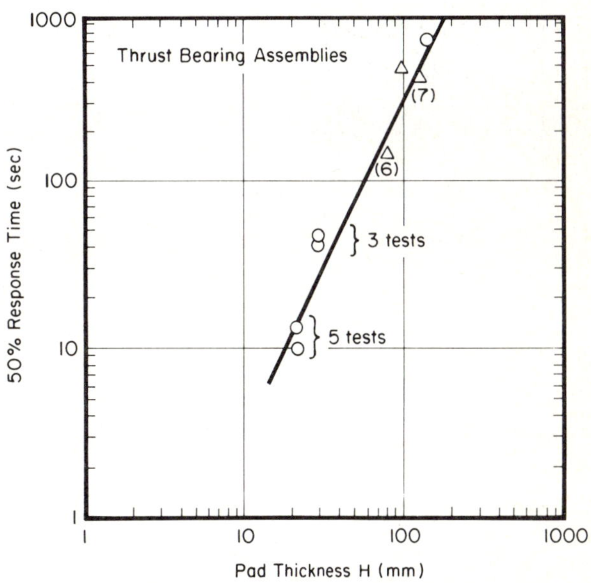

Fig.8 The 50% Response Time of Six Thrust Bearing Assemblies Plotted Against the Pad Thickness H. The open triangles are for hydrogenerator units. The numbers 6, 7 are references.

The reasonable concurrence of the data points in Figure 8 with H^2 must, to some extent, be fortitous. For the hydrogenerator bearings (shown as open triangles) the rotor thickness is difficult to estimate since the thrust runner ring is located against a massive shoulder on the shaft. In general, the rotor thickness can be taken as about twice that of the pads.

4.2 Journal Bearings

Figures 9a,b show the temperature increase at a point close to the film of a tilting pad journal bearing. These data are for a 76 mm bearing with five centrally pivoted pads, no preload and a clearance ratio of 0.00157 (δ = 1.57). The thermocouple was located at the 65% position in the hottest pad. The centre of the thermocouple junction was about 1.5 mm from the film. The bearing and test rig are described in [8]. The two results shown are for sudden speed increases from different initial conditions.

In the exponential expression

$$\frac{\Delta T}{\Delta T_F} = 1 - e^{-aF} \qquad (9)$$

the exponent a is 4.58 for Figure 9a and 5.73 for Figure 9.b. The radius of the shaft is used as the dimension in F. The concurrence of the experimental results with the exponential law is quite close.

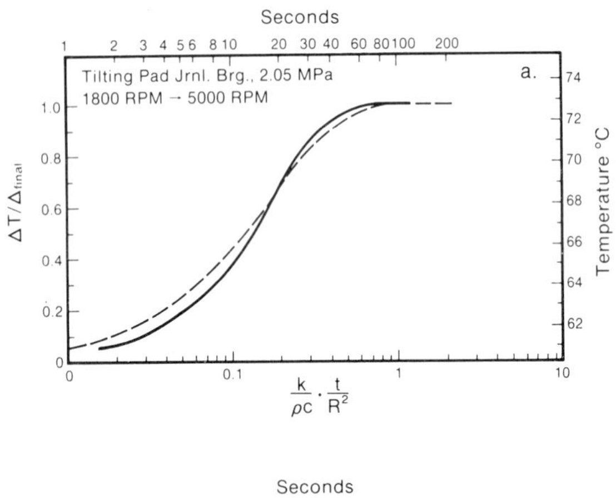

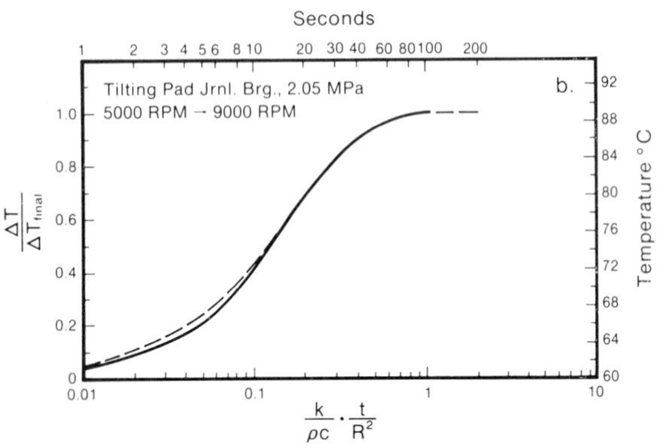

Fig.9 The Local Temperature in a 76 mm Tilting Pad Journal Bearing Following a Sudden Increase of Speed at Constant Load. (a) 1800 - 5000 rpm, (b) 5000 - 9000 rpm.

The same type of behaviour is shown for this bearing on rundown, Figure 10a, where a = 4.58. Agreement with the exponential law is less good for Figure 10b, that shows the rundown of a 305 mm journal bearing. The average temperature from several thermocouples close to the film is plotted against time and against Fourier number. For the dashed line shown, a = 9.21.

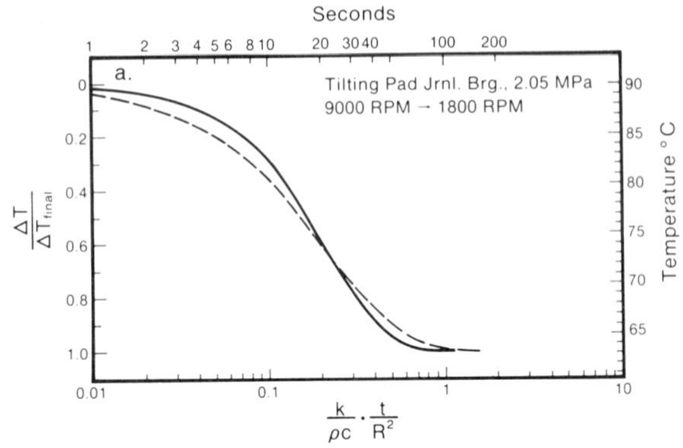

Fig.10a The Local Temperature in a 76 mm Tilting Pad Journal Bearing Following a Sudden Decrease of Speed at Constant Load

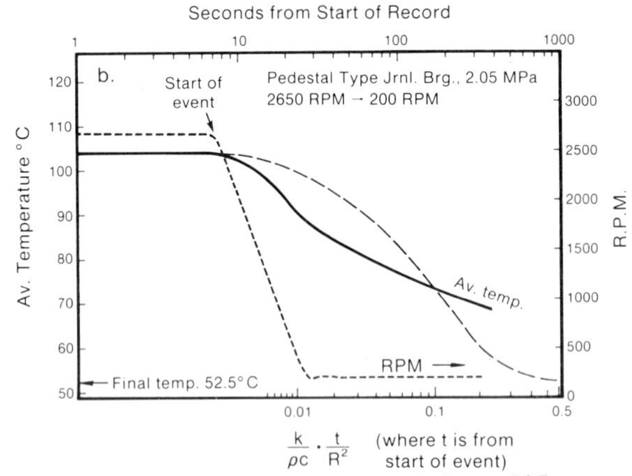

Fig.10b The Average Temperature in a 305 mm Journal Bearing Following a Sudden Decrease in Speed at Constant Load

5. DISCUSSION AND CONCLUSIONS

The decay time of thermal transients in a bearing assembly is of practical interest since

a) Errors can arise in laboratory or field tests if premature readings are made.
b) Temporary reductions in clearance (below the equilibrium value) can occur in journal bearings following a change to more severe conditions.
c) Thermal deflection of thrust shoes can occur that exceed the equilibrium value.

Some guidance as to elapsed time and transient thermal deformation may be found in the classical solutions of simple shapes. It is difficult to obtain quantitative data from these solutions, due to uncertainties in the effective Nusselt number. The Nusselt number directly affects the flux at the surfaces. Unfortunately the closed form solutions for Nu_f

for flow between plates can only give an upper bound value. In Figures 1 and 2 the upper bound of Nu is about 10 for a journal bearing. Practical values for journal bearing assemblies appear to be in the range 1 - 5.

These uncertainties do not arise in a numerical solution that couples the film, journal and bearing, although the complexity is increased. In particular cases, involving thermal ratchetting or unfavourable pad deformation, it may be worthwhile to develop a numerical model so that various options may be explored.

The numerical models developed here consider only the film and the two components bounding the film. It is assumed that the lubricant supply temperature (or chamber temperature) is unaffected by the change in operating conditions. This will not be the case in some self-contained bearing assemblies that include a sump with partial cooling or cooling by convection to ambient. In these cases a third thermal mass should be included (in addition to the two bounding components). In most cases it will be difficult to quantify the "third mass" effect, except by experiment.

In some of the data gathered for this paper, the cooling time for an assembly was (for some assemblies only) about twice as long as the heating time. This difference in heating and cooling rates was less for smaller step changes in conditions.

The majority of the experimental data could be matched fairly well with an exponential law of the form of Eq.(9). This implies that each fluid film bearing has a "time constant," and that the decay time for, say, a 90% change is unaffected by the magnitude of the change. change.

REFERENCES

1. Conway-Jones, J.M. and Leopard, A.J., "Plain Bearing Damage," Proc. Fourth Turbomachinery Symposium, Gas Turbine Laboratories, Texas A&M University, College Station, Texas, October 1976, 59-63.
2. Zerbe, Glenn H., "Transient Thermal Reflections of a Cantilever Beam," Senior Project, Dept. of Mech. Engr., Rensselaer Polytechnic Inst., Troy, NY, 1987.
3. Ezzat, H.A. and Rohde, S.M., "Thermal Transients in Finite Slider Bearing," ASME Jrnl. of Lubn. Tech., 96, July 1974, 315-321.
4. Czeguhn, K., "Variability with Time of Bearing Clearance Due to Temperature and Pressure," Proc. Inst. of Mech. Engrs., 1966-67, 187, Part 30, pp.216-223.
5. Ettles, C.M.M., "Transient Thermoelastic Effects in Fluid Film Bearings," Wear, 79, 1982, 53-71.
6. Chambers, W.S. and Mikula, A.M., "Operational Data for a Large Vertical Thrust Bearing in a Pumped Storage Application," Trans. STLE, 31, 1988, 61-65.
7. Nelson, D.V., Plummer, M.C. and McCulloch, R.C., "Overload Performance Testing and Evaluation of a Hydrogenerator Tilting Pad Thrust Bearing," Proc. American Power Conference, 46, 1984, 1057-1062.
8. Brockwell, K.R. and Dmochowski, V., "Calculation and Measurement of the Steady State Operating Characteristics of the Five Shoe Tilting Pad Journal Bearing," Proceedings of this Conference.

Paper VIII(iv)

Design procedures based on numerical methods for hydrodynamic lubrication

J. O. Medwell

SUMMARY

This paper presents a review of numerical schemes for predicting the performance of cylindrical bore journal bearings which have been developed under the initial guidance of F. T. Barwell at Swansea. The schemes are based on the finite element and finite difference methods which are used to solve the lubricant momentum and energy equations.

The effects of recirculatory flow, bush conduction and deformation, shaft expansion and misalignment have been considered.

NOMENCLATURE

B	=	bush thickness
C	=	radical clearance
C_p	=	specific heat
D	=	journal diameter
E	=	Young's Modulus
h	=	local film thickness
k	=	thermal conductivity
L	=	bearing length
M (x,y,z)	=	mesh coordinate point
p	=	local pressure
r	=	radial coordinate
R	=	journal radius
Re	=	film Reynolds number $[= \frac{\omega RC}{\nu}]$
T	=	temperature
u	=	velocity component in x direction
U	=	velocity vector ui + uj + wk
v	=	velocity component in y direction
w	=	velocity component in z direction
x,y,z	=	cartesian coordinates
θ	=	see Figure 1(a) or circumferential coordinate
∇^2	=	the Laplacian operator
α	=	see Figure 1(a)
β	=	see Figure 1(a)
ϵ	=	eccentricity ratio
μ	=	molecular viscosity
ν	=	kinematic viscosity
ν_x	=	turbulent viscosity
ν_z	=	turbulent viscosity
ν_E	=	Poisson's ratio
ξ,η,μ	=	parametric coordinate directions
ρ	=	lubricant density
φ	=	see Figure 1(a)
Φ	=	dissipation function
ω	=	rotational speed

INTRODUCTION

The theory of operation of hydrodynamic journal bearings was well established by Reynolds over a century ago. However, as pointed out by Barwell et al [1] the application of this theory to design was not simple because

(i) it is not possible to obtain a general analytical solution to the Reynolds equation

(ii) it is difficult to provide the correct value of viscosity for use in the equation because of its temperature dependence. Strictly, since large differences in temperature can arise the consequent variation of viscosity can only be determined by including the energy equation, containing the associated dissipative terms, in the design procedure. This usually involves using iterative techniques in order that the Reynolds and the energy equations can be solved simultaneously.

With the availability of modern day numerical and computational methods, this does not pose any particular problem. The numerical techniques which have been used most often are finite difference methods and, latterly, in the last decade or so, finite element methods. The former has been used in the generation of design charts and also forms the basis of software services such as provided by ESDU. However, although these may be used with confidence in normal journal bearing design calculations they can be suspect when off design conditions are met. Typical of these are: recirculatory flows that can occur in the highly eccentric lobes of profile bore bearings, shaft misalignment and thermo-elastic distortion in bearing bushes and shafts.

This paper will discuss how these conditions may be incorporated into standard numerical design procedures that have been developed, initiated by the late Professor F T Barwell, at Swansea.

In this respect most of the boundary conditions imposed in the solution procedures relate to the extensive experimental programmes that have been carried out in Swansea. These have been concerned with assessing the performance of standard cylindrical bearings which were fed by two diametrically opposed axial grooves at 90° to the load line as shown in Figures 1(a) and 1(b).

Wide ranges of speeds (both laminar and turbulent lubricant films) clearance ratios and load capacities were investigated (see Ref [2]). However, in the interests of brevity, detailed boundary conditions have been omitted and attention focused on the governing equations and solution strategies.

Design Procedures

The equations of motion and energy of an incompressible lubricant flow are based on the conservation laws of mass, momentum and energy. Generally, the three dimensional momentum equations may be written as

$$U\nabla . u = -\frac{1}{\rho}\frac{\partial p}{\partial x} + \nu\nabla^2 u \qquad [1]$$

$$U\nabla . v = -\frac{1}{\rho}\frac{\partial p}{\partial y} + \nu\nabla^2 v \qquad [2]$$

$$U\nabla . w = -\frac{1}{\rho}\frac{\partial p}{\partial z} + \nu\nabla^2 w \qquad [3]$$

and continuity

$$\nabla . U = 0 \qquad [4]$$

Similarly, the energy equation may be written as

$$\rho Cp\ U\nabla . T = k\nabla^2 T + \mu\Phi$$

In the above equations the lubricant film has been assumed to be laminar in motion and therefore the transport quantities (μ and k) and the viscous dissipation term Φ reflect the streamline nature of the flow. It is possible with extensive modification to cast the equations in a form suitable for turbulent motion. This can be accomplished by replacing the molecular properties with corresponding turbulent values, and then introducing appropriate models to describe them. The additional equations required generally increases the difficulties in obtaining a solution to such problems.

However, as mentioned earlier, in many practical applications the above equations can be simplified considerably by making appropriate assumptions. Typical of these are the neglect of cross stream velocities, all fluid advection terms, all conduction effects in the lubricant film. Thus the problem is reduced to solving the Reynolds equation and an energy equation which simply balances the generated heat with that convected through the film. Additionally, turbulence can be accounted for by replacing the molecular viscosity by expressions that modify it depending on the position considered in the film [3].

The above equations must be solved subject to the various imposed boundary conditions which will depend on the type of equations to be solved. For example, in their investigation of the role of inertia in lubricant films, Medwell et al [4] used the finite element method which, because of the elliptic nature of the equations of motion, required the specification six boundary values. These were either prescribed or cast in the form of gradients and, in some cases, updated during the solution procedure since they could not be stated a priori. As mentioned earlier, the solution procedure is necessarily iterative in form because of the dependence of the lubricant film profile on the bearing attitude angle and therefore requires continuous updating. In these investigations the three dimensional films profile was generated conveniently by scaling a unit cube (suitably discretized) to produce the appropriate film geometry through the relationships

$$M\ (x,\bar{y},\bar{z}) = P\ (\xi,\bar{\eta},\bar{\mu})\ \left[-\frac{L}{2}\right] \qquad [5]$$

$$M\ (\bar{x},y,\bar{z}) = P\ (\bar{\xi},\eta,\bar{\mu})\ \left[C\{1+\epsilon\ Cos(\theta+\alpha-\psi)\}\right] \qquad [5b]$$

$$M\ (\bar{x},\bar{y},z) = P\ (\bar{\xi},\bar{\eta},\mu)\ \left[R(\varphi-\theta)\right] \qquad [7]$$

where the overbar denotes a fixed coordinate.

This technique was first proposed by Gethin [2] and has been subsequently used successfully in analysing other performance aspects of journal bearing operation [5], [6] and [7].

In all analyses it is convenient to unwind the loaded part of the lubricant film because the radius of curvature of the film is very large in comparison with the film thickness Figures 1(a) and 1(b) shows the journal bearing geometry and the unwrapped film discretized in a form suitable for the finite element method of solution, using eight-node isoparametric elements.

If heat conduction in the journal bush is important it is necessary to include the appropriate form of the energy equation and boundary conditions into the solution

$$viz.\quad \nabla^2 T = 0 \qquad [8]$$

with the inclusion of conduction terms in the fluid film also.

A suitable discretization of the bush is also depicted on Figure 1(c).

So far, the equations that have been discussed are fully three dimensional in nature and reference has only been made to the finite element method of solution. This is because finite difference methods [8] and [9] solve parabolized form of the equations of motion with solution stability being maintained using upwind techniques. Where recirculatory flows occur the validity of these procedures becomes questionable.

The importance of recirculatory flows has been demonstrated for isothermal lubricant films [4] by solving the equations of motion in which only fluid advertion across the film was neglected. Although the analysis confirmed that lubricant inertia does not affect bearing performance significantly it does indicate that large areas of recirculation occur at high eccentricity ratios as shown on Figure 2. As pointed out earlier, this phenomena can be expected to appear in the lobes of profile bore bearings which can be operating in a highly eccentric condition even though the overall bearing eccentricity is modest.

The problem in the type of rigorous analysis mentioned above is that it can be expensive to compute and which can become prohibitive if thermal and elastic behaviour are included. Obviously, it is important to the designer to use methods which will compare favourably with both the rigorous analyses and experimentally determined performances and be economic computationally. An excellent exposition of this has been given by Gethin [5] who presented two models. One was based on the solution of the full equations including bush conduction (i.e. equations 1 - 4 and 8) and in the other the equations of motion were reduced to a generalized Reynolds equation, viz.

$$\frac{\partial}{\partial x}\left[G\frac{\partial p}{\partial x}\right] + \frac{\partial}{\partial z}\left[G\frac{\partial p}{\partial z}\right] = \omega R\frac{dh}{dx} - \omega R\frac{dF}{dx} \qquad [9]$$

where

$$G = \int_0^h \frac{y}{\mu}(y-F)dy; \quad F = \frac{F_1}{F_0}$$

where

$$F_0 = \int_0^h \frac{dy}{\mu} \text{ and } F_1 = \int_0^h y\frac{dy}{\mu}$$

A consequence of embodying the laws of conservation of mass and momentum into a Reynolds equation is the simplification of the boundary conditions which were now in the form of specified pressrues or gradients of pressure. The imposition of several different thermal boundary conditions was assessed and discussed fully while the viscosity dependence on temperature was 'inbuilt' into the solution procedure in the form of the Walther equation.

The strategy for the solution of both sets of equations was broadly the same consisting of the following basic steps:

(i) the bearing geometry and boundary conditions for the hydrodynamic equations were set up for an assumed value of attitude angle which was used to generate a film thickness profile for a specified eccentricity ratio.

The film thickness equation used was

$$h = C\{1 + \epsilon\cos(\theta + \alpha - \psi)\} \qquad [10]$$

(ii) the load was calculated and the new attitude angle was evaluated which was then used to update the film geometry.
This procedure was repeated until convergence was achieved and any parameters such as velocity gradients and/or pressure gradients that may be required for subsitution into the energy equation were calculated.

(iii) the boundary conditions for the energy equation were set up and this was solved to give temperature distributions.

(iv) the nodal values of viscosity were updated throughout the film and the steps (i), (ii), (iii) and (iv) repeated until either the velocity or pressure fields converge to a prescribed tolerance.

Some of the results from the two models are displayed on Figures 3 and 4 together with some experimental results. It can be seen that generally the more sophisticated model does not predict results which are significantly more accurate than those predicted by the method based on Reynolds equation. However, in terms of computing requirements the latter model used just 10% of the CPU time of the more rigorous analysis - a very significant saving.

The simultaneous solution of the Reynolds and energy equations using the above strategy forms the basis for many other journal bearing performance predictions using both finite element and finite difference techniques.

It was first used [10] to assess the influence of heat dissipation in high speed bearings where the lubricant film was in turbulent motion. Thus was done by using very crude turbulence models [3],

$$\text{viz. } \nu_x = \nu\,(1 + 0.000175\,Re^{1.06})$$

$$\text{and } \nu_z = \nu\,(1 + 0.000375\,Re^{1.06})$$

to replace molecular viscosity terms in the Reynolds equations. The energy equation employed neglected crossfilm and streamwise cnduction effects (the adiabatic condition) which results in the temperature of the lubricant film at any point being represented by a mean value.

Both equations were cast in conventional finite difference form based on central differences. The Reynolds equation was solved

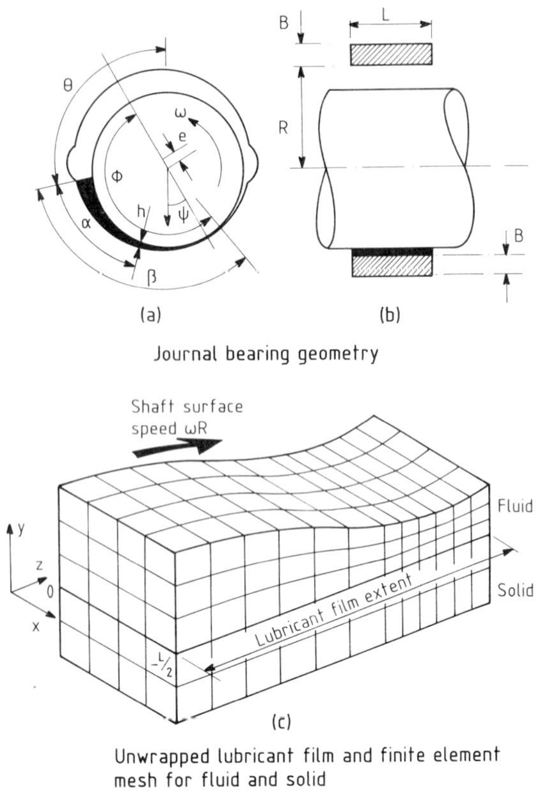

FIGURE 1

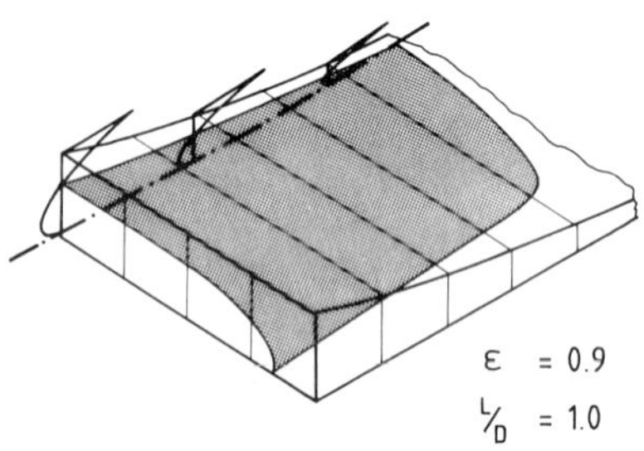

FIGURE 2 RECIRCULATING FLOW AT BEARING INLET

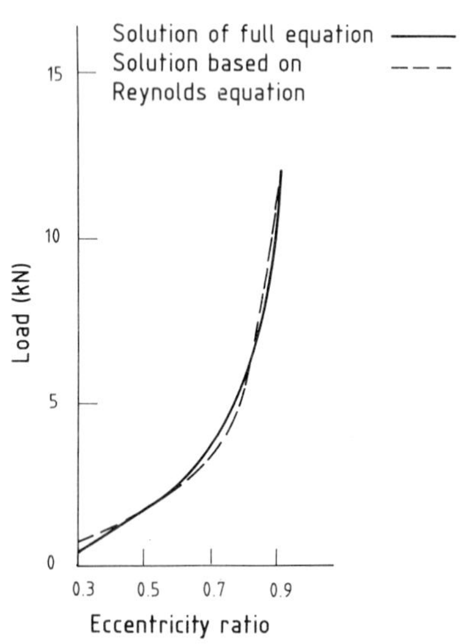

FIGURE 3 VARIATION OF LOAD WITH ECCENTRICITY RATIO
Bearing speed = 10,000 rev/min.
$C/R = 0.002$, $L/R = 1.0$

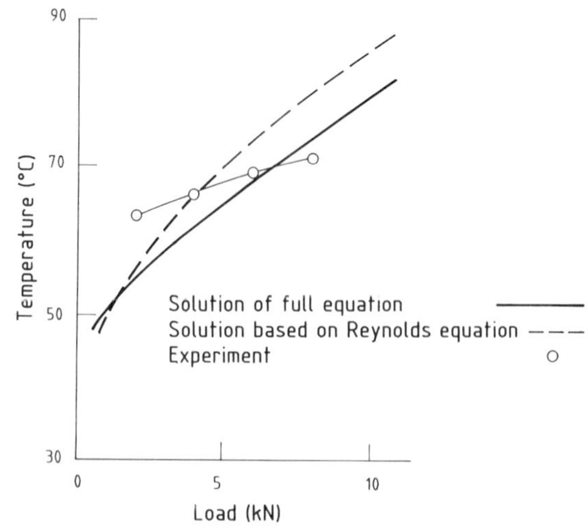

FIGURE 4 VARIATION OF LUBRICANT MAXIMUM TEMPERATURES WITH LOAD .
Bearing speed = 10,000 rev/min.
$C/R = 0.002$, $L/R = 1.0$

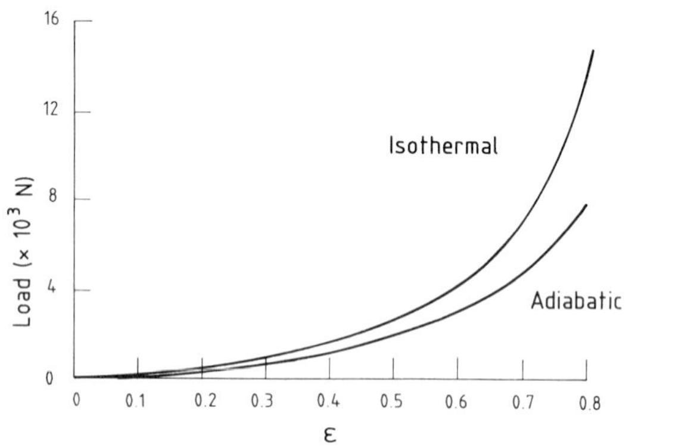

FIGURE 5 THEORETICAL VARIATION OF LOAD WITH ECCENTRICITY RATIO FOR ISOTHERMAL AND ADIABATIC FLOW.
SPEED = 30,000 rev/min.

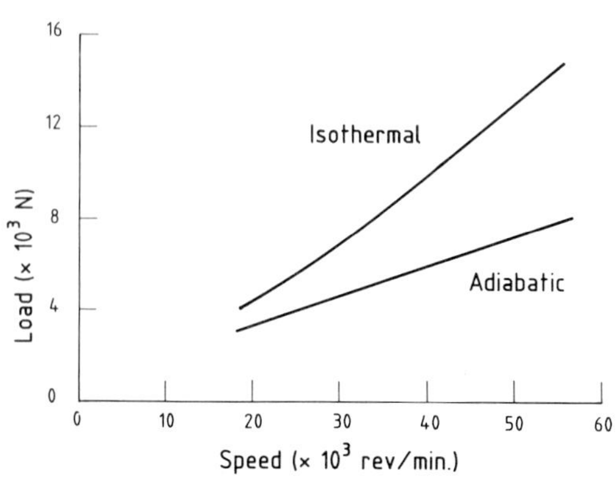

FIGURE 6 THEORETICAL VARIATION OF LOAD WITH SPEED FOR ISOTHERMAL AND ADIABATIC FLOW.
$\varepsilon = 0.7$

using a Gauss Seidel iterative technique while the simpler form of the energy equation allowed a straight through march to be used in obtaining a solution.

Some of the findings of the investigation are displayed on Figures 5 and 6.

The above method was extended by Medwell and Gethin [11] to account for off-design bearing operation such as misalignment. Identical Reynolds and energy equations were solved using the same models of turbulent viscosity and employing a modified film thickness equation which included an extra term to account for the misalignment. This is measured as the inclination of the shaft axis to that of the bush and is measured in the plane in which the shaft is loaded.

The main consequence of misalignment is shown in Figure 7 where maximum and bulk temperatures of the lubricant film are plotted as a function speed for an eccentricity of 0.65 and a misalignment angle of only 0.12°.

It can be seen that although the bulk temperatures exhibit modest rises over the inlet lubricant temperature the maximum temperatures generated are dangerously high and would certaily be a limiting parameter in such a bearing operation and which confirms the importance of misalignment effects in journal bearing applications.

Generally, when both finite element and finite difference methods are applied to similar problems as above, it is found that the difference methods are less demanding on computational requirements. However, as pointed out earlier, when recirculatory flow occurs the governing equations lose their parabolic form and the vality of upwind techniques introduced to achieve solution stability in the difference methods becomes questionable.

A further disadvantage of the finite difference method is its inability to cope efficiently with coupled problems such as those where bush conduction and thermo-elastic deformations are important. It is in these types of situations that the finite element method is particularly effective.

This derives from the capacity of the method to cope which property changes in adjacent elements either side of the fluid-solid interface that makes it extremely attractive to designers. An excellent example of this has been given by Gethin [12] in his theoretical investigation into the effect of thermoplastic bush deformation and shaft thermal expansion on journal bearing performance. He accounted for the distortion of the bush due to the temperature and pressure fields by introducing the relevant equations of solid mechanics. However, based on experimental evidence, computational economy was achieved by assuming that the distortion occurring on the bearing centre line only could be applied over the entire width of the bearing.

In polar coordinates the radical and circumferential components of strain were written in terms of a displacement matrix as

$$\begin{bmatrix}\varepsilon_r\\ \varepsilon_\theta\\ \gamma_{r\theta}\end{bmatrix} = \begin{bmatrix}\frac{\partial}{\partial r}, & 0\\ \frac{1}{r}, & \frac{1}{r}\frac{\partial}{\partial r}\\ \frac{1}{r}\frac{\partial}{\partial\theta}, & \frac{\partial}{\partial r} - \frac{1}{r}\end{bmatrix}\begin{bmatrix}u_r\\ v_\theta\end{bmatrix} \text{ or } \{\varepsilon^1\} = [B]\ \{\delta\} \qquad [11]$$

The stresses were related to strain by the constitutive equations

$$\begin{bmatrix}\sigma_r\\ \sigma_\theta\\ \tau_{r\theta}\end{bmatrix} = \frac{E}{1-\nu_E^2}\begin{bmatrix}\nu & 1 & 0\\ 1 & \nu & 0\\ 0 & 0 & \frac{1}{2}(1-\nu_E)\end{bmatrix}\begin{bmatrix}\varepsilon_r\\ \varepsilon_\theta\\ \gamma_{r\theta}\end{bmatrix}$$

$$\text{or } \{\sigma\} = [D]\ \{\varepsilon^1\} \qquad [12]$$

where ε_1, ε_θ and $\gamma_{r\theta}$ arise from the applied pressure and thermal loading. Applying an energy minimization principle and equations (11) and (12) an expression of the following form was obtained

$$\int_\Omega [B]^T [D]\ [B]\ \{\delta\}\ d\Omega = \int p d\Omega - \int q d\Gamma \qquad [13]$$

$$\text{or } [K]\ \{\delta\} = \{F\}$$

which can be solved for $\{\delta\}$, the displacement of the domain. In equation (13) the integral over the solid domain includes the thermal loading while the surface integral over Γ contains contributions arising from the imposed pressure that is at an elemental level.

$$\int_\Omega p d\Omega = \int_\Omega \alpha_\beta f(T) d\Omega$$

where f(T) is the temperature field over the element edge. Similarly

$$\int_\Omega q d\Gamma = \int_\Gamma f(p) d\Gamma$$

where f(p) is the pressure field over the element edge.

The applied pressure loading was obtained from solution of the hydrodynamic equations while the thermal loading was determined from solution of the energy equation in the bush.

A further assumption made in the analysis was that the bush was constrained in a rigid housing so that the computational domain was confined to the load carrying zone of the bearing.

The solution procedure follows that outlined earlier with the additional steps requried to calculate the distortion of the bush and shaft expansion which were then included in the calculation of the new film thickness.

Gethin presented a large amount of data to illustrate the effect of including distortion for a wide range of speeds (he assumed that the film was turbulent) and which highlight the success of his method is the maximum film temperature variation with shaft speed as shown on Figure 8.

CONCLUSIONS

This paper has been concerned with the use of finite element and finite difference methods in the prediction of journal bearing performances. Several aspects of journal bearing operation have been considered and it is concluded that the finite element method is a very powerful design tool which can cope with many aspects of bearing behaviour hitherto neglected.

REFERENCES

1. Barwell, F.T. and Medwell, J.O.; "Journal Bearing Calculations", Euromech Colloquium No. 124, Turin, Italy, 1979.

2. Gethin, D.T.; "Superlaminar flow in journal bearings" PhD Thesis, University of Wales, 1983.

3. Ng, C.N. and Pan, C.H.T.; "A linearised turbulent lubrication theory." Trans. ASME, J. Basic Engineering. Vol. 87.675.1965.

4. Medwell, J.O., Gethin D.T. and Taylor, C.; "A finite element analysis of the Navier-Stokes equations applied to high speed thin film lubrication." Trans. ASME J. Tribology Vol.109, 71, 1987.

5. Gethin, D.T.; "A finite element approach to analysing thermohydrodynamic lubrication in journal bearings. Tribology International, Vol.21, 67, 1988

6. Gethin, D.T. and Medwell, J.O.; "Analysis of high speed bearings operating with incomplete films." Tribology International Vol.18, 340, 1985.

7. Medwell, J.O. and Gethin D.T.; "A finite element analysis of journal bearing lubrication." Proc.10th Leeds-Lyon Symposium on Tribology (Eds. D Dowson et al), Butterworth Scientific, UK, 1983.

8. Launder, B.E. and Leschziner, M.A.; "An efficient numerical scheme for the prediction of turbulent flow in thrust bearings" Proc. 2nd Leeds-Lyon Symposium on Tribology (Eds. D Dowson et al) Mech Eng Publications, UK, 1975.

9. King, K.F. and Taylor, C.M.; "An estimation of the effect of fluid inertia on the performance of the plane inclined slider thrust bearing with particular regard to turbulent lubrication" Trans. ASME, J. Lub. Technology, Vol.99, 129, 1979.

10. Bowen, E.R. and Medwell J.O.; "A thermohydrodynamic analysis of journal bearings operating under turbulent conditions." Wear, Vol.51, 345, 1977.

11. Medwell, J.O. and Gethin, D.T.; "Synthesis of thermal effects in misaligned hydrodynamic journal bearings". Int. J. Num. Methods in Fluids, Vol.6, 445, 1986.

12. Gethin, D.T.; "An investigation into plain journal bearing behaviour inlcuding thermo-elastic deformation of the bush." Proc. Instrn. Mech. Engrs. Vol.199, 215, 1985.

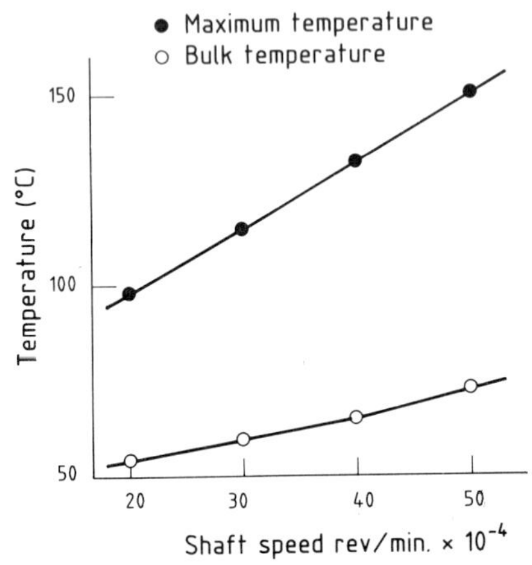

FIGURE 7 VARIATION OF MAXIMUM AND BULK TEMPERATURES WITH BEARING SPEED.
Inlet temperature = 40°C
Shaft misalignment = 0.12°
Eccentricity ratio = 0.65

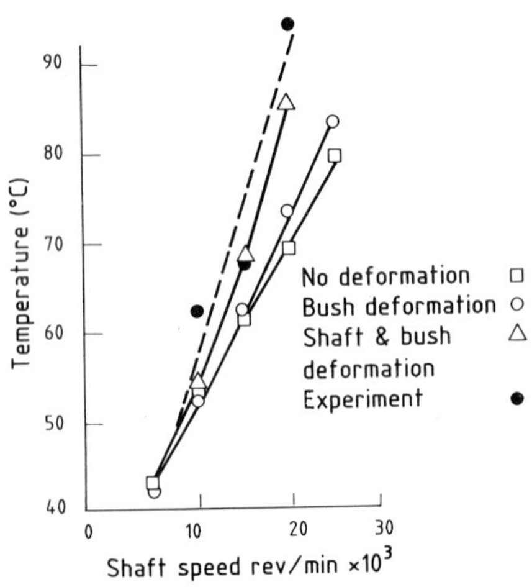

FIGURE 8 VARIATION OF MAXIMUM FILM TEMPERATURE WITH ROTATIONAL SPEED.
$C/D = 0.002$, $L/D = 0.5$, $B = 10mm$
Eccentricity ratio = 0.5

SESSION IX
WEAR

Chairman: Professor T.H.C. Childs

Paper IX(i)

The effect of residual stress and temperature on the fretting of bearing steel

M. Kuno and R. B. Waterhouse

An apparatus was designed to study the fretting wear of contact between a ball bearing and bearing steel flat (hardness 715VHN) over a range of slip amplitudes from 5μm to 50μm. The normal load was varied between 135 and 437N and the frequency was 50Hz. It was possible to subject the flat to a static tensile stress, and this was raised in steps up to 600MPa. The steel flat itself had a surface residual compressive stress of 800MPa, so the applied stress caused a reduction in magnitude and depth of influence of the residual stress. Two types of wear scar were produced in the test. Where there was partial slip ($T<\mu P$) an annular ring of damage was formed, the outer radius of which was constant, but the inner radius gradually decreased with increasing number of cycles. The second type of scar was produced when full slip occurred over the contact region ($T\geqslant\mu m$). The scar in this case was a full circle and its diameter increased with number of fretting cycles. When a tensile stress was applied cracks were generated in both types of fretting area. In the full slip regime these propagated to failure. The shortest time to failure was associated with a critical amplitude of 35μm which corresponded to a peak in the coefficient of friction of 0.72.

Raising the temperature to 200°C led to the development of a "glaze oxide" on the surfaces with a much reduced coefficient of friction. No cracks were generated in this case.

The presence of a residual compressive stress in the surface appears to have very little influence on the initiation of cracks by fretting. The development of fretting wear damage is discussed in relation to the mechanisms of fretting wear both when $T<\mu P$ and when $T\geqslant\mu P$.

1 INTRODUCTION

High carbon chromium bearing steel quenched and then tempered at low temperature is used as a material for rolling bearings. The microstructure is tempered martensite, and the hardness is approximately 720 VHN. The hardness is carefully controlled as there is a close correlation between the hardness and the rolling contact fatigue life. Hardness of the material is the most effective mechanical factor to prevent the raceways from surface damage. On the other hand, it is well known that there is a proportional relationship between brittleness and hardness. Hence, toughness of the material is decreased if hardness is increased. In fact, fracture of the rings arising from surface damage can often be seen in practice.

Fracture of the rings of rolling bearings is predominantly the result of heavy radial and axial loads and occurs even under lubricated conditions on the raceways. Surface damage should not arise as long as the rolling bearings are sufficiently well lubricated and the operating conditions are controlled. It has been observed that surface damage primarily occurs on the raceway, from which fracture is initiated. Therefore, in order to obviate fracture, the most effective countermeasure is to prevent the raceways suffering the initial surface damage. However, the circumstances where rolling bearings are used are extremely varied and becoming increasingly severe. In particular, miniature rolling bearings and solid lubricant rolling bearings are likely to be surface-damaged, because these bearings are not continuously operated for long periods. Hence, these bearings are often subject to fretting wear arising from vibration when they are nominally at rest.

Rolling bearings are usually subjected to a hoop stress as they are placed between shafts and housings. The hoop stress is large when the rolling bearings are used in machinery where accuracy is particularly required. The hoop stress of the outer ring is usually compressive, while that of the inner ring is usually tensile when the rolling bearing is installed as a machine component. For example, when the machine containing the rolling bearings is transported, vibration is likely to cause fretting damage on them. Fracture resulting from fretting damage tends to occur when the inner ring of the rolling bearing is subjected to a tensile hoop stress.

A new fretting wear test apparatus used in this study can apply tensile loads to the lower stationary specimens. The applied tensile load to the lower stationary specimens can be adjusted to equal the tensile hoop stress which is experienced by rolling bearing rings. Therefore, the fracture mechanism induced by fretting wear on the high carbon chromium bearing steel can be investigated with this test apparatus. The relationship between the factors which can be considered as affecting the fracture, i.e. applied tensile load, normal load (maximum Hertzian contact stress), slip amplitude and temperature, have been studied. In this study, the fretting wear tests have been carried out under unlubricated conditions. The specimen temperatures used were room temperature and 200°C, because 200°C is considered to be approximately the maximum temperature to which ordinary rolling bearings are subjected.

2 EXPERIMENTAL PROCEDURE

Fig. 1 shows a schematic diagram of the fretting wear test apparatus. The fretting contact consists of a sphere-on-plate arrangement. The plate is stationary and the upper spherical rider oscillates.

As shown in Fig 2, a tensile stress can be applied to the lower stationary specimen. Fig. 3 shows the geometry of the specimen used in this study. The upper spherical specimen is positioned at the centre of the lower guage length in the stationary specimen. A tensile stress is applied via a screw. One of the ends of the grips is linked with a load cell. The magnitude of the applied tensile stress can be adjusted by turning the nuts. The apparatus is described in detail in a previous paper. (1)

The upper moving specimen is a commercial 25.4mm diameter high carbon chromium bearing steel ball. The lower stationary specimen is also made of the high carbon chromium bearing steel. The lower stationary specimen was quenched (830°C → 60°C) and tempered (170°C for 1 hour) after machining, and was subsequently ground and polished with 1μm diamond polishing medium. The microstructure of the specimen is a typical tempered martensite, and the hardness is 715 VHN. The chemical composition is shown in Table 1.

The stress distribution below the surface of the lower stationary flat specimen has been measured by X-ray residual stress measurement, and is shown in Fig. 4.

The experimental conditions used in this study were as follows:

normal loads	: 135-437N
maximum Hertzian contact stresses	: 812-1200MPa
radius of static contact area	: 28-42μm
slip amplitudes	: 5-50μm
applied tensile stresses to the lower stationary specimens	: 0-600MPa
specimen temperatures	: in laboratory (17-22°C) 200°C

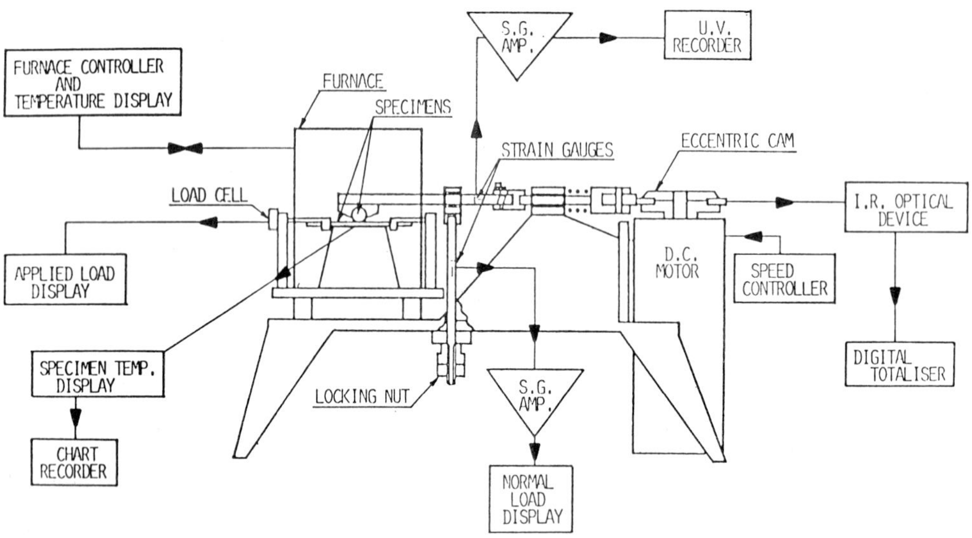

Fig 1 A schematic diagram of the fretting wear apparatus used.

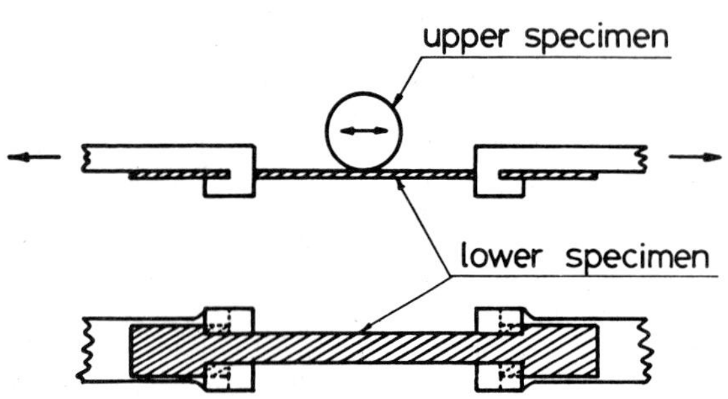

Fig 2 Fretting wear test arrangement.

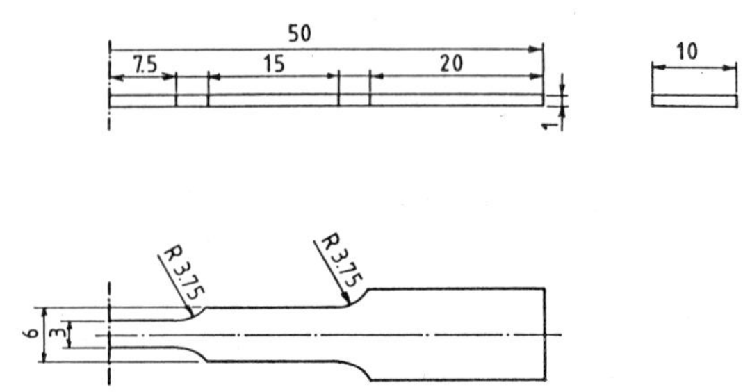

Fig 3 Geometry of the lower stationary specimen.

Table 1 Chemical composition of the material used

C	Si	Mn	P	S	Cr	Mo	Ni	Al	Cu	V
1.12	0.23	0.38	0.006	0.013	1.26	<0.03	0.04	0.02	0.06	<0.01

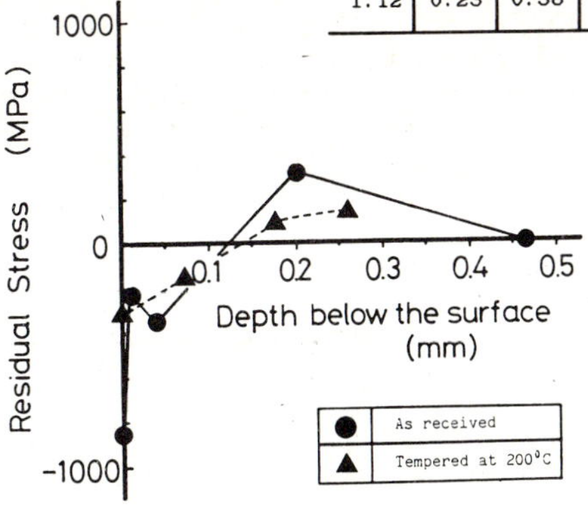

Fig 4 Stress distributions below the surface of the specimen.

The fretting wear tests at 200°C were carried out using the following method. After the specimen temperature had reached 200°C, it was held constant for 30 min., and then the upper and lower specimens were placed in contact. Therefore, the condition of the surfaces of the specimens is assumed to have reached equilibrium and to be chemically constant.

The fretting wear tests were individually carried out. Hence in each test the fretting contact situation at the beginning stage of the test is identical. Fretting wear scars produced on the lower stationary flat specimens without an applied tensile stress were examined to investigate the development of the fretting wear scar. The investigation was carried out by using two test conditions, which depend upon the numerical relationship between T and μP, $T<\mu P$ and $T \geqslant \mu P$. Here T represents the tangential force, μ the coefficient of friction and P the normal load. The test conditions used in this work were as follows,

		$T<\mu P$	$T \geqslant \mu P$
normal load	:	437N	135N
maximum Hertzian contact stress	:	1200MPa,	822MPa
radius of static contact area	:	42µm	28µm
frequency	:	50Hz	60Hz
amplitude	:	8µm	25µm

A tensile stress was applied to the lower stationary flat specimen during the fretting wear test at both laboratory temperature and 200°C (specimen temperature), as described above. In the test at 200°, the applied tensile stress was applied to the lower stationary specimen 30 min. after reaching the test temperature, and then the upper and lower specimens were placed in contact. Various slip amplitudes, normal loads and applied tensile stresses were used, as shown above. The test was predominantly carried out for up to 10^6 cycles, and if specimens had not failed at 10^6 cycles the test was discontinued. An example of a failure induced by fretting wear is shown in Fig. 5.

Both upper and lower specimens were connected with a constant current supply, and were supplied with a DC current of 1µA. The electrical contact resistance during fretting oscillation was recorded on a chart recorder.

3 EXPERIMENTAL RESULTS

Examples of the development of the fretting scar are shown in Fig, 6a and 6b. The outer circumferences of the fretting scars in Fig. 6a show the same diameter, although the width of the annulus of damage varies. On the other hand, the size of the scar significantly changes with increasing number of cycles in Fig. 6b. Fig. 6c shows the relationship between the number of cycles and the width of the annulus, (a-b).

An example of the electrical contact resistance is shown in Fig. 7. This fretting wear test was carried out under the following conditions,

normal load	:	135N
maximum Hertzian contact stress	:	822MPa
radius of static contact area	:	28µm
frequency	:	60Hz
amplitude	:	25µm
temperature	:	25°C
humidity	:	54%RH

The slip amplitude used in this test, 25µm is large enough for bulk sliding to occur with no non-slip area in the fretting contact area. Nevertheless, metal-to-metal contact can be recognised throughout the whole stage of the fretting wear.

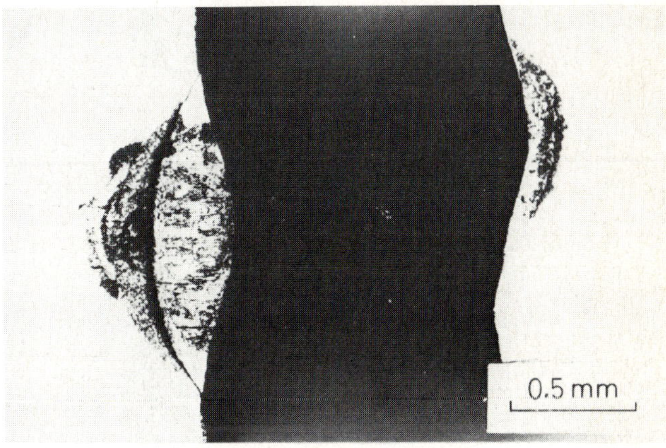

437 N (Pmax = 1200 MPa)
50Hz, 10 µm, fractured at 1.5×10^6 cycles
Applied tensile stress : 400 MPa

Fig 5 An example of a failure induced by fretting wear.

The fracture lives tested under various applied tensile stresses, normal loads and slip amplitudes at both laboratory temperature and 200°C are plotted on the diagrams shown in Fig.8. No fractures occurred in 7 x 10^4 cycles under various ranges of test conditions. In addition, the specimens did not fracture at 200°C although the test was carried out up to 10^6 cycles. According to Figs, 8a and b, short fracture life is the result of high normal load and occurs at a slip amplitude of 35μm in laboratory air.

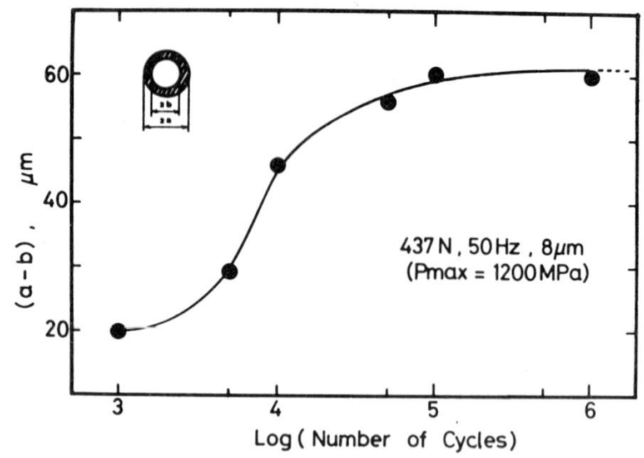

Fig 6c Relationship between the number of cycles and the width of the annulus of the scar, (a-b).

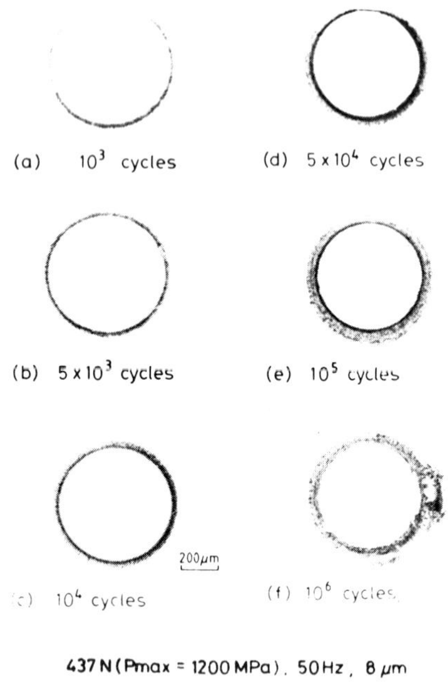

Fig 6a An example of the development of the scar when T<μP.

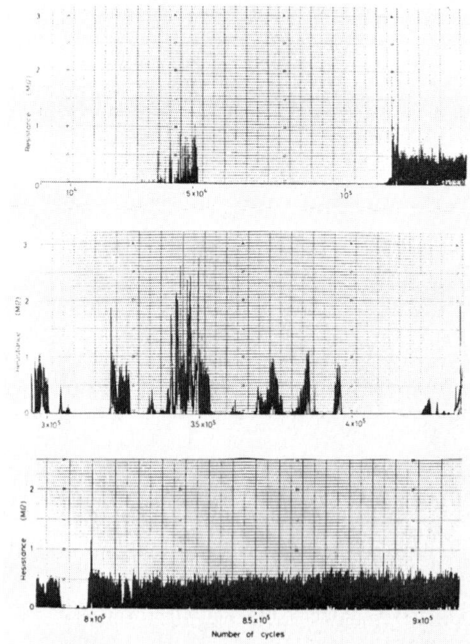

Fig 7 An example of the electrical contact resistance during fretting wear.

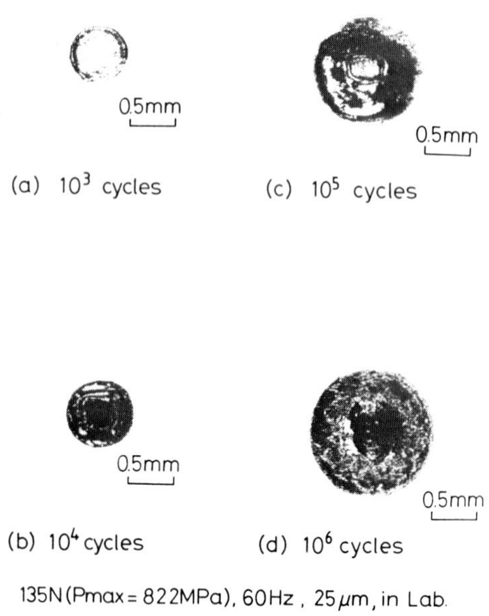

Fig 6b An example of the development of the scar when T≥μP.

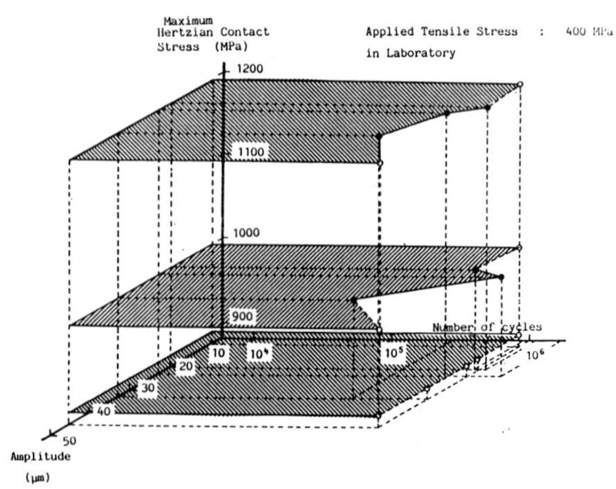

Fig 8a Relationship between the maximum Hertzian contact stress, slip amplitude and the number of cycles when the specimen is subjected to an applied tensile stress of +400MPa at room temperature.

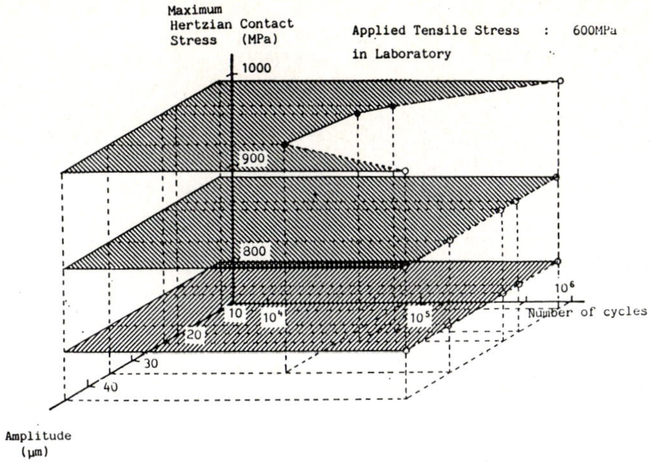

Fig 8b Relationship between the maximum Hertzian contact stress, slip amplitude and the number of cycles when the specimen is subjected to an applied tensile stress of +600MPa at room temperature.

4 DISCUSSION

According to Figs. 6a and 6b, it is obvious that the development process of the fretting wear scar is governed by the numerical relationship between T and μP. Fig 6a is a typical example when $T<\mu P$, while Fig. 6b is an example when $T\geqslant\mu P$.

When the slip amplitude is small enough for the fretting wear scar to contain a non-slip region (when $T<\mu P$), the fretting wear scar result from the following process. Initially, a small annular slip region is worn by microslip. The damage starts at the edge of fretting contact area as the relative slip is greatest there. Subsequently, the annular slip region is enlarged towards the centre of the annulus, while the outer diameter of the annulus remains constant. Sato et al (2) have also reported a similar wear process. Finally, the enlargement of the annular slip region ceases because the relative slip amplitude is zero at the inner boundary between the slip and non-slip regions, and the scar shows no further development. At this stage, little debris is produced.

A typical example of the relationship between the number of cycles and the width of fretting wear annulus when $T<\mu P$ is shown in Fig. 6c. According to this result, the development of the fretting wear annulus has settled down at 10^5cycles, and the fretting wear does not progress appreciably further from this stage.

On the other hand, when the contact region entirely slips (when $T\geqslant\mu P$), the size of the scar dramatically enlarges outwards, as shown in Fig 6b. The tangential shear strain is greatest at the periphery of the fretting contact area. The periphery of the contact area is subjected to a significant strain as the contact region entirely slips, Figs. 7 show the electrical contact resistance during the fretting oscillation when $T\geqslant\mu P$. It can be recognised that debris exists between the specimens during the fretting oscillation. However, metal-to-metal contact throughout the whole stage of the fretting wear also can be recognised. Metal-to-metal contact would predominantly occur at the periphery of the fretting contact area as the scar increases in size.

According to the results in Figs. 8, the normal load (maximum Hertzian contact stress), the applied tensile stress, the residual stress of the material and the slip amplitude are effective in fracture induced by fretting wear at room temperature. The existence of a critical slip amplitude can be recognised at approximately 35μm. Some researchers (3)-(7) have also found a critical amplitude in fretting fatigue. Nishioka et al (3)-(6) have obtained approximately 20μm for the critical slip amplitude using carbon steels. On the other hand, in Kudva et al's study (7), where the residual stress of the specimens was changed by heat treatments such as carburising, decarburising and quenching-tempering, they mentioned that the critical amplitude lies between 20 and 60μm and depends upon the residual stress in the specimen surface. In Figs 8a and 8b, the critical slip amplitude in this test is always 35μm although the surface residual stress is changed by the applied tensile stress to the specimen. This result does not agree with Kudva's result (7). Their results may include some other factors, for example the microscopic structures are different in the specimens used in their study. The relationship between coefficient of friction and slip amplitude obtained in this study in shown in Table 2. The coefficient of friction at 35μm of slip amplitude is approximately twice that at other amplitudes.

Table 2 Relationship between coefficient of friction and slip amplitude at room temperature.

Amplitude	10 μm	20μm	25μm	35μm	45μm
Coefficient of Friction	0.50	0.44	0.49	0.72	0.49

normal load : 437N (Pmax = 1200MPa)
frequency : 50Hz
applied tensile stress : 400MPa

coefficient of friction measured at 10^5 cycles

When $T \geqslant \mu P$, if the coefficient of friction increases, then the tangential force also increases under a constant normal load condition. Simultaneously, the tangential shear strain due to the tangential stress increases, and thus a high tangential shear strain energy is created. Consequently, the critical slip amplitude in fracture induced by fretting wear is obtained.

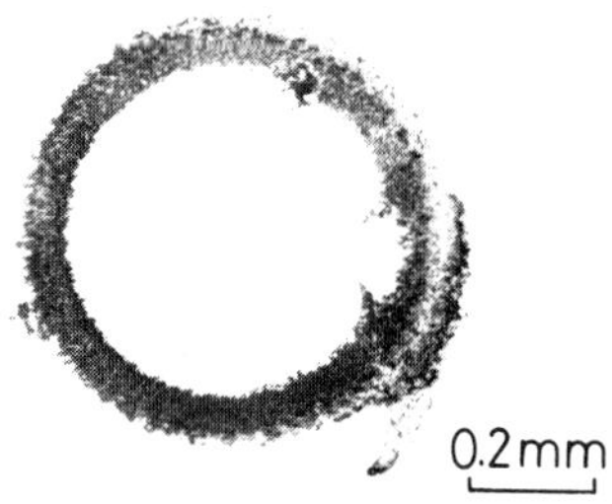

normal load : 437 N
maximum Hertzian contact stress : 1200 MPa
frequency : 50 Hz
amplitude : 8 μm
applied tensile stress : 0 MPa
number of cycles : 10^6 cycles
in laboratory air at R.T.
coefficient of friction : 0.5

Fig 9a An example of fretting wear scar at room temperature.

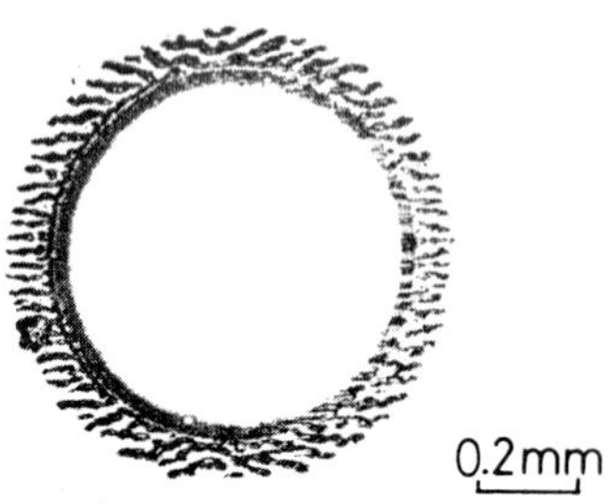

normal load : 437N
maximum Hertzian Contact stress : 1200MPa
frequency : 50Hz
amplitude : 8μm
applied tensile stress : 0MPa
number of cycles : 10^6 cycles
specimen temperature : 200°C
coefficient of friction : 0.21

Fig 9b An example of fretting wear scar at 200°C.

Figs. 9a and 9b show the scars obtained by fretting under the same condition but at different temperatures, at room temperature and 200°C (specimen temperature). The scar at 200°C (Fig 9b) shows a narrow annular ring with debris extruded from the contact region. It is very obvious that the width of the annular scar fretted at room temperature is much larger than that at 200°C. Using Mindlin's equation (8), coefficient of friction can be calculated,

$$T = \mu P \left(1 - b^3/a^3 \right),$$

where a represents the radius of the fretting wear scar and b the radius of the non-slip circular region. When the specimen shown in Fig. 9-a (at room temperature) was fretted, T was 118,52N. On the other hand, T was 24.69N in the specimen shown in Fig. 9-a. The coefficients of friction of these specimens are 0.5 and 0.21, respectively. This, therefore, confirms the effect of oxide film on fretting wear at an elevated temperature. From this result it can be seen there is no critical slip amplitude in the presence of the oxide film.

No specimen has been fractured at 200°C in these tests. This result indicates that coefficient of friction is a very effective factor in fracture induced by fretting wear. This may suggest an appropriate countermeasure for the prevention of this type of fracture.

As mentioned previously, the significant effect of the normal load on fretting fatigue has been pointed out (9)-(11). In the experimental rig used in this study, the slip amplitude can be controlled independently of the normal load. Hence, the results do not include the effect of reducing amplitude. Considering the crack initiation due to fretting oscillation, it is apparent that the tangential stress, and the tangential shear strain, which are subjected to a high contact stress lead to a high tangential strain energy, and consequently crack initiation is encouraged. Some researchers (9)-(11) have pointed out that a high normal load induces significant fretting damage and decrease of fretting fatigue life. It has also been reported that a high normal load impedes rapid crack propagation in the subsurface (12). It is very obvious that a high normal load leads to high tangential stress and strain. It appears that crack initiation is encouraged but crack propagation rate in the subsurface is reduced by high normal load. This effect of high normal load on crack initiation is of greater significance than its effect on crack propagation.

As mentioned above, Kudva et al (7) have found a critical slip amplitude in fretting fatigue tests. They have suggested that the critical slip amplitude depends upon the different surface treatments and stress states. A critical slip amplitude has also been found in this study. However, the critical slip amplitude shows a constant value which is approximately 35μm although the surface stress is mechanically changed. It therefore appears that the critical slip amplitude is independent of the stress state, but dependent upon thermo-chemical surface treatments such as carburising.

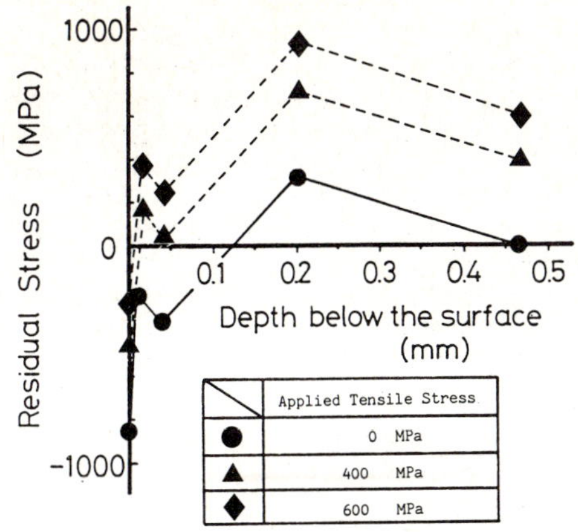

Fig 10 Stress distributions below the surface when tensile stresses are applied to the lower stationary specimen.

Fig 10 shows the stress distributions below the surface when the lower tensile specimen is subjected to various applied tensile stresses. Although the specimen is subjected to +600MPa, the residual stress on the specimen surface is still compressive. In contrast, the depth where there is a change-over to tensile residual stress is significantly dependent upon the applied tensile stress. Hence, the initiation of the cracks induced by fretting wear may be controlled by the tangential shear strain energy, but the propagation of the cracks may be governed by a combination of the tangential stress and the resultant subsurface tensile stress. It appears that the initiation of fretting wear cracks is little affected by the residual stress in the subsurface, but the subsequent crack propagation rate in fretting fatigue is very dependent upon the stress state. Nishioka et al (5)(6), Fair (3) and Mutoh et al (14) have also reported that residual stresses have little influence on crack initiation in fretting fatigue.
The residual stress in the surface of the specimen tempered at 200°C is -270MPa, as shown in Fig.4. The residual stress in the surface is +330MPa if a tensile stress of +600MPa is applied to the specimen. The specimen has not been fractured nor have cracks been found although the surface indicates a high tensile residual stress. As described above, the coefficient of friction at 200°C is much lower than that at room temperature. It, therefore, appears that coefficient of friction is more effective than surface residual stress in crack initiation.

5 CONCLUSIONS

The following conclusions are drawn from the investigations of fretting damage in rolling bearing.

(1) The development of the fretting scar depends upon the relationship between T and μP. When $T<\mu P$, the annular slip region grows towards the centre of the annulus. In contrast, when $T\geqslant\mu P$, the fretting scar grows outwards, and further metal-to-metal contact takes place at the outer edge of the scar during fretting oscillation.

(2) Fretting wear is dominated by the tangential shear strain. Cracks in fretting wear seem to initiate at the surface where the tangential strain is highest.

(3) A critical slip amplitude was found infracture induced by fretting wear. The coefficient of friction assumes a very high value at the critical slip amplitude. The critical slip amplitude was 35μm in this study.

(4) The oxide film produced at 200°C leads to a very low coefficient of friction. No specimen was fractured in the tests at 200°C. It is, therefore, obvious that the coefficient of friction is a very important factor in fracture induced by fretting wear. It is suggested that the critical slip amplitude in fretting fatigue is, under these conditions, nullified by the presence of oxide films.

(5) High normal load has little effect on crack propagation in fretting fatigue or fracture induced by fretting wear, but does influence crack initiation due to fretting oscillation by increasing the tangential stress and strain.

(6) Compressive residual stress in the subsurface of the fretted specimen has little influence on crack initiation, but has a great influence on crack propagation rate.

(7) Reduction in the coefficient of friction is the most effective way of improving performance in fretting wear and fretting fatigue as it is a parameter contributing to the tangential stress, which significantly affects the tangential shear strain.

6 ACKNOWLEDGEMENTS

The authors would like to express their appreciation to the NTN Toyo Bearing Co. Ltd., Osaka, Japan, for their financial support to carry out this investigation.

References

(1) Kuno M., Dinsdale K., Pearson B.R., Ottewell B. and Waterhouse R.B., A New Fretting Wear Test Apparatus, Journal of Physics E : Scientific Instruments, to be published.
(2) Sato J. Shima M., Igarashi J., Tanake M. and Waterhouse R.B., Studies of Fretting (Part 1), Japan Soc. of Lub. Eng. 26, 8 (1981) pp.555.
(3) Nishioka K. and Hirakawa K., Fundamental Investigations of Fretting Fatigue (Part 5), Bull. of Japan Soc. of Mech. Eng., 12, 52 (1969) pp.692.
(4) Nishioka K. and Hirakawa K., Fundamental Investigations of Fretting Fatigue (Part 2), Bull. of JSME, 12, 50 (1969) pp.180.
(5) Nishioka K. and Hirakawa K., FUndamental Investigations of Fretting Fatigue (Part 4), Bull. of JSME, 12, 51 (1969) pp.408.
(6) Nishioka K. and Hirakawa K., Fundamental Investigations of Fretting Fatigue (Part 6), Bull. of JSME, 15, 80 (1972) pp.135.

(7) Kudva S.M. and Duquette D.J., Effect of Surface Residual Stress on the Fretting Fatigue of a 4130 Steel, Residual Stress Effects in Fatigue (ASTM) (1982) pp.195.

(8) Mindlin R.D., Compliance of Elastic Bodies in Contact, J. Appl. Mech. 16 (1949) pp.259.

(9) Waterhouse R.B., Fretting Corrosion, Pergamon Press, Oxford, (1972) p.120.

(10) Sato K. and Fujii H., Crack Propagation Behaviour in Fretting Fatigue, Wear, 107 (1986) pp.245.

(11) Aronov V. and Mesyef T., Wear in Ceramic/Ceramic. Part 1, J of Tribology, ASME, 108 (1986) pp.16.

(12) Kaneta M., Murakami Y. and Okazaki T., Growth Mechanism of Subsurface Crack due to Hertzian Contact, J. of Tribology, ASME, 108 (1986) pp.134.

(13) Fair G.H., Effect of Shot-Peening on Fatigue and Fretting Fatigue of Aluminium Alloys, Ph.D. thesis (1988), University of Nottingham, U.K.

(14) Mutoh Y., Fair G,H., Noble B. and Waterhouse R.B. The Effect of Residual Stresses induced by Shot-Peening on Fatigue Crack Propagation in Two High Strength Aluminium Alloys, Fatigue Fract. Engng Material Structure, 10, 4 (1987) pp.261.

APPENDIX

Cracks were observed after 10^5 cycles in the fretting wear scars of the specimen tested under the conditions shown in Table 2. At slip amplitudes of less than 35μm no cracks were observed and only small amounts of debris was produced by the fretting oscillations. At 45μm, non-propagating cracks were observed within the fretting scar, the wear volume at this slip amplitude is, however, considerable and cracks may be worn away before they can propagate. The wear debris may also act as roller bearings as reported by Halliday et al [15]. At 35μm cracks were also observed within the fretting scar, initiation being enhanced by the high coefficient of friction, 0.72, at the metal-to-metal contact. The amount of wear debris produced at 35μm is, however, considerably less than that at 45μm, the cracks at 35μm are, therefore, not worn away and propagate to failure. Consequently, shorter lives were obtained at 35μm.

No specimens fractured at 45μm under these test conditions. On the other hand, the specimens eventually fractured at 20μm and 25μm, although no cracks were observed within the fretting scars after 10^5 cycles. It is apparent that the cracks are initiated even after 10^5 cycles within the fretting scars. As described above, the crack initiation and propagation are significantly governed by the production of wear debris. There is a critical amplitude where the alternating friction stress is high enough to initiate cracks , but where the production of wear debris is still very low [16], so that the fracture induced by fretting tends to occur at slip amplitudes where less wear debris is produced.

At 35μm crack initiation due to fretting is more dominant than the production of the wear debris. The effective surface stress of the specimens, tested under the conditions shown in Table 2, is -420MPa, as the specimens were subjected to an applied tensile stress of +400MPa. The predominance of either crack initiation or production of wear debris may depend upon the surface effective stress of the specimen. If the production of wear debris is by a surface fatigue process, e.g. delamination [17][18], then it could be expected that the level of the effective stress in the surface would have a significant influence on the process. It has been shown that a compressive residual stress reduces the fretting wear rate [19] and it is, therefore, likely that reducing the residual compressive stress will increase the wear rate. Also the coefficient of friction can be varied by varying the surface residual stress.[20] Therefore, the critical slip amplitude, indicating the highest coefficient of friction, is thought to be related to the surface effective stresses of the lower stationary specimens.

REFERENCES

[15] Halliday J.S. and Hirst W., The Fretting Corrosion of Mild Steel, Proc. Roy. Soc. A 236 (1956) pp.441

[16] Vingsbo O. and Söderberg S., On Fretting Maps, Proc. Conf. Wear of Materials, 1987, 5-9th April 1987, Houston, Texas, USA., pp.885

[17] Waterhouse R.B. and Taylor D.E., Fretting Debris and the Delamination Theory of Wear, Wear 29 (1974) pp.337

[18] Waterhouse R.B., The Role of Adhesion and Delamination Theory of Wear, Wear 45 (1977) pp.335

[19] Hills D.A., Ashelby D.W. and Waterhouse R.B., The Physical Influence of Surface Treatments on Wear, Proc. Conf. Wear of Materials, 1979, 16-18th April 1979, Dearborn, Mich.,USA., pp.369

[20] Bergman C.A., Cobb R.C. and Waterhouse R.B., The Effect of Shot-peening on the Friction and Wear of A Carbon Steel and An austenitic Stainless Steel in Fretting Conditions, Proc. Conf. Wear of Materials, 1987, 5-9th April 1987, Houston, Texas, USA., pp.23

Paper IX(ii)

The wear of hot working tools. Application to forging and rolling of steel

E. Felder

In hot working, two main types of tool damage occur: thermal fatigue and abrasion induced by the sliding velocity of the metal on the tool. First, by a mechanical and thermal analysis, we quantify the influence of friction, geometry of the operation and presence of scale on the workpiece, on these two types of damage. We then build a model of abrasion, which is then applied to rolling and forging; we obtain a good agreement with experiments performed on flat forging dies. Finally, in the case of forging, we model the evolution of die surface hardness, and give some experimental results about the influence of a nitriding treatment.

1 INTRODUCTION

The design of hot working processes requires models providing an estimation of tool life, which depends on various types of damage; the main ones (1) are abrasion, mechanical and thermal fatigue, and plastic deformation. This paper is devoted to the description of a model of abrasion and a discussion of some factors governing the thermal fatigue in the hot working of steels.

The prevailing damage type depends directly on the thermomechanical conditions at the tool-metal interface, mainly on the sliding velocity Δu of the metal on the tool. Two cases may arise:

- $\Delta u=0$ In this case, only contact thermal fatigue occurs: due to the periodic contact between the tool (temperature 40°C-200°C) and the steel workpiece (temperature 1100°C-1200°C), the tool surface undergoes cyclic heating and cooling; the thermal stresses thus generated produce a network of cracks, becoming deeper and deeper (2).
- $\Delta u \neq 0$ Thermal fatigue is still present, but moreover, oxyde particles carried into the contact by the workpiece induce abrasion and scratch the tool surface in direct proportion to Δu and the contact pressure p.

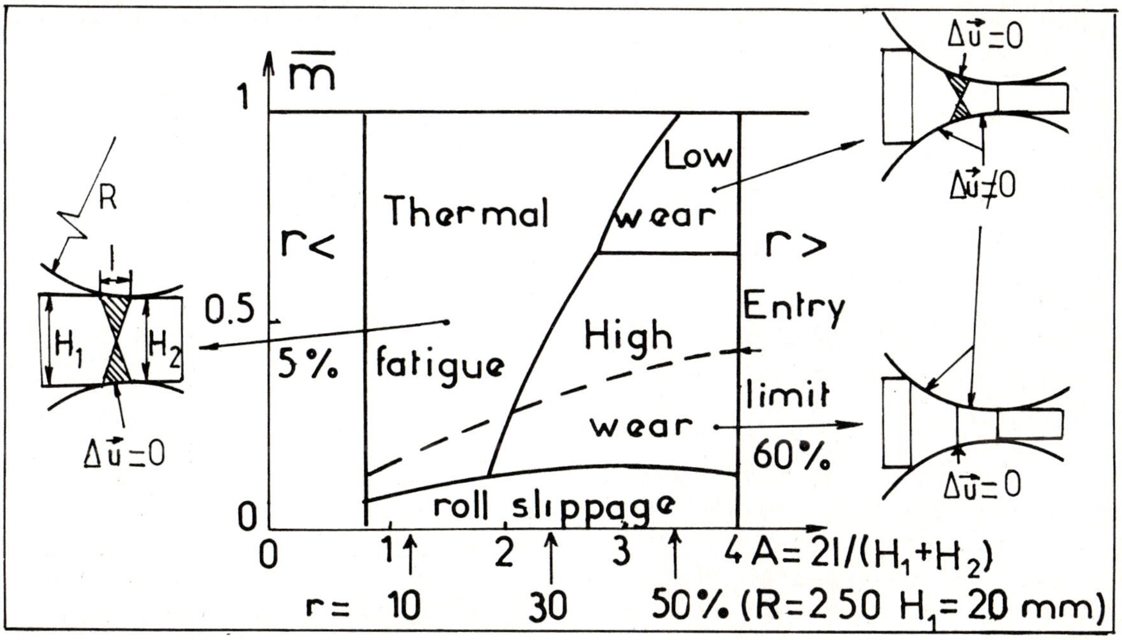

Fig 1 :damage map of flat rolling cylinders (4)

2 MECHANICAL ANALYSIS OF WORKING PROCESSES AND DIE DAMAGE MAPS

The mechanical analysis of the metal flow by the theory of plasticity (3) provides the values of Δu and p along the interface, which mainly depend on the friction stress τ, the geometry of the process and the tensile yield stress of the metal σ_0. Let us consider two simple, but instructive examples which can be analysed by elementary, classical methods.

2.1 Flat rolling

Figure 1 represents a wear map for flat rolling. The process is characterized by two parameters: the "TRESCA" friction coefficient $\overline{m} = \tau\sqrt{3} / \sigma_0 < 1$ (3) and the aspect ratio of the roll bite A. If A is below a critical value, which increases with $\overline{m}$, friction prevents any sliding and wear; above the critical value, sliding, hence abrasion, occurs. For given values of the roll radius R and the initial strip thickness H_1, the aspect ratio A increases with the reduction $r=1-H_2/H_1$ and we can define in the map of figure 1 five zones:

- too low or too high reduction for industrial application
- roll slippage, where friction is too low to carry the strip into the roll bite
- at low reduction, the field of thermal fatigue
- at high reduction, the zone of abrasion, the intensity of which depends on friction.

2.2 Axisymmetrical upsetting of cylinders

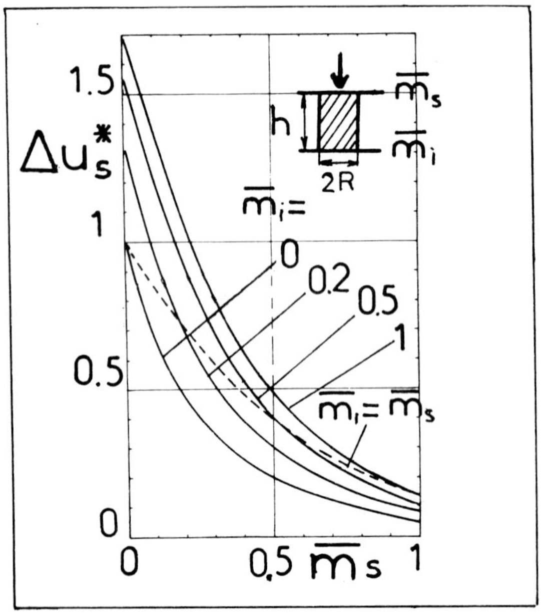

Fig. 2 : influence of the lower and upper die friction coefficient on the sliding velocity on the upper die for upsetting of cylinders (5) (diameter to height ratio: A=2R/h=0.75)

The interface conditions during hot forging depend on lubrication and temperature, and a dissymmetry between upper and lower die frequently occurs in industrial workshops. The sliding velocity on one of the dies is influenced by frictional conditions on both dies: for a given aspect ratio of the cylinder, figure 2 shows that the sliding speed on the upper die increases when friction $\overline{m}_s$ on the upper die decreases, or when the friction on the lower die $\overline{m}_i$ increases. Note that under very dissymetrical conditions ($\overline{m}_s << \overline{m}_i$), the sliding speed on the very well lubricated die can be greater than that obtained under symmetrical frictionless conditions ($\overline{m}_s=\overline{m}_i=0$); this is confirmed by observations on plasticine workpieces.

3 THERMOMECHANICAL BEHAVIOUR OF SCALE

The damage type and friction are mainly influenced by the effective hardness of the workpiece scale which depends on its temperature and its chemical composition. This is a very complex matter, since the high temperature scale on steel is composed of three iron oxydes FeO, Fe_3O_4 and Fe_2O_3. However wustite FeO is the main component; as its hardness is lower than that of steel above 900°C, scale can have a lubricating action if it cools down only slowly during each contact time Δt, which can vary between 5 ms (forging on drop hammer or final finishing rolling passes) and a few seconds (forging on hydraulic press or first roughing rolling passes). Figure 3, deduced from ring compression tests performed under dry conditions, demonstrates that the behaviour of the scale can be characterized by the ratio S of the scale thickness h_c and the depth of heat diffusion into scale (3):

$$S=h_c\,(6a_c\Delta t)^{-0.5} \qquad (1)$$

where a_c is the thermal diffusivity of scale ($=0.4 mm^2/s$).

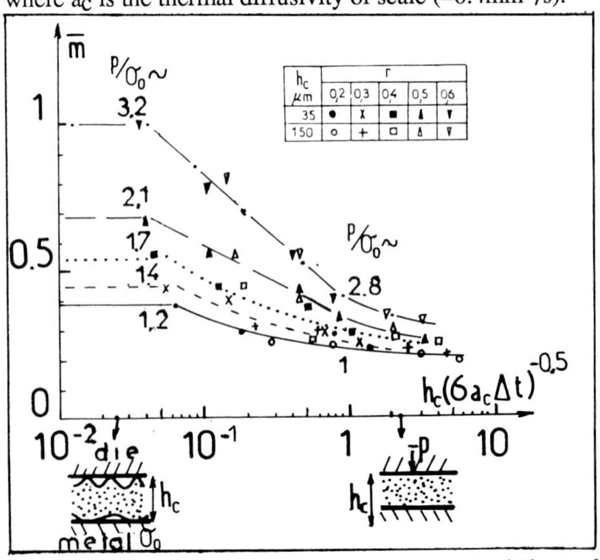

Fig. 3 : influence of the contact pressure p and thermal conditions on dry friction and behaviour of scale (6)

Three zones can be distinguished on figure 3:

- lubricating zone: S>2 the scale temperature remains high enough to induce a low TRESCA type friction, which is not very sensitive to p and Δt. As the real contact area approaches the apparent one, the scale can stick to the tools, but induces very low abrasion.
- wear zone: S<0.05 the scale is strongly cooled by the contact with the tool, and induces a higher, COULOMB-type friction: the friction stress is proportional to p and not very sensitive to Δt; the scale is harder than the workpiece, brittle, not very adherent, but induces high wear.
- intermediate zone: 0.05<S<2 friction and wear increase with both p and Δt.

4 MECHANICAL MODEL OF ABRASIVE WEAR

Tests on tribometers most of the time follow ARCHARD's law: wear volume δV is proportional to the normal load P and the sliding distance δl (during time δt):

$\delta V = k\, P\, \delta l$ (2)

Let us consider an apparent contact area A. Noting that contact pressure $p=P/A$, wear depth $\delta h=\delta V/A$, and sliding length $\delta l=\Delta u\, \delta t$, we can easily deduce from eq. (2) an estimate of the wear depth at a point M on the tool surface:

$\delta h = k \int p\, \Delta u\, dt$ (3)

We state precisely the value of wear coefficient k in paragraph 5 below.

4.1 Application to the metal working processes

The mechanical analysis of the process provides Δu and p (cf. paragraph 2). We first consider the rolling of a strip of length L in the high wear region ($\bar{m}<0.66$; A>3, cf. figure 1). For this stationary process, the rolls undergo uniform wear and integration of eq.(3) along the arc of contact provides the radial loss (4):

$$\Delta R=0.057\, k\, (H_1-H_2)^{1.5}\, (1+\bar{m}A/4)\, \sigma_o\, L/H_2R^{0.5} \quad (4)$$

The forging process is not stationary, and the analysis is more complex. As p and Δu vary with time and along the interface, δh depends on the number of forgings and varies along the interface. Figure 4 gives values of sliding length $\delta l=\int \Delta u\, dt$, the contact pressure p, die surface temperature Θ_S and wear depth profile δh for the upsetting performed under dissymetrical conditions on a drop hammer and a mechanical press.

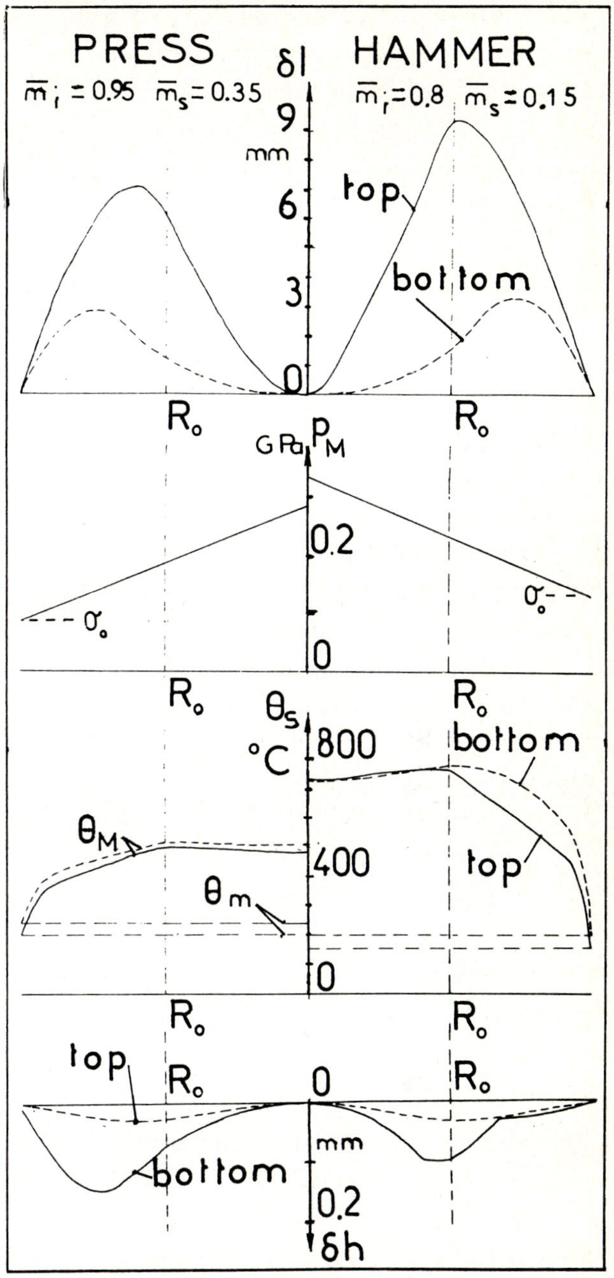

Fig. 4 : results of thermomechanical analysis of the upsetting of cylinders under dissymetrical conditions, and wear depth profile (7)

steel cylinder $2R_o$ = 30 mm ; h_o = 40 mm ; height reduction 75%; N=1000 forgings; Flat dies Z38CDV5 ; HV=4GPa

lower die (dry) upper die (lubricated)

4.2 The wear rate coefficient k

In order to describe the results obtained in forging (7), we assume that the wear rate coefficient k has the form:

$k = K_F\, K_W\, HV^{-2.1}$ (5)

First k decreases very markedly (exponent -2.1) as the Vickers hardness HV of the tool material increases, this is probably in close relation to the relatively small difference in hardness between the tool and the workpiece scale under the contact conditions (cf paragraph 3). Moreover, k depends on the microstructural features of the tool material (amount of carbides for instance); this effect is described by the factor K_W which can be written:

$$K_W = 1 + 5.1 \exp(-0.085\overline{W}) \qquad (6)$$
$$\overline{W} = (\%W) + 2(\%Mo) + 4(\%V) + 0.5(\%Cr)$$

$\overline{W}$ is called the tungsten equivalent.
Finally, the wear rate depends on the superficial films (lubricant film, transfer film formed by scale). The experimental values of the corresponding factor K are given in table 5.

	upper die (lubricated by graphitized oil)	lower die: dry; workpiece remains 1s before forging		
		Δt=0.1s	Δt=5 ms	
forgings	Δt=0.1s to 5 ms	S=0.015 to 0.05	centre	periphery S>0.2
K_F ($MPa^{1.1}$)	0.85	11	22.5	2.25
$\overline{m}$	0.4	0.15	0.95	0.8	

Table 5 : influence of contact time on K_F and friction

4.3 Experimental test of the model

In figure 6 we compare the theoretical and experimental wear depths profiles under various forging conditions. Under drop hammer forging conditions, we assume that the workpieces are randomly shifted horizontally by ±3mm (the positioning system is not perfect).We can note some interesting features from table 5 and figure 6:

- the lubricant diminishes drastically the wear rate at all contact times. Therefore, although lubrication increases the sliding distance (cf figure 4), lubricated wear is very low under our experimental conditions (figure 6); in industrial workshops where contact conditions are more severe, lubrication most of the time increases wear.
- for very small contact times such as with drop hammer (S=0.2), the scale of the lateral surface of the workpiece, which comes progressively in contact with the dies at the periphery, is not very abrasive (it remains hot). So at the periphery, under dry conditions, the wear depth is smaller on drop hammer than on press.
- on the contrary, in the central part of the dry lower die, the scale of the plane surface of the workpiece is cooled during the time lag before forging, and the wear rate is very high.

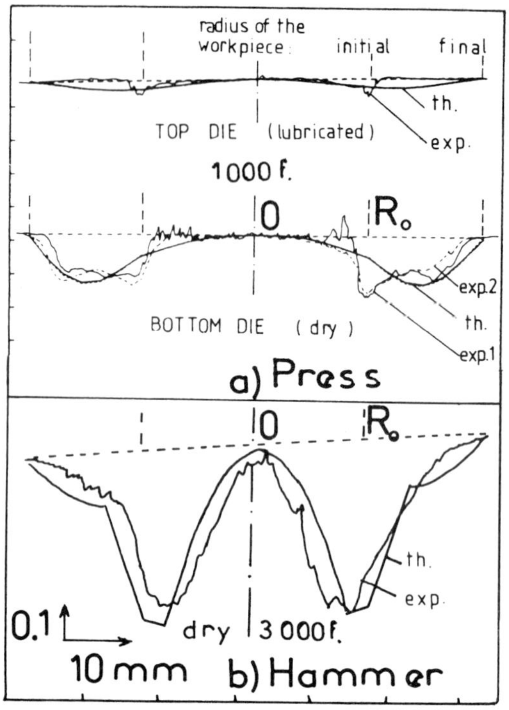

Fig. 6 : theoretical and experimental wear depth profiles in upsetting (conditions: same as figure 4).
Z38CDV5 flat dies (HV=4 GPa) (7)
a: mechanical press N=1000 forgings
b: drop hammer; bottom die

The values of the wear rate coefficient deduced from a few measurements performed in hot rolling with formula (4) are in good agreement with the values observed in hot forging.

5 EFFECTIVE HARDNESS OF FORGING DIES

The thermal solicitations of the forging dies are much more severe than those of the rolls of rolling mills, and can produce marked metallurgical, hence hardness changes in the superficial layers.Hence, the damage rate may greatly vary with the number of forgings performed. Depending on the maximal values of die surface temperature Θ_M (Fig. 7), superficial softening or hardening may occur.

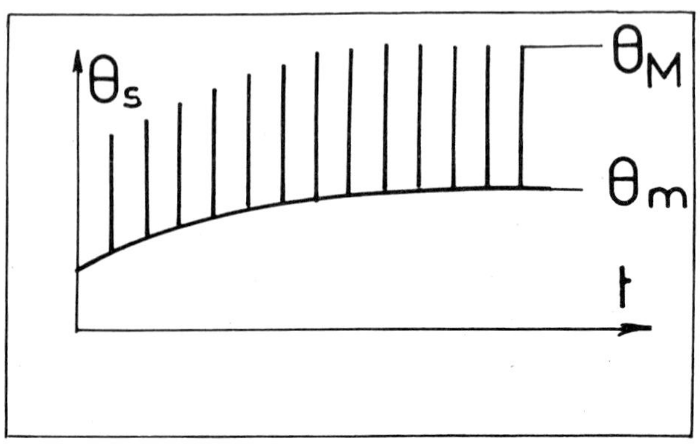

Fig. 7 : thermal cycle of the die surface

5.1 Surface softening

Let us first assume that the maximal surface temperature exceeds the tempering temperature of die material, but remains smaller than the temperature for start of austenitization AC1. The additional tempering can be calculated from the master tempering curve of the tool material (10,11). As the hardness decreases, wear rate can increase drastically according to formula (5). moreover, in the central part of the dies, where the sliding distance and wear remain small, thermal fatigue can increase drastically as demonstrated in figure 8. Due to the progressive reduction of the die flow stress, the plastic extension occuring after each forging increases steadily.

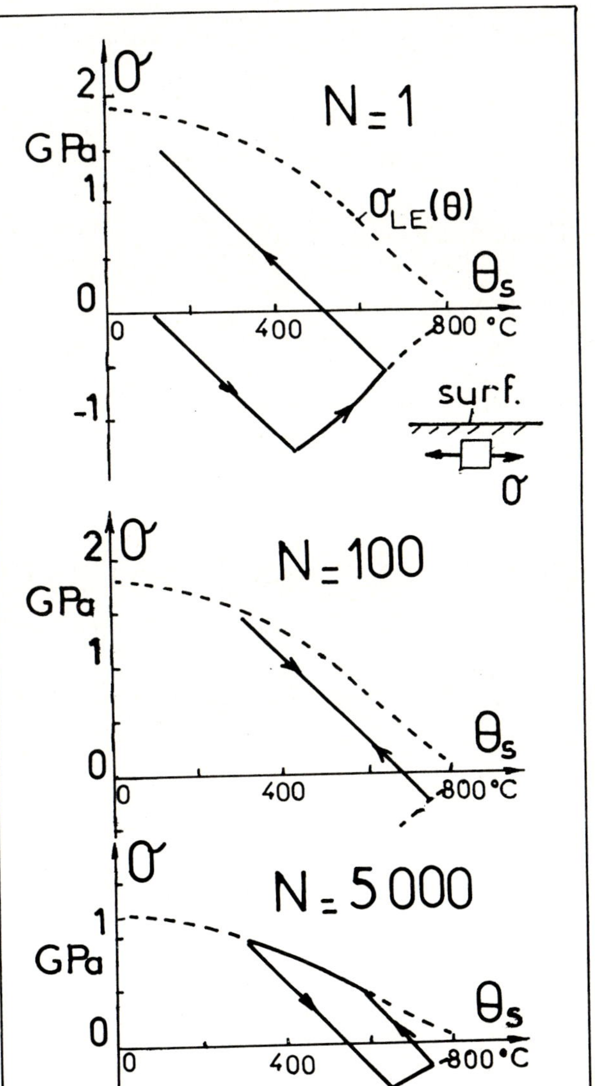

Fig. 8 : evolution of the superficial thermal and stress cycles with the number of forgings performed on Z38CDV5 dies (9)

5.2 Surface hardening

At the periphery of the drop hammer die, the sliding velocity can reach very high values (a few m/s), and induce very high die surface heating. If the maximal surface temperature Θ_M exceeds AC1, the hardening occurs by redissolution of the carbides in the matrix (all the more complete as Θ_M is higher) during heating followed by martensitic transformation (all the more complete as the bulk temperature Θ_m is lower) during cooling. We have built a model of this metallurgical change; the hardening occurs after a few forgings, increases with an increase in Θ_M and a decrease in Θ_m (figure 9).

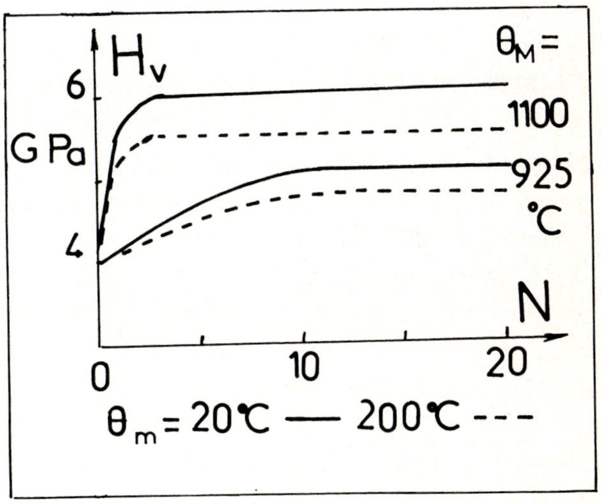

Fig. 9 : theoretical evolution of die surface hardness (6,7)

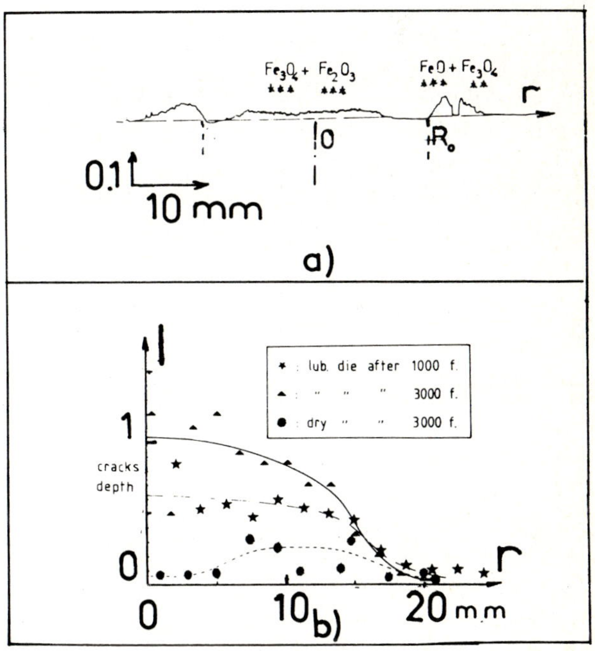

Fig. 10 : influence of nitriding treatment in hot forging with a mechanical press (conditions: same as fig. 6)

a: transfer film on dry die N=1000 forgings

b: maximal depth of cracks vs distance from centre

The high hardness of the "white layers" (6,11) so produced avoids abrasion, but they are brittle and most of the time appear cracked. It is interesting to notice that hardening of die surface by nitriding treatment has similar effects (figure 10): in press forging, the surface hardness of nitrided dies remain high enough (HV>8GPa) to prevent any wear, but promotes thermal fatigue (figure 10b). Under dry conditions, nitriding promotes formation of highly adherent transfer films which reduce the cracking rate in comparison with the lubricated die: lubricant hinders the adhesion of the scale and the thermal cycles are more severe.

6 CONCLUSION

The prediction of tool life in industrial practice clearly reveals that hot working is not easy: changes of some process parameters usually can modify the tool damage pattern. We have seen for instance that high tool surface hardness induced by thermal cycles or surface treatment hinders abrasive wear but promotes transfer films under dry conditions and thermal fatigue. Despite their simplicity, the models presented in this paper have been successfully applied to simple geometries. Work is presently in progress to introduce these models into Finite Element models of metalworking processes (12) to account for wear in the design of more complex operations.

REFERENCES

(1) THOMAS, A. : "wear of drop forging dies" in TRIBOLOGY IN IRON AND STEEL WORKS. Publ. by THE IRON AND STEEL INST. (LONDON) (1970), 135-141

(2) STEVENS,P.G., IVENS,K.P. and HARPER,P. : "Increasing work roll life by improved roll cooling practice." J. Iron Steel Inst. (Jan. 1971) 1-11

(3) BAQUE, P., FELDER, E., HYAFIL, J. and D'ESCATHA, Y.: MISE EN FORME DES METAUX. CALCULS EN PLASTICITE. Publ. DUNOD (PARIS) (1974)

(4) FELDER, E. : "Interactions métal-cylindre en laminage" CESSID-IRSID lectures (MAIZIERES-LES-METZ) (1985) , 85-171

(5) THORE, Y. and FELDER, E. : "A theoretical and experimental study of the interface conditions during the hot forging of steel using a dissymetrical flow model." J. Mech. Working Tech. 13 (1986) 51-64

(6) FELDER, E. and BAUDUIN, P. : "Usure et frottement en forgeage à chaud." Mécanique, Matériaux, Electricité (Jan. 1981) 4-15

(7) THORE, Y. : "Etude théorique et expérimentale du frottement et de l'usure par abrasion des matrices de forge à chaud des aciers. Influence d'une nitruration." Thèse Dr - Ing. (Ecole des Mines de Paris, CEMEF, 1984)

(8) FELDER, E., RENAUDIN, J.F. and THORE, Y. : "The tribology of a simple hot forging process: influence of lubrication and nitriding treatment." Proc. Conf. Advanced Techn. Plasticity (1984) Vol 1, 231-234

(9) BAUDUIN, P. , FELDER, E. , BATIT, G. and MONTAGUT, J.L. : "Study of the thermomechanical conditions and the wear of hot forging dies." Proc. 10th Int. Conf. Stamping (LONDON, JUNE 1980)

(10) PAYSON, P. : THE METALLURGY OF TOOL STEELS. Publ. J. WILEY and Sons (LONDON) (1962)

(11) FELDER, E. and COUTU, L. : "Aspects thermodynamiques de l'usure des matrices de forge à chaud de l'acier." Bull. Cercle Etude des Métaux 4, 14 (Dec. 1978) 153-176

(12) CHENOT, J.L. and ONATE, E. : MODELLING OF METAL FORMING PROCESSES. Publ. KLUWER (DORDRECHT) 1988

Paper IX(iii)

The influence of debris inclusion on the performance of polymeric seals in ball valves

B. J. Briscoe and P. J. Tweedale

This paper describes a simulation procedure which seeks to evaluate certain facets of the performance of polymeric valve seat in particle contaminated environments. The important feature of the simulation is the control of the rate of hard particle debris entrapment into a model contact and the subsequent effect on the magnitude of the friction. Two engineering polymers in common seal use (PTFE and GFPTFE) as well as several unproven materials (PEEK, a fluororelastomer modified PEEK and an aromatic polyester) are evaluated. The experimental results and their analysis indicate that by using the current simulation polymers may be effectively ranked in terms of their propensity to accumulate debris in particle contaminated environments. PTFE is shown to accumulate debris almost three times as rapidly as PEEK for example. The results of supplementary experiments correlate well with the ranking parameters provided by the simulation.

1. INTRODUCTION.

Many of the problems encountered with pipeline valves, and in particular ball valves, can be traced to the unsatisfactory performance of the valve seat materials. This subject has been researched and the type of service valve failure have been identified for the common valve types (1,2,3). A previous study of the catastrophic failure of ball valves by debris inclusion has been reported in the literature (4). Ball valves seats are generally fabricated from polymeric materials. The requirements of these seats are wide ranging. Ideally the valves should possess zero leakage rates at low and high differential pressures across the valve. They should have low operating torques to open and close, as well as providing good wear resistance. The engineering specification and design of effective seals invariably gives rise to a compromise in the main operating variables, such as operating torque and leakage, to yeild a practicable working design. Materials which are generally 'hard' have low creep compliance which allows for the onset of leakage to occur quickly due to poor sealing. A 'softer' material has better a creep compliance but tends to generate a higher operating torque particularly when the inclusion of debris from the fluid stream occurs.

The service requirements of all valves are that they should maintain their design characteristics over the prescribed lifetime. There are many field uses where such valves can be utilised. These include the relatively mild conditions in plant onshore, where short maintenance intervals may not be as important. At the other extreme, there are the hostile environments and the reliability demanded in offshore applications. This latter environment can be particularly demanding in that the fluid flow is invariably contaminated. This is the case on North Sea oil rigs where the operational temperature is ca. 80C and the fluid stream contains significant amounts of contaminants in the form of hard particles. This is a particularly frequent problem for ball valves as they are often chosen for this latter application due to their relative simplicity and light weight.

1.1 Valve Seat Materials.

It is useful to briefly review, in general terms, some the important features of the different types of valve seat materials. The ball is invariably constructed from a steel which has a hard steel coating.

Metallic seats have major advantages centred on their reliability. They have long service lifetimes and tolerance to wide ranging operating conditions. Their disadvantages are the difficulty in obtaining a satisfactory seal at low differential pressures and also the operating torques are often high. This later characteristic necessitates the use of larger actuators than are required for other types of seat.

Elastomeric seating materials possess very good sealing characteristics at low differential pressures with low operating torques. Such sealing materials are unfortunately susceptable to material production variability and also the deterioration in material properties in conditions of high dissolved gas content. The latter encompasses the so-called 'explosive decompression' phenomena where pressure cycling of gas into the seal can give rise to material failure. In addition, interaction with the gas (particularly hydrogen sulphide) can cause material modification. The combination of these factors has discouraged the use of these materials in certain applications especially offshore.

Polymeric materials as seats offer both good sealing and low operating torques, as well as being chemically inert to many of the fluids of common interest. Polymeric components are commonly included as one of the primary sealing members in ball valves. Usually the polymer elements are in the form of rings which are loaded against a metal ball which rotates about an axis which is parallel to the plane of the rings (see Figure 1). Normally two polymeric rings are used and the ball is fashioned with a diametral hole whose axis is orthogonal to the rotation axis. Flow control is effected by rotating this hole. This operation exposes the sealing element to the suspended particulate material invariably present in the fluids. These particles become embedded in the polymeric surface. The consequent deterioration in the valve seat characteristics can lead to failure due to the operating torque increasing significantly and eventually causing failure by seize. Alternatively, failure can occur by the sealing efficiency being compromised in a major way and catastrophic leakage occurring.

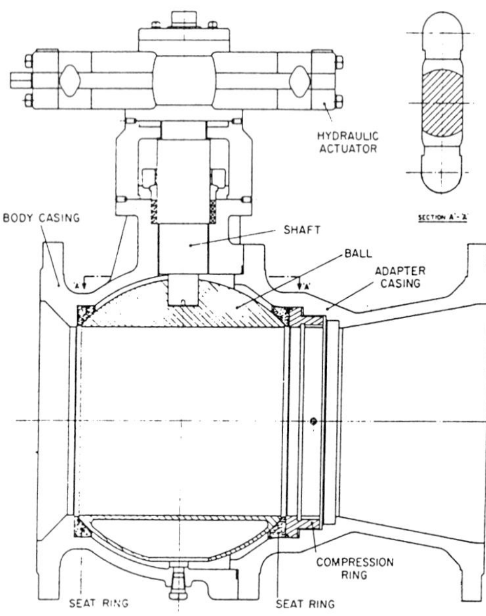

Figure 1. Diagram of a typical polymer seated ball valve.

1.2 Mechanism of Entrapment.

The mechanism of entrapment occurs by the debris being drawn into the sealing zone whilst the seal is in contact with the fluid flow and the ball is in motion. The particles are then slowly forced into the seal surface. The debris may thus progressively accumulate in the surface of the seal if it is not expelled when the contact is unloaded. Clearly the finest particles will tend to ingress most rapidly, although an analysis of failed seal components shows very substantial particles (some of the order of millimetres in diameter) can become trapped in the sealing zone during the lifetime of the part. The initial polymer/seal contact which is an effective seal and low friction contact is gradually transformed into, for example, a silica/steel contact which is not an effective seal and which has a much higher sliding and/or static friction.

1.3 Commercial Practice.

A wide range of organic polymers have been considered for use as valve seat materials. Amongst the most popular have been the fluorocarbons, in particular the PTFE's. These are attractive materials due to their chemical inertness, low friction and relatively low creep compliance. The combination of low sliding friction and rather high creep under normal load produces a material with an unusually effective intrinsic sealing capability. These materials however have rather poor wear resistance and debris accumulation characteristics. Also extrusion may occur at high operating pressures and temperatures. Another important class of polymers are the polyamides or nylons which are apparantly preferred for high pressure applications.

The testing of full scale valves in a reliable and systematic manner is an expensive and time consuming task, although this type of testing is carried out (1,3). The aim of the work described here was to understand at a fundamental level some of the material characteristics important in the inclusion of hard particulate debris into ball valve seats. This has involved the design and construction of a laboratory apparatus to simulate the process of particle inclusion and the measurement of the effect on the friction in metal/polymer couples.

2. EXPERIMENTAL PROCEDURE.

The contact configuration adopted for the simulation of debris accumulation into polymeric substrates was that of a metal roller loaded against a polymeric plate. The rolling action of the former trapped debris in the latter. The rolling speed defined the contact time of the debris in the contact zone. The debris retention and it's effect on the sliding friction was then measured by periodically clamping the roller and sliding it relative to the plate. This contact configuration was designed to simulate the typical operating conditions which occurs in ball valves seats; 3.5-14 MPa contact pressure. Selected sand particles were used as a contaminants at approx. 500ppm concentration in aqueous suspensions.

A series of model experiments were also undertaken to identify the critical material parameters affecting debris inclusions. The experimental configuration chosen here consisted of sliding rigid conical indentors over polymeric flats and measuring the friction and resulting scar widths (5). These studies sought to simulate the contact stresses generated at particle -polymer contacts.

An apparatus was constructed to perform the simulation described above. This is shown in Figure 2 and consisted of base plate upon which was mounted a loading arm and the experimental platform. The former consisted of compound arms to allow the application of the required forces to the roller from managable loading weights. The latter unit comprised of an immersion bath, connected to a reservoir equipped with a stirrer and circulator, mounted on two machine slideways. A motor drive attached to the lower slideway provided the motion in the contact, whilst the friction was monitored by a displacement transducer constructed from flexible beams connecting the roller holder to the loading arm. Analog/digital interfacing of the displacement transucer adjacent to the flexible beams allowed the experimental results to be logged in real time to a microcomputer.

The operation of the simulation experiment contained the following major elements:

(1) The roller was rolled backwards, forwards and backwards again over the sample before being locked into position. The locked roller was then slid across the surface and the frictional force developed was measured.

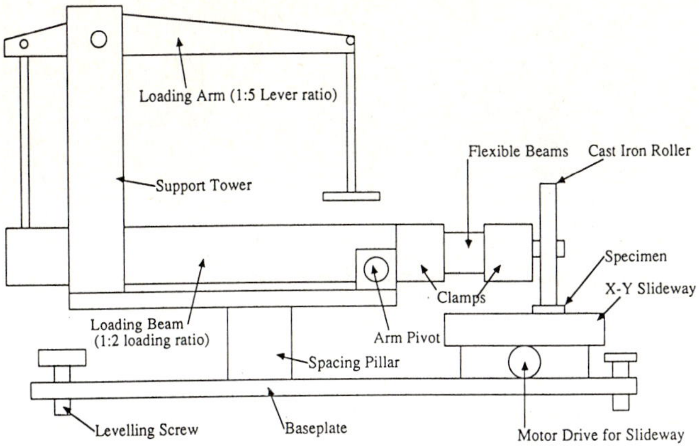

Figure 2. Side view of simulation apparatus.

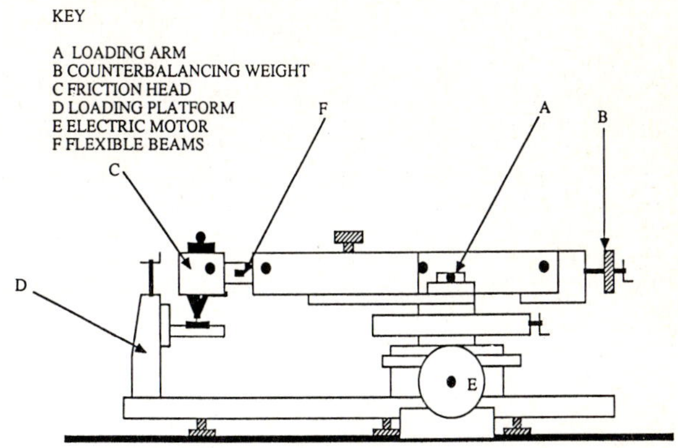

Figure 3. Side view of Eldredge apparatus.

(2) Three cycles of the operation described in (1) were undertaken in clean water in order to create a conforming contact and assess the friction in nominally clean conditions before introducing the contaminants.

(3) Debris accumulation cycles were then performed on a continuous basis with the debris contaminanted water being pumped into the accumulation bath. The friction force development with time was monitored in the manner outlined above until it plateaued or reached a value too high for the safe continuation of the experiment.

The loads imposed on the specimens were scaled by the ratio of their hardnesses in order to produce comparable contact pressures of ca. 14MPa. The speed of movement over the sample was 2.4 x 10-5 m/s.

The scratch hardness experiments were performed using a friction machine has been described in detail elsewhere (6). A general view is shown in Figure 3. The main functional part was a loading arm which is pivoted about the point (A), to the rear of this is a counter-balancing weight (B). Loads were applied directly to the friction head (C) by means of a dead weights once the arm has been balanced. The friction head was interchangeable and here is used to support a conical indentor. Beneath the friction head a loading platform (D) supported the polymer specimin. The loading arm was mounted on a pair of machine slideways at right angles to one another. The relative motion of the cone and specimen was achieved by a small electric motor (E), driving one of the slide-ways. The friction in the contact was monitored by strain gauge mounted on flexible beams connecting the friction head to the loading arm. The output was recorded on a chart recorder. The sliding velocity used was 4.2 x 10-5 m/s.

The experimental work was performed using various loads with one cone angle 60 degrees (relatively sharp) in order to simulate the contact characteristics generated by rough sand particles. The polymers studied were PTFE, GF-PTFE, a fluoroelastomer modified PEEK (Raychem V200) and an aromatic polyester (Celanese Vectra). Where susceptability to an aqueous environment was suspected the scratching experiments were carried out with a short contact time with water (of the order of the experiment duration), and also a longer contact time (with the samples immersed in water for approx. 72 hrs.). Experiments with complete saturation of the polymer matrix were not carried out due to the protracted times required for the fluid migration into the matrix. The frictional work was monitored. The track widths were measured immediately after scratching using a travelling microscope.

3. SIMULATION RESULTS.

The experimental results are given in Figures 4-8. The important features are:

(1) PTFE and GFPTFE results have a shape that reveals a significant increase in the friction initially. This then decreases to a lower value in the latter part of the experiment. The level of both of these features increases during the course of the simulation.

(2) The remaining polymers produce experimental results which are characteristically different from those of the PTFE systems. The initial large increase is absent with the frictional work being more stable at an overall level during the course of each cycle. The overall frictional level is higher initially but tends to increase more slowly than the PTFEs. This indicates the debris accumulation for these materials is significantly less than for the PTFE based systems. There is also some evidence of stick/slip behaviour.

The qualitative features of these results are presented schematically as in Figure 9. These results reflect those seen during the testing of full scale valves. A representative set of such data, for a PTFE based seal system which shows a dramatic rise in friction which is not typical, is shown in Figure 10. These differ from the results of the current simulation due to the geometrical changes in the ball valve contact in operation. However, the essential features of the rise in the frictional work of the contact with debris inclusion can be seen. The simulation therefore shows the essential features of ball valve operation in contaminated conditions.

4. ANALYSIS OF SIMULATION DATA.

A semiquantitative rationalisation of the experimental data can be made on the basis of a model originally used by Bowden and Tabor to predict the effect of the adsorption of boundary lubricants on the friction of metal sufaces (7). This approach has recently been adopted to interpret frictional phenomena in polymer boundary lubrication (8). Many other authors have undertaken

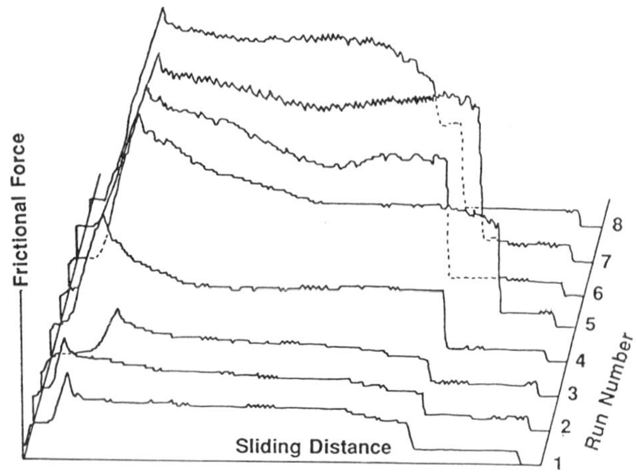

Figure 4. Simulation results for PTFE.

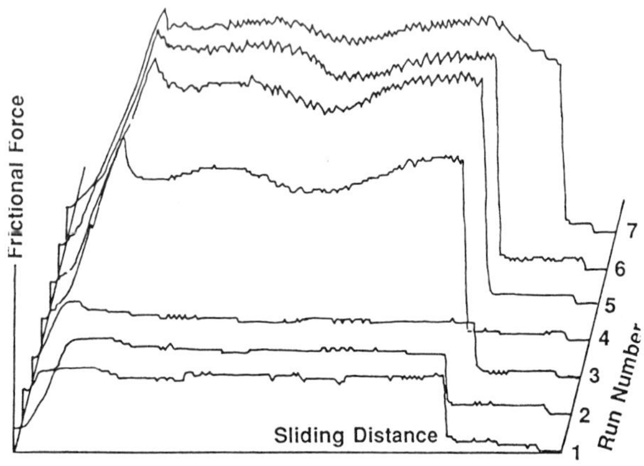

Figure 5. Simualtion results for GFPTFE.

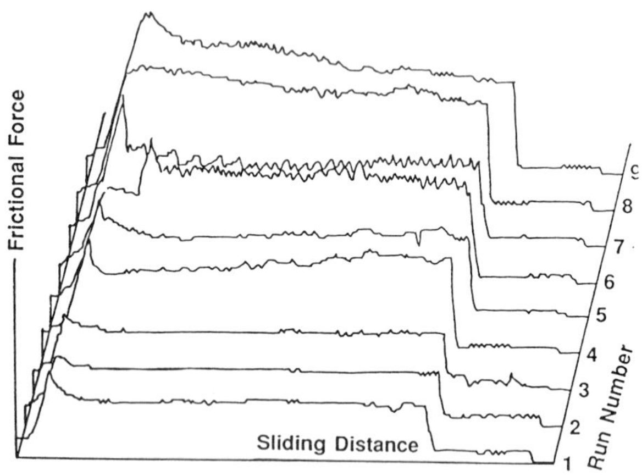

Figure 6. Simulation results for PEEK.

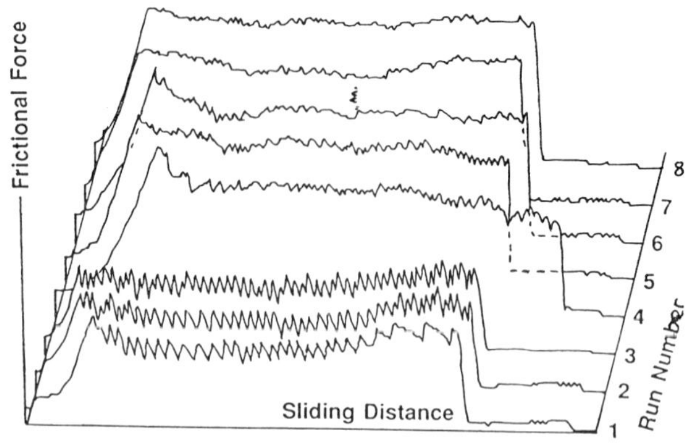

Figure 7. Simualtion results for Vectra.

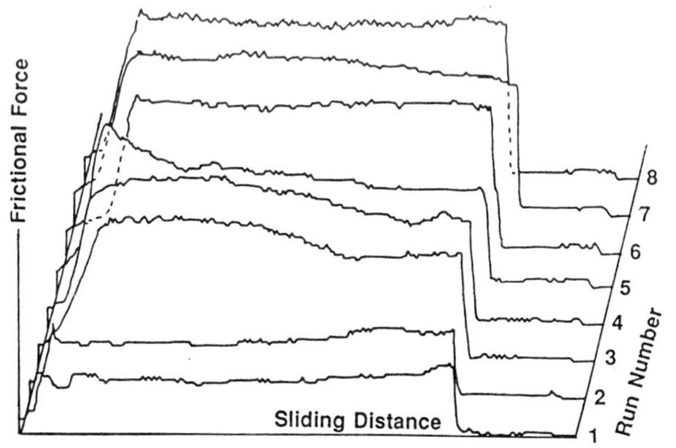

Figure 8. Simulation results for fluoroelastomer modified PEEK.

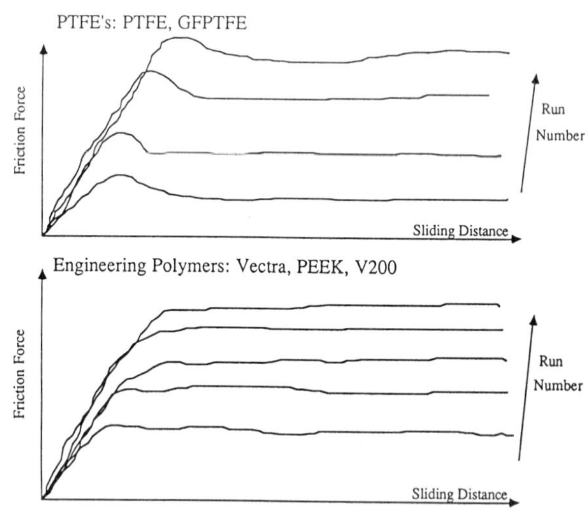

Figure 9. Characteristic behaviour patterns observed in simulation.

similar studies, and currently the idea is being applied to predict friction values in composite tribology (9). The original models consider the fractional coverage of a surface by surface active species and their effect on the measured frictional response. The effect of debris particles being included into a polymer surface can, as we shall see, be modelled in a similar manner. First we adopt the adhesion model of friction where F, the magnitude of the friction, is given by :-

$$F = A_T s_T .$$

Then we distinguish two sites on the surface;

$$F = A_p s_p + A_s s_s ,$$

one is the polymer and the other is the debris. Actually the problem is a more subtle one than just the coverage of the surface. The adhesion model of friction addresses areas of contact, and hence the specified sites are the contact sites. Then we have a total contact area AT, made up of Ap or As, the contact areas of polymer/steel and sand/steel respectively during sliding. Each type of contact zone has a different interface shear stress, sp and ss, for polymer/steel and sand/steel contacts respectively. The important point here is that

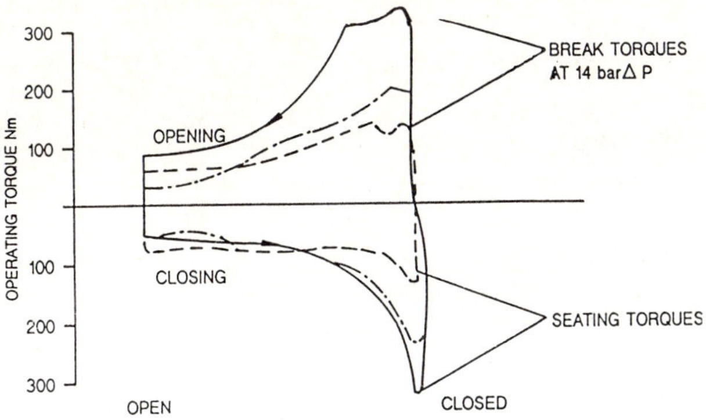

Figure 10. Typical torque trace/displacement curves from full scale testing of P"TFE seat supported ball valves.

$A_p/A_T = Ø$ may not be equal to fractional surface coverage deduced by optical examination of the surface of the debris encrusted polymer surface after unloading. Then

$$A_T = A_p + A_s \quad \text{therefore} \quad A_s = A_T - A_p$$

then, $$F = s_p A_p + (A_T - A_p) s_s$$

and $$F = A_T (s_s + (s_p - s_s) Ø)$$

$$F = A_T s_s (1 + Ø a) \qquad (1)$$

where a corresponds to $(s_p - s_s) / s_s$

In order to render this equation useful for the analysis of the experimental results, this expression for the frictional work during the simulation is subtracted from an estimate of the final frictional work. The latter was obtained from the assymptotic value that all surfaces approached when fully encrusted with sand. Equation (1) is a correct description of the problem but the application to the data presents several intractable problems. We assume that during each cycle of the simulation (N) a fraction (x) of the contact area, originally AT, is covered by particles. Then by enumerating a series of terms that may be written for the decrease in the contact area of polymer and the resulting accumulation of sand, Ø can be shown to be of the form (1 - x) N (see Apppendix 1). The expression for the sand accumulation, that is the increase in As, is a more complex summation of terms.

Since $$F_F = A_T s_s \quad \text{then}$$

$$F_F - F = A_T s_s - A_T s_s (1 + Ø a)$$

$$F = - A_T s_s (Ø a)$$

and $$F = - A_T s_s (1 - x)^N a \qquad (2)$$

or $$\log F = \log (- A_T s_s a) + N \log (1 - x)$$

The intercept of the resulting line will contain an estimate of the interfacial shear stress of the sand, whilst the slope will contain a parameter which describes the efficiency of inclusion of debris into the polymer contact as the term log(1 - x). The analysis of the experimental data according to this scheme are shown in Figure 11 for the polymers considered here. The slopes and intercepts of the

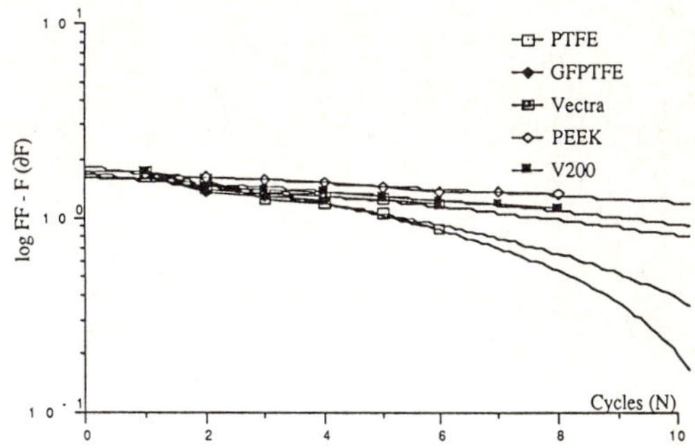

Figure 11. Graphical analysis results for polymers from simulation results (log F/cycle no.).

resulting plots fitted using linear regression analysis, are given in Table I.

Table I. Tabulation of material results from the simulation experiment analysis, and the consequent material parameters derived and the inclusion efficiency of debris into the polymer.

Material	Intercept	Slope	x
PTFE	1.851	-0.165	0.316
GFPTFE	1.72	-0.134	0.265
Vectra	1.62	-0.081	0.17
PEEK	1.74	-0.055	0.119
V200	1.66	-0.074	0.156

The estimate made of the frictional value for the fully sand encrusted surface, FF, effects the intercept values given, but this is common to all. The inclusion parameter, x, is however of considerable value as this is independent of the assumptions made with regard to the interpretation of these results. It may thus be regarded as a sensitive indicator for the propensity of the material in question to include debris into it's surface. The values of x are sensible and of the order expected. Thus for example, PTFE accumulates debris at a rate which is about three times faster than that of PEEK.

5. SCRATCH HARDNESS RESULTS.

The data are not presented in detail as a seperate entity but combined with data obtained from the simulation to test the value of two correlations. The scratch width data is regarded as a measure of the plastic stain capacity of the polymer surface and hence should correlate with the inclusion parameter, x. The friction coefficient, the work done in the experiment may or may not correlate with intercept parameter of the simulation depending upon the nature of the friction processes in the simulation. These correlations are presented in Figures 12 and 13 respectively. The former shows excellent correlation for PTFE, GFPTFE, PEEK and the aromatic polyester. The friction and intercept correlation is indifferent but follows the correct trend.

6. CONCLUSIONS.

We have demonstrated that it is possible using a

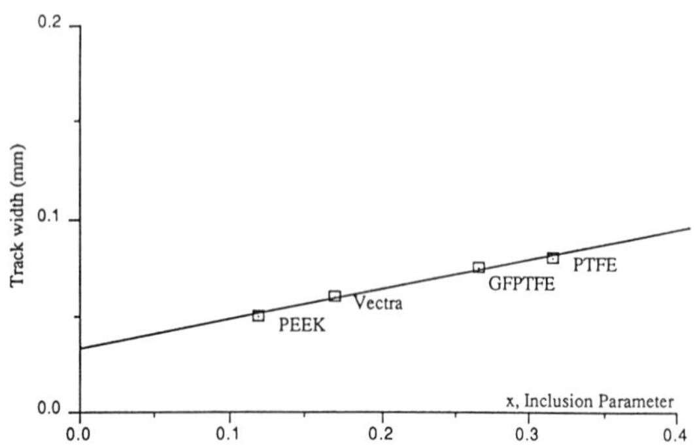

Figure 12. Correlation of track width in scratching and inclusion parameter (x) from simulation.

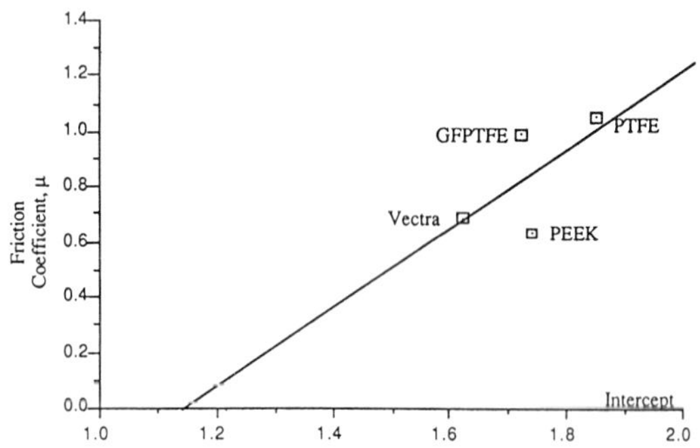

Figure 13. Correlation of friction coefficient in scratching and intercept parameter from simulation.

simple and relatively inexpensive experimental simulation to sensibly rank, in terms of their relative performance, the time dependent frictional behaviour of polymeric valve seat materials operating in particle contaminated aqueous media.

Systems based on PTFE show low initial sliding friction which increases rapidly as debris accumulates in the contact. They also show static friction values that are much higher than the kinetic values, even when debris accumulates in the contact. This is a common feature of certain types of PTFE contact (10, 11, 12), where significant reorientation occurs in the polymer surface.

The non PTFE based materials show a less interesting type of behaviour. They are characterised by higher initial friction which also increases with debris accumulation but less rapidly than for the PTFE systems. The static and dynamic coefficients of friction appear to be very similar.

The simple scratch hardness experiments provides two experimental quantities that can be measured easily, one of which shows a strong correlation with the inclusion parameter of the simulation.

Only the frictional characteristics of the valve seats with debris accumulation are addressed by this simulation. In the overall design other factors such as environmental stability, wear rates, sealing ability and to a lesser extent cost are additional considerations. However the simulation procedure described above does provide a means of quantifying one aspect of valve seat performance.

7. ACKNOWLEDGEMENT.

The authors would like to thank BP Research, Sunbury on Thames for the financial support for this work, and in particular Dr. D. Harrison for support and encouragement, and also providing Figure 10.

References.

1. BILLINGTON M J, HARRISON D, and VIVIAN B E. Paper J1 given at Intl. Conf. Developments in Valves and Actuators for Fluid Control, Oxford, (10-12 September 1985). 'The identification and evaluation of valve problems.'
2. VIVIAN B E. Process Engineering, 33, October 1985. 'Identifying and evaluating valve problems.'
3. BILLINGTON M J, HARRISON D, and VIVIAN B E. Paper given at 2nd. Intl. Conf. Developments in Valves and Actuators for Fluid Control, Manchester, (28-30 March 1988). 'Performance testing to identify reliable valves.'
4. CONNELL R A, SUMMERS G G, SHEPHERD J P, and SHONE E B. Proc. 3rd. Leeds-Lyon Symposium on Tribology, (1976), Paper IX(iv). 'The use of a filled PTFE as a ball valve seat material.'
5. BRISCOE B J, EVANS P D, and LANCASTER J K. J. Phys. D: Appl. Phys., 20, 346-353, (1987). 'Single point deformation and abrasion of gamma irradiated PTFE.'
6. BOWDEN F P, and TABOR D. 'Friction and Lubrication of Solids.' Part 2. Clarendon Press, Oxford, 1964.
7. BOWDEN F P, GREGORY J N, and TABOR D. Nature, 156, (1945), 97-101. 'Lubrication of metal surfaces by fatty acids.'
8. SKELCHER W L. in Microscopic Aspects of Adhesion and Lubrication, pp. 719-728, Ed. J.M.Georges. 'A proposed adhesive mechanism of boundary lubrication for polyphenylene oxide and silicone fluid.'
9. BRISCOE B J, and TWEEDALE P J. Proc. Intl. Conf. ICCM V, held at Imperial College, London, July 1987. 'A critical reveiw of the tribology of polymer composites.'
10. POOLEY C M, and TABOR D. Proc. Roy. Soc., A329, 251, (1972). 'Friction and molecular structure: The behaviour of some thermoplastics.'
11. BRISCOE B J, and STOLARSKI T A. Wear, 104, 121, (1985). ' Transfer wear of polymers during combined linear motion and load axis spin.'
12. BRISCOE B J. Phil. Mag. A43, 511-527, (1981). 'Friction and wear of organic solids and the adhesion model of friction.'

APPENDIX 1.

DERIVATION OF DECREASE IN POLYMER SURFACE AREA WITH SAND INCLUSION.

For a contact originally composed of polymer of surface area, Ap, at the start of the experiment this will be the total surface area, AT; if the contact is then covered by sand particles a fraction, x, of the available area during each cycle of the simulation, N. Then the following expressions for the amount of free polymer surface may be written down

Cycle Number	Term for Polymer Area.	Equivalent Expression.
Start	1	-
1	1 - x	(1 - x)
2	1 - x - x(1 - x)	
	(1 - x) - x(1 - x)	
	(1 - x)(1 - x)	$(1 - x)^2$
3	1-x-x(1-x)-x(1-x-x(1-x))	
	(1-x)-x(1-x)-x((1-x)-x(1-x))	
	(1-x)((1-x)-x(1-x))	
	(1 - x)(1 - x)(1 - x)	$(1 - x)^3$
N		$(1 - x)^N$

SESSION X
ROLLING ELEMENT BEARINGS (2)

Chairman: Dr E Ioannides

PAPER X(i)	Tribological Characteristics of Needle Bearings
PAPER X(ii)	Power Loss Prediction in Ball Bearings
PAPER X(iii)	The Effect of Roller End-Flange Contact Shape Upon Frictional Losses and Axial Load of the Radial Cylindrical Roller Bearing
PAPER X(iv)	The Study of Roller End and Guiding Shoulder Construction of Roller bearings

Paper X(i)

Tribological characteristics of needle bearings

S. Blair and W. O. Winer

A Needle Bearing Test Rig has been constructed which allows the measurement of frictional torque, axial thrust, needle axial skew angle, and needle complement velocity for an applied radial load of up to 3500 N and a speed of up to 7000 rpm. Either shaft rotation or cup rotation can be accommodated. Preliminary data indicate that frictional torque is greater for full-complement bearings than caged bearings, that torque is higher for cup rotation than shaft rotation, and that torque is minimized when the needles (full-complement) are axially aligned with the shaft in the load zone. A correlation was found between needle skew angle and axial thrust. Less than one degree change of needle skew is required to go from no thrust to the maximum thrust developed (about 5% of the radial load). Full complement bearings tend to operate with a preferred needle skew direction and attending thrust direction with occasional spontaneous reversals. Operation without thrust and needle skew is not stable.

1 INTRODUCTION

Needle roller bearings are a special case of cylindrical roller bearing in which the roller has a high length to diameter ratio for reduced radial space requirements. They are used in both the caged and full-complement variations.

Empirical estimates of roller bearing frictional torque have been made (Ref [1,2]) and are frequently used for design purposes. Roller skew is known (Ref [2]) to occur in roller bearings and was measured for a single roller through one revolution of a roller bearing by Nypan (Ref [3]). Axial forces were measured for imposed needle skew angles by Ulezelski et al. (Ref [4]). Needle skew is believed to produce axial force (Ref [4]). A force transducer on the needle bearing shaft of an automotive roller tappet produced the axial force data in Figure 1 using a simulator described in Ref [5]. The axial force is seen to occur at the frequency of needle passage and can be positive or negative with magnitude of as much as 140 N.

A Needle Bearing Test Rig was designed and assembled in the Georgia Tech Tribology and Rheology Laboratory to provide measurements of frictional torque, axial thrust, needle skew angle, and needle complement velocity. Details of the machine design and some preliminary data are reported here.

2 NEEDLE BEARING RIG

2.1 Construction

The Needle Bearing Simulator is shown in Figure 2 in the shaft rotation mode. The cup may also be rotated with the shaft fixed in another mode. The shaft spans three support blocks. The left and center block have identical openings so the drive module and the torque module may be interchanged among them. The center block contains a load diaphragm and load cell for radially loading the test bearing which is always contained in the center block. Regulated gas pressure is supplied to the diaphragm at the top fitting. When the drive module containing the test bearing is in the center block, the cup rotation mode is available. In Figure 2 the shaft is supported in the right block by a journal bearing. For cup rotation, the journal bearing sleeve is replaced by a special ball bearing which allows both rotation and translation of this right shaft journal. Also at the right block is a ball thrust bearing, thrust cell and micrometer for measurement of the axial force (thrust) applied to the shaft by the needles. The micrometer, pushing through the thrust cell and thrust bearing, preloads the shaft against the drive module (or alternatively, the torque module) at the left block. A flexure on the left end of the shaft increases the axial compliance of the shaft at the left support relative to the axial compliance at the thrust cell end. A preload at the micrometer of 0.076 mm is used for axial force measurements, resulting in 92 N preload on the thrust cell.

The drive module spool is rotated by a 3 mm V-belt, the pulley for which serves as an encoder for a rotational velocity pick-up. Optical probes (Figure 2) of 0.49 mm diameter active element are positioned at the left and right top (load zone) ends of the test bearing by a pair of 3-axis micropositioners (not shown in Figure 2).

2.2 Torque Transducer

The torque module (Figure 2) provides a housing for the specimen bearing for shaft rotation mode and support for the shaft in cup rotation mode while allowing low friction freedom of rotation of the inner spool. Friction torque may be measured by determining the magnitude of the force required to resist rotation of the inner spool at the force cell link in Figure 2. However, it is necessary to correct for the frictional torque of the bearings which support the inner spool. The inner spool is supported by two ball bearings which are housed inside an intermediate spool which, in turn is supported by two ball bearings. Jogging of the intermediate spool or the inner spool causes jogging (oscillating rotation) of the support bearings while the specimen (or the shaft in cup rotation mode) is stationary. If support bearing friction is symmetric with respect to rotation direction, jogging causes a noise on the torque signal which if averaged gives the true test bearing torque.

2.3 Needle Optical Probes

Two fiber optic probes (MTI Fotonic Sensors) transmit light to and receive reflected light from the trunnions (Figure 3) at both ends of the needles. The 0.9 mm diameter trunnion ends are lapped (600 grit paper) for this purpose. The test bearing cup flanges are machined such that the flange inside diameter is at the needle center line. This way, the flange still axially locates the needles but exposes enough of the trunnions to be optically detected. The needle skew angle changes the phase relationship of the optical probe signals as shown in Figure 3.

2.4 Data Acquisition System

Six channels of data were sampled concurrently. They included signals from the thrust, load, and torque charge amplifiers, the drive module speed pickup and the two needle optical probes. Sampling rate was 2 to 10 kH_z for 8192 points per channel. Raw data was stored on disc as ASCII files for later reduction.

3 EXPERIMENTAL PROCEDURE

In this work, the measurement of absolute frictional torque was a result of averaging several revolutions and the thrust and needle angle measurements were transient measurements with no averaging. For this reason, two test procedures were used. Test bearings were cleaned in acetone in a "sonic" cleaner before being pressed into the torque or drive module.

3.1 Absolute Friction Measurements

For all absolute friction tests, the inner spool support bearings in the torque module were jogged at a frequency of 1 H_z and an amplitude of 11° for a maximum velocity of 13 rpm. Each test bearing was lubricated once with a 10W-40 commercial engine oil. Oil was added by syringe until it overflowed from the horizontal bearing. The test procedure follows:

1) Without load, the bearing was brought to speed and jogging of the torque module begun.
2) The load was applied for about five (unless noted) seconds. Averaged data of this period of time yielded the torque level.
3) The load was removed and the drive motor braked.
4) The drive module was jogged by hand several times. Data of this period of time were averaged to give the zero friction level.
5) Voltage levels from data averaged during (4) were subtracted from those averaged during (2) to give absolute frictional torque.

The thrust signal level was found to be sensitive to normal load, possibly due to deflection of the shaft which would shorten its effective length. For this reason, the thrust signals that were recorded during these tests are not reported as the zero thrust signal level is not known. However, any transients in thrust which occurred are stored on disc. If the inner spool support bearing friction is symmetric, averaging data while jogging these bearings removes this source of error.

3.2 Needle Angle/Thrust Measurements

Needle angle/thrust tests were performed in the shaft rotation mode without jogging of the inner spool. A grease (NLGI reference system "A") was chosen for the lubricant for these tests to reduce the possibility of oil drops interfering with the optical path between the probes and the needle ends.

A Gould Waveform Recorder was used to sample data with the trigger set for a ± 20% window on the thrust range. Pretrigger is set at 50% so that the trigger event occurs in the center of the data file. Sample rate should provide at least 10 points per needle passage. The test procedure follows:

1) Bring the shaft to speed then apply the load.
2) Adjust the position of the two optical probes to maximize their signals as viewed on an analog scope.
3) Zero the load, thrust, and torque charge amplifiers.
4) Arm the recorder. The recorder triggers when the thrust level rises or falls by 20% of the recorder range.
5) When the record is complete, brake the motor.
6) Jog the drive by hand with normal load still applied and the thrust signal level being sampled by the Nicolet scope.
7) Graphically average the thrust signal level during (6) to be supplied as the zero thrust signal level to software later.
8) Transfer the 48 k bytes of data to the computer and store on disc.

4 TEST RESULTS

4.1 Friction Torque Tests

The friction coefficient, μ, can be defined as

$$\mu = \frac{T}{r_s W}$$

where T is frictional torque, r_s is the radius of the shaft and W is the radial load on the bearing.

Friction coefficients for shaft rotation and 5 seconds of operation with 10W-40 motor oil are plotted vs. shaft speed for four loads in Figure 4. The test bearing is B1212, an unmodified full-complement 19.05 mm bore drawn cup needle bearing with 27 needle rollers. Similar data for the same bearing and cup rotation is plotted in Figure 5, along with results for a caged bearing. The caged bearing was J1212, a 19.05 mm bore drawn cup bearing with 17 retained needles.

In all cases at lower loads the friction coefficient is sensitive to speed with values that increase with speed. At high load, the friction coefficient is reduced from that at low load and is relatively insensitive to speed. These effects can be attributed to the predominance of viscous losses at light load (speed sensitive) and Coulomb type friction at high load (speed insensitive). To check, these tests might be repeated for a low viscosity oil or a grease.

The frictional torque of the caged bearing was less than half that of the full-complement version for the same load. This can be attributed to the replacement of a roller/roller contact of infinite slide to roll ratio (for which hydrodynamic lubrication is difficult to achieve) with a cage/roller contact with slide to roll ratio of 2. The cage also guides the needles to reduce axial skew and attending friction (see later section).

The rotating cup consistently showed higher friction than the rotating shaft mode for the full-complement bearing (the caged bearing was only used in the cup rotation mode). At low load the ratio was about 2 to 1 and for high load about 1.5 to 1. Since the affect of rotation mode is greatest at light load, viscous drag may be the cause of the difference. The rotating cup would tend to retain oil in the cup while the rotating shaft would not.

4.2 Needle Angle/Thrust Tests

All results in this section are for shaft rotation in a full-complement bearing lubricated with grease. Figure 6 illustrates typical behavior when a reversal (sign change) in thrust occurs. The needle bearing appears to operate somewhat stably with a non-zero preferred thrust which will occasionally and unpredictably reverse. In fact, most attempts to record a spontaneous thrust change on this rig were unsuccessful. Often, acceleration of the drive brought on a thrust change. This observation is in agreement with Ref [4] where it was concluded that the needles in a loaded full-complement bearing operate skewed. Note the similarity between Figures 1 and 6.

Figures 7-8 present relative needle angle, absolute thrust, relative frictional torque, and needle complement to shaft speed ratio as functions of time. Needle angle is calculated from the phase difference between the signals from the two optical probes at needle ends (in the bearing load zone) and the average complement velocity for the entire record and the length of a needle. Needle angle is defined in Figure 3. The frictional torque which is plotted is relative since the zero torque signal level is undetermined for these tests. Also the support bearings are not jogged so that the attenuation of the signal due to support bearing friction is not known. The torque data is presented here, however, since the transients that are plotted not only show the trend of the torque response but also represent the lower limit of torque variations during a record. The needle complement speed to average shaft speed ratios plotted are for both the measured complement speed (calculated from the time required for 27 needles to pass the optical probes) and the complement speed calculated for the no slip condition from the measured shaft speed using

$$\left.\frac{N_{compl}}{\bar{N}_{shaft}}\right|_{\text{no slip}} = \frac{r_s}{2(r_s+r_n)} \frac{N_{shaft}}{\bar{N}_{shaft}}$$

where the over-bar indicates an average over the data record and r_s and r_n are shaft and needle radii respectively. This presentation was selected because it shows the acceleration of the shaft during a record. Measured complement speed was consistently about 1.5% below the no slip speed.

It should be noted that data was only collected when a significant change in axial thrust was encountered. In Figures 7-8 there is apparently a correlation between needle angle and thrust and torque. For the data of Figure 8 frictional torque and thrust were plotted as functions of needle angle in Figure 9. The record was divided into one hundred equal time intervals. Angles, torques, and thrusts were averaged over each period before being plotted. Therefore, in Figure 9 each point represents about six needle passages. The thrust vs. needle angle curve clearly shows the characteristic "S" shape of traction vs. side slip angle (Ref [6]). The traction coefficient saturates at about 0.05. Of interest also is the minimization of frictional torque at the relative needle angle for zero thrust. It can be argued that the relative angle at which thrust is zero and torque is minimized should correspond to a needle angle of zero, i.e., the needles are axially aligned. Then, for the data of Figure 9, the needle angle is equal to the relative angles less 0.8°. The reduction in frictional torque when the needles are axially aligned is 0.014 N·m which for the load of 860 N gives a change of friction coefficient of $\Delta\mu = 1.7 \times 10^{-3}$ or about 25% of the friction coefficient shown in Figure 5. As this $\Delta\mu$ value has likely been attenuated by support bearing friction, the actual torque reduction may have been much greater.

In Figure 10, the angle data of Figure 7 are plotted as individual needle angles for one-fourth of the data record, centered about the middle of the record.

5 CONCLUSION

Frictional torque is greater for full-complement bearings than caged bearings, torque is higher for cup rotation than shaft rotation, and torque is minimized when the needles (full-complement) are axially aligned with the shaft in the load zone. A correlation was found between needle skew angle and axial thrust. Less than one degree change of needle skew is required to go from no thrust to the maximum thrust developed (about 5% of the radial load). Full complement bearings tend to operate with a preferred needle skew direction and attending thrust direction with occasional spontaneous reversals. Operation without thrust and needle skew is not stable.

In many full-complement needle bearing applications a means of reacting the thrust generated by the needle bearing must be provided , e.g., thrust bearing surfaces on the faces of transmission planet gears. If a geometry (needle aspect ratio, clearance, etc.) can be found which minimizes needle skew and thereby thrust, this requirement may be relaxed and limiting speed would probably be higher. Frictional torque would be reduced. All of which would increase the applications for these full-complement bearings.

6 ACKNOWLEDGEMENT

The authors gratefully acknowledge the support of The Torrington Company and the assistance of Michael Brauer, Dr. Y. P. Chiu and Robert Lugosi in the conduct of this research.

References

(1) Tedric A. Harris, Rolling Bearing Analysis, John Wiley and Sons, 1984, p. 425-427.

(2) Eschmann, Hasbargen, Weigand, Ball and Roller Bearings, John Wiley and Sons, 1985.

(3) L. J. Nypan, "Roller Skewing Behavior in Roller Bearings," ASME, JOLT, Vol. 104, July 1982.

(4) Ulezelski, Evans, Haka, Malloy, "Needle Bearing Axial Thrust Study," Society of Automotive Engineers, 830568, 1983.

(5) Bair, Winer and Griffioen, "The Tribological Behavior of an Automotive Cam and Lifter System," Trans. ASME, JOT, 108, 1986.

(6) Johnson, K. L., and Tevaarwerk, J. L., "Shear Behavior of Elastohydrodynamic Oil Films," Proc. Roy Soc. of London, 365A, 1977.

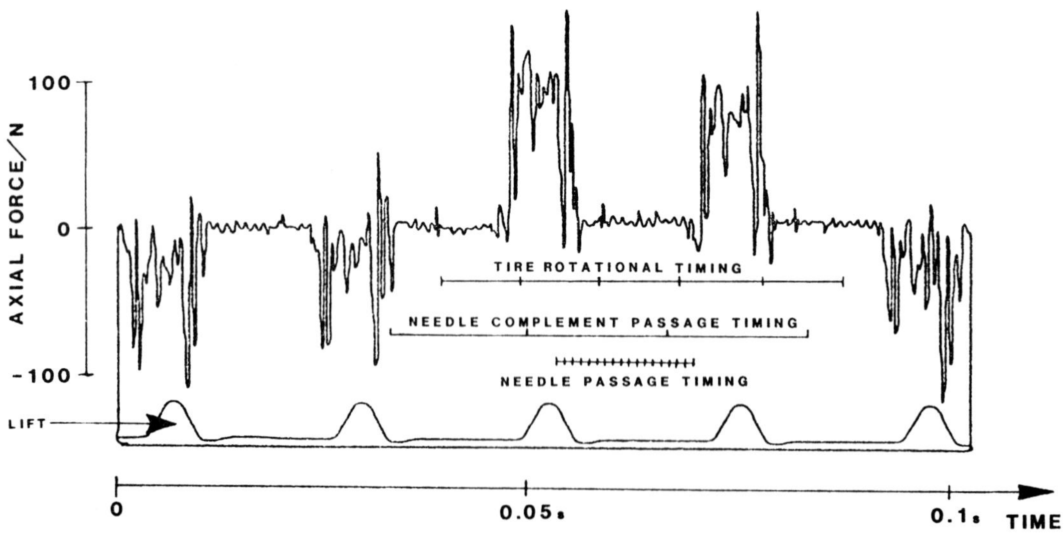

Figure 1. Shaft Axial Force for a Needle Roller Tappet at 2700 rpm of cam

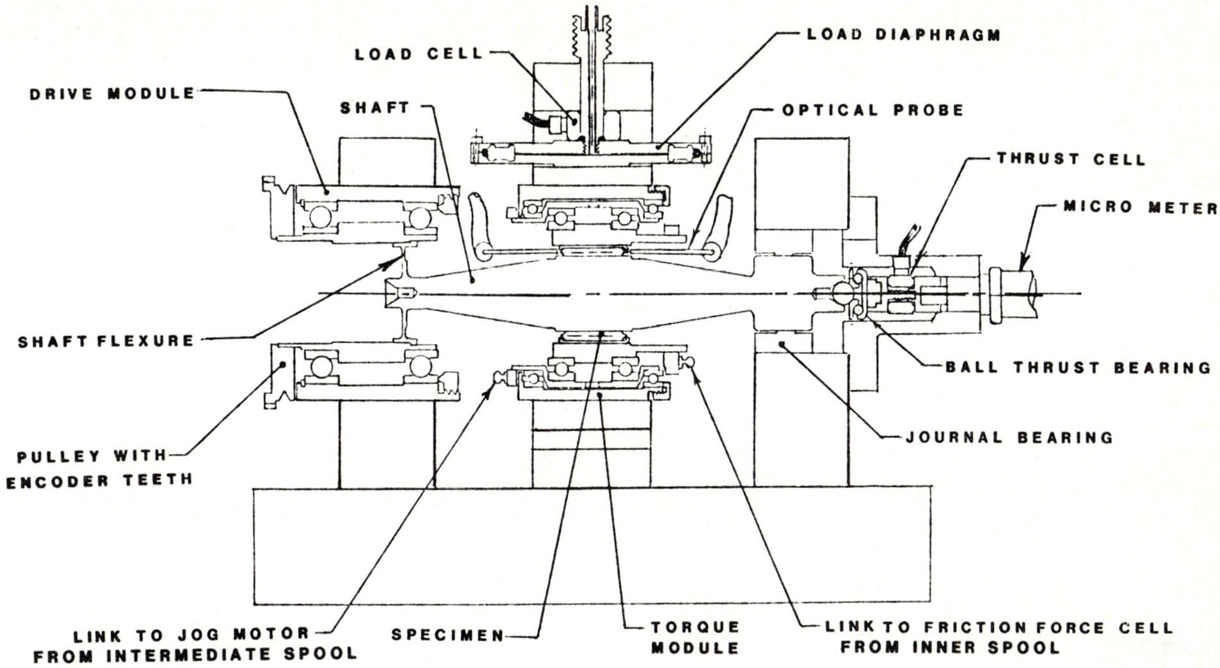

Figure 2. Needle Bearing Simulator

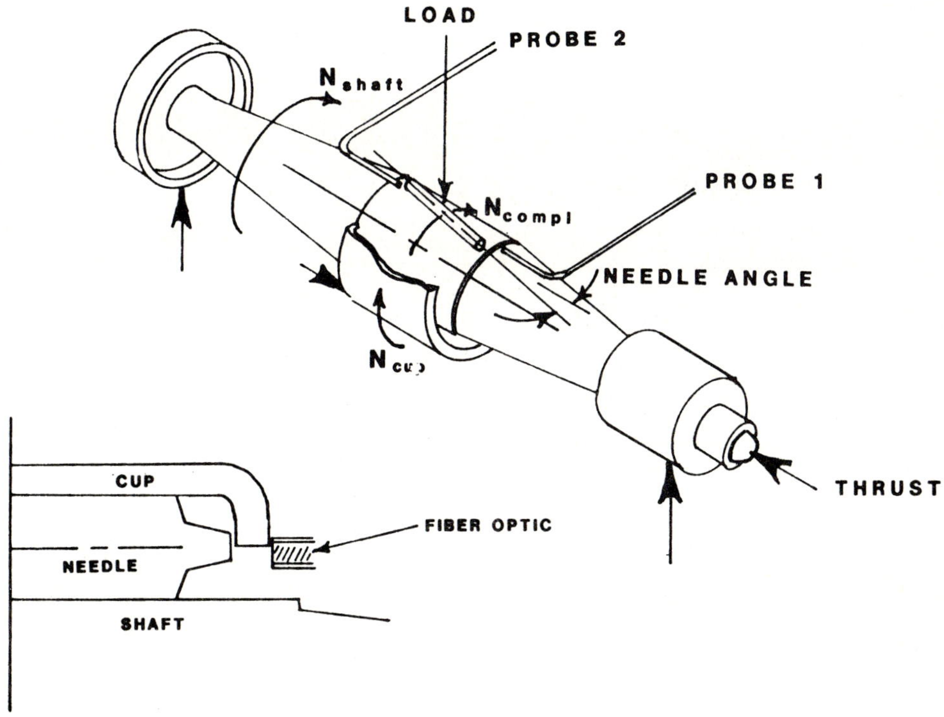

Figure 3. Definitions of Rotation Directions, Load, Thrust and Needle Angle and Optical Probe Detail

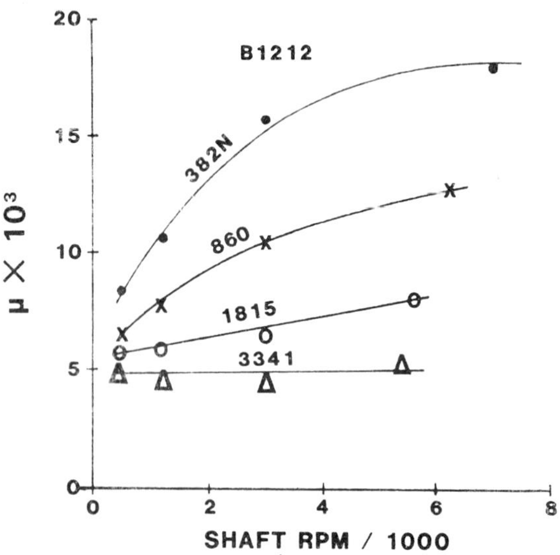

Figure 4. Friction Coefficient, μ, for shaft rotation, 5 sec running, 10W-40 Motor Oil. Radial load on shaft as noted.

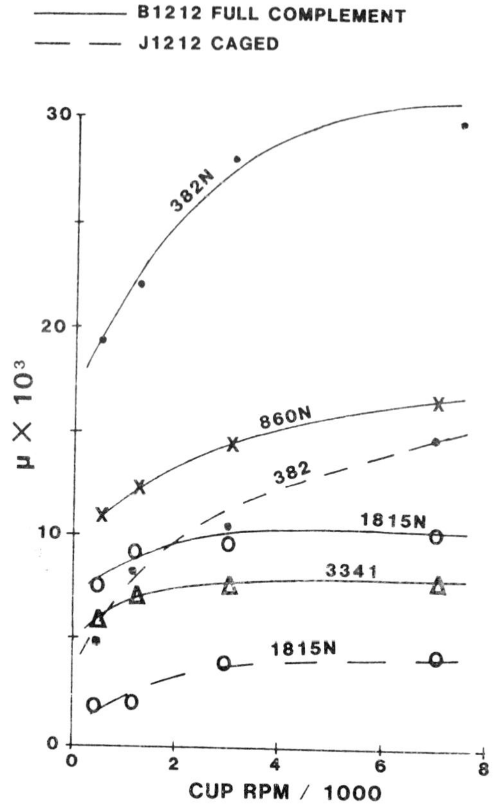

Figure 5. Friction Coefficient, μ, for cup rotation, 5 sec running, 10W-40 Motor Oil

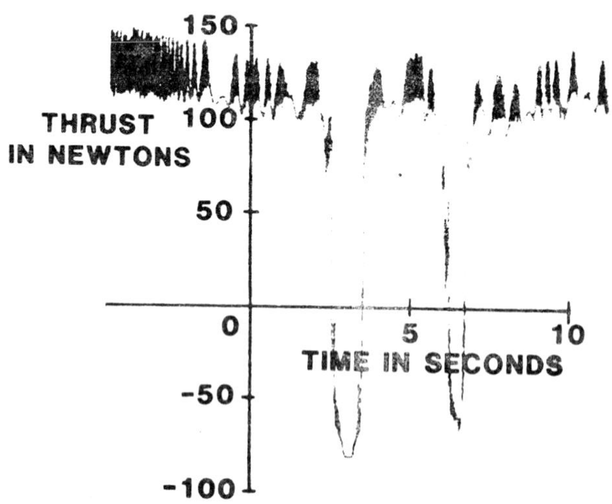

Figure 6. Axial Thrust for B1212 at 3000 rpm of Shaft and 1815 N Load, Oil Lubricated.

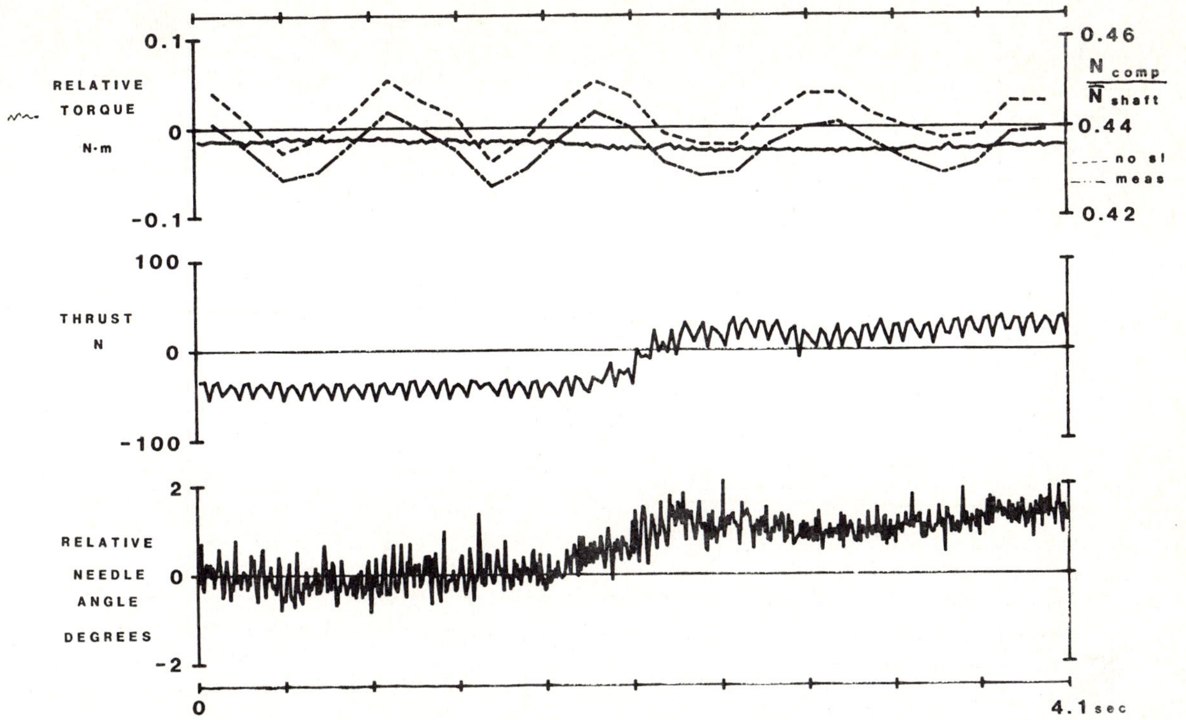

Figure 7. Run D, $\overline{N}_{shaft}$ = 841 rpm, 860 N load, grease

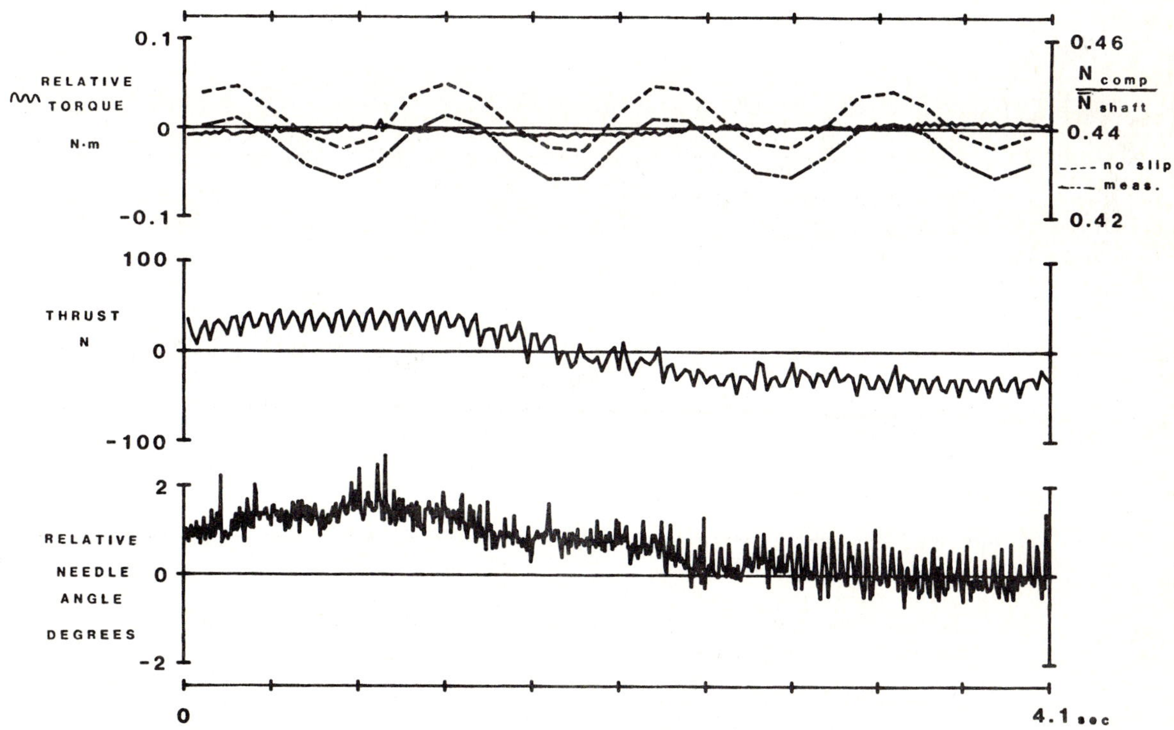

Figure 8. Run E, $\overline{N}_{shaft}$ = 850 rpm, 860 N load, grease

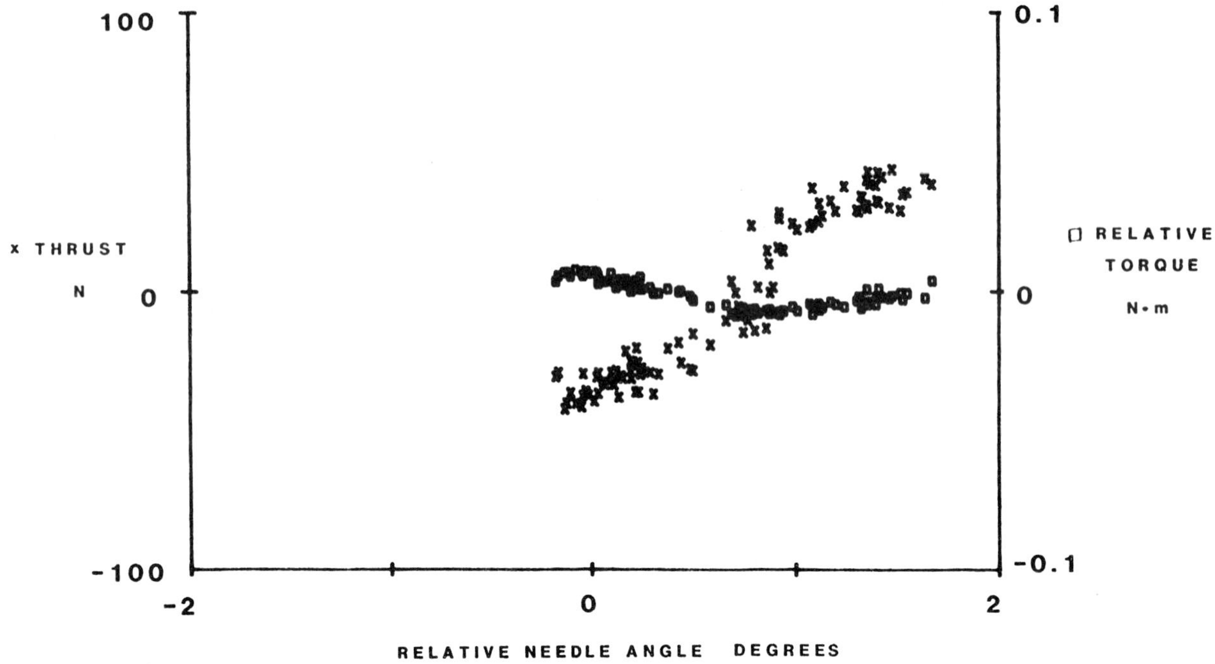

Figure 9. Run E, $\overline{N}_{shaft}$ = 850 rpm, 860 N load, grease

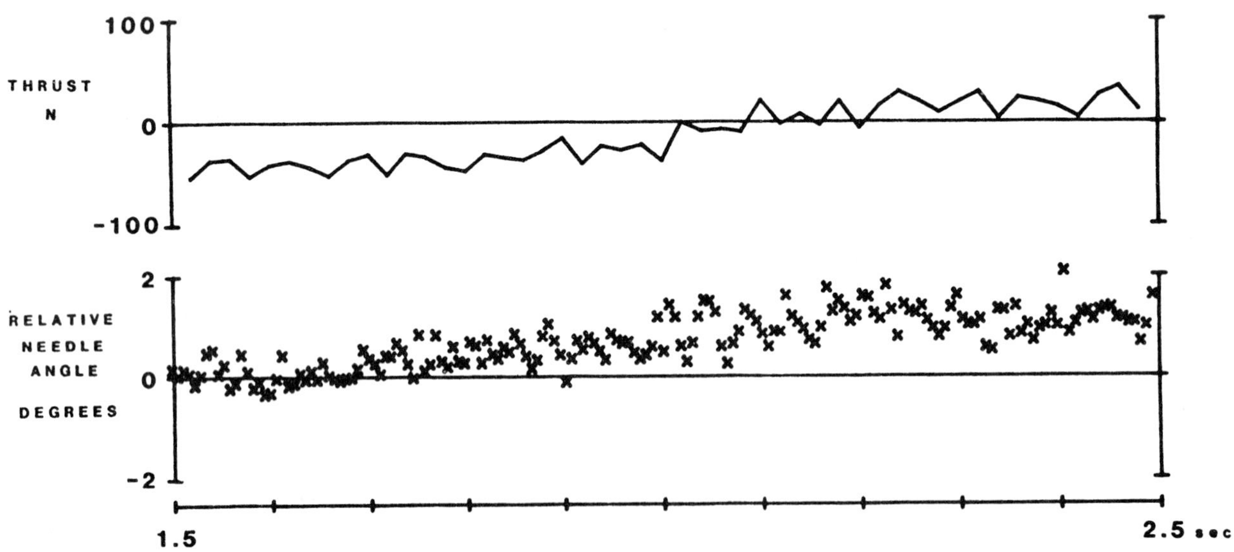

Figure 10. Run D, $\overline{N}_{shaft}$ = 841 rpm, 860 N - Individual Needle Angles and Corresponding Thrust

Paper X(ii)

Power loss prediction in ball bearings

R. J. Chittenden, D. Dowson and C. M. Taylor

SYNOPSIS

The determination of power losses and traction forces in modern ball bearings typical of aerospace applications is a complex task. The high rotational speeds commonly experienced by such bearings cause inertia effects to be important, and this inturn loads the balls heavily against the outer raceway of the bearing. The resulting disparity in contact angle tends to introduce a spin component to the motion of the inner raceway/ball contacts on top of the usual sliding and rolling motion. A complete analysis of the contacts would have to include the elastohydrodynamic analysis of heavily loaded contacts, and conjunctions with continuously variable surface velocities. The iterative solution required for such conditions, despite the use of powerful, modern numerical methods, still requires many tens of minutes processing on a large mainframe computer. The computations required for a bearing having 20-30 rolling elements (40-60 contacts) is therefore unreasonable, and even if this were possible its incorporation into the overall dynamic analysis of the bearing would again introduce many more levels of difficulty.

For many years the design of bearings for aerospace applications has been based on knowledge gained experimentally, but with the slow introduction of various theoretical work to aid the designer in such calculations as minimum lubricant film thickness. The ever increasing cost of experimental work and the need to make the initial design the best possible, however, provide a requirement for much improved theoretical methods, and hence the need for developments of the type described in this paper. The purpose of the work has been to unite an analysis of bearing dynamics with realistic elastohydrodynamic pressure and film thickness profiles, and then to calculate realistic values for power loss and power transfer within the bearing together with the provision of information on the contact traction forces. A well tried analysis was adopted for calculation of the bearing dynamics, and the use of a simple extrapolation allowed existing solutions to the elastohydrodynamic lubrication problem to be introduced. Calculations of power loss and traction force were then undertaken with the fluid assumed to be either Newtonian or Non-Newtonian, firstly for a single contact and secondly for the whole of an example bearing. The results obtained were then compared with other computations for the same bearing which were known to agree well with the available experimental evidence, and finaly the sensitivity of the analysis to various lubricant parameters was examined. The conclusions drawn from the work were that realistic power loss calculations could be made with a Non-Newtonian fluid model, and that the most important lubricant parameters were the pressure-viscosity coefficient and the limiting-shear-stress-pressure coefficient.

1 INTRODUCTION

It has been known for many years that bearing systems may generate considerable amounts of heat which must be removed to ensure the correct operation of the associated mechanical system. The requirement is particularly demanding in aerospace applications where the trend has been to seek bearings which are able to operate in ever more arduous conditions. The design process is far from simple since the amount of additional weight that can be introduced must be minimised, as must the size of the installation, and hence the bearing is often closely incorporated within the surrounding engine components. In order to satisfy all of these restrictions successfully at the design stage, therefore, reliable knowledge of the heat generation and heat transfer between components is essential. In the past this information has often been gained by incorporating an experimental test programme into the design effort and arriving at a satisfactory result by a trial and error method. Owing to the complexity of a modern gas turbine and the cost of running an associated experimental programme the initial design is critical to the success of an engine. It is important, therefore, that designers must have access to improved methods of calculating parameters such as the power loss associated with viscous effects in the bearing chamber.

The starting point for analytical methods developed to predict the power loss within the bearing chamber must be a dynamic analysis of the whole bearing, or a representative ball where conditions are such as to result in a purely axial load. Compared to many areas of tribology, however, the amount of attention paid to this type of analysis of high speed bearings has been small. This is partly explained by the complexity of the analysis required to determine the load, surface velocities, and geometry associated with each conjunction within the bearing, a task most readily undertaken by an iterative procedure on a modern high speed digital computer. In reality there must be an interaction between the traction forces acting upon the conjunctions and the dynamic modelling of the bearing, but by neglecting the lubricant and assuming a coefficient of friction over the contact area Jones(1,2) managed to simplify the analysis and the resulting procedure has been used as a standard by engineers for many years. More recent studies were able to include some of the influences of a lubricant into the calculations by employing experimentally derived expressions for such features as traction forces and viscous drag. (For example the work of Gentle and Boness (3) and Pasdari and Gentle (4)).

A more complete analysis, in which the traction forces and power losses are determined from the shear forces acting on the surfaces of the contacting solids poses many problems. The high loads, particularly at the outer raceway/ball contacts, where inertia forces are significant, mean that the local elastic deformation of the solids can be large when compared with the resulting lubricant film thickness, and hence the appropriate type of analysis for these contacts, as well as those between inner raceway and ball, is that of elastohydrodynamic (E.H.D.) lubrication. The solution procedure for these contacts requires the simultaneous solution of the expressions governing the local elastic deformation and the associated Reynolds' equation for fluid film lubrication, which must include the effects of pressure (and perhaps temperature) upon the lubricant density and dynamic viscosity. The typical contact between a ball and a raceway may be represented by a "line contact" (Dowson and Higginson (5)) for the calculation of minimum film thickness, but the evaluation of power loss requires the possibility of spin to be included into the analysis. This can only be accomplished by the use of the more complex elastohydrodynamic "point contact" analysis of the type presented by Dowson and Hamrock (6,7,8,9) and further developed to allow the influence of typical conjunction loads to be examined (Evans and Snidle (10)), and by Chittenden et al (11). Recent developments in numerical techniques (Lubrect et al (12,13)) can also be introduced with advantage. The modifications required to allow for spin were made by Xu (14), whose work also indicated the need to consider the lubricant to be Non-Newtonian as shear rates could reach levels where a Newtonian analysis would be unrealistic. This feature had been indicated in earlier rheological studies, Sandborn and Winer (15) and Adams and Hirst (16). The effort to merge all these effects into a single analysis would clearly be immense, especially when it is remembered that a single E.H.D. solution can still consume many tens of minutes of computer time on a large, high-speed, mainframe computer. At the outset it was recognised that by making several simplifying assumptions it would be possible to evaluate power loss and traction force with a fair degree of accuracy, and it is this work that is described in this paper. The major steps in the analysis are indicated below and, with the exception of the dynamic analysis of the bearing, are detailed in subsequent sections:-

(a) Determine the conjunction loads and surface velocities by the use of a standard dynamic analysis of the unlubricated bearing.

(b) Estimate the elastohydrodynamic lubrication features of each contact within in the bearing by extrapolating from existing solutions obtained for similar geometrical configurations.

(c) Compute traction force and power loss for each contact based upon the extrapolated elastohydrodynamic pressure and film thickness profiles, and hence the total power loss and power transfer for the whole bearing by summation. [A Non-Newtonian lubricant could also be used at this stage.]

Having set out the assumptions embodied in the analysis and dealt with the operation of its components, for which a flow diagram is provided as an overview, the procedures are applied to a representative bearing. Single contacts are first considered to show the effects of considering a Non-Newtonian fluid, and then by expanding to include the whole bearing the influences of several lubricant parameters are examined.

1.1 Notation

C_1-C_3	constants used for film thickness and pressure extrapolations
d	ball diameter (m)
D	distance between raceway curvature centres (m)
E'	reduced elastic modulus (Pa)
F_a	applied axial load (N)
F_{app}	load applied to a particular conjunction (N)
F_{int}	load obtained from a numerical solution (N)
F_r	applied radial load (N)
$g_{\tau o}$	pressure-representative shear stress coefficient (Pa^{-1})
G_∞	elastic shear modulus of the lubricant (Pa)
G^*	fluid shear modulus (Pa)
H_{cen}	dimensionless central film thickness
H_j	dimensionless film thickness at node "j" in the computational zone
H_{min}	dimensionless minimum film thickness
H_o	dimensionless film thickness constant used for definition of elastohydrodynamic film thickness
H_∞	dimensionless central deformation due to Hertz pressure distribution
H_1	largest dimensionless film thickness within the computational zone
L_a	axial distance between ball centre and outer raceway curvature centre (m)
L_r	radial distance between ball centre and outer raceway curvature center (m)
p	fluid pressure (Pa)
P_j	dimensionless pressure at node "j", $P = p/E'$
$S(x,y)$	geometrical separation of the contacting bodies
$W(x,y)$	elastic deformation
x,y	coordinate directions (see text)
α	pressure-viscosity coefficient (Pa-1)
β_f	free contact angle (degrees)
β_i	inner race contact angle (degrees)
β_o	outer race contact angle (degrees)
δ_a	axial displacement of inner raceway (m)
δ_r	radial displacement of inner raceway (m)
δ_x, δ_y	elemental dimensions of the computational area associated with each node
$\underline{\eta}$	lubricant dynamic viscosity (Pas)
$\bar{\eta}$	dimensionless lubricant viscosity
η_o	lubricant dynamic viscosity at conjunction inlet (Pas)

$\dot{x}$ contact strain rate (s^{-1})
τ shear stress (Pa)
τ_o representative shear stress, indication the point at which the fluid becomes appreciably non linear (Pa)

2 DESCRIPTION OF THE PROBLEM

(a) Assumptions

In order to undertake the computations presented in this paper it was necessary to make use of a few additional assumptions. These significant additional assumptions are listed below:-

(i) That an existing solution of the elastohydrodynamic lubrication problem could be used as the basis of an extrapolation to yield pressure and film thickness profiles for similar contact geometries within the bearing.
(ii) That the extrapolated profiles may be used as an approximation to a contact which is spinning and/or sliding in addition to rolling, the only surface motion of the original solution, and that a non-Newtonian lubricant model could be imposed on these conjunction features.
(iii) That the extrapolated isothermal conjunction characteristics would not be significantly changed by the heat generation within the contact associated with its spinning, sliding or rolling motions.

The details of the extrapolation process are given in a later section, however, use of these procedures was found to result in good agreement between the small number of solutions available for testing. (A good extrapolation being deemed to be one where there were no observable differences between the isometric projections of the extrapolated film thickness or pressure profiles and those of an existing solution with the extrapolated characteristics.). At the outset of the project knowledge of the likely effects of spin and non-Newtonian lubricant characteristics was limited, but recent work at Leeds has allowed a clearer picture to be obtained. Xu (14) investigated the elastohydrodynamic lubrication of contacts with spin and found that with spin speeds typical of those in rolling element bearings the film thickness and pressure profiles were changed only slightly. A similar situation has been found by Esfahani (17) concerning the influence of a non-Newtonian lubricant model. This work found that with a slide-roll ratio twice that typically found in ball bearings the computed elastohydrodynamic solution differed from the Newtonian situation by approximately 10%. The third assumption is based upon the knowledge that conditions within the contact are largely determined by situation at the conjunction inlet and provided that these are specified correctly the effect of a rise in fluid temperature within the conjunction should not be significant.

(b) Pressure Profile Extrapolation

The extrapolation of pressure, and of film thickness, is based on existing elastohydrodynamic results which define a solution in a normalised form with respect to a regular grid structure specified with respect to the centre of contact.

In order to extrapolate the pressure profile it was necessary to determine the values of conjunction load and geometry from the bearing dynamics analysis. These parameters then allowed the semi-major and semi-minor axes of the Hertzian contact ellipse to be evaluated by the one point iteration technique of Hamrock and Anderson (18). With these parameters and the reduced elastic modulus if the contact known it was then possible to redefine the existing, normalised, elastohydrodynamic solution and to evaluate a real conjunction load as:-

$$F_{int} = \sum_j P_j E' \delta_x \delta_y \qquad (1)$$

A pressure multiplier was then computed from the ratio of applied load to integrated load, and the specified load obtained by multiplying all the pressures forming the integrated load by this constant.

$$P_{j_{new}} = P_{j_{old}} \frac{F_{app}}{F_{int}} \qquad (2)$$

It should be noted that this simple form of extrapolation takes no account of the changes in pressure profile likely to result from the differences in operating conditions between the inner and outer raceway/ball contacts. This would be particularly noticeable for some of the relatively high speed, low load conditions on the inner raceway where the extrapolated profile might be a poor approximation to a specific E.H.D. calculation.

(c) Film Thickness Extrapolation

The development of a procedure for the extrapolation of the film thickness profile proved to be somewhat more difficult than that for pressure. Minimum and central film thickness were calculated for the conjunction operating conditions by means of the formulae presented by Chittenden et al. (1985). Initial attempts to extrapolate the film thickness were based on the linear relationship:-

$$H_{j_{new}} = C_1 H_{j_{old}} + C_2 \qquad (3)$$

C_1 and C_2 being determined such that the extrapolated profile had the predicted values of minimum and central film thickness set into it. Observation of the resulting film profile, however, indicated that as H_j increased above the predicted central film thickness value the extrapolation became more and more unrealistic. This problem was particularly severe around the first point on the solution grid (H_1, the greatest distance from the centre of contact.), and hence this node was chosen as the third film thickness value that should be specified in the extrapolation procedure, along with H_{min} and H_{cen}.

The standard approximation for the film thickness at node "j" in the elastohydrodynamic solution process is:-

$$H_j = H_o + S(x,y)_j + W(x,y)_j \qquad (4)$$

In order to allow H_1 to be determined it was assumed that the elastic deformation ($W(x,y)_1$) was negligible at the first grid node, and that the film thickness constant (H_o) could be approximated by the predicted central film thickness minus the deformation at the centre of contact. Hence the film thickness at (H_1) may be written as:-

$$H_1 = H_{cen} - H_{oo} + S(x,y)_1 \quad (5)$$

The film profile was then extrapolated by the use of the following expression:-

$$H_{j_{new}} = C_1 H^2_{j_{old}} + C_2 H_{j_{old}} + C_3 \quad (6)$$

where

$$C_1 = \frac{\left[\dfrac{H_{1_{new}} - H_{cen_{new}}}{H_{1_{old}} - H_{cen_{old}}} - \dfrac{H_{min_{new}} - H_{cen_{new}}}{H_{min_{old}} - H_{cen_{old}}}\right]}{H_{1_{old}} - H_{min_{old}}}$$

$$C_2 = \frac{H_{min_{new}} - H_{cen_{new}}}{H_{min_{old}} - H_{cen_{old}}} - C_1 (H_{min_{old}} - H_{cen_{old}})$$

$$C_3 = H_{min_{new}} - C_2 H_{min_{old}} - C_3 H^2_{min_{old}}$$

(d) Shear Stress Calculations

The shear stresses associated with each node in the solution grid were determined for the driving surface (z=0) and the driven surface (z=h) by means of the following expressions:-

$$\delta F_{x_{z=0}} = \left[\frac{-h}{2}\frac{\partial p}{\partial x} - \frac{\eta(u_{lower} - u_{upper})}{h}\right]\delta_x \delta_y \quad (7)$$

$$\delta F_{x_{z=h}} = \left[\frac{-h}{2}\frac{\partial p}{\partial x} - \frac{\eta(u_{lower} - u_{upper})}{h}\right]\delta_x \delta_y \quad (8)$$

where u_{lower} and u_{upper} are the surface velocities at z=0 and z=h respectively, and δ_x, δ_y define the area of each grid element. The pressure gradients were specified by a finite difference approximation, and the viscosity was evaluated using the Barus pressure-viscosity expression - ($\eta_j = \eta_o e^{\alpha p_j}$). It may be seen from these two expressions that for an increasing load (increasing viscosity, pressure gradiants and reduced central film thickness) and/or increasing sliding speed the Newtonian lubricant has no limit to the possible values of local shear force. The prediction of unrealistic values is therefore possible and in order to keep them within reasonable bounds a non-Newtonian model must be introduced.

(e) The non-Newtonian Models

It may be seen from equations (7) and (8) that the local shear forces have been represented by a term involving pressure gradient, and another involving velocity gradient. The influence of these terms is a function of position since the pressure gradients are greatest around the edge of the Hertzian region whilst the velocity gradient term would tend to be more significant within the Hertzian region where the film thickness was smallest, and the viscosity highest. In order to facilitate the inclusion of Non-Newtonian effects it was assumed that the two components could be evaluated separately, and that the sliding component of shear force could be replaced by a term developed from either of the following Non-Newtonian expressions:-

Stress-Strain relationship from Johnson and Tevaarwerk (19).

$$\dot{\gamma} = \frac{1}{G^*}\frac{d\tau}{dt} + \frac{\tau\tau_o}{\tau_e \eta}\sinh\left(\frac{\tau_e}{\tau_o}\right) \quad (9)$$

Stress-Strain relationship from Gecim and Winer (20).

$$\dot{\gamma} = \frac{1}{G_\infty}\frac{d\tau}{dt} + \frac{\tau_L}{\eta}\tanh^{-1}\left(\frac{\tau}{\tau_L}\right) \quad (10)$$

In cases where spinning caused the local slip direction to be displaced from the rolling direction the appropriate resolved component was used, and for production of the results shown in this paper the only stress-strain relationship used was that according to Johnson and Tevaarwerk.

3 COMPUTATIONAL PROCEDURE

The computational scheme is illustrated by means of a flow chart, Figure 1. If a new bearing was to be analysed then the first stage was to run the bearing dynamics analysis. This followed the analysis of a high speed bearing presented by Jones (1,2) as set out by Harris (21). The analysis assumed values of the axial and radial displacement of the inner raceway (δ_a,δ_r) together with bearing misalignment (θ), which was taken to be fixed at zero degrees for the computations presented in this paper. Expressions governing the bearing geometry and equilibrium of forces were then solved for the secondary unknowns of axial and radial ball centre/outer raceway curvature centre distance (L_a,L_r) together with the deflections at the inner and outer ball/raceway contacts. The primary unknowns could then be altered by checking for axial and radial equilibrium and the iteration was continued until the change in the primary unknowns was less than a pre-defined limit. (A full analysis of the bearing should also consider equilibrium of moments, but because of the difficulty of specifying the resistance to misalignment of the bearing on its own this feature was not included. During the initial testing of the analysis misalignment was permitted to such an extent that the residual moment from the fully aligned case was reduced by 50% and for this situation very little change was observed in the computed power losses.)

The power loss/traction force analysis involved calculations for each contact in the bearing starting with the inner raceway and then dealing with the outer raceway. The operating conditions for each contact were used to extrapolate an existing elastohydrodynamic solution to represent the specified conditions (Section 2(b)-(c)). The local shear stresses and associated shear forces were then evaluated with non-Newtonian effects incorporated if required. (The term $\frac{1}{G^*}\frac{d\tau}{dr}$ being approximated by $\frac{1}{G^*} U_s \frac{d\tau}{dx}$). The calculation of local shear stress could also be repeated for a Hertzian pressure profile, in which case the Hertzian flat was separated by the predicted central film thickness.

The summation of the local shear forces was carried out to give the total shear force and moment acting on the two surfaces, and the spin power loss was then easily determined:-

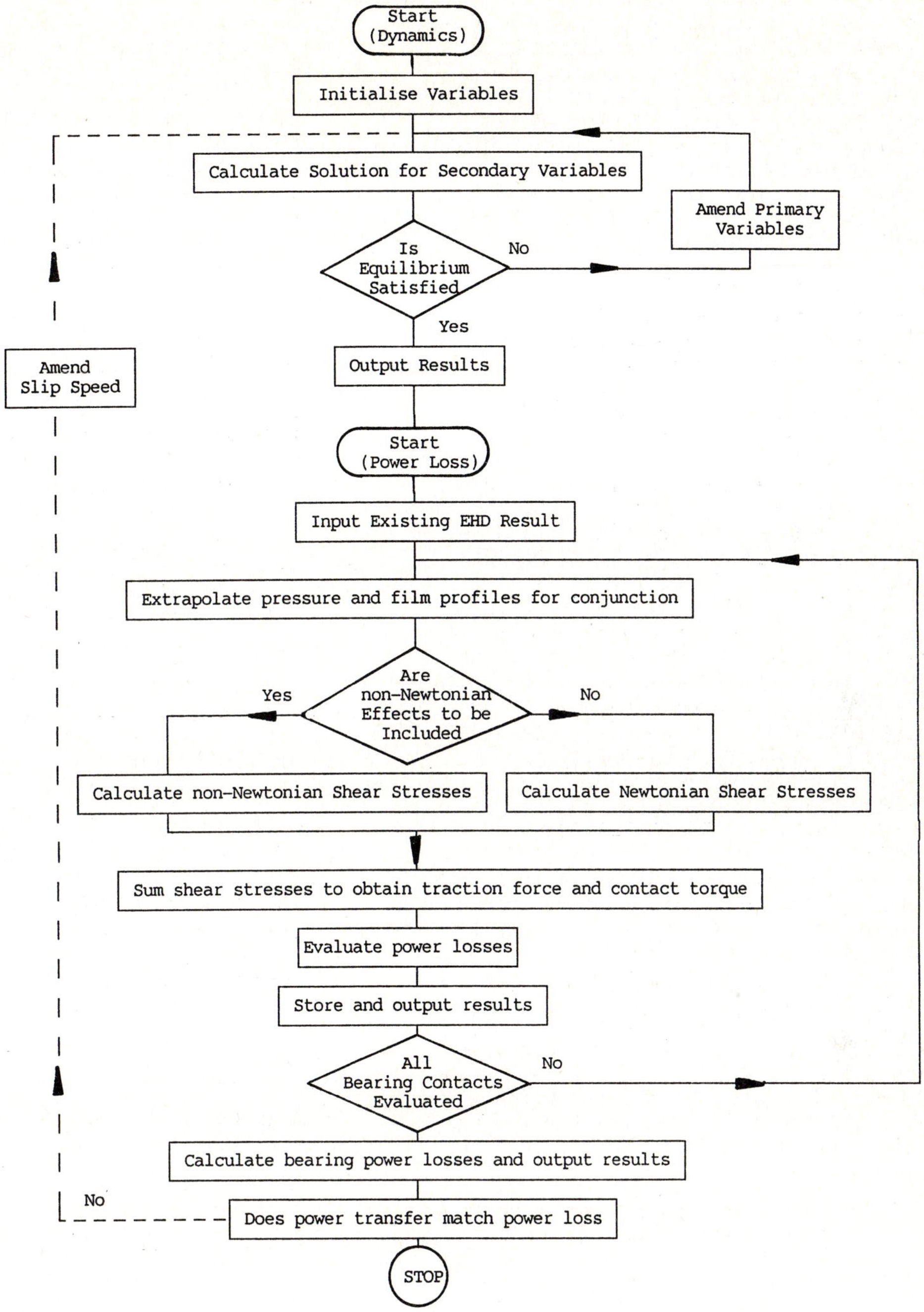

Figure 1 Flow diagram for calculation procedure

$$W_{spin}=\Omega\left(\sum_j \tau_j \sin\phi\,\delta_x\,\delta_y\,.x+\sum_j \tau_j \cos\phi\,\delta_x\,\delta_y\,.y\right) \quad (11)$$

(Ω denoting spin speed).

Separation of the rolling and sliding power losses was undertaken by specifying the rolling power loss to be that expected for a contact without spin or sliding,

$$W_{roll}=2_{upper}\sum_j\left(\frac{h_j}{2}\frac{\partial p_j}{\partial x}\right) \quad (12)$$

The sliding power loss was the rolling/sliding power loss less the rolling power loss.

$$W_{slide}=U_{lower}F_{T_{lower}}-U_{upper}F_{T_{upper}}-W_{roll} \quad (13)$$

4 RESULTS

To illustrate the use of the above analysis a typical aerospace bearing of 250mm diameter running at 10,000rpm was taken as an example. The basic operating conditions of the bearing are set out in Table 1, whilst typical conjunction parameters are provided in Table 2 together with the changes in these values around the bearing. The bearing dynamics were then determined for these conditions, and although not specified to the analysis the raceway control was that of the outer raceway. To allow the power loss/traction force analysis to be undertaken it was necessary to specify a number of lubricant constants, see Table 3, taken from manufacturers data or estimated from the work of Johnson and Tevaarwerk (19) or Evans and Johnson (22). The bearing and its operating conditions were chosen so that they could be compared with others analysed by Rolls-Royce using a heat transfer analysis, (Nicholson (23)), which were also known to have good agreement with experimental measurements. The bearing dynamics were first evaluated, assuming 3% inner raceway slip and before computations commenced on the whole bearing, changes in sliding speed were examined for a conjunction on the inner raceway. The results show the difference in power loss predictions between the Newtonian and non-Newtonian fluid models, Figure 2.

Results for the whole bearing are shown in Table 4. This contains values for inner raceway rolling, spinning and sliding power loss together with power transfer across to the outer raceway contacts, together with the predicted outer raceway power loss for four different analysis techniques:-

(i) Rolls-Royce calculation procedure. This analysis is based upon the calculation of heat balance for the bearing and, as noted above, is detailed in Nicholson (1986).

(ii) Elastohydrodynamic analysis, (Newtonian lubricant). Here, the film thickness and pressure profiles obtained according to the extrapolations described in Sections 2(a) and 2(b) are used in conjunction with the local shear stress expressions given as equations (7) and (8).

(iii) Elastohydrodynamic analysis, (non-Newtonian lubricant). In this situation the same film thickness and pressure profiles are used as in case (ii). The Newtonian sliding shear stress (denoted by the right hand term within the brackets of equations (7) and (8)) is replaced by a non-Newtonian expression derived from equation (9).

(iv) Hertz approximation. This analysis employs the well known Hertz pressure profile and dry contact surface shape. The latter, however, is separated by the predicted elastohydrodynamic central film thickness to represent the typical lubricant film over the majority of the contact area.

The contribution to the overall power loss by each contact is illustrated in Figure 3 for both inner and outer ball/raceway contact with the elastohydrodynamic analysis (non-Newtonian lubricant).

As may be seen from this table there is a noticeable difference between the Rolls-Royce results and those obtained by the analytical techniques presented in this paper. To investigate the sensitivity of the analysis to changes in the lubricant characteristics, computations were carried out with the elastohydrodynamic film and pressure profiles and a Non-Newtonian fluid model for five-fold changes in the following parameters compared with those given in Table 3:-

conjunction viscosity at inlet (η_o)
lubricant pressure/viscosity coefficient (α)
critical shear stress (τ_o)
critical shear stress/pressure coefficient ($g_{\tau c}$)

Axial load	4500 N
Radial load	1000 N
Rotational speed	10,000 rpm
Slip speed	3%
Bearing Pitch Diameter	0.25 m
Number of Balls	26
Ball Diameter	0.022m

Table 1 Operation Conditions for the Example Bearing

Reduced Elastic Modulus for Ball Raceway Contacts (E')		217×10^9 Pa
Inner Raceway Ball Contacts	Side leakage radius of curvature (R_s)	3.9×10^{-1}m
	Entraining radius of curvature (R_e)	9.8×10^{-3} m
	Rolling velocity (v_{ri})	46 m/s
	Sliding velocity (v_{si})	3.6 m/s
	Spin speed (ω_{si})	6000 rad/sec
	Ball load (F_{Bi})	210 N
Outer Raceway Ball Contacts	Side leakage radius of curvature (R_s)	8.5×10^{-1}
	Entraining radius of curvature (R_e)	1.3×10^{-2}
	Rolling velocity (v_{ro})	82.5 m/s
	Ball load (F_{Bo})	2920 N

Table 2 Conjunction Parameters for a Typical Ball

Inlet dynamic viscosity (η)	1.65×10^{-5} Pas
Viscosity-pressure coefficient (α)	1.02×10^{-8}
Representative shear stress (τ_o)	3.0×10^{6} Pa
Shear-stress-pressure coefficient (s)	6.0×10^{-3}
Lubricant shear modulus (G^*)	2.5×10^{8} Pa
Shear-modulus-pressure coefficient (m)	4.0

Table 3 Lubricant Parameters

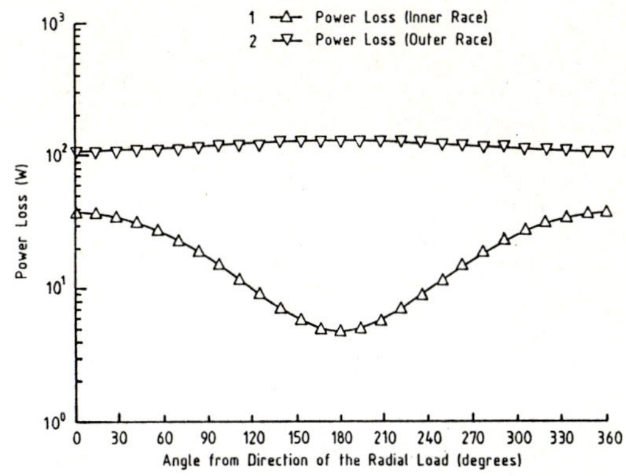

Figure 3 Predicted Power Loss from the Bearing. Non-Newtonian, Barus Pressure Viscosity Expression

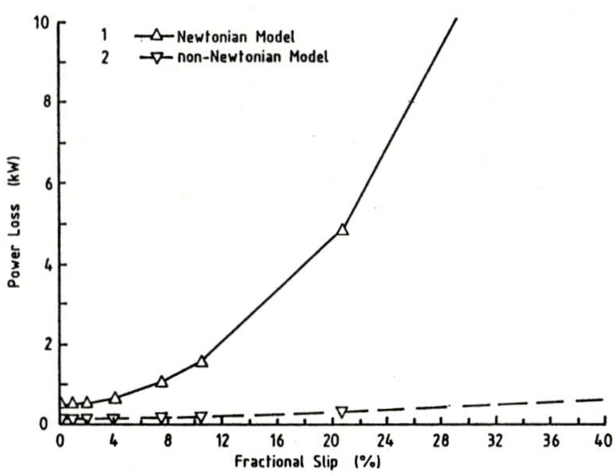

Figure 2 Total Power Loss for a Single Conjunction

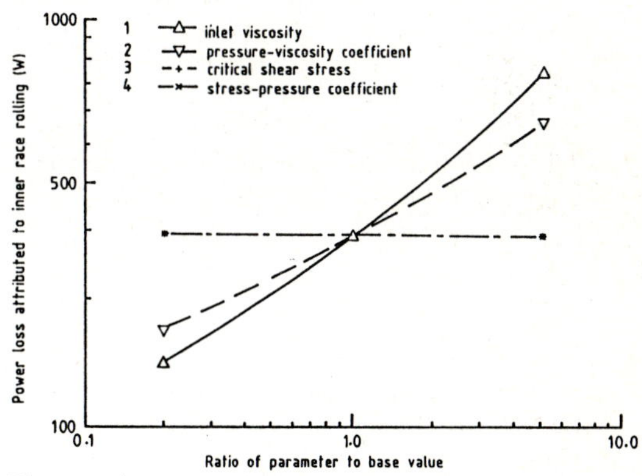

Figure 4 The Effect of Changes in the Lubricant Parameters upon the Inner Race Rolling Power Loss

		Rolls-Royce	Newtonian	non-Newtonian	Hertz
Inner Race	Rolling	1200	290	290	14
	Spinning	220	100	63	88
	Sliding	1300	220	140	160
	Transmitted Power	5200	2600	1500	2100
Outer Race	Rolling	5200	3100	3100	110
Total Power Loss		7290	3710	3593	372

Table 4 Power Losses for the Different Models (All results in Watts)

The results of these sensitivity tests are displayed in Figures 4 to 8, which illustrate inner raceway rolling power loss, inner raceway sliding power loss, inner raceway spinning power loss, power transfer and outer raceway power loss respectively.

5 DISCUSSION

The results shown in Figures 2 clearly show the need to introduce the non-Newtonian fluid model, since as the fractional slip increases the shear stresses generated within the contact reach the stage where the power loss predicted by a Newtonian analysis is totally unreasonable. Indeed if the bearing operating characteristics were such that it was operating towards its design limit of perhaps 10,000 N axial load, then the power losses given by the Newtonian analysis would be many orders of magnitude greater than the corresponding Non-Newtonian analysis, and therefore beyond the level which would be considered realistic for these situations.

The influence of radial load on the power loss at each conjunction, shown in Figure 3, also shows the large difference between inner and outer raceway power loses caused by the effects of inertia acting on the rotating elements. The slight increase in the outer raceway power loss, from a minimum at 0 degrees to the radial load to a maximum at 180 degrees, may be attributed to the change in contact surface velocity rather than to any change in the loading conditions.

The results presented in Table 4 show that for the particular conditions considered there is only a small difference between the Newtonian and non-Newtonian results for full elastohydrodynamic profiles. The overall level of agreement is not good, with the Hertz approximation yielding very unsatisfactory results for the rolling power losses. It is a little surprising that the Newtonian results fall short of those obtained by Rolls-Royce, but although the model was simple, neglecting asperity interaction and thermal effects, it is likely that these differences can be explained by modest changes in the parameters defining the lubricant characteristics. The powerful effect of these values is examined shortly.

The way in which power loss is ascribed to the various components should not be used to choose the best method for calculation, however, since correlation with experimental work can be made only in terms of the total power loss. Two further points must be noted regarding this table, the first being that the rolling power loss for a Hertzian situation employing both positive and negative pressure gradients should be zero. The non-zero value in the table was attributed to numerical errors associated with the finite difference approximation of pressure gradients which were induced by the slight displacement of the solution grid in the entraining direction coupled with the application. Secondly, during early testing of the computer codes the local shear stresses were checked and it was found that large pressure gradients around the edge of the conjunction could produce local traction

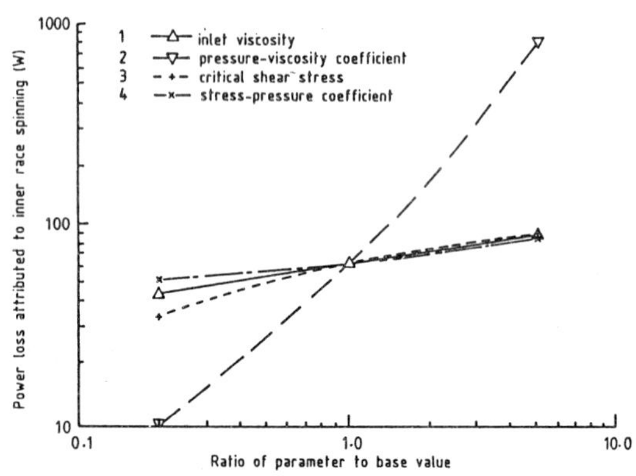

Figure 6 The Effect of Changes in the Lubricant Parameters upon the Inner Race Spin Power Loss

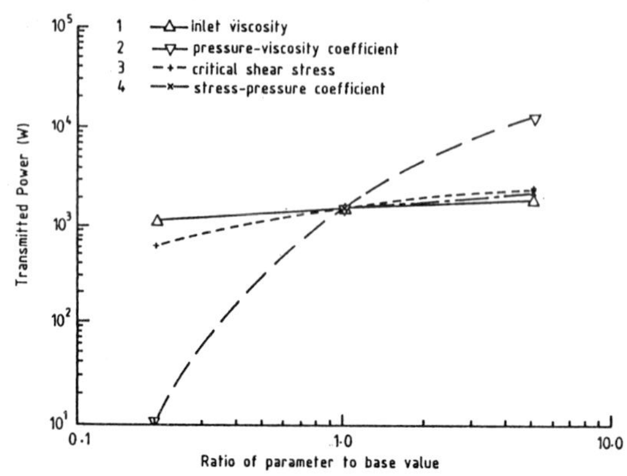

Figure 7 The Effect of Changes in the Lubricant Parameters upon the Power Transmitted from the Inner Race

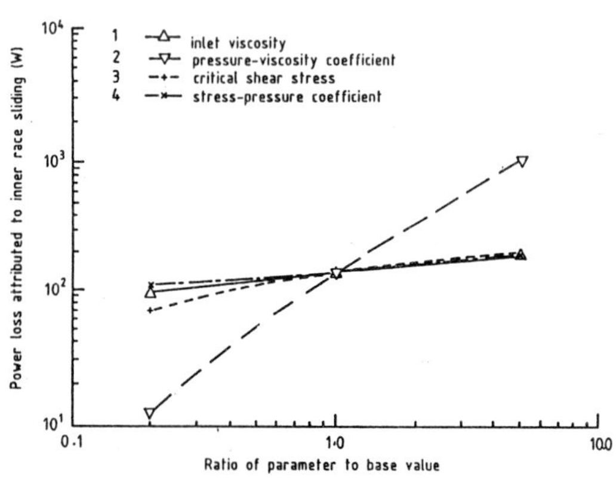

Figure 5 The Effect of Changes in the Lubricant Parameters upon the Inner Race Sliding Power Loss

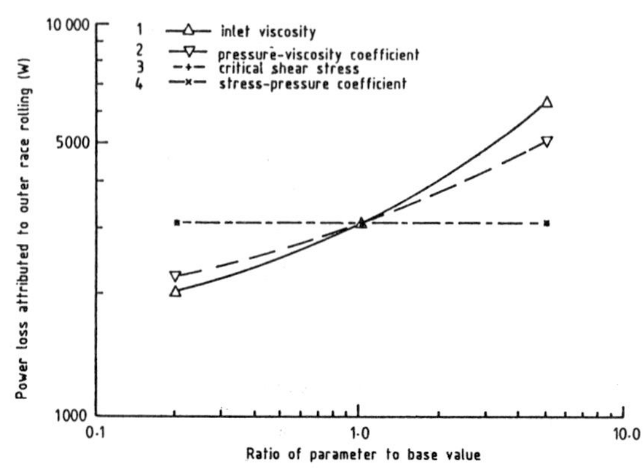

Figure 8 The Effect of Changes in the Lubricant Parameters upon the Outer Race Rolling Power Loss

coefficients well in excess of unity. To restrict the local traction forces at these points to within reasonable limits a limiting local traction coefficient of 0.7 was imposed. (This procedure was applied to all computations and was influential in the calculation of the non-zero Hertzian rolling power loss when coupled with point one, above).

An interesting feature observed from the calculations with the full elastohydrodynamic film profiles was the production of a side force acting on the contact, shown in Figure 9. This feature resulted from the shape of the elastohydrodynamic conjunction in which the side force at a point some distance from the transverse centre line is not balanced by that of a point at an equal distance on the other side of the centre line. The size of this force, however, is some two orders of magnitude less than the corresponding traction force.

The influence of the lubricant characteristics on inner race rolling power loss, shown in Figure 4, may be seen to be most significant for the inlet viscosity value, which acts through the film thickness formulae to affect the local shear stresses by being raised to a greater power than the pressure-viscosity coefficient. As neither the critical shear stress nor critical shear stress-pressure coefficient appear in these formulae or are included in the pressure extrapolation, these value have no effect on the rolling power loss. The inner race spin power loss, Figure 5, is most dependent upon the pressure viscosity coefficient, and comparison of the y-axis scale of this figure with that of Figure 4 shows that the changes produced are much more marked. An increase in this parameter also increases the predicted film thickness and hence decreases the local sliding shear stresses. The associated increase in viscosity due to the Barus expression, however, more than makes up for this deficiency. The comments made for Figure 5 also apply to the transmitted power shown in Figure 6. The effect of changes in the critical shear stress is also noticeable in this figure. Its influence is by setting the level at which Non-Newtonian effects are important and hence is a measure of the extent to which Newtonian effects still hold. The final illustration in this set, Figure 8, exhibits the same effects as Figure 4 although the magnitudes are slightly different.

The last item of Table 4 that has not yet been commented upon is the balance between the power transfer and the power lost at the outer raceway/ball contacts. Since the outer race was taken to be stationary, the power transferred should match the power lost by these contacts. To ensure this balance the missing link, shown as the dashed line in Figure 1, would be required. An estimate of the necessary inner race slip ratio, however, was obtained by additional computations for slip values of 0.6% and 15% (a value rather higher than typical in aerospace bearings). These results are recorded on Figure 10, and from this it may be seen that the required balance point is at approximately 7% slip.

6 CONCLUSIONS

The results of this study can be summarised as follows:-

- A method has been developed for the calculation of power loss in deep groove ball bearings based purely upon theoretical considerations. Simple extrapolations have allowed realistic elastohydrodynamic film thickness and pressure profiles to be included and a non-Newtonian lubricant model has been incorporated into the computations.

- Predictions of power loss in a representative aero-engine bearing have been made and, with the exception of sliding power loss, found to be of similar magnitude to those predicted by Rolls-Royce using a different approach. The latter was found to agree well with experimental measurements. (The total power loss being in agreement to almost a factor of 2 although the constituent parts do not show the same correlation).

- The influence of the lubricant characteristics upon the predicted components of power loss has been examined. This indicated that lack of precise knowledge of these values, particularly the pressure-viscosity coefficient, could account for much of the difference between the power loss predicted by the computations detailed in this paper and those obtained by Rolls-Royce.

Finally, the missing link in the interaction between the bearing dynamics analysis and the power loss calculations was approximated by undertaking calculations for

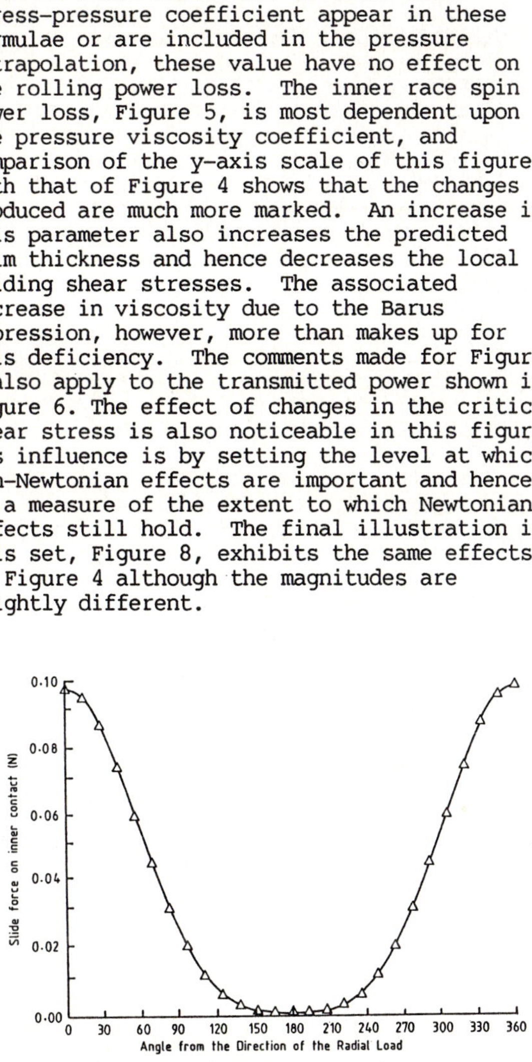

Figure 9 The Side Force on the Contact due to Spinning at the Ball/ Inner Raceway Contact

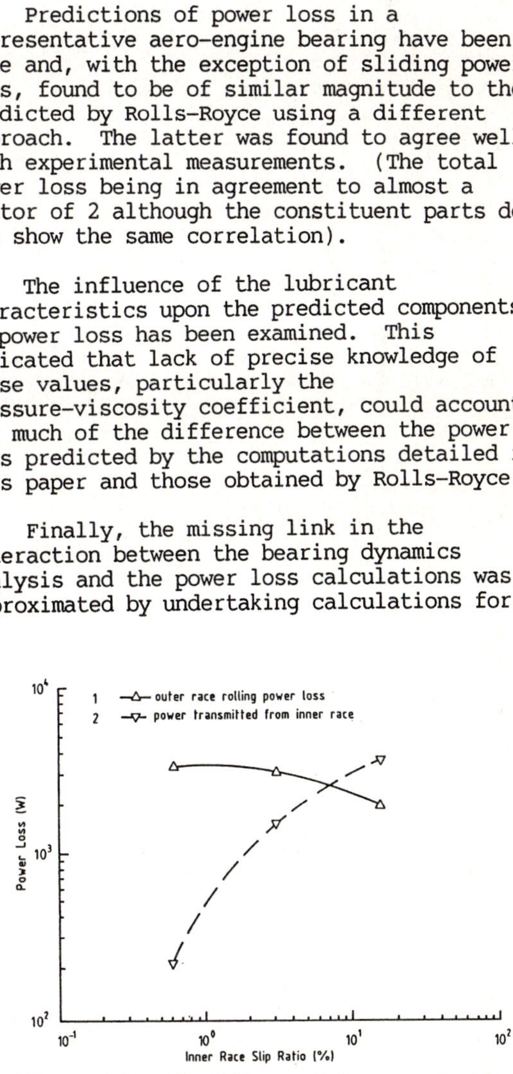

Figure 10 The Effect of Changes in the Inner Race Slip Ratio upon Power Transfer and Outer Race Rolling Power

different assumed slip ratios. This then allowed the slip ratio required to balance the power transfer and outer race power loss to be estimated.

7 ACKNOWLEDGEMENTS

The authors would like to thank Rolls-Royce plc. and the S.E.R.C. for the funding of this project as part of the Roll-Royce/S.E.R.C. co-funded scheme, and in particular Dr J. Dominy and Mr R. Nicholson of Rolls-Royce for their invaluable help and comments.

References

(1) JONES A.B. (1959) 'Ball Motion and Sliding Friction in Ball Bearings', Trans. A.S.M.E., J. Basic Eng., 1-12.
(2) JONES A.B. (1960) 'A General Theory for Elastically Constrained Ball and Roller Bearings Under Arbitrary Load and Speed Conditions', Trans. A.S.M.E., J. Basic Eng., 309-320.
(3) GENTLE C.R. and BONESS R.J. (1976), 'Prediction of Ball Motion in High-Speed Thrust Loaded Ball Bearings', J. Lubr. Technol., 98(3), 463-471.
(4) PASDARI M. and GENTLE C.R. (1986), 'Computer Modelling of a Deep Groove Ball Bearing With Hollow Balls', Wear, 111, 101-114.
(5) DOWSON D. and HIGGINSON G.R. (1965), 'Elastohydrodynamic Lubrication, The Fundamentals of Roller and Gear Lubrication', Pergamon, Oxford.
(6) HAMROCK B.J. and DOWSON D. (1976a), 'Isothermal Lubrication of Point Contacts, Part I, Theoretical Formulation', J. Lubr. Technol., 98(2), 223-229.
(7) HAMROCK B.J. and DOWSON D. (1976b), 'Isothermal Lubrication of Point Contacts, Part II, Ellipticity Parameter Results', J. Lubr. Technol., 98(3), 365-378.
(8) HAMROCK B.J. and DOWSON D. (1977a), 'Isothermal Lubrication of Point Contacts, Part III, Fully Flooded Results', J. Lubr. Technol., 99(2), 264-276.
(9) HAMROCK B.J. and DOWSON D. (1977b), 'Isothermal Lubrication of Point Contacts, Part IV, Starvation Results', J. Lubr. Technol., 99(1), 15-23.
(10) EVANS H.P. and SNIDLE R.W. (1981), 'Inverse Solution of Reynolds' Equation of Lubrication Under Point Contact Elastohydrodynamic Conditions', J. Lubr. Technol. 103(4), 538-546.
(11) CHITTENDEN R.J., DOWSON D., DUNN J.F. and TAYLOR C.M. (1985), 'A Theoretical Analysis of the Isothermal Lubrication of Concentrated Contacts, Part II, General Case, with Lubricant Entrainment along either Principal Axis of the Hertzian Contact Ellipse or at Some Intermediate Angle', Proc. R. Soc. Lond., A397, 271-294.
(12) LUBRECT A.A., TEN NAPEL W.E. and BOSMA R. (1987), 'Multigrid, and Alternative Method of Solution for Two-Dimensional Elastohydrodynamically Lubricated Point Contact Conditions', J. of Tribology, 109(2), 437-443.
(13) LUBRECT A.A., VENNER C.H., TEN NAPEL W.E. and BOSMA R. (1987), 'Film Thickness Calculations in Elastohydrodynamically Lubricated Circular Contacts, Using A Multigrid Method', J. of Tribology, 110(3), 503-507.
(14) XU H. (1988), PhD. Thesis, Department of Mechanical Engineering, Leeds University.
(15) SANDBORN D.M. and WINER W.O. (1971), 'Fluid Rheological Effects in Sliding Elastohydrodynamic Point Contacts with Transient Loading: 1- Film Thickness', J. Lubr. Technol., 93(2), 262-271.
(16) ADAMS D.R. and Hirst W. (1973), 'Frictional Traction in Elastohydrodynamic Lubrication', Proc. R. Soc. Lond., A332, 505-525.
(17) ESFAHANI M.Z. (1988) Private Communication.
(18) HAMROCK B.J. and ANDERSON W.J. (1973), 'Analysis of Arched Outer-Race Ball Bearing Considering Centrifugal Forces', J. Lubr. Technol., 95(3), 265-276.
(19) JOHNSON K.L. and TEVAARWERK J.L. 'Shear Behaviour of Elastohydrodynamic Oil Films', Proc. R. Soc. Lond., A356, 215-236.
(20) GECIM B. and WINER W.O. (1980) 'Lubricant Limiting Shear Stress Effect on EHD Film Thickness', J. Lubr. Technol., 102(2), 213-221.
(21) HARRIS T.A. (1984) 'Rolling Bearing Analysis', John Wiley and Sons, New York.
(22) EVANS C.R. and JOHNSON K.L. 'The Rheological Properties of Elastohydrodynamic Lubricants', Proc. Inst. Mech. Engrs., 200(C5), 303-312.
(23) NICHOLSON R. (1986), 'The Prediction of Operating Temperatures in High Speed Angular Contact Bearings', Paper XXI(ii), Fluid Film Lubrication- Osbourne Reynolds Centenary, Proceedings of the 13th Leeds-Lyon Symposium on Tribology. Editors D. Dowson, C.M. Taylor, M. Godet and D. Berthe, Elsevier.

Paper X(iii)

The effect of roller end-flange contact shape upon frictional losses and axial load of the radial cylindrical roller bearing

H. Krzemiński-Freda and B. Warda

The subject of the consideration is the radial cylindrical roller bearing of the NJ type with conical flanges and spherical ends rollers. In the paper, theoretical bases have been presented for selection of the roller end radius and the flange inclination angle resulting from the compromise between an attempt to minimizing frictional losses in the bearing and the necessity to guarantee an appropriate shape of the oil gap conducive to formation of an oil film at the roller end - flange contact.

1 INTRODUCTION

The radial cylindrical roller bearing of the NJ type can also carry, apart from radial load, some axial load. This load is carried by the bearing roller end - flange contacts.

Normal forces occurring at the contacts and friction forces accompanying them during the bearing operation, cause the roller tilt in two reciprocally perpendicular planes intersecting its axis (Fig. 1). The tilt in the plane parallel to the bearing axis, referred to as the skew (angle φ_1), causes a slide at the roller - main race contacts and, as a result, an increase in the frictional losses in the bearing. The tilt in the plane intersecting the bearing axis (angle φ_2) leads to an occurrence of unfavourable distributions of pressure at the roller - main race contacts, manifesting themselves as a pressure increase in one of the ends of the contact area. It is one of the factors reducing bearing durability.

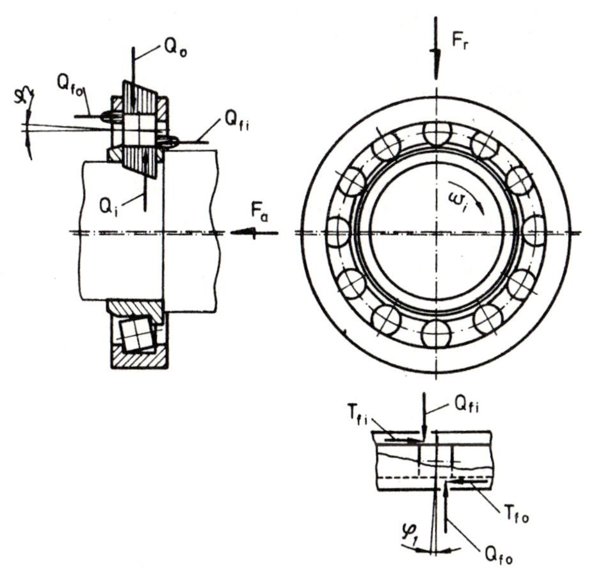

Fig. 1. Skew and tilt of the roller in the radial cylindrical roller bearing subjected to combined load.

However, the roller end - flange mating conditions have a decisive effect on the durability of a radial cylindrical roller bearing loaded with an axial and a radial force. Due to slides and accompanying friction, at the roller end - flange contacts a large amount of heat is generated, which causes fast wear of the mating surfaces.

The amount of heat generated in the bearing can effectively be reduced by such modification of the shape of the contact surfaces, which will guarantee a decrease in the frictional losses. One should aim at the following:

- ensuring such a shape of the oil gap at the roller end - flange contact that would be conducive to formation of an oil film,
- minimizing pressures at the contacts,
- minimizing the skew angle of the roller.

Numerous experimental and theoretical investigations (1,2,3,4,5,6) have shown that the most favourable mating conditions of the roller end and the flange are created by conical flanges and spherical roller ends. However, despite many achievements in this field, the operating conditions of the radial cylindrical roller bearing subjected to combined load have not been investigated sufficiently.

In the paper, an attempt was made to carry out a complex analysis of the roller balance conditions, considering all the most important factors affecting this balance. The subject of the analysis is the bearing with conical flanges and spherical ends rollers (Fig. 2).

Its aim is to create theoretical bases for selection of the most advantageous - from the viewpoint of frictional losses in the bearing - values of the angle of inclination of flanges (β) and the roller end radius (R_w). A more detailed discussion of the problems dealt with the paper can be found in the work (7).

1.1 Notation

a	- major semiaxis of the contact ellipse
b	- minor semiaxis of the contact ellipse
b_1, b_2, b_3	- parameters determining the position of the roller end-flange contact area centre
C	- dynamic basic load rating

D	- outside diameter of the bearing
$D_{i(o)}$	- diameter of the main race
D_w	- roller diameter
d	- bore diameter
e	- distance of the roller end - flange contact area centre from the main race
H	- convexity of the generatrix of the roller or main race
k	- number of the contact line division elements
l_a	- length of the generatrix of the roller
u	- slide velocity
v	- linear velocity
α	- angle determining the position of the contact area centre on the flange
δ^*	- dimensionless contact deformation
μ	- coefficient of friction
τ	- friction force per the unit of the contact line length
ψ	- angle determining the roller position in the bearing
ω	- angular velocity
χ	- angle determining the direction of action of the force Q_r.

Subscripts
i - inner ring
o - outer ring
f - flange
w - roller

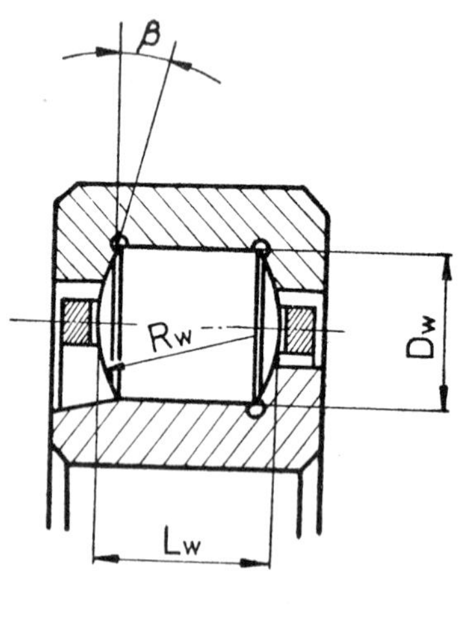

Fig. 2. Inner geometry of the modified radial cylindrical roller bearing.

2 THE ANALYSIS OF THE ROLLER BALANCE CONDITIONS

The analysis of the roller balance conditions in the radial cylindrical roller bearing subjected to combined load was carried out with the following assumptions:

1. The bearing material is ideally elastic and isotropic.
2. The surfaces in contact have ideal shapes.
3. The pressure distributions at the contacts have been assumed according to hertzian theory.
4. The roller - bearing race mating takes place in the conditions of fluid friction.
5. The mass forces, the effect of the cage on the rollers and the losses of the rolling friction in the bearing have been neglected.
6. The bearing rings can displace relative to one another only in radial and axial directions, but they cannot till.

Under the action of the radial load (F_r) and axial load (F_a) at the roller - bearing race contacts pressures appear, accompanied by surface deformations (Figs. 3a, c).

The pressures and deformations at the roller end - flange contacts were determined acc. to hertzian formulae for the point contact:

$$(1)\qquad p_{Hfi(o)} = \frac{3Q_{fi(o)}}{2\pi \cdot a_{i(o)} \cdot b_{i(o)}},$$

$$(2)\qquad \delta_{fi(o)} = 2.782 \cdot 10^{-4} \cdot \delta^*_{fi(o)} \cdot Q_{fi(o)}^{2/3} \cdot \Sigma\varrho_{fi(o)}^{1/3} \quad [\mathrm{mm}],$$

$Q_{fi(o)}$ in N, $\Sigma\varrho_{fi(o)}$ in mm^{-1}. The latter formula is right for the materials of modulus of elasticity $E = 2.08 \cdot 10^5$ MPa and of Poisson´s ratio $\nu = 0.3$. $\Sigma\varrho_{fi(o)}$ is the sum of curvatures at the point being the centre of the contact area of the roller end and the flange of the inner or the outer ring.

Pressure distributions at the roller - main race contacts were determined using Lundberg´s formula (8), in which the value of force per contact line length unit is dependent on the deformations resulting from interpenetration of the surfaces of the roller and the races.

The deformation within the range j + 1 of the contact line (Fig. 3c) can be expressed by the formula:

$$(3)\qquad \Delta_{i(o)(j+1)} = (-4(H_w + H_{i(o)})/l_a^2) \cdot r_j^2 + (4(H_w + H_{i(o)})/l_a - \tan\varphi_2) \cdot r_j + \Delta_{i(o)1},$$

$$(4)\qquad r_j = l_a \cdot j/k \qquad j = 0,1,\dots k.$$

The deformation $\Delta_{i(o)(j+1)}$ depends on the reciprocal approach of the bearing rings in the radial direction:

$$(5)\qquad \delta_{io} = \delta_i + \delta_o \approx \Delta_{i(k/2+1)} + \Delta_{o(k/2+1)}.$$

The distribution of forces per the contact line length unit (Fig. 3b) is defined by the formula:

$$(6)\qquad q_{i(o)(j+1)} = 78200 \cdot \Delta_{i(o)(j+1)}^{1.11} \cdot l_a^{-0.11} \quad [\mathrm{N/mm}],$$

while the distribution of the maximum pressure (Fig. 3a) by the dependence:

$$(7)\qquad p_{Hi(o)(j+1)} = 190.73 \cdot (q_{i(o)(j+1)} \cdot \Sigma\varrho_{i(o)})^{1/2} \quad [\mathrm{MPa}],$$

$\Delta_{i(o)(j+1)}$ and l_a in mm, $\Sigma\varrho_{i(o)}$ in mm^{-1}. $\Sigma\varrho_{i(o)}$ is the sum of the curvatures at the roller - main race contacts.

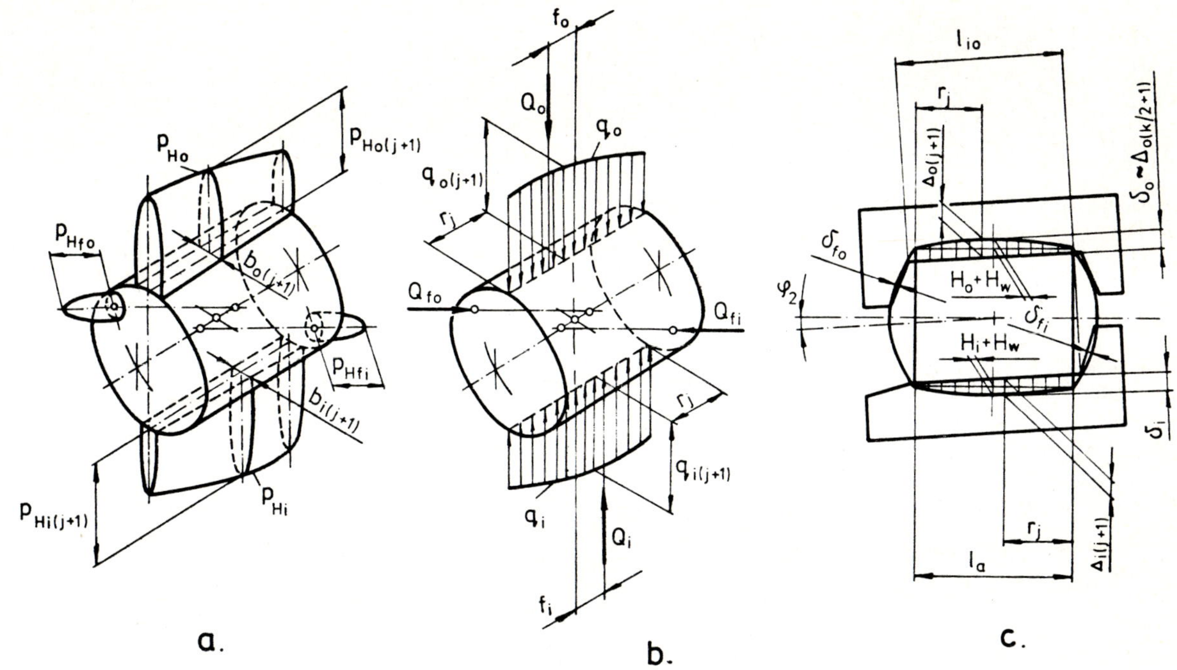

Fig. 3. a - Pressure distributions at the roller - flange contacts.
b - Resultant normal forces at the roller - race contacts; distributions of forces per the contact line length unit.
c - Deformations at the roller - race contacts.

In the analysis of the balance conditions, continuous pressure distributions were raplaced by concentrated forces:

- Q_{fi} and Q_{fo} - at the roller end - flange contacts,
- Q_i and Q_o - at the roller - main race contacts.

Normal forces Q_i and Q_o and the points of their application (Fig. 3b) were determined by integrating the distributions of forces per the contact line length unit:

$$(8) \qquad Q_{i(o)} = \int q_{i(o)}\, dr \approx \sum_{j=1}^{k} q_{i(o)(j+1)} \cdot l_a / k \,,$$

$$(9) \qquad f_{i(o)} = l_a/2 - \left\{ \int q_{i(o)}\, r\, dr \right\} / Q_{i(o)} \approx \approx l_a/2 - \left\{ \sum_{j=1}^{k} q_{i(o)(j+1)} \cdot r_j \right\} \cdot \frac{l_a}{Q_{i(o)} \cdot k} \,.$$

An analysis of the inner geometry of the bearing, taking the skew (φ_1) and the tilt (φ_2) angles into consideration, provided information about the position of the centres of the roller end - flange contact areas, while an analysis of the bearing kinematics made it possible to determine the directions of slide velocities and the values of relative slides at the contacts (Fig. 4).

The relative slides at the roller - main race contacts are defined by the formula:

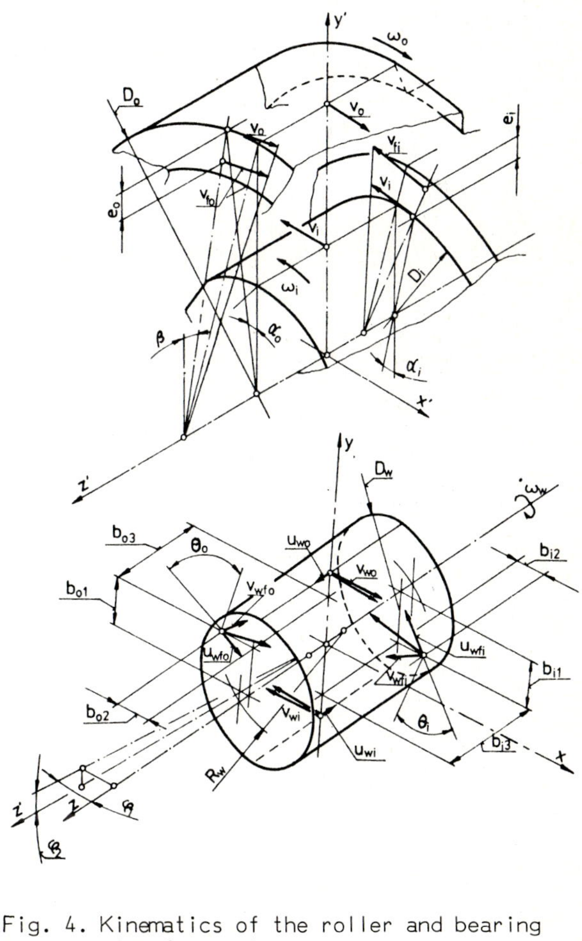

Fig. 4. Kinematics of the roller and bearing ring fragments.

$$(10)\quad s_{i(o)} = u_{wi}/(v_i + v_{wi})\cdot 100\% = \sin\varphi_1/(1+\cos\varphi_1)\cdot 100\%;$$

the direction of the slide velocities is in accordance with the roller axis direction.

The relative slides at the roller end - flange contacts were determined from the dependence:

$$(11)\quad s_{fi(o)} = u_{wfi(o)}/(v_{wfi(o)} + v_{fi(o)})\cdot 100\%,$$

where:

$$(12)\quad u_{wfi(o)} = v_{fi(o)}^2 + v_{wfi(o)}^2 - 2\cdot v_{fi(o)}\cdot v_{wfi(o)}\cdot \cos\Theta_{i(o)},$$

$$(13)\quad v_{fi} = (1 + 2e_i/D_i)\cdot v_i,$$

$$(14)\quad v_{fo} = (1 - 2e_o/D_o)\cdot v_i,$$

$$(15)\quad v_{wfi(o)} = 2\cdot v_i\cdot \cos\varphi_1\cdot (b_{i(o)1}^2 + b_{i(o)2})^{1/2}/D_w,$$

$$(16)\quad v_i = \omega_i\cdot D_i/2.$$

The slide velocity directions are determined by the angles θ_i and θ_o which are the functions of the roller skew and tilt angles and the inner bearing geometry.

The coefficient of friction was presented in the function of the relative slide (s in%) and the maximum pressure at the contact (p_H, Fig.5). The results of the experiments described in the papers (9,10,11) and the dependence given in the paper (12) were used.

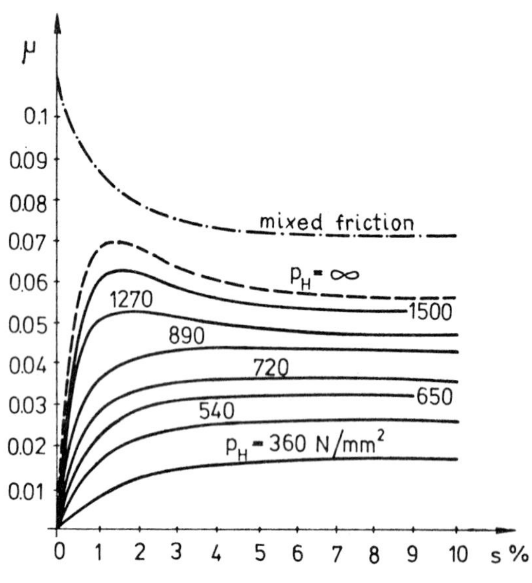

Fig. 5. The characteristics of the coefficient of friction.

The resultant friction forces at the roller end - flange contacts are described by the dependence:

$$(17)\quad T_{fi(o)} = \mu_{fi(o)}\cdot Q_{fi(o)}.$$

The friction force directions are in accordance with the slide velocity directions at the points which are the centres of the roller end - flange contacts areas (Figs. 4 and 6).

The resultant moments of friction at the contacts with the flanges were determined according to the formula:

$$(18)\quad M_{fi(o)} \approx \frac{3}{16}\pi\cdot \mu_{fi(o)}\cdot Q_{fi(o)}\cdot a_{i(o)}.$$

The resultant friction forces at the roller-main race contacts are determined by the formula:

$$(19)\quad T_{i(o)} = \int \tau_{i(o)}dr \approx \sum_{j=1}^{k} \mu_{i(o)(j+1)}\cdot q_{i(o)(j+1)}\cdot l_a/k,$$

and the position of points of application of these forces by the dependence:

$$(20)\quad t_{i(o)} = l_a/2 - \left\{\int \tau_{i(o)} r dr\right\}/T_{i(o)} \approx l_a/2 - \left\{\sum_{j=1}^{k} \tau_{i(o)(j+1)}\cdot r_i\right\}\cdot \frac{l_a}{T_{i(o)}\cdot k}.$$

The equilibrium state of the roller and the bearing outer ring fragment is presented in Fig. 6.

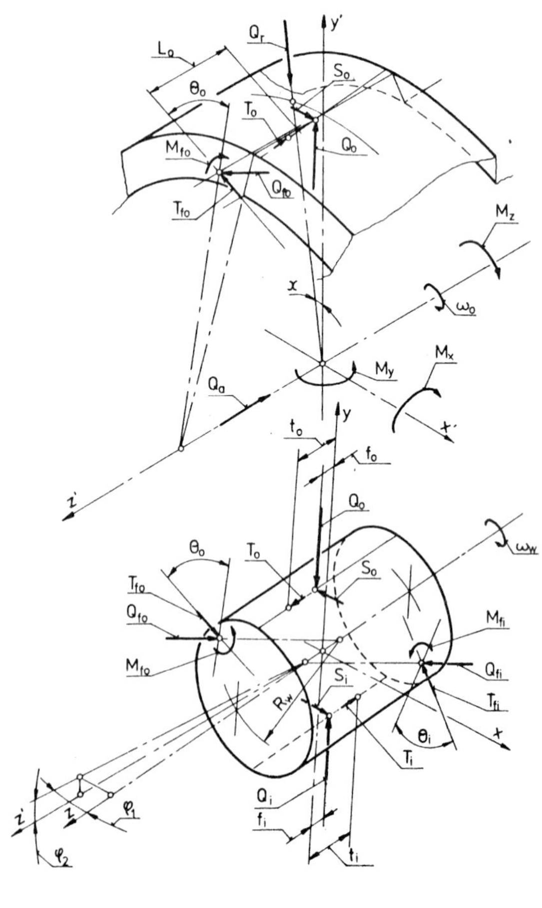

Fig. 6. Ilustration of the equilibrium state of the roller and the bearing outer ring fragment.

The adequate balance equations have the following form:

$$(21)\ \Sigma Q_x = Q_{fox} + T_{fox} - Q_{fix} - T_{fix} + S_o - S_i = 0,$$

$$(22)\ \Sigma Q_y = Q_{fiy} + T_{fiy} - Q_{foy} - T_{foy} + Q_i - Q_o = 0,$$

$$(23)\ \Sigma Q_z = Q_{fiz} - T_{fiz} - Q_{foz} + T_{foz} - T_i + T_o = 0,$$

$$(24)\ \begin{aligned}\Sigma M_x &= (Q_{fiy} + T_{fiy})\cdot b_{i3} + (Q_{foy} + T_{foy})\cdot b_{o3} - (Q_{fiz} - T_{fiz})\cdot b_{i1} \\ &- (Q_{foz} - T_{foz})\cdot b_{o1} + (T_i + T_o)\cdot D_w/2 + Q_i\cdot f_i + Q_o\cdot f_o - M_{fix} - M_{fox} = 0,\end{aligned}$$

$$(25)\ \begin{aligned}\Sigma M_y &= (Q_{fix} + T_{fix})\cdot b_{i3} + (Q_{fox} + T_{fox})\cdot b_{o3} - (Q_{fiz} - T_{fiz})\cdot b_{i2} \\ &- (Q_{foz} - T_{foz})\cdot b_{o3} - S_i\cdot f_i - S_o\cdot f_o + M_{foy} + M_{fiy} = 0,\end{aligned}$$

$$(26)\ \begin{aligned}\Sigma M_z &= (Q_{fiy} + T_{fiy})\cdot b_{i2} + (Q_{foy} + T_{foy})\cdot b_{o2} - (Q_{fix} + T_{fix})\cdot b_{i1} \\ &- (Q_{fox} + T_{fox})\cdot b_{o1} + (S_i + S_o)\cdot D_w/2 + M_{fiz} + M_{foz} = 0,\end{aligned}$$

$$(27)\ \Sigma Q_{x'} = Q_r\cdot \sin\varkappa - Q_{fox'} - T_{fox'} - T_{ox'} + S_{ox'} - Q_{ox'} = 0,$$

$$(28)\ \Sigma Q_{y'} = Q_{foy'} + T_{foy'} + Q_{oy'} - T_{oy'} - Q_r\cdot \cos\varkappa = 0,$$

$$(29)\ \Sigma Q_{z'} = Q_{foz'} - T_{foz'} - T_{oz'} - S_{oz'} - Q_{oz'} - Q_a = 0,$$

$$(30)\ \begin{aligned}\Sigma M_{z'} &= (Q_{foz'} + T_{fox'})\cdot(D_o/2 - e_o)\cdot\cos\alpha_o \\ &- (Q_{foy'} + T_{foy'})\cdot(D_o/2 - e_o)\cdot\sin\alpha_o - (S_{ox'} - T_{ox'} \\ &- Q_{ox'})\cdot D_o/2 - T_{oy'}\cdot t_o\cdot\sin\varphi_1 + Q_{oy'}\cdot f_o\cdot\sin\varphi_1 \\ &- M_{foz'} - M_z = 0.\end{aligned}$$

The balance equations were solved numerically, using Hook-Jeeves method.

3 THE ANALYSIS OF THE ROLLER MATING CONDITIONS IN THE BEARING SUBJECTED TO COMBINED LOAD

The analysis comprised a number of bearings of different sizes and different dimension series. Calculations were made for a few values of angles of inclination of the flanges, $\beta \approx 0.25^\circ \div 4^\circ$.

The solution of the balance equations made it possible to determine a number of parameters describing the roller balance state for definite inner geometry of the bearing and set values of the forces Q_r and Q_a falling to one roller. The most important of them are the following:

- angles of skew and tilt of the roller,
- position and dimensions of the roller end - flange contact area,
- pressure distributions at the contacts,
- magnitude of frictional losses resulting from the axial load,
- reciprocal position of the bearing rings in the axial (l_{io}, Fig. 3c) and radial direction (δ_{io}).

Fig. 7 presents the curves of dependence of the forces Q_a and Q_r on the parameters l_{io} and δ_{io} for the NJ 2316 E bearing and the angle $\beta = 0.9550^\circ$. These diagrams make it possible to determine distributions of load on particular rollers for any radial and axial load (F_r, F_a) and any radial clearance of the bearing (g).

Fig. 8 presents distributions of radial and axial load on the rollers for the NJ 2316 E bearing and the angle $\beta = 0.9550^\circ$, the radial clearance g = 0.025 mm, the radial load $F_r \approx 0.11.C$ and two cases of the axial load $F_a/F_r = 0.15$ and $F_a/F_r = 0.6$. It can be seen that the radial load distribution is similar to the distribution corresponding to the action of the pure radial load (F_r), denoted by broken line in Fig. 8. The area of the loaded rollers is then defined by the angle ψ_ε. However, due to the tilt of the rollers caused by the axial force action, a greater number of rollers are loaded radially. The angle defining the area of the loaded rollers (ψ_l) is the greater the bigger the F_a/F_r ratio.

A characteristic feature of the axial load distribution is almost identical load of the rollers in the wide range of the angles ψ, virtually up to the angle ψ_ε. It causes the ratio Q_a/Q_r for the most loaded roller to be about twice as small as the ratio F_a/F_r.

The position and dimensions of the roller end - flange contact area are decisive of the shape of the oil gap. The position of the contact area depends on the skew and tilt angles of the roller. Figures 9 and 10 present characteristics of the angles φ_2 and φ_1 for the most loaded roller ($\psi = 0^\circ$) in the function of the ratio F_a/F_r and the angle β, for the bearings under investigation. They were made, as the characteristics of the other parameters, for the constant value of the ratio $F_r/C \approx 0.11$. Such radial load corresponds to the average, most often found operating conditions of radial cylindrical roller bearings. It results form these diagrams that the angle of inclination of the flanges in its practical selection range has a negligible effect on the roller tilt (φ_2), thereby on the pressure distributions at the roller - main race contacts, while it has a substantial effect on the value of the skew angle (φ_1). Yet, for the angles $\beta < 1^\circ$, the value of the angle φ_1 does not exceed 2, and the phenomena related with the skew have a negligible effect on the bearing operation, especially upon the frictional losses.

Fig. 11 presents changes of position and dimensions of the theoretical roller end - inner ring flange contact area and the oil gap shapes corresponding to them in the function of the angle β. As the angle β decreases, the theoretical contact ellipse encompasses a greater and greater fragment of the roller end - flange mating area. The actual contact area occupies still greater part of this area. In Fig. 11, the boundaries of the actual contact area and the shape of the oil gap corresponding to it for the angle $\beta = 0.2386^\circ$ have been marked with broken line. Small angles of inclination of the flanges ($\beta \approx 0.25^\circ$) may lead to a considerable reduction of the inlet zone, and in the extreme case, at great axial forces, even to its disappearance, which will make formation of the oil film at the roller end - flange contact impossible.

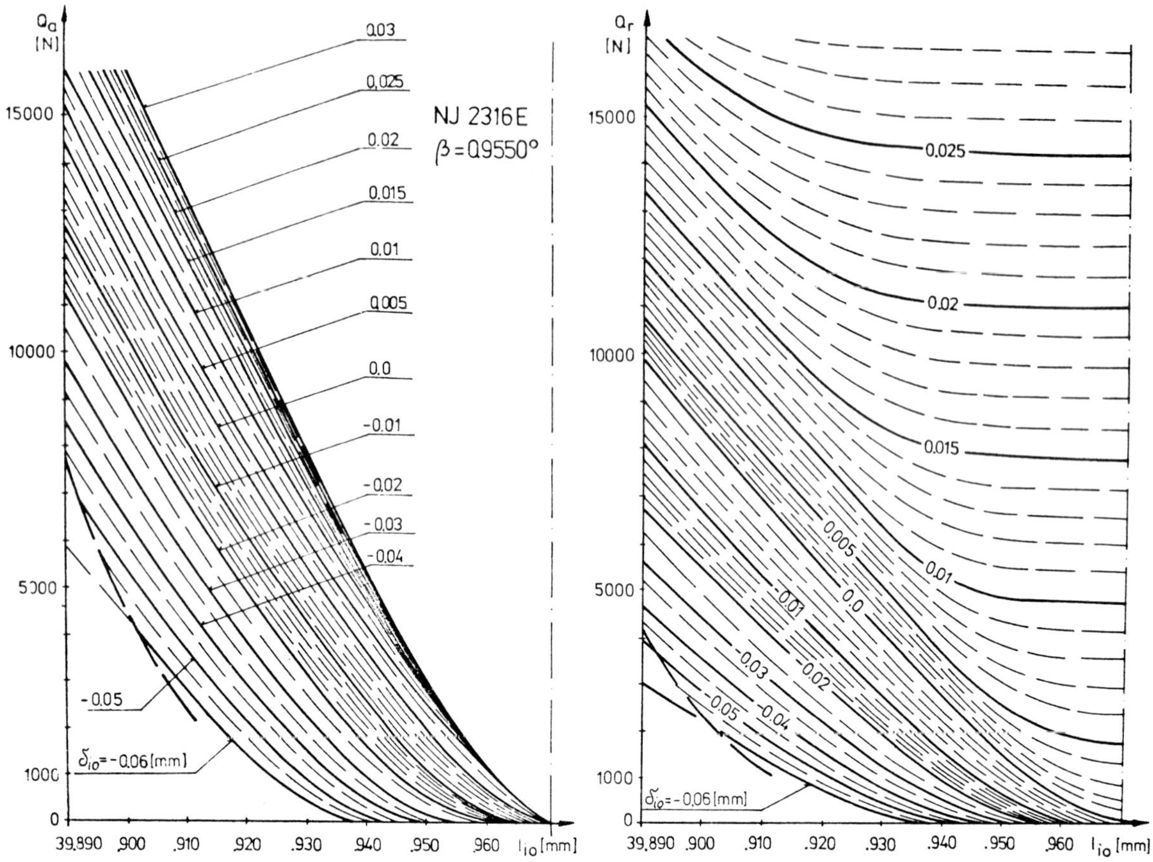

Fig. 7. The characteristics of the forces Q_a and Q_r in the function of the parameters l_{io} and δ_{io}.

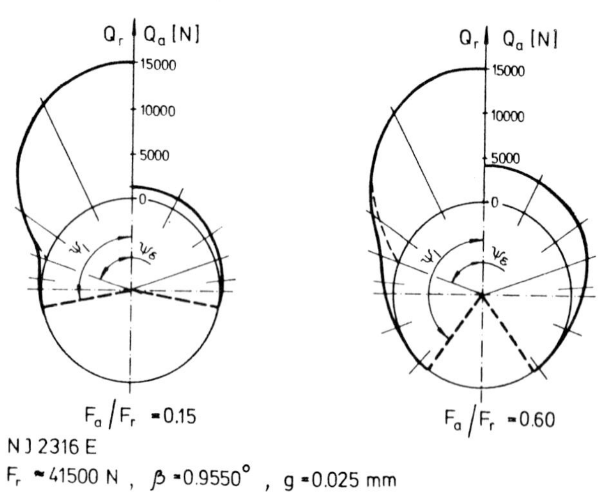

Fig. 8. Distributions of radial and axial load on the rollers.

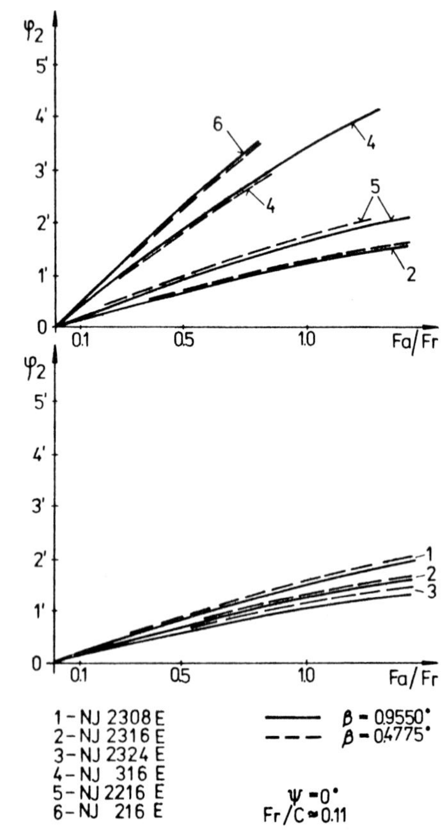

Fig. 9. Effect of the angle β on the angle of tilt of the most loaded roller.

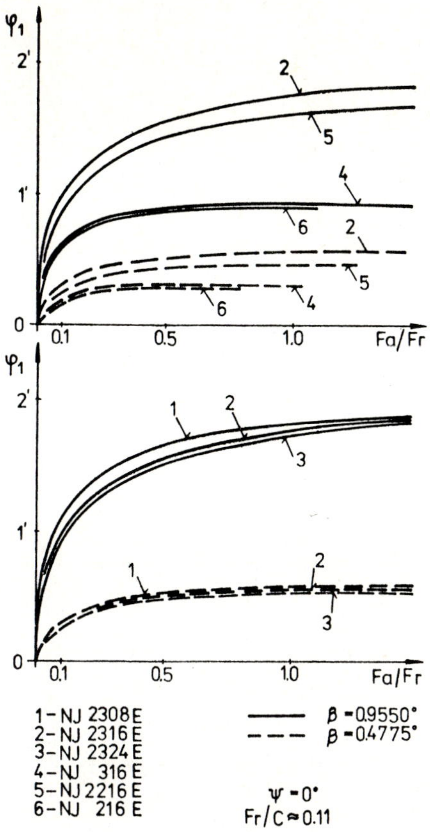

Fig. 10. Effect of the angle β on the angle of skew of the most loaded roller.

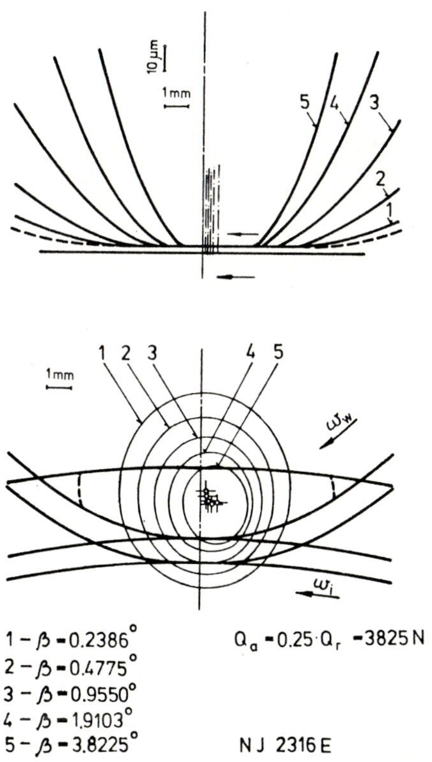

Fig. 11. Effect of the angle β on the oil gap shape and on the position and dimensions of the theoretical roller end - inner ring flange contact area.

Smaller angles β are accompanied by smaller pressures at the roller end - flange contacts (Fig. 12), and as a results, by smaller frictional losses in the bearing. For the angle $\beta = 0.5°$, the coefficient of moment of friction resulting from the axial load (f_2 , Fig. 13) for a wide range of axial loads is 0.001 ÷ 0.0015. It is about six times smaller than the coefficient f_2 for the bearing with flat roller ends and with flat flanges. The moment of friction can be determined from the formula:

$$M_2 = f_2 \cdot F_a \cdot d_m , \tag{31}$$

where $d_m = (D + d)/2$ is the mean bearing diameter.

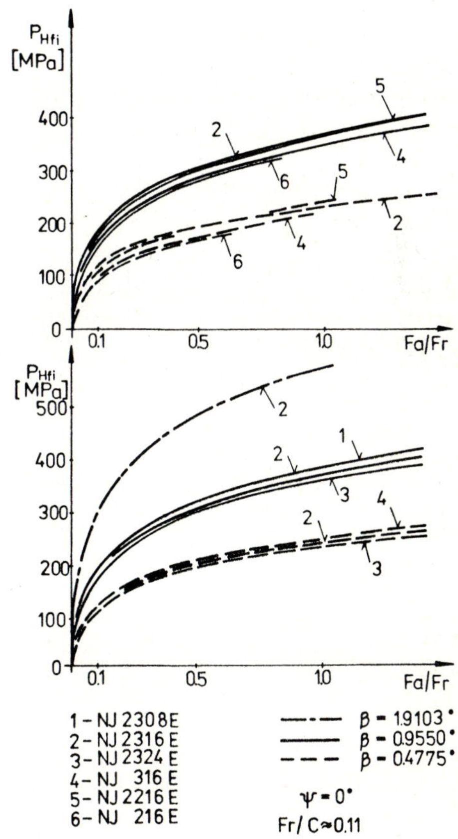

Fig. 12. Effect of the angle β on the value of the maximum pressure at the roller end - inner ring flange contact.

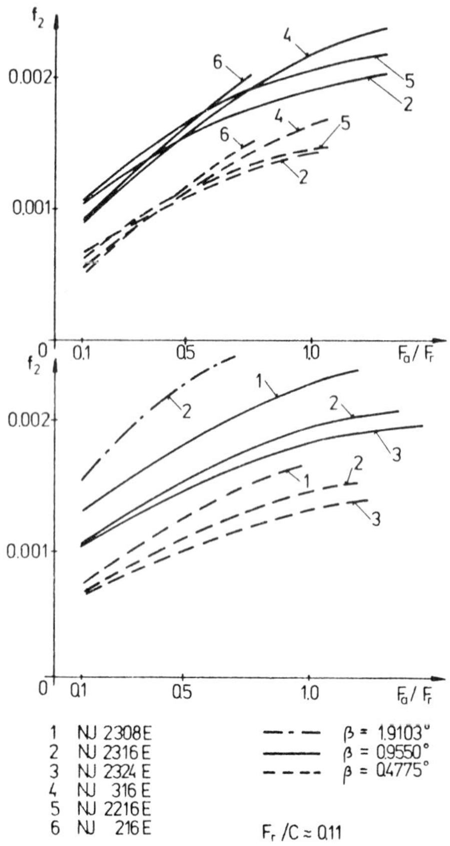

Fig. 13. Effect of the angle β on the value of coefficient of moment of friction f_2.

4 SELECTION OF THE INCLINATION ANGLE OF FLANGES AND THE ROLLER END RADIUS

Small angles of inclination of the flanges ensure the most advantageous operating conditions of the bearing. Small pressures at the roller end - flange contacts, small angles of skew of the rollers and consequently small frictional losses in the bearing correspond to them. The angle of inclination of the flanges, however, must be big enough to guarantee an appropriate shape of the oil gap at the roller end - flange contact. The factors that limit the smallest permissible value of the angle β, apart from the dimensions of the contact area, are also the tolerances of the angle β and the roller end radius. In the case of very small angles ($\beta < 0.25°$) and, in this connexion, of big radii of the roller end, even slight deviations of both these quantities from the nominal values may cause unfavourable changes in the oil gap shape. In particular, it may lead to the roller end - flange edge contact, which will render formation of the oil film difficult or impossible.

Basing on the above observations, the range of inclination angles have been established, which make it possible to achieve a compromise between an attempt to minimizing frictional losses in the bearing and the necessity to guarantee a favourable shape of the oil gap conducive to formation of an oil film at the roller end - flange contact. The values of angles proposed range within $\beta = 0.4° \div 0.6°$, depending on the size and the dimension series of the bearing.

The value of the roller end radius is related with the value of the angle of inclination of the flanges by the dependence:

$$R_w = (D_w/2 - e_{in})/\sin\beta. \tag{32}$$

The roller end radius should be so selected as to make the distance of the roller end - inner ring flange contact point from the main race fulfil the equality:

$$e_{in} = (1/3 \div 1/2)\cdot h_{ie} + h_i - h_{ie}, \tag{33}$$

h_{ie} is the effective height of the inner ring flange (Fig. 14).

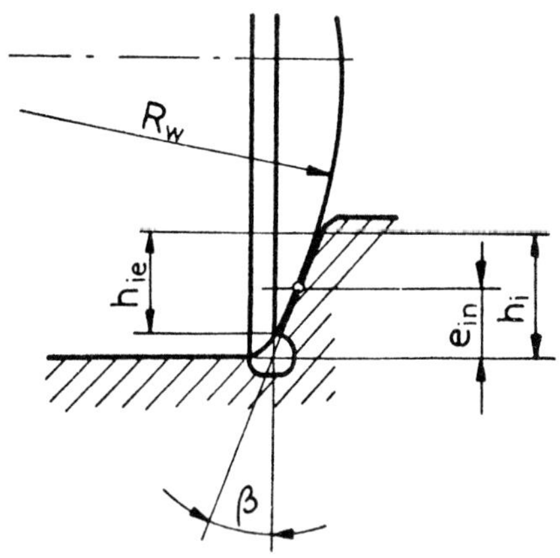

Fig. 14. Selection of the roller end radius R_w.

5. CONCLUSIONS

1. The analysis made allows one to establish the range of the angles of inclination of the flanges ensuring a favourable shape of the oil gap conducive, with appropriate selection of lubricant, to formation of an oil film at the roller end - flange contacts. The values of these angles range within $\beta = 0.4° \div 0.6°$.
2. Small frictional losses in the bearing correspond to the values of the angles β proposed.
3. In view of very small values of the skew angles of the rollers (for the given range of the angles β, $\varphi_1 < 1'$) the phenomena accompanying the skew have a negligible effect upon the bearing operation.

References

(1) PURDY, G. T. 'Rolling bearings,' U.S. Patent 3, 268, 278, 1966.

(2) KORRENN, H. 'The axial load-carrying capacity of radial cylindrical roller bearings', Trans. ASME, Series F, Vol. 92, (1970), 129-137.

(3) HARRIS, T. A. 'The endurance of a thrust loaded double row radial cylindrical roller bearing', Wear, Vol. 18, (1971), 429-438.

(4) MORRISON, F.R., PIRVICS, J. and CRECELIUS, W.J. 'A functional evaluation of a thrust carrying cylindrical roller bearing design', Trans. ASME, Series F, Vol. 101, (1979), 164-170.

(5) KISPERT, K. 'Das SU-Lager - ein axial hochbrlastbares Radial-Zylinderrollenlager', VDI-Z, Vol. 123, (1981), 31-34.

(6) KOUDELA, V. and MANHALTER, P, 'The cylindrical roller bearing for combined load', In: Loziska 83. Brno 1983, 137-147 (in Czech).

(7) WARDA, B. 'The effect of the roller end - flange contact shape of the radial cylindrical roller bearing upon ability to transfer axial load'(doctoral thesis), Łódź: Technical University of Łódź, 1987, (in Polish).

(8) PALMGREN, A. 'Grundlagen der Walzlagertechnik,' Stuttgart: Francklische Verlagshandlung 1964.

(9) SCHOUTEN, M.J. 'Einfluss Elastohydrodynamischer Schmierung auf Reibung, Verschleiss und Lebensdauer vor Getrieben' (doctoral thesis), Eindhoven: Technische Hogeschool Eindhoven, 1973.

(10) TRACHMAN, E.G. and CHENG, H. S. 'Thermal and non-newtonian effects on traction in elastohydrodynamic contact', In: Proc. Elastohydrodynamic Symp. 1972, Institution of Mechanical Engineers, London, 142-148.

(11) KUZMIN, N. F. 'O koefficente trienija v tjażełonagrużennom kontakte', Vestnik Mashinostroenija, 5, (1954), 18-26.

(12) RACZYŃSKI, A. 'The effect of the working surface shape of the spherical roller bearing upon the collaboration condition of the rollers with the guide ring' (doctoral thesis), Łódź: Technical University of Łódź, 1980 (in Polish).

Paper X(iv)

The study of roller end and guiding shoulder construction of roller bearings

M. Li and S. Wen

This paper deals with the roller end and guiding shoulder construction of roller bearings on the basis of EHL theory and experimental analysis. An orthogonal test dealing with the main effect factors of friction torque and temperature increasing has been done in a wide scope. Optimized roller end and guiding shoulder construction which can improve the lubrication condition has been obtained, it is shown that the axial load carrying capacity of the new construction can be increased by 5-10 times, compared with the usual ones.

1 INTRODUCTION

Roller bearing has several advantages, say, the biggest radial load carrying capacity, the high limit rotating speed, and the simplicity of assemble and adjustment. But, its axial load carrying capacity is very small, which severely limits its usage. In general, in the case the load, rotating speed, and varying of temperature are high, additional thrust bearings have to be used together with it. In order that the roller bearings can be used alone in a great radial load combined with a certain axial load condition, some improvement must be made in the construction of this type of bearing.

2 THEORETICAL ANALYSIS

2.1 Possible good shapes of roller end and guiding shoulder

For roller bearing, the axial loads are carried by roller ends and guiding shoulders. The roller end and guiding shoulders of commercial products are flat in shape. Theoretically, EHL film can not exist on the surfaces of roller end and guiding shoulder in this construction. To obtain good lubrication conditions, several new designs have been made recently (1,2).

From lubrication point of view, for a given axial load, line contact state between roller ends and guiding shoulders is favourable, which can be understood by the minimum film thickness formula given by Hamrock and Dowson (3)

$$H_{min} = H_{minl}\left[1-1.6e^{-0.62(R_y/R_x)^{2/3}}\right] \quad (1)$$

where H_{min} is the dimensionless minimum EHL film thickness in an elliptical contact, H_{minl} is the dimensionless minimum film thickness in a line contact, R_y and R_x are the equivalent curvature radii of contact solids. It is shown from equation (1) that, when $R_y/R_x > 6$, H_{min} is near H_{minl}, but when $R_y/R_x = 1$, H_{min} is only 14 percent of H_{minl}. Thus the most favourable roller end and guiding shoulder construction should get a value of R_y/R_x as large as possible. Therefore, the possible best shape of roller end and guiding shoulder is cone (See Fig. 1). Another reasonable good construction is that the roller ends are spherical in shape but the guiding shoulders are conical in shape.

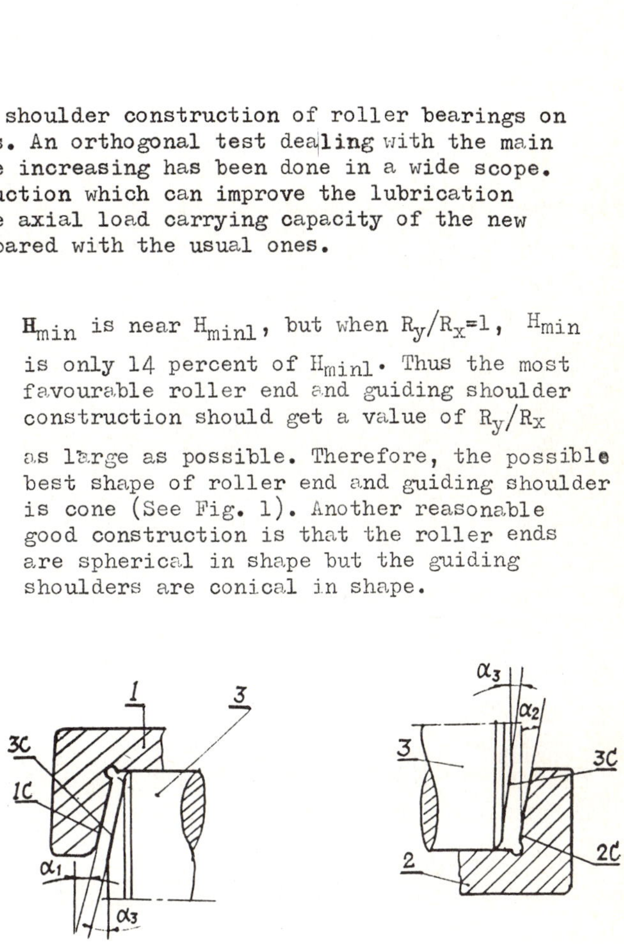

Fig. 1 Conical-to-conical construction

1. outer race, 2. inner race, 3. roller,

1c. conical surface of outer race,
2c. conical surface of inner race,
3c. conical surface of roller end.

2.2 Contact stability

The stability of roller for the constructions shown in Fig.1 should be noticed. For a flat end roller, the direction of M_a, the moment of supporting forces, is always contrary to that of M_f, the moment of friction force on the axial load carrying roller(See Fig.2), but, for conical end roller, sometimes the direction of M_a and M_f are identical, this will increase the skewness of the axial load carrying roller (See Fig.3). We call the former case the stable one and the latter case the unstable one.

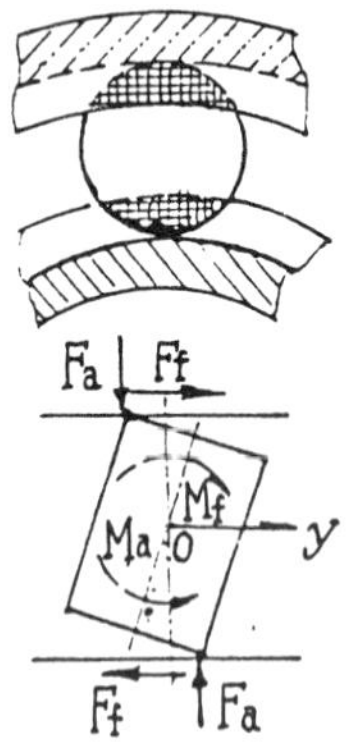

Fig.2 Stable case

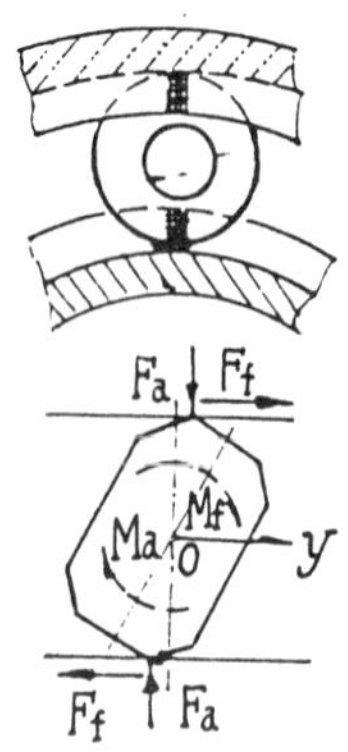

Fig.3 Unstable case

2.3 The location of contact point

The location of contact point on surfaces of guiding shoulders is important to lubrication state. The best location of contact point can be selected by the following principle: the contact state to line contact state is as near as possible, the sliding between surfaces is as small as possible, the entrainment velocity of surfaces is as great as possible. From kinetic analysis of roller (See Fig.4), the optimum location of contact point on the rotating inner race shoulder is point O, the optimum location on the fixed outer race shoulder is somewhere between point C_V and C3.

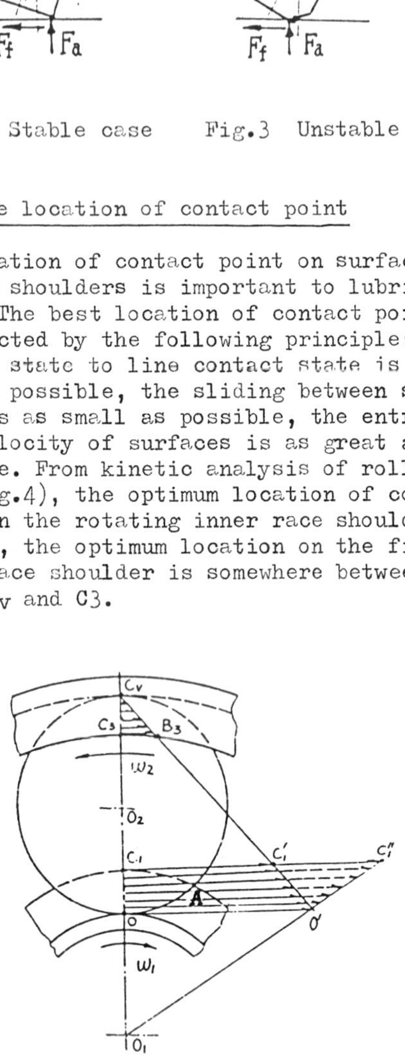

Fig.4 Kinetic analysis diagram of roller

2.4 Inclination of guiding shoulder

The inclination angle α' of guiding shoulders is the main factor affecting the axial load carrying capacity of roller bearings. To make clear the effect of α' on lubrication state, numerical analysis has been made to find EHL solutions by solving Reynolds equation and elastisity equation simultanneously, for α' = 15', 45', 120', and 900', respectively(4).

In numerical analysis, the roller end is assumed conical and spherical in shape. The paramete r h_{VH} which stands for the location of contact point (See Fig.5) is taken as 1.2 mm and 1.6 mm respectively.

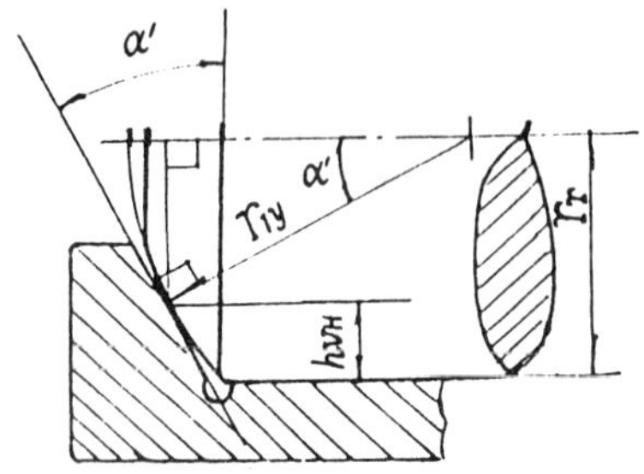

Fig.5 Location of contact point

From numericall solution, the varying of P_{max}, the maximum film pressure; h_{min}, the minimum film thickness in contacts, and μ, the friction coefficient; versus α', the inclination angle of guiding shoulders, are given in Fig.6.

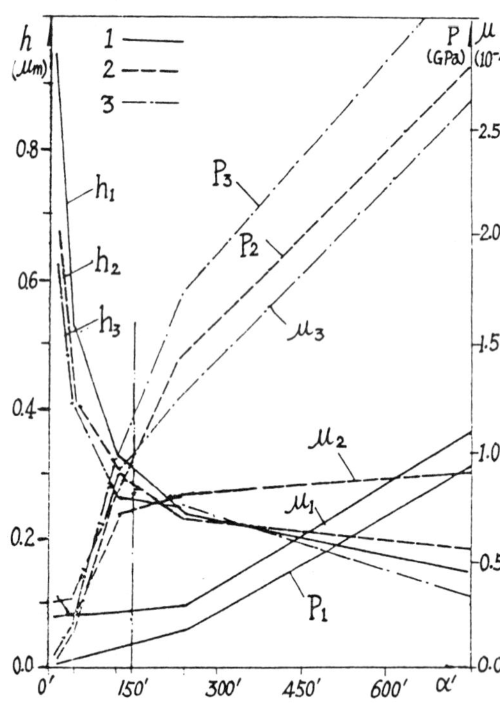

Fig.6 **Varying** of P_{max}, h_{min}, and μ versus α'

1. conical-to-conical, 2. conical-to-spherical, h_{VH}=1.2 mm, 3. conical-to-spherical, h_{VH}=1.6 mm.

From Fig.6, it is known that, when α' is increased, P_{max} and μ are increasing, but h_{min} is decreasing, thus the inclination angle α' should not be very great. It seems that the good range of α' is 15'-45'.

It is also known from Fig.6 that the conical-to-conical construction is much better than the conical-to-spherical construction, For spherical-to-conical construction alone, the lubrication state is better when h_{VH} =1.2 mm than when h_{VH} =1.6 mm.

3 EXPERIMENTAL ANALYSIS

The factors which affect the axial load carrying capacity of roller bearings are complicated. In order to obtain the optimum construction parameters and the practical axial load carrying capacity, experiments have been made in a wide scope.

3.1 Arrangement of effect factors

The factors which affect the load carrying capacity of roller bearings have been selected and arranged according to the orthogonal principle. These factors are:

1. Inclination angle of guiding shoulder, α'. For orthogonal test, four degrees are selected, namely, 1: α'=15', 2: α'=45', 3: α'= 120', and 4: α'=900'.

2. Shapes of roller end. About this factor, two degrees are selected, 1: conical end, this degree is represented by "<"; the inclination angle of roller end is identical and 2: spherical end, this degree is represented by "(" and the radius of spherical surface which is determined from the inclination angle α', and the location of contact point, h_{VH}, see Fig.5.

3. Location of contact point h_{VH}. Constrained by construction, this factor can be varying only in the scope 0.8-1.8 mm, thus only two degrees are selected about this factor, 1: h_{VH} =1.2 mm, 2: h_{VH} =1.6 mm.

Making use of orthogonal table L8(4×2^3), the arrangements of aformentioned factors for test are determined, which is given in Table 1.

TABLE 1 Arrangement of test factors

Sample	α'	Shape	h_{VH}
1	1: 15'	1: <	2: 1.6mm
2	3: 120'	2: (	2: 1.6mm
3	2: 45'	2: (	2: 1.6mm
4	4: 900'	1: <	2: 1.6mm
5	1: 15'	2: (	1: 1.2mm
6	3: 120'	1: <	1: 1.2mm
7	2: 45'	1: <	1: 1.2mm
8	4: 900'	2: (	1: 1.2mm

Eight sets of bearings have been produced in Luoyoung Bearing Research Institute corresponding to every factor arrangement in Table 1. Fig.7 shows the assemble diagram of bearing samples.

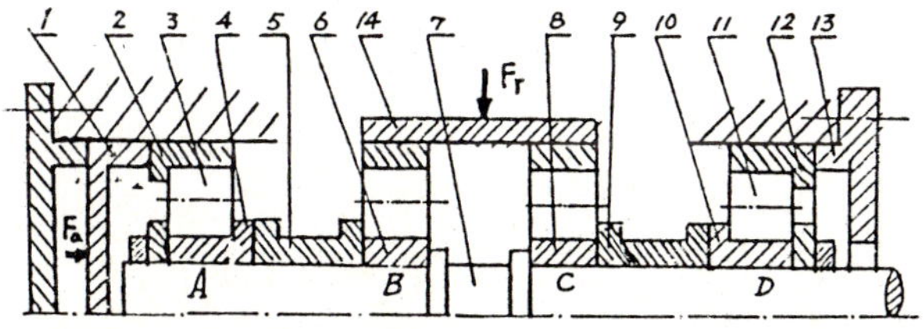

Fig.7 Assemble diagram of bearing sample

3.2 Apparatus and testing conditions

The rotating speed of the mainshaft of test machine can be changed continually in the range of 0-3000 rpm. The outer ring of tested bearing can move axially but can not move tangentially in the bore of testing machine, so that the axial load can be applied on the rollers. The contact surface temprature on outer race guiding shoulder is measured with thermo-couples, the body temperature of outer ring is measured with four mercury thermometers which contact directly with the outer ring. The friction torque is measured with a torque transducer. In order to ensure the comparibility of testing results, the testing conditions for any bearing sample are identical and the radial load applied in every test is F_r=6000 N, the lubricant is mineral oil, $\nu_{20°C}$=125 cSt, $\rho_{20°C}$=0.89 g/cm^3, $\nu_{50°C}$=30 cSt, $\rho_{50°C}$=0.88 g/cm^3.

4 EXPERIMENTAL RESULTS

The results are shown in Fig.8, where F_a is the axial load, M is the friction torque, T_f is the surface temperature on outer race guiding shoulder, T_b is the body temperature of outer ring.

It is known from Fig.8(a) to (f) that, all the friction torque M, the body temperature T_b and the contact surface temperature T_f of the bearing sample 7 are lower than those of others. Thus the best construction among the eight bearing samples is the construction of sample 7. The performance of sample 2 and sample 6 are also good. Experiments have been repeated by sample 7 and sample 2, which shows excellent repeatability.

The results show that the good range of inclination angle α' is 40'-130', its possible optimum value is 100'. But for conical-to-conical construction alone, the optimum value of α' is about 50', whereas for spherical-to-conical construction alone, the optimum value of α' is about 110'.

The surface topography of roller ends and guiding shoulders have been measured by a talysurf profilometer before and after experiment. The measurement traces show that for small values of α', the contact between roller ends and the fringe of guiding shoulder of outer race takes place when very slight skewness of rollers occurs, this contact makes indents on roller ends. This phenomenon is more evident for spherical-to-conical construction (sample 5), see (4).

For great value of α', the effective curvature radii at contact point are small, contact stress there is great, thus, wear is easy to take place there. Moreover, excessive α' will shorten the roller, so great skewness will occur while running (sample 8), see (4).

For α' =45' (sample 7 which is conical-to-conical construction) and for α' =120'(sample 6 which is spherical-to-conical construction), only slight change of surface topography can be observed after experiment, see (4).

Comparing experiment has also been done between sample 7 and commercial flat-to-flat construction bearing. It is shown that, under the same axial load condition, both the temperature increasing and the friction torque of sample 7 are much lower than those of commercial bearing. For the same temperature increasing, say, 50 °C–60 °C, the axial load carrying capacity of sample 7 can be increased by 5–10 times, compared with the commercial bearing.

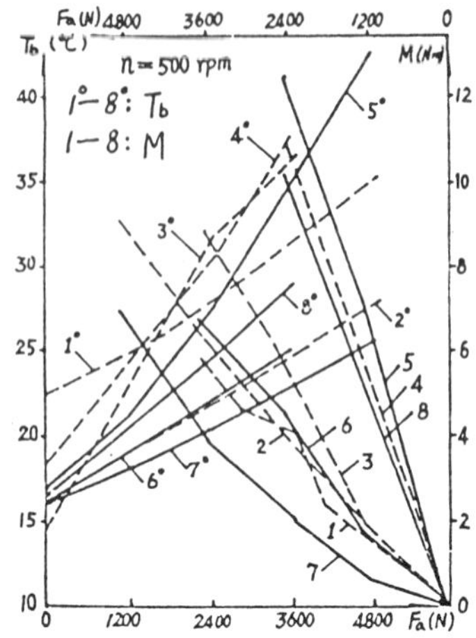

Fig.8 (a)

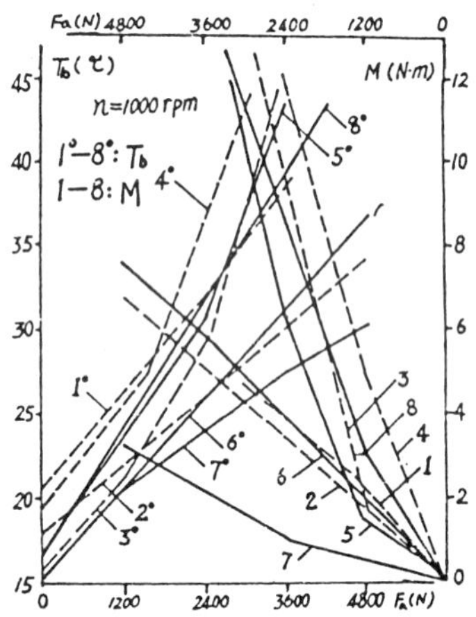

Fig.8 (b)

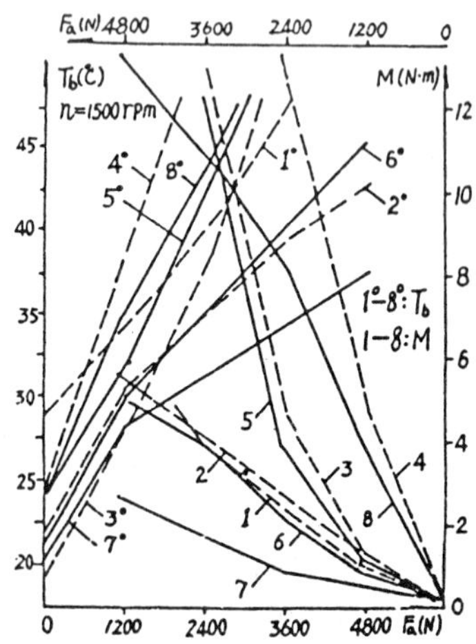

Fig.8 (c)

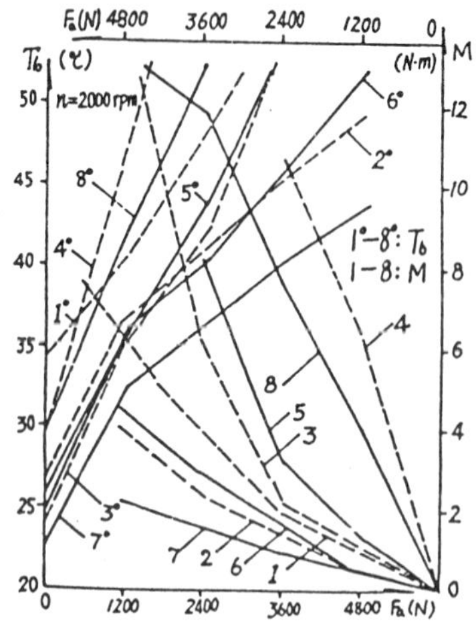

Fig.8 (d)

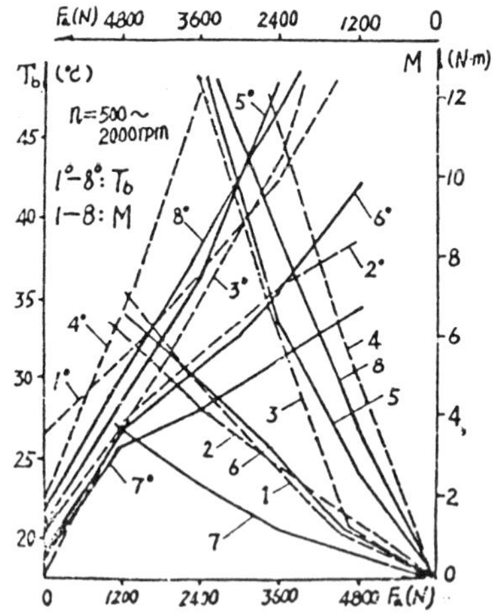

Fig.8 (e)

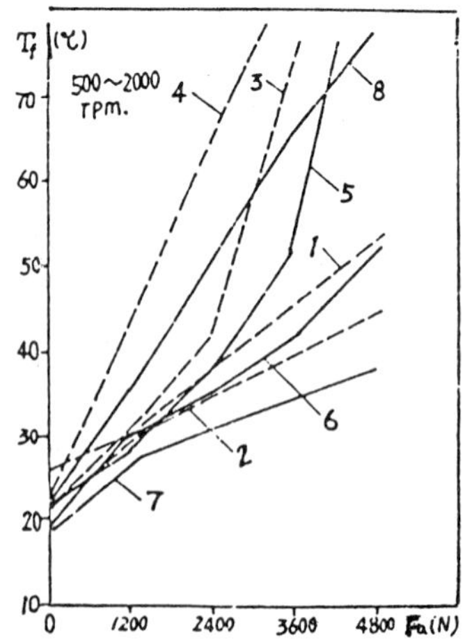

Fig.8 (f)

Fig.8 Experiment results

5 CONCLUSION

The shapes of roller end and guiding

shoulders are the main factors affecting the axial load carrying capacity of roller bearings. The new conical-to-conical construction proposed in this paper is better than that of spherical-to-conical or usual flat-to-flat constructions.

The inclination angle α' of guiding shoulders is the key factor affecting axial load carrying capacity of this type of bearings. To conical end rollers, the optimum α' is 45'-55' for outer race shoulder, and 40'-50' for inner race shoulder. To spherical end rollers, 110'-120' for outer race shoulder, 105'-115' for inner race shoulder.

The location of contact point on shoulder surface is also an important factor, it ought to locat at about one third the shoulder height from race, never locat at where above two third the shoulder height.

Using the best construction proposed in this paper,the axial load carrying capacity of roller bearings can be increased by 5-10 times, compared with usual bearings.

References

(1) Koudela Vojtech, Czechoslovakia Patent, 1983.3.15-8616-78. (12.20,1978) Int. Cl. F16C 19/26, 33/10, 33/34.

(2) Nakamura, United States Patent 4,318,574 May 9, 1982.

(3) Hamrock, B.J. 'Isothermal elastohydrodynamicl lubrication of point contact, part II--Ellipticity Parameter Results' Journal of Lubrication Technology, Trans. ASME, Vol.98, No.3, 1976, pp.375-383.

(4) Li Mingzhang, 'The analysis and experiment of EHL for the rib contact of roller bearings', Tsinghua University, Beijing, China, 1987.

SESSION XI
PLAIN BEARINGS (2)

Chairman: Dr G M Hamilton

Paper XI(i)

Dynamically loaded journal bearings: a modal approach to EHL design analysis

A. Kumar, J. F. Booker and P. K. Goenka

A new computational model, using the finite element method for both fluids and solids, is presented for transient elastohydrodynamic lubrication analysis.

The relative (differential) displacements of lubricated surfaces are represented by linear combinations of selected modeshapes, which form an acceptably adequate basis for both relative rigid body displacements and relative elastic deformations.

The model combines both elastostatic (solid) and hydrodynamic (fluid) aspects in a single coupled initial value problem of relatively low order.

A case study of an idealistic bearing structure is presented to explain fully the detailed workings of the new computational model, which is subsequently applied to a realistic automotive connecting rod to demonstrate its versatility and accuracy.

1 INTRODUCTION

The modern study of elastohydrodynamic lubrication (EHL) in dynamically loaded journal bearings was initiated by Blok [1975]. However, quantitative application of his concept of "flapping action" was precluded by the lack of computational power available at the time.

The finite element method (FEM) was first applied to the problem by Oh and Goenka [1985]; LaBouff and Booker [1985] proposed a similar model and provided a comprehensive review of relevant prior literature. (The motivation for the FEM lies in its flexibility in handling complex geometries and boundaryconditions and in its conceptual ease in coupling elastostatic and hydrodynamic analyses.) Such "complete" models require very intensive computation, though they have been applied to design optimization by Goenka and Oh [1986b].

Faster (though much less "complete") computational models have been proposed by Fantino, Godet and Frêne [1983], van der Tempel, Moes and Bosma [1985], and Goenka and Oh [1986a]. These models all incorporate serious simplifications in elastostatic and/or hydrodynamic representation.

No one existing computational model combines acceptable speed and reliability with adequate accuracy and generality; none provides suitable insight for design in the presence of EHL "flapping action".

We now proceed to outline a novel simulation procedure which is essentially a variable (higher) order generalization of the 2nd order (rigid body translation) procedures used by Goenka [1984] and by LaBouff and Booker [1985]. This model can be optimized for each application (depending on accuracy requirements and available computational resources) and can provide unique physical insight in the process.

1.1 Notation

scalars (integer)

n	nodal order	
m	modal order	
k	rigid body freedom order	

scalars (real)

x,y,z	film coordinates	
X,Y,Z	system coordinates	
t	time	

vectors (real)

$\mathbf{h} = \{h^z\}$	thickness	(nx1)
$\mathbf{c} = \{c^z\}$	clearance	(nx1)
$\mathbf{d} = \{d^z\}$	relative displacement	(nx1)
$\mathbf{v} = \{v^z\}$	relative velocity	(nx1)
$\mathbf{p} = \{p\}$	pressure	(nx1)
$\mathbf{b} = \{b^z\}$	residual force	(nx1)
$\mathbf{r} = \{r^z\}$	internal force	(nx1)
$\mathbf{f} = \{f\}$	external force/moment	(kx1)

matrices (real)

$\mathbf{A} = [A]$	area	(nxn)
$\mathbf{C} = [C]$	damping	(nxn)
$\mathbf{K} = [K]$	stiffness	(nxn)
$\mathbf{B} = [B]$	elastic equilibrium	(nxk)
$\mathbf{G} = [G]$	static equilibrium	(nxk)
$\mathbf{T} = [T]$	transformation	(nxm)
$\mathbf{I} = [I]$	unit	(*x*)
$\mathbf{0} = [0]$	null	(*x*)

2 PROBLEM FORMULATION (ANALYTICAL)

2.1 Physical relations

Figure 1 shows a journal bearing sleeve and lubricant film with their fixed rectilinear system coordinates X,Y,Z and curvilinear film coordinates x,y,z. Though the geometry here is cylindrical, the analysis below is equally applicable to others (e.g., planar or spherical).

The sleeve is restrained as shown; the journal is free to move under load.

The sleeve is assumed to be elastic and may be either regular or irregular; the journal is assumed rigid and regular (though much of the analysis would apply if it were neither). Both journal and sleeve are assumed to be massless.

Consider a finite element meshing of the lubricant film with n nodes. As suggested schematically by Figure 2, nodal displacements and nodal forces are normal to journal/sleeve surfaces; tangential displacements are ignored, and tangential forces are neglected.

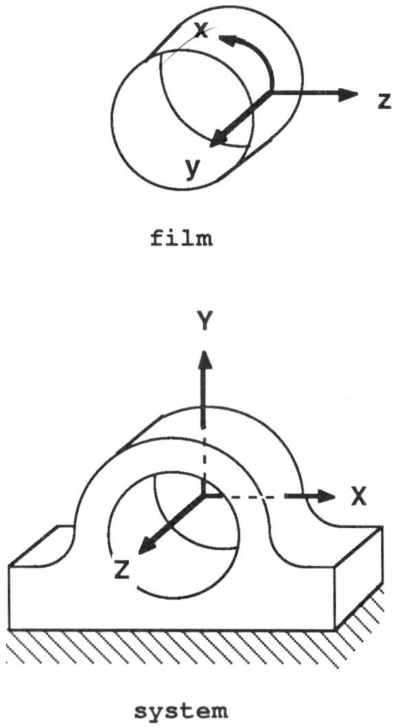

Fig. 1 Coordinates

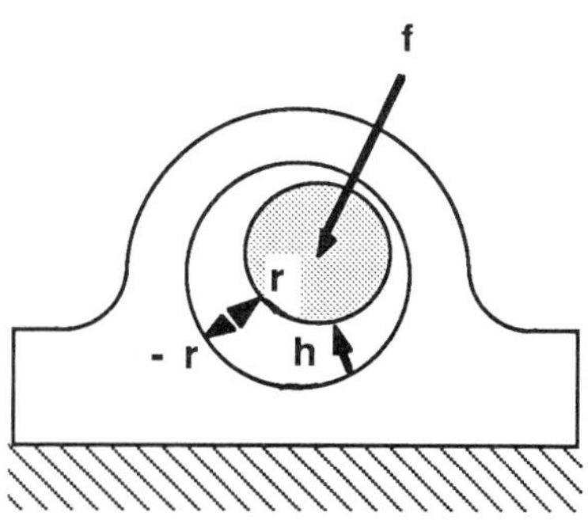

Fig. 2 Forces and Displacements

Nodal film thickness, clearance, and relative displacement are measured from sleeve to journal and must satisfy the n kinematic relations

$$\mathbf{h} = \mathbf{c} + \mathbf{d}$$

while nodal pressures and nodal forces acting on the journal (and oppositely on the sleeve) must necessarily satisfy the n hydrostatic (equivalence) relations

$$\mathbf{A}\,\mathbf{p} = \mathbf{r}$$

where the area matrix depends only on geometry. It is symmetric, nonsingular and negative definite (as here defined).

Conventional 2-D finite element analysis applied to the lubricant film (with clearance, boundary conditions, and cavitation configuration specified) gives the n hydrodynamic relations

$$\mathbf{C}(\mathbf{d},t)\ \mathbf{v} + \mathbf{b}(\mathbf{d},t) = -\mathbf{r}$$

with relative velocity

$$\mathbf{v} = d\mathbf{d}/dt$$

where the damping matrix and residual force vector depend on fluid film geometry and viscosity. Their computation is (in principle) a straightforward application of thin film fluid mechanics (as reviewed by Booker [1987]).

Conventional 3-D finite element analysis of the journal/sleeve structure (for fixed restraint boundary conditions) under k≤4 external forces and moments acting on the k≤4 journal freedoms along and about the system X,Y axes gives the n elastostatic relations

$$\mathbf{K}\,\mathbf{d} + \mathbf{B}\,\mathbf{f} = \mathbf{r}$$

where the (relative) stiffness and elastic equilibrium matrices depend on both elasticity and geometry. Their computation, which is outside the scope of this paper, is a straightforward application of structural mechanics. The stiffness matrix is always symmetric, singular, and positive semi-definite (since the structure is conservative but permits rigid body relative motion).

Incorporated in the elastostatic relations are the k equilibrium relations

$$\mathbf{G}^T\,\mathbf{r} = \mathbf{f}$$

where the static equilibrium matrix depends only on geometry.

Comparison of the elastostatic and equilibrium relations shows that

$$\mathbf{G}^T\,\mathbf{K} = \mathbf{0}$$

$$\mathbf{G}^T\,\mathbf{B} = \mathbf{I}$$

which demonstrate the interdependence of the 3 system matrices and the singularity of the symmetric stiffness matrix.

2.2 Modal relations

The central and essential approximation of the present procedure is replacement of n physical (nodal) displacements by m<<n modal (generalized) displacements through the n transformation relations

$$\underline{d} = \underline{T}\,\underline{d}'$$

where the m≥k columns of the indicated modal transformation (constraint) matrix are chosen quite arbitrarily to represent both rigid body motion and the essential features of elastic displacements.

Note that orthogonality of the transformation $\underline{T}$ would allow quite simple determination of m modal displacements $\underline{d}'$ which correspond (in a least-squares "best fit" sense) to n physical displacements $\underline{d}$ through use of the inverse transformation

$$\underline{d}' = (\underline{T}^T\,\underline{T})^{-1}\,\underline{T}^T\,\underline{d}$$

which is not otherwise required.

Applying the direct transformation gives m (modal) elastostatic relations

$$\underline{K}'\,\underline{d}' + \underline{B}'\,\underline{f} = \underline{r}'$$

where

$$\underline{K}' \equiv \underline{T}^T\,\underline{K}\,\underline{T}$$

$$\underline{B}' \equiv \underline{T}^T\,\underline{B}$$

$$\underline{r}' \equiv \underline{T}^T\,\underline{r}$$

and m (modal) hydrodynamic relations

$$\underline{C}'(\underline{d}',t)\,\underline{v}' + \underline{b}'(\underline{d}',t) = -\underline{r}'$$

where

$$\underline{C}' \equiv \underline{T}^T\,\underline{C}\,\underline{T}$$

$$\underline{b}' \equiv \underline{T}^T\,\underline{b}$$

(In practice, matrix $\underline{C}'$ and vector $\underline{b}'$ can be determined for fixed $\underline{d}'$ and t through correlation of output $\underline{r}'$ with input $\underline{v}'$ for a series of m+1 trials. The modal damping matrix thus found is symmetric, nonsingular, and negative definite.)

Combining the transformed (modal) elastostatic and hydrodynamic relations gives the set of m (modal) coupled equations of motion

$$\underline{C}'(\underline{d}',t)\,\underline{v}' + \underline{K}'\,\underline{d}' + \underline{b}'(\underline{d}',t) = -\underline{B}'\underline{f}(t)$$

Specification of m appropriate initial (typically arbitrary) modal displacement (state variable) values completes the problem formulation. (Inverse transformation from n physical to m modal initial conditions could utilize orthogonality as discussed above; in practice the issue seldom arises.)

2.2.1 Infinite stiffness

Though the underlying physical model requires finite stiffness, the modal model can be made to accommodate infinite stiffness through application of the transformation matrix

$$\underline{T} = \underline{G}$$

so that the modal matrices

$$\underline{K}' \equiv \underline{T}^T\,\underline{K}\,\underline{T} = \underline{0}$$

$$\underline{B}' \equiv \underline{T}^T\,\underline{B} = \underline{I}$$

are based entirely on system geometry (and not elasticity).

The result is the set of m=k coupled equations of (rigid body) motion

$$\underline{C}'(\underline{d}',t)\,\underline{v}' + \underline{b}'(\underline{d}',t) = -\underline{f}(t)$$

based only on geometry and hydrodynamics.

2.2.2 Transformation matrix choices

Wide latitude exists in the selection of transformation matrix columns (modes).

It is convenient (but not necessary) to scale these modes so as to normalize the modal (generalized) displacements, with unity corresponding to local contact under a uniform clearance distribution. For example, following this convention, the modal displacements corresponding to rigid body translations become the ordinary eccentricity ratio components.

The convenience of orthogonality in the inverse transformation of initial conditions has been noted above, though it is not at all clear that orthogonality is really necessary (or even desirable) in every case.

Problem-independent orthogonal modes can be chosen intuitively to represent any physically allowable rigid body or elastic displacements, such as either circumferential (trigonometric) waviness or purely axial (polynomial) variation.

Problem-dependent orthogonal modes can also be chosen quite systematically from among the n orthogonal eigenvectors of the physical stiffness matrix; the corresponding modal stiffness matrix will be diagonal.

The applications below illustrate the latter (problem-dependent) approach.

2.3 Constraints

Discussion of constraints is included for the sake of analytical completeness; they have little (if any) real effect on computation except in extreme cases.

The transformation on which the present work is based can be thought of as imposed by nodal forces of constraint acting together with nodal fluid forces on the elastic structure. Because such physical forces of constraint do no work, no counterparts appear in the modal equations of motion.

Owing to boundary and/or cavitation conditions, some of the n nodal pressures are specified, thus forming a second type of constraint. It is thus not possible to specify independently initial conditions for n state variables.

It is convenient here to distinguish between n_1 "active" or "free" nodes (where pressure is to be determined) and n_2 "passive" or "fixed" nodes (where pressure is already specified). Through the hydrostatic equivalence and elastostatic equilibrium relations these same n_2 pressure specifications can be expressed as n_2 constraints on physical displacements and/or velocities.

When analysis is couched in terms of physical displacements, it is convenient (and possible) to specify independently initial conditions for only some n_1 "active" physical displacements (as demonstrated by Booker & Shu [1984]).

When analysis is couched in terms of modal displacements, it appears equally convenient (and possible) to specify independently the initial conditions for an arbitrarily-chosen set of n_1 modal displacements. (Thus the apparent problem of constraint can be ignored in practice so long as a low-order state description is employed with $m \ll n_1$.)

3 PROBLEM SOLUTION (NUMERICAL)

Numerical formulation and solution of the initial-value problem above is effected by a comprehensive program package centered on a dynamic simulation program, FEMEHL, as documented in more detail in reports by Booker [1987] and Kumar[1987]. Commercial program packages and libraries used (NASTRAN, CAEDS, IMSL, and LINPACK or the equivalents) are widely available.

Preprocessors generate appropriate 2- and 3-dimensional element meshes, perform static elastic analysis (using programs CAEDS and/or NASTRAN), any necessary 3-D/2-D condensation and absolute/relative displacement shift, (using the LINPACK program library) eigenvalue/vector extraction and optional combination (using the IMSL program library), transformation matrix selection and application where possible, etc. Choices of meshes and solid element types are problem-specific and are noted below as appropriate.

Dynamic simulation program FEMEHL is the heart of of the program package. Table 1 provides a procedural outline for a simple version using Runge-Kutta order M integration; other integration methods involve similar time/state tests. Program FEMEHL for this paper uses the "stiff system" subroutine DGEAR from the IMSL program library. Selection of (variable) time step is automatic.

Program FEMEHL also makes use of the IMSL and LINPACK program libraries for matrix decomposition/equation solution.

The hydrodynamic lubrication portion of FEMEHL uses 3-node triangular elements as described in detail by Booker and Huebner [1972]).

Computation of the viscous damping matrix and force vector is carried out numerically in modal coordinates through correlation of input velocity with output force for a series of m+1 numerical trials.

The cavitation algorithm of FEMEHL (implemented by the only iteration loop in the simulation procedure of Table 1) is equivalent to that of Oh and Goenka [1985] and/or LaBouff and Booker [1985] (accounting for a published sign error). In fact, the entire simulation procedure is a variable order m generalization of the order 2 (rigid body translation) procedures used by Goenka [1984] and by LaBouff and Booker [1985], to which it reduces for m=2.

Postprocessors perform appropriate data reduction and plotting functions as required.

```
set fixed boundary conditions
set initial time and state
until cycle finished
   for each time step
      set time step size
      for test 1 to M
         choose (test) time and state
         find time-dependent duty
         find time-dependent boundary conditions
         find film thickness
         assemble system
         choose initial cavitation partition
         set cavitation boundary conditions
         until cavitation closure complete
            merge boundary conditions
            modify system
            for trial 1 to 1+m
               choose (trial) state rate
               solve system for pressure
               find resultant force
            find state rate for equilibrium
            find pressure, flow and density rate
            check cavitation closure
            if closure incomplete
               change cavitation partition
         if test 1
            output instantaneous results
      find approximate average state rate
      update time and state
output cycle results (extrema, means, sums)
stop
```

Table 1 Dynamic simulation procedure (Runge-Kutta order M)

4 APPLICATIONS

A case study with an idealistic bearing is provided to explain the model, which is then applied to a realistic automotive connecting rod to demonstrate its versatility and accuracy.

4.1 Idealistic bearing

Figures 3 and 4 show an idealized journal bearing sleeve and an associated finite element meshing of a lubricant film (shown in its planar development). The film mesh contains n=48x11=528 nodes. (The circumferentially symmetric geometry and unrealistic material combination are deliberately chosen to demonstrate features of the computational model.) The sleeve is completely restrained over its entire outer diameter.

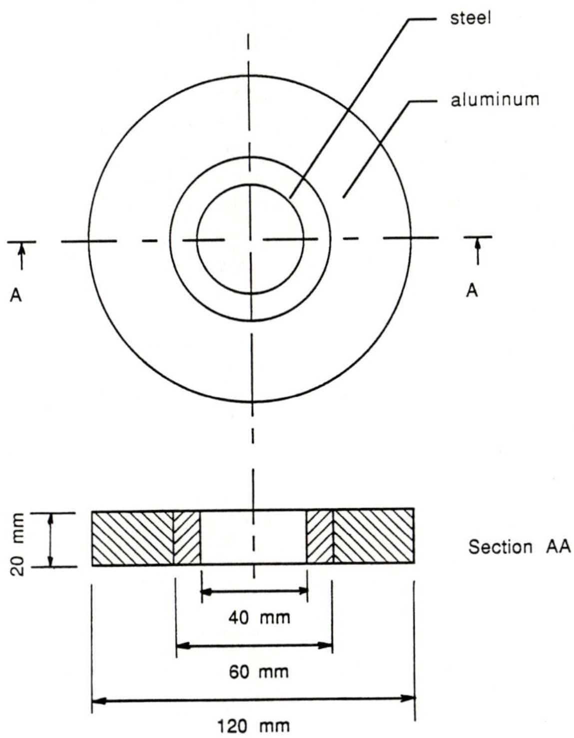

Fig. 3 Idealistic bearing - sleeve geometry

4.1.1 Hydrodynamic(fluid) parameters

Table 2 provides geometric and material properties for the lubricant film. This information is required for the damping matrix and residual force vector.

length	(mm)	20
radius	(mm)	20
clearance	(μm)	20
viscosity	(mPa·s)	7

Table 2 Idealistic bearing - fluid geometric/material properties

4.1.2 Elastostatic(solid) parameters

Figure 3 and Table 3 provide geometric and material properties for the sleeve; the journal is assumed perfectly rigid. This information completely defines the stiffness and elastic equilibrium matrices, which were found using 8-node solid elements and the general program package CAEDS.

material	steel	aluminum
Young's modulus (GPa)	206.8	71.0
Poisson's ratio (-)	0.290	0.334

Table 3 Idealistic bearing - solid material properties

4.1.3 Modeshapes

Table 4 and Figure 5 describe some members of an orthogonal set of n=528 eigenvectors of the stiffness matrix arranged in the order of increasing eigenvalue (and general complexity).

Owing to the symmetry of the stiffness matrix, such an orthogonal set of eigenvectors always exists.

Owing to the singularity of the stiffness matrix, eigenvectors describing rigid body relative displacement arise with null eigenvalues.

Owing to the circumferential symmetry of this idealized structure, eigenvectors describing relative elastic displacement with any circumferential variation arise in pairs with a repeated (single) eigenvalue.

Table 4 provides interpretive (and sometimes approximate) descriptions of circumferential and axial variation of the eigenvectors shown in Figure 5. As usual, "even" and "odd" functions are respectively symmetric and antisymmetric about their origins. Other common terms such as "rigid" and "elastic" have their usual meanings.

Modes 1 and 2, exactly represent relative rigid body <u>translation</u> along system axes X and Y, respectively.

Modes 3 and 4, exactly represent relative rigid body <u>rotation</u> about system axes Y and X, respectively.

Modes 5,6, 9,10, 13,14, and 19,20 approximate undulating elastic <u>expansion</u> with successively larger wave numbers.

Modes 11,12, 15,16, 17,18, and 21,22 approximate undulating elastic <u>torsion</u> with successively larger wave numbers.

Mode 7 approximates a perfectly <u>conical</u> elastic deformation.

Mode 8 approximates a uniformly <u>cylindrical</u> elastic expansion.

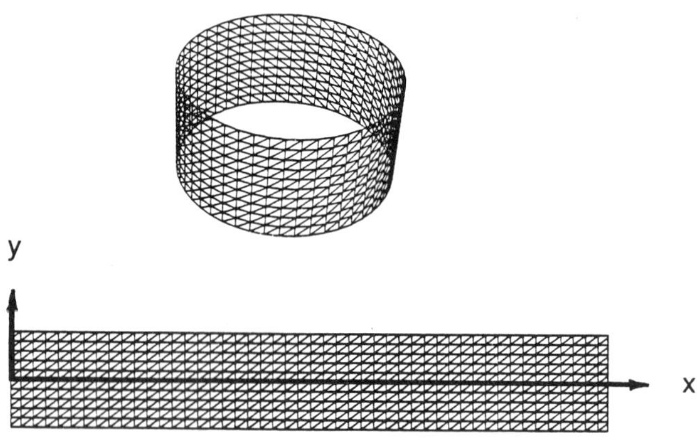

Fig. 4 Idealistic bearing - film mesh

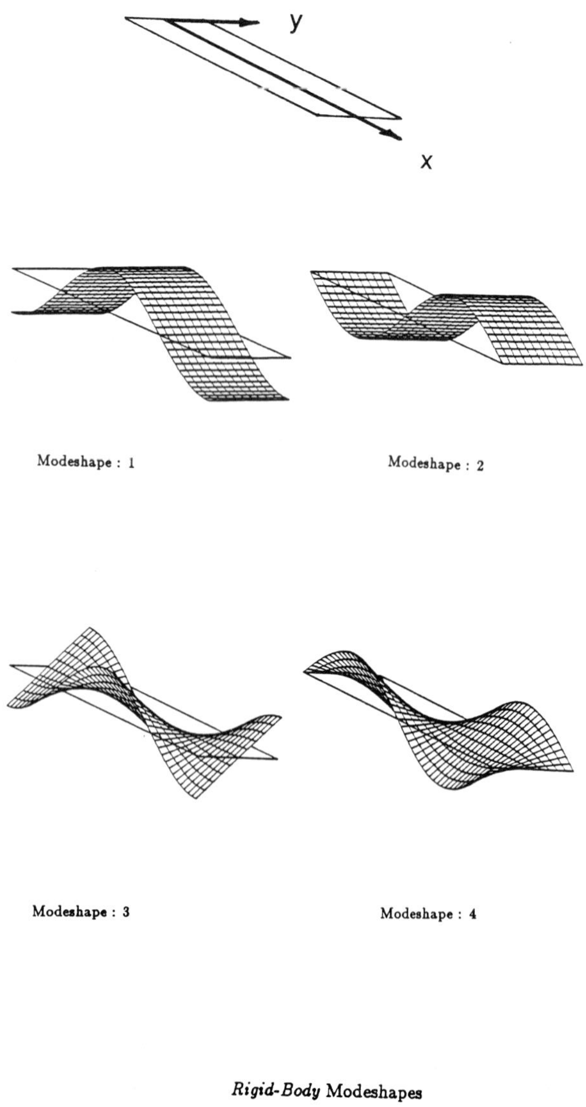

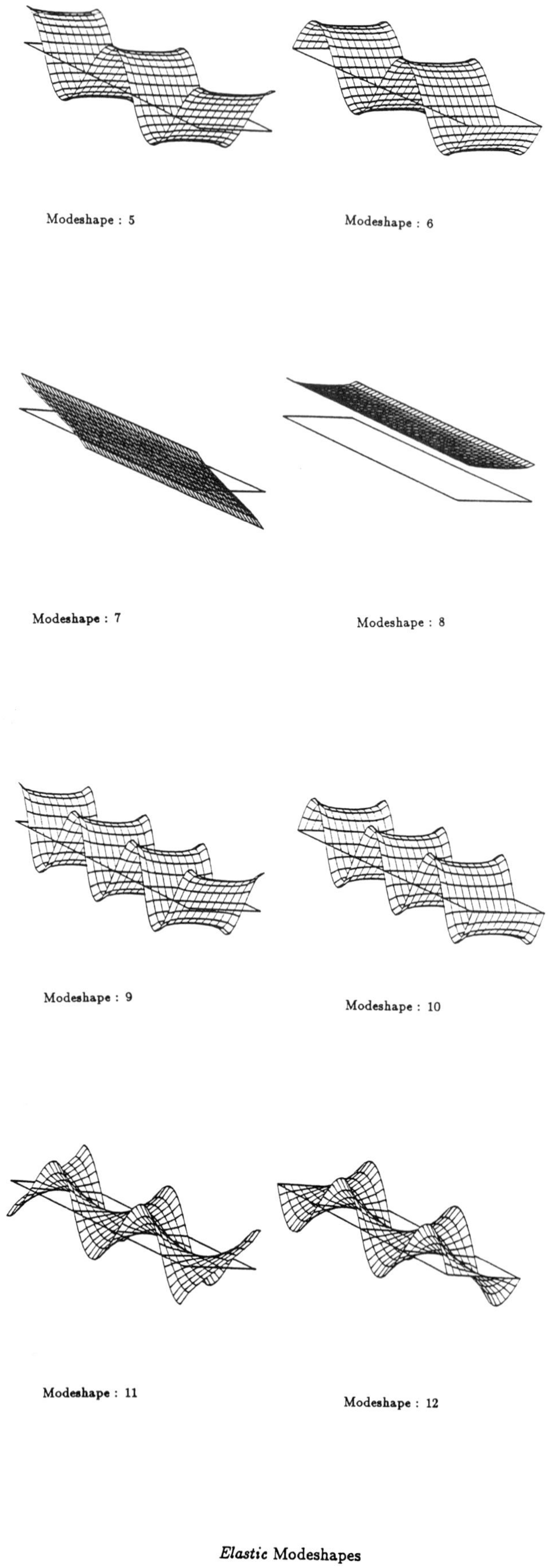

Fig. 5 Idealistic bearing - eigenvector mode set

Rigid & Elastic Modes			
Mode	Eigenvalue	Eigenvector x (wave)	y (order)
1	*0.*	*even (1)*	*even (0)*
2	*"*	*odd (")*	*" (")*
3	*"*	*even (")*	*odd (1)*
4	*"*	*odd (")*	*" (")*
5	23 467 200.	even (2)	even (~0)
6	"	odd (")	" (")
7	27 217 746.	even (0)	odd (~1)
8	32 206 422.	even (0)	even (~0)
9	38 056 573.	even (3)	even (~2)
10	"	odd (")	" (")
11	52 553 122.	even (2)	odd (~3)
12	"	odd (")	" (")
13	61 675 211.	even (4)	even (~2)
14	"	odd (")	" (")
15	65 746 385.	even (3)	odd (~3)
16	"	odd (")	" (")
17	82 267 879.	even (4)	odd (~3)
18	"	odd (")	" (")
19	87 584 647.	even (5)	even (~2)
20	"	odd (")	" (")
21	99 518 319.	even (5)	odd (~3)
22	"	odd (")	" (")
.	.	. .	. .
.	.	. .	. .
.	.	. .	. .
528	611 535 730.	even (0)	even(~10)

Table 4 Idealistic bearing eigenvector mode set

mode	m = 2	4	6	8	10
1	•	•	•	•	•
2	•	•	•	•	•
3		•	•	•	•
4		•	•	•	•
5			•	•	•
6			•	•	•
7					
8					
9				•	•
10				•	•
11					
12					
13					•
14					•

Table 5 Idealistic bearing - eigenvector mode subsets

4.1.4 Transformation (Modal) Matrix

Table 5 introduces 5 subsets of modes (with m=2,4,6,8,10) to demonstrate the consequences of such systematic choices. Though the selection of subset content is arbitrary,it is also quite deliberate. Because loading is axially-symmetric, axially-even modes are generally included (except for exclusion of expansion mode 8 for improved numerical stability), and axially-odd modes are generally excluded (except for inclusion of rigid-body rotation modes 3 and 4 as a check).

These 5 subsets of modes provide 5 transformation matrices for comparison in the 2 very simple studies below, demonstrating the most salient features of the transient solution scheme.

4.1.5 No load or rotation: initial conditions

The first simple study follows transient decay of one modal displacement under no external load and no journal rotation.

The choice of transformation matrix is arbitrarily taken as modal subset m=6 (4 rigid body and 2 elastic modes normalized as above).

In particular, mode 5 approximates an elliptically cylindrical deformation, with major diameter on the system X axis. Mode 6 is essentially identical though rotated 45 degrees about the Z axis.

Initially, modal displacement 5 is given a value of 0.9; all other modal displacements are initially null.

Figure 6 shows a portion of the ensuing response. The abscissa is time; a single ordinate suffices for all m=6 state variables (modal displacements).

A complete lack of modal coupling is apparent in Figure 6. In particular, the axially-odd rigid body modes 3 and 4 are not excited, despite the evident lack of axial symmetry in the mesh of Figure4.

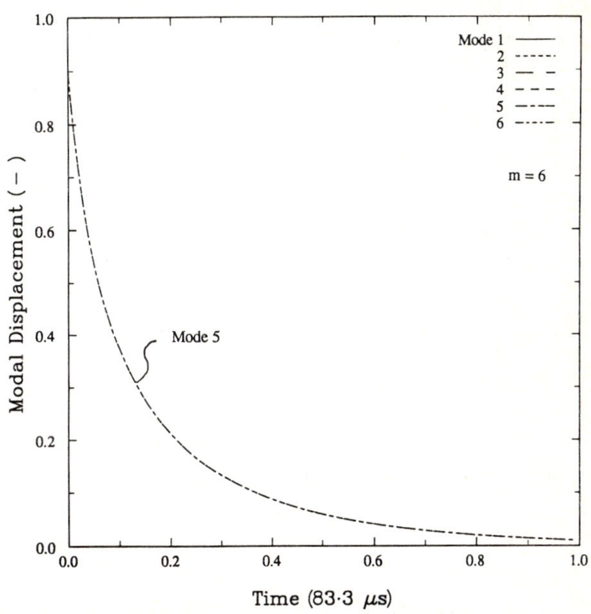

Fig. 6 Idealistic bearing - (no load or rotation) transient modal response

4.1.6 Steady load and rotation: final equilibrium conditions

The second study follows the transient approach to a final equilibrium state under the combined steady external load and journal rotation given in Table 6.

This time the transformation matrix is arbitrarily taken as modal subset m=10 (4 rigid body and 6 elastic modes normalized as above).

All m=10 modal displacements are initially null.

Figure 7 shows a portion of the ensuing response. The abscissa is time (or shaft angle); a single ordinate suffices for all m=10 state variables (modal displacements).

Figure 7 shows significant response in several of the m=10 modes present. (Axially-odd rigid body modes 3 and 4 are still not significantly excited, however, despite the axial asymmetry in the mesh of Figure 4.)

Figure 8 shows spatial distributions of film thickness and pressure computed at the "final" equilibrium state.

Figure 9 shows transient variation of film thickness and pressure extrema computed using modal subsets m=2,6,8,10. (Predictably, results for m=2 and m=4 are essentially identical.)

Figure 9 shows clearly the familial effects of increasing model complexity, as well as the (possibly) beneficial effects of elasticity on film thickness and pressure extrema.

Studies of idealistic applications such as this can provide rough guidance in trial selection of model complexity for particular realistic applications such as that below.

axis		X	Y	Z
load	(kN)	0	-10	0
rotation	(rev/s)	0	0	33.3

Table 6 Idealistic bearing - steady load and rotation

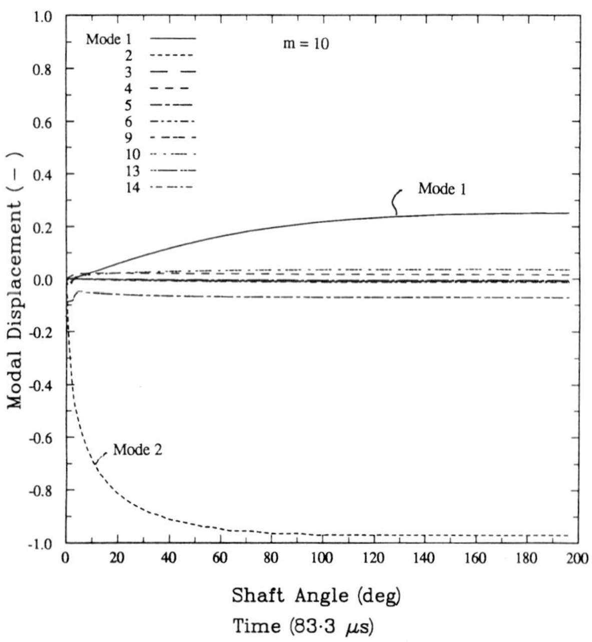

Fig. 7 Idealistic bearing - (steady load and rotation) transient modal response

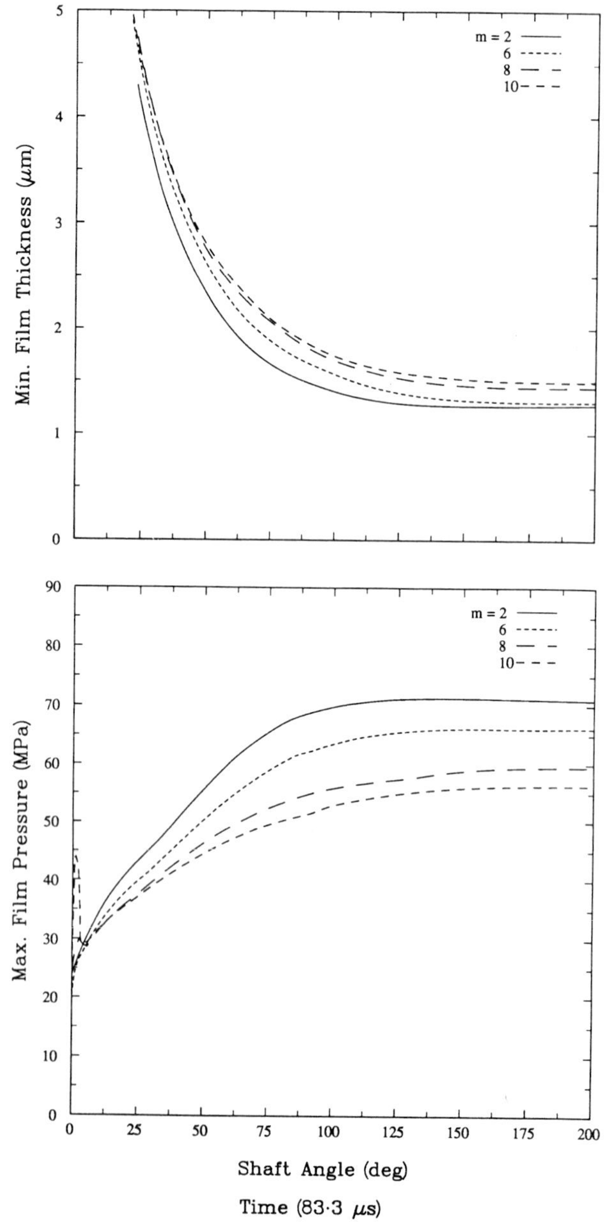

Fig. 9 Idealistic bearing - (steady load and rotation) film thickness and pressure transient extrema

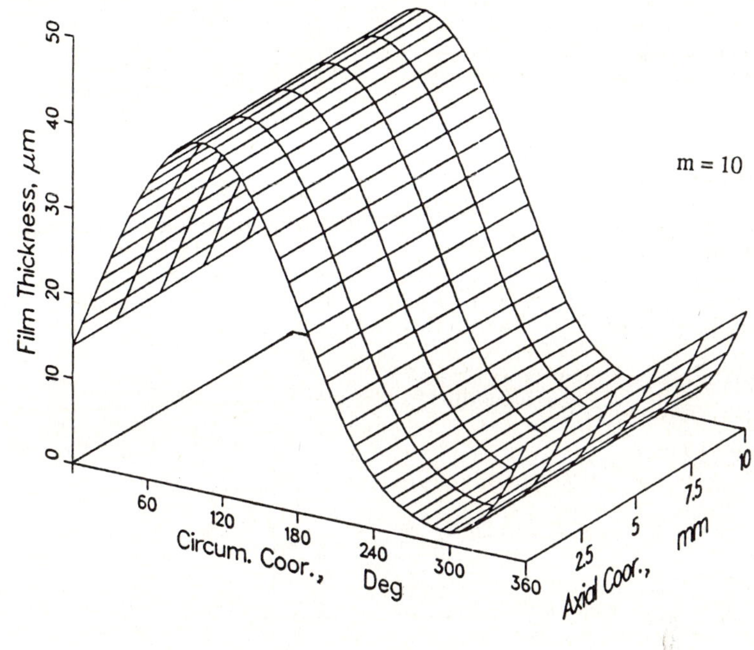

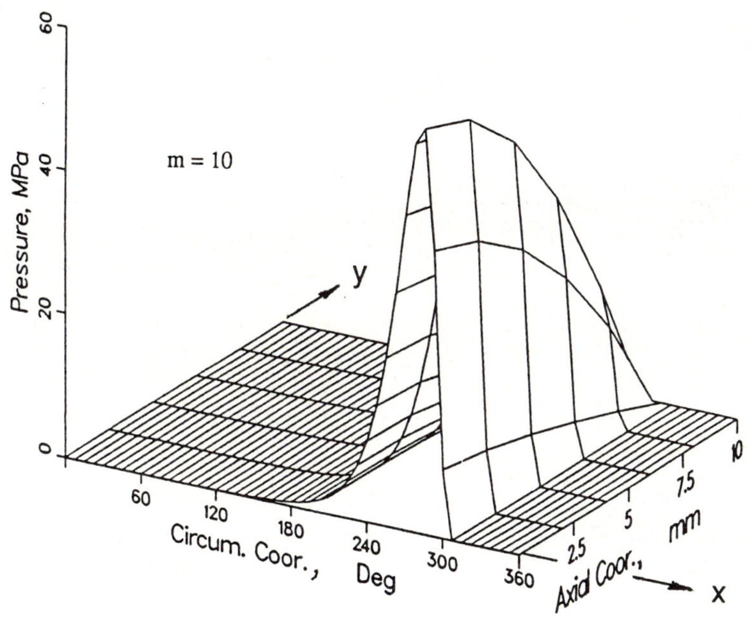

Fig. 8 Idealistic bearing - (steady load and rotation) film thickness and pressure equilibrium distributions

4.2 Realistic bearing

Figure 10 shows half of a connecting rod bearing, chosen to demonstrate both the versatility and accuracy of the present modal approach, and studied previously (as 'Design 1') by Goenka and Oh [1986b]. The connecting rod is fully restrained at the column cross-section indicated.

Owing to symmetry, only a half-film with n=6x48=288 nodes must be analyzed. Stiffness andelastic equilibrium matrices are found using 8-node solid elements with the program package NASTRAN. Orthogonal eigenvector modes are found as above using the program library IMSL.

Since the problem is axially-symmetric, only axially-symmetric modes need be included in the modal transformation matrix. For this example we have included only 2 rigid body modes and 4 simple elastic modes, for a total of m=6 modes.

Lubricant film properties and load/rotation duty cycle are as provided by Goenka and Oh [1986b].

Initial conditions are selected arbitrarily, and the numerical simulation is continued until solution periodicity is achieved (usually well within 2 duty cycles).

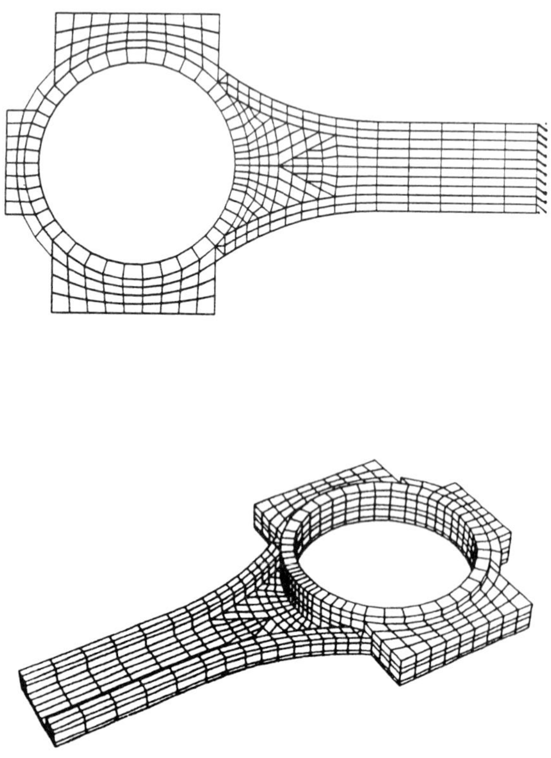

Fig. 10 Realistic bearing mesh (Goenka & Oh [1986b] 'Design 1')

Figure 11 compares predictions of instantaneous minimum film thickness and maximum film pressure for present and previous approaches for both rigid and elastic bearings. (Such a local comparison is a particularly rigorous test of such a global approximation.) For the rigid case, present and previous results are (predictably) identical; for the elastic case, results are qualitatively similar but quantitatively different.

The differences, which are most noticeable near the cyclical extrema, should be largely eliminated by inclusion of additional modes in the transformation matrix. Still, the present agreement is remarkable, considering the simplicity (m=6) of the modal subset used presently (and consequent gains in computing time over the more exact previous solution).

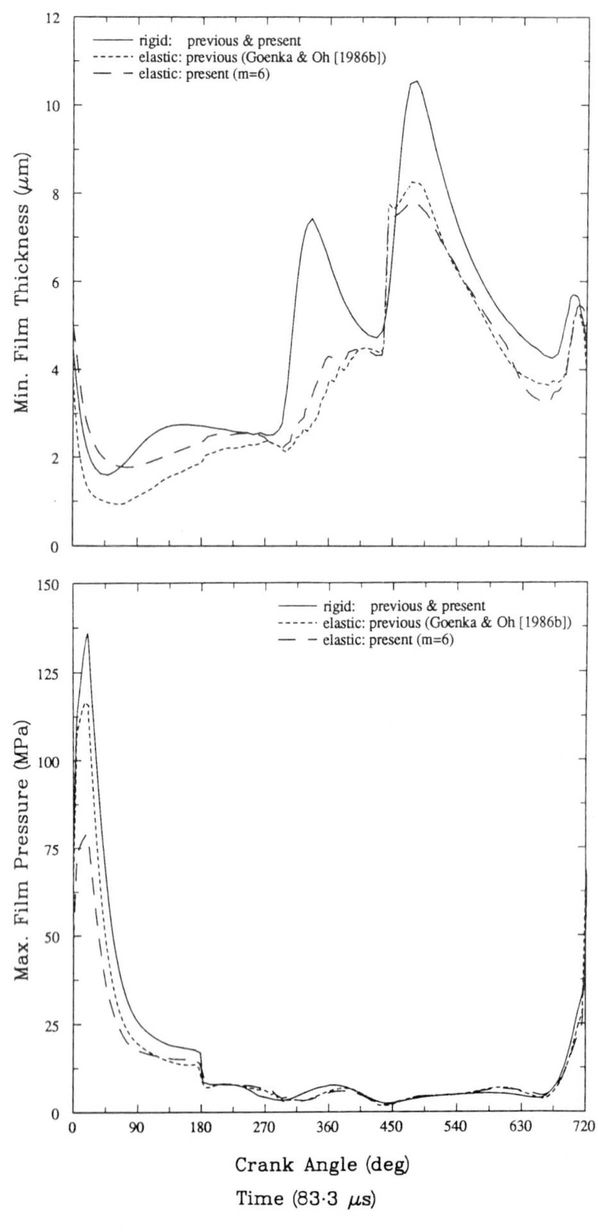

Fig. 11 Realistic bearing - film thickness and pressure transient extrema

5 SUMMARY AND CONCLUSIONS

We have outlined above in some detail a new computational model for transient elastohydrodynamic lubrication analysis of dynamically loaded journal bearings.

It is a variable (higher) order generalization of successful 2nd order (rigid body translation) procedures, introducing no additional nonlinearities (or associated additional iterations), and its modal order is very much less than the film discretization nodal order (so limiting displacement complexity); it is thus flexible, robust, and fast.

Its consistent use of relative (rather than absolute) displacements contributes to both numerical accuracy and ease of interpretation.

As described throughout this paper, the representation is essentially global, rather than local. It should thus be best for calculation of global phenomena (such as power loss and shaft dynamics), and worst for very local phenomena (such as thickness and pressure extrema), for which its accuracy should be adequate if an adequate modal subset is utilized.

The modal basis of the new model should provide unique physical insight for optimal bearing structural design in the presence of elastohydrodynamic "flapping action".

The new computational model can be optimized for particular applications (depending on accuracy requirements and available computational resources). Its application to a particular bearing will be compared in considerable detail with application of a "complete" model in a future publication by the authors.

6 ACKNOWLEDGEMENT

This research used the facilities of the Integrated Mechanical Analysis Project in the School of Mechanical Engineering at Cornell University.

REFERENCES

1972 BOOKER, J.F., and HUEBNER, K.H., 'Application of finite-element methods to lubrication: an engineering approach', Trans. ASME, J. Lubrication Technology, 1972, 94, 313-323. Errata, 1976, 98, 39.

1975 BLOK, H., 'Full journal bearings under dynamic duty: impulse method of solution and flapping action', Trans. ASME, J. Lubrication Technology, 1975, 97, 168-179. Errata, 1977, 99, 223.

1983 FANTINO,B.,FRENE,J., and GODET,M.M., 'Dynamic behavior of an elastic connecting rod bearing - theoretical study', Studies of Engine Bearings and Lubrication (SAE/SP-539), Society of Automotive Engineers, 1983, 23-32.

1984 BOOKER, J.F., and SHU, C.F., 'Finite element analysis of transient elasto-hydrodynamic lubrication', Developments in Numerical and Experimental Methods Applied to Tribology (Proceedings, 10th Leeds-Lyon Symposium on Tribology, Lyon, France, September 6-9,1983), Eds. D. Dowson, C.M. Taylor, M.M. Godet, and D.Berthe, Butterworths, London, England,1984, 157-163.

1984 GOENKA, P.K., 'Dynamicallyloaded journal bearings: finite element method analysis', Trans. ASME, J. Tribology, 1984, 106, 429-439.

1985 OH, K.P. and GOENKA, P.K., 'The elastohydrodynamic solution of a journal bearing under dynamic loading', Trans. ASME, J. Tribology, 1985, 107, 389-395.

1985 LABOUFF, G.A., and BOOKER, J.F., 'Dynamicallyloaded journal bearings: a finite element treatment for rigid and elastic surfaces', Trans. ASME, J. Tribology, 1985, 107, 505-515.

1985 VAN DER TEMPEL, L., MOES, H., and BOSMA, R. 'Starvation in dynamically loaded flexible short journal bearings',Trans. ASME, J. Tribology, 1985, 107, 516-521.

1986a GOENKA, P.K., and OH, K.P., 'An optimum short bearing theory for the elastohydrodynamic solution of journal bearings', Trans. ASME, J. Tribology, 1986, 108, 294-299.

1986b GOENKA, P.K., and OH, K.P., 'An optimum connecting rod design study - a lubrication viewpoint', Trans. ASME, J. Tribology, 1986, 108, 487-493.

1987 BOOKER, J.F., 'Basic equations for fluid films with variable properties', ASLE/ASME Tribology Conference,San Antonio,TX, October 5-8, 1987, Paper 87-Trib-3, Trans. ASME, J.Tribology. In press.

1987 BOOKER, J.F., 'Program FEMEHL: Theoretical Guide', Report No. MSD-87-14, revised November 1987, Sibley School of Mechanical and Aerospace Eng'g, Cornell University, Ithaca, NY.

1987 KUMAR, A., 'Program FEMEHL:Programmer's Guide', Report No. MSD-87-15, revised December 1987, Sibley School of Mechanical and Aerospace Eng'g, Cornell University, Ithaca, NY.

1988 KUMAR, A., 'Elastohydrodynamic Lubrication Analysis of Dynamically Loaded Journal Bearings: A Modal Approach', MS Thesis, January 1988, Cornell University, Ithaca, NY.

Paper XI(ii)

Shape defects and misalignment effects in connecting-rod bearings

P. Maspeyrot and J. Frene

The analysis of the behaviour of a big-end bearing for an internal combustion engine is presented. The minimum film thickness, the friction torque and the axial flow are obtained for a bearing with a conic or a parabolic axial profiles and for a misalignment of the journal. Moreover the effect of the radial clearance, in a perfect bearing under dynamic loading, on the minimum film thickness is shown. The technique used for bearing characteristic evaluation is an extension of the mobility method of Booker.

1 INTRODUCTION

Numerous studies deal with the transient analysis of dynamically loaded journal bearings. Many of the major assumptions used in the earliest studies were not realistic but they permitted to obtain a model compatible with the computing capabilities. The computing power increasing, the assumptions made were not so restrictive and a summary of the models obtained can be found in the paper of Campbell et al. |1|. The mobility method of Booker |2, 3| had been often applied and it had been a great advance in the analysis of dynamically loaded bearings. The journal motion is solved in two components, squeeze and whirl, which permit a very fast calculation of the journal centre orbit.

On the last fifteen years, works have been carried out on more realistic bearings. So, oil film history |4|, inertia |5| or roughness |6| effects, oil-feed holes or grooves |7|, shape defects of the bearing |8| were be taken into account. More and more the elastic deformations in the bearings are studied. These studies can be divided in two groups ; in the first, only local elastic deformations occuring at the bearing surfaces are considered |9|, and global deformations of the connecting-rod bearing characterize the second group |10, 11, 12|, |13|.

Misalignment in bearings had been often studied in the case of steady load and for particular case of misalignement |14, 15, 16|. Among them, Nicolas |17| studied both theorically and experimentally the general problem of misaligned bearing under steady load. Recently, Goenka |18| developped a finite element formulation for transient analysis of journal bearings, this method can be used for the study of misaligned bearing.

1.1 Notation

C Radial clearance (m)

D Bearing diameter (m)

F Applied load (N)

$H = h/C$ Non-dimensional film thickness

L Bearing length (m)

M Mobility factor

M_ε, M_ϕ Mobility components

$P = PDL/F$ Non-dimensional pressure

Q Axial flow (m^3/S)

R Bearing radius (m)

T_s Shaft friction torque (mN)

U_s, V_s, V_b Linear velocities (m/s)

d Projected length of journal centre-line on mid-plane bearing (m)

e Eccentricity (m)

e_o Mid-plane eccentricity (m)

h Fluid film thickness (m)

p Pressure (Pa)

t Time (s)

Z Axial coordinate (m)

$\bar{Z} = Z/L$ Non-dimensional axial coordinate

OXYZ Coordinate system

α Mobility angle (rd)

β Misalignment angle (rd)

$\delta = d/C$ Non-dimensional projected length

$\varepsilon = e/C$ Relative eccentricity

$\varepsilon_o = \varepsilon_o/C$ Mid-plane relative eccentricity

ϕ Attitude angle (rd)

μ Dynamic viscosity (Pa.s)

$\bar{\omega} = (\omega_s - \omega_b)/2 - \dot{\psi}$ Relative angular velocity (rd/s)

ω_b, ω_s Angular velocities (rd/s)

θ Angular coordinate (rd)

θ_2 Crankshaft angle (rd)

$(\ )_b$, $(\ )_s$ Related respectively to the bearing and to the shaft

2 THEORY

For the analysis presented here, the following assumptions have been made :

- Isothermal conditions
- Laminar flow
- Newtonian lubricant
- Rigid bearing surfaces.

2.1 Film geometry equation : Fig. 1-2

For any axial position, the film thickness may be calculated from :

$$h = C + e_o \cos\theta + f(\theta, Z) \qquad (1)$$

where $f(\theta, Z)$ is a function depending on the shape defect :

- Perfect bearing : $f(\theta, Z) = 0$
- Tapered bearing : $f(\theta, Z) = (R-r)\frac{Z}{L}$
- Bearing with a parabolic axial profile : $f(\theta, Z) = (R_1-R)\,(2\frac{Z}{L})^2$
- Misaligned bearing : $f(\theta, Z) = C\delta\frac{Z}{L}\cos(\theta-\beta)$

where R and r are the radius at $Z = \pm\frac{L}{2}$ and the reference point is assumed to be at $Z = 0$.

There will be a moment acting on the journal due to the axially non symmetric pressure profile, this moment is not considered.

2.2 Reynolds equation

In non-dimensional form, the Reynolds equation is given by :

$$\frac{\partial}{\partial\theta}\left[H^3\frac{\partial P}{\partial\theta}\right] + \left[\frac{R}{L}\right]^2\frac{\partial}{\partial Z}\left[H^3\frac{\partial P}{\partial Z}\right] = \frac{12\,\mu\,L\,D}{F\left[\frac{C}{R}\right]^2}\left\{(\bar{\omega}-\dot{\phi})\frac{\partial H}{\partial\theta} + \dot{\varepsilon}_o\cos\theta\right\} \qquad (2)$$

where $\bar{\omega} = \frac{\omega_s - \omega_b}{2} - \dot{\psi}$ and $H = \frac{h}{c}$

Following mobility method, ε_o and ϕ can be expressed as the two components M_ε and M_ϕ of the mobility vector $\vec{M}$.

$$\left.\begin{aligned} \varepsilon_o &= \frac{F(\frac{c}{R})^2}{\mu L D}M_\varepsilon \\ \varepsilon_o(\dot{\phi}-\bar{\omega}) &= \frac{F(\frac{c}{R})^2}{\mu L D}M_\phi \end{aligned}\right\} \qquad (3)$$

where the components of the mobility vector can be written as follows :

$$\left.\begin{aligned} M_\varepsilon &= M\cos\alpha \\ M_\phi &= -M\sin\alpha \end{aligned}\right\} \qquad (4)$$

In these relations, α is the angle between the mobility vector and the eccentricity direction. Introducing the non-dimensional pressure $P = P/M$ and using relations (3) and (4), equations (2) becomes :

$$\frac{\partial}{\partial\theta}\left[H^3\frac{\partial P}{\partial\theta}\right] + \left[\frac{R}{L}\right]^2\frac{\partial}{\partial Z}\left[H^3\frac{\partial P}{\partial Z}\right] = 12\left[\cos(\theta+\alpha) - \frac{\delta\bar{Z}}{\varepsilon_o}\sin\alpha\sin(\theta-\beta)\right] \qquad (5)$$

Equation (5) is the classical mobility form of the Reynolds equation when $\delta = 0$. So, in the cases of taper or parabolic axial profiles, the only effect in the Reynolds equation of these defects is on the film thickness expression. At mid-plane ($\bar{Z} = 0$), equation (5) is reduced also to the classical mobility form of the Reynolds equation.

Solution of equation (5) can be obtained by an iterative method (Fig. 3) : choosing a given value of the mobility angle α, the pressure distribution is calculated using finite differences with the Gauss-Siedel method and the Reynolds boundary conditions. The non-dimensional load is then calculated by integration which gives both a value for the attitude angle ϕ and a new value for the mobility angle α. Thus the new value of ϕ for the next step is obtained by linear interpolation. This procedure is pursued until convergence is achieved.

2.3 Friction torque

The friction torque T_s is calculated on the shaft for a full journal bearing without cavitation by the relation (6) :

$$T_s = \int_{-L/2}^{L/2}\int_0^{2\pi}\left\{R\,\frac{h}{2}\frac{\partial p}{\partial\theta} + \frac{\mu\,R^2}{h}(U_s - U_b)\right\}d\theta dz \qquad (6)$$

For a perfect bearing this equation gives an analytical result :

$$T_s = \frac{C\,\varepsilon\,F\sin\phi}{2} + \frac{\mu\,L\,R^3}{C}\frac{2\pi}{\sqrt{1-\varepsilon^2}}(\omega_s-\omega_b) \qquad (7)$$

In the other cases, h depends upon θ and Z and the friction torque is obtained numerically.

2.4 Axial flow

The axial flows through the ends of the bearing are :

$$\left.\begin{aligned} Q_1 &= -\int_0^{2\pi}\frac{h^3}{12\,\mu}\frac{\partial p}{\partial Z}R\,d\theta \quad (Z = -L/2) \\ Q_2 &= -\int_0^{2\pi}\frac{h^3}{12\mu}\frac{\partial p}{\partial Z}R\,d\theta \quad (Z = L/2) \end{aligned}\right\} \qquad (8)$$

The total axial flow is : $Q = |Q_1| + |Q_2|$.

3 RESULTS AND DISCUSSION

Numerical results are obtained for the big-end connecting-rod bearing, without circumferential groove, of the Ruston-Hornsby 6 Veb-X M_K III diesel engine whose characteristics are given table 1, and the load diagram is given in fig. 4. For a perfect bearing a value of 8,78 μm has been predicted for the minimum film thickness by Goenka |17|. Our computations give a value, with the same conditions, equal to 8,5 μm.

3.1 Radial clearance

The effect of radial clearance on the minimum film thickness for a perfect bearing under dynamic loading conditions is shown fig. 5. The

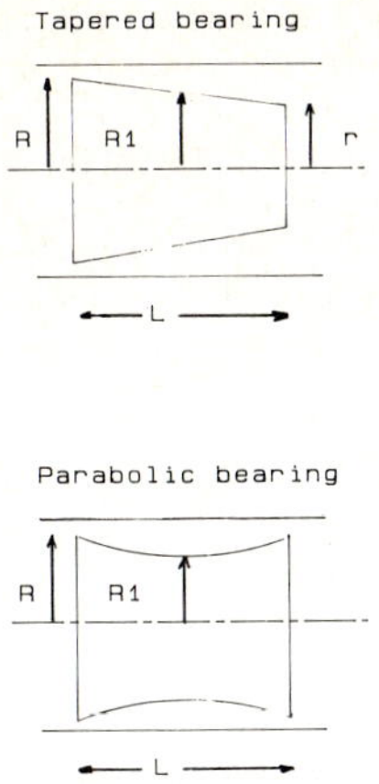

Fig 1 : Representation of a non-perfect bearing

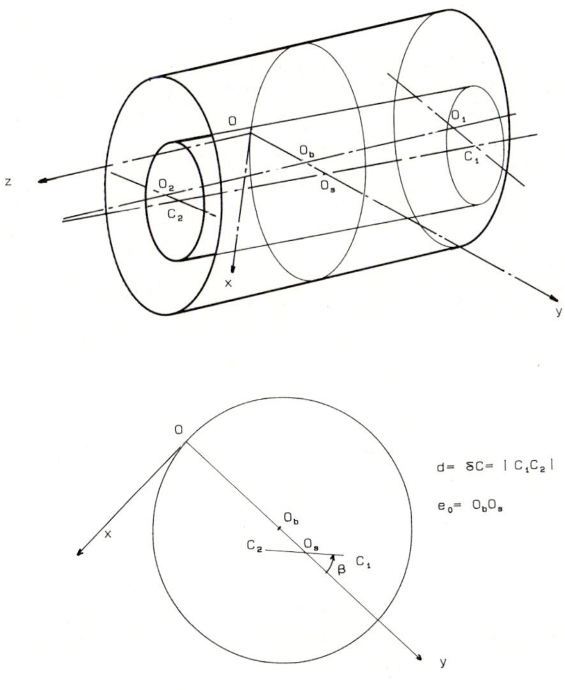

Fig 2 : Representation of the misaligned bearing

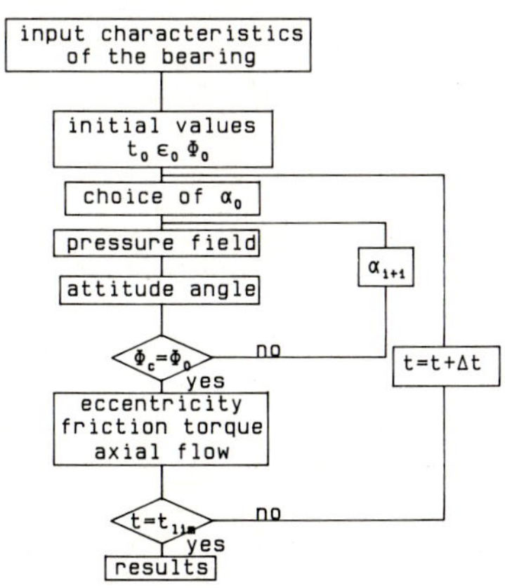

Fig 3 : Flow chart of solution technique

Bearing length	: L = 0.127 m
Crankshaft arm length	: L2= 0.184 m
Connecting-rod length	: L3= 0.782 m
Bearing diameter	: D = 0.203 m
Radial clearance	: C = 82.55E-6 m
Crankshaft rotational speed	: N =600 rpm
Oil viscosity	: μ = 0.015 Pa.s

Table 1: Characteristics of the bearing

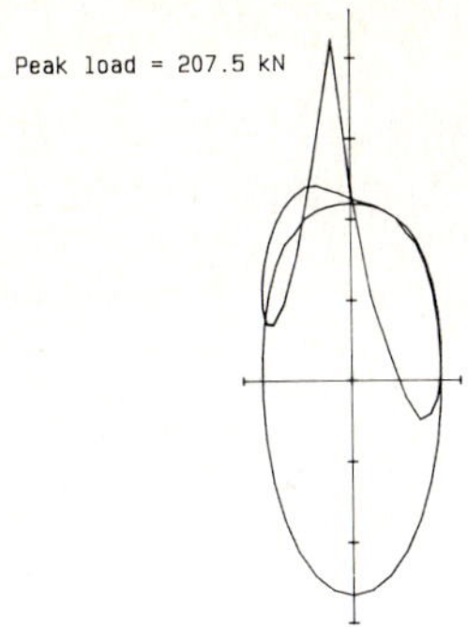

Fig 4 : Load diagram

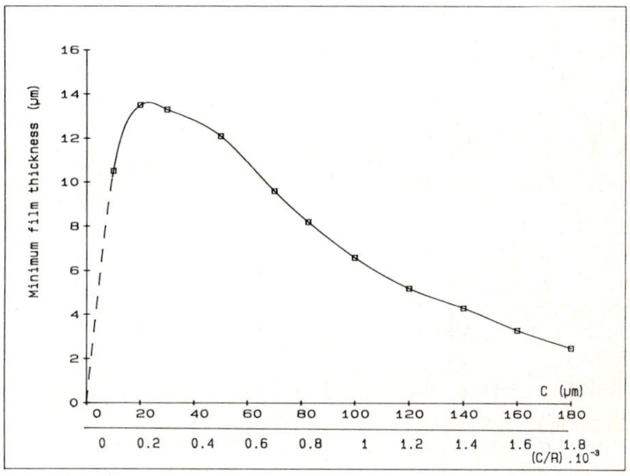

Fig 5 : Effect of radial clearance on the minimum film thickness

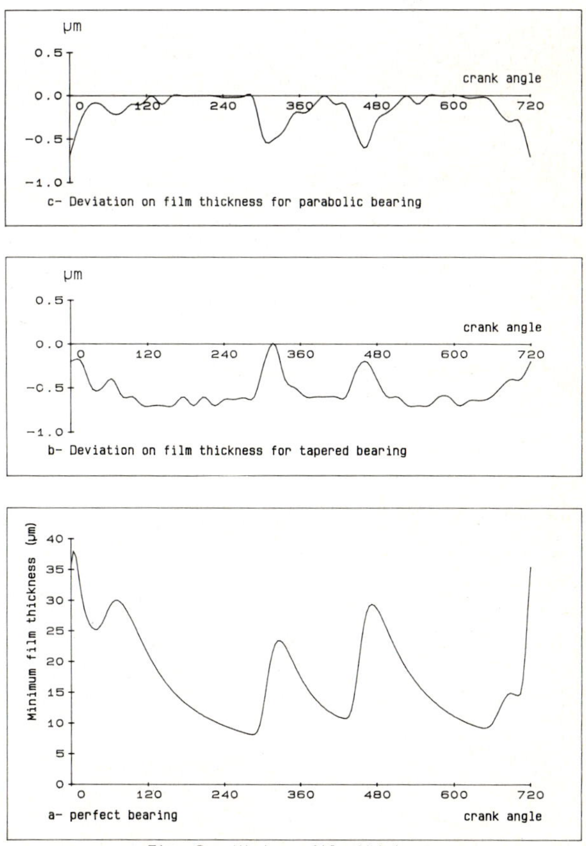

Fig 6 : Minimum film thickness

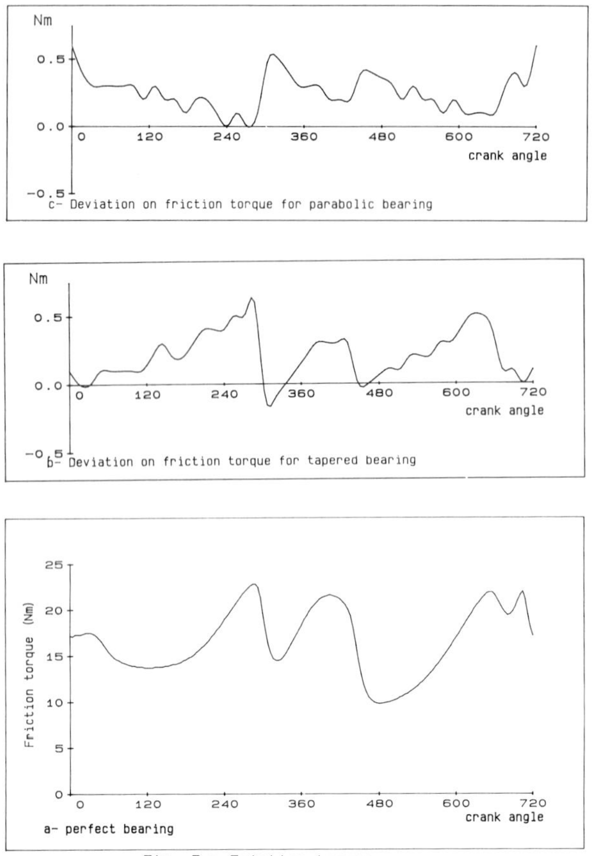

Fig 7 : Friction torque

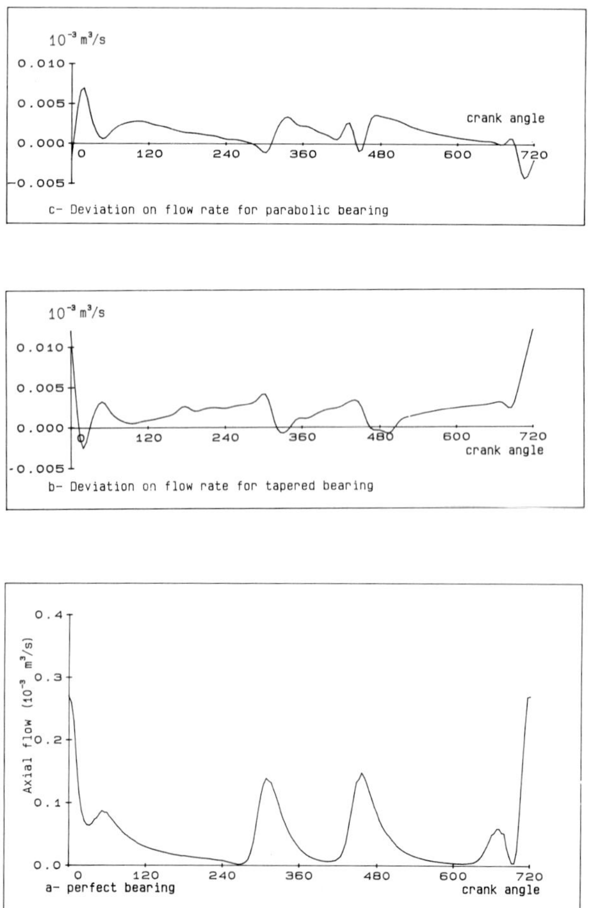

Fig 8 : Axial flow

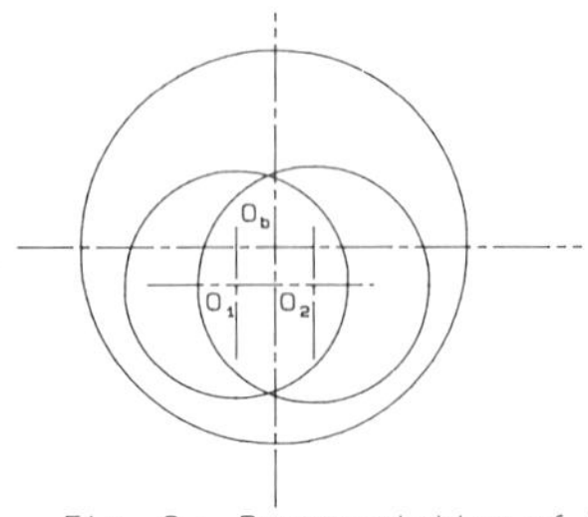

Fig 9 : Representation of horizontal misalignment

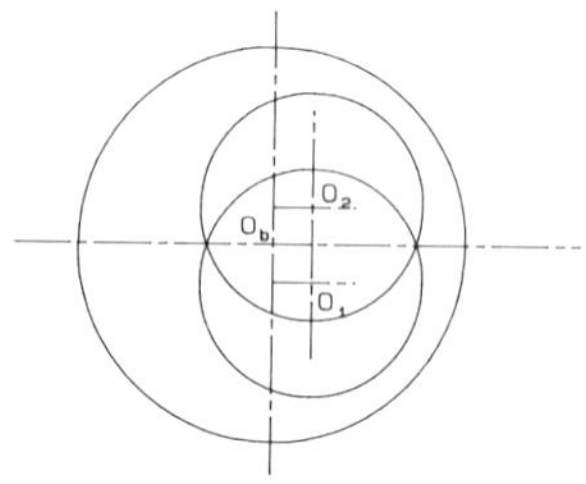

Fig 10 : Representation of vertical misalignment

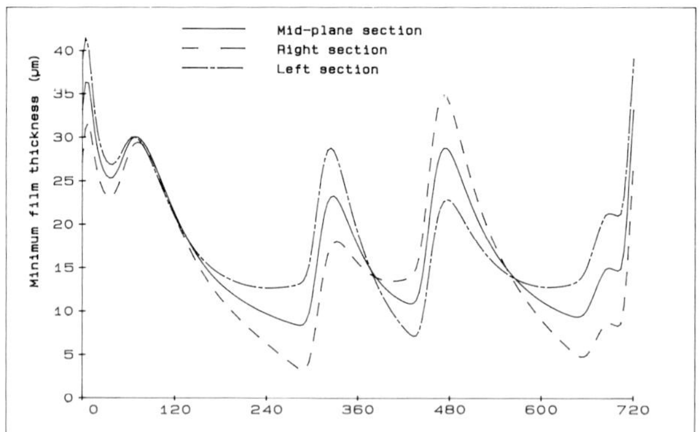

Fig 11 : Minimum film thickness for horizontal misalignment in three sections of the bearing

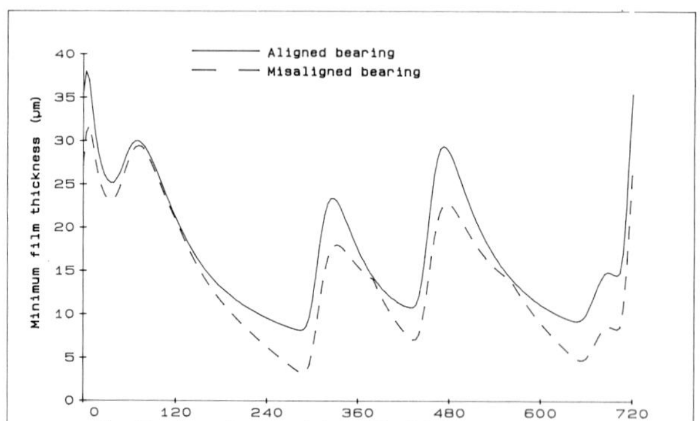

Fig 12 : Comparison on minimum film thickness between aligned and horizontal misaligned bearing

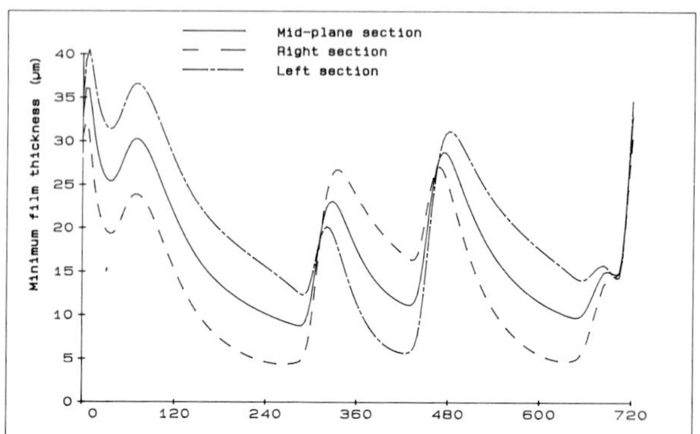

Fig 13 : Minimum film thickness for vertical misalignment in three sections of the bearing

shape of the curve obtained has a similar aspect than for a bearing under steady state conditions. For that bearing the optimal value of the radial clearance, i.e. for the maximum value of the minimum film thickness, is about 22 µm. This value is very small compared to the actual value (82,55 µm) and corresponds to a value hard to achieve. If other assumptions had been made on the operating conditions, i.e. taking account of thermal effects or elastic deformations of the bearing surfaces, it would be possible to find an optimum value of the radial clearance nearly 82,55 µm.

3.2 Tapered and parabolic bearings

The minimum film thickness as a function of the crank angle is shown fig. 6 for the tapered and parabolic bearings. For the tapered bearing the defect is R-r = 8 µm and for the parabolic bearing $R-R_1$ = 8 µm. The results are obtained for a mean constant radial clearance. Fig. 6-a-presents the film thickness curve for a perfect bearing and Fig. 6-b give the difference between the minimum film thickness for a tapered bearing and the minimum film thickness for a perfect bearing. Fig. 6-c gives the same difference for a parabolic bearing. The same representation is used for the friction torque (fig. 7-a-b-c) and the axial flow (Fig. 8-a-b-c). Tapered and parabolic bearings reduce the film thickness almost throughout the cycle. The maximum decrease is about 0,7 µm for the two shape defects which represents 8 % of the minimum value of the minimum film thickness.

These geometric shape defects have little effects on the friction torque and the axial flow, respectively increasing about 3 and 5 percent.

3.3 Misalignment

The horizontal misalignment represents a constant misalignment perpendicular to the rod axis Fig. 9, and vertical misalignment along the rod axis Fig. 10. In these two cases the misalignment is fixed to d = 12 µm and the journal is assumed to rotate about its centreline C_1 C_2.

Fig. 11 gives the minimum film thickness with respect to the crankshaft angular rotation for three sections of the bearing ; right section ($Z = \frac{L}{2}$), mid-plane section (Z = 0) and left section (Z = - L/2) for the horizontal misalignment. Fig. 12 compares the minimum film thickness for both aligned and misaligned bearings. The minimum film thickness for misaligned bearing is obtained taking the minimum value of the three sections. The minimum film thickness is reduced by about 55 percent due to horizontal misalignment. Fig. 12 can be compared to the curve obtained by Goenka |17| concerning a study of a misalignment perpendicular to the rod axis, the shape of the curves are similar.

The same kind of results are presented Fig. 13 and 14 for the vertical misalignment. The curve are slightly differents, the vertical misalignment giving more important differences between the two sides of the bearing than horizontal misalignment.

Fig. 15-16 give a comparison for the two cases of misalignment for the friction torque and Fig. 17-18 for the axial flow. Due to these misalignments the friction torque increases about 5 % in the two cases and the axial flow increases about 8 % for horizontal misalignment and 10 % for vertical misalignment.

4 CONCLUSION

An analysis of the behaviour of a big-end bearing with misalignment and shape defects has been presented. Based on a technique using the mobility method of Booker, we have determined the bearing characteristics of a diesel engine. The effect of the radial clearance on the minimum film thickness under dynamic loading gives a shape of curve similar to the steady state condition case.

The friction torque and the axial flow are slightly modified by the shape defects or the misalignments, the increasing is less than 10 percent.

The conic or the parabolic profiles reduce the minimum film thickness by about 8 % which is not very important compared to the horizontal or vertical misalignments which reduce the minimum film thickness by about 55 and 45 %.

5 ACKNOWLEDGEMENT

This work was supported by the European Economic community, Brite contract n° RI 1B-0139-F (CD) and by RNUR, PSA, FIAT, and VOLVO Companies.

References

(1) CAMPBELL, J., LOVE, P.P., MARTIN, F.A. and RAFIQUE, S.O. 'Bearings for Reciprocating Machinery'. A review of the present state of theoritical, experimental and service knowledge. Inst. of Mech. Eng. Vol. 182. 1967.

(2) BOOKER, J.F. 'Dynamically loaded journal Bearings : Mobility Method of solution'. Journal of Basic Eng., ASME, Series D, Sept. 1965.

(3) BOOKER, J.F. 'Dynamically loaded Journal Bearings : Numerical Application of the mobility method'. JOLT-ASME, Series F, Janv. 1971.

(4) JONES, G.J. 'Crankshaft Bearings : Oil Film History'. Proc. of the 9th Leeds-Lyon Symposium on Tribology. Sept. 1982.

(5) MARTIN, F.A. and BOOKER, J.F. 'Influence of Engine Inertia Forces on Minimum Film Thickness in Con-rod- Big-end Bearings' Proc. Inst. Mech. Engrs, Vol. 181, 1966-67.

(6) BERTHE, D., FANTINO, B., FRENE, J. and GODET, M. Surface Roughness on the Hydrodynamics of Lubricated Systems'. Journal of Mechanical Engineering Science. IME, Vol. 16, n° 3, 1974.

(7) JONES, G.J., LEE, C.S. and MARTIN, F.A. 'Crankshaft Bearings : Advances in Predictives Techniques Incorporating the Effects of Oil Holes and Grooving'. A.E. Tech. Symposium (April 1982).

(8) GOENKA, P.K. and BOOKER, J.F. 'Spherical Bearings : Static and Dynamic Analysis via the Finite Element Method'. ASME-JOLT, Vol. 102, n° 3, July 1980.

(9) OH, K.P. 'The Numerical Solution of Dynamically loaded Elastohydrodynamic Contact as a Nonlinear Complementary Problem.' ASME-JOLT, Vol. 106, Jan. 1984, p. 88-95.

(10) FANTINO, B., FRENE, J. and GODET, M. 'Dynamic Behaviour of an Elastic Connecting Rod Bearing. Theorical Study'. Studies of Engine Bearings and Lubrication, SP. 530, SAE, Feb. 1983.

(11) LABOUFF, G.A. and BOOKER, J.F. 'Dynamically Loaded Journal Bearings : A Unified Finite Element Approach to Hydrodynamic and Elastohydrodynamic Lubrication'. ASME, paper n° 84.

(12) OH, K.P. and GOENKA, P.K. 'The Elastohydrodynamic Solution of Journal Bearings under Dynamic Loading'. ASME Paper n° 84, Trib. 20.

(13) Van Der TEMPEL, L., MOES, H. and BOSMA, R. 'Starvation in dynamically loaded flexible short journal bearings'. ASME, Oct. 1985.

(14) ASANABE J., AKANOSHI, M. and ASAI, R. 'Theorical and Experimental Investigation of Misaligned Journal Bearing Performance'. Tribology conventian 1971. Inst. Mech. Eng., 1972.

(15) SMALLEY, J. and Mc CALLION, N. 'The effect of Journal Misalignment on the performance of a Journal Bearing under steady Running conditions'. Proc. Inst. Mech. Eng. 1966-67, Vol. 181.

(16) PINKUS, O. and BUPARA S.S. 'Analysis of Misaligned Grooved Journal Bearings'. JOLT-ASME, Oct. 1979.

(17) NICOLAS, D. and FRENE, J. 'Tilting torque Permissible in Plain Bearings. Theory, Experimental Results and Application to Machine Design'. First European Tribology Congress, 1973, p. 353-360.

(18) GOENKA, P.K. 'Dynamically Loaded Journal Bearings : Finite Element Method Analysis'. ASME, JOLT, Vol. 106, Oçt. 1984.

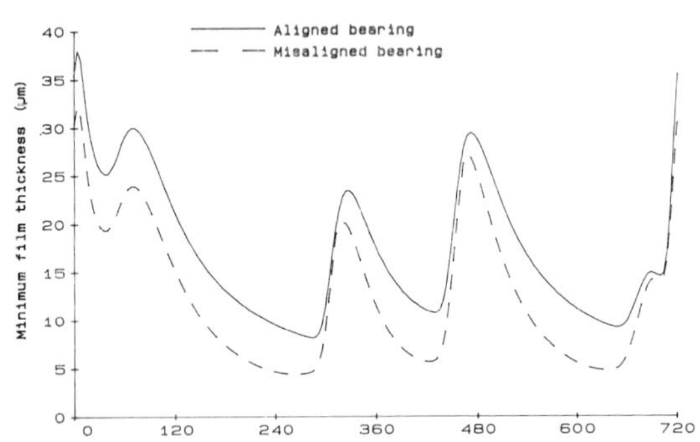

Fig 14 : Comparison on minimum film thickness between aligned and vertical misaligned bearing

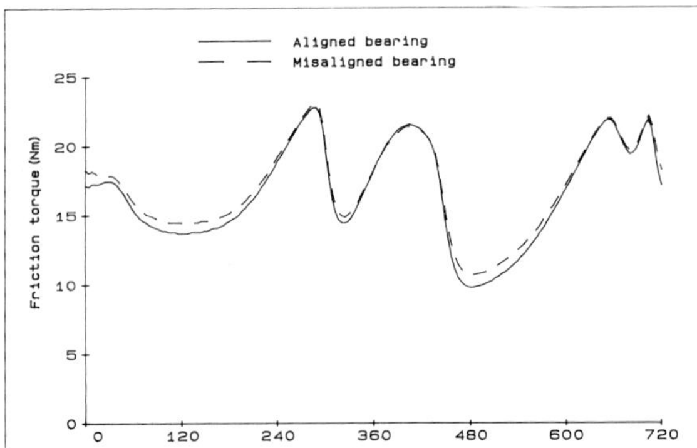

Fig 15 : Comparison on friction torque between aligned and horizontal misaligned bearing

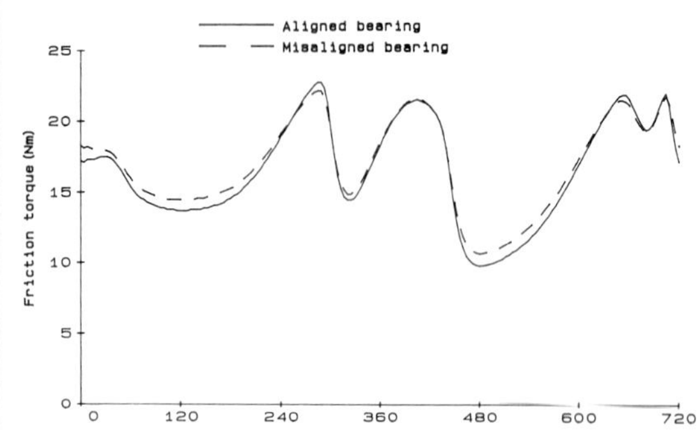

Fig 16 : Comparison on friction torque between aligned and vertical misaligned bearing

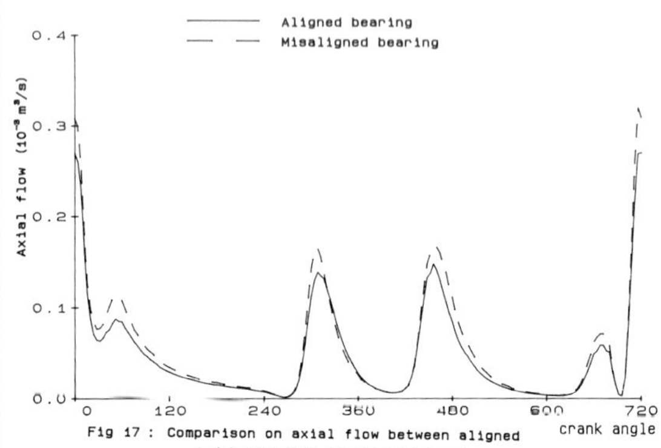

Fig 17 : Comparison on axial flow between aligned and horizontal misaligned bearing

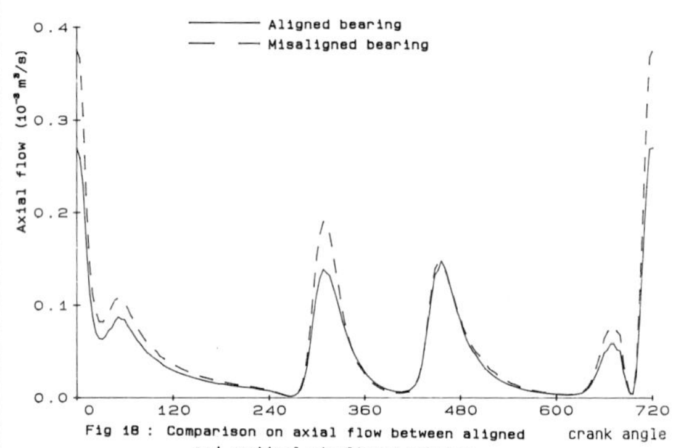

Fig 18 : Comparison on axial flow between aligned and vertical misaligned bearing

Paper XI(iii)

Transient dynamics of engine bearing systems

S. Boedo and J. F. Booker

A realistic single-cylinder engine model is formulated for investigation of bearing response to transient duty cycle variation. Connectingrod and main bearing duty cycles are calculated by either of two distinct approaches, based respectively on either specified kinematics (crankshaft rotation) or specified dynamics (crankshaft external torque). Journal displacement response is subsequently determined through application of the (short-bearing) mobility method. Three numerical examples compare the two approaches. The resulting bearing minimum film thickness under transient variations in the engine duty cycle is shown to be substantially less than would otherwise be predicted under the usual steady-state assumptions.

1 INTRODUCTION

The analysis and design of engine journal bearings can be viewed as two sequential steps. The first step requires the determination of appropriate bearing duty (i.e., loads and kinematics) based upon a representative engine model. The second step employs the resulting bearing duty with an appropriate journal bearing model (and bearing analysis method) to determine critical design parameters (e.g. minimum film thickness, maximum film pressure) that characterize bearing life and durability.

Considerable improvement has been made in the second design analysis step in recent years. Hydrodynamic (Booker [1965, 1971], Childs [1977], Goenka [1984a, 1984b]) and more recent elasto-hydrodynamic (Fantino, Frêne, and Godet [1983], Oh and Goenka [1985], LaBouff and Booker [1985]) lubrication models have progressively improved the ability to predict the behavior of dynamically-loaded journal bearings under arbitrary loading.

However, little improvement has been made in the first design analysis step; as a result, the required bearing duty used in the second step has not been completely representative of the total dynamic response of the engine (and also, as we shall see, of its dynamically-coupled environment). Traditionally, bearing loads and kinematics have been calculated from such classic assumptions as specified (though usually constant) crankshaft angular velocity and periodic cylinder pressure forcing functions (Den Hartog [1934]). Even recent elastic crankshaft formulations (Booker and Stickler [1982], Welsh and Booker [1983]) have utilized the classic approach. A shortcoming of these assumptions is reflected in the observation that a real reciprocating engine, even at "steady-state", operates with a fluctuating crankshaft angular velocity. Moreover, the transient bearing response due to aperiodic load variations, such as aperiodic cylinder pressure variation (e.g., "throttling") has not been generally addressed.

As a first step toward a more realistic bearing duty calculation, the present work describes a general rigid-body treatment of engine and bearing analysis for the single-cylinder reciprocating engine. This analysis starts with the development of an engine and environment model dynamically equivalent to the corresponding actual rigid-body engine and actual engine application, respectively. General expressions for bearing duty are then derived based upon this model. Bearing response is obtained through application of the resulting bearing duty to an appropriate bearing analysis method. Comparisons to the classic analysis approach are illustrated with three numerical examples based upon a typical production engine.

1.1 Nomenclature

θ	crankshaft rotation angle	(-)
p	cylinder pressure	(FL^{-2})
α	cylinder pressure attenuation	(-)
A	piston cross-sectional area	(L^2)
F_c	connectingrod bearing film load	(F)
F_m	main bearing film load	(F)
c_{ext}	external viscous damping constant	(FLT)
J_{ext}	external rotational inertia	(FLT^2)
T_g	general torque forcing function	(FL)
T_{ext}	external torque	(FL)
e_c	connectingrod journal eccentricity	(L)
e_m	main journal eccentricity	(L)
z	state	(various)
r	crank radius	(L)
λ	connectingrod length	(L)
g	gravitational acceleration	(LT^{-2})
M_1-M_7	point masses	$(FL^{-1}T^2)$
ζ	counterweight attitude angle	(-)
J_r	residual connectingrod inertia	(FLT^2)
J_f	flywheel rotational inertia	(FLT^2)
x,y,z	inertial frame coordinates	(L)
t	time	(T)
$(\dot{\ })$	time derivative	(T^{-1})
$(\_)$	vector quantity	

2 PROBLEM FORMULATION

The simplest prototypical engine and bearing system is the single cylinder reciprocating engine, shown schematically in Figure 1. The moving parts of this engine include piston, connectingrod, crankshaft, and flywheel components. Cylinder gas force acts upon the piston to produce reciprocating motion. The resulting reciprocating piston motion is converted into rotary motion via the connecting-rod. Connections include a pistonpin bearing (at the piston end of the connectingrod), a connectingrod bearing (at the crankshaft end of the rod) as well as two main bearings supporting the crankshaft. The flywheel provides enough rotational inertia so that the resulting crank-shaft angular velocity is nearly uniform under steady-state operating conditions (even in the absence of external inertia).

Several conventional design assumptions are introduced which simplify the analysis yet retain desired generality. The engine is symmetrically designed so that main bearing loads and motions are identical. The piston is assumed constrained by cylinder walls so that its motion parallels the cylinder axis. The pistonpin bearing is approximated by a kinematically-ideal pin joint, since the radial clearance of this bearing is small relative to main and connectingrod bearing clearances.

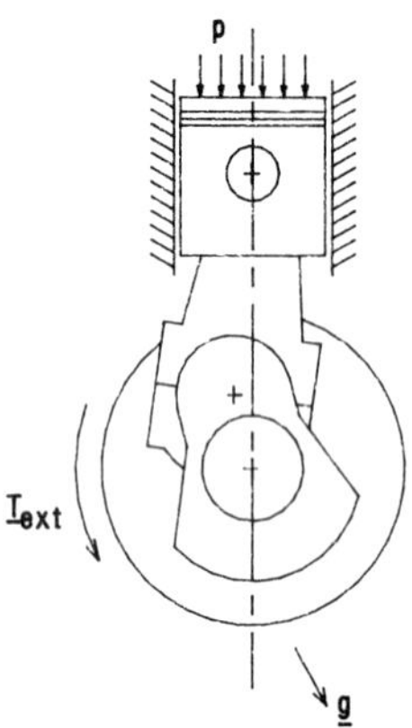

Fig. 1 Engine schematic

For the purpose of bearing duty calculation <u>only</u>, main and connnectingrod bearings are approximated as pin joints. Except under extraordinary circumstances, this decoupled approach of engine and bearing dynamics has been demonstrated to be sufficiently accurate (Boedo [1986]).

Piston, connectingrod, crankshaft, and flywheel components are assumed perfectly rigid; hence, all components can be modelled as mass-less rigid bodies embedded with discrete point masses (as well as residual inertia). See Appendix A for a more detailed explanation.

The resulting engine model and associated geometry are idealized in Figure 2. The fixed (inertial) x,y coordinate system is the primary reference frame used in this study. This coordinate system is fixed to the assumed rigid and fixed engine block, and its origin is coincident with the main bearing sleeve centers, also fixed to the engine block. Main and connectingrod journal eccentricites are exaggerated for clarity; bearing clearance spaces are indicated by dashed circles. The cylinder axis is coincident with the x-axis; thus, "offset" engines are not considered. Gravitational acceleration $\underline{g}$ is shown in a general direction anticipating the effect of cylinder bank angling.

External loads are applied to the model in the form of cylinder pressure p acting on the piston and external torque T_{ext} applied to the flywheel end.

The variation of cylinder pressure p with crankangle θ is assumed to be specified. To complete the problem, the determination of bearing duty variation rests with the choice of one of the following approaches:

1 Crankangle θ is specified as a function of time t; the variation of crankshaft external torque T_{ext} must be determined. This is the basis of classic design analysis approaches and is designated as Problem I below.

2 External crankshaft torque T_{ext} is specified as a function of time t (based upon a model representing the external environment); the variation of crankangle θ must be determined. This is the basis of the proposed design analysis approach and is designated as Problem II below.

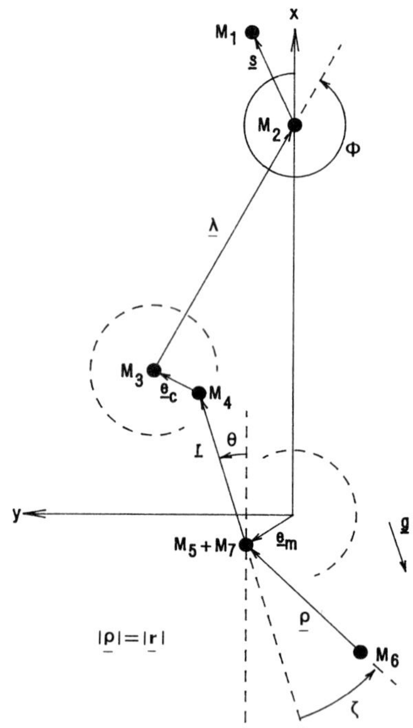

Fig. 2 Engine idealization

2.1 <u>Engine Dynamics</u>

2.1.1 Problem I: Specified Rotation

Suppose that crankshaft rotation is explicitly specified as a function of time only; i.e.

$$\theta = \theta(t) \tag{1}$$

For the purpose of bearing duty calculation only, pin joint approximations for connectingrod and main bearings require

$$\underline{e}_c \equiv \underline{0}, \quad \underline{e}_m \equiv \underline{0} \tag{2}$$

Since $\theta(t)$ is specified, the engine system motion is completely determined as an explicit function of time only.

As shown elsewhere (Boedo [1986]), components of connectingrod and main bearing film loads $\mathbf{F}_c$ and $\mathbf{F}_m$ as functions of time are

$$F_c^x = -(M_1 + M_2 + M_3)g^x + p(\theta)A - (M_1 + M_2)(r\dot{\theta}^2\cos\theta + \lambda\dot{\phi}^2\cos\phi + r\ddot{\theta}\sin\theta + \lambda\ddot{\phi}\sin\phi) - M_3 r\dot{\theta}^2\cos\theta - M_3 r\ddot{\theta}\sin\theta \quad (3a)$$

$$F_c^y = -M_3 r\dot{\theta}^2\sin\theta + M_3 r\ddot{\theta}\cos\theta - M_3 g^y + (M_3 r\dot{\theta}^2\cos\theta + M_3 r\ddot{\theta}\sin\theta + M_3 g^x + F_c^x)\tan\phi - J_r\ddot{\phi}\lambda^{-1}\sec\phi \quad (3b)$$

$$2F_m^x = (M_4 + M_5 + M_6 + M_7)g^x - F_c^x + M_4 r\dot{\theta}^2\cos\theta - M_6 r\dot{\theta}^2\cos(\theta + \zeta) + M_4 r\ddot{\theta}\sin\theta - M_6 r\ddot{\theta}\sin(\theta + \zeta) \quad (3c)$$

$$2F_m^y = (M_4 + M_5 + M_6 + M_7)g^y - F_c^y + M_4 r\dot{\theta}^2\sin\theta - M_6 r\dot{\theta}^2\sin(\theta + \zeta) - M_4 r\ddot{\theta}\cos\theta + M_6 r\ddot{\theta}\cos(\theta + \zeta) \quad (3d)$$

where $\dot{\phi}$ and $\ddot{\phi}$ are found from

$$\dot{\phi} = -r\dot{\theta}\cos\theta \,/\, (\lambda\cos\phi) \quad (3e)$$

$$\ddot{\phi} = (r\dot{\theta}^2\sin\theta + \lambda\dot{\phi}^2\sin\phi - r\ddot{\theta}\cos\theta)/(\lambda\cos\phi) \quad (3f)$$

The torque T_{ext} required to maintain the kinematic constraint (1) is found from

$$T_{ext} = (M_4 g^x - F_c^x)r\sin\theta - M_6 g^x r\sin(\theta + \zeta) - (M_4 g^y - F_c^y)r\cos\theta + M_6 g^y r\cos(\theta + \zeta) + (M_4 + M_6)r^2\ddot{\theta} + J_f\ddot{\theta} \quad (4)$$

where film loads (and $\ddot{\theta}$) are given above.

2.1.2 Problem II: Specified Torque

As a simple example, suppose that the single cylinder engine model is torsionally coupled to an external world model represented by

$$T_{ext} = -J_{ext}\ddot{\theta} - c_{ext}\dot{\theta} + T_g(t) \quad (5)$$

where J_{ext}, c_{ext}, and $T_g(t)$ represent external rotational inertia, linear torsional viscous damping constant, and general time dependent torque forcing function, respectively.

As in Problem I, connectingrod and main bearing pin joint appoximations for bearing duty calculaton require

$$\mathbf{e}_c \equiv \mathbf{0}, \quad \mathbf{e}_m \equiv \mathbf{0} \quad (6)$$

This reduces the engine model to a one-degree-of-freedom system. The state of the engine system is represented by a two-dimensional state vector

$$\mathbf{z} = \begin{Bmatrix} z_1(t) \\ z_2(t) \end{Bmatrix} = \begin{Bmatrix} \theta(t) \\ \dot{\theta}(t) \end{Bmatrix} \quad (7)$$

It is now desired to determine state rate $\dot{\mathbf{z}}$ with initial conditions

$$\mathbf{z}(0) = \begin{Bmatrix} z_1(0) \\ z_2(0) \end{Bmatrix} = \begin{Bmatrix} \theta(0) \\ \dot{\theta}(0) \end{Bmatrix} \quad (8)$$

As shown elsewhere (Boedo [1986]), the external load equation (5) and the kinematic relations (Appendix B) are substituted into the dynamic equilibrium equations (Appendix B), resulting in state rates

$$\dot{z}_1 = z_2 \quad (9a)$$

$$\dot{z}_2 = V_2 \,/\, V_3 \quad (9b)$$

where

$$V_2 = V_1\cos\phi - (H_1\cos\phi + \alpha_1 I_2)r\sin\theta + [(H_1\cos\phi + \alpha_1 I_2)\tan\phi - H_2\cos\phi + \alpha_1 I_4]r\cos\theta \quad (9c)$$

$$V_3 = I_1 r^2\cos\phi\sin^2\theta - (I_1\sin\phi - I_3\cos\phi + I_2)r^2\sin\theta\cos\theta + (I_2\tan\phi + M_3\cos\phi + I_4)r^2\cos^2\theta + J_1\cos\phi \quad (9d)$$

$$V_1 \equiv -M_4 g^x r\sin\theta + M_6 g^x r\sin(\theta+\zeta) + M_4 g^y r\cos\theta - M_6 g^y r\cos(\theta+\zeta) - c_{ext}\dot{\theta} + T_g(t) \quad (9e)$$

$$H_1 \equiv (M_1 + M_2 + M_3)g^x - p(\theta)A + (M_1 + M_2 + M_3)r\dot{\theta}^2\cos\theta + (M_1 + M_2)\lambda\dot{\phi}^2\cos\phi \quad (9f)$$

$$H_2 \equiv (M_3 r\dot{\theta}^2\cos\theta + M_3 g^x)\tan\phi - M_3 r\dot{\theta}^2\sin\theta - M_3 g^y \quad (9g)$$

$$\alpha_1 \equiv r\dot{\theta}^2\sin\theta + \lambda\dot{\phi}^2\sin\phi \quad (9h)$$

$$I_1 \equiv M_1 + M_2 + M_3 \quad (9i)$$

$$I_2 \equiv (M_1 + M_2)\sin\phi \quad (9j)$$

$$I_3 \equiv M_3\tan\phi \quad (9k)$$

$$I_4 \equiv J_r\lambda^{-2}\sec\phi \quad (9l)$$

$$J_1 \equiv (M_4 + M_6)r^2 + J_f + J_{ext} \quad (9m)$$

$$\sin\phi \equiv -r\sin\theta/\lambda \quad (9n)$$

$$\cos\phi \equiv (1 - \sin^2\phi)^{1/2} \quad (9o)$$

$$\dot{\phi} \equiv -r\dot{\theta}\cos\theta \;/\; (\lambda\cos\phi) \quad (9p)$$

The state rate equations (9) with initial conditions (8) completes specification of a second-order initial-value problem.

Numerical integration of (9) with initial conditions (8) is now performed in a routine manner (e.g. Euler, Runge-Kutta, etc.) to determine state $\underline{z}$ at desired times.

Subsequent connectingrod and main bearing reaction film loads $\underline{F}_c$ and $\underline{F}_m$ are given as functions of state $\underline{z}$ (and time) by equations (3) with $\ddot{\theta}$ given by equation (9b).

2.2 Bearing Dynamics

Given bearing duty (translational loads and angular kinematics as functions of time) calculated as in either Problem I or II, it is now possible to study bearing dynamics as a subsequent (or decoupled) problem. The mobility method (Booker [1965], [1971]) based upon a short-bearing, circumferentially symmetric, cavitated π-film model has been chosen for its familiarity and computational efficiency; recent methods based upon finite element models (Goenka [1984a],[1984b], LaBouff and Booker [1985]) are equally applicable.

The result is a second-order initial value problem for each bearing. State variables are x and y components of journal eccentricity (in the reference frame of Figure 2). For the connectingrod bearing, the state rate equation becomes

$$\dot{\underline{e}}_c(t) = \underline{f}_c(\underline{e}_c, t) \quad (10)$$

with initial eccentricity vector given by

$$\underline{e}_c(0) = \underline{e}_{c0} \quad (11)$$

Similarly, for the identical main bearings

$$\dot{\underline{e}}_m(t) = \underline{f}_m(\underline{e}_m, t) \quad (12)$$

with initial eccentricity vector given by

$$\underline{e}_m(0) = \underline{e}_{m0} \quad (13)$$

The exact functional form of equations (10) and (12) is described in detail elsewhere (Booker [1965]).

Numerical integration of equations (10) and (12) is performed in a routine manner to determine connectingrod and main journal eccentricity "orbits" at desired times.

3 APPLICATIONS

Table 1 represents a typical set of engine and bearing specifications for the single-cylinder engine model. These model parameters reflect representative physical data for production single-cylinder engines (Boedo [1986]).

3.1 Example I: Specified Rotation

This example illustrates the classic analysis approach of Problem I. Assume crankshaft angle

$$\theta(t) = 60\pi t \text{ rad} \qquad t \ge 0 \quad (14)$$

Table 1 Typical Engine/Bearing Specifications

Inertia Distribution

M_1	=	0.4	kg	piston/pistonpin
M_2	=	0.1	kg	connectingrod
M_3	=	0.2	kg	
J_r	=	0	$kg\cdot m^2$	
M_4	=	1.5	kg	crankshaft
M_5	=	0.5	kg	
M_6	=	2.0	kg	
ζ	=	0	rad	
M_7	=	8.0	kg	flywheel
J_f	=	0.06	$kg\cdot m^2$	

Engine Configuration

A	=	5900	$(mm)^2$	piston cross-sectional area
r	=	30	mm	crank radius
λ	=	100	mm	connectingrod length
g^x	=	-9.81	m/s^2	gravity x-component
g^y	=	0	m/s^2	gravity y-component

Journal Bearing Parameters

		Rod	Main		
D	=	40	(50)	mm	journal diameter
L	=	20	(25)	mm	bearing length
C	=	20	(25)	μm	radial clearance
μ	=	6	(6)	mPa·s	oil viscosity

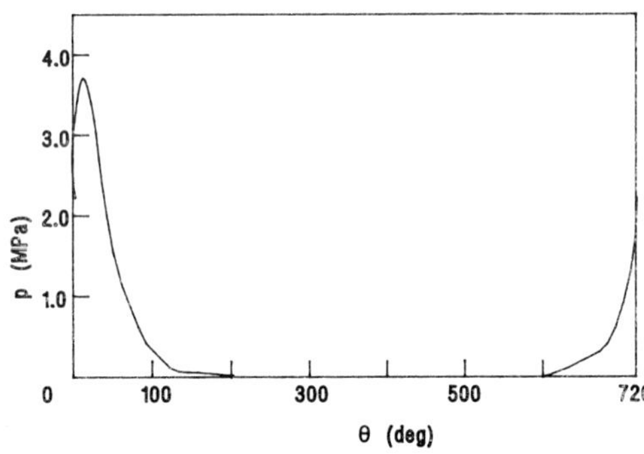

Fig. 3 Cylinder pressure

and the (periodic) cylinder pressure variation shown in Figure 3.

Once the periodic duty response is calculated by equations (3), periodic time response histories of main and connectingrod journal eccentricity are subsequently obtained (Figure 5). Numerical integration of bearing equations (10) and (12) is performed using a fourth-order Runge-Kutta method at fixed time step intervals of 46.295 μs (corresponding to 0.5 degree crankshaft rotation). Each bearing is started concentrically, and the simulation is continued until periodic orbits are apparent. (Transient contributions to bearing response due to arbitrary initial conditions at startup persist for approximately two engine cycles.)

3.2 Example IIA: Specified Torque - Periodic Response

This example illustrates the proposed design analysis approach of Problem II. Assume that the external world is represented by external torque T_{ext} of the form (5), with[1]

$$J_{ext} = 0.0028 \quad kg \cdot m^2 \tag{15a}$$

$$c_{ext} = 0.1113 \quad N \cdot m \cdot s \tag{15b}$$

$$T_g(t) \equiv 0 \tag{15c}$$

and that the (periodic) cylinder pressure variation shown in Figure 3 is specified.

Bearing duty is calculated through numerical integration of state rate equations (9) with initial conditions

$$\theta(0) = 0 \quad rad$$

$$\dot{\theta}(0) = 60\pi \quad rad/s \tag{16}$$

at time step intervals of 46.295 μs until periodic state response is apparently obtained. (Approximately 10 engine cycles are required.)

The subsequent bearing analysis is performed using the bearing duty determined above with the bearing analysis procedure outlined in Example I.

Figures 4 and 5 show periodic response of crankshaft angular velocity and bearing eccentricity orbits, respectively. Note that the steady-state crankshaft speed is no longer constant. Comparison of Examples I and IIA shows that, although the engine speed is fluctuating about an average value of 60π rad/s, the effect of this fluctuation on bearing response is negligible. Thus, from a periodic orbit analysis viewpoint, the assumption of a constant angular velocity is quite satisfactory.

3.3 Example IIB: Specified Torque - Transient Response

This example is identical to Example IIA except that the cylinder pressure forcing function is modified during the simulation to reflect engine "throttling".

The engine simulation is started with (arbitrary) initial conditions given by equations (16) with indicated "full-throttle" cylinder pressure provided in Figure 3, modified by (multiplying) by the normalized throttling attenuation factor α of Figure 6. As shown in Figure 6, the simulation is continued with the "full-throttle" cylinder pressure history until periodic state response has been obtained. The pressure history is then reduced by 50 percent in magnitude at approximately 360 degrees into the combustion cycle, when the cylinder pressure is zero. The simulation continues with this "half-throttle" pressure history until periodic state response is once again obtained. Then (again at 360 degrees into the combustion cycle), the cylinder pressure is restored to the full-throttle condition. The simulation is terminated when periodic state response again appears to have been obtained.

Figure 7 shows resulting crankshaft angular velocity as a function of time for the entire simulation. It is observed that approximately 20 engine cycles are required to reach periodicity at the reduced cylinder pressure, and 25 additional cycles are needed to regain full-throttle periodic response. The periodic average angular velocities at full- and half-throttle positions are found to be approximately 180 and 90 rad/s, respectively.

Given the resulting duty variation (film loads and angular kinematics from the engine simulation), time response histories of journal eccentricity are sequentially obtained for each bearing as a subsequent separate problem. Each bearing is started concentrically, and the state rate equations (10) and (12) are solved using fourth-order Runge-Kutta numerical integration with a time step identical to that used in the engine simulation. The simulation is continued for the entire duty response history.

Figure 8 shows connectingrod and main journal eccentricity ratio at the throttle-down cycle. The indicated full-throttle orbit has attained periodicity. The transient half-throttle orbit results from the reduction of cylinder pressure at the indicated throttle-down point (at 360 degrees into the orbit cycle).

Figure 9 shows connectingrod and main journal eccentricity ratio at the throttle-up cycle. The indicated half-throttle orbit has attained periodicity. The transient full-throttle orbit results from the restoration of full-throttle cylinder pressure at the indicated throttle-up point (again at 360 degrees into the orbit cycle.)

Figures 8 and 9 show that, although minimum film thickness changes considerably, the orbits retain their basic qualitative characteristics.

Figure 10 shows connectingrod and main journal cyclically-minimum film thickness values for each engine cycle of the simulation. It is observed that, at the point of engine throttle-down, the minimum film thickness initially <u>increases</u> by about 40 percent of full-throttle periodic value. At steady-state, however, the resulting half-throttle minimum film thickness values are very similar to full-throttle values. At the point of throttle-up, the peak value initially <u>decreases</u> by 60 percent of full-throttle but approaches the full-throttle value at steady-state.

In the case of transient loading, the use of steady-state analysis for predicting minimum film thickness has thus been shown to be extremely non-conservative.

A final note pertaining to periodicity is that the "true" average angular velocity (after 25 load cycles) at full-throttle is found to be about 8 percent less than the 60π rad/s result obtained in Example IIA (after only 10 cycles). Hence, one drawback of this design analysis method is that a considerable number of engine cycles is required to obtain duty cycle periodicity; however, this difference appears to have little or no effect on resulting steady-state bearing eccentricity orbits.

---

[1]For comparison to Example I, this damping constant c_{ext} is determined through several trial runs so that the average crankshaft speed is approximately 60π rad/s after 10 engine cycles. In real practice, the damping constant can be calculated directly or measured experimentally.

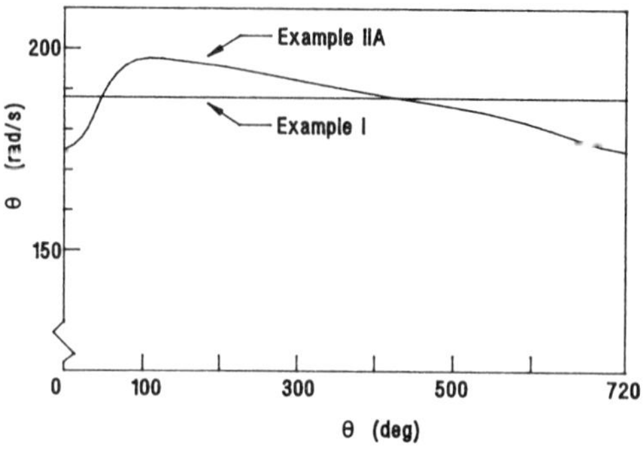

Fig. 4 Periodic crankshaft speed response (Examples I, IIA)

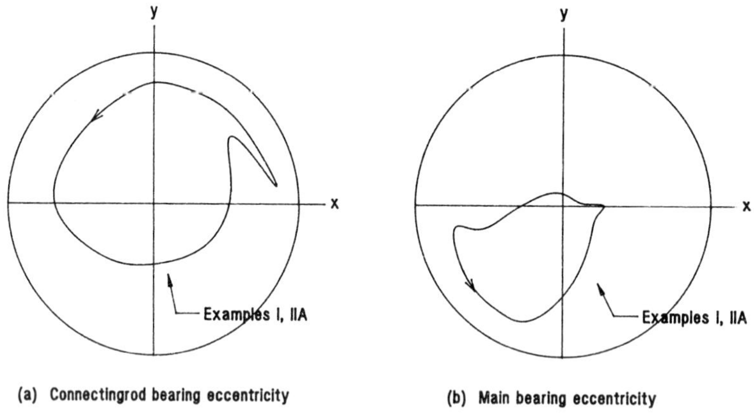

Fig. 5 Periodic duty response (Examples I, IIA)

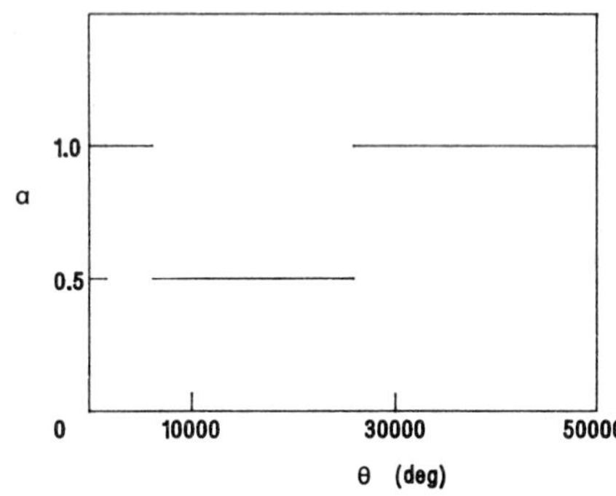

Fig. 6 Cylinder pressure attenuation factor (Example IIB)

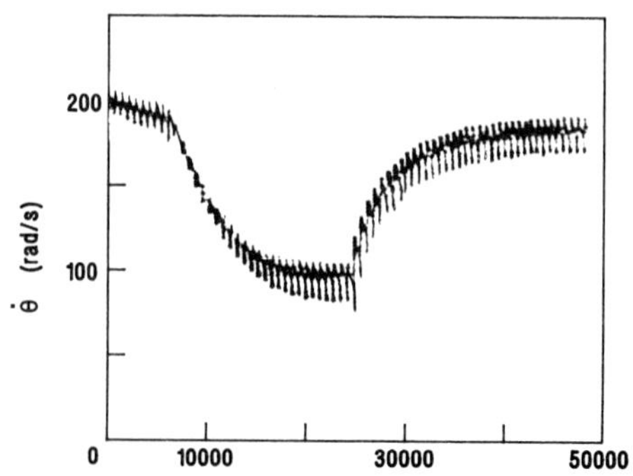

Fig. 7 Transient crankshaft speed response (Example IIB)

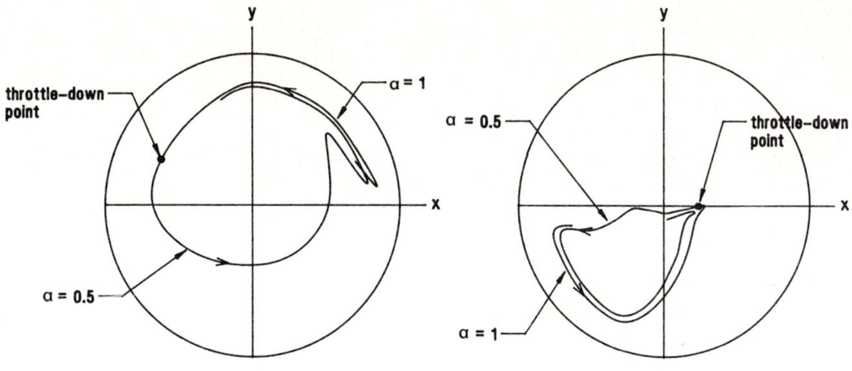

(a) Connectingrod bearing eccentricity (b) Main bearing eccentricity

Fig. 8 Transient "throttle-down" duty response (Example IIB)

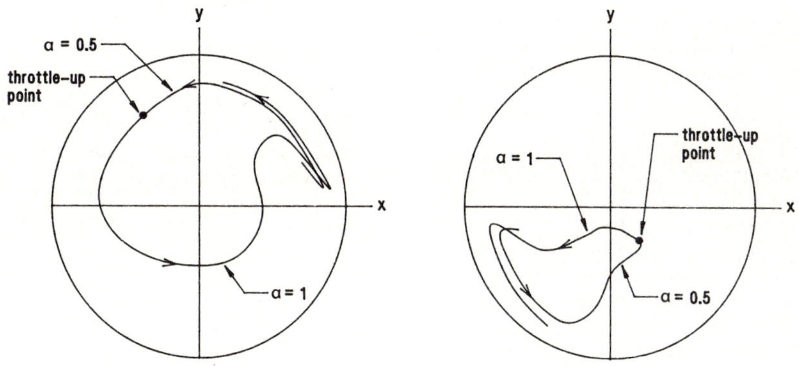

(a) Connectingrod bearing eccentricity (b) Main bearing eccentricity

Fig. 9 Transient "throttle-up" duty response (Example IIB)

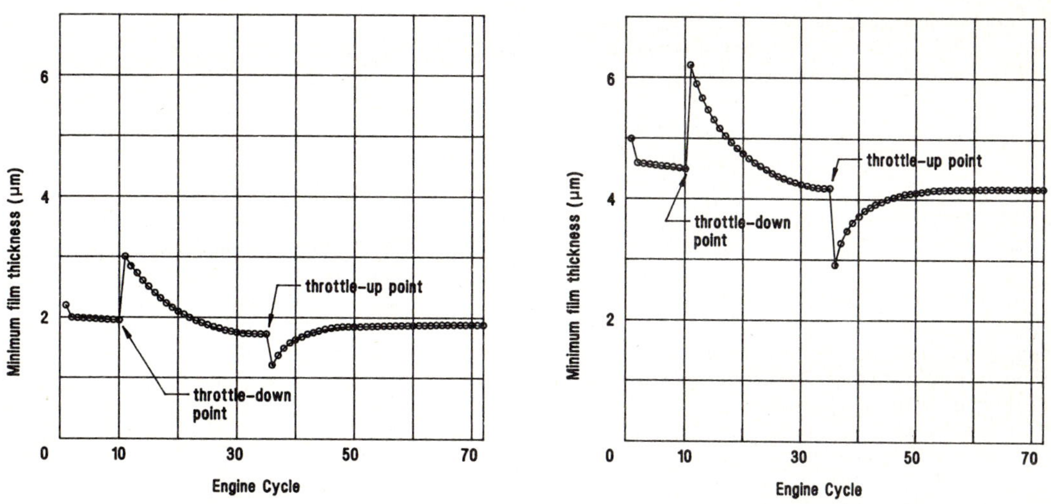

(a) Connectingrod bearing (b) Main bearing

Fig. 10 Cyclically-minimum film thickness (Example IIB)

4 SUMMARY AND CONCLUSIONS

A realistic engine model formulation has been developed with the intent of incorporating such usually-ignored effects as engine speed variation, flywheel and crankshaft mass distributions, and general external loading. Two analysis procedures (Problems I, II) for determining bearing duty response have been examined based upon specified crankshaft rotation and specified torque, respectively. Subsequently, bearing displacement response is calculated by the short-bearing mobility method.

Three numerical studies (Examples I, IIA, IIB) reveal that periodic bearing and engine response produce similar results through either analysis method. However, the power of the specified torque method rests with the ability to investigate transient response in a relatively realistic and straightforward manner (Example IIB). Aperiodic cylinder pressure variation shows that bearing minimum film thickness varies very considerably after engine "throttle-up" or "throttle-down"; moreover, these transient effects persist for several engine cycles.

The potential severity of such transient excursions seems to justify the (possibly substantial) computation time required for their prediction.

5 APPENDIX A: DETERMINATION OF MODEL PARAMETERS

Under the assumed kinematic constraints of plane motion, dynamic equivalence requires that the total mass, mass center location, and polar moment of inertia of actual and model components be identical.

Figure A1 shows actual and model components disassembled from the corresponding engine (Figure 1) and model (Figure 2).

For the piston/pistonpin model, dynamic equivalence requires

$$M_1 = M_p \qquad (1)$$

where M_p is the measured total mass of the actual piston/pistonpin assembly.

For the connectingrod, model small-end mass M_2, big-end mass M_3, and residual ring inertia J_r (Paul [1979]) are found from

$$M_2 = (M_r x_r) / \lambda \qquad (2a)$$

$$M_3 = M_r(1 - x_r/\lambda) \qquad (2b)$$

$$J_r = J_{rod} - M_r x_r \lambda \qquad (2c)$$

where λ, M_r, x_r, and J_{rod} are measured length, total mass, mass center location, and polar moment of inertia, respectively, of the actual connectingrod. Lengths x_r and λ are measured from the big-end rod center, and polar moment J_{rod} is taken about the same center.

For the crankshaft, model masses M_4, M_5, M_6, and attitude angle ζ are chosen so that

$$\cos\zeta = (1 - K^2) / (1 + K^2) \qquad (3a)$$

$$M_4 = [J_c(1 - K^2) - rM_c x_c^{\xi}(1 + K^2)]/(2r^2) \qquad (3b)$$

$$M_5 = (M_c r^2 - J_c) / (r^2) \qquad (3c)$$

$$M_6 = [(rM_c x_c^{\xi} + J_c)(1 + K^2)]/(2r^2) \qquad (3d)$$

where M_c, J_c, x_c^{ξ}, x_c^{η} are measured total mass, polar moment of inertia, and mass center coordinates[2], respectively, of the actual crankshaft, and K is given by

$$K = (rM_c x_c^{\eta}) / (J_c + rM_c x_c^{\xi}) \qquad (3e)$$

Polar moment J_c is measured about the crankshaft axis of rotation.

For the flywheel, model parameters M_7 and J_f are chosen so that

$$M_7 = M_{fly} \qquad (4a)$$

$$J_f = J_{fly} \qquad (4b)$$

where M_{fly} and J_{fly} are flywheel mass and (mass center-based) polar moment of inertia, respectively, of the actual component. It is implicitly assumed that the flywheel is dynamically balanced about the axis of rotation.

6 APPENDIX B

6.1 Kinematic Relations

Let x_i and y_i denote the x and y position coordinates, respectively, of mass point i shown in Figure 2. From this figure, a general set of kinematic relations are[3]

$$\sin\phi = -(e_m^y - e_c^y + r\sin\theta) / \lambda \qquad (1)$$

$$\cos\phi = (1 - \sin^2\phi)^{1/2} \qquad (2)$$

$$\dot{\phi} = -(\dot{e}_m^y - \dot{e}_c^y + r\dot{\theta}\cos\theta)/(\lambda\cos\phi) \qquad (3)$$

$$\ddot{\phi} = -(\ddot{e}_m^y - \ddot{e}_c^y - r\dot{\theta}^2\sin\theta - \lambda\dot{\phi}^2\sin\phi + r\ddot{\theta}\cos\theta) / (\lambda\cos\phi) \qquad (4)$$

$$\ddot{x}_1 = \ddot{e}_m^x - \ddot{e}_c^x - r\dot{\theta}^2\cos\theta - \lambda\dot{\phi}^2\cos\phi - r\ddot{\theta}\sin\theta - \lambda\ddot{\phi}\sin\phi \qquad (5)$$

$$\ddot{y}_1 = 0 \qquad (6)$$

$$\ddot{x}_2 = \ddot{x}_1 \qquad (7)$$

$$\ddot{y}_2 = \ddot{y}_1 \qquad (8)$$

$$\ddot{x}_3 = \ddot{e}_m^x - \ddot{e}_c^x - r\dot{\theta}^2\cos\theta - r\ddot{\theta}\sin\theta \qquad (9)$$

$$\ddot{y}_3 = \ddot{e}_m^y - \ddot{e}_c^y - r\dot{\theta}^2\sin\theta + r\ddot{\theta}\cos\theta \qquad (10)$$

$$\ddot{x}_4 = \ddot{e}_m^x - r\dot{\theta}^2\cos\theta - r\ddot{\theta}\sin\theta \qquad (11)$$

$$\ddot{y}_4 = \ddot{e}_m^y - r\dot{\theta}^2\sin\theta + r\ddot{\theta}\cos\theta \qquad (12)$$

$$\ddot{x}_5 = \ddot{e}_m^x \qquad (13)$$

$$\ddot{y}_5 = \ddot{e}_m^y \qquad (14)$$

---

[2]Referred to the ξ-η coordinate frame fixed to the crankshaft and centered at the crankshaft axis of rotation (Fig. A1).

[3]The imposed kinematic constraints on the piston require $\dot{\underline{s}} \equiv \underline{0}$, $\ddot{\underline{s}} \equiv \underline{0}$

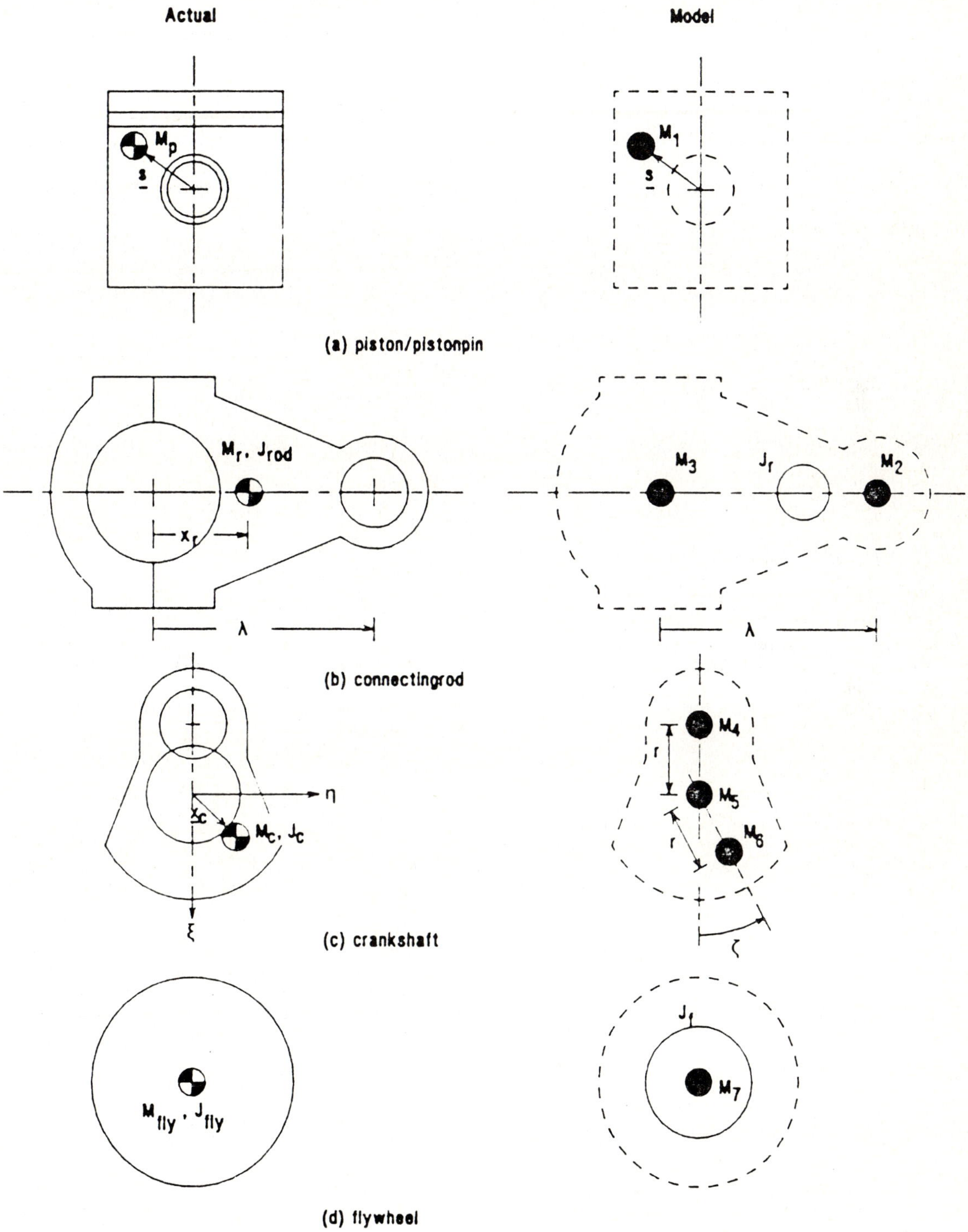

Fig. A1 Dynamic equivalence of actual and model rigid bodies

$$\ddot{x}_6 = \ddot{e}_m{}^x + r\dot{\theta}^2\cos(\theta + \zeta) + r\ddot{\theta}\sin(\theta + \zeta) \quad (15)$$

$$\ddot{y}_6 = \ddot{e}_m{}^y + r\dot{\theta}^2\sin(\theta + \zeta) - r\ddot{\theta}\cos(\theta + \zeta) \quad (16)$$

$$\ddot{x}_7 = \ddot{x}_5 \quad (17)$$

$$\ddot{y}_7 = \ddot{y}_5 \quad (18)$$

6.2 Dynamic Relations

As shown elsewhere (Boedo [1986]), dynamic equilibrium of the idealized engine of Figure 2 requires

$$(M_1 + M_2 + M_3)g^x - pA + F_c{}^x - M_1\ddot{x}_1 - M_2\ddot{x}_2 - M_3\ddot{x}_3 = 0 \quad (19)$$

$$(M_1 + M_2 + M_3)g^y + N + F_c{}^y - M_3\ddot{y}_3 = 0 \quad (20)$$

$$-(M_3\ddot{x}_3 - M_3g^x - F_c{}^x)\lambda\sin\phi + (M_3\ddot{y}_3 - M_3g^y - F_c{}^y)\lambda\cos\phi - J_r\ddot{\phi} = 0 \quad (21)$$

$$(M_4 + M_5 + M_6 + M_7)g^x - F_c{}^x - 2F_m{}^x - M_4\ddot{x}_4 - M_5\ddot{x}_5 - M_6\ddot{x}_6 - M_7\ddot{x}_7 = 0 \quad (22)$$

$$(M_4 + M_5 + M_6 + M_7)g^y - F_c{}^y - 2F_m{}^y - M_4\ddot{y}_4 - M_5\ddot{y}_5 - M_6\ddot{y}_6 - M_7\ddot{y}_7 = 0 \quad (23)$$

$$-(M_4g^x - F_c{}^x - M_4\ddot{x}_4)r\sin\theta + (M_4g^y - F_c{}^y - M_4\ddot{y}_4)r\cos\theta + (M_6g^x - M_6\ddot{x}_6)r\sin(\theta + \zeta) - (M_6g^y - M_6\ddot{y}_6)r\cos(\theta + \zeta) + T_{ext} - J_f\ddot{\theta} = 0 \quad (24)$$

where N is the reaction force transmitted from the cylinder wall to the piston, p is the specified cylinder pressure applied to the piston head, and A is the piston cross-sectional area.

7 ACKNOWLEDGEMENT

This research was conducted using the Cornell National Supercomputer Facility, a resource of the Center for Theory and Simulation in Science and Engineering (Cornell Theory Center), which receives major funding from the National Science Foundation and IBM Corporation, with additional support from New York State and members of the Corporate Research Institute.

REFERENCES

[1934] DEN HARTOG, J.P. 'Mechanical Vibrations', 1934 (McGraw-Hill Book Co., Inc., New York, N.Y.).

[1965] BOOKER, J.F. 'Dynamically-Loaded Journal Bearings: Mobility Method of Solution', Trans. ASME, Journal of Basic Engineering, 87, 1965, p. 533.

[1971] BOOKER, J.F. 'Dynamically-Loaded Journal Bearings: Numerical Application of the Mobility Method', Trans. ASME, J. Lub. Tech., 93, 1971, p. 168. Errata: No. 2, 1971, p. 315.

[1977] CHILDS, D.W., MOES, H., van LEEUWEN, H.J. 'Journal Bearing Impedance Descriptions for Rotor-Dynamic Applications', Trans. ASME, J. Lub. Tech., 1977, p. 198.

[1979] PAUL, B. 'Kinematics and Dynamics of Planar Machinery', 1979 (Prentice-Hall, Englewood Cliffs).

[1982] BOOKER, J.F., STICKLER, A.C. 'Bearing Load/Displacement Determination for Multi-Cylinder Reciprocating Machinery', ASME 2nd International Computer Engineering Conference, San Diego, California, August 15-19, 1982.

[1983] FANTINO B., FRENE, J., GODET, M. 'Dynamic Behavior of an Elastic Connecting-rod Bearing- Theoretical Study', SAE/SP-539-- Studies of Engine Bearings and Lubrication, Paper 830307, Feb. 1983, pp. 23-32.

[1983] WELSH, W.A., BOOKER, J.F. 'Dynamic Analysis of Engine Bearing Systems', SAE International Congress, Detroit, Michigan, Feb. 28 - March 4, 1983, Paper 830065.
'Tribology of Reciprocating Engines', Eds. D. Dowson et al., (Butterworths, London, England, 1983), pp. 29-36.

[1984a] GOENKA, P.K. 'Analytical Curve Fits for Solution Parameter Studies of Dynamically-Loaded Journal Bearings', Trans. ASME, J. Tribology, 106, 1984, p. 421.

[1984b] GOENKA, P.K. 'Dynamically-Loaded Journal Bearings: Finite Element Method Analysis', Trans. ASME, J. Tribology, 106, 1984, p. 429.

[1985] OH, K.P., GOENKA, P.K. 'Elasto-hydrodynamic Solution of Journal Bearings Under Dynamic Loading', Trans. ASME, J. Tribology, 107, 1985, p.389.

[1985] LaBOUFF, G.A., BOOKER, J.F. 'Dynamically-Loaded Journal Bearings: A Finite Element Treatment for Rigid and Elastic Surfaces', Trans. ASME, J. Tribology, 107, 1985, p. 505.

[1986] BOEDO, S. 'Dynamics of Engine Bearing Systems: Rigid Body Analysis', M.S. Thesis, Cornell University, Ithaca, N.Y., 1986.

Paper XI(iv)

Thermal considerations in engine bearings

G. A. Clayton and C. M. Taylor

In the analysis of dynamically loaded journal bearings, such as those occurring in the internal combustion engine, the prediction of shaft orbit and the minimum oil film thickness during the cycle has attracted much of the attention of designers. Relatively little effort has been expended in developing techniques to give a reliable estimation of the effective lubricant temperature through thermal analysis. The present study has been directed towards a thermal analysis of the Ruston and Hornsby 6VEB-X Mk III big-end bearing which has become a benchmark test case. Predictions of effective temperature based on the estimation of lubricant flow rate and power loss by simple techniques have been compared with the results of a more rigorous approach incorporating consideration of oil film history. Good agreement between the two analyses was obtained for the particular bearing considered.

1 INTRODUCTION

The mobility method of dynamically loaded journal bearing analysis (1,2) has over a period of some twenty five years firmly established itself as a practical design tool. The main thrust of the method is to determine the orbit of the shaft centre during an operating cycle. With this established the minimum lubricant film thickness during the cycle may be calculated and the magnitude of this has been used as an important design criterion. The technique involves a series of assumptions and approximations. Whilst these lead to a simple and robust approach to orbit specification through computer analysis, it must be recognised that the resulting data can only be used effectively in a comparative sense as opposed to being considered precise. That is, designers would hope that the analytical technique would give correct trends in relation to film thickness variation as parameters are altered but would accept that absolute magnitudes can be called into question.

It is appropriate to note some of the explicit and implicit assumptions of the mobility method,

- Short journal bearing analysis (usually)
- Circumferential symmetry (no supply holes)
- No elastic or thermal distortions
- Newtonian lubricant behaviour
- No misalignment
- Perfectly circular components
- Isothermal lubricant film

Bearing in mind such a catalogue of approximations it is clearly evident why the method is applied in a comparative manner, with the use of parametric studies and widespread reference to practical operating experience.

The most commonly used form of Mobility analysis adopts the short journal bearing approximation in which circumferential pressure gradients are considered to be small with respect to axial pressure gradients. Taken with the assumption that negative gauge pressures may be neglected, giving a cavitation region always extending over 180° (half the developed bearing surface), this results in an elegant and flexible design procedure to step out the journal movement.

It is perhaps surprising, at first, to find that few researchers and designers have developed a thermal analysis to complement the orbit prediction technique. The latter relies upon a specification of the fixed effective temperature in the lubricant film leading to a constant dynamic viscosity. This is commonly determined on the basis of experience of experimentally determined operating temperatures but may in effect be a guess. The purpose of the present paper is to explore a simple, theoretical, thermal analysis approach and to extend this to more complex considerations.

1.1 Notation

b	bearing width (one land)
c_d	diametral clearance
d	bearing diameter
F	bearing load
H_{SH}	shear power loss
H_{SQ}	squeeze power loss
J_1^{00}	$\int_{\theta_1}^{\theta_1+\pi} \frac{d\theta}{(1+\varepsilon\cos\theta)}$
M	mobility scalar
p_f	groove supply
Q_H	hydrodynamic flow rate
Q_p	pressure flow rate
V	shaft centre velocity
β	angle between load line and shaft centre velocity vector
ε	eccentricity ratio

η dynamic viscosity

θ_1 angular location of pressure build up

ω_b bearing angular velocity about its centre

ω_j journal angular velocity about its centre

2 THEORETICAL ANALYSIS - SHORT BEARING THEORY

The curve fitted valves of mobility number according to short journal bearing analysis have been obtained by considering an isothermal lubricant film and a cavitation region extending over half the bearing circumference (1). The location of the 180° portion of the film where superambient pressures are generated, with respect to the line of journal and bush centres, depends upon the ratio of squeeze to entraining velocities. The method takes no account of the influence of lubricant supply pressure. Thus for the Ruston and Hornsby 6VEB-X Mk III big-end bearing to be considered, which has a full central circumferential groove, it is a relatively straightforward matter to calculate what will be called the hydrodynamic flow (rate). At any instant this depends upon the surface velocities and has been termed velocity induced flow rate by some. This is given by (3),

$$Q_H = \frac{FM(c_d/d)^2 c_d}{2\eta} \qquad (1)$$

and represents the flow from both sides of a bearing. Determination of lubricant flow rate by this equation neglects the contribution of the lubricant supply pressure. It has been a widespread engineering practice to determine a pressure induced flow rate taking all surface velocities to be zero, and to sum the hydrodynamic and pressure flow components to obtain the total side leakage. For a full central circumferential groove the pressure induced flow rate may be shown to be (3,4),

$$Q_P = \frac{0.0327\ p_f c_d^3}{\eta} \left(\frac{d}{b}\right) (1 + 1.5\varepsilon^2) \qquad (2)$$

where (b) is the width of one land and the flow rate (Q_P) is from one side of the bearing only. Recent studies have been undertaken by the authors for a **steadily loaded** central circumferential grooved bearing incorporating coupled surface velocity and supply pressure influences on flow rate together with a consideration of film rupture and reformation. This work, which has yet to be published, has shown that the **total** flow rate from one side of the bearing may be given to a good approximation by,

$$Q_P = \frac{0.0315\ p_f c_d^3}{\eta} \left(\frac{d}{b}\right) (1 + 1.5\varepsilon^2) \qquad (3)$$

It will be recognized that this is independent of rotational velocity. This demonstrates the sound engineering judgement adopted in the ESDU design item for plain journal bearings (4). Here for central circumferential groove bearings, in the absence of reliable data, the use of equation (2) was recommended to predict the total flow rate. The decision to do this was based upon the nature of the individual Couette and Poiseuille flow rates from the groove with no consideration of cavitation.

The determination of the total flow rate at any instant by summing the hydrodynamic and pressure flow components for a dynamically loaded bearing according to equations (1) and (2) has been shown to be inaccurate by Jones (5). He undertook analysis of engine bearings with various groove arrangements and a consideration of oil film history. That is, a detailed account was taken of continuity of flow at the rupture and reformation boundaries of the cavitation region during the transient process. The incorporation of oil transport effects demonstrated that the simple addition of the hydrodynamic and pressure flows overpredicted the more precise calculation. This is in accord with the remarks made previously in relation to steadily loaded bearings. Jones showed that for a variety of grooving arrangements the pressure induced flow rate alone gave better agreement with the more rigorously determined side leakage, and indeed with experimental measurements. He also commented that the oil film history model was particularly demanding on computing resources, such that it was only fleasible to use it for critical bearing analyses.

The prediction of power loss in dynamically loaded bearings has only received the attention of research workers relatively recently. Martin et al (6) detail expressions for instantaneous power loss due to shear (neglecting pressure flow influence) and what is termed squeeze effects. These are based on rotational and translational velocities respectively with the shear power loss given by,

$$H_{SH} = \frac{\eta d^3 b\ (\omega_j - \omega_b)^2 J_1^{00}}{4c_d} \qquad (4)$$

and that due to 'squeeze' by,

$$H_{SQ} = FV\cos\beta \qquad (5)$$

A typical thermal balance calculation to determine the effective operating temperature of a bearing (4) would take the viscous dissipation, or some proportion of it, and estimate the temperature rise due to this being convected away by the lubricant. Thus according to the theoretical approach outlined above, the instantaneous predictions of flow rate (equations 1 and 2) and power loss (equations 4 and 5) would be integrated or summed around a full cycle and combined to calculate a temperature rise from the lubricant supply condition. If the new effective film temperature did not coincide with that adopted to determine the journal orbit used for the calculation, an iterative process would be required to achieve convergence.

Such thermal calculations in the past have been known to give substantially incorrect values of operating temperature. This, it has been recognized, may have been due to the summation of hydrodynamic and pressure flow rate values to determine total flow rate, but could also be associated with imprecise determination of power loss, for example due to inadequate specification of the cavitation region. The approaches adopted by the authors will now be outlined.

3 THERMAL ANALYSIS

Calculations have been carried out for the big-end bearing of the Ruston and Hornsby 6VEB-X Mk III diesel engine. Some important specifications adopted for the determination of the performance of this full circumferential grooved bearing are given below,

Overall bearing width	127.0mm
Width of one land	57.15mm
Bearing diameter	203.2mm
Diametral clearance	0.165mm
Engine speed	10Hz (600 rpm)
Lubricant dynamic viscosity	0.095Pas at 40°C
	0.0097Pas at 100°C
Oil supply pressure	$276kN/m^2$

Very full details of all data required to calculate the orbit of the crankpin within the big-end housing are presented in (7), including the loading to be carried resulting from cylinder pressures and inertial effects. For the computations carried out in reference (7) a fixed operating oil viscosity of 0.015 Pas, corresponding approximately to an effective temperature of 81°C, was used. This effective viscosity will vary for the thermal calculations carried out here. The stated oil supply pressure to the groove was 276 kN/m^2 (2.76 bar), although this did not impinge upon the analysis in reference (7) to predict shaft orbit. In the present work the effect of varying supply pressure will be considered.

3.1 Method of Selective Summation for Oil Flow Rate Prediction

An improved but still simple approach to predict flow rate has been examined as part of the study. This has been termed selective summation. In Figure 1 the variation of instantaneous oil flow rate components as a function of crank angle has been plotted. The separate hydrodynamic and pressure flow elements have been summed to show the combined or total flow and the integrated flow rates over the four stroke cycle are indicated. These calculations were carried out on the basis of the analysis presented in section 2 and for an effective lubricant viscosity of 0.015 Pas and a supply pressure of 276 kN/m^2.

It has already been stated that on the basis of his oil film history analysis, Jones (5) found that the total flow rate was better approximated by the pressure flow component alone. To be more precise hydrodynamic and pressure flow components substantially over estimated the total flow rate, whilst the pressure flow element itself somewhat underestimated this value. Here, the hydrodynamic and pressure flow components have been calculated at each time instant considered during the cycle and the **larger** value used in integrating or summing to obtain the side leakage during two revolutions of the crankshaft. This 'selective summation' of flow components is a simple approach which might be expected to give better agreement with the total flow rate determined by more precise means than by talking either (Q_H + Q_P) or (Q_P) alone. The results to be presented later will bear out this contention for the bearing under consideration.

3.2 Present Simplified Oil Film History Technique

The authors have developed a solution procedure incorporating oil film history effects and an iterative thermal balance. Jones (5) did not describe the basis of his film history technique in great detail, however, the approaches appear to be similar. It is not proposed to describe here the details of the present oil film history analysis.

Jones did comment upon the substantial computer central processing unit (CPU) time necessary for his analysis technique. This is consequent upon the necessity of having to solve Reynolds' equation, with the correct determination of film rupture and reformation boundaries, perhaps several thousand times to obtain a correctly converged journal orbit. The imposition of a thermal balance iteration over and above this would suggest unacceptable computer run times may be required even by present day standards. The following procedure was therefore adopted to circumnavigate this problem. It would not be expected that the incorporation of oil film history effects would significantly affect the journal orbit prediction for a given effective dynamic viscosity **for the bearing under consideration.** This is because oil film pressures and hence load capacity are only marginally influenced. Thus it is reasonable to determine the journal orbit for an assumed effective viscosity (or temperature) using the mobility method either according to short bearing (1) or finite width (2) analytical approaches. Jones showed the modest effect that consideration of oil film history had on the journal orbit for a **full circumferential groove bearing.**

With the orbit determined the shaft centre velocity at any instant during the cycle is known. At each successive position Reynolds' equation is solved with the cavitation region adjusted according to whether the oil available can fill the new control volume at each nodal point. There is no iteration to determine shaft centre velocity or orbit and hence the solution time is drastically reduced. Each successive pressure distribution presents a suitable starting point for the next position. It will be noted that as well as satisfying flow continuity, the finite difference approximation technique used for solution of the governing Reynolds' equation enable the supply pressure at the groove to be incorporated into the analysis.

This procedure was repeated until the total lubricant flow rate calculated at the same position on successive cycles was within a specified tolerance (2%). This ensured that the cavitation region was the same according to oil film history requirements. After the convergence the total flow rate and total power loss for the 720° cycle were determined by numerical integration of their components determined at each time step.

A thermal balance procedure could now be undertaken. By assuming that the **total** power dissipation in the film was convected away by the lubricant, the effective temperature of the oil film could be determined according to the temperature rise from the specified inlet condition. If this effective temperature (and hence viscosity) did not agree within a specified tolerance (0.25°C) to that assumed for the orbit calculation, then the process was repeated. The flow chart for the analysis is shown in Figure 2.

3.3 Calculation Details

An important aim of the present study was to

investigate effective simplifications which might be undertaken in the thermal analysis of dynamically loaded journal bearings. It is a straightforward matter to write down in all their complexity the governing equations for the problem and the associated boundary conditions. However, for some time yet the full solution will be impractical and designers will have recourse to approximate techniques. The previous two sub-sections have described simplified aspects adopted here but it is also important to clarify some more detailed points to facilitate understanding and to enable comparison with the work of others.

(i) For the results described in this paper the mobility data of Goenka (2) for finite width bearings was used. Computations carried out using the Booker short journal bearing curve fits for mobility (1) lead to precisely the same conclusions. Where hydrodynamic flow rate was calculated according to equation (1), the Goenka mobility data was used although the equation is strictly only valid for a 180° film according to short bearing theory. The difference involved in flow prediction for the example considered was only a few percent (see Table 1).

(ii) The calculation of power loss may be quite complex involving three components (a) Couette and Poiseuille contributions associated with viscous shear, (b) an element due to the translation of the journal through the oil film. The latter component may be thought of as the work required to push the journal around its orbit. There is some confusion of terminology in the literature. For the oil film history solutions presented later the contribution of all three components have been incorporated in the analysis. It should be noted that the Couette viscous shearing is the dominant element responsible for around 92% of the power loss. The viscous shear has then been determined from the Couette considerations only neglecting pressure on Poiseuille effects (see (8)). The translation term has also been included in this '2PI FILM SHEAR' approach, but its influence on power loss is small (5%) and could be neglected.

4 RESULTS AND DISCUSSION

The results presented in Figures 3 and 4 are not for a thermal analysis. Each relates to an assumed effective dynamic viscosity of 0.015 Pas. Figure 3 shows the variation of oil flow rate as a function of crank angle, for a supply pressure of 276 kN/m^2 and according to three analytical approaches. These are the full film history analysis after Jones (5) and the selective summation and simplified oil film history technique developed by the authors.

The excellent agreement between the predictions of flow rate by the three analyses is encouraging. This is true for both instantaneous lubricant flow rates and the cyclic value. Indeed the oil flow rates during the 720° cycle agree to better than 6%. The significance of this in terms of computational time to undertake an analysis is worthy of note. Typically the simplifed oil film history approach of the authors needed about 300 times the central processing unit requirement of the selective summation technique (the latter taking about 1 second CPU on an Amdahl 570 machine). The full oil film history considerations of Jones might be estimated to consume an order of magnitude more in run time than the authors approximate method. Thus the selective summation scheme of oil flow prediction based on a shaft orbit determined by the mobility method could be judged to reduce computational time by a factor of some 3000.

It must be borne in mind that the comments above are made in respect of an analysis of only one bearing having a full central circumferential groove. The work of Jones (5) has shown that for such a grooving arrangement the predicted journal centre orbits, with and without oil film history considerations, are indeed similar. This reflects the adequate lubricant supply situation afforded by a full circumferential groove. The situation will be less satisfactory for a partial groove or supply hole. Table 1, presented below, summarizes the prediction of cyclic lubricant flow rate for the Ruston and Hornsby big-end bearing according to the various analysis approaches which may be adopted. The figure in relation to the oil film history considerations of Jones has been estimated from his data.

ANALYSIS APPROACH	PREDICTED OIL FLOW RATE (m^3/s) $x10^6$
JONES(5) OIL FILM HISTORY	46.9
SIMPLIFIED OIL FILM HISTORY	
(a) Booker Mobility Data	51.5
(b) Goenka Mobility Data	49.5
SELECTIVE SUMMATION	
(Goenka Mobility Data)	48.0
COMPONENT FLOWS	
(Goenka Mobility Data)	
(a) Hydrodynamic (Eqn. 1)	27.1
(b) Pressure (Eqn. 2)	38.9
(c) Hydrodynamic + Pressure	66.0

Table 1 SUMMARY OF OIL FLOW RATE PREDICTIONS

The influence of groove supply pressure is indicated in Figure 4. Oil flow rate is shown for both the authors simplified oil film history analysis and the selective summation approach. The agreement is within 3% over the supply pressure range of 160 to 370 kN/m^2. Perhaps the most notable feature is the almost linear dependence of total flow rate on oil supply pressure. This might be expected from the selective summation analysis, when it is noted from Figure 1 that for a substantial proportion of the total cycle the instantaneous pressure flow component is larger than the hydrodynamic one.

The remaining Figures (5,6 and 7) are all concerned with results from thermal analysis in which an iterative process has been undertaken to converge to the effective oil temperature. The lubricant supply pressure was fixed at

276 kN/m² and typically five cycles were needed to achieve a satisfactory solution. However, this did depend upon the accuracy of the original estimated effective temperature.

Power loss, oil flow rate and effective oil temperature are plotted in Figures 5, 6 and 7 respectively according to the predictions of the authors' simplified oil film history analysis. Their variation as a function of lubricant inlet temperature from 40 to 100°C is presented. From Figure 5 it is evident that an increase in oil inlet temperature results in an increase in effective temperature and hence a fall in dynamic viscosity, thus causing a drop in power loss. Over the inlet temperature range 40 to 100°C, the power loss falls from about 1900 to 800 watts, a 58% reduction. Also shown in Figure 5 are the power loss, predictions according to what is termed a '2PI FILM SHEAR' analysis. Here the viscous component of the power loss has been calculated by taking the oil film to be full at every instant during the cycle (i.e. no cavitation) and determining the Couette or velocity shear loss only neglecting pressure flow influences. The instantaneous power loss for such an approach taking the film to extend over 2π radians is given by (e.g. 8),

$$H_{SH} = \frac{\pi}{2}\,\frac{\eta d^3 b(\omega_j - \omega_b)^2}{C_d(1-\varepsilon^2)^{1/2}} \qquad (6)$$

Martin (8) has shown that such a simplification is very effective in predicting the trends of cyclic power loss variation and the overall agreement (to within about 7%) shown in Figure 5 bears out his contention.

The variation of oil flow rate according to lubricant inlet temperature is substantial as evidenced by Figure 6. An increase in inlet temperature from 40° to 100°C is predicted to increase the cyclic flow rate by 142% from 36×10^{-6} m³/s to 87×10^{-6} m³/s (130 to 313 litres/hour). Such an increase is consequent upon the reduction in effective dynamic viscosity because of the temperature increase. The range of situations taken may be considered to correspond to those from a near cold start to fairly hot running.

The change in effective oil film temperature with increase in oil inlet temperature is depicted in Figure 7. It can be seen that over the inlet temperature range examined, the effective lubricant temperature is predicted to change from 69°C to 107°C. For the 40°C inlet temperature the rise to the effective temperature is 29°, whilst at 100°C inlet the increase is restricted to 7°. This reflects the variations in power loss and oil flow rate shown in Figures 5 and 6. As the inlet temperature of the lubricant increases, power loss drops and oil flow rate increases and thus the temperature rise will fall significantly.

The following Table 2 shows the predicted values of effective oil film temperature, oil flow rate, power loss and minimum oil film thickness according to the present oil film history solution for inlet oil temperature variation. It will be recalled that Campbell et al (7) carried out orbit calculations with an effective dynamic viscosity corresponding to a constant oil film temperature of about 81°C.

Inlet Temp. (°C)	Effective Film Temp. (°C)	Oil Flow Rate ($m^3/s\times10^6$)	Power Loss (kW)	Minimum Thickness (μm)
40	69	36	1.88	6.7
60	78	44	1.44	5.3
80	91	59	1.10	3.9
100	107	87	0.80	2.9

Table 2 Influence of Inlet Oil Temperature

The importance of carrying out a thermal iterative balance to determine the effective operating temperature will be apparent. The errors in predicting important performance characteristics are substantial if the wrong effective film temperature is chosen.

One final, important point remains to be highlighted. The authors have presented performance data for the Ruston and Hornsby 6VEB-X Mk III big-end bearing incorporating a thermal analysis with a simplified oil film history consideration. It has been shown (Figures 3 and 4) that the approximate (but simple) approach to calculating oil flow rate which has been called 'selective summation' gives excellent agreement with the predictions of more precise analyses carried out. Further, the '2π FILM SHEAR' method of determining power loss is likewise an effective simlification compared with the more exact analysis (Figure 5). It follows, therefore, that for the specific example dealt with acceptable predictions of operational characteristics of the bearing with thermal considerations can be made by an analysis incorporating both simple approaches. The computer processing unit time with such a technique will be very much less than for the more exact analysis, making it a sound engineering approach for industrial design. As has been indicated such a conclusion will probably not be valid for engine bearings with partial grooves or oil supply holes.

5 CONCLUSIONS

(i) A study of the thermal analysis of dynamically loaded journal bearings has been undertaken. The well known Ruston and Hornsby 6VEB-X Mk III big-end bearing has been chosen as a test case. This bearing has lubricant supplied through a full central circumferential groove. The thermal modelling has assumed the total power dissipation in the lubricant film to be convected away by the flow. No consideration of heat conduction to the bounding solids has been undertaken.

(ii) An important feature in thermal considerations is the assessment of oil flow rate. The precise prediction of this will be influenced by oil film history and is complicated by detailed consideration of lubricant supply grooves. For the bearing considered here results with a full oil film history analysis have been determined by others (3). Two simplified approaches to oil flow rate prediction were examined,

(a) Selective summation, in which the greater of the analytical predictions for hydrodynamic and pressure flow was taken to represent the instantaneous value for flow rate.
(b) A simple oil film history analysis, in which the orbit of the journal was assumed to be satisfactorily predicted by a mobility analysis, and the oil film history and thermal analyses were carried out with this orbit.

(iii) The predictions of oil flow rate by the two simpler approaches agreed closely with the result of Jones. The computer central processing unit time for the 'selective summation' method was, however, orders of magnitude less than for the oil film history techniques. It is recognized that the same conclusion as regards oil flow rate prediction could probably not be drawn for bearings with partial grooves or oil supply holes.
(iv) The power loss in the bearing was determined by a full analysis of the components arising from rotational and translational velocity components but also by a simple approach (after Martin (8)) in which only Couette shearing was considered for a full, non cavitating, film. The agreement between the two predictions of power loss was excellent. The prediction of oil flow rate and power loss enabled a thermal iteration to be carried out and detailed predictions have been made for different assumed values of the oil inlet temperature at the groove. The implications in relation to predicted performance is shown to be significant.
(v) Because of the effective simplifications possible for the prediction of oil flow rate and power loss, it may be concluded that a reliable thermal analysis may be carried out for a full central circumferential groove bearing without recourse to complex and time consuming oil film history analysis. This may well not be true for other grooving arrangements.

6 ACKNOWLEDGEMENT

One of the authors (GAC) was a postgraduate research student in the Department of Mechanical Engineering at the University of Leeds whilst the work described in this paper was undertaken. The support of the Science and Engineering Research Council (SERC) for provision of a Cooperative Award in Science and Engineering (CASE) studentship is acknowledged. The cooperating body was Ricardo Consulting Engineers Ltd and their financial contribution and technical advice of staff are acknowledged with thanks. In particular the support and comments of Mr. Adrian R. Heath and Mr. David J. Lacy are gratefully recognized.

References

(1) BOOKER, J.F. 'Dynamically loaded journal bearings : mobility method of solution', Trans. ASME, Journal of Basic Engineering, Sept., 1965, pp.537-546.

(2) GOENKA, P.K. 'Analytical curve fits for solution parameters of dynamically loaded journal bearings', Trans. ASME, Journal of Tribology, Oct., 1984, Vol 106, pps 421-428.

(3) MARTIN, F.A. 'Developments in engine bearings', Proc. of the 9th Leeds-Lyon Symposium on Tribology - "Tribology in Reciprocating Engines", 1983, Butterworths, pps 9-28.

(4) 'Calculation methods for steadily loaded pressure fed hydrodynamic journal bearings', ESDU design item 66023, 1966, pp. 70.

(5) JONES, G.J. 'Crankshaft bearings : oil film history', Proc. of the 9th Leeds-Lyon Symposium on Tribology - "Tribology in Reciprocating Engines", 1983, Butterworths, pps 83-88.

(6) MARTIN, F.A., LO, P.M. and BOOKER, J.F. 'Power loss in connecting rod bearings', Proc.I.Mech.E. International Conference - "Tribology-friction lubrication and wear fifty years on", Vol. II, 1987, pps 701-708.

(7) CAMPBELL, J., LOVE, P.P., MARTIN, F.A. and RAFIQUE, S.O. 'Bearings for reciprocating machinery : a review of the present state of theoretical, experimental and service knowledge', Proc. I.Mech.E. Conference on Lubrication on Wear, London, Vol. 182, Pt 3A, 1968, pp. 34.

(8) MARTIN, F.A. 'Friction in internal combustion engine bearings', I.Mech.E. Conference - 'Combustion Engines - Reduction of Friction and Wear', Publication 1985-3, March 1985, pps. 1-18.

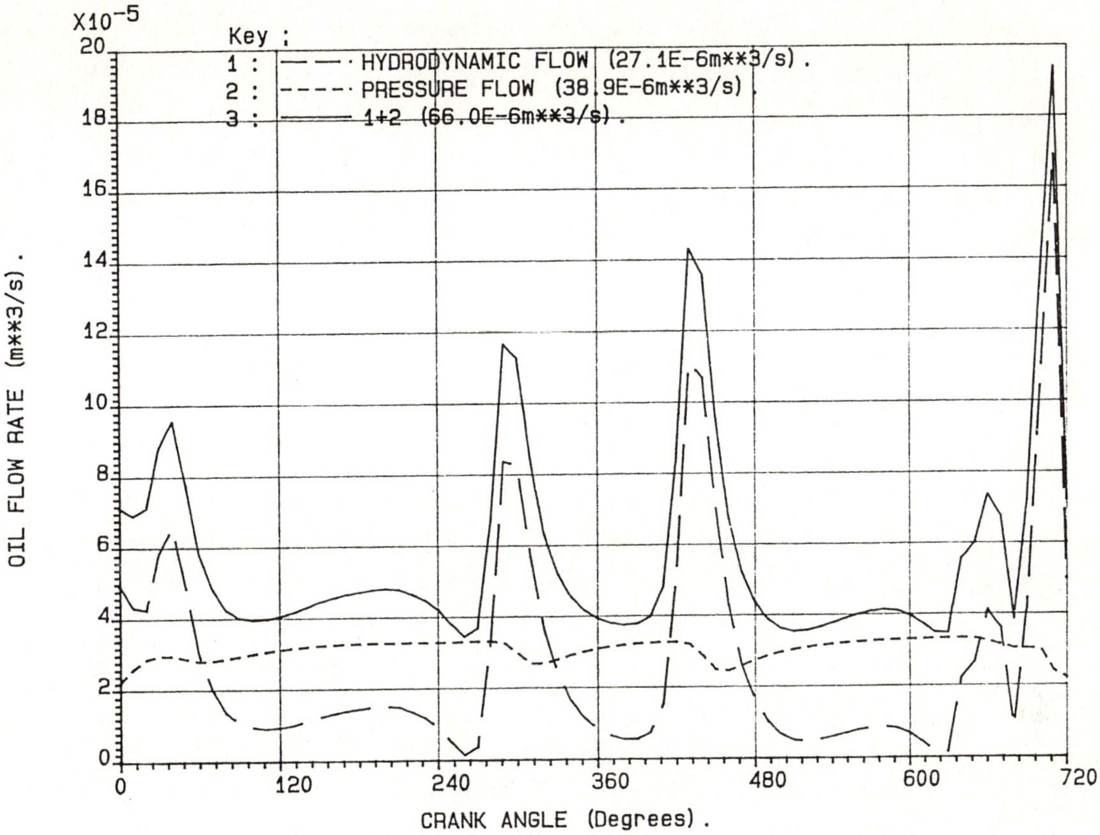

FIGURE 1. OIL FLOW RATE VARIATION WITH CRANK ANGLE.

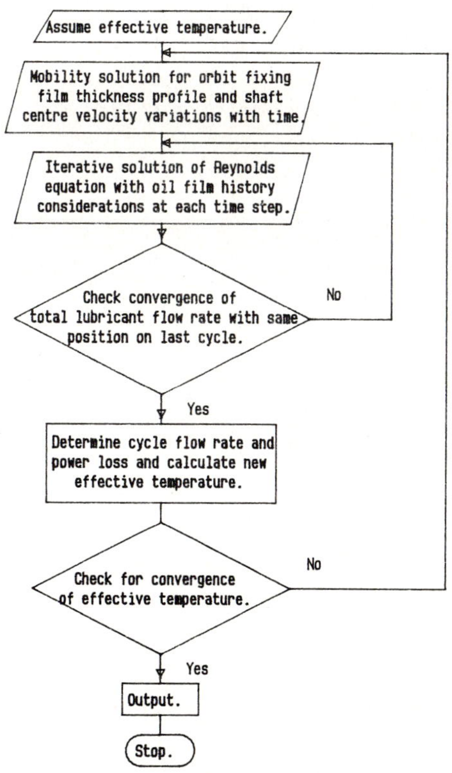

FIGURE 2. FLOW CHART FOR OIL FILM HISTORY THERMAL SOLUTION.

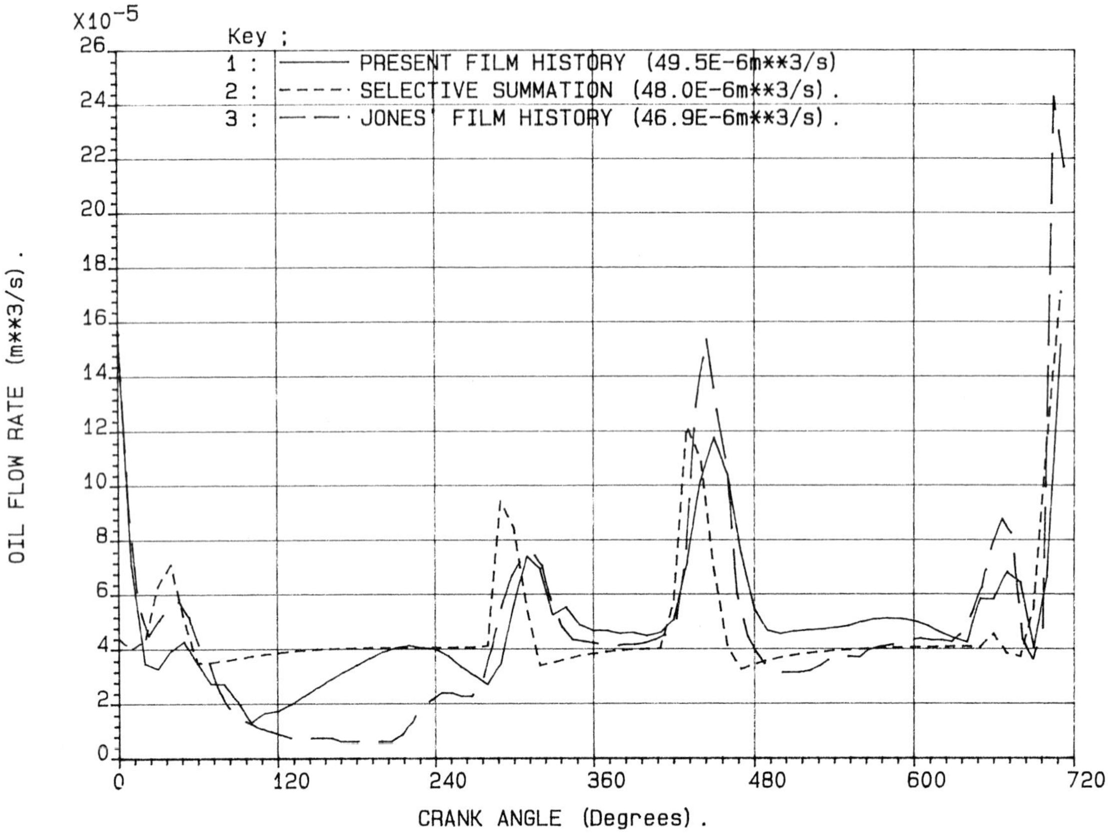

FIGURE 3. OIL FLOW RATE VARIATION WITH CRANK ANGLE.

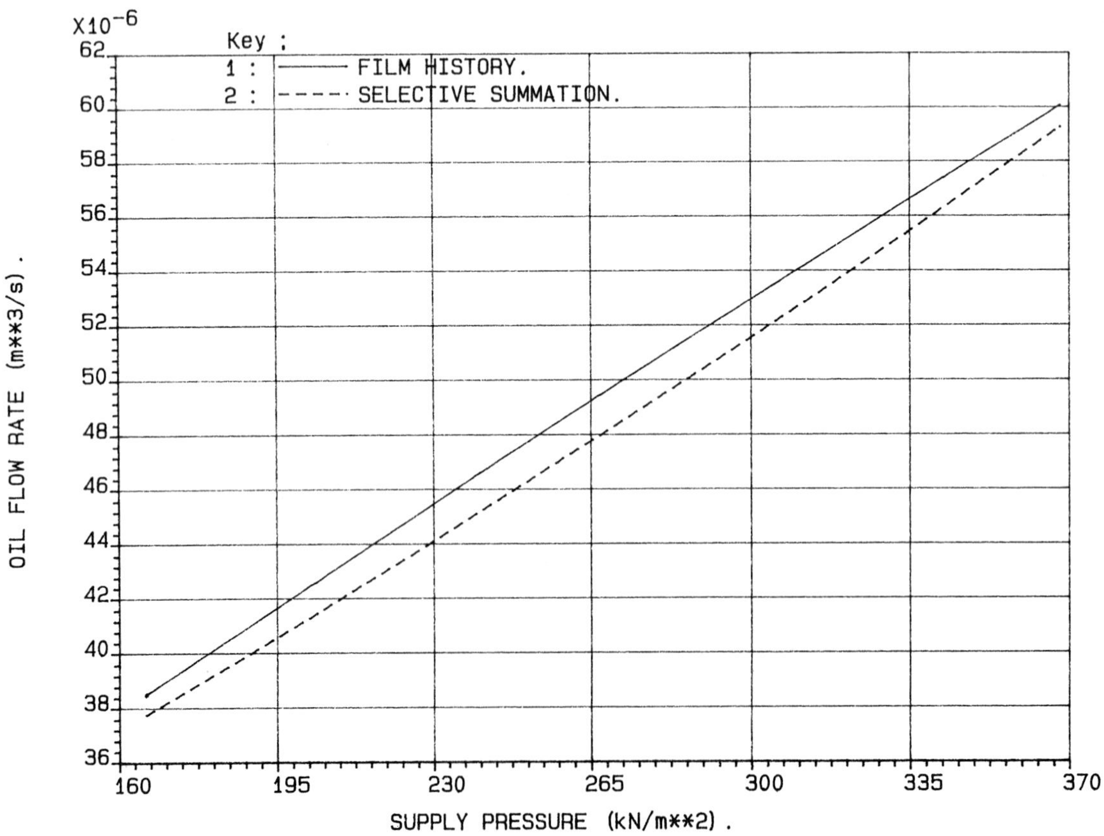

FIGURE 4. OIL FLOW RATE VARIATION WITH SUPPLY PRESSURE (Note false origin) .

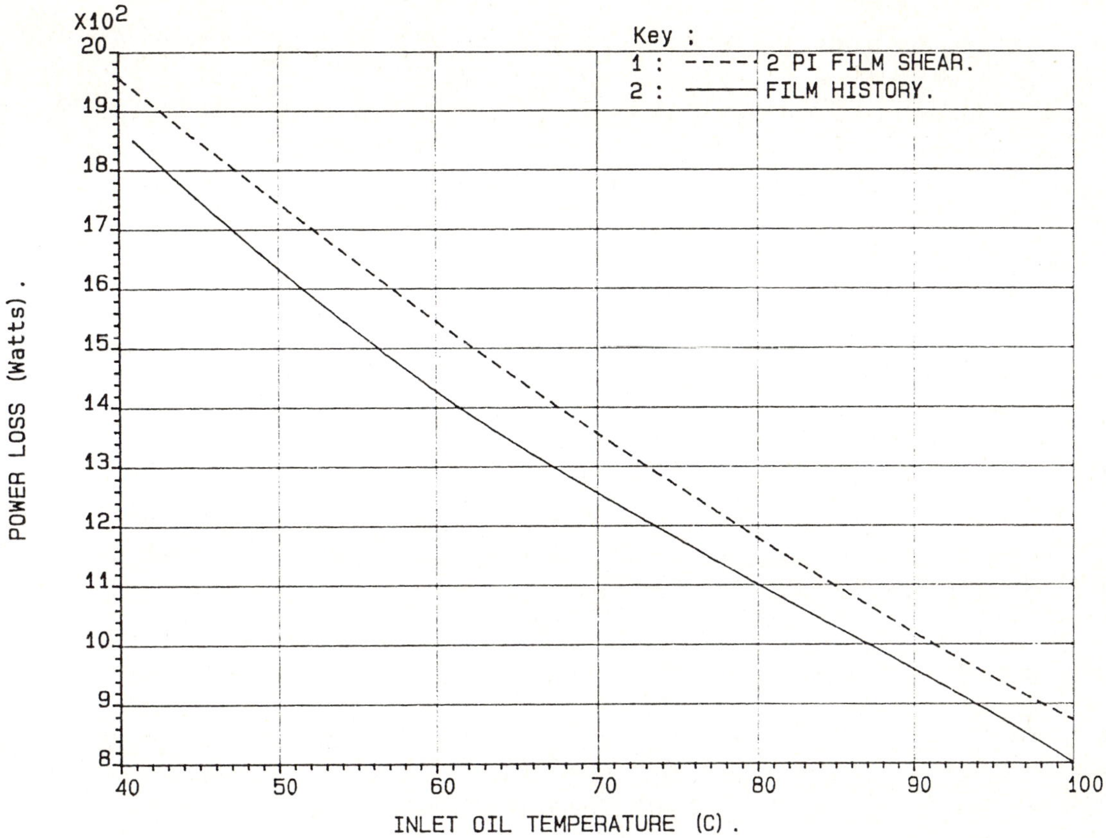

FIGURE 5. POWER LOSS VARIATION WITH INLET OIL TEMPERATURE.
(Note false origin).

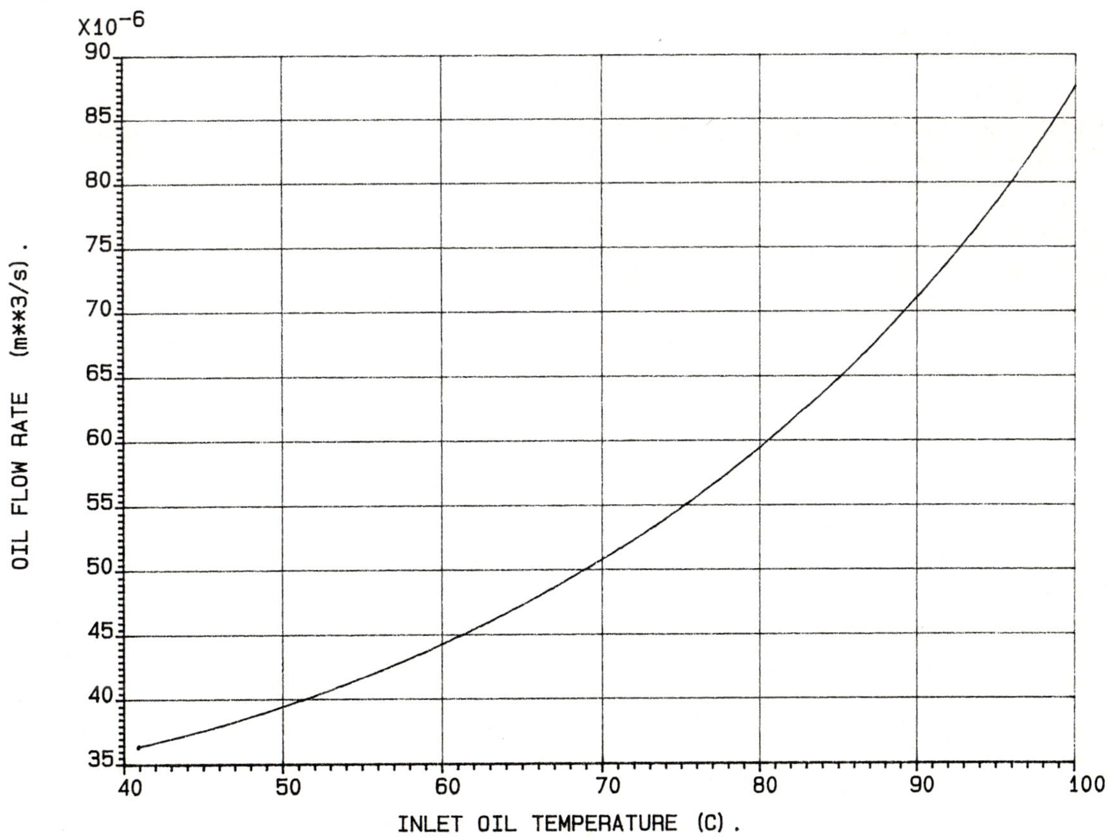

FIGURE 6. OIL FLOW RATE VARIATION WITH INLET OIL TEMPERATURE.
FILM HISTORY SOLUTION (Note false origin).

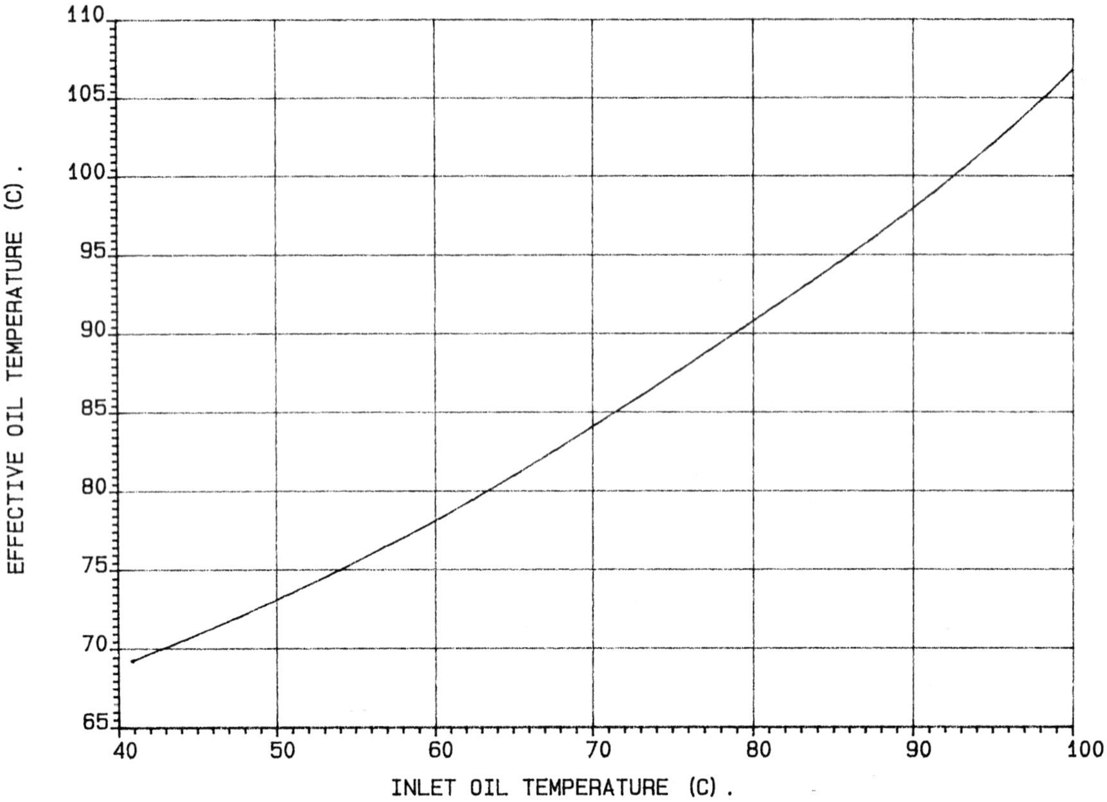

FIGURE 7. EFFECTIVE OIL TEMPERATURE VARIATION WITH INLET OIL TEMPERATURE-FILM HISTORY SOLUTION (Note false origin).

SESSION XII
CERAMICS

Chairman: Dr J C Bell

Paper XII(i)

Design requirements of ceramic sliding contacts

R. J. Gozdawa and T. A. Stolarski

The main aim of this paper is to discuss, using design case study, factors controlling the performance and affecting the utilization of engineering ceramics, particularly the silicon carbide, as a material for the components of a sliding bearing. The bearing system, comprising combined radial and thrust tilting pad bearings and designed for Framo Engineering, was destined for a special pump used in North Sea oil exploration.

1 INTRODUCTION

The science of interacting surfaces in relative motion has arisen, principally, from experience with metallic and organic materials, but advances in the technology of non-metallic inorganic materials now open the way to use them as the bearing materials as well.
The advantages seem to be quite significant promising lower friction, longer lives, high temperature capability and relaxation of the lubricant specification (1). However, it is recognized that the use of refractory, inorganic non-metallic compounds and the advantages their unique combination of such properties as high hardness and chemical inertness bring to bearing surfaces should be considered in the light of strict precautions which must be observed in their application (2). It is especially true when the ceramics are used as inserts and liners or structural and load bearing components in their own right.
Although ceramics offer a significant potential advantages for sliding contact applications there is insufficient amount of information regarding their behaviour, performance and requirements related to the design of sliding contact components. The majority of research on friction, wear and load carrying capacity has been restricted to specimen testing (3,4,5) and there is very little information about real systems behaviour or design recommendations in the form of suitable design rules exploiting the obvious advantages at acceptable level of risk.

2 DESIGN CONSIDERATIONS

The bearing system designed for Framo Engineering, Norway, consisted of radial and thrust tilting pad bearings combined together in one unit. The system is shown, schematically, in Fig.1. The design brief required that the system, supporting vertically positioned shaft of the pump, should be lubricated with the working fluid which was a mixture of a cooling oil and hard, abrasive particles. This requirement excluded a number of traditional bearing materials and put ceramics at the top of the list. Additional requirements of specific loads up to 13 MPa, rotational speed of

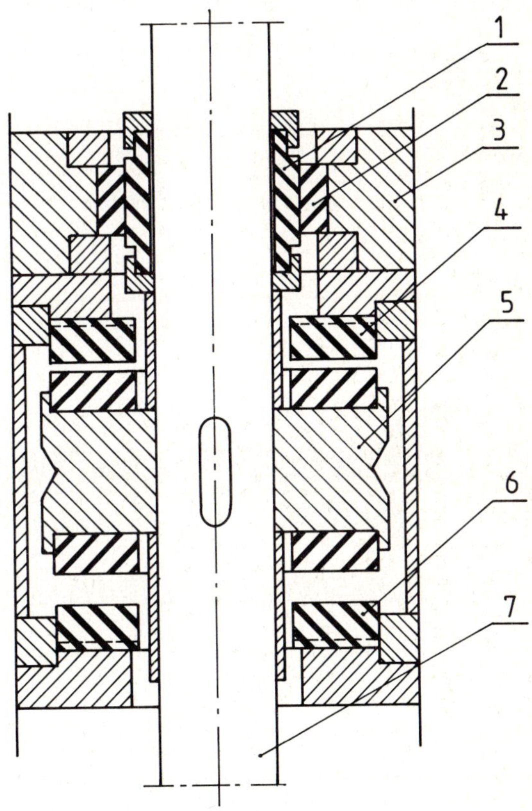

Fig.1 Bearing system: radial/thrust tilting pad 1 - journal sleeve assembly, 2 - journal pads, 3 - journal housing, 4 - inactive thrust pads, 5 - thrust collar assembly, 6 - active thrust pads, 7 - shaft.

6000 rpm and the temperature of the working fluid of up to 130°C provided further support for ceramics as the only suitable type of material for this particular application. Silicon carbide (reaction sintered) was finally chosen for bearing elements.
Having decided on ceramic-to-ceramic contact, a number of essential factors have to be considered in the final design of the ceramic component and its associated metal counterpart. For example, the attachement or anchoring of ceramic components

to the metal structure must be carefully considered and the possible differences in coefficients of thermal expansion between two materials ought to be taken into account. Moreover, high tensile stress concentrations must be rigorously avoided.
During the design stage of the bearing system in question and subsequent testing of a prototype the following problems were addressed and carefully considered:

(i) load capacity of actual ceramic bearing components to the point of failure in order to establish the prevailing mode of failure and to derive satisfactory safety factors,
(ii) ways to minimize and diffuse tension forces and any stress concentration,
(iii) means to attach brittle materials to themselves, and to ductile materials with the possible utilization of compressive loads acting away from edges,
(iv) design and fit of mating surfaces since it is not possible to use force in the assembly of components,
(v) the importance of close tolerances and high quality surface finish in contact stress reduction,
(vi) design ways to accomodate differential thermal expansion,
(vii) the significance of creep and relaxation over extended periods of time as press fits were used.

Some of the above problems could only be properly investigated on experimental way. Therefore, it was decided that extensive testing of a prototype system should precede the final implementation of it.

3 TESTING OF PROTOTYPE

3.1 Test procedure and conditions

A full scale prototype of the bearing system was tested in a specially constructed test rig in order to establish the limits of operating conditions safe for a long service life.
A schematic lay-out of the rig is shown in Fig.2. The axial load on thrust bearing was generated by a hydraulic ram and was at two levels that is 25 and 40 kN. Radial load on the journal bearing was applied by means of a load screw turned manually and was kept constant at 2000 N.
A load cell was used to measure the radial load while the axial load was assessed by measuring the pressure of oil in the hydraulic ram. The position of the radial bearing was such that the load was acting between two adjacent pads and was equally shared by them. Thermocouples, located near the contact zone between a shaft and the pads, were used to monitor the temperature in the bearing. Temperature of the lubricant at the inlet to the bearing was the main variable used to assess the performance of the bearing system. By changing the inlet temperature of the lubricant it was possible, through the change in its viscosity, to affect the lubrication conditions to a considerable extent. A diesel oil and Tellus T-37 were used as lubricants. Five different oil inlet temperatures were used, namely: 80, 90, 100, 110 and 120°C. Inlet temperature was thermostatically controlled but, nevertheless, there were some difficulties in keeping it constant throughout a given test. Lubricant flow rate for all tests was constant at 38 l/min. All tests were carried out at the speed of 6000 rpm.

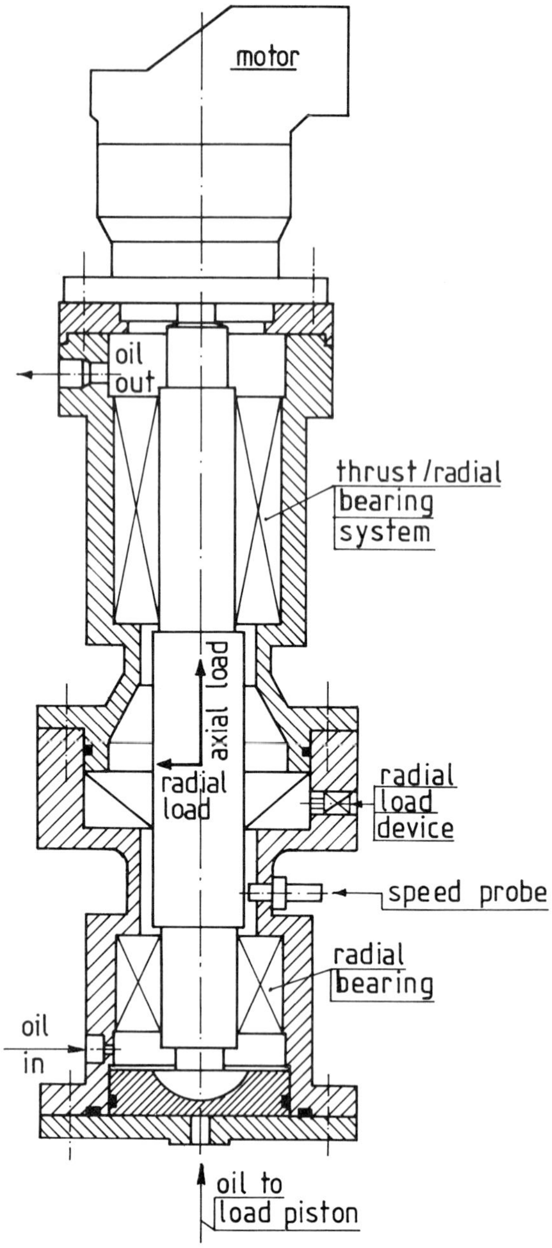

Fig.2 Lay-out of the test rig

3.2. Discussion of results

The results can best be expressed in terms of bearing temperature rise over the oil inlet temperature. The results of prototype testing are shown in Fig.3-6.
During the course of testing, two failures occurred. First failure of the thrust bearing pad took place under specific load of 12.75 MPa and speed of 6000 rpm. The lubricant was a diesel oil. Figure 7 shows the scanning electron micrograph of the failure surface. The silicon carbide face of the rotating thrust collar was also slightly damaged, mainly in the form of deep scratches, by loose particles generated by fracture of the silicon carbide pad at the pivot point. This failure was apparently caused by excessive loading which was approximately three times greater than the design load.
The second failure of the thrust bearing pad

occurred at much less specific load of 8 MPa and after a relatively short period of time of 4.5 hours. A hydraulic oil, Tellus T-37 was used as a lubricant. The test programme leading to the failure consisted in the progressive oil inlet temperature increase of $10^{\circ}C$ every 60 minutes starting from $80^{\circ}C$. At $120^{\circ}C$ oil inlet temperature, the bearing was run for only 30 minutes when the failure occurred. After failure, an inspection of the pads showed that carbide-to-carbide contact had occurred on some pads presumably due to the loss of the oil film. Unfortunately, there was no way of measuring the film thickness during running. However, taking into account high specific loads and low lubricant viscosity caused by elevated temperature, it is reasonable to say that boundary lubrication was a prevailing mode of lubrication prior to failure.
Inspection of the failed pad also suggested brittle fracture. This is clearly seen in Fig.8 in the form of scanning electron micrograph. As in the previous case, loose particles produced by fracture process scratched the surface of the rotating thrust collar.

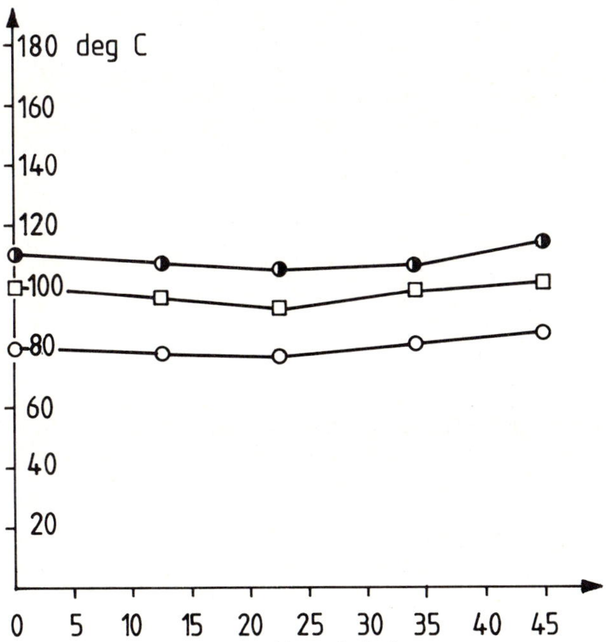

Fig.3 Results of prototype testing.
Test conditions: oil inlet temperature - $80^{\circ}C$, speed - 6000 rpm, thrust load - 25 kN, radial load - 1 kN.

Further visual inspection of the bearing after the second failure suggested that the load was not equally shared by all the pads and was probably caused by the variations in pad thickness. This point to the importance of precision during the manufacture of ceramic elements as they do not deform sufficiently to absorb any eventual inaccuracies.
As it can be seen in Fig.3-6, temperature rise in the bearing was consistent with oil inlet temperature. For the oil inlet temperature from 80 to $120^{\circ}C$, corresponding temperature

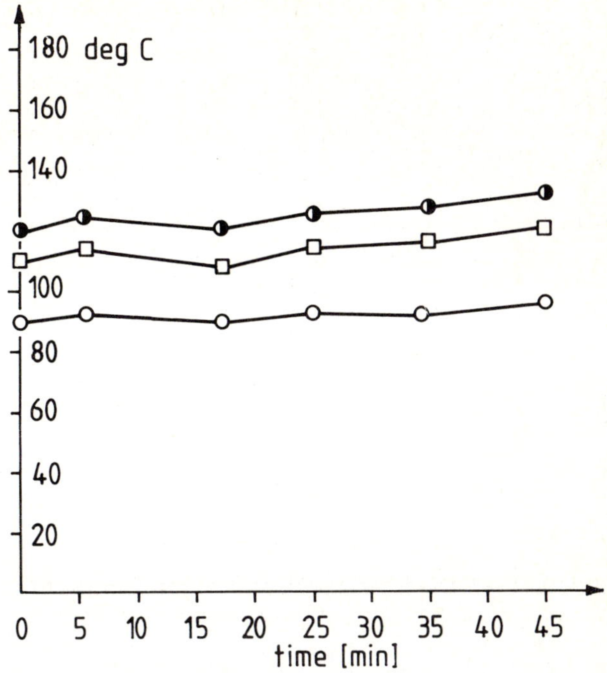

Fig.4 Results of prototype testing.
Test conditions: oil inlet temperature - $90^{\circ}C$, speed - 6000 rpm, thrust load - 25 kN, radial load - 1 kN.

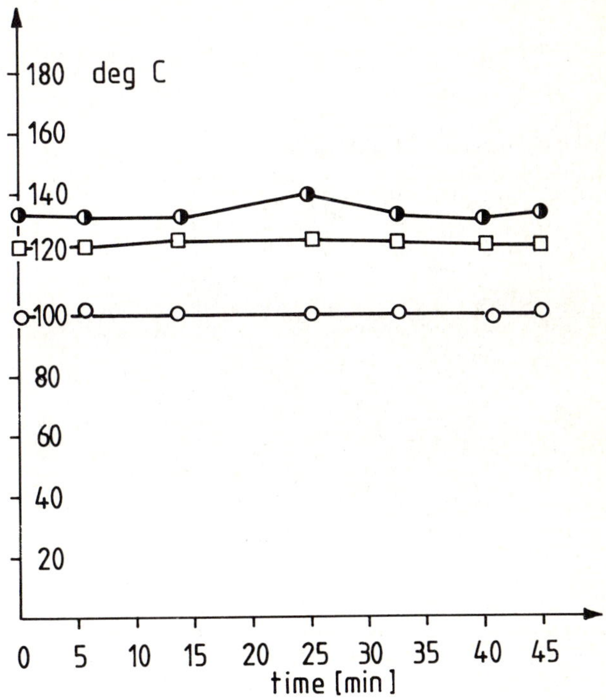

Fig.5 Results of prototype testing.
Test conditions: oil inlet temperature - $100^{\circ}C$, speed - 6000 rpm, thrust load - 25 kN, radial load - 1 kN.

rise in the radial bearing was in the range of 20 to 30° C respectively. Temperature rise in the thrust bearing, however, was roughly constant and equal to, approximately, 34° C regardless oil inlet temperature.

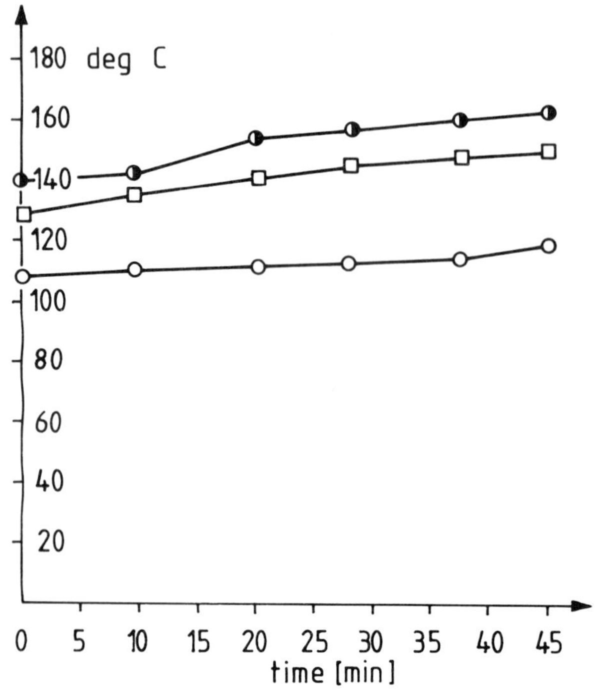

Fig.6 Results of prototype testing.
Test conditions: oil inlet temperature - 110° C, speed - 6000 rpm, thrust load - 25 kN, radial load - 1 kN.

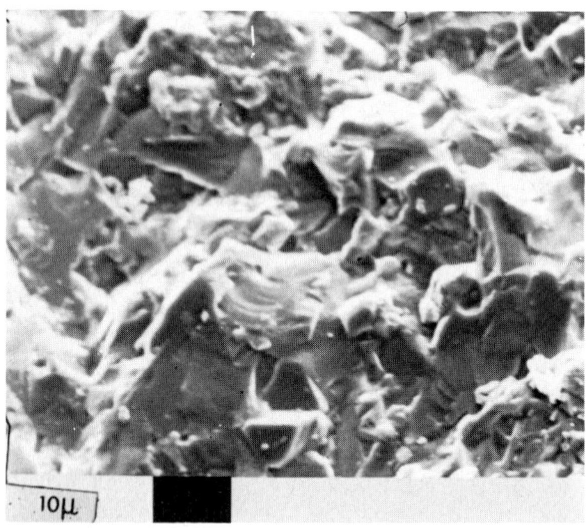

Fig.7 Micrograph of the failure surface of a thrust pad. Failure occurred under specific load of 12.75 MPa and speed of 6000 rpm. The lubricant was a diesel oil.

4 CONCLUSIONS

The main conclusion to be drawn from data accumulated during prototype testing is that the silicon carbide tilting pad readial/thrust bearing system seems to offer a good solution for the conditions specified in the design brief. The first pad failure was presumably due to high loading and excessive bending stress resulting from it. The pad fracture originated at the pivot point. Specific loads applied were very much higher than that normally used

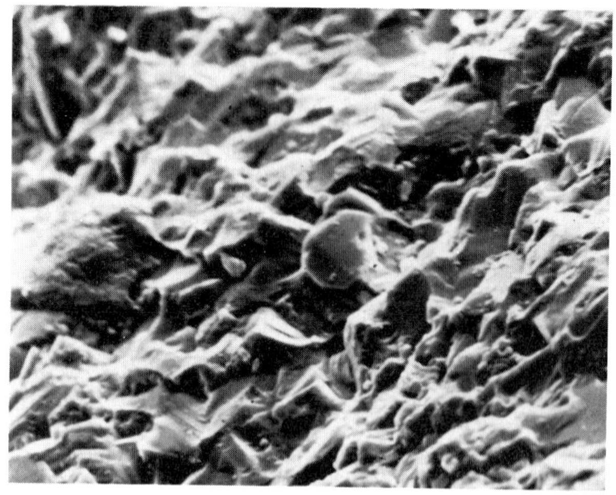

Fig.8 Micrograph of the failure surface of a thrust pad. Failure occurred at specific load of 8 MPa and inlet oil temperature of 120° C. The lubricant was Tellus T-37 and speed was 6000 rpm.

for traditional bearing materials.
A dimensional check of an assembled bearing revealed only a small variations. The maximum variation in inner and outer diameter was 0.022 mm and the maximum circumferential variation in thickness was 0.025 mm. Individual components showed even smaller variations. Pad thickness varied in the range of 0.006 mm while the backing plate thickness variation at the pad pivot point was only 0.018 mm.
The second pad failure emphasizes the importance of proper lubrication of ceramic sliding bearing as it is quite sensitive to any disruption in lubrication. Ceramic sliding bearing failure is usually sudden and occurs without any clear prior warning.
It is now quite clear that sliding bearings with ceramic components can be operated at significant and considerably higher unit loads than that used for conventional materials. However, to ensure long service life a proper and effective lubrication must be secured.

5 ACKNOWLEDGEMENT

The authors would like to thank Mr V.A.Lumpkin Chief Engineer, Framo Engineering for making prototype test data available.

References

(1) CRANMER, D.C. 'Ceramic tribology - Needs and opportunities', Tribology Trans, 1988, vol.31, 164-173
(2) GARDOS, M.N. 'On ceramic tribology', Lubrication Engineering, 1988, May, 400-407
(3) SCOTT, D.and BLACKWELL, J. 'Hot-pressed silicon nitride as a bearing material, Wear, 1973, vol.24, 25-30
(4) FISHER, T.F.and TOMIZAWA, H. 'Interaction of tribochemistry and microfracture in the friction and wear of silicon nitride', Wear, 1985, vol.105, 124-131.
(5) MOROLIKE, B.L. 'The frictional properties of carbides and borides at high temperatures Wear, 1960, vol.3, 234-241.

Paper XII(ii)

Unlubricated wear and friction behaviour of alumina and silicon carbide ceramics

G. Kapelski, F. Platon and P. Boch

Ceramics are brittle materials in which fracture develops when the stress field reaches a critical state. In the case of wear and friction tests with high contact pressures, cracking of the friction track can develop during the very beginning of the test. The extent of damage and the production of wear debris are sensitive to such cracking. This means that the entire life of a tribological couple can be affected by its initial instants. This work was thus focused on the study of the damage which can occur during the first instants of the life of a ceramic-to-ceramic couple, tested using a ball-on-disc configuration. Several ceramics were studied and in particular silicon carbide and alumina. Friction tests were compared to indentation tests, in order to determine the correlation between the different fracture strengths which can be defined.

1 INTRODUCTION

Ceramics behave as brittle materials at low and medium temperatures. Fracture suddenly develops when the stress field reaches a critical state. In the case of wear and friction tests with high contact pressures, cracking of the friction track can develop during the very beginning of the test. The extent of damage of the contact surfaces as well as the production of wear debris are very sensitive to the presence of cracks. This means that the entire life of a tribological couple can be affected by its initial instants, and that the wear and friction behaviour of a ceramic-to-ceramic couple is very dependent on an accidental overload.

The critical state for the stress field above which fracture develops depends on many parameters. The critical values of the three principal stresses depend on whether the stress field is uniaxial or multiaxial. Therefore, it is sometimes difficult to correlate the damages which are observed in tribological studies of ceramics - which correspond to multiaxial stress states - with the damages which are observed during fracture tests. In particular, the "tribological strength", understood as the yield value above which cracking develops, is generally higher than the macroscopic strength (as can be determined using, for instance, a uniaxial bending test). This work has been focused on the study of the damage which occurs during the first instants of the life of a ceramic-to-ceramic couple, for a ball-on-disc configuration. Different ceramic materials were studied and in particular silicon carbide and alumina. The friction tests, called "dynamic" because they involve the displacement of the ball on the disc, were completed by indentation tests, called "static" because they simply correspond to loading a disc by a ball.

2 INDENTATION TEST AND DYNAMIC TEST

For the static test of a ball (radius R) loaded on a slab by a normal load (F), and assuming that materials behave in a perfectly elastic manner, the value of the radius of the contact area is (1) :

$$a = (0.75\ k\ R\ F/E)^{1/3} \qquad (1)$$

with :

$$k = (1-\nu^2) + (1-\nu'^2)E/E'$$

where E, E', ν, and ν' are Young's modulus and Poisson's ratio of the materials of the slab and ball, respectively. The principal tensile stress within the slab (σ_{1s}) reaches its maximum at the surface, on the circle of radius "a" :

$$\sigma_{1s} = (1-2\nu)(0.75\ k\ R/E)^{-2/3}\ F^{1/3}/(2\ \pi) \qquad (2)$$

Cracks develop along a circular cone when σ_{1s} exceeds the static fracture strength (σ_{1sf}).

For the dynamic tribological tests, Hamilton & Goodman (2) and Lawn (3) have taken into account the tangential force at the surface. The maximum of the principal tensile stress (σ_{1d}) is now located at the rear of the friction contact, and its value depends on the friction coefficient :

$$\sigma_{1d} = (\sigma_{1s})(1+k_v\ f) \qquad (3)$$

with $k_v = 3\pi\ (4+\nu)/(8(1-2\nu))$

Correspondingly, cracks now initiate at the rear of the friction contact, and they develop along a cone section.

3 EXPERIMENTAL

3.1 Static indentation tests

Indentation tests were carried out using a laboratory-made device, loaded by a tensile machine working in its compression mode. Either a JJ-M30K model or an Instron-1121 model was used.

3.2 Dynamic tests ; tribometer

The tribometer was a laboratory-made model (4). It allows the use of two tribological configurations, namely i) ball-on-disc or ii) disc-on-disc. Trapping of debris is minimized in the former case, whereas it is maximized in the latter case. The experimental device is located inside a chamber with a controlled atmosphere (e.g. dry air, or air with a fixed humidity). The chamber is surrounded by an electric furnace, which allows the tests to be performed from room temperature to approximately 1000°C.

Most of the experiments were carried out using the ball-on-disc configuration, with loads varying from 5 to 50 N, at room temperature, and with humidity ranging from 45 to 55%. Sliding velocities were 0.01 or 0.1 m/s.

4 MATERIALS

The discs were made of silicon carbide (SiC) or alumina (Al_2O_3). Balls were generally made of alumina. However, ball-bearing steel, silicon nitride (Si_3N_4), and "SiAlON" ceramics were used in some instances. Table 1 gives the main characteristics of the various materials (with E Young's modulus, ν Poisson's ratio, ρ and ρ' the density and the apparent density, respectively, and σ_{3p} the 3-point flexural strength).

Mat.	SiC	Al_2O_3	SiAlON	Si_3N_4	Steel
E GPa	420	313	310	310	210
ν	.14	.22	.31	.28	.30
ρ g/cm^3	3.15	3.71	3.20	3.20	7.88
ρ' (%)	98	93	>99	>99	100
σ_{3p} MPa	450	285	-	-	-

Table 1 Main characteristics of the materials used in this study (SiC from Céramiques et Composites, Tarbes, France, Al_2O_3 from C.I.C.E., Montreuil, France).

5 RESULTS AND DISCUSSION

5.1 Static indentation tests

Indentation tests were carried out using alumina balls on SiC and Al_2O_3 discs. Initial tests were performed to determine the range of loads leading to cracking, then subsequent tests were performed to determine the mean critical load ($F_{s.av}$) which is required to crack 50% of the disc batch.

For SiC, cracks can be easily imaged by optical microscopy. However, a chemical etching improves the optical contrast (Figure 1). The cracks exhibit a regular, nearly perfectly circular shape. For Al_2O_3, the cracks do not exhibit such a regular shape. They form a diffuse circular array, composed of connected microcracks. A dye penetrant liquid helped us to visualize the extent of cracking.

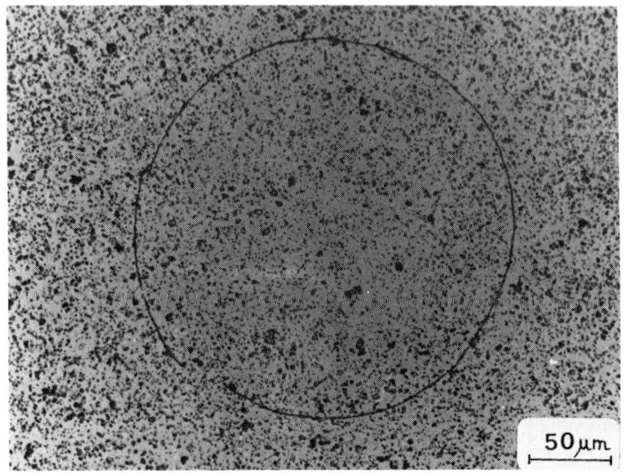

Figure 1 Crack array in a silicon carbide disc loaded by an alumina ball (F = 500 N).

Relations (1) and (2) are used to evaluate the Hertz radius (a) and the indentation strength (σ_{isf}), in function of the load (F) :

$$a = A\ F^{1/3} \text{ and } \sigma_{is} = B\ F^{1/3}$$

Table 2 gives A and B, and the corresponding values of the radius of the contact area (a), of the mean critical load ($F_{s.av}$), and of the strength (σ_{isf}).

Materials	A µm/N$^{1/3}$	B N$^{2/3}$ m^2	$F_{s.av}$ N	a µm	σ_{isf} MPa
Al_2O_3 on SiC	27	154	434	206	1164 ± 9
Al_2O_3 on Al_2O_3	28	108	1250	305	1160 ± 5

Table 2 A, B, a, $F_{s.av}$, and σ_{isf} values for static indentation tests.

The values of $F_{s.av}$ are different for Al_2O_3 (1250 N) and SiC (434 N). However, strength values appear to be nearly the same (approximately 1160 MPa) for both materials. This "indentation strength" values are high in comparison with bending strength values. In the case of SiC, for instance, σ_{1sf} is 1160 MPa whereas σ_{3p} is only 450 MPa. Three complementary reasons can explain this discrepancy :

(1) Indentation tests yield a biaxial stress state, with a compressive component (σ_2) superimposed on the tensile component ($\sigma_1 = -\sigma_2$). The presence of a compressive component could increase the fracture strength (5).

(2) The stress value is over-evaluated, because the real diameter of the crack zone is not "a" but a_f, with $a_f > a$. The corrected stress value (S_{1sf}) is :

$$S_{1sf} = (1-2\nu)F(a/a_f)^2/2\pi a^2$$

For SiC, it was found $a_f = 235 \pm 5$ µm, which yields $S_{1sf} = 895 \pm 50$ MPa. For alumina, a similar derivation gives $S_{1sf} = 1050 \pm 65$ MPa.

(3) The stressed volume is approximately 6000 times larger for the flexural test than for the indentation test (in the case of silicon carbide) and approximately 80000 (for the case of alumina). Therefore, the Weibull probability of finding a critical flaw of a given severity is much lower in the former case than in the latter. It can be shown (6) that :

$$(V_{3p}/V)^{1/m} = S^*_{1sf}/\sigma_{3p}$$

where m is Weibull's modulus, and V_{3p}, V, σ_{3p}, and S^*_{1sf} are the volumes and strengths corresponding to the flexural and the indentation tests, respectively. For SiC, "m" is about 12, which yields a corrected value of σ_{3p}, S^*_{1sf}, of 945 MPa. This corrected value is in a good agreement with the experimental result which is $S_{1sf} = 895 \pm 50$ MPa. For Al_2O_3, "m" is about 9, which yields a corrected value of σ_{3p} of 1000 MPa, in very good agreement with the value of S_{1sf} (1050 ± 60 MPa).

5.2 Dynamic tribological tests

5.2.1 Al_2O_3 ball on SiC disc

For the first tests, the ball was loaded on the disc for only one revolution. The sliding velocity was low. Actually, the results were independent on the sliding velocity, at least for the range of velocities (0.01 to 0.1 m/s) which was studied here.

The crack array on the wear track was observed after chemical etching (Fig.2). Most of the cracks are 100 to 250 µm long. However, the crack array is not exactly the same at the beginning of the track compared to its end. At the beginning, the cracks are smaller, their curvature is more pronounced, and they are more closely packed. At the end, the cracks are longer, their curvature is less pronounced, and the mean distance between them is larger.

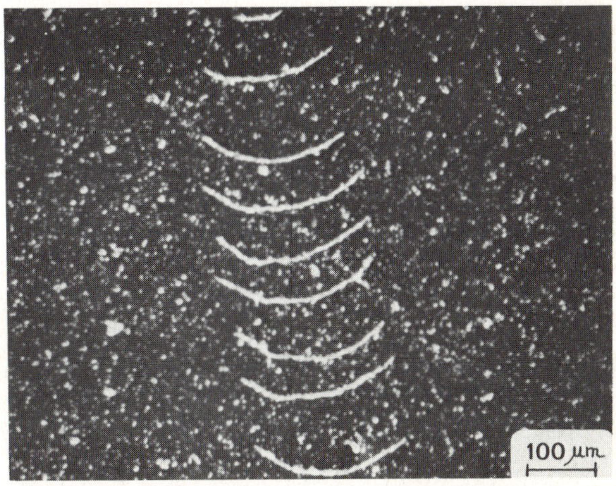

Figure 2 Crack array for a alumina ball-on-SiC disc test (F = 10 N).

The extension of cracking below the surface was determined on specimens cut along a plane tangential to the wear track (Fig.3). For a load of 10 N, the mean extension is 30 to 35 µm, with some cracks extending to 45 µm. Cracks are nearly perpendicular to the surface, with a slight inclination towards *the front* of the contact. This result disagrees with some literature data (3) where cracks are said to be initiated perpendicularly to the surface with a subsequent inclination towards *the rear* of the contact.

Cracking is very sensitive to the value of the coefficient of friction. In some cases, a load of 10 N is not enough to allow cracking to develop on the entire wear track. The zones where the coefficient of friction is higher are cracked, whereas the zones where it is lower are uncracked. Fig.4 shows a typical chart of the coefficient of friction (f) vs. the distance of friction (l). The cracked zones correspond to places where "f" exceeds a critical value, f_c. It was found that $0.36 < f_c < 0.43$. Such a value is in good agreement with the theoretical value which can be derived from the relation :

$$F_{d.av} = F_{s.av}/(1+k_v f_c)^3$$

where $f_c = 0.37$. The "tribological dynamic strength" (σ_{1df}) which can be derived from the given values of parameters (i.e. $F_{d.av} = 10$ N and $f_c = 0.4$) is 1230 ± 72 MPa. It is in a good agreement with the "tribological static strength" ($\sigma_{1sf} = 1164$ MPa). It must be pointed out that, in the tribological test, it is not possible to determine the difference between Hertz's radius (a) and the radius of the cracked zone (a_f). Therefore, it was not possible to calculate a corrected value (say : S_{idf}) which would have been the counterpart of S_{isf}. This means that the comparison was carried out between the σ_{idf} and σ_{isf} values.

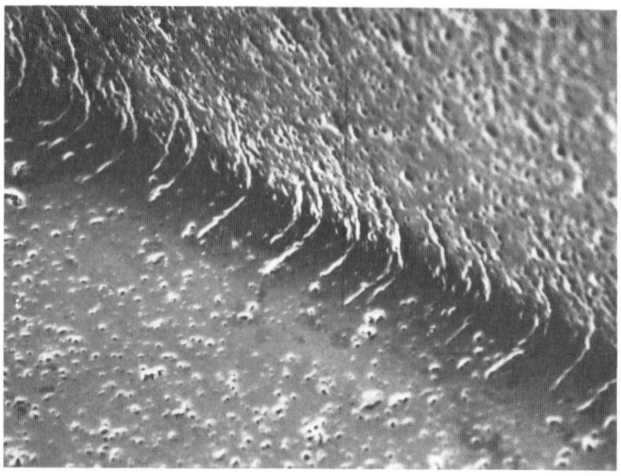

Figure 3 View of a section of a SiC disc (cut perpendicularly to the surface and tangentially to the wear track).

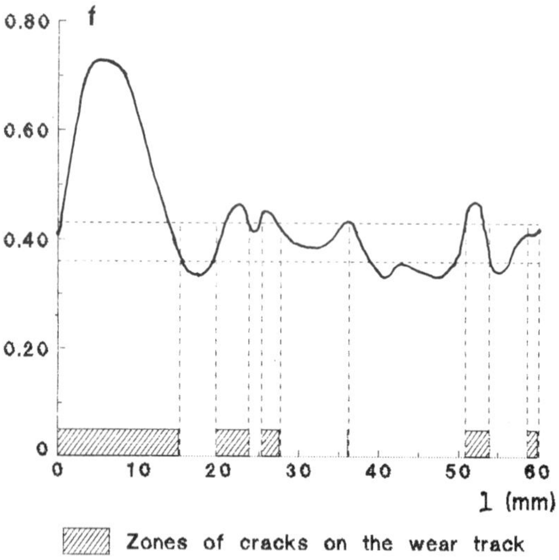

Figure 4 Coefficient of friction vs. distance of friction (alumina ball on silicon carbide disc).

The coefficient of friction (f) is approximately 0.4 in the case of dry friction, with no lubrication. It can be decreased below 0.2 by lubricating the disc with distilled water. For f = 0.15 and F = 10 N, σ_{id} is only 760 MPa, and no cracking is observed. It is necessary to increase the load up to 40 N (where f = 0.22) to observe cracking. The critical value of the coefficient of friction is 0.2, which corresponds to σ_{idf} = 1240 ± 94 MPa.

Complementary tests were carried out to study the evolution of the crack array when the number of revolutions (n) increases. Observations were made for n = 2, 10, 100, and 1000. The results are as indicated below :

(1) n = 2. New cracks develop. They do not constitute a new array but they connect with the initial array (Fig.5).

(2) n = 10. New cracks of small size develop between the former cracks (Fig.6).

(3) n = 100. Several new crack arrays develop and superimpose on the initial array. The mean length of cracks is smaller than that of the initial cracks. (Fig.7).

(4) n = 1000. The wear track is densely cracked, with numerous, connected cracks of small length (10 µm). This microcracking favours the formation of wear debris (Fig.8).

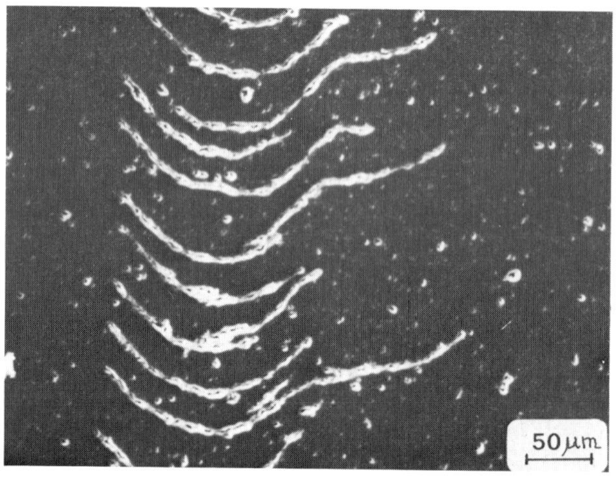

Figure 5 Crack array after 2 revolutions (Al_2O_3 ball on SiC disc).

The evolution of cracking is associated to the evolution of the tribological configuration. The diameter of the worn area on the ball increases as the number of revolutions increases. Therefore, debris are trapped in a more efficient manner, which leads to a "third body" effect and modifies the stress field.

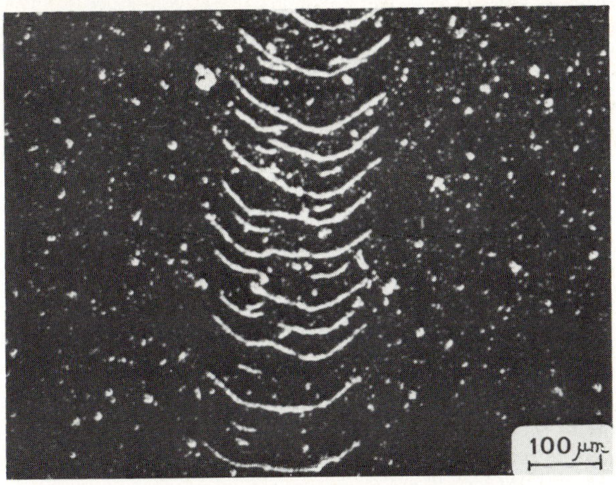

Figure 6 Crack array after 10 revolutions (Al_2O_3 ball on SiC disc).

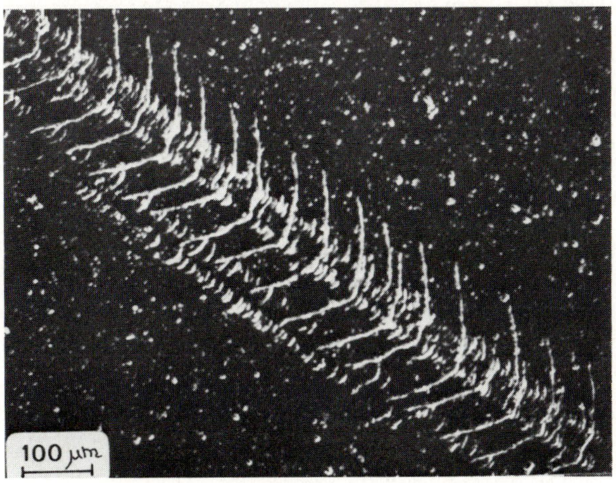

Figure 7 Crack array after 100 revolutions (Al_2O_3 ball on SiC disc).

5.2.2 Alumina disc

The alumina ball-on-alumina disc configuration was studied first. Under a load of 10 N, the coefficient of friction is only 0.22-0.25. This yields a maximum stress within the disc of 770 ± 60 MPa, which is well below the tribological strength, and no cracking was observed. Since the stress intensity increases as the cubic root of the load, cracks were not expected to develop unless very high loads were used. However, such high loads were forbidden by the design of the tribometer. Therefore, the alumina ball was replaced by balls of other materials (silicon nitride, SiAlON, and ball-bearing steel), in the hope of increasing the coefficient of friction up to values sufficient to yield cracking.
Table 3 gives the mean value and the maximum values of the coefficient of friction, for one revolution under a load of 10 N, and the corresponding stress values. It can be seen that they continue to be low, which explains why no cracking was observed.

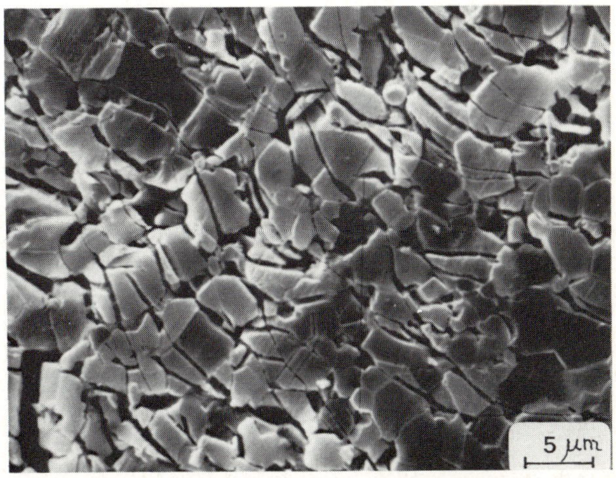

Figure 8 Crack array after 1000 revolutions (Al_2O_3 ball on SiC disc).

Ball	f_{av}	σ_{1dav} MPa	f_{max}	σ_{1dmax} MPa
Al_2O_3	0.22	708	0.25	770
Si_3N_4	0.36	1015	0.42	1115
SiAlON	0.32	1090	0.36	1190
Steel	0.41	980	0.48	1113

Table 3 Mean and maximum coefficient of friction, and stress values, for tests of one revolution under a load of 10 N.

Figure 9 shows the diagram of σ_{1d} vs. f, for different ball materials. It must be pointed out that the diameter of the SiAlON ball was 7.9 mm, instead of 10 mm for the other balls. Therefore, SiAlON balls yield to the highest stress, although they do not yield the highest coefficient of friction. Tests were carried out with loads of 30 N, which gives f = 0.42. In this case, some cracking was observed. However, cracks were very difficult to observe, and it was not possible to determine the critical value of the coefficient of friction. Only the maximum stress was evaluated, from data on the maximum of the coefficient of friction. It was found that σ_{1df} = 1920 MPa.

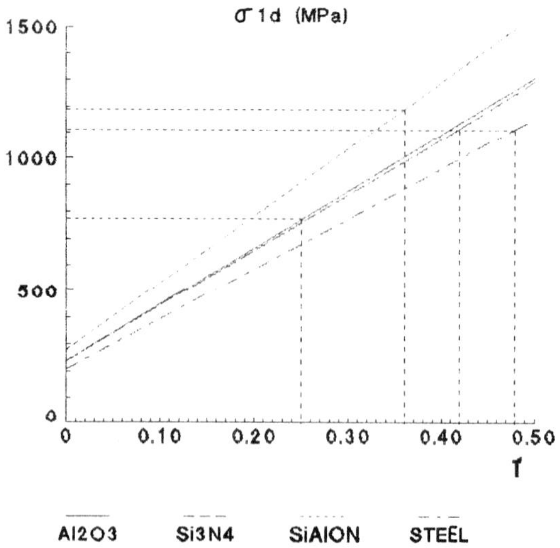

Figure 9 σ_{1d} value vs. friction coefficient for various materials.

6 CONCLUSIONS

Two sorts of tests have been carried out. Static (indentation) tests and dynamic (sliding) tests, for a ball-on-disc configuration. Attention was focused on the cracking of ceramic disc. For silicon carbide, the static and the dynamic strength values are nearly the same (σ_{sf} = 1160 MPa and σ_{df} = 1230 MPa). The coefficient of friction is an important parameter. For a given load, there is a critical value, f_c, beyond which tribological cracking develops. When the number of cycles increases, the crack array becomes more and more dense, which favours the formation of debris to feed the "third body". Data for alumina are less conclusive, because lower coefficients of friction did not allow us to observe cracking easily. However, results are qualitatively the same as those for SiC. This indicates that special care must be taken to avoid accidental overloads when ceramic parts are concerned.

References

(1) JOHNSON K.L. 'Hertz theory of elastic contact', 90-104, 'Pressure applied to a circular region', 56-63, in Contact Mechanics, the Press Syndicate of the University of Cambridge Publ.,(1985).(2) HAMILTON G.M. and GOODMAN L.E., 'The Stress Field Created bya Circular Sliding Contact', in J. Appl. Mech., 371-374, (1966).

(3) LAWN B.R., 'Partial cone crack deformation in brittle material loaded with a spherical indenter', in Proc. Roy. Soc., A299, 307-316, (1967).

(4) PLATON P. and BOCH P. 'Standard Tribological Test for Engineering Ceramics", in Ceramics Materials and Components for Engines, Deutsche Keramische Ges. Publi., 735-757, (1986).

(5) WRONSKI A. Comment on 'Fracture Criterion for Alumina Ceramics subjected to Triaxial Stresses", in J. Am. Ceram. Soc. 70(8), C-201 (1987).

(6) CREYKE W.E.C., SAINSBURY I.E.J. and MORRELL R., in 'Design with Non-Ductile Materials', Applied Science Publishers, London (1982).

Paper XII(iii)

The effects of surrounding atmosphere on the friction and wear of ceramics

S. Sasaki

The friction and wear behaviors of ceramics, Al_2O_3, ZrO_2, SiC and Si_3N_4, have been studied with the three kinds of pin-on-disk machines under the following four conditions; in vacuum condition at pressure ranged from 10^{-6}Pa to 10^5Pa in either nitrogen or oxygen, in nitrogen with organic compound vapors of C_2H_5OH, CH_3COOH, $(CH_3)_2CO$, C_6H_{14} and C_6H_6, in humidity controlled air and in deionized water. Based on the experimental results, discussion is made for mechanisms of the effects of surrounding atmosphere on friction and wear of ceramics.

1. INTRODUCTION

In order to examine the possibility of using ceramics for sliding parts, it is essential to comprehend the behavior of friction and wear of various kinds of ceramics under hostile conditions. From the industrial point of view, lubrication with process fluids is also of important.

However, the mechanical properties of ceramics are more sensitive to the environments than metals. This fact sometimes make comprehension of phenomena sophisticated since ceramics are highly affected by various atmospheric factors which closely interact with each other. For example, the effects of humidity on friction of Si_3N_4 depends largely on the kind of sintering aids, manufacturing process and friction conditions[1].

This report covers the study on the behavior of friction and wear of four kinds of ceramics (Al_2O_3, ZrO_2, SiC and Si_3N_4) in vacuum, nitrogen with various organic compound, humid air, and water as shown belows. Discussion is also presented for mechanisms to understand how such conditions affect the friction and wear of the ceramics.

2. EXPERIMENTS

The following four conditions were tested in this experiment using the three kinds of pin-on-disk machines (Fig.1(a)-(c)), which have been specially designed to be operated in vacuum, ambient gases and fluids:

(a) vacuum condition --- pressure ranged from 10^{-6}Pa to 10^5Pa in either nitrogen or oxygen.
(b) nitrogen with organic compound vapor --- nitrogen saturated with organic compound vapors of C_2H_5OH, CH_3COOH, $(CH_3)_2CO$, C_6H_{14}, and C_6H_6 at temperature of 20 C, then raised up to 30 C.
(c) humid air --- premixed dry air with moisture-saturated air at certain level of relative humidity.
(d) water --- deionized water.

After the test, weight loss of both pin and disk were measured, then the total specific wear rate was calculated. When the weight loss was very small, the planimeteric wear of pin and disk were calculated.

In order to examine the wear behavior, the observations and analysis of worn surface and wear debris were conducted using a SEM(Scanning Electron Microscopy) equipped with EPMA(Electron Probe Microanalysis), ESCA(Electron Spectroscopy for Chemical Analysis) and other analytical tools.

Sintered-Al_2O_3, s-partially stabilized ZrO_2(PSZ), s-SiC, s-Si_3N_4 and hot isostatic pressed-Si_3N_4, used in this study, were manufactured by Kyocera Co.Ltd. Their mechanical properties are shown in Table 1. Figure 2 shows the geometry of pin and disk specimens: the size of the disk specimen was 25 mm outside diameter and 20 mm wear track diameter, whereas the pin specimen was 4 mm outside diameter and 30 mm long with 4 mm radius round shape at both edges. The sliding surfaces were prepared by polishing at a surface roughness of 0.03 um Rz. The details of the test conditions are tabulated in Table 2.

(a)

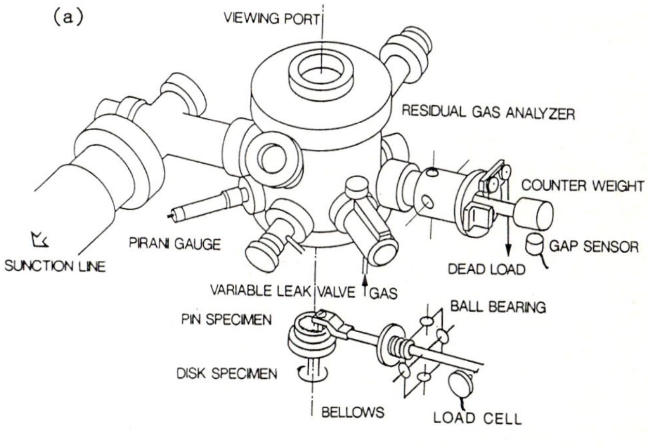

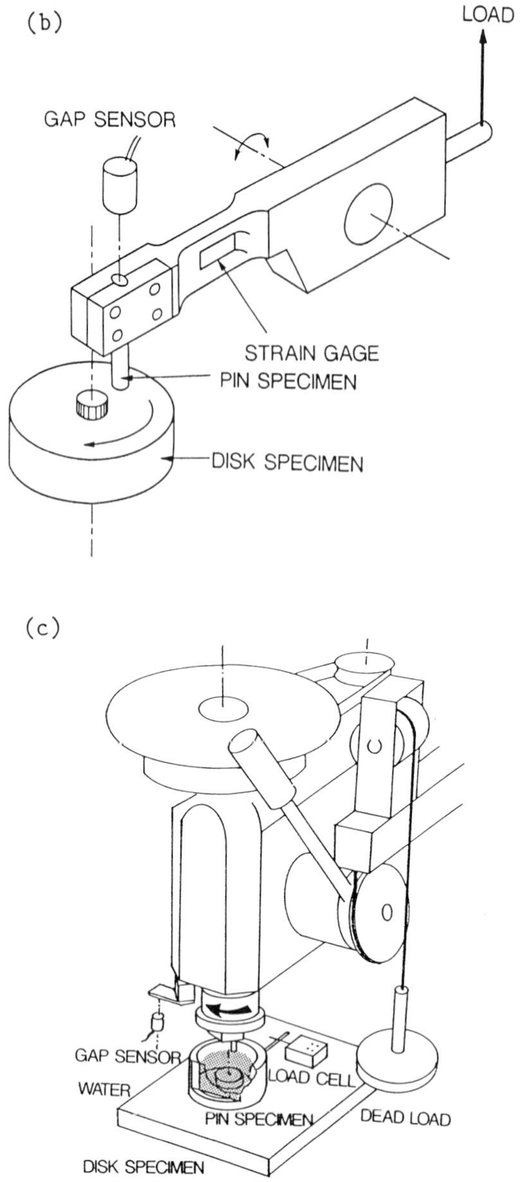

Fig.1 Sliding testers operated (a)in vacuum, (b)in fluids, and (c) in ambient gases

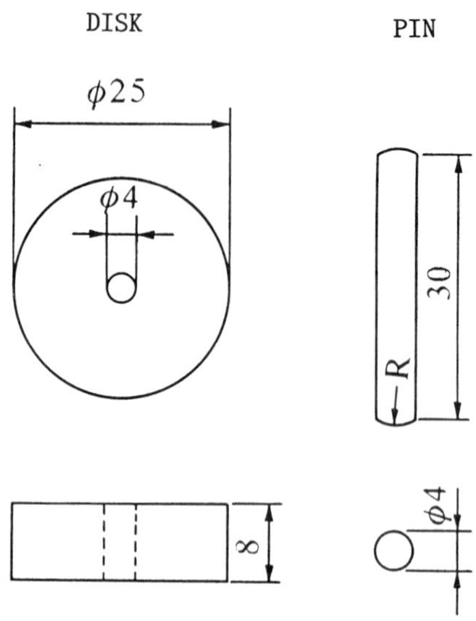

Fig.2 Geometry of pin and disk specimens

Table 2 Sliding condition

ENVIRONMENT	LOAD (N)	VELOCITY (mm/s)	WEAR DISTANCE (m)
VACUUM	3	100	--------
N_2+ORGANIC COMPOUND VAPOR	10	400	1256
HUMID AIR	10	400	1256
WATER	50	100-800	1256

Table 1 Mechanical properties of test specimens

		s-Al_2O_3	s-PSZ-1	s-PSZ-2	s-SiC	s-Si_3N_4	hip-Si_3N_4
Bulk Density		3.8	5.9	6.0	3.2	3.2	3.25
Young's Modulus	$kg/cm^2(x10^6)$ (*1)	3.7	2.1	2.2	4.4	2.9	3.1
Poisson's Ratio		0.23	0.31	0.31	0.16	0.28	0.27
Hardness	Hv (HRa)	1800	1250	1250	94(HRa)	92(HRa)	1530
Flexural Strengh	kg/cm^2	3300	10000	10000	5500	6800	10000
Coefficient of linear Thermal Expansion	$1/C(x10^{-6})$ (*2)	7.9	11.0	11.0	4.4	3.1	3.0
Thermal conductivity	cal/cm·s·C (*3)	0.23	0.009	0.009	0.15	0.06	0.07

(*1) Load: 500g, (*2) 40-800 C, (*3) 20 C, s-: sintered hip-: hot isostatic pressed

3. RESULTS

3.1 Frictional behavior in vacuum condition

Using the high vacuum sliding tester as shown in Fig.1(a), the effects of gaseous pressure on friction coefficient were examined at a sliding velocity of 0.1 m/s and a load of 3 N.

The gaseous pressure was varied in the range of 10^{-6}Pa to 10^{5}Pa by adjusting the suction line consisting of a turbo-molecular pump backed up with rotary pump and the introduction line with a variable-leak valve. Oxygen or nitrogen were introduced in order to examine the differences of effect of adsorption between them. The steady state friction coefficients u were determined at each level of pressure from 10^{-6} to 10^{5}Pa.

The results are summarized in Fig.3(a)-(d). For Al_2O_3 as shown in Fig.3(a), no significant change in friction coefficient was recorded up to 10^{-2}Pa in either nitrogen or oxygen. The decrease in friction, then, occurred at around 10^{-1}Pa. In comparison between nitrogen and oxygen at the pressure range of 10^{2} to 10^{5}Pa, the friction curves show different patterns.

For ZrO_2(PSZ-2) as shown in Fig.3(b), similar curves were obtained in both nitrogen and oxygen. These curves were also similar to that of Al_2O_3 in oxygen. This may be attributed to non-reactivity of oxygen to these oxides and also non-reactivity of nitrogen to ZrO_2, but to Al_2O_3. Therefore, the behavior of the friction of those combinations merely depended on the adsorption of gases on the surface of the oxides ceramic.

The friction coefficient of SiC as shown in Fig.3(c) was changed up to 10^{2}Pa in nitrogen, then it began to drop. The lowest u value of 0.4 and highest u value of 0.65 were obtained at 10^{4} and 10^{5} Pa, respectively. In oxygen, the friction coefficients dropped slightly at around 10^{-2}Pa and the lowest u value of 0.4 was recorded at around 10^{2}Pa, then they began to rise as the pressure increased.

For Si_3N_4 as shown in Fig.3(d), the friction coefficients at high vacuum range were around 0.3, which were the lowest values among the four kinds of ceramics. As the pressure approached the atmospheric pressure, the friction coefficients began to rise rapidly and finally reached the value as high as 0.8. Particularly in oxygen, the fluctuation of the curve was more significant than in nitrogen. This is attributed to the formation of SiO_2 on the surface by a tribochemical reaction. This fact was confirmed by ESCA analysis as described later.

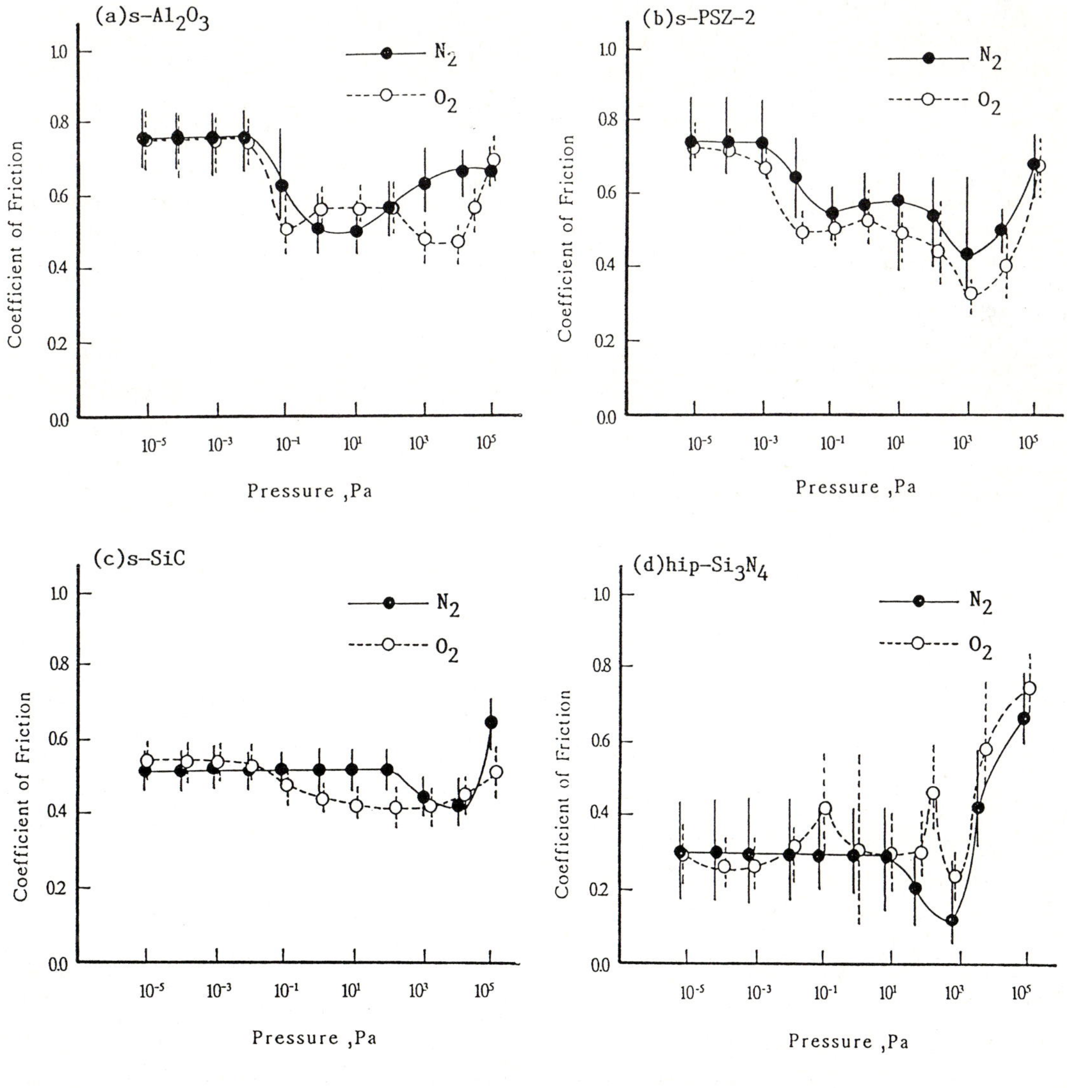

Fig.3 Friction behavior at oxygen and nitrogen reduced pressure

3.2 Friction and wear behavior in nitrogen with organic compound vapor

Using the wear tester as shown in Fig.1(b), the friction coefficient and specific wear rate were measured in nitrogen gas contained with C_2H_5OH, CH_3COOH, $(CH_3)_2CO$, C_6H_{14}, C_6H_6; Organic compound vapors were saturated in nitrogen at temperature of 20 C, then raised up to 30 C. Table 3 shows the saturated vapor pressure at 20 C and its relative vapor pressure at 30 C. As a comparison, the tests were also conducted in both dry and humid nitrogen without any organic compound. The test was conducted at a load of 10 N, a sliding velocity of 400 mm/s and a total wear distance of 1256 m.

Table 3 Vapor pressure of organic compound

ORGANIC COMPOUND	VAPOR PRESSURE,mmHg		RELATIVE VAPOR PRESSURE,%
	20°C	30°C	
H_2O	17.5	31.8	55.0
C_2H_5OH	44.7	79.6	56.2
CH_3COOH	11.5	20.6	55.8
$(CH_3)_2CO$	186.9	286.5	65.2
C_6H_{14}	122.6	189.0	64.9
C_6H_6	76.1	120.7	63.0

Figure 4(a)-(d) show the friction and wear of the four kinds of ceramics under organic vapor-contained nitrogen. In general, the friction coefficient and wear rate of Al_2O_3 in these gases were smaller than in dry nitrogen gas, except in N_2+CH_3COOH vapor. In $N_2+C_6H_{14}$ and C_6H_6, some reaction products like friction polymer were observed on the sliding surfaces as shown in Fig.5.

As shown in Fig.4(b), ZrO_2(PSZ-1) showed severe wear in the humid air and C_2H_5OH, both of which have an -OH in their molecules. This can be attributed to the occurrence of stress corrosion, which will be discussed later.

In the case of SiC as shown in Fig.4(c), the wear rates were insensitive to organic compound vapors in nitrogen, whereas the friction coefficient was rather sensitive to those vapors. In $N_2+C_6H_{14}$, which gave the lowest friction coefficient of 0.1 among the organic compounds, some reaction products like as friction polymer were observed on sliding surface.

For Si_3N_4 as shown in Fig.4(d), the friction coefficients in $N_2+(CH_3)_2CO$, C_6H_{14} and C_6H_6 were as low as 0.1-0.2 and the wear rates were negligibly small.

As summarizing remarks, the effect of organic compound on the behavior of friction and wear varied with the kinds of ceramics. For example, C_2H_5OH exhibited an excellent lubricating capability for Al_2O_3, but did not for ZrO_2, it caused unacceptable wear. These facts may suggest that the organic compound vapors not merely form a simple adsorbed layer on the surface, but forms some kinds of chemical reactivity on the surface.

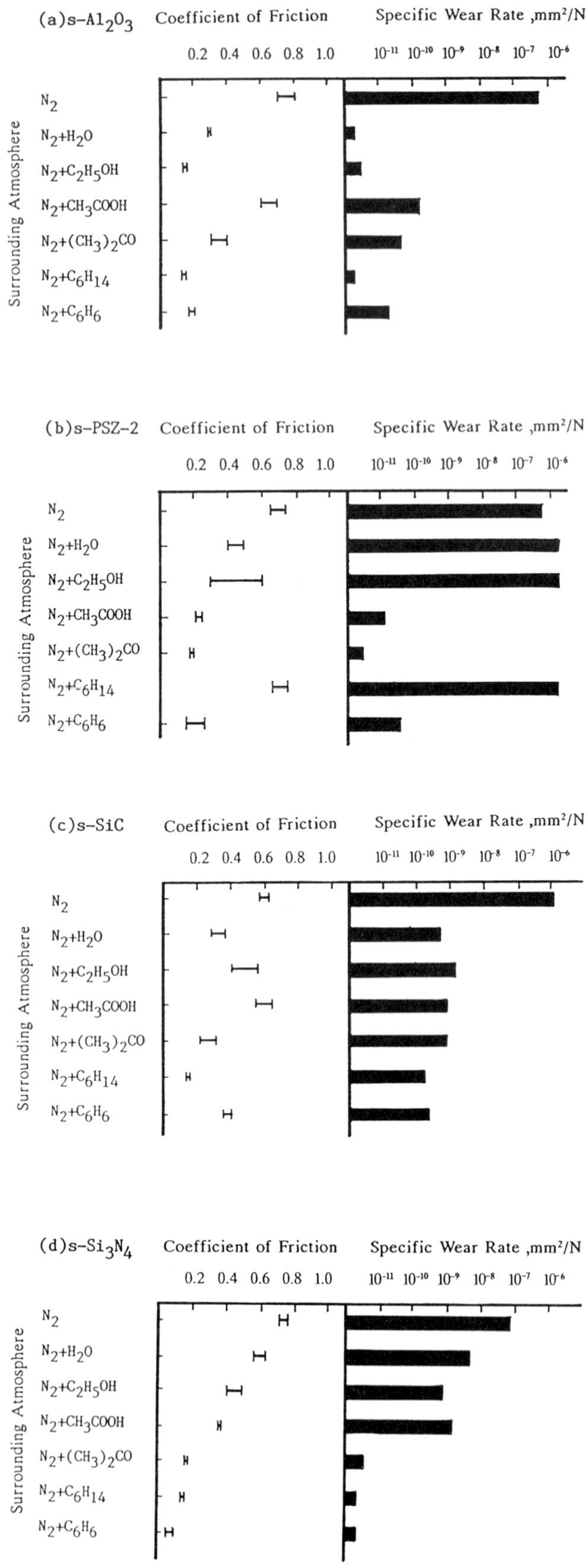

Fig.4 Friction and wear behavior in N_2 with various kinds of organic compound

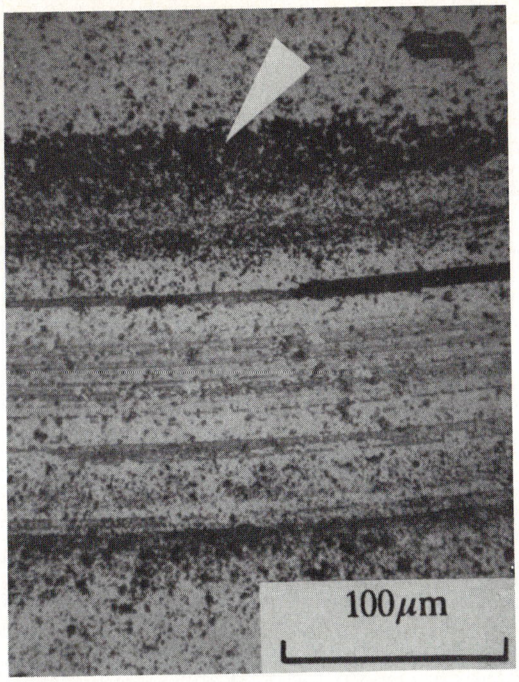

Fig.5 Optical micrograph of Al_2O_3 in N_2+C_6H_6, showing the formation of friction polymer indicated by an arrow

3.3 Effects of humidity

The effect of humidity on friction and wear was examined using the sliding tester as shown in Fig.1(b) at a load of 10 N, a sliding velocity of 400 mm/s and total sliding distance of 1256m.

Figure 6 shows the results. The friction coefficient decreases with an increase in relative humidity for all ceramics. But the wear rates did not show such a definite tendency. For example, the wear rates of Al_2O_3 and SiC decreased as relative humidity raised up. But in ZrO_2(PSZ-1), the highest wear rate appeared at 20% R.H. humidity . As discussed later, the wear of ZrO_2 caused by stress-induced corrosion cracking, which plays a major role especially at low relative humidity range around 20% R.H. As the atmosphere becomes more humid, however, the lubricative effect of water-adsorption layer begins to dominate the behavior of the friction and wear of ZrO_2.

In the case of Si_3N_4, the lowest wear rate appeared at 50% R.H. humidity. The increase in humidity accelerates a tribochemical reaction of Si_3N_4 with water[2)], resulting in the formation of reaction products, SiO_2 xH_2O. In this case, however, the reduction of friction coefficients caused by the increase in the reaction products was relatively small, so that a high degree of wear resulted in the high humidity range.

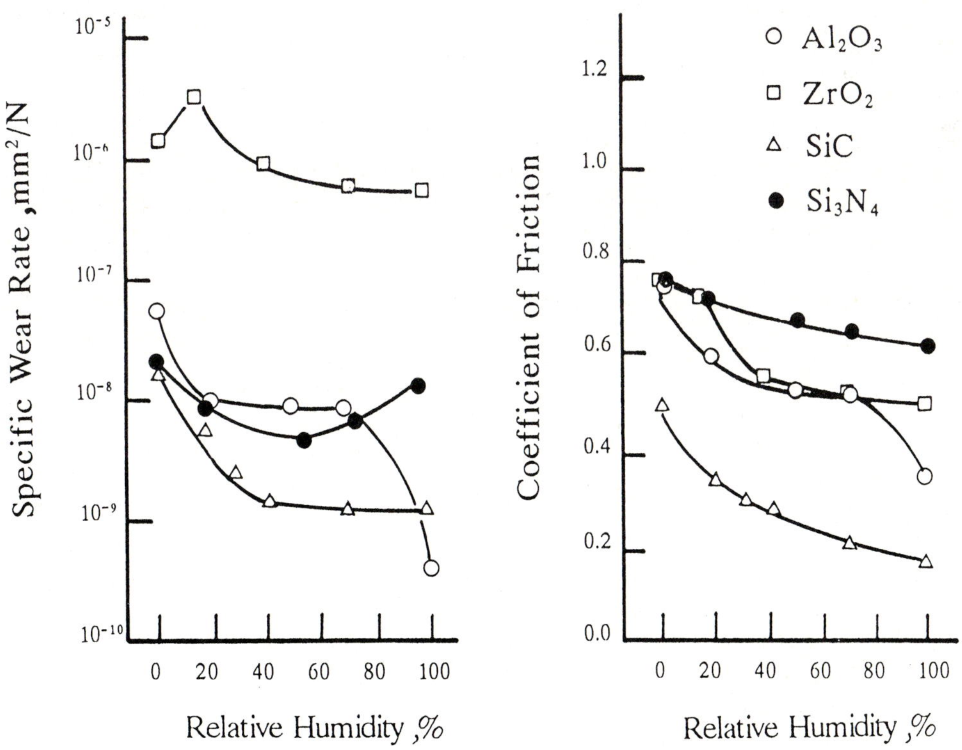

Fig.6 Friction and wear behavior as a function of relative humidity in air

3.4 Effects of water

The effect of sliding velocity on friction coefficients and wear rates with the four ceramics in water was examined using the friction tester as shown in Fig.1(c) at a load of 50 N, a sliding velocity of 100-800 mm/s and a total sliding distance of 1256 m.

As shown in Fig.7, the friction coefficient of Al_2O_3 was constant at around 0.3 in the range of 100-800 mm/s sliding velocity. The wear was smallest among ceramics tested. It decreased with an increase in sliding velocity. The similar tendency was observed for SiC and Si_3N_4. The friction coefficients of those materials, however, reached as low as 0.01 for whole range of sliding velocity for SiC and above 400 mm/s for Si_3N_4. This may be due to tribochemical wear as suggested by Fisher and Tomizawa[3]. The tribochemical wear is a phenomena in which the soft hydroxides are formed on the sliding surface by the reaction of ceramics and water(in the case of Si_3N_4,the products are $SiO_2\ xH_2O$). These products may act as a deaccelerator of friction and a accelerator of wear[4] caused by the slight removal of the product from the sliding surface within the range of this experiment. When this tribochemical wear begins to dominate on the surface, the surface becomes smooth and a good lubricative state is generated in conjunction with the effect of hydrodynamic lubrication.

For ZrO_2(PSZ-1), both friction and wear were the highest among ceramics tested: the wear rate increases as sliding velocity increase. The friction coefficient kept as high as 0.5 for the whole velocity range tested. This fact may be due to the results that shear stress and friction work at higher velocity become higher which results in higher rate of corrosion cracking as described later.

3.5 Analytical Results

Figures 8-11 show the secondary electron(SE) images on the worn surfaces observed using SEM. The worn surfaces of Al_2O_3, SiC and Si_3N_4 in water(Fig.8(c), 10(c), and 11(c)) were observed to be very smooth. This was caused by the above mentioned tribochemical wear.

Figures 9(c) and (d) show a SE image and corresponding cathodo luminescence(CL) image of worn surface of PSZ in water. By comparing the two images, it is understood that the luminescence originates from some smoothed area where the applied load was carried on. A strong CL intensity observed in CL image of PSZ was probably come from m-ZrO_2[5]: phase transformation from t-ZrO_2 to m-ZrO_2 might be occur at these luminescent area on sliding surface.

Figure 12 shows the analytical results of the worn surface of Si_3N_4 using the SEM-EPMA. It is noted that oxygen atoms replaced the nitrogen atoms on the sliding areas. This is attributed to the oxidation of Si_3N_4 to SiO_2.

In the case of Si_3N_4 in water, the PH value of the water shifted to the alkaline side after the wear test. The ammonium ion was detected by using a ion-sensitive test paper. For the other ceramics, however, no ammonium ion was detected after the wear tests. This fact indicates that ammonia was formed by the tribochemical reaction of Si_3N_4 with water[6].

Figure 13 shows the analytical results on the worn surface of Si_3N_4 in oxygen at pressure of 10^5 Pa using ESCA. By comparing the worn surface with the unworn site, an energy peak probably caused by Si-O bond was observed. This agrees with the above analytical results and proves the presence of an oxidation of Si_3N_4. In conclusions, Si_3N_4, which is a non-oxide silicone of ceramics, forms SiO_2 and $SiO_2\ xH_2O$ by the following tribochemical reactions:

$$Si_3N_4+3O_2 \longrightarrow 3SiO_2+2N_2$$
$$Si_3N_4+6H_2O \longrightarrow 3SiO_2+4NH_3$$
$$SiO_2+XH_2O \longrightarrow SiO_2\ xH_2O$$

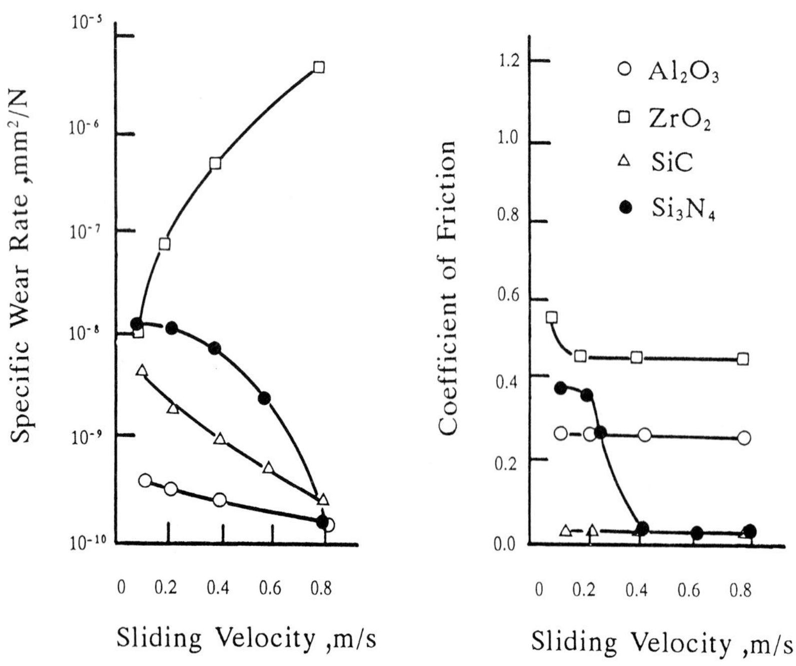

Fig.7 Friction and wear behavior as a function of sliding velocity in water

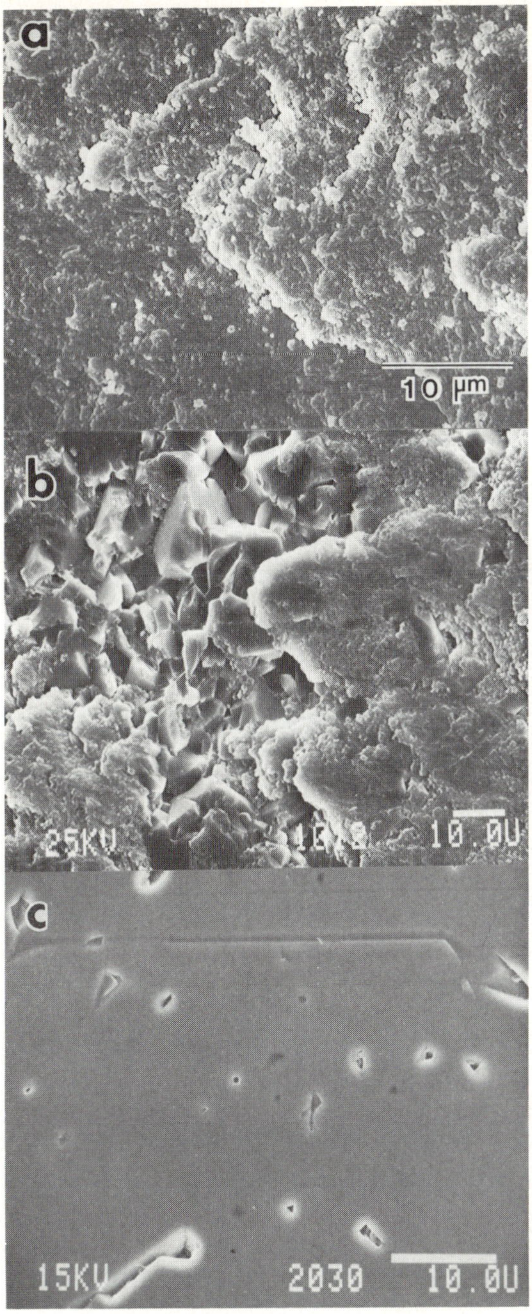

Fig.8 SEM micrograph of worn surface of s-Al_2O_3
(a) in Vacuum (10^{-6}Pa).
Load:3N Velocity:100mm/s
(b) in Humid air (R.H.50%).
Load:10N Velocity:100mm/s
(c) in Water.
Load:50N Velocity:400mm/s

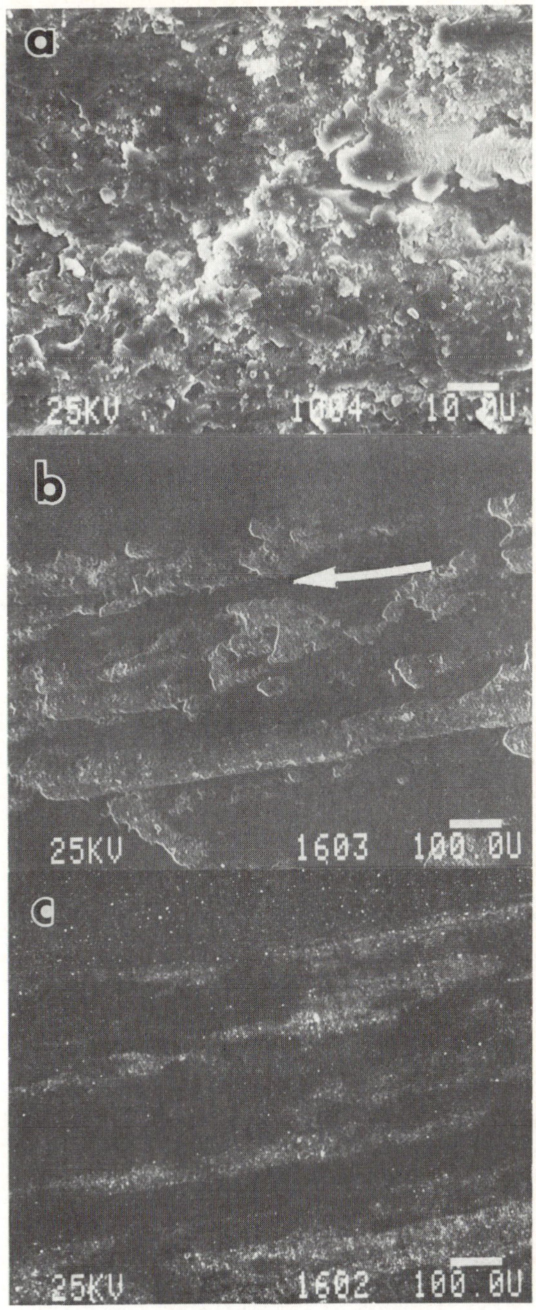

Fig.9 SEM micrograph of worn surface of s-PSZ
(a) in Humid air (R.H.50%) (PSZ-2)
Load:10N Velocity:400mm/s
(b) in Water (PSZ-1)
Load:50N Velocity:100mm/s
(c) CL image comparing to (c)

4.DISCUSSION

Based on the above results, discussion is made for mechanisms of the behavior of friction and wear which take place in four kinds of ceramics. There are three factors to be considered for the effects of environments.

(1) Formation of adsorption layer
(2) Change of mechanical properties by adsorption.
(3) Formation of products caused by tribochemical reaction.

4.1 Formation of adsorption layer

The formation of adsorption layer on the friction surface avoids direct contact between the solids, leading to the reduction of friction. In Fig.3, the drop of friction coefficients is observed in all cases at pressure range 10^{-3}-10^{3} Pa. This phenomena is related to the formation of adsorption layer which is rather difficult in high vacuum atmosphere, rather than in the low vacuum atmosphere.

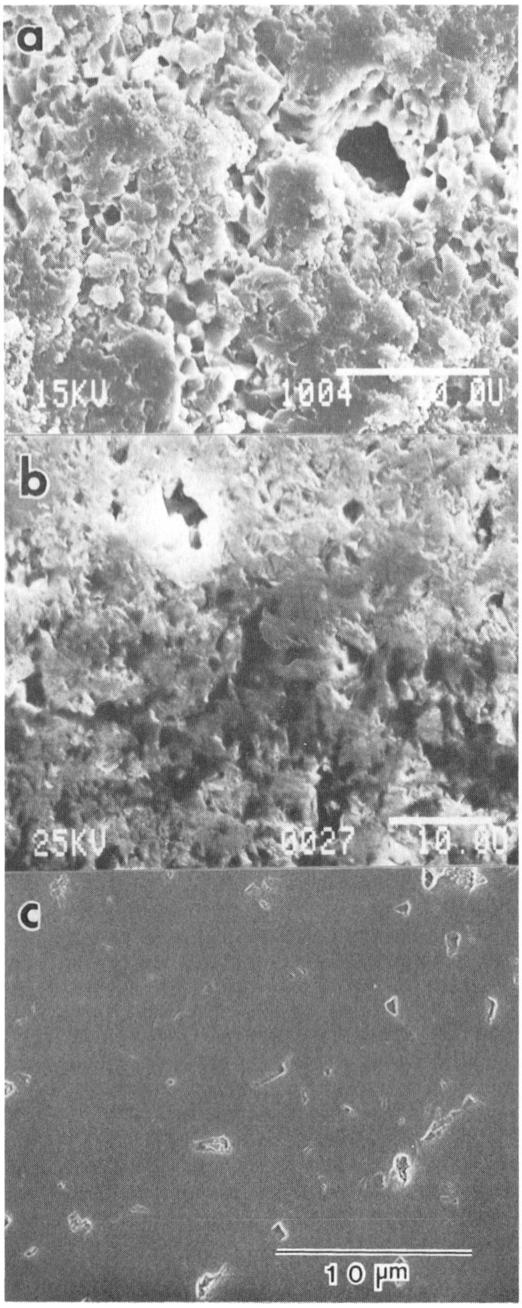

Fig.10 SEM micrograph of worn surface of s-SiC
(a) in Vacuum (10^{-6}Pa).
Load:3N Velocity:100mm/s
(b) in Humid air (R.H.50%).
Load:10N Velocity:400mm/s
(c) in Water.
Load:50N Velocity:400mm/s

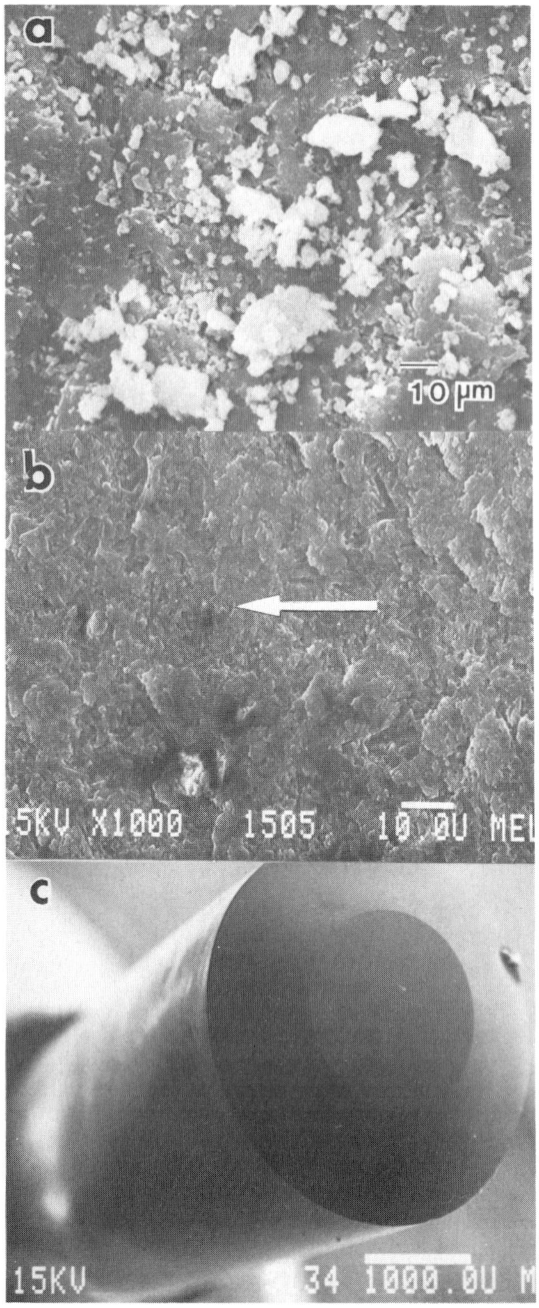

Fig.11 SEM micrograph of worn surface of hip-Si_3N_4
(a) in Vacuum (10^{-6}Pa).
Load:3N Velocity:100mm/s
(b) in Humid air(R.H.50%).
Load:10N Velocity:400mm/s
(c) in Water.
Load:50N Velocity:400mm/s (Pin specimen)

4.2 Change of mechanical properties by adsorption

The adsorption of the molecules in the atmosphere on the sliding surface changes its surface energy or causes stress corrosion. This sometimes results in change of mechanical properties on the surface of ceramics.

4.2.1 Decrease of surface energy

As is well known, the surface energy of solids decreases due to the adsorption of various molecules on the surface[7], depending on what the molecules are and how much of such reactive molecules are contained in the atmosphere. Particularly in such brittle materials as ceramics, there is a close relationship between the surface energy and mechanical strength, i.e. as the surface energy deceases, the mechanical strength decreases[8]. In order to clarify those mechanisms, the following discussion is presented. When strong polar compounds such as water are adsorbed on the friction surface, the surface energy decreases, resulting in the reduction in mechanical strength.

Here, highly stressed region of real contact area may easily flow due to this effect,

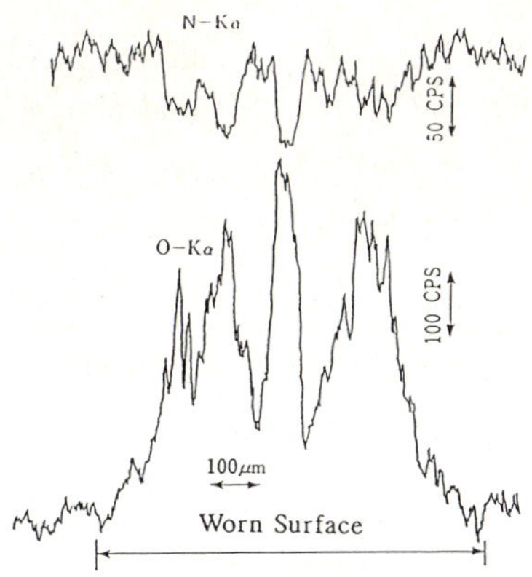

Fig.12 Line profile of N-Kα and O-Kα across the worn surface of s-Si_3N_4 by EPMA analysis.

In humid air (R.H.20%).
Load, 10N. Sliding velocity, 400mm/s.

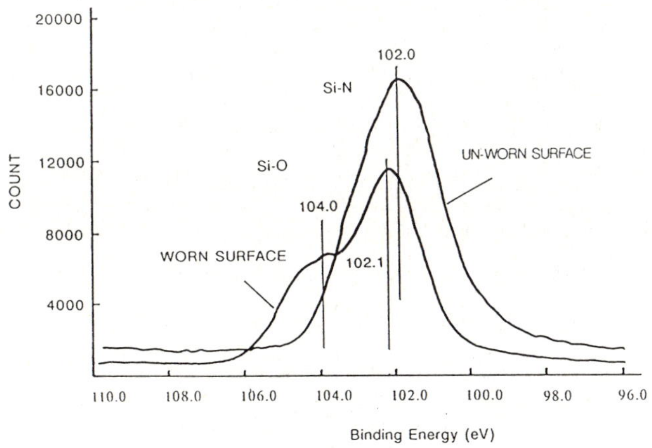

Fig.13 ESCA analysis of Si_{2P} peak of hip-Si_3N_4.

At oxygen reduced pressure (10^4Pa).
Load, 3N. Sliding velocity 100mm/s

which results in reduction of the friction shear stress. These deformed zones also protect the sliding surfaces from severe damage. This mechanism also applied to the explanation of the frictional behavior of Al_2O_3 in humid air and N_2+C_2H_5OH as shown in Fig.6 and Fig.4(a). In fact, some plastic flow observed on sliding surface of Al_2O_3 as can be seen in Fig.8(a).

4.2.2 Stress Corrosion

Stress corrosion is defined as the acceleration of crack development velocity taking place on the material surface under stress due to adsorption of molecules having lone electron pairs adjacent to the hydrogen atom[9]. This type of corrosion enhances to take place the surface crack and leads to destructive wear. In the case of PSZ, the serious wear occurs by its phase transformation[10]. The mechanism is shown in Fig.14.

PSZ remains tetragonal phase, which is stable at elevated temperature, partially in theirs structures. When a crack occurs, the tetragonal phase immediately transforms at the crack tip to monoclinic phase (martensite type phase transfer) and absorbs the destructive energy simultaneously, which results in high intensity and toughness. Such stress induced phase transformation is enhanced by taking place stress corrosion to break Zr-O-Zr bond. In this manner, tiny cracks are formed and accelerated to lead severe wear. These effects are significant in N_2+C_2H_5OH or H_2O, in humid air and in water, The occurrence of phase transfer to m-ZrO_2 was also confirmed by CL analysis as shown in Fig.9(d)

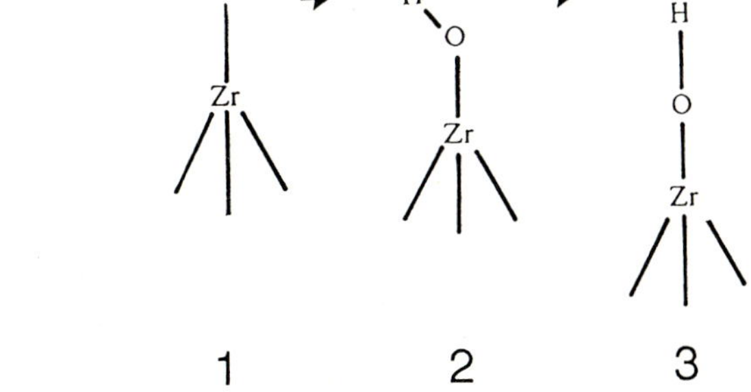

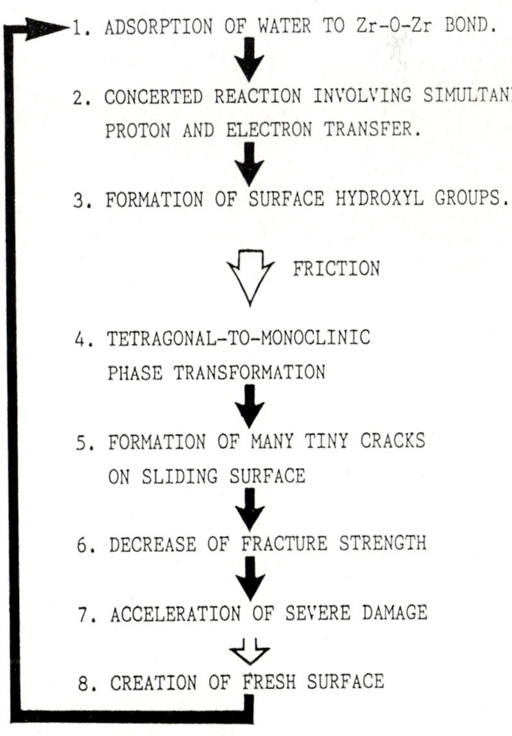

Fig.14 A model of stress-induced corrosion cracking of PSZ

4.3 Formation of products caused by tribochemical reaction

The products of tribochemical reaction are classified into the following two categories: First type of reaction products is formed by reactions of ceramics with atmospheric molecules and ceramics, for instance SiO_2 formed by the reaction of Si_3N_4 with H_2O or O_2. In the first type, when molecules in the atmosphere and ceramics react to form the reaction products, the wear proceeds by the removal of these products. However, this removal is very small, so that it makes the surface smooth. This phenomena frequently obtains a good lubrication in concert with the effect of hydrodynamic lubrication. The typical case of this type is low friction and low wear of Si_3N_4 and SiC in water.

Second type is characterized by reactions of adsorbed molecules on sliding surface. The typical case is the friction polymer-like substance observed on the sliding surface of the four kinds of ceramics in N_2+C_6H_6. The tribochemical reaction products act as a lubricant to reduce the friction coefficients and also protect the sliding surface from destructive damage.

5.CONCLUSION

(a) Under reduced pressure of oxygen and nitrogen, no change was observed in friction coefficients at the pressure range of 10^{-6}-10^{-3} Pa. On the other hand at pressure above 10^{-3} Pa, the friction minimum appeared. The difference in friction behavior between nitrogen and oxygen was noted for Al_2O_3, SiC and Si_3N_4.

(b) In the N_2 + organic compound vapor, each compound affected the friction and wear behavior of ceramics in different ways. For example, C_2H_5OH exhibited a good lubrication characteristic for Al_2O_3, but not for ZrO_2; it brought the increase in wear. In C_6H_6 and sometimes in C_6H_{14} and $(CH_3)_2CO$, a friction polymer-like compound was formed on the sliding surface of four kinds of ceramics. This product gave a good lubricative property.

(c) In humid air, the increase of relative humidity reduced the friction coefficients of all ceramics used in this study. The wear rate of Al_2O_3 and SiC decreased with the increase of relative humidity. In ZrO_2, the maximum wear appeared at around 20% R.H. humidity, whereas in Si_3N_4, the minimum wear appeared at around 50% R.H. humidity. The analysis of worn surface of Si_3N_4 revealed the formation of SiO_2 and its hydration as a result of tribochemical reactions of Si_3N_4 with oxygen or water.

(d) In water, it was clear that the wear rates of Al_2O_3, SiC and Si_3N_4 decreased with the increase of sliding velocity. Particularly in the case of SiC and Si_3N_4, the friction coefficients dropped to as low as 0.01. SEM observation proved the formation of a very smooth surface. On the contrary for ZrO_2, severe wear occurred and the wear rates increased when the sliding velocity increased. CL analysis indicated a phase transformation from t-ZrO_2 to m-ZrO_2 on the friction surface

6.ACKNOWLEDGMENT

The author thanks Y.Enomoto for helpful discussions and K.Saito for scanning electron microscopy.

REFERENCE

(1) H,Shimura and Y.Tsuya,"Effects of Atmosphere on the War Rate of Some Ceramics and Cermets",Wear of Materials, 1977, K.Ludema, ed., Amer.Soc. Mechanical Eng., 1977, p.452.
(2) T.E.Fisher and H.Tomizawa,"Interaction of Tribochemistry and Microfracture in the Friction and Wear of Silicon Nitride", Wear,105,(1985)29.
(3) H.Tomizawa and T,E.Fisher,"Friction and Wear of Silicon Nitride and Silicon Carbide in Water: Hydrodynamic Lubrication at Low Sliding Speed obtained by Tribochemical Wear",ASLE Transaction,30,(1987)41.
(4) T.Sugita, K.Ueda and Y.Kanemura,"Material Removal Mechanisms of Silicon Nitride during rubbing in water", Wear,97,(1984)1.
(5) J.T.Czernuszka and T.F.Page, "Cathodoluminescence: A Microstructural Technique for Exploring Phase Distributions and Deformation Structures in Zirconia Ceramics", J.Am.Ceram.Soc.,68,8,(1985)C-196.
(6) Y.Kanno, K.Suzuki and Y.Kuwahara,"NH_3 Formation Caused by the Presence of H_2O in the Wet Grinding of Silicon Nitride Powder", YOGYO-KYOKAIHI, 91, (1983) 386. (Japanese)
(7) G.I.Gaines and D.Tabor, Nature (London), 178, (1965)1304.
(8) M.L.Hammond and S.F.Ravitz, J.Am.Ceram.Soc., 46,(1963)329.
(9) T.A.Michalske and S.W.Freiman, " A Molecular Mechanism for Stress Corrosion in Vitreous Silica",J.Am.Ceram.Soc.,66,4,(1983)284.
(10)T.Sato, S.Ohtaki and M.Shimada, " Transformation of yttria partially stabilized zirconia by low temperature annealing in air",J.Mater.Sci., 20,(1985)1466

Paper XII(iv)

Wear performance of materials for ball screw and spline applications in Candu reactor fuelling machines

P. E. Dale and R. Tristani

Several materials have been evaluated under sliding conditions in air to asses their tribological performance for potential replacement of cobalt containing Stellite balls in CANDU nuclear reactor fuelling machine ball screws and splines. Stellite was found to have the best friction and wear performance of the materials tested although Si_3N_4 offers the best replacement potential under sliding conditions.

1 INTRODUCTION

1.1 Fuelling machine

In the CANDU nuclear reactor, there are two fuelling machine heads, each one services an end of the horizontal fuel channels. They work in conjunction to move fuel in and out of the core during on-line refuelling.

The Bruce fuelling machine design uses ball screw rams to push fuel into the fuel channels. As a ball nut is rotated, the screw travels over the length of the assembly and is restrained from rotation by a ball spline attached to the screw.

1.2 Ball screw and spline

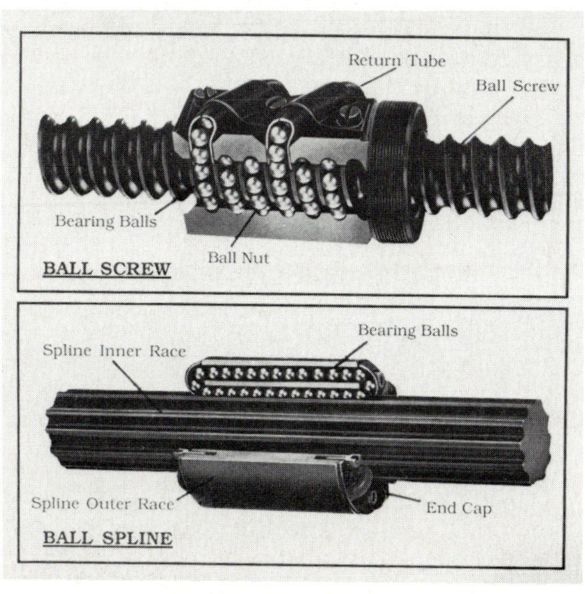

Figure 1 Sectional view of ball screw and spline mechanisms. *(Saginaw)*

Ball screws and splines, fig 1, are similar to their conventional counterparts except that their sliding components are separated by rows of ball bearings. Efficiency and reliability are improved by replacing the high sliding friction of a conventional screw or spline with the rolling friction of a ball bearing. Rolling balls, which run in matching hardened steel grooves, are diverted from the trailing end to the leading end of the nut or race by guide tubes.

1.3 Service conditions

Ball screw and spline components work extremely well under normal lubricated conditions. However, the fuelling machine application calls for ph10 heavy water (D_2O) lubrication at temperatures ranging from 90 to 300 C. As a result, wear of the screws, splines and balls is higher than would be expected for oil lubricated operation.

The fuelling machine is an important component in the operation of the nuclear station where unscheduled shut-downs are extremely expensive due to the cost of replacement power. Replacement of balls at regular intervals to avoid premature failure is a much less expensive option but uses up valuable radiation exposure allowances for maintenance personnel. In addition, Stellite No.6 balls used in the ram ball screw and splines, contain 56% cobalt. Cobalt 60 results from exposure of the cobalt 59, contained in stellite wear debris, to the high neutron flux in the reactor core. Cobalt 60 is a strong gamma radiation emitter and has a relatively long half life which means that cobalt containing wear debris is a serious contaminant in the heat transport system.

A program was initiated at Ontario Hydro Research to look for a replacement material for Stellite No.6 balls. The work reported here is the first stage in a plan to evaluate alternative materials.

1.4 Failure mechanisms

There are a number of possible ball screw and spline failure mechanisms; fatigue of the balls, wear of the balls due to sliding and corrosion. This test program is concerned only with the abrasive and adhesive wear failure mechanisms resulting from ball sliding, rather than rolling, contact.

Sliding occurs when the external forces on the balls exceed the friction forces in the load bearing contacts. These external forces are the result of a combination of possible mechanisms, fig 2, :-

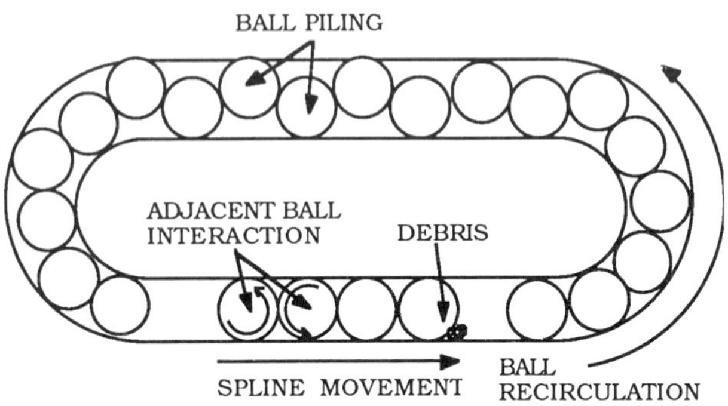

Figure 2 Mechanisms leading to ball sliding in a ball spline application.

- Surfaces of adjacent balls move in opposite directions which produces sliding contact between the balls. This can be minimised by the use of slightly smaller diameter spacer balls which carry no load but accommodate the relative velocities.
- Debris in the load contact will cause the balls to stall and result in ball sliding.
- Balls in the return tubes are not under load and are forced back into the loaded zone by balls entering the return tubes. This can cause "piling" of the balls in the return tube due to the high forces required to return the balls to the loaded zone. This can cause ball sliding in the return tube but a severe case could stall all the balls in the track.

2 TEST PROCEDURE

2.1 Materials

Three candidate ceramic materials were selected for evaluation, Aluminium Oxide (Al_2O_3), MgO bonded hot pressed Silicon Nitride (Si_3N_4) and Zirconia partially stabilized with MgO (PSZ). Si_3N_4 was the most promising due to previous application as a bearing material and the reported long fatigue life of the material compared to existing bearing steels[1]. In addition, a metallic ball material, Cenium ZNZ which contains only .085% Cobalt, was included for evaluation. These alternatives were compared to the existing Stellite No.6 alloy ball material. Typical properties for all the ball materials are given in table 1.

Material	Hardness Hv	Fracture Toughness $MPa\ m^{0.5}$	Density $g.cm^{-3}$
Al_2O_3	1365	4	3.84
Si_3N_4	2000	6	3.26
PSZ	2100	8.1	5.71
Stellite No.6	390	---	8.71
Cenium ZNZ	736	---	8.06
S17400	361	---	------
S45500	467	---	------

DISK	C	Mn	Si	P	S	Cr	Ni	Mo	Cu	Ta	Ti	N
S45500	.005	.06	.03	.006	.004	11.54	8.42	.03	2.14	.25	1.23	.004
S17400	.036	.54	.55	.017	.002	15.13	4.29	.02	3.22	.22	-	-

Table 1 Ball and disk material properties and analysis.

Two materials have been used in the fuelling machine for ball screws and splines, S17400 and S45500, the latter being used more extensively. These are both precipitation hardening martensitic stainless steels and their analysis and significant properties are given in table 1. Both materials were included in this test program.

2.2 Test conditions

As the intent was to compare the sliding wear behavior of alternative materials to those of Stellite No.6, it was considered acceptable to run the samples in air for the first stage of the work. Under the sliding conditions of load and speed encountered in the fuelling machine, the water has limited effectiveness as a lubricant. Test parameters were selected from design and operational data to be:-

Load	155 N
Speed	5.25 cms^{-1}
Sliding Distance	750 m

Each test was run for a duration of 4 hours or until 1mm of linear wear was detected by a displacement transducer on the test machine. All tests were performed in air at ambient temperature with humidity controlled to 40% R.H. ± 10%.

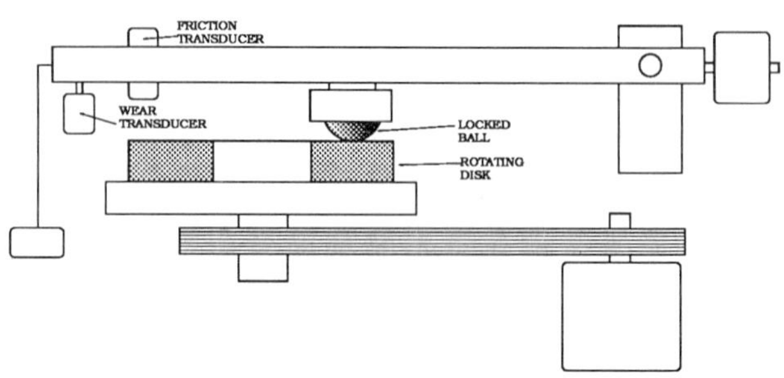

Figure 3 Schematic of ball on disk test machine.

Tests were conducted on a basic fixed pin on rotating disk test machine, fig 3. The ball was clamped in a cup fixture to prevent rotation and loaded on a flat disk. Wear was continuously measured by a linear variable differential transformer (LVDT) type transducer and the friction force, by load cell. Data from the tests was collected by a desk top computer interfaced to the displacement transducer and friction load cell. Balls were tested against each of the two disk materials. Si_3N_4 ball tests were repeated to confirm apparent anomalies.

Prior to each test, the samples were degreased with acetone and trichloroethane in an ultrasonic bath, oven dried and weighed on an analytical balance. After testing, samples were cleaned and reweighed. Several samples were examined using a scanning electron microscope (SEM).

3 RESULTS

3.1 Coefficient of friction

Friction coefficients for tests with S45500 disks are slightly lower than for S17400 disks with the same ball material, fig 4. Although most of the material combinations have friction coefficients between 0.6 and 0.7, two combinations are outside this range. Stellite No.6 has a friction coefficient of .55 when sliding against S45500 which may lead to wear due to ball sliding in the ball spline and ball screw mechanisms where friction is required to cause the ball to roll. If the external forces on the ball are greater than the friction between the ball and track, then the balls will slide producing wear.

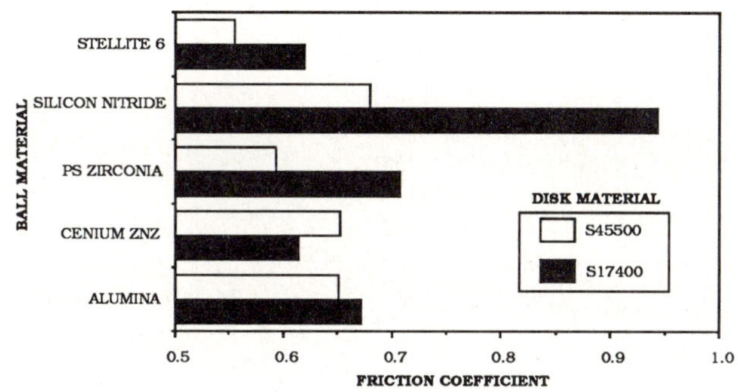

Figure 4 Friction coefficients for ball sliding wear tests.

The coefficient of friction between Si_3N_4 and S17400 is very high, .94 compared to .68 for Si_3N_4 sliding against S45500. This evidence suggests a different wear mechanism for the two Si_3N_4 combinations.

3.2 Combined ball and disk wear rates

Wear rate results are expressed in g/m of sliding and represent the average over 750 m. For this reason, they do not separate running-in from equilibrium wear.

With Stellite and PSZ balls, the combined ball and disk wear rates for both disk materials are similar whilst for Al_2O_3, Cenium ZNZ and Si_3N_4 balls, wear is significantly higher with S45500 disks, fig 5.

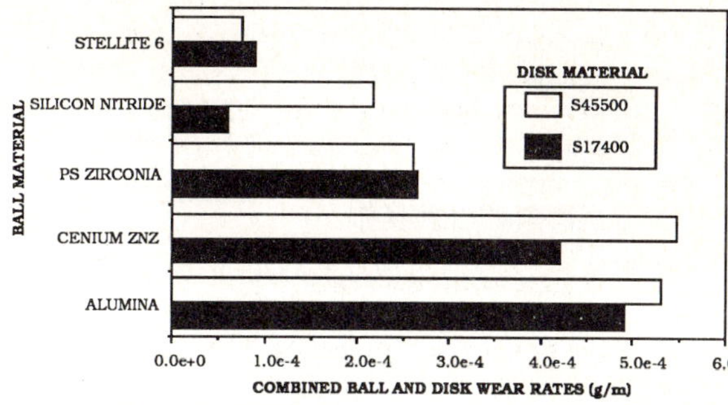

Figure 5 Combined ball and disk wear rates for ball sliding tests.

Tests run with the Stellite balls show the lowest combined ball and disk wear rates. Si_3N_4 ball tests result in substantially different wear (and friction) performance when sliding against S45500 than against S17400. This difference in wear rate, a factor of 3.5, is also indicative of a different wear mechanism than when sliding against the two disk materials. With Si_3N_4 balls, wear rates are roughly a factor of 1.5 worse than for Stellite due to the poor performance of Si_3N_4 when sliding against S45500. Al_2O_3 and Cenium ZNZ ball tests exhibit similar combined ball and disk wear rates, a factor of 6 worse than with Stellite balls, although the average for both disk materials shows Al_2O_3 to be slightly worse. A combination of PSZ balls with either disk shows slightly inferior wear performance to Stellite by about a factor of 3.

3.3 Individual ball and disk wear rates

Examination of the individual ball and disk wear rates, figs 6 and 7, indicates that, for all but Al_2O_3 balls, the majority of wear during the tests occurs on the disks. This is not surprising as the ceramic materials typically have hardnesses in the order of 1300 to 2000 Hv compared to the disk hardness of 360 to 470 Hv.

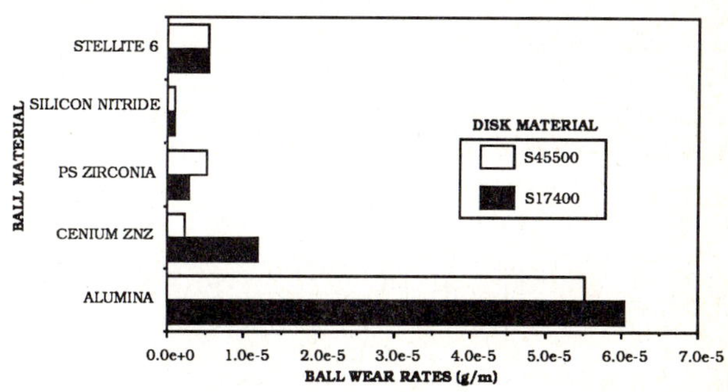

Figure 6 Ball wear rates for ball sliding wear tests.

In addition, the wear performance of both Stellite and Cenium ZNZ depend on a very hard carbide network in their microstructure which have hardnesses in excess of 1000 Hv.

Si_3N_4 shows the lowest ball wear rates and Al_2O_3, the highest for both disk materials tested.

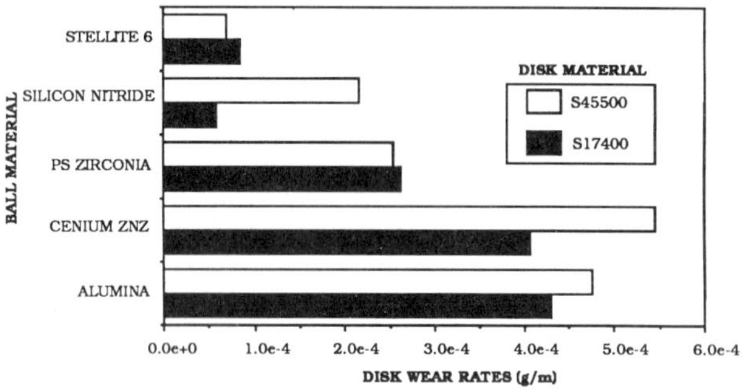

Figure 7 Disk wear rates for ball sliding wear tests.

4 DISCUSSION

None of the alternative ball materials tested have superior tribological performance to the Stellite No.6 ball currently in use. Post test SEM examination of Stellite No.6 ball sliding against S17400 disk reveals a very smooth ball wear surface, fig 8.

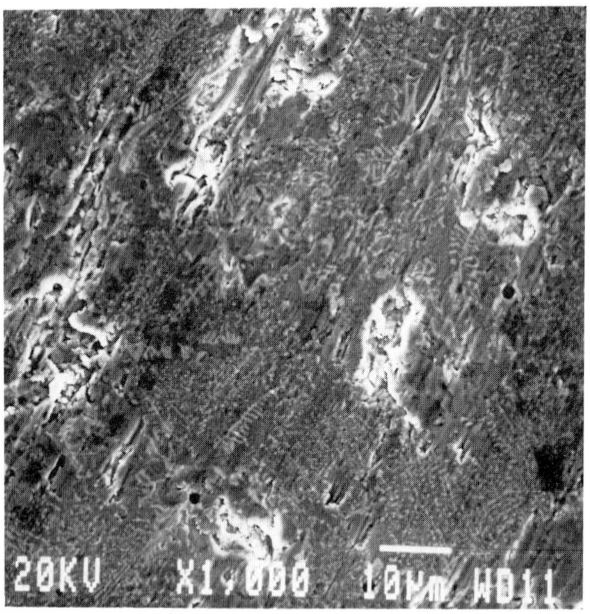

Figure 8 Typical surface of Stellite No.6 ball material when sliding against S17400 disk material.

The presence of a dendritic structure combined with a smooth, almost polished surface, suggests abrasion to be the dominant wear mechanism on the ball. However, there is some evidence of delamination and localized adhesive wear on the ball surface. In contrast, the disk contact surface, fig 9, has undergone extreme plastic deformation which does not appear to be consistent with the ball surface.

Figure 9 S17400 disk surface when sliding against Stellite No.6 ball material.

The adhesive wear mechanism which produced the deformed disk surface does not result in significant permanent material transfer to the ball. This is confirmed by the limited evidence of adhesion on the ball surface.

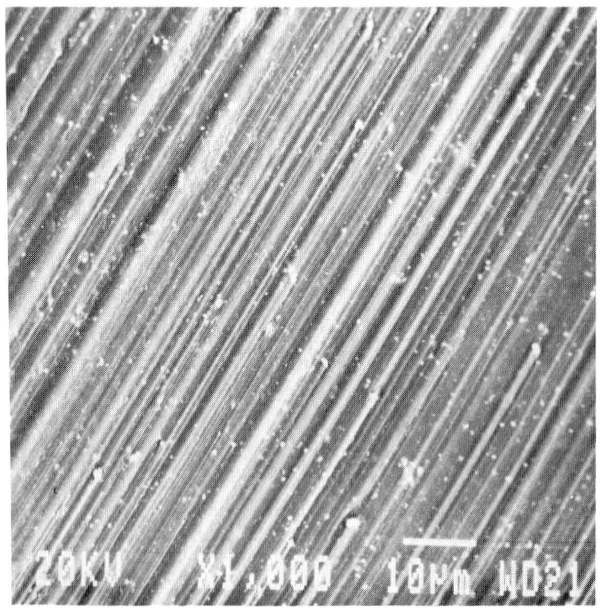

Figure 10 Si_3N_4 ball surface when sliding against S17400 disk material.

For the S17400 disk test, the ball surface shows a grooved morphology, fig 10. Close examination shows evidence of abrasion caused by loose debris still on the surface. Although the high hardness of ceramic materials would seem to preclude abrasion, abrasive wear of ceramics has been identified as a possible failure mechanism[2].

From examination of the disk counter surface, fig 11, it would appear that the ball has been sliding on compacted wear debris and is not in contact with the disk surface. It is the compacted debris which produces abrasion on the ball surface.

In comparison, SEM examination of the S45500 disk test shows different characteristics to the S17400 disk material sliding against Si_3N_4. The ball, in this case, has a layer of deformed material on the rubbing surface, fig 12, while the disk surface is relatively smooth with evidence of abrasive wear, fig 13.

Figure 11 Typical S17400 disk surface when sliding against Si_3N_4 ball material.

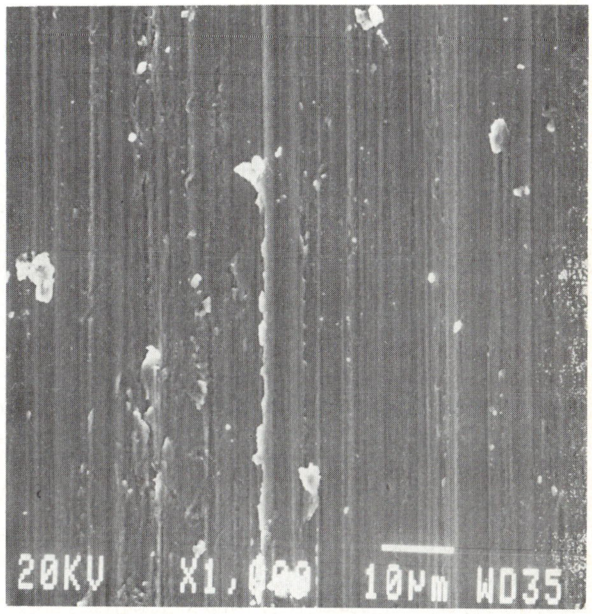

Figure 13 Typical S45500 disk surface when sliding against Si_3N_4.

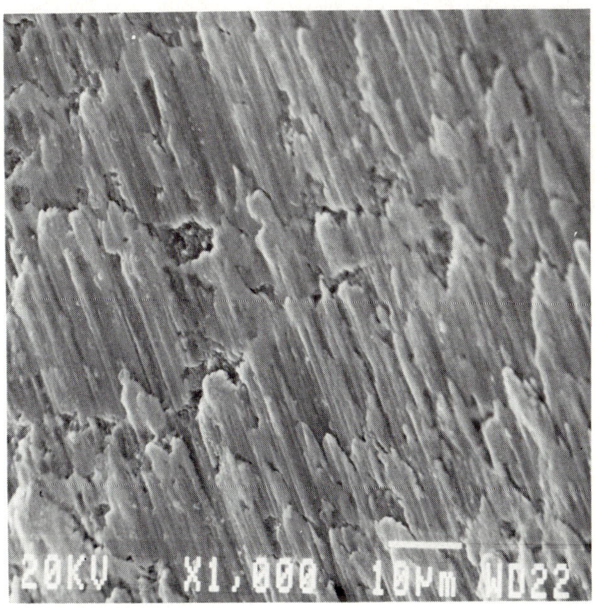

Figure 12 Typical Si_3N_4 ball surface when sliding against S45500 disk material.

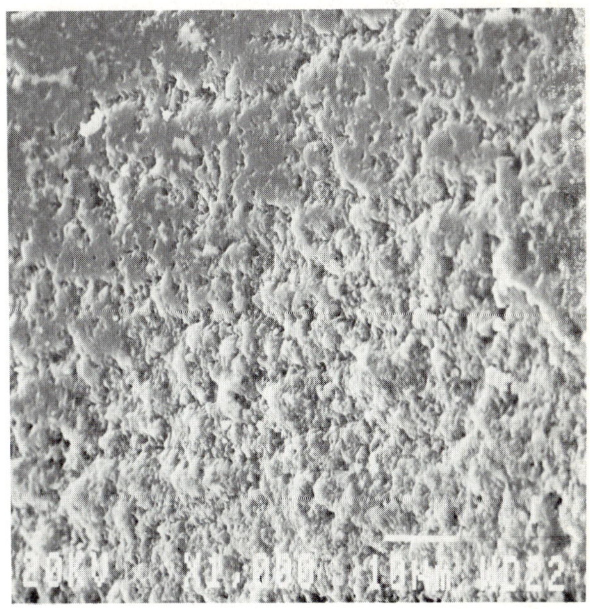

Figure 14 Si_3N_4 ball surface near edge of scar when sliding against S45500 disk material.

This ball surface is similar to plastic deformation identified on Al_2O_3 when sliding against AISI 4340 steel[3]. At the exit edge of the ball wear scar there is evidence of a possible microfracture wear mechanism on the ball surface, fig 14, which was not apparent when sliding against S17400.

The difference in wear rate between the two disk materials and Si_3N_4 is because the S17400 disk surface is protected by a layer of compacted wear debris and the ball is hard enough to resist subsequent abrasive wear. With S45500 disk material, the ball becomes the abrader for which the disk has no protection and results in a higher disk wear rate. SEM examination of the ball and disk surfaces has confirmed a difference in wear mechanism, indicated by the difference in both friction coefficients and wear rates for the the two Si_3N_4 combinations tested. The reason for this tribological difference is not completely understood. The sequence leading to the equilibrium wear mechanism requires more investigation.

Friction coefficients for Si_3N_4 on both S17400 (.94) and S45500 (.68) are much higher than the 0.15 reported by Bhushan and Sibley[1] for Si_3N_4 sliding on M50 bearing steel. However, this result was obtained at a slow sliding speed (3 $mm.s^{-1}$) and low load (20 N) in dry air which was thought to result in a thin SiO_2 film on the surface. It is possible that a coherent oxide film does not form under the present test conditions resulting in a higher friction coefficient. Si_3N_4 offers the best potential for replacement of Stellite No.6 balls in the fuelling machine. The ball wear rates are very low, fig 4, and disk wear rates are close to those of Stellite.

Of the other materials, PSZ is close to the Si_3N_4 in terms of its' sliding wear performance and it has lower friction coefficient than Si_3N_4 when sliding against either disk material. However, Si_3N_4 offers the best potential for replacement of the Stellite balls due to its' lower ball wear rate when sliding against both disk materials. Both Cenium ZNZ and Al_2O_3 combinations have inferior wear rates and do not appear to be a useful alternative to the Stellite balls in this application

In view of the reported reactions of Si_3N_4 with water[4], it is extremely important to evaluate the Si_3N_4 material in the aqueous environment encountered in the fuelling machine. The presence of only 1% water by volume in a base oil has been shown to substantially increase the wear rate of self mated Si_3N_4[5]. In contrast other work has indicated a reduction in wear in the presence of water[4]. Any benefit in the ceramic/metal contact of this application would depend strongly on the adherence of any layer or protective film.

5 CONCLUSIONS

(i) Si_3N_4 is the most promising replacement material for the cobalt containing Stellite No.6 in the fuelling machine ball screw and spline application.

(ii) Both Cenium ZNZ and Al_2O_3 balls are subject to severe wear under these sliding conditions and are therefore considered unsuitable for ball screw and spline applications.

(iii) The wear rate and friction coefficient of Si_3N_4 is dependant on the wear mechanism which has been shown to be substantially different for two similar precipitation hardening stainless steels. The reasons for these tribological differences are not fully understood.

(iv) Ball sliding is only one of the possible failure mechanisms in the ball screw and spline. More laboratory simulation tests have to be performed, therefore, before a full fuelling machine evaluation of alternative materials could be contemplated.

References

[1] Bhushan, B., and Sibley, L.B., "Silicon nitride rolling bearings for extreme operating conditions," Trans. ASLE 1982, vol. 25, No.4, 417-428

[2] Buckley, D.H. and Miyoshi, K., "Friction and wear of ceramics," Wear, 100 (1984), 333-353

[3] Mehrotra, P.K., "Mechanisms of wear in ceramic materials," Proc. Intl. Conf. on Wear of Materials, April 11-14, 1983. 194-201.

[4] Tomizowa, H. and Fischer, T.E., "Friction and wear of silicon nitride and silicon carbide in water: hydrodynamic lubrication at low sliding speed obtained by tribochemical wear", Trans ASLE 1986, vol 30, No. 1, 41-46

[5] Habeeb, J.J., Blahey, A.G. and Rogers, W.N., "Wear and lubrication of ceramics", I.Mech.E, C132/87, 555-564

SESSION XIII
REVIEW PAPERS

Chairman: Professor W O Winer

PAPER XIII(i)	Tribological Design - The Power Generation Industry
PAPER XIII(ii)	Tribological Design and Assessment - The Nuclear Industry
PAPER XIII(iii)	Tribological Design - The Spacecraft Industry
PAPER XIII(IV)	Tribological Design - The Electronics Industry

Paper XIII(i)

Tribological design - The power generation industry

P. G. Morton

ABSTRACT

Conventional power generation plant covers a very wide field, many of the machines being common to other engineering plant. The paper concentrates on those machines that are unique to power generation and outlines past and present problems with a tribological basis. The scope is limited to a selection of typical problems mainly relating to large steam turbosets, although sets driven by water turbines and gas turbines are mentioned.

1. INTRODUCTION

This paper deals only with conventional power plant since nuclear plant is being covered in another paper in these Proceedings (1). The term 'Power Engineering' could be taken as applying to the complete path from energy source to the consumer of electricity. This would include consideration of such areas as fuel handling, boilers, condensers, transmission networks, switchgear etc. There are certainly tribological considerations in the design of equipment used in these areas, widely disparate examples being wear in fuel pulverizers and tube damage in heat exchangers. Power generation plant also utilizes a wide range of machines common to other applications such as small and medium sized electric motors, fans, pumps, compressors and gearboxes. The components unique to power generation plant, which present the greatest challenge to the tribological designer, are however those in the prime mover/generator line-out and some large auxiliaries, therefore the paper will concentrate mainly on tribological problems in this area.

The power and mass of equipment often has a bearing on the design problems encountered. Powers available from single shaft line-outs range from 5 MW to 1300 MW. Diesel sets are limited to a maximum of around 40 MW. the higher end of the gas turbine range is typically around 100 MW Water turbine sets have been made with outputs exceeding 500 MW, but this is exceptional. The highest powers are available from steam turbine sets. Maximum outputs of 1300 MW have been reached although sets with outputs of 660 MW running at line frequency are currently the most numerous.

In terms of the combination of physical size, mass and rotor speed, the steam driven turboset with a two pole generator, is undoubtedly the most impressive rotating machine in the engineering field. The water turbine generator set is also very massive although it runs at much lower speeds. Some of the problems discussed in this paper demonstrate clearly how increases in physical size, mass and power throughput have brought to the fore new aspects of tribological design.

2. SCOPE

Tribology is about interacting surfaces in relative motion (2). Over the years, tribologists have broadened their original scope to include interaction mechanisms where the surfaces are not necessarily in close juxtaposition. Indeed, recently even the subject of electromagnetic attraction seems to have been included in the tribologist's remit! (3). Tribology has always of course covered a wider area than friction and wear; the influence of tribological processes on machine dynamics has always been an area shared with dynamicists. In presenting examples of problem areas in the design of elements in power engineering machinery, this paper will use the broadest definition of tribology. Rather than generalize over the whole power generation field, the paper will concentrate on a few specific examples of how tribology impinges on the design process of the larger machines.

3. PROBLEMS IN STEAM TURBOSETS

It is convenient to divide the turboset into three zones; the non-rotating components, the rotating components and the interface.

3.1 Non Rotating Components

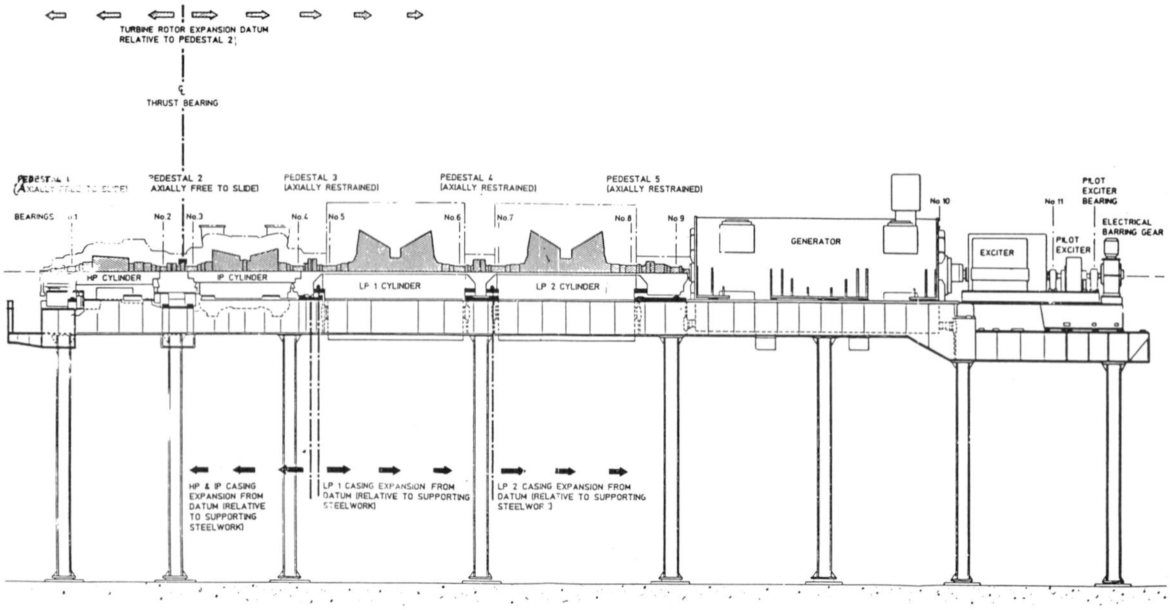

Fig. 1. Support and Expansion Arrangements, Elevation.

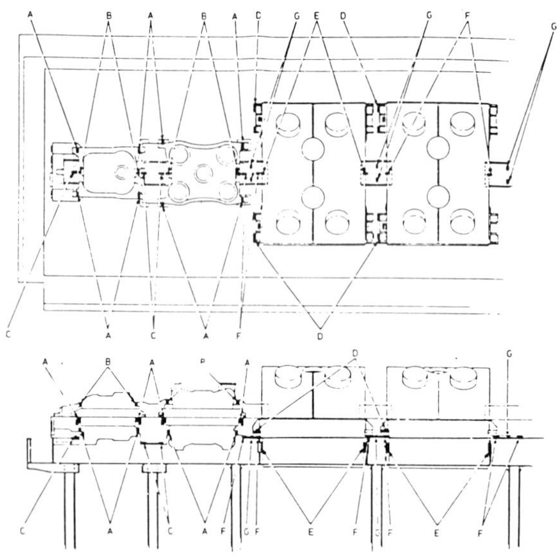

Fig. 2. Support and Expansion Arrangements, Plan.

The static parts of a typical large set comprise a high pressure (H.P.), turbine cylinder, an intermediate pressure (I.P.) cylinder and two or three low pressure (L.P.) cylinders, generator and exciter casings. Fig. 1 shows the arrangement of these components. The H.P. and I.P. turbine cylinders are constructed from two or more shells in order to minimize the thermal stressing. Differential movement is achieved internally by allowing for sliding in the joints. The cylinder assembly is rigidly coupled axially. Pedestals are provided which transfer the bearing loads from each turbine to the foundations. The generator bearing is integral with its casing. The total mass of a large turboset could reach 1200 Tonnes including the rotor and is reacted by the foundations at a series of pads. The general requirement is that all interfaces between the set and earth should be capable of reacting operating loads in a determinate manner, and guiding the set so that misalignments do not occur. Temperatures vary through the structure from admission steam temperatures of 565°C to ambient and all interfaces have to allow for expansion. Figs. 1 and 2 shows that the arrangement is complex but basically axial guidance is provided at the pedestals, the H.P. and I.P. cylinders are allowed to expand laterally by sliding on the pedestals and the L.P. cylinders by sliding on the foundations. Vertical guidance is also provided for all cylinders. The generator casing is also supported on pads and is separately located. The whole assembly is around 60 m in length. The supports usually comprise a packer plate mating with the surfaces themselves, or key blocks attached to those surfaces.

The supports and guides are subjected to a combination of sliding, pressure, temperature and agressive environment. This particular combination presents a tribological problem for two reasons. First, if the friction forces are not symmetrical about the axial centreline, slewing or 'crabbing' may occur resulting in severe misalignment problems. Second, any significant wear between planned maintenance periods could result in internal clearances between rotor and stator being bridged, causing rubbing or damage within the turbine or malfunctions of the bearings. The permitted level of misalignment is discussed in (4). The degree of wear depends on the choice of material pairs, the lubrication strategy, the environment, the product of pressure and distance moved and the speed of movement. Gross movements occur at start up and between significant load changes. Small movements due to vibration also produce wear. Low (5) computes typical total movements of between 1 km and 15 km per year and axial thermal movements of up to 5 cm at start up. Much of the aggregate movement at the pedestal and casing locations, arise from running frequency vibrations. Pressure between sliding contacts varies but it could be as high as 4 MN/m^2. In the past, systems based on grease lubrication were used, having steel, iron or leaded bronze packers mating with low alloy steel components and problems due to high temperature and poor grease distribution were experienced. For more than 20 years however, designers have used a variety of dry bearing materials, problems occurring with older plant being dealt with as they arise, either by maintenance or substitution for newer materials. Due to the use of multiple H.P. and I.P. turbine casings, and in some case integral condensers around the L.P. supports, sliding in a steam environment occurs. As far at high temperature steam is concerned, the formation of an oxide which is self healing is beneficial but the surfaces have to be protected from galling prior to this formation. Various hard coatings have been found to be effective in this respect.

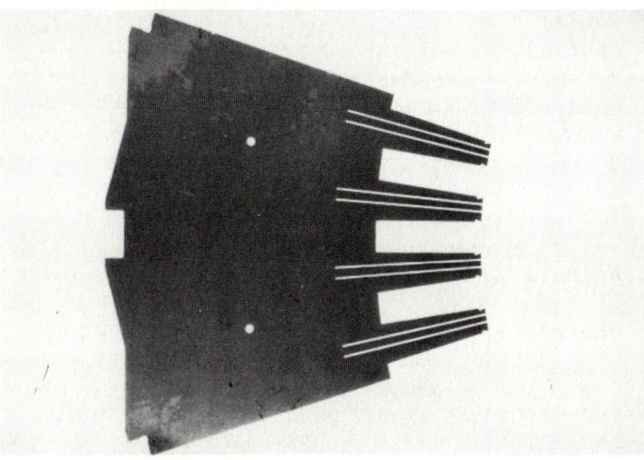

Fig. 3. Typical Stator Core Lamination.

The generator is common to all power plant and always has a stator core built up from laminations. In a large turbogenerator there are around 150000 laminations similar to the one shown in Fig. 3. The laminations are coated with insulation and assembled by overlapping and clamping to form a tube 6 m long and 3.3 m bore diameter. The whole assembly of laminations in held together by friction maintained by axial pressure applied by end plates clamped with bolts. Tribological problems hardly occur nowadays in the main body of the core due largely to the care taken in selection and assembly of the laminations. Some fretting problems have occurred on the end laminations when the clamping was not uniform, this has been rectified by improving the clamping plate design.

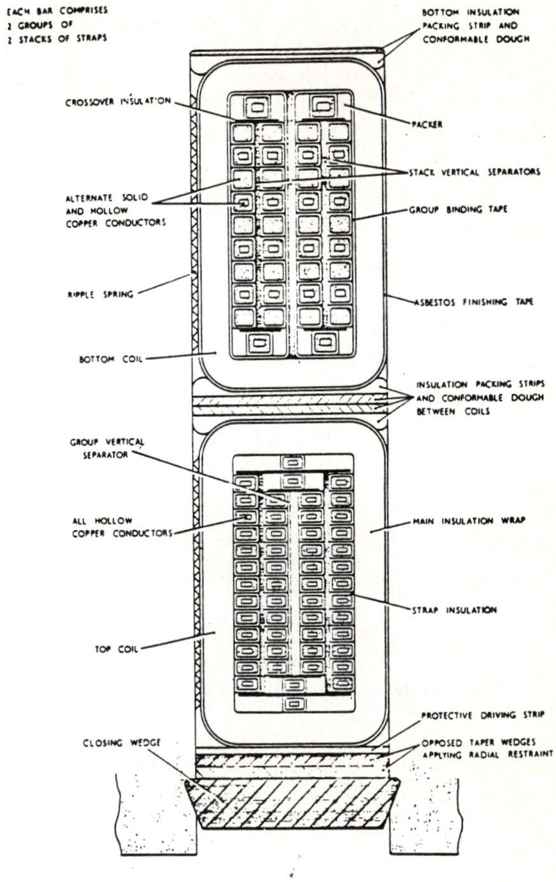

Fig. 4. Stator Coil.

Stator coils are shown in Fig. 4. The replacement of bituman based insulating varnish by epoxy or polyester resin bonded mica, has virtually eliminated damage due to thermal expansion and contraction. Consolidation techniques incorporating taper wedges and epoxy glass ripple springs, have prevented conductors, 'bouncing' in their slots under the influence of magnetic forcing.

3.2 Rotating Components

Large steam turbosets have either four-pole or two-pole generators rotating at half line frequency or line frequency respectively. The latter have a much more slender rotor, and thus pose greater problems, so we will concentrate solely on this type of design. All of the rotors in the line-out are solidly coupled to make a continuous shaft more than 60 m long and weighing up to 300 Tonnes. Tribological design problems in this rotating assembly arise from the nature of the operating conditions which are:-

a) Steady base load operating.

*b) Load and speed cycling during 'two-shifting' operating.

---

* A U.K. requirement in which the set is unloaded and rapidly reloaded after a shutdown period of 4-6 hours.

c) Operation at barring speed (2 to 30 rev/min.

The following table from Ref. 6 gives the loading cycling to be expected during a plant design life of 40 years.

Load Variation	No of Times
Shutdowns of up to 10 hours duration	4000
Shutdowns of up to 72 hours duration	1000
Cold starts	200
Routine overspeed tests	165
Overspeeds after load reflection	30

Considerable periods of barring i.e. running at very slow speed to equalize rotor temperatures, is implied in the above table.

Centrifugal force, particularly in line frequency sets can be beneficial. The large forces on some L.P. turbine blades for instance can minimize the possibility of fretting at root fixings and lacings. Protracted periods of barring where the blades flap from side to side under gravity as they rotate has to be taken into account by the designer.

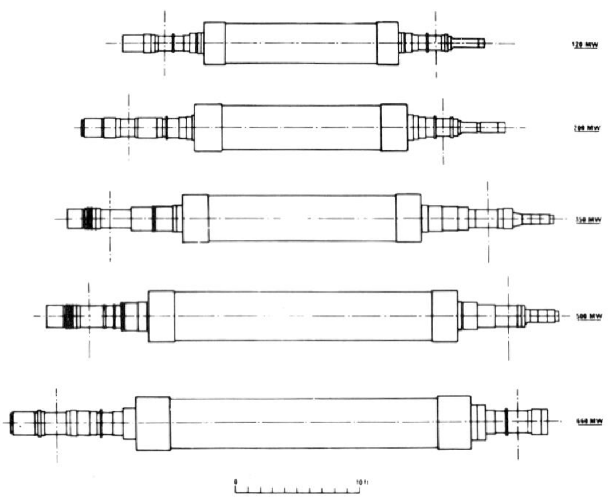

Fig. 5. Generator Rotor Size Increases.

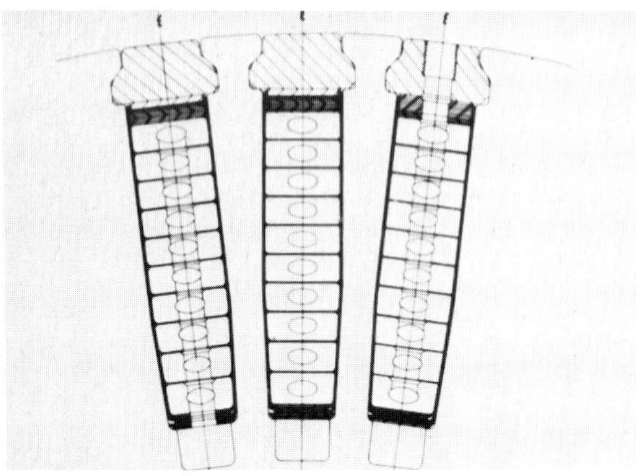

Fig. 6. Rotor Coil Assembly.

In general, the greatest design problems on the rotating system arise from speed and load changes. Perhaps the most sensitive areas of the complete rotor which weighs up to 300 Tonnes lie in the generator rotor. Over three decades these rotors have grown in size and mass as Fig. 5 shows and the sag under self weight of a 660 MW rotor is around 5 mm. The rotor is deeply slotted to accommodate windings, the arrangement of copper insulation and wedges being shown in Fig. 6.

Relative movements between conductors, wedges and slots arise from two sources viz differential expansion and self weight rotor bending. Regarding differential expansion between the windings and both rotor and wedge, this depends on the load conditions which are related to winding current. A typical differential strain would if unrestrained be 6×10^{-4} according to Marlow (7). The effect of friction centrifugally imposed is to lock up a proportion of the winding thus inducing stress. In the past, this frictionally induced stress sometimes caused plastic deformation which at each stop/start cycle shortened the copper; eventually this ratchetting mechanism caused failure. This problem was cured by the introduction of silver bearing hard-drawn copper (8). With the present day flexible rotor, the designer has to face a new problem. If the coefficient of friction varies between coils, bending moment arising from different tractions due to different slip lengths will cause a rotor bend to be locked in under load, thereby causing appreciable unbalance which can be magnified at certain speeds. The cure is to provide a determinate slip layer between, winding and wedge using for instance resin glass and P.T.F.E. formulations.

Cyclic movements due to self weight bending occur at barring, maximum levels being of the order of 0.5 mm peak to peak. At full speed, relative motion occurs over the end 75 mm and is of the order of 0.008 mm. As has already been remarked, the time spent in run-ups and in barring is not insignificant, so that this self-weight induced fretting could well result in unacceptable wear at the ends of the windings. Due regard has to be made to this factor when selecting insulating materials.

Fig. 7. Rotor Endwinding and Retaining Ring.

Windings leave the rotor body at the ends as shown in Fig. 7 and are supported by retaining rings shrunk onto the rotor. These rings expand and distort with speed and relative motions of the end windings occur. Suitable insulation material is required to prevent excessive wear at operating temperatures. In the past, asbestos reinforced material was used, but this has been discontinued. New materials have been introduced to give the required properties.

Wedges retaining the windings used to be made from relative short lengths of steel (around 300 mm). As cyclic microstrain at the ends of these wedges became larger with increasing rotor flexure, the initiation of fretting cracks became more probable. In addition, larger self weight bending stress together with stresses due to thermal gradients, increased the possibility of fretting fatigue in the rotor. Ref. 6 shows an example of fretting damage due to a short rotor slot wedge. Designers have overcome the problem by making the wedges from aluminium alloy, which reduces the traction forces by reducing Young's Modulus by a factor of three. Furthermore fretting initiation is not as likely between aluminium alloy and steel as between steel and steel. The wedges are also now made continous over the full length of the rotor so that the relatively small area of microslip is located at longitudinal stress-free parts of the rotor, i.e. the ends. Even if fretting cracks occurred they would have no stress field in which to propagate. The end rings which are an interference fit on the rotor end at low speed, and a light fit at running speed also require special attention to prevent fretting damage.

A component on large generators which has been the subject of a great deal of development in the past, is the slip ring, which transfers current from brushes into the rotor winding. The current density under the brush is around 0.1 A/mm^2 and the surface speed can be as high as 75 m/sec. The rings used to be bronze but are now made of tool steel (1.0% C). Graphite impregnated with synthetic lubricant has been developed to provide a good compromise between brush wear and contact resistance. It is necessary to cool the rings to an average surface temperature between 75°C and 85°C and in very dry atmospheres it has been found necessary to use humidifiers to limit brush wear - the ideal is 60-90% relative humidity.

Water droplet erosion in the last stages of the L.P. turbine blading was recognized as a problem were than 50 years ago by Parsons. The erosion mechanism was studied extensively by Gardner (9). The problem occurs at the leading edges of the L.P. blades near the tips where the peripheral speed can by in excess of 600 m/sec. Alleviation is possible by water extraction but this is not widely adopted. Due to changes in thermodynamic design the phenomenon is now confined to the last row blade. Several techniques of reducing erosion have been used, one being local hardening of the blades which are usually an alloy steel of around 13% Cr. Other techniques involve the deposition of hard coatings. The most popular design however was brazed-on shields made from cobalt based alloys although, sometimes shields are made from tool steel.

Having dealt with some wear problems on the rotor it would be useful to look at indirect consequences of slip. It has been known for sixty years that any rotating structure which has internal slip has the potential for instability (10). The effect of internal microslip in the generator for instance, is to inject negative damping into the system. Provided the system has enough positve damping from other sources this is alright. In conjunction with other destabilizing phenomena (which will be dealt with later) the aggregate effect could be to introduce sub-synchronous vibrations. Lund discusses this destabilizing phenomenon in ref. (11).

3.3 Interface Problems

3.3.1 Bearings

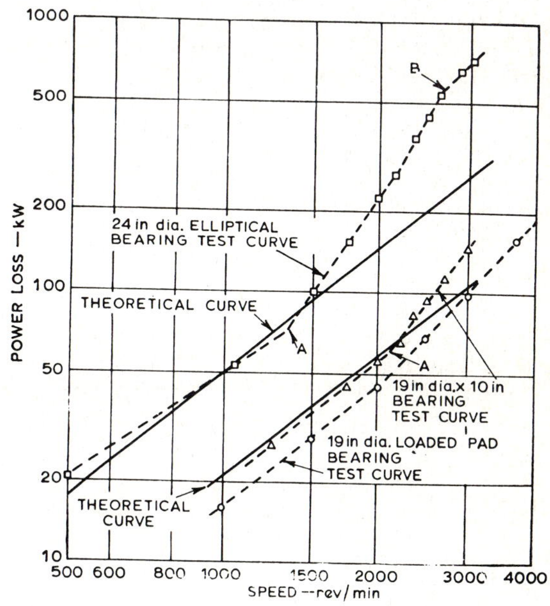

Fig. 8. Losses in a Large Generator Bearing.

The bearing is the universal interface in all rotating machines and in the case turbomachinery we are concerned with oil lubricated bearings both for thrust and journal applications. Typically there could be 14 journal bearings and a single thrust bearing located between H.P. and I.P. rotors. The journal bearing diameters are fixed by the shaft sizes necessary to cater for torque transmission. Specific pressures have to be limited to 2 MPa to avoid excessively thin oil films. Bearing diameters in the line frequency sets run from 610 mm at the generator end to 350 mm at the H.P. turbine with L/D ratios of 0.7 to 0.5. The sets running at half line frequency have larger bearings, the maximum diameter being above 760 mm. In most cases because the rotor has to be barred and run up safely, high pressure jacking oil is supplied during these operations. The bearings run in the superlaminar regime and the losses are large. Fig. 8 taken from Ref. 12 shows losses of 1/2 MW in typical generator bearings. It would be desirable to reduce bearing losses, first because of the considerable savings in energy (13) and second because the wasted energy increases the operating temperature of the bearing thus reducing safety margins. It is of course obvious that reducing the swept area occupied by oil in the bearing, while maintaining the oil film thickness over that area would undoubtedly reduce power loss. The partial arc, single feed bearing is a good starting point and the addition of a pressure feed to make the bearing a hybrid one may offer some improvement (14). Further research is necessary into methods of reducing the losses in large bearings.

A good example of a tribological problem which was cured by material changes, is provided by the 'wirewool' shaft damage experienced some 25 years ago (15). Failures occurred both on the test bed and in service within a few hundred hours, on turboset ratios and gas turbine compressors at the journals. The symptom was the production of a hard iron carbide 'scabs' embedded in the white metal of the bearing with hardnesses over 1000 V.P.N. Once formed, these scabs machined away the journal at a very high rate, e.g. 3mm in 10 minutes. It was eventually found that two conditions were necessary in order to produce the problem. First, particles in the oil had to be large enough to bridge the thinnest oil film. Second the journal had to be made from medium or high chrome steel; most of the journals affected were made from 3 Cr Mo steel. The cure was to make rotors from steel with a very low chromium content.

In general the oil lubricated bearing is reasonably well behaved from the tribology view point provided the oil is kept clean, and the peak white metal temperature and pressure are kept within reasonable limits. It is very necessary for good alignment, both lateral and angular to be maintained in order to prevent excessive local pressures. It also necessary for the designer to be able to predict with some certainty the pressure and temperature profiles as well as the dynamic (stiffness and damping) properties. The foundations of hydrodynamic theory were laid down by Reynolds a century ago (16) and significantly improved by Christopherson in 1941 (17), when he introduced practical ways of modifying the theory to include thermal effects. More recently, others have introduced oil heating and conduction in the bush and journal into the heat balance e.g. (18) (19). In large turbo type bearings, even these improved techniques have failed to predict with any real accuracy the characteristics of the bearing. Recent work, both theoretical and experimental (20), (21) has clearly demonstrated that it is necessary to take into account the distortion of the bush in any mathematical model of a large heavily loaded journal bearing. It has been shown that pressure and temperature profiles, journal attitude and bearing dynamic properties are all significantly affected by thermoelastic behaviour. In a large set where the alignment may change it is also necessary to predict the effect of these changes (22). Severe misaligment may overload the bearing or by unloading it, reduce its stability.

Bearings are capable of modifying the turboset dynamics profoundly. The most common form of vibration in turbomachinery is the once per rev or synchronous vibration, due in the main to unbalance in the rotor. All hydrodynamic bearings will reduce this vibration. Unfortunately, such bearings also have the ability to inject negative damping into the system under certain operating conditions, when forced or tuned into frequencies below one half of their rotational speed. Parameter changes that mitigate one property, aggravate the other, so that a compromise has to be sought. It is generally supposed that tilting pad bearings provide some sort of solution to the dilemma. This is not so - they achieve stability at the expense of synchronous damping, and this is important, since we may expect very high amplification factors, at the generator in particular, as the machine runs through its lower critical speeds. A variety of bearing designs have been proposed, but the limiting constraint is the minimum oil film thickness. At some level of size and therefore flexibility and critical speed the two pole turbogenerator rotor will have to rely on some external device, either active or passive, to achieve acceptable dynamic characteristics. What output power can be reached before this situation occurs, depends on the ingenuity of the electrical and mechanical designers. Much work has been done on squeeze film bearings, which might be applied to the power generation industry, but there is need for more research into the practicability of active control applied to large machines.

3.3.2 Seals

Seals are a necessary evil in all turbomachinery and the large turbine generator set is no exception. There is no area on the rotor where permanent metal to metal contact is made with any static part. Seals vary from simple non-ferrous knife edge scrapers and wipers to complex labyrinth arrangements. The simple scrapers do not present any problem. Their function is to prevent low pressure oil from escaping along the shaft and they wear themselves clear with the action of the shaft eccentricity. Some seals control or prevent the flow of steam through or from the turbines. Other seals prevent the escape of hydrogen from the generator casing.

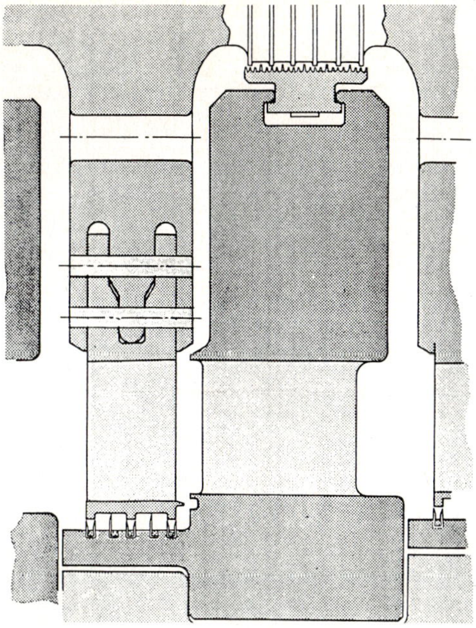

Fig. 9. A Disc and Diaphragm Turbine Stage.

Steam seal problems occur mainly in the H.P. turbine where the pressure drop and therefore the flow velocities are high. The steam path through a stage of the H.P. turbine is shown in Fig. 9 and includes the gaps at the shroud of the turbine blades as well as the inter-stage seals. The diaphragm or gland seals are constructed in six segments and spring backed so that during misalignment and shaft run-out hard contact will be prevented. It will be seen that the inter-stage and tip seals are of the labyrinth type in order to impede the steam passage through them. When fluid flows through any annulus, a quasi-spring and damping characteristic will be produced. As the rotating member becomes eccentric to the stationary one the flow pattern is modulated to produce not only restoring forces, but 'cross coupling' coefficient (23). The velocity of the lateral movement will also result in damping forces. The net effect of this force vector is to produce an aerodynamic instability known as 'steam whirl'. The destabilization is dependent on flow velocity and therefore upon turbine power. The effect of steam whirl, which in a sub-synchronous orbiting of the rotor, is therefore to limit the power available from a turboset. Much research has been carried out on the flow of compressible fluids through labyrinths, and it has been found that the effect of swirl is of great importance (24). If the swirl is reduced, the destabilising cross coupling coefficients are greatly reduced. In some cases the introduction of axial guides to the steam flow into the glands has proved effective in curing this problem.

Another kind of problem occurs at seals where contact with the shaft is made repetitively as the shaft rotates. If the shaft is whirling synchronously with running speed, the temperature at the contact point will rise above that of the other side of the shaft thus producing thermal bending. It can be shown that under certain circumstances the shaft vibrations will grow in an unstable manner due to the coupling between increasing vibration level and shaft temperature (25). Solid contact is not always necessary. Floating oil lubricated radial seals, such as are used as hydrogen seals in generators, if sufficiently restricted in their housing can produce the same differential thermal effect (26). The possiblity of unstable spiralling vibration of this type depends on the dynamic magnification of the system, the differential energy input at the seal and axial position of the seal on the shaft. Of these only the seal frictional characteristics can be dealt with tribologically. In the case of solid contact, seal materials that clear themselves by material removal or seals with low coefficients of friction may be helpful, but the only certain cure is to achieve adequate clearance. In hydrodynamic/ hydrostatically lubricated radial seals, these should be free to float and the oil flow in the annular clearance should possess a strong centralising property. A type of hydrogen seal which has been extensively used is the face type seal. Some designs started off as an annular, white metalled thrust ring with radial grooves onto a collar on the generator shaft. Oil was fed into the bearing and flowed radially inwards to prevent gas leakage. Axial thrust was maintained on the ring by hydrogen and oil pressure and reacted by the hydrostatic/ hydrodynamic film. The earlier designs were found to be unsuitable as the oil film thickness was low enough to cause overheating and white metal fatigue. As hydrogen pressures increased, the bearing profile was changed to introduce 20 to 30 taper/parallel lands which resulted in a thicker hydrodynamic oil film. The seal ring which was non-rotating moved axially to follow up shaft expansion and this necessitated the development of appropriate seals from 'O' rings through 'D' rings to lip seals.

3.4 Boiler Feed Pumps

A steam plant auxiliary whose design involves some quite important tribological considerations is the boiler feed pump, a machine whose power can exceed 10 MW (27). It is typically a high speed machine with maximum speeds of over 6000 revs/min. It can be driven by steam turbines, squirel cage induction motors, or more recently by variable frequency synchronous motors. Problems involving flow through the water passages have occupied designers over many years. Impeller-diffuser flow can introduce several sub-synchronous phenomena, including of course rotating stall at around 80% of the running speed (28). Perhaps the most heavily researched area is the study of flows through the annular clearance spaces in impeller neck ring seals and balance drums. The flow is turbulent, clearance ratios being about 1% and the existence of both stiffening and destabilizing forces was first discussed by Black (29).

4. GAS TURBINE GENERATORS

There are of course many tribological problems in gas turbine design. Many of these are not unique to the power generation industry. Seals are of great importance in the gas turbine and most of the important seals comprise lands on the rotating parts mating with honeycombs. The hot gas environment has to be taken into account in the design of seals, which can either be abradable or non-abradable. In the former case appropriate surface hardening has to be provided on the seal lands to prevent pick-up. The possibility of fretting of components in the hot gas path has also to be taken into account. Seal rubbing can sometimes occur causing thermal bending just as in the steam turbine case. The environment is not ideal for the hydrodynamic bearings, since housing temperatures are high and ovalling of the bush can occur. The avoidance of such problems is the designer's task and good mechanical design practice as well as the correct choice of materials is the key. Many gas turbine generators are used for 'peak-lopping' and have to be run up quickly on demand. The turbine may have to be decoupled from the generator and thus instead of solid couplings, gear or spline type couplings are used. The internal friction problem, mentioned earlier, assumes greater importance here, since larger slipping amplitudes are involved particularly with large and flexible assemblies. Subsynchronous unstable or limit cycle oscillations are possible and must be suppressed by building adequate damping into the system by careful bearing design.

5. HYDROELECTRIC PLANT

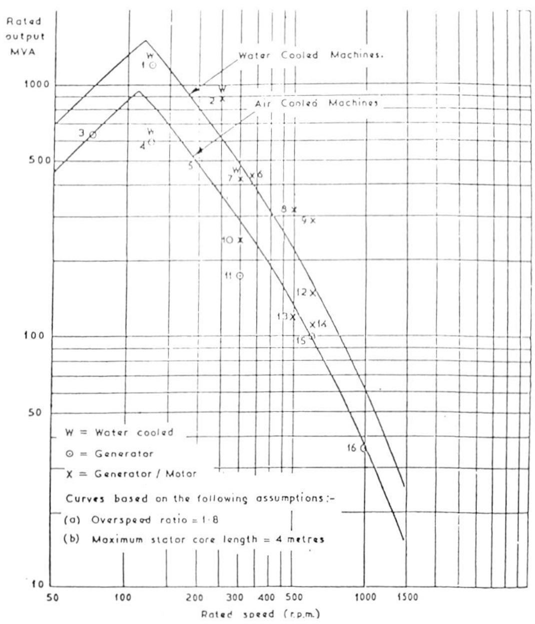

Fig. 10. Output Limits for Hydroelectric Machines.

Although horizontal (axial flow) arrangements are in use, the vertical shaft machine is preferred for large power outputs. The design of hydro-generators is influenced by the high runaway speed possible when a turbine governer failure occurs (this condition is reckoned to be 1.8 times rotational speed). Relationships between speed and power for large hydro-electric sets is given in Fig. 10, taken from Ref. (30), the choice of speed depending partly on the available head. In the water turbine, erosion problems can occur due to the presence of quartzite sand particles. This problem has been virtually overcome by the use of erosion resistant materials such as 13% Cr steel in the water passages. Cavitation is another form of wear which can occur in low pressure areas of water turbines (31). Once again suitable materials (13% Cr 18/8 Cr Ni or aluminium bronze) have been used to combat the problem. In some turbines the design of runner seals requires special care, owing to the high relative speeds and the danger of abrasive particles.

Fig. 11. Kingsbury Type Thrust Bearing.

Probably however, one of the most sensitive component in the large vertical set is the thrust bearing. This component has to react to the static weight of the rotor as well hydraulic forces. Peak loadings of 7 MN for a 500 revs/min set operating at over 300 MW are typical. The preferred position for the thrust bearing is underneath the rotor and the use of tilting pads immersed in an oil bath is universal. The oil in the bath is often directly water cooled, and on some occasions the bearing pads themself have cooling passages. The pads can be solidly pivoted with a line contact, or a spherical, in which case some load equalization technique is required, or they can be supported on a spring mattress (Fig. 11). With pump storage machines the shaft rotation is reversible, requiring symmetric pad supports. The bearing has also to be capable of coping with rapid run-up from barring to full speed.

The design of heavily loaded thrust bearings is complex, requiring attention to thermal and elastic effects (32), (33). Elastic effects which are worst during jacking and low speed operation result in a concave surface. Thermal effects make the pad convex. Direct water cooling which is supposed to compensate for thermal effects complicates the thermoelastic behaviour further. The fact that the pads are submerged does not prevent the oil carry over mentioned by Ettles (34). The runner which transfers the load to the shaft also distorts thermally and elastically. Centrifugal effects also complicate the oil flow in the bearing by supplying a radial component. Finally transient effects are important because load changes happen much faster than the thermal time constants of the bearing. Due to the high specific load, oil films are very thin and even small distortional changes are important. There have been several examples of damage to white metalled pads particularly on the outer corners. This is due to the fact that concave distortion, produces an oil film profile incapable of maintaining hydrodynamic lift. The use of a large number of smaller pads in which elastic distortions are much reduced, can be helpful in such cases. Profiling is also a way of improving the slow speed behaviour of the pads. The generator rotor is usually a large diameter (5 m typically) laminated rim with laminated salient poles held on with 'V' or 'T' keys. As long as the laminations are securely clamped, and the pole fixings are tightly secured, the risk of fetting damage is minimal.

6. DIESEL SETS

The design of diesel sets is a highly specialized field and a vast amount of literature has been published on the many tribological problems in this type of machine (32). The very large diesel - generators may run at speeds a little over 100 rev/min and use residual fuel, whose contents aggravate ring and liner wear. It is inevitable that special attention has to be paid to cylinder lubrication by injected and metered oil. The high loads carried by the cross-head bearings, under a two stroke duty, require attention to oil feeds. Detailed comment on these and other problems is beyond the scope of this paper.

7. CONCLUDING REMARKS

The power generation field is enormous and this paper has, due to restrictions in length, only scratched the surface of the subject. A few examples have been chosen to demonstrate the nature of the tasks designers have had to face as power levels have increased. Their work has been assisted by the large volume of tribological research both in universities and in industrial R and D laboratories, the results of which have been available for many years and virtually waiting for problems to arise. Mathematical techniques and material data are necessary adjuncts to the innovative skills of the designer and the intuitions of the 'trouble-shooter'. After all, we are still using the work of Reynolds after a century. In reviewing the few problems presented in this paper it is apparent that temperature effects have a major influence on tribology in machine elements. Since temperature and power are connected, it is inevitable that as power generation machinery becomes bigger and more powerful, new problems will continue to emerge.

REFERENCES

(1) Chivers, T.C. 'Tribological design - the nuclear industry', 15th Leeds-Lyon Symposium on Tribology 1988.

(2) Jost, H.P. et.al. 'Lubrication (Tribology) education and research', H.M.S.O. 1966.

(3) Chan, H.M. 'Active magnetic bearing design methodology - a conventional rotordynamics approach', 15th Leeds-Lyon Symposium on Tribology 1988.

(4) Hashemi, Y. 'Alignment changes and their effects on the operation and integrity of large turbine generators', Instn. Mech. Engrs. Conference Paper C 3/83, 19-30.

(5) Low, M.B.J. 'The sliding of high temperature components in steam turbines', Proc. Instn. Mech. Engrs., 1980, 194, 171-180.

(6) Heathcote, C., Petty, D.J. and Smith, R.J. 'Lifetime capability of turbo-generators to withstand vibration and other cyclic effects', CIBRE International Conference on Large High Voltage Electrical Systems, 1988.

(7) Marlow, B.A. 'The mechanical design of turbogenerators', Proc. Instn. Mech. Engrs., 1986, 200, No. 135.

(8) Juhlin, G.A. 'Deformation of turbo-alternator windings due to temperature rise', J. Instn. Elect. Engrs. 139, 85.

(9) Gardner, G. 'Events leading to erosion in the steam turbine', Proc. Instn. Mech. Engrs., 1963-64, 178, 593.

(10) Newkirk, B.L. 'Shaft rubbing', Mech. Eng., 1926, 48, 830.

(11) Lund, J.W. 'Destablisation of rotors from friction in internal joints with micro-slip', IFToMM Int. Conf. on Rotordynamics, Tokyo 1986, 487.

(12) Duffin, S. and Johnson, B.T. 'Some experimental and theoretical studies of journal bearings for large turbine generators', Symposium of Journal Bearings for Reciprocating and Turbo-Machinery paper 4, 30, Nottingham 1966, (Instn. Mech. Engrs.).

(13) Jost, H.P. and Schofield, J. 'Energy saving through tribology: a techno-economic study.

(14) Rowe, W.B., Xu, S.X., Chong, F.S., Weston, W., 'Hybrid journal bearings with particular reference to hole entry configurations', Tribology International 1982, 339-347.

(15) Dowson, PH. and Fidler, F. AEI Engineering Vol 2. No. 2.

(16) Reynolds, O. 'On the theory of lubrication and its application to Mr. Beauchamp Tower's experiments, including an experimental determination of the viscosity of olive oil', Phil. Trans. R. Soc. Lond. A, 1886, 177, 157-234.

(17) Christopherson, D.G. 'A new mathematical method for the solution of film lubrication problems', Proc. Instn. Mech. Engrs. 1941, 146, 126-135.

(18) Lund, J.W. and Thomsen, J. 'A calculation method and data for the dynamic coefficients of oil-lubricated journal bearings', In Toprics in fluid film bearing and rotor bearing system design and optimization, 1978, pp 1-28, Am. Soc. Mech. Engrs.

(19) Ferron, J. Frêne, J. and Boncompain, R. 'A study of the thermo-hydrodynamic performance of a plain journal bearing: comparison between theory and experiments', J. Lubric. Technol., 1983, 105, 422-428.

(20) Morton, P.G. and Keogh, P.S. 'Thermoelastic influences in journal bearing lubrication', Proc. R. Soc. A 1986, 403, 111-134.

(21) Morton, P.G., Johnson, J.H. and Wale, G.D. 'The effect of thermal distortion on the characteristics of large hydrodynamic bearings', Proc. Instn. Mech. Engrs., 202, No. C3, p 219-225.

(22) Ettles, C., Wells, D.E., Stokes, M. and Matthews, J.C. ' Investigation of bearing misalignment problems in a 500 MW turbo-generator set', Proc. Instn. Mech. Engrs., 1974, 18835.

(23) Smith, D.M. 'Dynamic characteristics of turbine journal bearings' proc. Lubric. and Wear Conf. 1963, paper 8, 72, (Instn. Mech. Engrs.).

(24) Banckert, H. and Wachter, J. 'Flow induced spring coefficients of labyrinth seals for application in rotor dynamics', NASA C.P. 2133, 1982, 205-222.

(25) Dimargonas, A.D., Newkirk, B.L. 'Effect: thermally induced dynamic instability of high speed rotors, ASME Paper No. 73-GT-26.

(26) Kellenberger, W. 'Spiral vibrations due to seal rings in turbo-generators. Thermally induced interaction between rotor and stator', ASME Paper No. 79-DET-61.

(27) Brown, R.D. 'Vibration phenomena in boiler feed pumps originating from fluid forces', Int. Conf. Rotordynamic Problems in Power Plants, Proceedings, Rome 1982, 497.

(28) Black, H.F. 'Advisory notes on boiler feed pump', University of Virginia, Dept. of Mechanical and Aerospace Engineering 1979.

(29) Black, H.F. 'Effects of hydraulic forces in annular pressure seals on the vibrations of centrifugal pump rotors', Jounral of Mech. Eng. Sci. 1969, 11, No. 2, 206-213.

(30) Foster, E.N. and Parker, F.J., 'Hydro-electric machines', Proc. IEE 1333, 1986, Part No. 3, Paper No. 4B76C, 126-136.

(31) Sharp, R.E.B. 'Cavitation of hydraulic turbine runners', Trans. ASME, October 1940.

(32) Ettles, C.M. and Cameron, A. 'Thermal and elastic distortions in thrust bearings', Instn. Mech. Engrs. Lubrication and Wear Convention 1963, 60-71.

(33) Ettles, C.M. 'The analysis and performance of pivoted pad journal bearings considering thermal and elastic effects', J. Lub. Tech. ASME, 1980, series F, 102, 182-192.

(34) Ettles, C.M. 'Hot oil carry-over in thrust bearings', Proc. Instn. Mech. Engrs, 1969-70, 184, pt 3L, 75-81.

(35) Lilly, L.C.R. 'The diesel engine reference book', Butterworth, 1984, Chapter 27.

Paper XIII(ii)

Tribological design and assessment - The nuclear industry

T. C. Chivers

The safe and economic production of electrical power from nuclear reactors requires high engineering standards. Many of the engineering components include tribological elements and it is with their design that this paper is concerned. In general the environments encountered are specific to the nuclear industry, and frequently the contacts are unlubricated. In addition some components may require lives in excess of thirty years, whilst others will have a target maintenance period of some eight years. In the absence of a basic theory for the assessment of wear processes the paper discusses how modelling is performed to underwrite the integrity of components whilst avoiding excessive margins and the associated economic penalties.

1 INTRODUCTION.

The successful safe and economic production of electricity from Nuclear Power Reactors puts considerable demands on engineering design. The environment encountered is dictated by the nuclear physics of the system. Hence not only can temperatures be high, but material selection is restricted. For tribology these considerations exclude lubricants from most areas and also materials which may be regarded as benevolent. Where lubricants can be employed additives may well be prohibited. As a consequence conventional design codes will be violated. Further, components that may well have an established satisfactory performance in another field cannot simply be transplanted into a nuclear environment. Many of the problems encountered will be unique to the nuclear industry. Under these circumstances the industry has to develop its own design methods. This is particularly true in a subject like tribology where the surface processes are so strongly influenced by environmental effects.

As with many capital intensive plants, outage costs associated with the loss of output from a nuclear power plant are high. Unscheduled outages associated with component failure or malfunction can result in expenditures far in excess of the value of the item concerned. It is essential, then, to maintain very high design standards whilst also avoiding excessive capital or revenue penalties.

Despite the restrictions on materials and lubricants it is necessary to have within these reactors tribological elements which perform some mechanistic function, or are able to accommodate, for example, relative movement. The design of such elements in the absence of established codes requires a pragmatic approach in which component validation plays a significant part.

1.1 <u>Notation</u>

Ed Energy dissipated

K Specific wear rate

L Interfacial load

S Sliding distance

u Friction coefficient

2 STRATEGY

The tribological elements encountered in nuclear reactors fall into two broad categories. Firstly, there are the permanent components which have to survive the life of the reactor unless they are to become life limiting features. Examples in this category are boilers and thermal insulation. Here sliding junctions are required to accommodate thermal movement whilst transmitting system stresses. They do not have to perform a mechanical function. In the second category are the more conventional tribological elements associated with such things as mechanical drives to control devices of one form or another. This latter category of elements can be serviced, but as with all maintenance a financial penalty can accrue if the intervals are either too long or too short.

In the absence of a relevant history, and if the penalty associated with malfunction is significant, then components in the second category are subjected to design validation by undergoing simulated operational trials. This requires considerable financial investment in environmental facilities and testing. In some instances full environmental conditions will be employed and the duty cycle replicated. This is only feasible where operational requirements are intermittent and the degradation processes are cycle rather than real time dependant. Because of the cost of high temperature facilities this approach is usually (but not exclusively) restricted to components subjected to relatively low temperatures (less than some 100C). When full environmental simulation is not feasible then the duty cycle will be replicated and scaling requirements assessed seperately.

For the permanent components, where a full reactor lifetime may exceed 30 years, degradation processes are a function of time. Here validation will be based on calculation or model testing. Where modelling of structural components is concerned the philosophy is restricted to determining movements, loads etc. and hence formulating small scale test programmes. These tribology tests are conducted in enviromental rigs and can model a wide range of reciprocating, impacting, and impact-slide conditions. Continuous slide is not a wear mode commonly encountered in nuclear environments.

The approach to the design assessment of these two categories of plant items will now be discussed beginning with those components that are unlubricated and requiring long life.

3 UNLUBRICATED WEAR.

3.1 Theoretical Background

The designer of what is to be a commercial reactor requires a means of assessing a design to ensure that it will have an adequate life without carrying undue conservatisms and the associated economic penalties. Neither must safety be compromised. However the theory of unlubricated wear processes is not well developed in any predictive sense. Many theories have been advanced and certainly a qualitative understanding exists. A major difficulty is that the models advanced are not of universal applicability; each relates to a specific subset of the wear process, although recently attempts have been made to map more clearly the areas of relevance appropriate to a particular wear mechanism (1). Aspects on which these theories are agreed is that wear should proceed linearly with sliding distance and should be proportional to load. Given the absence of a comprehensive theory a simple approach is adopted and that is to use a model based on Archard's approach (2) viz:-

$$\text{Wear volume} = K.L.S \qquad [1]$$

where K is termed the specific wear rate, L is the interfacial load, and S the relative sliding distance.

The simplicity of this presentation masks the complexity of the processes. For example, Childs (3) when reviewing unlubricated wear showed that K could vary by up to 7 orders of magnitude. Much of this variability is associated with the wear process undergoing a transition from an initial 'severe' metallic wear to a more benign 'mild' mode. The initial contact processes can result in considerable surface deformation, hardening and roughening, and a high initial wear rate. As the process proceeds oxides can become incorporated into the surface layers, further increasing the hardness, the surfaces will smooth and wear rate will drop. It appears that many parameters such as material pair and operational and/or environmental conditions influence wear behaviour and it needs to be recognised that their effects are interactive.

To deal with these wear transitions two parameter Archardian equations have been used (see, for example (4)). In general this merely adds another element of complication and their role is currently restricted to one of interpretation of experimental data rather than design assessment. Hence here such considerations will be ignored, and the strategy to be developed is based on defining a single value for K that adequately defines the total process. The way to proceed therefore is to ensure that design assessments recognise this variability in specific wear rate and that the appropriate values are employed. The determination of the appropriate 'K' value is addressed in Section 3.2.2.

Interfacial loads can generally be assessed with a reasonable degree of accuracy. Again a complication can arise in that wear transitions can be influenced by small changes in interfacial load.

Sliding distances are not always easily determined. In modern power reactors high power densities are employed and this can result in vibration and consequential fretting wear. It follows that the total sliding distance in fretting is dependant upon the frequency and amplitude of oscillation and its duration. In some circumstances energy levels may exceed the constraining forces and impacting can occur. Evidence exists to show that the specific wear rate is influenced by amplitude and frequency of vibration. Time can also be important in that transitions in wear behaviour are often observed.

A considerable body of experimental data now exists showing that a number of parameters influence wear rate. For example, as temperature or time increases so wear rate is generally found to decrease. If frequency or amplitude of oscillation increases so too can K. Other variables (e.g. contact geometry, material pair, fluid chemistry, interfacial load) influence wear rate in a more complex manner and generalisations are not possible.

If a relevant wear rate is known for a design condition, and if the load and sliding distances are defined then adequate wear margins can be accommodated into a bearing design. However in certain features excess material can have severe economic penalties and there is a need to reduce conservatisms. For example to thicken a boiler tube to increase wear margins will adversely affect heat transfer and increase building costs. In many instances all of the inputs into Equation [1] will not be known at the time of the initial design and some experimental investigation will be required. It is this category which will now be considered.

3.2 Evaluation of Parameters

3.2.1 Relative movement and load

For mechanism elements associated with precision components, relative movement is not difficult to determine. When, however, it is necessary to accommodate thermal movements and unlubricated conditions an element of freedom at bearings is essential. Unless clearances are adequate seizure or enhanced loading can result from surface roughening and galling during the wear process and/or oxidation. Because of this extra freedom which must be incorporated the resultant relative movement is not simply dictated by the mechanical function of the mechanism; it is influenced by the total forces acting at the joint and these could well be fluctuating as a result of periodic excitation e.g. by fluid dynamic forces. The dynamic behaviour of vibrating structures is difficult to predict accurately. This is particularly true when frictional damping is involved and when excitation arises from flow induced forces. Under such conditions modelling is required to determine the damping and excitation forces. To model complex systems is always costly. However to look at excitation/response characteristics does not require full environmental simulation. For example the dependance upon material properties is sufficiently well understood not to require full temperature simulation, and dynamic excitation may be modelled using, for example, different fluid densities etc. If the environmental conditions are not simulated then such a test cannot be used directly for wear assessments. However, an indirect assessment may be made if, in addition, relevant scaling data are obtained covering the difference between model and prototype conditions. It may be necessary for the model tests to run for long periods to yield meaningful tribology data. How, then, can the vibration characterising tests produce data that will permit wear life assessments?

Equation [1] may be rewritten as

$$\text{Wear volume} = K.Ed/u \qquad [2]$$

where Ed = u.L.S and is a measure of the energy dissipated.

Several approaches to wear assessment have been made using such a model and these are reviewed in (5). A difficulty here is that it is easier to assess the energy being fed into a vibrating system than it is to measure that dissipated by friction; assuming that all energy is dissipated in this way introduces an element of pessimism. This pessimistic or worst case approach is acceptable as a starting position. However the problem then is to distribute the input energy among the various wear sites. If it is assumed that the wear occurs uniformly then this clearly introduces an unacceptable optimism into the analysis. Hence complex structures need to be extensively instrumented to obtain information on the distribution of movements etc.

An alternative strategy is available based on this wide deployment of accelerometers on vibration assessment models. For all systems it will be possible to bound the loads that will be operating at the intefaces. Similarly it will be possible to analyse any joint and estimate its behaviour under a set of forces (including assessed frictional components) and excitation. Whilst precision may not be possible here realistic bounds can be determined. Double integration of acceleration readings will then give information on vibration amplitude and hence can be converted to slip. However, in most cases vibration behaviour is complex and double integration raises two problems. Firstly the analysis is performed by considering specific frequency bandwidths and dividing by frequency squared. This can produce large errors at low frequencies. Secondly the result is an amplitude-frequency spectrum which has to be summed. The first problem can be minimised by using a single integration to yield velocity data, although filtering and cut-off of the data still needs to be considered. Combining this data with an analysis of the mechanism permits sliding distance at any joint to be calculated as proportional to the mean velocity and time. The problem of variable amplitude remains, and this will be returned to later, but information is now available on the spread of activity and the most vulnerable areas can be identified.

3.2.2 Small Scale Tribology Modelling

Having established estimates of movements and loads the appropriate specific wear rate now needs to be determined. As identified above, K is influenced by a large number of both intrinsic and extrinsic variables. Problems associated with such variability are minimised by reproducing in the test facilities the correct operating conditions. For example the correct material pair, environment, temperature, and movement type will be employed (i.e. reciprocating motion, sliding or impact fretting etc). It will not always be possible to replicate the frequency-amplitude spectrum. Test duration will normally be much shorter than componentt life leading to a need for considerable extrapolation. Load and contact geometry also must be considered, and wherever possible full scale contact geometry and load is employed in model experiments, see for instance (6) (7) (8) and (9).

However, in some situations it is not possible to use full scale contact geometries in test rig simulations. Again the absence of a definitive theory is a handicap as no guidance is available on scaling. Conformal geometry is the most difficult to replicate in test rigs, and thus it is perhaps fortuitous that most contact geometries of interest in long term modelling can be reasonably well represented by dome or cylinder on flat. Geometries such as pin in bush appear more frequently in plant subject to maintenance. In one such instance assessment trials showed significant differences in wear rate with subtle changes in geometry and specific model tests were required to assess palliatives (10).

The use of full scale load in model tests is justified by the simple contact theory used in most wear analyses. This theory states that the real contact area is proportional to the applied load and inversely proportional to the material hardness. Questions related to relevant hardness and work hardening can be avoided by using the prototype material in the correct condition. However, use of the prototype load in model tests can lead to problems, e.g. it may over stress the test rig and hence lower loads must be used. Under these conditions the notional contact pressure is modelled, although this cannot be rigorously defended. An empirical approach, therefore, is necessary in support of data acquisition. This is to explore the effect of contact pressure or load on the wear processes to identify potential wear transition boundaries. If a relevant transition were to be found then there would be no alternative to larger scale testing unless the most pessimistic data were acceptable in terms of component life.

Also explored are the roles of other parameters. Temperature and slip amplitude can be important variables and inter-related. This is demonstrated in Figure 1. What can be of significance here is the relative importance of time spent in different operating conditions. For example under some conditions wear rates can

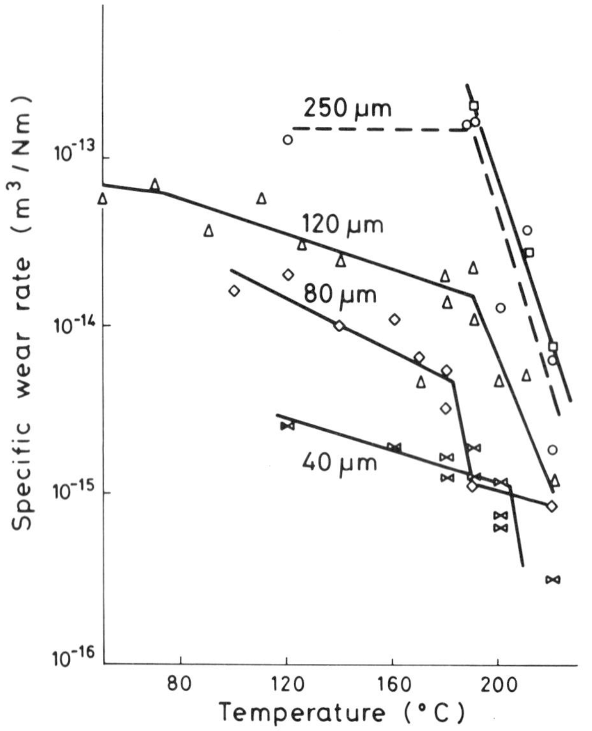

Fig. 1. Specific wear rate of mild steel against stainless steel at various strokes plotted against temperature.

be 100 times lower at 225C than at 175C. A corollary is that if a wear allowance of a component is deemed just acceptable for a 30 year life at a normal operating temperature in excess of 200C then problems are likely if time is spent at lower temperatures. This has considerable implications for commissioning of nuclear plant where power levels and temperatures are incremented upwards. Such procedures have to be defined early in the design stage so that the consequences can be assessed.

Most experiments are conducted under conditions that are notionally invariant with time. This is not normally the case on operating plant where a range of conditions are encountered. A strategy employed under these circumstances is to identify regions under which conditions will be psuedo-constant and to treat these separately (11). i.e.

$$\text{Total wear volume} = (\text{Wear vol})1 + (\text{Wear vol})2 + \text{etc.} \quad [3]$$

Such a formulation has not been rigorously validated to date. Mixed frequency - amplitude spectrums have, however been assessed (9).

A further aspect relating to scaling is wear scar morphology. Large structures can contain massive wear scars: small specimens cannot. When investigating plant problems large and deep wear scars have been seen (11). Some of these scars were larger than specimens employed in test rigs to investigate the problem. With forensic work it is possible to compare, for example, wear rates, morphology, oxides etc. For design purposes it is not possible to do this and in some instances large scale testing on a limited scale may have to be undertaken to define or validate model testing.

Having explored the variability of wear rate with contact and environmental parameters reasonably pessimistic values are selected for use in wear volume assessments. It needs to be appreciated that for a given set of conditions the specific wear rate will be subject to statistical variation. The nature of this distribution is not adequately defined although it can appear to be logarithmic (see for example (12)). For practical purposes it is usually adequate to consider the upper bound as arithmetic mean times 3.

3.2.3 Failure Criterion.

Equation [1] is written in terms of wear volume removed. In most bearing geometries it is the depth loss which is going to be most critical. For example in the case of pressurised tubes it is thining of the walls that will lead to failure. In other instances stress concentrations associated with scar morphology may determine the acceptable material loss. There is no way of determining wear volume-depth or morphology relationships other than by experiment.

The establishment of data bases has yielded valuable information on such aspects, and in many instances pessimistic assumptions can be made. Currently no hard and fast rules can be made on the issue of wear volume-depth relationships and, once again, in critical applications recourse to experimentation is necessary.

3.3 <u>Assessments.</u>

The preceeding sections have outlined how the various parameters in the Archard equation are assessed and a predicted wear volume evaluated. This is then compared with the critical wear volume for failure as defined above. It may be that the volume calculated exceeds that

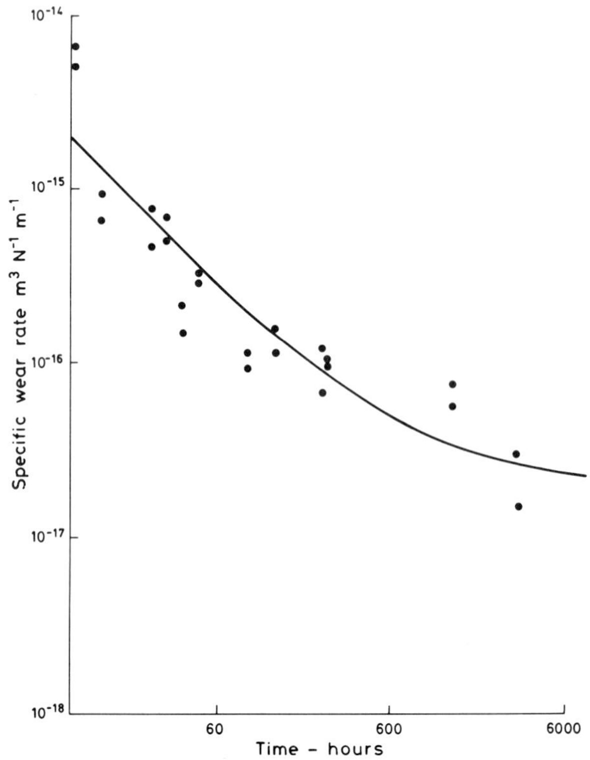

Fig. 2 Specific wear rate data for a hard coating plotted against time.

permissable. In these circumstances it is necessary to iterate, and to look for pessimisms in the route followed. For example in assessing vibration behaviour or the response to such movements, simplifying assumptions may have been made in the analysis because they were known to result in a conservative answer. An important consideration here is the effect of time. It will be recalled that we are addressing complex phenomena using a very simple wear equation. In many cases transitory behaviour is evident prior to establishment of the equilibrium wear process. An example is shown in Figure 2. It is seen that as time increases so K, as defined in equation 1, reduces,and only approaches the equilibrium value at long times. To extend tests times is therefore a valid strategy in component assessments. Whilst this is not a cheap option it is certainly a preferable option to plant modification. Established data bases give a feel for the achievable. Having established values for load, movement, and volume it is a simple matter to calculate an acceptable wear rate. Should this value be less than some E-17m3/Nm then the requirement would be regarded as difficult to demonstrate reliably. If the value was greater than E-12m3/Nm then this would be regarded as readily achievable and perhaps not worth investigating. Only when such routes have been exhausted is design change seen as a requirement. This strategy is summarised in Figure 3. Here design change must be interpreted broadly. For example given the non-linear relationships involved in vibrations and wear it

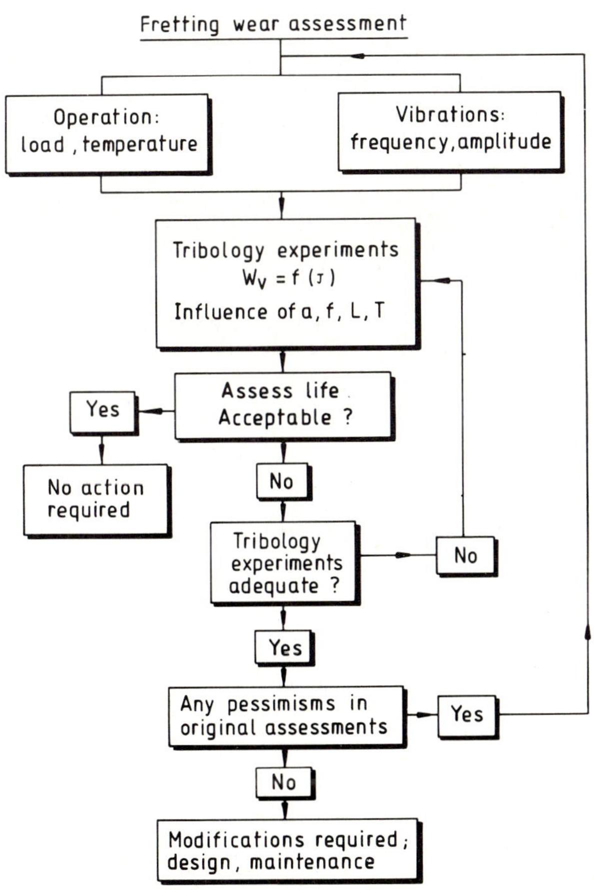

Fig. 3. Simplified decision tree for assessment of component life as limited by fretting

may be demonstrated that life will be adequate if a certain narrow spectrum of operation is avoided. Provided this carries no unacceptable penalties both in terms of safety and economics then this will be seen as a valid conclusion.

4 NON CONTINUOUS OPERATION

The above has considered the situation with regard to plant items that are subject to near continuous movement. Other components are only required to undergo relative movement at intervals and it may be possible to simulate the full duty cycle for assessment purpose on a contracted time scale. An assumption would be that behaviour is distance and not time dependant. Components that would be subjected to full duty testing would be those that were essential to the safety of the plant or those whose failure would lead to loss of capacity and excessive replacement energy costs.

4.1 Control Rod Actuator Mechanisms

An example of intermittent motion devices are the control rod actuator mechanisms. Their function is to regulate the reactivity of the core by insertion or removal of a steel rod. There is also another requirement and this is to drop the rods into the reactor to shut it down. This may arise from normal operation or in an emergency. Their early testing revealed problems associated with the adoption of conventional design codes for various grease lubricated rolling element bearings. Experimental investigation reproduced the failure which was attributed to environmental effects (the atmosphere was moist CO2 rather than air) (13). By running a lengthy test programme, a statistically sound data base of results was established for the new environment. This showed where traditional design codes were deficient, and also where they were probably valid. Hence a revised code was developed and, after review, adopted for the mechanism. Whilst the approach may seem radical for safety related hard-ware, there was considerable redundancy in the system, and deliberate pessimisms in the assessments. Also, whilst the mechanisms were designed for a maintenance interval of about 3 years, to avoid a large and disruptive workload after this time, the maintenance cycle actually started from day one thus giving early warning of any unanticipated failure processes. Following the success of this approach, maintenance intervals were increased on 2 power stations, and are currently being revised at two others. This has significant cost savings without compromising safety. More experimental work is in hand in order to extend the arguments to further stations.

4.2 Valves

Valves play a crucial role in reactor operation. This is particularly true for the Pressurised Water Reactor where outage costs associated with valve malfunction and maintenance have been high. A further difficulty arises in that cobalt containing alloys are frequently employed in hard facings on the valves and seats. Any debris released into the circuit and passing through the core can be activated to Co60, and this contributes significantly to the dose rate received by maintenance workers. To try and minimise these difficulties work is underway throughout the world to improve valve reliability and to seek cobalt free hard facing alloys having equivalent performance.

The strategy adopted in the U.K. on reliability is to subject designs to simulated service conditions (14). Here the duty cyle is accelerated to compress timescales and valves either pass or fail the requirements in terms of the mechanical integrity of actuators, leakage across seats and glands etc. All of these tests are performed in co-operation with the manufacturers, and constructive dialogues follow to ensure maximum benefit accrues from the test work. Wear assessments are also made in an attempt to measure material loss and any possible contribution to the Co60 inventory.

Programmes on Cobalt replacement have yet to be satisfactorily concluded. Whilst Cobalt containing alloys appear to fulfill all the requirements of valve components the alternatives seem to have one or more weaknesses whether in ease of deposition, or resistance to galling, wear or erosion.

5 INDUSTRIAL PROCEDURES

The preceding sections have discussed design processes that have been evolved within the Nuclear Power industry because relevant Standards do not exist. Under these circumstances it is necessary to have established routes for the approval of such internal codes as this aids their acceptance by the Licencing Authority which regulates the operation of nuclear plant.

The basic route followed for the evaluation of tribology assessment methods and input data for the U.K. Nuclear Power industry is as follows. A group of tribologists and other experts representing the Research and Development side of the industry will consider a specific aspect of design and reach conclusions on the assessment route that should be pursued and the data to be employed. Having reached agreement these recommendations go forward for endorsement by a Parent or Steering Body. At this higher level, industry wide representation again exists with a requirement that the views of Engineers and Designers are sought on the practicalities of the proposals. If approval is obtained then the procedure will be the adopted assessment route and any deviations must be justified.

Peer review is also regarded as a valuable adjunct to these procedures. This is promoted by the publication of Papers in the open literature and presentations at Conferences and Seminars of ideas on modelling and data assessments.

6 CLOSURE

This paper has discussed tribological assessment procedures for the validation of components within CEGB nuclear power plant. It is clear that as far as tribology is concerned these proceedures currently rely on experimentally determined input. The explanation is simple. Adequate design codes do not exist. The reasons for this are two fold. Firstly, environmental considerations invalidate most standard codes. Secondly the basic theory of unlubricated wear processes requires considerable development before it can be of value in design.

The first problem is very specific to the nuclear industry and can be circumvented by the acquisition of data and the formulation of its own codes. However the second problem, the absence of a theoretical base, is much more pervasive. Whilst the development of data bases is leading to a reduced requirement for continuing experimentation, the need to incorporate pessimisms in extrapolations is undesirable. Many industries operate mechanism elements under unlubricated conditions. Mineral extraction, food processing, printing, aerospace are just a few. Economic pressures are such that industry is trying to maximise its returns on investments. Attempts to increase throughputs by increasing speeds have often come to grief because of the non-linear relationship between wear and velocity. Greater understanding of wear processes can result in significant economic returns.

7 ACKNOWLEDGEMENTS

The contents of this paper rely on contributions from colleagues throughout the industry, not just the CEGB; their efforts are gratefully acknowledged. This paper is published by permission of the Central Electricity Generating Board.

References

(1) LIM, S.C. and ASHBY, M.F. Acta. Metall. 1987, Vol 35(1), pp1-24
(2) ARCHARD, J.F. and HURST, W. Proc. R. Soc., 1956, A236, pp397-410.
(3) CHILDS, T.H.C. Tribology Int., 1980, pp285-293.
(4) SKINNER, J. J.Lub.Techn.,1981, Vol 103, pp228-235.
(5) KO, P.L. ASME Pressure Vessel and Piping Conference, Chicago, 1986, Paper 86-pvp-1.
(6) MILLMAN, R.S. SLEIGHTHOLME, A. and PARRY, A.A. Wear Vol 106, 1985, pp77-96.
(7) LEVY, G. and MORRI, J. Wear Vol 106, 1985 pp97-138.
(8) JONES, D.H. NEHRU, A.Y. and SKINNER, J. Wear Vol 106, 1985 pp139-162.
(9) LEHEUP, E.R. MILLMAN, R.S. and PARRY, A.A. I.Mech.E. Conference, Tribology-Friction, Lubrication and Wear, London, 1987, Paper C162/87.
(10) RADCLIFFE, S.J. Tribology Int. 1981, pp263-269.
(11) CHIVERS, T.C. GORDELIER, S.C. ROY, J and WHARTON, M. BNES Conf. Vibration in Nuclear Plant, Keswick, 1978.
(12) CHIVERS, T.C. Tribology Int., 1987, pp225-233.
(13) RADCLIFFE, S.J. I.Mech.E Tribology Conv., Durham, 1976.
(14) AIREY, D.R. RICHARDS, D.J.W. HILSLEY, D.E. and BRYANT, S. BHRA Int Conf on Valve Performance, Oxford, 1985.

Paper XIII(iii)

Tribological design - The spacecraft industry

R. A. Rowntree, E. W. Roberts and M. J. Todd

Modern spacecraft abound with mechanisms where relative movement between contacting surfaces occurs. Abnormal performance of such mechanisms, arising from high friction or unexpected wear, can rapidly lead to loss of mechanism function; for critical items this can lead to the complete failure of the spacecraft mission. Tribology, therefore, has an essential role in the modern spacecraft industry and is, or should be, an integral part of the design process. It is not a process to be added when the design is complete. This paper reviews the major tribological design issues of current spacecraft mechanisms.

1 INTRODUCTION

On 4th October 1957, Sputnik 1, man's first spacecraft achieved orbit around the earth and, it can be said, the spacecraft industry was founded. Early spacecraft, in the 1960s, had few moving parts and only short lives, typically 3 to 6 months at maximum. Mechanisms were relatively simple and conventional terrestrial vacuum lubricants, or simple derivatives thereof, were initially adequate. However, several spacecraft failures were attributed to the then short life of dry or solid-film lubricants.

Modern spacecraft of the 1980s employ many diverse mechanisms which require relative movement of contacting surfaces. Most of these mechanisms, such as solar array drives, momentum wheels and deployment systems, are fundamental to the operation of the spacecraft (these mechanisms will be discussed later). Spacecraft lifetimes of 7 to 10 years are now commonplace. Spacecraft in low earth orbit can now, via the Space Shuttle, be serviced in orbit to achieve extended operation; others by design or by chance have been, or will be, reused. Thus the modern spacecraft designer has to cope with more critical and complex mechanisms, yet achieving longer lifetimes in a thermal vacuum environment, and with extreme reliability. As spacecraft manufacturing costs alone are now measured in tens of millions of pounds, there is a vital role and need for sound tribological design practice in the spacecraft industry.

From the European perspective, the importance of spacecraft tribology was recognised in the 1970s by the newly formed European Space Agency (ESA). In 1972 the European Space Tribology Laboratory (ESTL) was created to research and develop space lubricants, and to perform thermal vacuum testing of space hardware. The benefit of this approach has been to integrate tribological developments directly into the expanding European spacecraft industry. The European Space Tribology Symposium (1), the Space Tribology Workshops (2,) and the European Space Mechanism and Tribology Symposia (3-5) highlight the technology transfer achieved between ESA-supported research activities and the spacecraft industry.

The approach of this review paper is to consider the environmental and operational requirements peculiar to spacecraft, analyse the essential properties and merits of solid and liquid space lubricants, describe some of the fundamental spacecraft mechanisms where sliding or rolling contacts occur, and consider their tribological design issues.

2 THE SPACECRAFT ENVIRONMENT

The environmental conditions in which an orbiting spacecraft must operate and their influence on mechanism design are discussed below.

2.1 Vacuum

In low earth orbits, approximately 200 km in height, the atmospheric pressure is of the order of 10^{-8} mb; in geostationary or geosynchronous orbits, approximately 36,000 km in height, the pressure is of the order of 10^{-12} mb. The predominant gas species are hydrogen and helium ions. However, the local external environment around a spacecraft is at higher pressure due to outgassing of water vapour and other desorbed vapours, and the presence of station-keeping propellant products. The local external pressure is, therefore, typically in the range 10^{-8} to

10^{-10} mb. Within spacecraft mechanisms themselves, the pressure may be even higher (up to 10^{-6} mb) because of the limited ability to pump internal volumes through small gaps. These higher pressures decrease with the time in orbit and eventually stabilise at around 10^{-8} mb, dependent on the mechanism location and venting arrangements (6).

While pressures down to 10^{-6} mb are relatively 'benign', at pressures significantly below 10^{-7} mb, metal surfaces can rapidly weld or at best exhibit high friction and wear. This is because the normal repair and junction limiting-limiting process of oxidation and physical adsorption is not operative. Progressive oxide film attrition produces high friction and severe wear, for materials which would otherwise be protected in air. A further consideration is that any lubricant, introduced to prevent such severe wear, must have a vapour pressure below that of the environment to restrict evaporation losses, or that means of sealing and replenishment are provided.

2.2 Temperature

Because of the electromagnetic radiation from the sun, the eclipsing of the spacecraft by the earth's shadow, spacecraft orientation and the absence of convection, temperature extremes of -100°C to +150°C are common for earth-orbiting spacecraft. The temperature changes within the body of a spacecraft can be small due to its relatively high thermal capacity and use of thermal control systems. However, those mechanisms attached to the exterior of the spacecraft have, in general, low thermal capacities and rely on passive thermal control. As a consequence such mechanisms can experience, in addition to large temperature extremes, large temperature differentials between stationary and rotating components (±55°C is not uncommon). The use of cryogenically cooled IR detectors on spacecraft further necessitates the operation of tribomechanisms at -269°C. Larger temperature extremes can be found on spacecraft for deep space missions or for solar studies, but from a design viewpoint these are uncommon.

The imposed temperature range radically affects the choice of materials and lubricants, but, because of thermal movements, can also lead to high and sometimes unforseen loads on moving surfaces.

2.3 Particle radiation, micrometeorite impact and atomic oxygen erosion

Spacecraft are subject to proton and electron bombardment, and additionally, in low earth orbits, to atomic oxygen attack. The proton and electron flux in geostationary orbit is equivalent to a dose rate of about 10^9 rad/yr, though most is stopped by the exterior of the spacecraft. The electron flux is more penetrating and dose rates of 10^5 rad/yr are estimated for mechanism internals (6). At present atomic oxygen erosion appears to be limited to the external surfaces of spacecraft. From the design viewpoint, materials and lubricants must withstand such radiation doses in vacuum. PTFE is one material which undergoes chain scission and its properties degrade at relatively low absorbed doses. However the damage is less severe in vacuum. Micrometeorite impact is considered unlikely to effect tribological components. For example, most of the mechanisms on Giotto, the spacecraft that flew though the dust cloud surrounding the nucleus of Halleys Comet were operational after encounter.

2.4 'Zero gravity'

The near-zero gravity or micro-gravity of a spacecraft has, for a spacecraft designer the beneficial effect of eliminating the weight imposed on the tribological surfaces of a mechanism. However the generation of any wear debris from sliding surfaces may not be constrained and can adversely affect nearby surfaces or components, for example by the shorting of electrical insulation or contamination of optical surfaces. The distribution and containment of liquid lubricants is essentially governed by their surface energy and rarely by gravitional forces.

3 THE OPERATIONAL REQUIREMENTS OF SPACECRAFT

Prior to, or after attainment of orbit, mechanisms on a spacecraft must meet various operational requirements, the major ones being summarised below.

3.1 Launch

The high vibration during launch (typically 20g to 40g) places severe mechanical loads on spacecraft tribocomponents. The prevention of bearing brinelling, and the fretting and adhesion of highly loaded contacts must be avoided. Design techniques include the offloading of bearings to provide by-pass load paths and gear unmeshing. In practice all mechanisms are subject to vibration testing and performance checks, prior to spacecraft integration.

3.2 Lifetime

Mechanisms can be broadly classified into two groups; those which are required to perform only a very few operations or perform intermittently (eg deployment actuators), or those which operate continually throughout the spacecraft lifetime (eg momentum wheels). In the former, required lifetimes can range from a few seconds to a few hours of operation in total; in the latter lifetimes may be 7 to 10 years, or even longer.

In practice both have tribological problems. In the former, sound tribological practice is sometimes forgotten in the mistaken belief that it can be taken for granted. Also, it is not uncommon for the designed operational life to be exceeded even before launch, because of the unforeseen need for additional testing. The vacuum lubricant may then have been exhausted before launch, yet this may not be apparent from its performance in ground testing in air.

In the case of continually operating mechanisms, then actual life testing (either accelerated or real-time) may be the only way to qualify a design and the tribosystems used therein. The use of dry lubricants on spacecraft mechanisms allows, in practically all cases, tribologically-valid accelerated life testing. This is not true when fluid lubricants are used, because of changes in lubrication regime. Other time- and temperature-dependent effects include corrosion, fluid creep, fluid degradation and in certain designs, fluid supply. An increasing tribological 'dilemma' for the spacecraft industry is that the only reliable way to perform valid life tests on mechanisms, which use liquid lubricants, is in real time. This is rarely compatible with spacecraft manufacturing timescales or with allocated launch dates. Besides which, advances in technology or simply non-availablity of lubricant in the future, can subsequently negate the value of the test.

3.2 Reliability

For unmaned spacecraft, a reliabiliy factor of >0.99 is demanded. The tribological implications of this alone implies the use of redundant systems though, historically, redundant bearings, by their added complexity, have not enhanced reliability. In practice, therefore, rolling element bearings are often single point failure items within spacecraft mechanisms and the attention these receive, in terms of design, traceability, precision, performance and lubrication are commensurate with this.

The prediction of tribological behaviour for some designs of space mechanisms is simply not accurate or well enough understood to meet this reliability level. Thus in designing a mechanism, it is prudent to avoid making its operation dependent on the close control of the friction coefficient. A safety factor of three is common.

3.3 Minimum mass and power

To achieve minimum spacecraft mass, the majority of housings, shafts, etc. for mechanisms are made from aluminium or titanium alloy, or beryllium. None are good tribological materials, unless they are under very low contact stresses, and have appropriate surface treatment and lubrication.

The particular advantage of beryllium is its high strength to weight ratio, coupled to its similar thermal expansion coefficient to bearing steels. Fibre-reinforced materials and other lightweight composites used for the structure of the spacecraft, find limited applications with mechanisms. The use of ceramics is increasing, to date their major tribological use being either as coatings or as monolithic balls in rolling element bearings.

Spacecraft power is always limited and tends to decrease during life because of solar array and internal battery degradation. For example, the total power generated by solar arrays in geostationary spacecraft is currently in the range 1.5 to 8 kW. However, a typical power budget for a mechanism is 15 W (or less). Spacecraft tribological design is directed at achieving low and stable friction coefficients for maximum efficiency. Conversely power consumption implies heat generation and rejection, which can be important, for example in cryogenic mechanisms or for the internal temperature distribution within a mechanism.

3.4 Contamination

All materials used on spacecraft have to meet acceptable outgassing criteria; in Europe this is known as the micro-vcm test (7). These criteria specify a limit for the amount of volatile contaminant material which may be released and condense on a cooled plate of standard dimensions. The objective is to ensure the minimum of cross-contamination for space hardware. Because of their vapour pressure and their ability to creep over surfaces, liquid lubricants are perceived as a major potential source of contamination. Creep can arise from surface energy differences, from the presence of scratches - leading to capillary flow, by thermal gradients and by fractionation - the Marangoni effect (8). Fluid lubricants can also be contaminated by other materials with detriment to their lubricity characteristics. Hence where contamination control is important ie the nearby location of sliding electrical contacts, mirrors, telescopes or other sensors, then either solid lubricants are employed or special measures taken to seal off potential contaminants.

The above review summarises the major environmental and operational 'drivers' or constraints which influence spacecraft mechanism design. As the design process identifies, via a trade-off study, the preferred concept and the material and components, then other constraints will emerge. These 'secondary' design drivers include: type of movement, precision of axis of rotation, torque uniformity, allowable preload or contact stress, electrical resistance and noise, operation and testing in air, storage, corrosion and most importantly the lubricant choice and application technique.

4 THE CHOICE OF LUBRICANT TYPE

Fundamental to the design of a spacecraft mechanism, where relative motion occurs, is the choice of lubricant. While an example will be given later, it is rare to rely soley on the low adhesion between sliding surfaces, treated or otherwise, without a lubricant being present to give low friction and wear in vacuum. Two basic approaches have been applied in the spacecraft industry. These are to employ either liquid lubricants with low vapour pressures or thin solid films of dry lubricants. The choice of type and application technique are important in the design of a spacecraft mechanism and its ultimate performance. It is beyond the scope of this review paper to assess individual lubricant properties, some examples of types and uses will be given later. Table 1, however, summarises the principal relative merits and properties of these basic lubrication techniques.

Table 1. Relative merits of solid and liquid space lubricants

Dry lubricants	Wet lubricants
Neglible vapour pressure, no contamination	Finite vapour pressure, oil loss & contamination
Wide operating temperature with constant friction coefficient	Viscosity, creep & vapour pressure all temperature dependent
Negligble migration	Seals normally required to slow effusion loss
Accelerated testing possible	Speed and time-dependent effects can invalidate accelerated testing
May have short life in laboratory air (50% RH)	Usually insensitive to air or vacuum environment
Debris causes frictional 'noise'	Low frictional 'noise' in most lubrication regimes
Friction is often speed independent	Friction (viscous) is speed dependent
Life determined by lubricant wear	Life determined by lubricant degradation
Poor thermal characteristics	Increases thermal conductance between surfaces
Electrically conductive	Electrically insulating

5 SPACECRAFT MECHANISMS

It is possible to identify 5 broad types of mechanism on spacecraft, dependent on function. A short resume, to illustrate their specific tribological design issues and lubricant choice, is given here.

5.1 Attitude control

Positional and attitude control of 3-axis stabilised spacecraft is obtained by the use of reaction-control thruster jets or by using the angular momentum vector of a large rotating wheel, or by a combination of both techniques. In the rotating wheel approach either momentum wheels, of fixed speed, are used which are connected either directly to the spacecraft or via a gimbal arrangement, or reaction wheels, of variable speed, are used.

A current design of reaction wheel for the OLYMPUS spacecraft is shown in Fig 1, manufactured by British Aerospace (9,10). For the purposes of this review, the essential tribological features are similar to those of momentum wheels. Low torque noise is an key requirement. In this design, the wheel rotates in a low pressure atmosphere - either air or helium - to minimise windage losses and contamination problems. Hence the bearing system is not subject to all the constraints specific to space usage. Rotational speeds are in the range $\pm$3500 rpm. For simplicity, reliability and weight saving, a conventional 'soft' preloaded (ie low axial stiffness) angular-contact bearing pair was chosen. A load by-pass system ensures the bearings are protected from the launch loads, whilst ensuring a low friction torque during operation.

Because of the relatively low temperature excursions experienced and the enclosed nature of the mechanism (ie minimal contamination risk), a refined mineral oil, KG80, was chosen as the bearing lubricant. This oil is widely used in the gyro industry, it contains an oxidation inhibitor

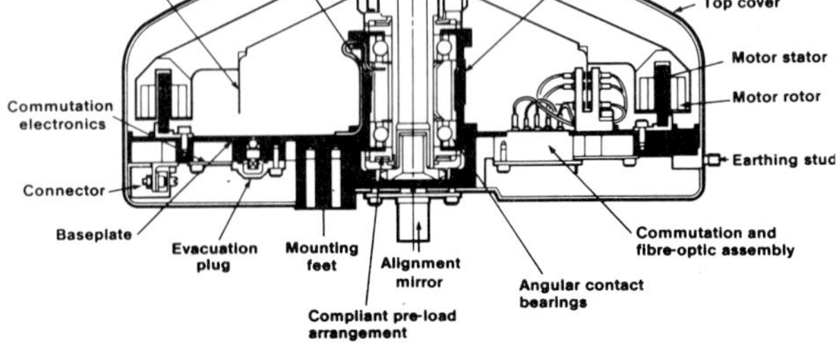

Fig. 1 10 Nms reaction wheel (BRITISH AEROSPACE).

and the boundary lubricant tri-cresyl phosphate. The amount of oil added to each bearing is carefully designed to give full EHL conditions when at speed but without causing starvation nor excessive viscous losses. In practice, 10 microlitres of oil is employed as the initial charge for each bearing. For lubricant replenishment a porous nylon reservoir, impregnated by the oil, is placed between the bearings. Lubricant transfer is by molecular flow and creep from the reservoir, and, if needed, a foil heater is wrapped around the reservoir to accelerate fluid transfer.

Other vital tribological design features include the use of labyrinth seals and creep barriers to minimise lubricant loss, a cage design which avoids instabilities and hence torque excursions, and the precision requirements of the bearings to achieve the allowable torque noise and shaft runouts.

5.2 Pointing mechanisms

Pointing of antennae, telescopes etc. can be achieved by various techniques dependent on spacecraft type. For spin-stabilised spacecraft, for example GIOTTO, the pointing of the high gain antenna was achieved by the de-spinning of this part of the spacecraft, at the precise reverse speed. For 3-axis stabilised spacecraft, techniques include gimbal arrangements, screw-jacks, linear actuators and swash plate designs. Perhaps the most common design in Europe, at this time, is the swash plate arrangement and an example is shown in Fig 2.

In this design there are four main structural elements: a spacecraft interface, an antenna interface and two wedges which can be rotated independently (10,11). Between the wedges and interfaces, there are three large-diameter thin-section bearings. By suitable rotation of the wedges, the antenna interface can be orientated to any specified direction within the combined wedge angle range (in this case, a pointing range of 8.6 deg, with a pointing accuracy of ± 0.015 deg). A flexible bellows ensures the torsional

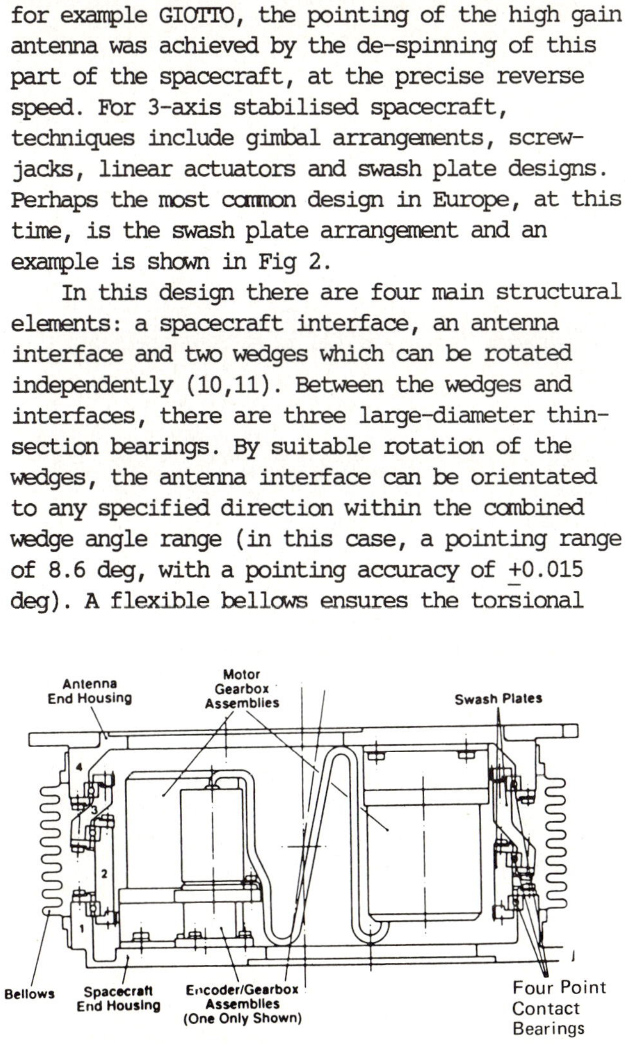

Fig. 2 General arrangement of BRITISH AEROSPACE Antenna Pointing Mechanism.

stiffness of the design. The main bearings rotate at a maximum speed of 0.3 deg/s and the motion is largely oscillatory.

The major tribological design issues are the bearing performance and the actuators used to rotate the wedges. The bearings are a key structural element of this design. Not only are they exposed to space vacuum, they are required to have high thermal conductance (to minimise thermal gradients and hence torque changes), high stiffness to radial, axial and moment loads, but also have high precision to maintain pointing accuracy. Thin-section four-point-contact bearings were chosen after a trade-off study of allowable preload (and hence bearing torque) versus required stiffness. Bearing lubrication is by a perfluorinated grease (Braycote 601), the base oil of which has a vapour pressure of 10^{-12} mb at 20 deg C. Even when exposed to the vacuum of deep space, such a low vapour pressure fluid will not (theoretically) evaporate for many tens of years. Antimigration barriers are used to reduce fluid loss by creep.

Wedge actuation is achieved by using a multi-pass spur train of ratio 70:1 from motor shaft to wedge ring gears. To maintain high pointing accuracy and avoid changes in backlash, minimum gear wear is essential. All gears are plasma nitrided to reduce wear and further lubricated with Braycote 601. For both bearings and gears, a critical tribological issue was the performance of the perfluorinated grease under boundary conditions and any lubricant interaction with the surrounding materials.

5.3 Solar array drive mechanisms

Sun-orientation of solar arrays, for 3-axis stabilised spacecraft, is achieved by Solar Array Drive Mechanisms or SADMs. They are the rotating joint between spacecraft and array, they also transfer electrical power and signal information. Their rotational speeds are low; in geo-stationary orbit the required rate is 1 revolution per day. In addition they must exhibit low contact resistance and electrical noise during electrical transfer, and impose low disturbance torques on the spacecraft.
An advanced design of SADM, manufactured by British Aerospace, is shown in Fig 3.

In this design (10,12) a pair of back-to-back flexibly preloaded bearings is used to support the shaft carrying the solar array (not shown). In a similar manner to the reaction wheel bearing arrangement, a dual compliance disc spring is used which avoids the need for a separate launch lock. There is a need, however, to allow the inner race of the inboard bearing to slide on its shaft. To avoid fretting and adhesion with the beryllium shaft, a chrome plated sleeve, lubricated with sputtered MoS_2 is used. Because of the temperature range of operation and the need to avoid contamination of the slip-rings,

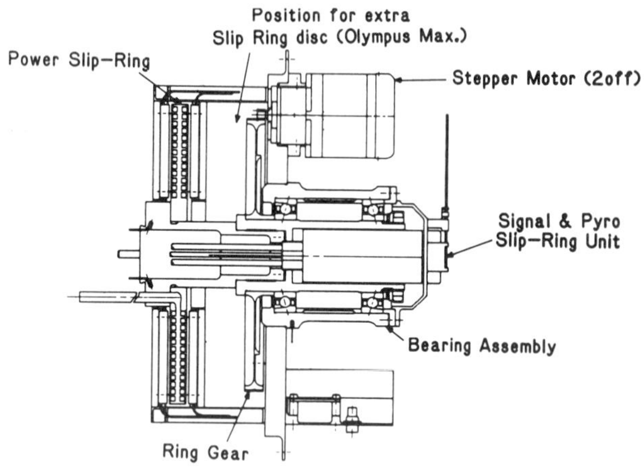

Fig. 3 BRITISH AEROSPACE Solar Array Drive Mechanism for Olympus 1.

dry lubricants are used throughout this mechanism. The main bearings are lubricated with a thin film (approximately 0.3 microns) of ion-plated lead, with a cage manufactured from lead bronze. This lubricant combination is particularly suitable for high stress, low speed applications in space and where torque noise is of lower concern. A substantial advantage of lead and other dry lubricating films is that when applied to bearings, the resulting torques are independent of temperature and rotational speed (13). The performance of ion-plated lead films has been extensively proved in ground simulation tests, while in orbit nearly two million hours of operational experience has been achieved to date.

The number of power slip-ring circuits needed and size limitations, resulted in the choice of a pancake arrangement, where the slip-rings are mounted concentrically on the faces of a disc. The drawback of this approach is the increased friction torque and sliding distance of the brushes, compared to cylindrical designs. The choice of brush material, ring material and applied load is a compromise between acceptable wear, low friction and contact resistance. The present preferred choice for power brushes is a composite of silver (82.5%), copper (2.5%) and MoS_2 15%); and for the ring gold plated copper. Because of their low contact loads and short sliding distances, single wire brushes are used in signal slip-rings, manufactured from tarnish resistant materials.

This design of SADM uses a single-pass external-spur geared drive, as opposed to a direct drive, for shaft rotation. The wheel gear is manufactured from nitralloy steel, the pinion from 440C; both are plasma nitrided to confer wear resistance. Lubrication is also by ion-plated lead.

5.4 Deployment mechanisms

There are many diverse designs of deployment mechanism, from simple spring-actuated devices through to complex motor gearhead arrangements. The example shown in Fig 4 is that to be used on the Hubble Space Telescope for the primary deployment of the solar arrays (14). It essentially consists of a planetary gearbox driving a crank-link arrangement through 90 deg. The approach to the lubrication of this mechanism is typical of deployment mechanisms.

Because of the cleanliness requirement of this spacecraft, dry lubricants were preferred. In the case of the gears these are manufactured from aluminium and anodised. With the duty cycle of this mechanism and low operating contact stress, no lubricant is applied to the gear teeth. The bearings rely on the use of TiC coated WC balls, with races manufactured from SAE 52100, to avoid adhesion. A lead bronze cage is used, but the number of operations performed by the mechanism is insufficient to develop a transfer lubricant film. Sucessful performance depends on the surface treatments rather than lubrication. This approach is only justified for this type of light-duty, short-life mechanism and where thorough qualification and life testing proves mechanism performance.

Fig. 4 CONTRAVES Space Telescope Solar Array Primary Deployment Mechanism.

5.5 Scanning mechanisms

The use of infra-red scanning mechanisms for either attitude sensing or earth observation is increasing. Dependent on type, their rotational speeds can be in the range 1 rpm to around 500 rpm, at either a constant or variable speed. At the higher and more constant rotational speeds, fluid lubrication is possible, provided sealing and means of lubricant replenishment are available. In most other cases dry lubrication is preferred. Fig 5 shows, as an example, the Infrared Horizon Scanning Sensor, manufactured by Sodern for the SPOT satellite (15). An important requirement was the ability to operate the sensor for long periods (800 hrs) in air for mechanism checks.

The scanning shaft rotates at a constant speed of 60 rpm, with a required lifetime of 5 years. For simplicity, no offloading arrangement is employed and the bearings are sized to withstand the launch loads. A brushless DC motor drives the shaft via 16:1 reduction gear train. PTFE is one of the few dry lubricants that performs equally as well in both air and vacuum, and lubrication of both gears and bearings relies on PTFE transfer. The bearings employ a cage manufactured from Duroid (a PTFE/MoS_2/glass fibre composite), while the gear train uses a Duroid vs stainless steel combination. The wear rate and stress limitations of this material, when used as a bearing cage, have been studied and design information is available (16). Thus the bearing preloads and gear tooth loads were designed to match the required life. A simulated life test has confirmed the performance of the lubricant approach, though increasing shaft jitter was found. This is a consequence of the cage wear and the unavoidable presence of cage debris in the bearings.

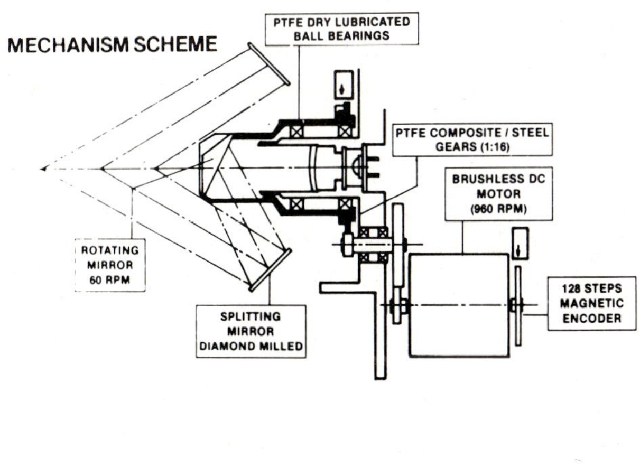

Fig. 5 Schematic of SODERN Infrared Horizon Scanning Sensor.

6 CLOSING REMARKS

This paper has reviewed the critical role of tribological design in the spacecraft industry. The unique environmental and operational requirements of spacecraft has necessitated the development of specialised thin-film solid and low vapour pressure liquid lubricants, as well as surface treatments to minimise wear. Some examples of mechanism types and their sucessful tribological solutions have been presented. Even so, the spacecraft currently under design for the 1990s place further demands in terms of life, operating temperature range, contact stress, noise characteristics and materials on the spacecraft industry. The development of space and the continued expansion of the spacecraft industry is, therefore, synonymous with advances in space tribology.

References

(1) Proc. 1st Space Tribology Symposium, Fracscati, Italy, April 1975, ESA-SP-111.

(2) Proc. 2nd Space Tribology Workshop, ESTL, Risley, UK, Oct 1980, ESA-SP-158

(3) Proc. 1st European Space Mechanisms & Tribology Symposium, Neuchatel, Switzerland, Oct 1983, ESA-SP-196.

(4) Proc. 2nd European Space Mechanisms & Tribology Symposium, Meersberg, Germany, Oct 1985, ESA-SP-231.

(5) Proc. 3rd European Space Mechanisms & Tribology Symposium, Madrid, Spain, Sept 1987, ESA-SP-279.

(6) ROBBINS, E.J. 'Tribology tests for satellite application: simulation of the space environment', Proc. 1st Space Trib. Symp., 1975, 101-114.

(7) ESTEC Spec. ESA-PSS-01-702, 'A thermal vacuum test for the screening of space materials.

(8) ESTL Technical Bulletin No. 5, 'The prevention of oil surface migration (creep) in spacecraft mechanisms', 1986.

(9) STURTIVANT, G.J. 'Results of a 10 Nms reaction wheel test programme, including tests on a nylon pore lubrication system', Proc. 2nd Eur. Space Mech. & Trib. Symp., ESA-SP-231, 1985, 273-279.

(10) SHEPPARD, J.S. 'Some tribological problems in space mechanisms', Instn. Mech. Engrs. Seminar 'Tribology in aeropsace', 1985, 9.1-9.16.

(11) LOVETT, P.J. 'Development and qualification of the Olympus antenna pointing mechanism', Proc. 3rd Eur. Space Mech. & Trib. Symp., ESA-SP-279, 1987, 273-280.

(12) SHEPPARD, J.S. 'Design and development of an advanced solar array drive mechanism', Proc. 1st Eur. Space Mech. & Trib. Symp.,ESA-SP-196, 1983, 19-26.

(13) TODD, M.J. 'Solid lubrication of ball bearings for spacecraft mechanisms', Tribology International, 1982, 15(6), 331-338.

(14) VEIT, A. and CHANDLER, D. 'Space Telescope - solar array deployment mechanism', Proc. 1st Eur. Space Mech. & Trib. Symp., ESA-SP-196, 1983, 27-31.

(15) POCHARD, M. and WALLIS, G.'High lifetime mechanisms implemented in infrared horizon scanning sensors', Proc. 3rd Eur. Space Mech. & Trib. Symp., ESA-SP-279, 1987, 183-189.

(16) STEVENS, K.T. and TODD, M.J. 'Parametric study of solid-lubricant composites as ball bearing cages', Tribology International, 1982, 15(5), 293-302.

Paper XIII(iv)

Tribological design - The electronics industry

E. A. Muijderman, A. G. Tangena, F. Bremer, P. L. Holster and A. v. Montfoortand

Tribological research and development work within Philips is mainly performed in the Philips Research Labs. and at the Philips Centre for Manufacturing Technology (CFT). This work is largely concentrated in Eindhoven, the Netherlands. Design is mostly performed within the Product Divisions (PD). We will eludicate how we transfer tribological knowledge from research to PD and how we acquire such knowledge.

Full-film lubrication
Externally pressurized air and oil bearings: In the research attention is focussed on the dynamic behaviour and on the construction of optimally (infinitely) stiff bearings.
Spiral-Groove Bearings (SGBs) Grease-lubricated SGBs for the scanner motor of video recorders have been used in mass production since 1978.
Foil bearings This research investigates the tape-scanner system of video recorders, where the tape acts as an air-lubricated foil.

Mixed & boundary lubrication
For the selection of lubrication fluids we have a well-established routine. An important research topic is precision machining, which includes the machining process itself, machine dynamics, measurement and control, and their interactions.

Dry-running systems
In this field studies are made of material characterization, contact mechanics (for use in wear models like the zero-wear model and Archards model), coatings and magnetic recording. Part of this work has the nature of research, while some other part is more development-oriented.

There is an enormous amount of tribological R&D work which is of interest to Philips.
- a universal, reliable, user-friendly program package to calculate full-film bearing performance
- dynamic analysis of full-film bearing behaviour
- analysis of transition diagrams, so that research results can be transferred to practical applications
- interaction of lubricants with metals, plastics and ceramics
- fundamental understanding of wear processes.

1 INTRODUCTION

For nearly a century Philips have occupied a special position in the markets of the world. From our earliest days we have had to export our products, because our home market was too small to support economic lamp production. Today Philips' activities go far beyond making lamps, and the company has national organizations in more than sixty countries. In 1987 the national organizations operated a total of 420 factories and employed 340,000 people. In sales (22 billion ECU in 1987) Philips comes about fifth in the world list of electronic companies.

Philips are a high-technology company - because of the nature of our products and our research and development efforts. We spend 8.3% of our net sales on R&D activities, and employ a total of 40,000 people in research and development. The corresponding figures for research alone are about one tenth of these. This R&D effort is the mainspring for the rapid technological development and product innovation for which Philips are known.

Our nine Product Divisions sell to a total market that can be divided into five groups:
- Lighting and batteries
- Consumer electronics
- Domestic appliances and personal care products
- Products and systems for professional applications
- Industrial supplies including electronic components and ICs.

Philips' product policy has gradually moved towards 'horizontal diversification' - supplying world markets with the fullest possible range of related products and services. There are only a few companies in the world offering this wide range of products.

To be competitive our products must offer high performance and a long lifetime. Since most Philips products are small, product designers tend to think that loads and also friction and wear are negligible. But small contacts could imply high contact pressures with the danger of large wear. The situation is even worse since mostly the amount of wear is related to the dimensions of the product. So that whenever there are moving parts, the tribology must be considered from the earliest design stages. Products like the Compact Disc (CD) player and the video recorder demand the highest possible mechanical accuracy during their entire lifetime in millionfold mass production. Our production and measuring machines must meet the highest standards with virtually no wear. The accuracy of "useful wear processes" like precision machining is approaching the nanometre level. The increased use of "mechatronics" demands precisely defined mechanical systems with a well-defined frictional behaviour. The production of IC

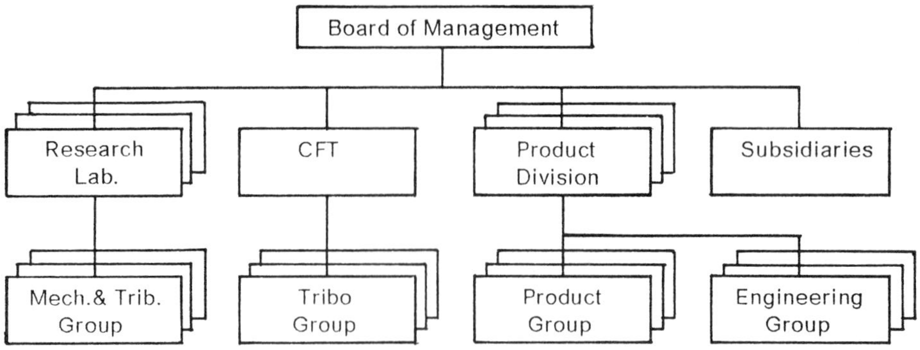

Figure 1: Organizational structure of Philips

chips is critically vulnerable to dust contamination. One of the major sources of dust in clean rooms is debris from wear processes. So it is clear that for an electronic company like Philips tribology is important.

Figure 1 gives an overview of the organizational structure of Philips. Our research is carried out by 4200 people at eight laboratories in six countries. The Research Labs. do not depend on Product Divisions for their financing, their budget is determined by the Board of Management. Research topics are divided between the laboratories. The individual research workers have considerable influence on the research program. Drawing on their scientific expertise and their appreciation of industrial requirements, they propose research topics. Research management sets priorities. Of course there is frequent consultation with development laboratories of Product Divisions.

The Philips Research Lab. in Eindhoven has 2400 employees, working in some 50 research groups. Two of these groups are mechanical research groups. In one of them there is a Tribology section. Approximately 10 researchers are engaged on tribology, doing fundamental research.

Besides the Research Labs. there is a Centre for Manufacturing Technology (CFT). This centre operates as a second knowledge centre within Philips. The CFT gives technical support, undertakes turnkey projects, and gives advice. In many cases it is the link between Research and Product Divisions. The financing of the CFT is partly dependent on paid projects for Product Divisions. Around 1000 engineers are working here in about 80 groups. In the mechanics and mechanisms department there is a tribology group of 9 persons.

Of course there is also a lot of knowledge in the field of tribology available inside the different Product Divisions and subsidiaries.

Inside a large organization like Philips it is essential to have efficient communication of knowledge and know-how. We have organized the transfer and acquisition of tribological knowledge in a number of ways:

Transfer of knowledge

- Internal courses: Engineering
  Constructing for friction and wear
  Lubrication
- Publications: Reports (CFT, Research Labs.) (internal)
  Magazines (external)
- Advice: by telephone or in writing
- Symposia and conferences (internal and external)
- Investigations for Product Divisions
- Expert system and computer programs

Acquisition of knowledge

- Our own investigations (Research Labs, CFT and Product Divisions)
- Publications, magazines, books and leaflets
- Attending symposia and conferences
- Contacts with: universities, TNO (the governmental research organization in the Netherlands) and other companies.

As a company with wide interests in the field of tribology Philips is always interested in coopering with other parties if this contributes to our research goals.

Table 1: Comparison of properties of radial SGBs, ball-bearings, sintered bearings and externally pressurized bearings in consumer products (mass production):

Property	SGB		ball-bearing	sintered bearing	ext. press. bearing	
	Air	Grease			Air	Oil
Price	- -	+	-	+ +	- -	- -
load carrying capacity	-	+	+	+	+	+ +
smooth running	+ +	+ +	+	-	+ +	+ +
running friction	+	-	+ +	-	+ +	+
starting torque	-	+	+ +	+	+ +	+ +
noise	+ +	+ +	-	-	+ +	+ +
life	+	+ +	+	-	+ +	+ +
vibration	+ +	+ +	- -	- -	+ +	+ +
suited for production	+	+ +	-	+ +	- -	- -
speed range	+	-	+ +	-	+ +	+

+ + positive → - - negative

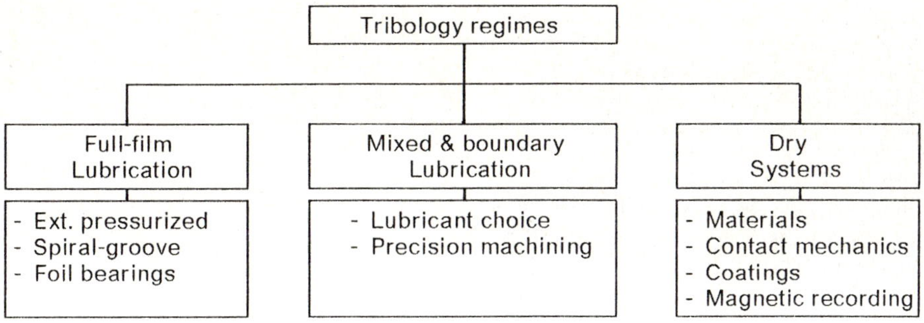

Figure 2: Tribology classification within Philips.

2 TRIBO ACTIVITIES WITHIN PHILIPS

For every tribological problem the designer has to consider four aspects: construction, lubrication, materials and environment. He can use a 9-point checklist to see whether or not his construction is sound from a tribological point of view. Small PC-based programs are used to assist the designer. We can get an impression of the type of information used in the checklist if we look at table 1. Table 1 defines the suitability of four different bearing types for application in consumer products at a nominal speed of revolution.
Recently an expert system, developed at the CFT was introduced. We started with an expert system that functions as a diagnosis/repair system. In future it will also function as an aid in the design stage. We regard such a system as an extremely helpful tool for transferring specific knowledge from the tribologist to the product designer. We will not discuss the checklist and the expert system in detail, since it will be the subject of a separate contribution to this Leeds-Lyon conference.
From a research point of view it is easier to classify tribology in lubrication regimes (see figure 2). In this paper we will use this classification. For every regime we will discuss the available techniques for achieving a good construction and we will indicate the materials and lubricants used.

2.1 Full-film lubrication

In research and development full-film lubrication is mainly studied for externally pressurized bearings, spiral-groove bearings (SGB) and foil bearing applications. We distinguish between bearings with a compressible medium like air and those with an incompressible medium like oil (grease in SGBs), because this determines the bearing design (e.g. recess volume, supply pressure). Magnetic (video)recorders, where the tape is often running on an airfilm, use a special class of bearings, called foil bearings.

Externally pressurized bearings

We prefer to design full-film bearings ourselves because optimal designs require integration of the bearing in the construction. Besides, when bearings do not function properly it is easier to rely on Philips experts. Nevertheless several machines with non-Philips bearings are in use and also separate bearing components like grinding spindles, slides etc. are built in.

Oil:
Externally pressurized oil bearings are mainly applied in machining centres for high precision, mostly in combination with a hydraulic drive [1,2]. Bearing design is based on analytical formulas plus graphs. For radial bearings we have a finite element (FE) program that can calculate the influence of tilt, geometry deviations and hybrid effects. For step bearings there is a finite difference (FD) program.
Typically we construct MDR (variable flow restrictor) bearings [3,4] with infinite stiffness and integrated bearings (with restrictor in the bearing gap) [5]. In the mid sixties a small series of completely hydraulic lathes was manufactured (see figure 3) [6].

Figure 3: The all-hydraulic lathe [6].

Air:
Air bearings are mainly used in measuring devices (see figure 4) [7] both in production [8] and in development [9,10], in machining equipment (see figure 5)[11] as grinding spindles, fiber-connector turning lathes [12] etc. and in special applications like wafer-steppers [2,13], mastering for CDs and chip assembly. In 1958 we used our first air bearing in a production line for adjusting the resistance value of a carbon resistor by the grinding process. Now we have proven designs for infinitely stiff thrust and journal bearings.
The design of air bearings starts with simple formulas [14] supplemented with MTI graphs [15]. There is a program for (infinitely stiff) tapered gap thrust bearings (see figure 6)[16,17] and FE programs for tapered gap journal bearings and for porous bearings. We preferably design bearings with inherently compensated orifices to avoid pneumatic hammer. Recently we succeeded in analysing stiffness and damping coefficients for tapered gap thrust bearings to good accuracy (see figure 7).
For externally pressurized bearings it is important that the dimensions of the bearing gap remain the same at all times. Since the materials do not make contact the

Figure 4: Measuring set-up used in the production of VCR units. Sensing and detection are performed at 70 locations in a 300*200*100 mm space with an accuracy of 10 μm.

Figure 5: Aerostatic slides in operation during production of VCR scanner heads.

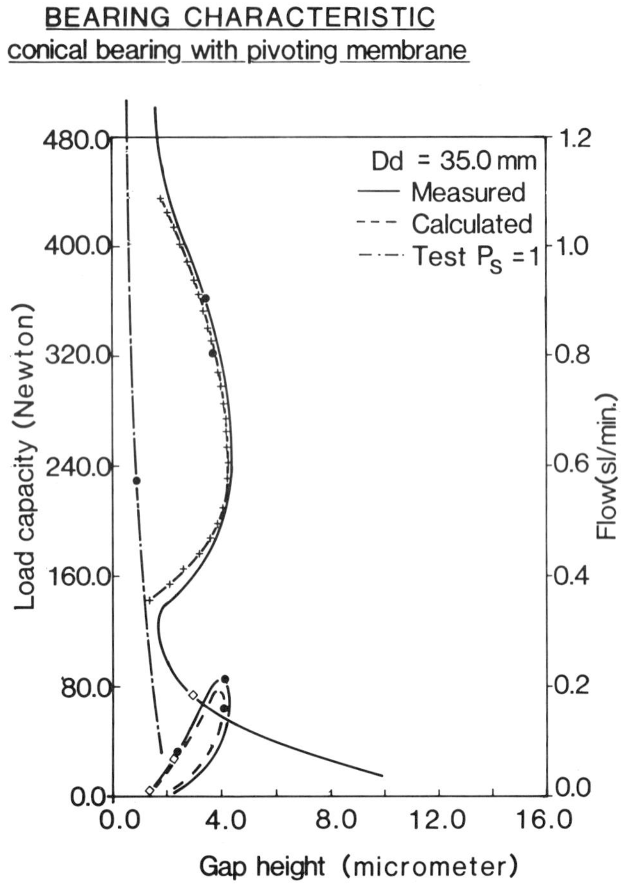

Figure 6: Calculated (solid lines) and measured (dashed lines) results for an infinitely stiff circular thrust bearing pad [17].

Figure 7: Measuring set-up for dynamic measurements of conical pad thrust bearings.

most stringent demands are that the materials must be thermally stable (expansion) and thermodynamically stable (corrosion) [18].

Spiral-groove bearings (SGBs)

Spiral-groove bearings are a special type of hydrodynamic bearings [19], used in professional applications, and nowadays chiefly in consumer products where high precision is required. Their main features are excellent running accuracy, virtually no wear and no need for an external pressure source [20]. Moreover they prove to be well suited for mass production (see figure 8) and for special designs [21].

Oil and grease:
The important consumer application is currently in video scanners (grease lubricated radial SGB) [22]., Future applications will include DAT scanners (digital audio tape) (grease-lubricated radial and axial SGB) and ESP (electronic still picture). The extreme durability of SGBs is demonstrated by oil-lubricated SGBs for satellite flywheels [23,24,25]. Such bearings have shown excellent performance for almost 20 years now, without any maintenance.
Calculations on thrust bearings are done with Muijderman's equations [26], using his correction method for the end effects. Recently an FE program was written that can evaluate the influence of geometrical errors and also calculate squeeze forces.
The work of Reinhoudt (FEM)[27], Bootsma (FDM and perturbation method) [28], and of Tielemans [29] and Sastra (perturbation method) is the long-standing basis for the analysis of thrust and journal bearings. The computer programs are based on an infinitesimal number of grooves and the effects of free edge and cavitation are included. There is also an FE program for SGBs with a fixed number of grooves, but this program does not include the effects of free edge and uses a primitive

Figure 8: Some motor casings with SGBs.

Figure 9: Axial and radial air-lubricated SGBs in the Philips Synscan Laser scanner. Development Polymotor Italiana Spa. Via Avossa, 94-16015, Casella (Genoa Italy).

Figure 10: A compressor equipped with SGB that uses the process medium as lubricant.

cavitation condition. For radial bearings we can calculate the influence of the geometrical parameters (including geometrical errors) on the load-carrying capacity, the friction and the stiffness and damping coefficients of the bearing.

In conclusion there exists a large capacity for analysis of SGBs but unfortunately not all programs are readily available or complete, and some are too far away from the product designers. Furthermore when we design an SGB all working conditions are taken worst case. These working conditions include loads (magnitude and direction), tolerances, temperature (maximum for stiffness and load-carrying capacity, minimum for friction), consumer use (position of the apparatus) and the condition of the lubricant (broken down grease). In all combinations of those circumstances the bearing must fulfil demands for minimum stiffness or load-carrying capacity and limited driving torque. Because not all conditions occur simultanously we usually end up with a bearing that is better than necessary, even though we have limited experience on how these conditions interact.

We use three different greases developed specially for our bearings (Vermulst, Remmers [30,31,32]). In practice we mostly use steel-bronze for grease-lubricated SGBs.

Air or gas:

We believe that the increasing capabilities of precision grinding are opening up a promising future for SGBs with compressible lubricants. We succeeded in developing a laser scanner (see figure 9) (air-lubricated radial and axial SGB) and we expect great advantages for a rotating free piston (see figure 10) (radial SGBs) [33].

For designing journal bearings we used the results of Fleming [34] and Hamrock [35] and for thrust bearings we still use the incompressible theory as far as allowed by the local compressibility number. Both procedures have been outlined by Muijderman [36]. There are also some FE programs that can calculate specific bearing configurations (see figure 11), but much remains to be done to convert them into a universal design tool.

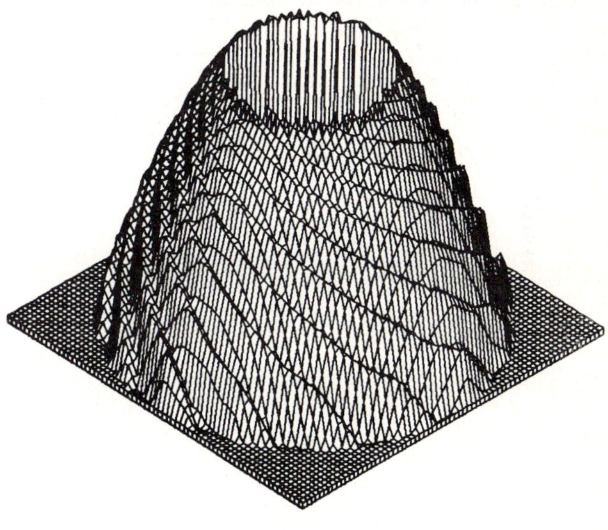

Figure 11: Pressure distribution inside an SGB calculated using the FEM.

The succes of gas-lubricated SGBs depends to a great extent on the choice of material. This choice is determined by the start-stop behaviour under dry-running conditions. Such conditions not only cause wear, but sometimes also a too high starting torque. For the laser scanner we advised Cr_2O_3 coatings on steel running against itself, with excellent results. Bulk ceramics are also potential possibilities although they require expensive grinding procedures and their friction behaviour is sometimes disappointing. We gained experience with the choice of gas-bearing materials by ranking them on our so-called ring-ring test apparatus.

Foil bearings

The running of a magnetic tape around a video scanner relies on an airfilm between the tape and the scanner.

This situation is made more complicated by the fact that the video heads have to penetrate through the airfilm and to contact the tape, which changes the air gap between the tape and the scanner. We are investigating both dynamic and static aspects of such contacts using the FE for theoretical analysis and a very sophisticated measuring apparatus for the experimental verification (see figures 12, 13, 14)[37]. The material used for the head drum is dictated by the thermal and thermodynamical stability of the whole system and ease of machining.

Figure 12: The sophisticated set-up for measuring the air film between tape and scanner and the protrusion of the head in the tape.

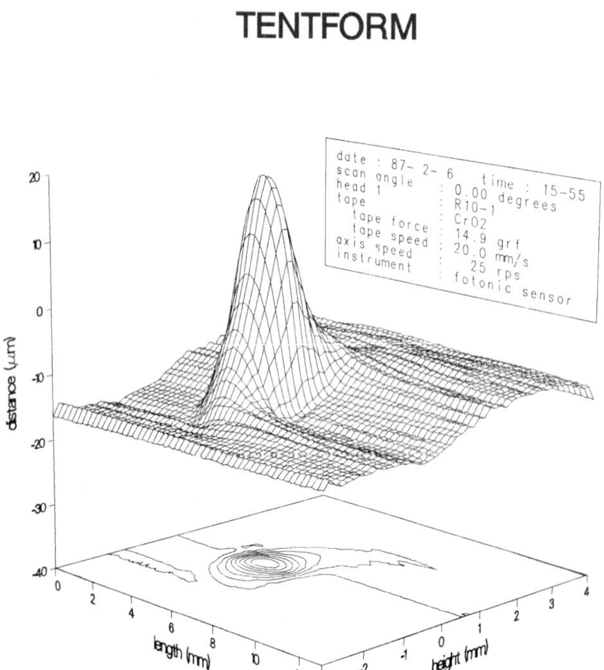

Figure 13: An example of the results obtained on the set-up of figure 12. The protrusion of the head in the tape can clearly be seen, as well as the air film between tape and scanner.

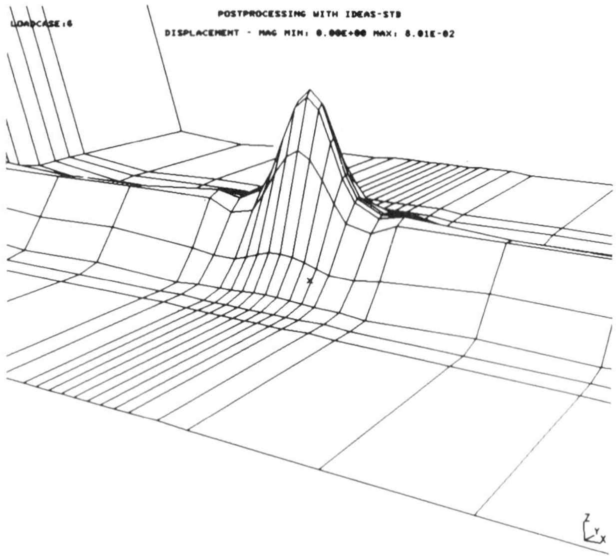

Figure 14: Calculated deformations (FEM) of a tape, when a video head is making contact.

2.2 Mixed & boundary lubrication

For the mixed & boundary lubrication regime and also for the dry-running systems design rules and materials choice are not as well-defined as for the full-film regime. The complexity of this regime and also the dry-running regime led us to design our expert system with focus points in these regimes.

Lubrication

Within Philips there is a committee which prescribes the choice of lubricants for maintenance of machinery. The lubricants are tested in practical conditions. The results are worked out in the form of lists of preferred lubricants for certain applications. This activity is mainly the task of the maintenance departments.

In addition to this committee there is an engineer at the CFT who deals with specific requests. The choice of lubricant is mostly based on information from leaflets of lubrication companies and on our own experience. A start has been made to map the diffuse area of lubricant interaction with surface (checking the compatibility of lubricants with plastics), environment etc. This work might lead to a small Philips expert system on lubricant choice.

At the CFT many lubricants have been tested in a pen/ring configuration. In this test the load and the speed at which severe wear occurs are examined. The results are interpreted in terms of so-called transition diagrams.

Stribeck curves of different bearing types have been measured as well (see figures 15, 16). This work will be continued. We have also investigated porous bearings [38].

The research work mentioned has led to several personal computer (PC) programs for the mechanical designer. There is a program that indicates if there is a lubricant film possible and in what regime it operates. There is also a database for k-values (wear factors) for the different lubrication regimes and material combinations. This database incorporates our own research work as well as some work from the literature.

Some fundamental work on lubrication during metal deformation has been done at the Research Labs [39,40,41].

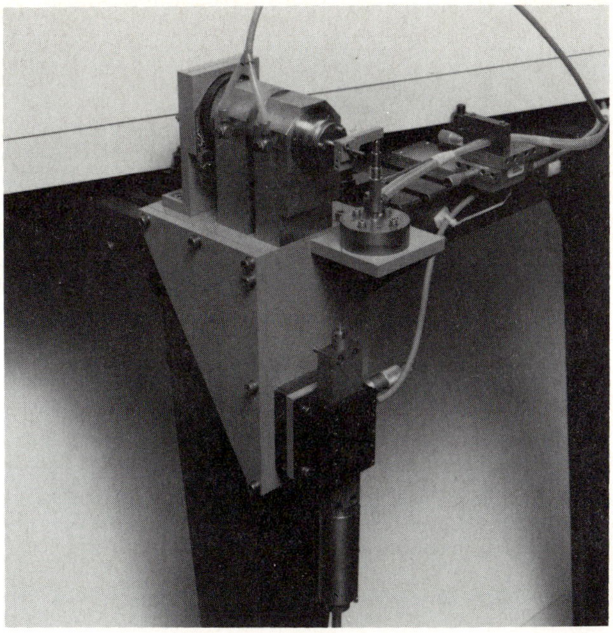

Figure 15: The fully airborne measuring set-up for testing different bearing types.

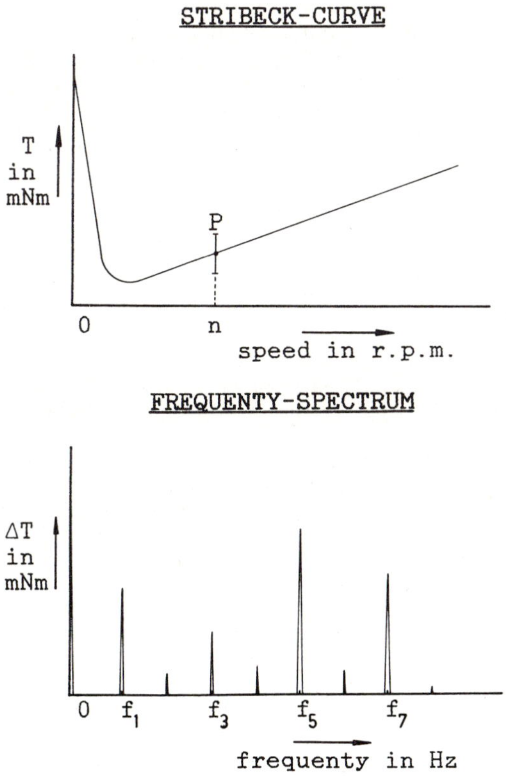

Figure 16: Results obtained on the set-up of figure 15. The top part of the graph represents a Stribeck curve, the bottom part the frequency spectrum for one rotational speed. We remark that for one point of operation of the Stribeck curve the frequency spectrum gives additional information about for instance the unroundness of the axis of the bearing.

Material removal processes (precision machining)

Precision machining is very important for a company like Philips [1,11]. Parts used in Compact Disc players and video recorders, and also in appliances like shavers, must meet the highest tolerances. Precision machining is the subject of a multidisciplinary investigation at the Research Labs. Since it involves many different disciplines it is a joint effort with metallurgists, control theory experts, mechanical engineers, chemists and tribologists participating. The processes involved are grinding, turning, lapping and polishing. The investigation focusses on the machining process itself, machine dynamics [42], measurement and control, and their interaction.

As for the machining process we investigate the generation of chips, the influence of machining fluids on plastic deformation and fracture processes and the influence of parameters like turning speed and feed rate on the exact position of individual scratches [43].

In machine dynamics we look for the static and dynamic behaviour of bearings and modal analysis of machine components [44].

In measurement and control sophisticated measuring equipment is used to generate parameters that are used by the control system to steer the machining process.

The results of this investigation will provide insight into machining processes and will provide guidelines on how to obtain the higher accuracies required in future products.

2.3 Dry-running systems

Although the division between boundary lubrication and dry-running is an arbitrary one, since most systems do have some medium, be it water, in between the two contacting surfaces, we will use the term dry-running for systems that are not intentionally lubricated.

Many contact situations occurring in Philips products are dry-running. Examples are tape guides, many electrical contacts and most plastic bearings.

Material characterization

At the CFT many new materials have been tested to check their usefulness as wear-resistant materials, mostly in a pin-disc configuration (see figure 17) [45]. The results of these tests (k-values and friction coefficients) are filed in a PC database program, from which they can readily be retrieved when needed. The emphasis in these tests is on plastics and ceramics. For plastic-metal combinations there is a database with pv and pt values.

A special class of wear materials are the tools. In their lifetime they interact with new surfaces all the time. Their wear behaviour therefore differs considerably from that found in for instance pin-disc tests. For such conditions a special test set-up was made in the CFT.

Contact mechanics

Contact mechanics is an important tool to check the feasibility of a certain material combination for wear appli-

Figure 17: Sophisticated pin-on-disc test machine. Note the fully air-borne arm for the test pin.

cations [46]. For solid materials the contact pressure together with the k-value obtained from the database are used in the Archard wear equation. For applications where no wear at all is allowed (IC manufacturing) we use the zero wear model [47].
There are PC programs available that calculate contact parameters in case of an elastically deforming contact of solid materials (according to Hertz). The influence of friction on these parameters is obtained from formulas of Hamilton [48]. Although in principle plastic deformation must be avoided, in some cases like layered systems some plastic deformation of the substrate can be allowed without plastic deformation of the coating. In such cases we use FE to obtain the stresses in the layered system. Up to now most calculations were done on 2-D or very special 3-D situations, but we are working with FE on the description of deformations in more general 3-D situations, that include plastic deformation and friction (see figure 18).

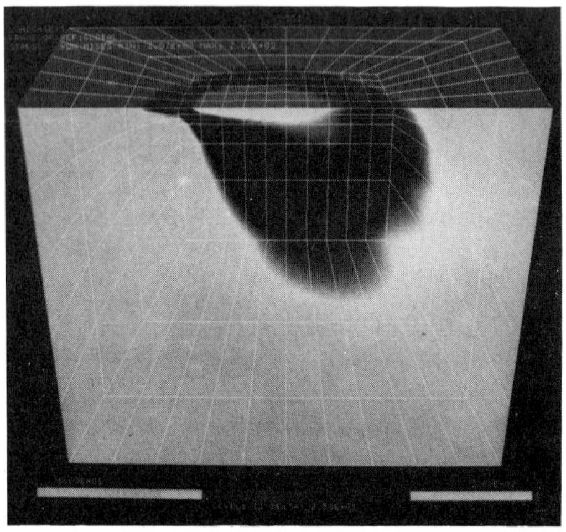

Figure 18: 3-D calculation of the von Mises stresses under a surface when a sphere is sliding over it with a friction coefficient of 0.5 (elastic calculation).

At the Research Labs. the influence of surface asperities on frictional behaviour was simulated with a FE program [49,50], and also contact phenomena in metal forming were studied extensively [51,52,53].

Coatings

Coatings present an economical way to change the properties of surfaces. Within Philips there is a large choice of galvanic, plasma-sprayed, diffused, physically or chemically vapour-deposited or ion implanted coatings. In a number of leaflets and technical notes the CFT has evaluated the most important ones.
New types of wear-resistant, low friction coatings are being developed in our Research Labs. in Hamburg. A special research group there is investigating methods to coat surfaces at lower temperatures than conventionally used in deposition processes, so that bulk properties are not affected. The emphasis is on metal-carbon, boron-nitride and MoS_2 layers. Also being studied are the use of direct ion implantation and the enhancement of adhesion by ion implantation (ion beam mixing) [54,55,56,57,58,59].
At the Research Labs. in Eindhoven we are trying to understand the wear behaviour of coatings. Here we consider the influence of the substrate, coating, geometry, normal load and friction on the mechanical stresses occurring in the coating using FE calculations. These mechanical stresses and especially the von Mises stress are then correlated to the wear occurring in the system. This systematic approach to the wear of layered systems has been used successfully to describe the wear of electrical contact systems (see figure 19) [60,61,62,63,64].

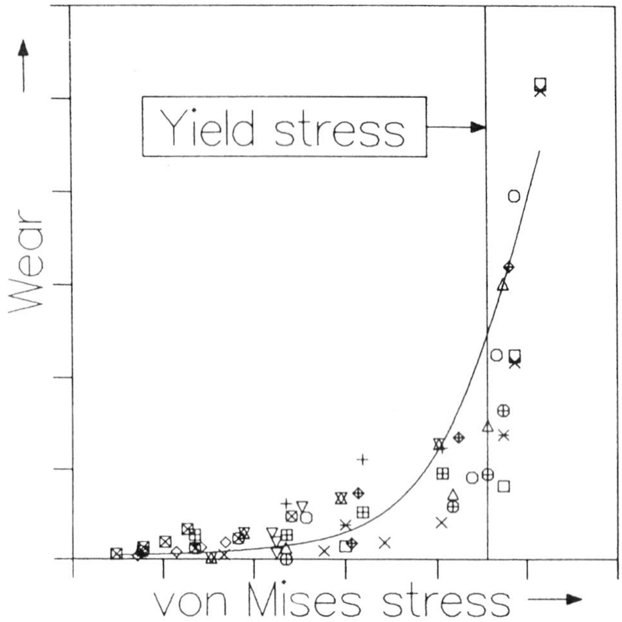

Figure 19: The correlation between wear and the von Mises stress in the top layer of a coated system. For many normal loads, layer thicknesses and substrate hardnesses there is a good correlation. The solid line represents a low-cycle fatigue model.

Magnetic recording

Tape-head interaction is a special problem occurring in magnetic recording media [65,66,67,68]. Requirements for these systems are very severe. Over an extremely long wear path only a few microns of wear depth can be tolerated in the video head. In the Research Labs. the wear mechanism of different head materials running against different tapes is investigated [69,70]. A special test, the Sphere On Tape (SOT) test was developed to quickly rank tape-head combinations [71]. The choice of good combinations requires special expertise in view of the interaction between magnetic and tribological demands. In this field experts are also engaged within the Consumer Electronic Products Division.
At the Research Labs we are investigating the running behaviour of a tape through a recorder. A start has been made on analysing the capstan-tape interaction.

3 DESIRABLE FUTURE ACTIVITIES

Recently an extensive list of future directions in tribology research was provided by Jahanmir [72]. He indicates necessary work on the nature of friction, the mechanism and processes of wear, lubricant chemistry, the modelling of tribological processes and tribo-systems, work on tribo-materials and lubricants, and research for novel experimental and diagnostic techniques. We particularily endorse the remark he makes in his summary: "The one area that clearly stands out is the establishment of models for the design and failure prediction of tribological components and systems". So Jahanmir stresses that real fundamental knowledge is needed. For technologists literature and / or university studies are not the primary source of knowledge. Mostly they rely on information from suppliers and manufacturers. In our opinion the primary task of universities is to provide fundamental knowledge, so that results of for instance pin-disc tests can be transferred to practical situations. Technological testing is best done by industry itself.
In order not to break with the general outline of this pa-

per, we will follow the outline of chapter 2, summarized in figure 2. We shall refer in this chapter to research that will not necessarily be performed by Philips in the coming years, but that closely fits in with our interests. In some cases our present activities are already mainly concentrated on fundamental research (e.g. foil bearings, precision machining, magnetic recording). Such subjects are omitted in this chapter.
For the designer it is important that functional requirements like life, running accuracy or minimum friction should be translated into specific demands upon the tribo system like type of contact, dimensions of the contact or contact materials. An expert system might be the solution for such a problem.

3.1 Full-film lubrication

Externally pressurized bearings

Theoretical modelling of externally pressurized bearings would be simplified and the application of such bearings extended if a program package were available that could calculate precisely the effect of tolerances, bearing dimensions (pneumatic hammer), non-conventional layout and hybrid bearings. Such a program could be set up in much the same way as FE packages for mechanical behaviour. At present we are trying to create such a package ourselves, since as far as we know it is not available on the market. We would be glad however if we could buy such a package or could start a joint effort.
For gas bearings we would like to know more about the static and dynamic behaviour of inherently compensated restrictors, discharging into tapered gaps and the transition to pure orifices. Furthermore it seems desirable to investigate the influence of roughness and geometry tolerances on the running accuracy.
For making porous bearings we lack a material with good permeability and low porosity. Also a very smooth (granite) surface would be useful for inherently compensated bearing pads.

Spiral-groove bearings

In recent years the use of SGBs has increased not only within Philips but also outside our company.
For theoretical modelling we need a program package like the one wanted for externally pressurized bearings. For SGBs rotor-bearing dynamics are also important. At the CFT and the Research Labs. such programs are being developed.
To obtain higher precision in dynamic mechanical systems further investigations of individual components and their interactions are necessary. The dynamic behaviour of SGBs is not yet fully understood, especially their stiffness and damping behaviour in relation to cavitation and free boundary.

3.2 Mixed & boundary lubrication

The mixed and boundary lubrication regime is one of the most difficult ones to study, since it involves many interacting processes. Such a study therefore requires the cooperation of many disciplines. The need for collaboration is greatest in this field.

For mixed & boundary lubricated and dry-running systems an important parameter for judging the performance of the system is the surface temperature in the contact. An easy-to-use computer program for heat transfer would be very helpful.

Lubrication

For most experimental set-ups the transfer of results to practical situations is difficult. Most results are highly dependent on the test chosen. For instance the transition diagrams as advocated by the International Research Group of the OECD are obtained in a pin-ring configuration. There is an obvious need to translate the results of such transition diagrams into a more or less dimensionless form so as to give them a more universal character.
The interaction of lubricants with metals and plastics and lubricant behaviour under extreme pressures in dies needs further study.
As for the lubricants themselves we feel the need for lubricants whose viscosity depends less on temperature than present ones and are acceptable in real designs. Silicon oils spread too much. We would also like lubricants that are electrically conductive and special lubricants for ceramics.

3.3 Dry-running systems

An important step towards controlling wear behaviour is to understand the process of wear particle generation. If we understand this process better, we will be able to devise zero or controlled wear situations. Perhaps the best way to tackle this problem is to classify wear processes through an organized form like wear maps, which indicate a range of effective operating conditions for each wear mechanism. We think there is a need to study processes like surface fatigue, repeated plastic deformation and adhesion.

Material characterization

Some applications of new materials like ceramics, which have in principle great potential, are hampered by the lack of knowledge of their tribological behaviour and because the materials are extremely difficult to machine. Further research seems necessary.

Contact mechanics

As a support to the above-mentioned understanding of wear particle generation it is necessary to know precisely the mechanical stress field when such a particle is generated. 3-D FE calculations including friction and plastic deformation can give such results. At present such calculations are still hampered by the amount of computer memory and the long computing times necessary.

Coatings

The basic wear processes and failure modes for coated systems are more complicated than for solid materials. It is more difficult to define precisely the circumstances and the mechanical stresses in the coated system at the time of failure. But since coatings are increasingly used it is of great economic importance to know their mode of failure. It seems especially desirable to examine the relevance of existing test methods for coatings like the scratch test.
Research for new types of coatings that have excellent tribological properties while not affecting bulk substrate properties remains desirable.

REFERENCES

(1) GIJSBERS T.G., 'COLATH a numerically controlled lathe for very high precision', Philips Techn. Rev., Vol 39 (1980), pp. 229-244.

(2) BROUWER A.G.,BRUEL R.H.,VAN HEEK H.F., KLOSTERMAN F.T., MANNETJE 'T J.J.,' The

OPTHYCOGRAPH', Philips Techn. Rev., Vol 34 (1974), pp. 257-269.

(3) DE GAST J.G.C., 'A new type of controlled restrictor (M.D.R.) for double film hydrostatic bearings and its application to high precision machine tools', Proceedings of the 7th International Machine Tool Design and Research conference, Pergamon Press, New York, Sept. 1966.

(4) DE GAST J.G.C., 'Dynamic behaviour of double film hydrostatic bearing with variable flow restrictor', Paper 69-WA/Lub-12, ASME.

(5) KRAAKMAN H.J.J., 'Hydrostatische lagers met verhoogde stijfheid III Slot', PT-Werktuigbouw, 32 (1977), p.690.

(6) KRAAKMAN H.J.J.,DE GAST J.G.C., 'A precision lathe with hydrostatic bearings and drive', Philips Techn. Rev., Vol 30 (1969), p. 117.

(7) HOLSTER P.L., JACOBS J.A.H.,SASTRA B.'Measuring the radial error of precision air bearings', Philips Techn. Rev., Vol 41 (1983/84), pp.334-337.

(8) BREHM R. DRIESSEN J.C., GROOTEL P. van, HEIJNEMANS W.A.L.,'Low thermal expansion materials for high precision measurement equipment', Precision Engineering, Vol 7 (1985), pp 157-160.

(9) SASTRA B., DOHMEN G.M., 'High precision numerically controlled measuring system for cylindrical surfaces', Precision Engineering, Vol 6, No 1, Jan 1984, pp 12-16.

(10) ENGELEN G.A.J.van, HAGEN J.L.M., HEIJNEMANS W.A.L., ' An equipment for measuring the magnetic fields of television deflection coils', Philips Techn. Rev., Vol 39 (1980), pp 277-282.

(11) GIJSBERS T.G., 'Some aspects of precision machining within Philips in 1984', SPIE vol 508, Production aspects of single point machining optics (1984), pp 26-30.

(12) KHOE G-D., LEEST H.F.M. VAN, LUIJENDIJK J.A., 'Single-mode fiber connector using core-centered ferrules', IEEE Journal of Quantum Electronics, Vol QE-18, No 10, October 1982, pp 1573-1580.

(13) KLOSTERMAN F.T., ' A step- and repeat camera for machining photo masks for integrated cicuits', Philips Techn. Rev., Vol 30 (1969), pp 57-70.

(14) HOLSTER P.L.,'Reliable and easy to handle design formulae for externally pressurized gas thrust and journal bearings', Paper 2 of the 3rd Gas Bearing Symposium in Southampton, 25-28 April 1967.

(15) 'Design of gas bearings', MTI design course held in Southampton 1969, Vol.1 and Vol. 2.

(16) HOLSTER P.L.,JACOBS J.A.H., 'Theoretical analysis and experimental verification on the static properties of externally pressurized air-bearing pads with load compensation', Presented at the Gas Bearing Symposium in Washington 1986.

(17) HOLSTER P.L., JACOBS J.A.H., 'Theoretical analysis and experimental verification on the static properties of circular air-bearing pads with load compensation', Tribology International, October 1987, Vol 20, No 5, pp. 276-289.

(18) MUIJDERMAN E.A., TANGENA A.G., 'Vollfilmgeschmierte (Spiralrillen)-Lager und die Verwendung von Keramikwerkstoffen', Proc. 1. Fachsymposium: Ingenieurskeramik für hochbeanspruchte Reibsysteme, Frankfurt 1988, Technik + Kommunikation Verlags GMBH: Praxis-Forum, Berlin.

(19) MUIJDERMAN E.A.,'New possibilities for the solution of bearing problems by means of the spiral groove principle', Proc. Instn. Mech. Engrs 180 (1965-66), p. 174.

(20) MUIJDERMAN E.A., 'Bearings', Scientific American 214, March 1966, p. 60.

(21) VOLMAN H.J.W.M., 'The 'push-pull' spiral-groove bearing - a thrust bearing with self-adjusting internal preloading, Philips Techn. Rev., 35 (1975/76), p.11.

(22) MUIJDERMAN E.A., REMMERS G., TIELEMANS L.P.M., 'Grease-lubricated spiral-groove bearings', Philips Techn. Rev., 39 (1980/81), p. 184.

(23) MUIJDERMAN E.A., 'Grease lubricated spiral-groove bearings for space machinery', April 1975, European Space Tribology Symposium, Frascati, E.S.R.O., Italy.

(24) WAL VAN DE U., 'Application of spiral-groove bearings to spacecraft', April 1975, European Space Tribology Symposium, Frascati, E.S.R.O., Italy.

(25) REINHOUDT J.P., 'A flywheel for stabilizing space vehicles', Philips Techn. Rev., 30 (1969), p. 2.

(26) MUIJDERMAN E.A., 'Spiral-groove bearings', Thesis Technological University Delft, The Netherlands, March 1964, 197 pages.

(27) REINHOUDT J.P., 'On the stability of rotor-and-bearing systems and on the calculation of sliding bearings, Thesis Technological University of Eindhoven, 1972.

(28) BOOTSMA J., 'Liduid-lubricated spiral-groove bearings', Thesis Technological University Delft, The Netherlands, 1975, 190 pages.

(29) BOOTSMA J.,TIELEMANS L.P.M., 'Conditions of leakage-free operation of herringbone grooved journal bearings', Paper Nr. 76 - Lub 27, ASME Lubrication Conference Boston, Oct. 1976, 8 pages.

(30) REMMERS G., 'Grease-lubricated spiral-groove bearings for a straight-through shaft, Philips Techn. Rev., 27 (1966), p. 107.

(31) REMMERS G., 'Grease lubricated spiral-groove bearings of plastic, full-film lubrication for consumer products', Presented at I.M.E. Tribology Convention 1976, Durham, Great Britain.

(32) REMMERS G., 'Grease-lubricated helical-groove bearings of plastic, Philips Techn. Rev., 34 (1974/75), p. 103.

(33) BREMER F., MUIJDERMAN E.A., HOLSTER P.L., 'Influence of pressure difference and axial velocity on a spiral groove bearing for a moving piston', Proc. Leeds-Lyon Symposium on Tribology, D. Dowson, ed., Elsevier, The Netherlands 1987.

(34) FLEMING D.P., HAMROCK J.H.,'Optimization of

self-acting herringbone -grooved journal bearings for maximal stability', NASA TN D-7803, 1974.

(35) HAMROCK J.H., FLEMING D.P.,'Optimization of self-acting herringbone journal bearings for maximum radial load capacity', NASA TN D-6351, 1971.

(36) MUIJDERMAN E.A.,'Luftgeschmierte Spiralrillenlager', Chapter 5 of the book 'Luftlagerungen', Expert Verlag, 1982, (Kontakt & Studium, Band 78), ISBN 3-88508-618-2.

(37) FIJNVANDRAAT J.G.,'The effect of finite width in foil bearings: Theory and experiment', to be published in Leeds-Lyon Symposium on Tribology, Leeds 1986.

(38) BRAUN A.L., 'Porous bearings', Tribology International, Oct. 1982, p. 235-241.

(39) NOORDERMEER L.J., BRAUN A.L., TANGENA A.G., 'Investigation on lubricants for the extrusion of aluminium using the pin-on-ring test', Proc. North American Metalworking Research Conference VII, Michigan 1979, p. 159.

(40) TANGENA A.G., HURKX G.A.M., GALENKAMP H., 'The testing of boundary lubricants in metalworking processes', Proc. 8th Leeds-Lyon Symposium on Tribology, D. Dowson ed., Butterworth London 1981, p. 230.

(41) HEIJNINGEN G.J.J. VAN, HURKX G.A.M., TANGENA A.G., 'The analysis of plastic deformation processes using finite element calculations and experiments on a twist compression tester', Proc. 10th Leeds-Lyon Symposium on Tribology, D. Dowson ed., Butterworth London 1984, p. 307.

(42) MUIJDERMAN E.A.,'Algebraic formulas for the threshold and mode of instability and the first critical speed of a simple flexibly supported (overhung) rotor-bearing system', Presented at the International Conference on Rotor Dynamics, Sept. 14-17 1986, Tokyo, JSME, IFToMM.

(43) FRANSE J., JONG DE G.J., 'Roughness generation in grinding of small optical surfaces', SPIE Vol. 803: Micromachining of elements with optical and other submicrometer dimensional and surface specification, The Hague, 1987, p. 43.

(44) SCHRAMA R.J.P., FRANSE J., 'The precision cutting process as a non linear closed loop system', to be published in Precision Engineering, 1988.

(45) VERBEEK H.J., 'Tribological systems and wear factors', Wear 56 (1979), p. 81-92.

(46) TANGENA A.G., HURKX G.A.M., 'Calculations of mechanical stresses in electrical contact situations', IEEE Trans. Comp. Hybr. Manuf. Techn., vol. CHMT-8, No. 1, March 1985, p. 13.

(47) MC GREGOR C.W., Ed., 'Handbook of Analytical Design for Wear', Plenum Press New York, 1964.

(48) HAMILTON G.M., 'Explicit equations for the stress beneath a sliding spherical contact', Proc. Instn. Mech. Engrs., Vol. 197C (1983), p. 53.

(49) TANGENA A.G., WIJNHOVEN P.J.M., 'Finite element calculations on the influence of surface roughness on friction', Wear, 103 (1985), p. 345.

(50) FRANSE J., TANGENA A.G., 'The influence of asperity deformation and surface roughness on the coefficient of friction during dry sliding', Proc. 12th Leeds-Lyon Symposium on Tribology, D. Dowson ed., Butterworth London 1986.

(51) BAAIJENS F.P.T., VELDPAUS F.E., BREKELMANS W.A.M., 'On the numerical simulation of contact problems in forming processes', Proc. NUMIFORM '86 Conference / Gothenburg / 25-29 August 1986, p. 85.

(52) BAAIJENS F.P.T., BREKELMANS W.A.M., VELDPAUS F.E., STARMANS F.J.M., 'A constitutive equation for frictional phenomena including history dependency', Proc. NUMIFORM '86 Conference / Gothenburg / 25-29 August 1986, p. 91.

(53) WIJNGAARDEN H. van, BAKEL B.L.M. van, VERWEY A.D., MEERSHOEK J.E.A., 'Numerical simulation of thermocompression bonding', Proc. NUMIFORM '86 Conference / Gothenburg / 25-29 August 1986, p. 157.

(54) KOBS K., DIMIGEN H., LEUTENECKER R., RYSSEL H., 'Adhesive and abrasive wear of nitrogen implanted steels', Proc. 6th International Conference on Ion and Plasma Assisted Techniques, Brighton, UK, 27-29 May 1987 and Proc. Materials Modification by High Fluence Ion Beams (NATO Advanced Study Inst.), Viana do Castelo, Port., 24 Aug.-4 Sept. 1987.

(55) DIMIGEN H., KOBS K., 'Wear resistance of nitrogen-implanted steels', Materials Science and Engineering, 69 (1985) 181-190.

(56) KOBS K., DIMIGEN H., HÜBSCH H., TOLLE H.J.,' Improved adhesion of solid lubricating films with ion-beam mixing', Philips Techn. Rev., 44, No. 1, 24-29.

(57) KOBS K., DIMIGEN H., HÜBSCH H., TOLLE H.J., LEUTENECKER R., RYSSEL H., 'Enhanced endurance life of sputtered MoS_x films on steel by ion beam mixing', Materials Science and Engineering, 90 (1987) 281-286.

(58) KOBS K., DIMIGEN H., HÜBSCH H., TOLLE H.J., LEUTENECKER R., RYSSEL H., 'Improved tribological properties of sputtered MoS_x films by ion beam mixing', Appl. Phys. Lett. 49 (9), 1 Sept 1986, 496-498.

(59) TOLLE H.J., KOBS K., 'Untersuchungen tribologischer Schichten mit SIMS', Frezenius Z Anal Chem (1987) 329:391-394.

(60) TANGENA A.G., HURKX G.A.M., 'The determination of mechanical properties of thin metal films using indentation tests', Trans. ASME, J. Eng. Mat. Techn., July 1986, vol. 108, p. 230.

(61) WIERENGA P.E., FRANKEN A.J.J., 'An ultramicroindentation apparatus for the mechanical characterization of thin films', J. Appl. Phys., 55 (1984), p. 4244.

(62) TANGENA A.G., WIJNHOVEN P.J.M., 'The correlation between mechanical stresses and wear in a layered system', Wear, 121 (1988), p. 27.

(63) TANGENA A.G., 'Tribology of thin film systems', Ph. D. Thesis, Eindhoven University of Technology, April 1987.

(64) TANGENA A.G., WIJNHOVEN P.J.M., MUIJDERMAN E.A., 'The role of plastic deformation in wear of thin films', accepted for publication in: Trans. ASME, J. Tribology.

(65) WIERENGA P.E., SCHAAKE R.C.F., 'The effect of mechanical and chemical surface properties on the friction and wear of magnetic tapes', Wear, 119 (1987) 29-50.

(66) VELZEN VAN P.N.T., WIERENGA P.E., SCHAAKE R.C.F., 'Surface characterization of particulate video tapes by static SIMS, submitted to ASLE Trans.

(67) BRONDIJK J.J., FEEKES E.E., SPRANGERS W.J.J.M., WIERENGA P.E., 'Roughness and deformation aspects in calendering of particulate magnetic tape, IEEE Trans. Magn., MAG-23 (1987) 146-150.

(68) TANGENA A.G., WIJNHOVEN P.J.M., BRONDIJK J.J., WIERENGA P.E., 'Calendering of magnetic tape: A comparison between elastoplastic calculations and experimental results', in: Tribology and Mechanics of Magnetic Storage S, STLE Spec. Publ. SP-22, 1987.

(69) BROESE VAN GROENOU A., KADIJK S.E., 'Slip patterns by sphere indentations on single crystal MnZn ferrite', submitted to Acta Metallurgica.

(70) KADIJK S.E., BROESE VAN GROENOU A., 'Cross-slip patterns by sphere indentations on single crystal MnZn ferrite, submitted to Acta Metallurgica.

(71) BROESE VAN GROENOU A., KADIJK S.E., 'Sliding sphere wear test on NiZn and MnZn ferrites, Wear, 126 (1988) 91-110.

(72) JAHANMIR S., 'Future directions in tribology research', Trans. ASME, J. Tribology, April 1987, vol. 109, p. 207.

SESSION XIV
HYDROSTATIC BEARINGS

Chairman: Professor H S Cheng

PAPER XIV(i)	Optimum Design and Automatic Drawing of Recessed Hydrostatic Bearings
PAPER XIV(ii)	Computer Aided Design of Externally Pressurized Bearings
PAPER XIV(iii)	A Theoretical Investigation of Hybrid Journal Bearings Applied to High Speed Heavily Loaded Conditions Requiring Jacking Capabilities
PAPER XIV(iv)	Behaviour of a High Speed Hydrostatic Thrust Bearing with Recess Inserts and Grooved Lands
PAPER XIV(v)	An Experimental Comparison Between the Performance of a 'Total Cross Flow' and an Equivalent Conventional Design Hydrostatic Journal Bearing

Paper XIV(i)

Optimum design and automatic drawing of recessed hydrostatic bearings

S. Xu and B. Chen

The optimum design and automatic drawing of the hydrostatic bearing with any number of recesses are studied. For the analysis and calculation of the dynamic effects on the lands the Reynolds' equation is directly solved by the finite difference method with inequal steps or the finite element method.

In the optimum design of the bearing, there are some optimization methods and several objective functions to be offered to the users. Because several technical approaches are employed, it become practicable to do the optimization of bearings by solving the Reynolds' equation directly.

To draw the bearing draught automatically the interface between the standard graphic software AUTO-CAD and the programming language is worked out.

Nomenclature

$\overline{A_f}$	Effective dimensionless friction area $(=A_f/D^2)$
B	Flow factor of the bearing at the concentric condition
B_r	Flow factor of a recess at the concentric condition $(B_r=B/N_o)$
C_d	Diametral clearence $(=2h_o)$
D	Bearing diameter
H	Dimensionless Film thickness $(=h/h_o)$
EP	Eccentricity
EPA	Eccentricity of the thrust bearing
G	Bearing stiffness
H_p	Pumping power $(=p_s \cdot Q)$
H_f	Frictional Power
H_t	Total power $(=H_f+H_p)$
K	Power ratio $(=H_f/H_p)$
L	Bearing length
L_T	Axial flow land width
L_A	Circumferential flow land width
N	Rotational speed
P_s	Supply pressure
$\overline{Q}$	Dimensionless flow rate $(=Q \cdot \frac{12\eta}{p_s \cdot h_o^3})$
N_o	Number of recesses
S_H	Speed variable in Hybrid bearing
S_{Ho}	Optimal speed variable in hybrid bearing
W	Radial bearing load
W_A	Axial bearing load
W_o	Static bearing load
β	Concentric pressure ratio $(=P_r/P_s\|_{\varepsilon=0})$
η	Dynamic viscosity
φ	Attitude angle

1. INTRODUCTION

The structure of the recessed hydrostatic bearing without axial grooves (Fig.1) is simple and it is convenient to manufacture. This type of recessed bearing has larger effective supporting area because of no axial grooves. Since the hydrodynamic effects ,the load capacity and stiffness of this kind of bearing is greater than those of the similar bearing with axial grooves. Also, this type of recessed bearing has small flow rate and can prevent the influence on performances of the bearing from the air sucked into the recess through the grooves at high speed by cavitation .

There are a lot of research work on hydrostatic bearings , but only a little work on its optimum design has been made. Although the optimum design of hydrostatic and hybrid bearings was discussed , the dynamic effects were not taken into account and only some curves and graphics based on the relative formulas (mainly simple formulas) to be used for comparasion and selection [1][2][3] . In 1980 , Author derived the flow continuity equation and power representation of the recessed hydrostatic bearings and got a series of computational equations by means of the matrix analysis and then got the optimal parameters with the optimization method . After that , the optimum design of slot-entry hybrid bearing and hole-entry hybrid bearing was made by solving the Reynolds' equation directly using the finite difference method [4][5][6].

The authors have already worked out the program about the analysis , calculation ,optimum design and automatic drawing of the recessed hydrostatic bearings .For the analysis and calculation of the bearing , the Reynolds' equation is directly solved by the finite difference method with inequal steps or the finite element method .In this optimization program , the size and performance constraints of the bearing are considered and several bearing performances, such as load capacity, stiffness, flow rate, total power loss and total power loss/load capacity can be regarded as the objective function respectively at user's desire.

In optimization, the analysis program should be as simple and available as possible on the condition of the desired accuracy in order to save the computer time , therefore , the benefit of the finite difference method is apparent, e.g.,its computation formula is simple, its required memory capacity is rather small and the efficiency is high enough , so that it is the most suitable for the optimization . If the solved domain is made discretization reasonable and the step of the mesh is inequal , both accuracy and efficiency of the computation can be ensured as desired .

To draw the bearing draught automatically the interface between the standard graphic software AUTO-CAD and the programming language is worked out.By this, the drawing of bearings can be made out automatically with plotter or printplotter.It is interactive and easy to manage and the user's interface is transparent. With many tests and application, it has been shown that this software system is realiable and efficient and can be offered to the designers directly.

2. THE MATHEMATICAL MODEL OF THE OPTIMUM DESIGN

In the recent years , with the application of the computer and the development of the science and technology , the optimum design has become one of the important techniques in the modern design theory . Using the optimum design , we can consider the various complicated factors and find the optimum design values, subject to the prescribed conditions.

For a bearing design, the diameter of the bearing D and the load W at a certain speed N is usually prescribed . The eccentricity EP and the radial clearence h_o should be known to meet the demands for the manufaturing accuracy and the bearing stiffness , and in general the number of recesses is arbitrary .

These known conditions for the design of a recessed bearing without axial grooves are as follows :

$$D,\ N,\ W,\ W_o,\ W_A,\ N_o,\ EP,\ EPA,\ H_o$$

The following independent dimensionless parameters which mainly affect the bearing performance are considered as the design variables in optimization.

$$X=\{L/D, L_A/L, L_T/D, \beta, K\}$$

It is essential that some bearing performances (e.g. load capacity) can be regarded as an objective function at user's desire . In general, with the speed and power increasing in machine, the engry comsumption and the temperature rise become a critical problem, so that the ratio of total power dissipation to load capacity is usually regarded as the objective function* and its minimum value has to be found.

$$F=(H_t/W) \longrightarrow \min. \qquad (1)$$

It has to subject to these following constraints:

(a). $1.2 >= L/D >= 0.5$

(b). $0.5 >= L_A/L >= 0.1$

(c). $\pi/(2\cdot N_o) >= L_T/D >= 0.$

(d). $0.9 >= \beta >= 0.$

(e). $12. >= K >= 0.$

(f). $WNO >= W_o$

(g). $P_s\max >= P_s >= P_s\min$

(h). $\eta\max >= \eta >= \eta\min$

3. THE ANALYSIS OF THE BEARING AND ITS PROGRAMMING

In the hydrostatic and hybrid bearings , a dimensionless form of the Reynolds' equation can be stated as follows:

$$\frac{\partial}{\partial X}\left(H^3\frac{\partial \bar{P}}{\partial X}\right) + (D/L)^2\frac{\partial}{\partial Z}\left(H^3\frac{\partial \bar{P}}{\partial Z}\right) = 24\pi S_H\frac{\partial H}{\partial X} \qquad (2)$$

where, $X= x/D,\quad Z= z/L,\quad H=h/h_o,$

$P= p/P_s,\quad S_H=\frac{\eta N}{P_s}(D/C_d)^2,$

$C_d=2\cdot h_o$

Now , the speed variable S is introduced , which is a synthetic parameter of the bearing performance . Then , it is so convenient for the computer programming that the supply pressure P_s and dynamic viscosity η are not needed to know first.All we have to do is to select the suitable values P_s and according to the solved factor S_H at last.

It is easy to obtain the following expressions:

The flow factor of the bearing at the concentric condition is

$$B=12\cdot\beta\cdot\bar{B} \qquad (3)$$

$$\text{where } \bar{B}=\frac{\pi D}{6L_A} \qquad (4)$$

The optimal speed variable in hybrid bearing is

---

* Also, the other objective functions , such as minimization of flow rate of bearing , minimization of temperature rise , can be used here , only the subprogram of the objective function should be changed.

$$S_{Ho} = \frac{1}{4\pi}\sqrt{\frac{B}{12\bar{A}_f}} \quad (5)$$

$$\bar{A}_f = \frac{A_f}{D^2} \quad (6)$$

$$A_f = (A_L + \frac{A_R}{4}) \cdot N_o \quad (7)$$

where, A_R is the area of a recess, and A_L is the area of the land around a recess

The speed variable in hybrid bearing is

$$S_H = S_{Ho} \cdot \sqrt{K} \quad (8)$$

After the dimensionless load capacity $\bar{W}$ is known, variables P_s and η will be determined as follows :

$$P_s = \frac{W}{\bar{W} \cdot L \cdot D} \quad (9)$$

$$\eta = \frac{P_s \cdot S_H}{N}\left(\frac{2h_o}{D}\right)^2 \quad (10)$$

3.1. The finite difference method

The solution domain of the bearing is divided into the differential mesh lines with inequal steps and the general five-point difference formula is used. (see Fig.2)

$$A_{ij} \cdot \bar{P}_{i,j+1} + B_{ij}\bar{P}_{i,j-1} + C_{ij}\bar{P}_{i+1,j} + D_{ij}\bar{P}_{i-1,j} + E_{ij} = F_{ij}\,\bar{P}_{i,j} \quad (11)$$

where,

$$A_{ij} = \left(\frac{\Delta X_1 + \Delta X_2}{2}\right)\left(\frac{D}{L}\right)^2 \frac{H^3_{i,j+\frac{1}{2}}}{\Delta Z_2}$$

$$B_{ij} = \left(\frac{\Delta X_1 + \Delta X_2}{2}\right)\left(\frac{D}{L}\right)^2 \frac{H^3_{i,j-\frac{1}{2}}}{\Delta Z_1}$$

$$C_{ij} = \left(\frac{\Delta Z_1 + \Delta Z_2}{2}\right)\frac{H^3_{i+\frac{1}{2},j}}{\Delta X_2}$$

$$D_{ij} = \left(\frac{\Delta Z_1 + \Delta Z_2}{2}\right)\frac{H^3_{i-\frac{1}{2},j}}{\Delta X_1}$$

$$E_{ij} = -24\pi S_H \left(\frac{\Delta Z_1 + \Delta Z_2}{2}\right)(H_{i-\frac{1}{2}} - H_{i+\frac{1}{2}})$$

$$F_{ij} = A_{ij} + B_{ij} + C_{ij} + D_{ij}$$

The pressure distribution of the bearing can be obtained by solving both the Reynolds' equation and the flow continuity equation. the traditional solution is that suppose the variables P_s and η are known, the flow through the bearing and the restrictor can be calculated respectively and got equilibrium by iteration. In this paper, a more effective solution is taken successfully. The dimensionless design parameters of the restrictors are used and according to the corresponding formulas given by the fluid mechanics, the flow continuity equation can be carried out as follows.

The dimensionless flow rate through the restrictor is

$$Q_c = \delta(1-\bar{P}_r) \quad (12)$$

or $\quad Q_o = \lambda\sqrt{1-\bar{P}_r}$

Where

$\delta = \dfrac{3\pi\, d_c^4}{32\, h_o^3 \cdot l_c}$ is the dimensionless design parameter of a capillary restrictor,

and $\lambda = \dfrac{3\sqrt{2}\cdot \alpha d_o^2 \cdot \eta}{h_o^3\sqrt{\rho P_s}}$ is the dimensionless design parameter of an orifice restrictor.

Now, according to its expression at concentric condition, we can obtain

$$\delta = \frac{Br}{1-\beta}, \qquad \lambda = \frac{Br}{\sqrt{1-\beta}} \quad (13)$$

The flow rate out of a recess consists of two components, i.e. axial flow rate and circumferential flow rate. Using the differential quotient ,their formula can be obtained easily. The flow rate feeding a recess is equal to the flow rate through a restrictor, so that the flow continuity equation can be written simply as

$$\bar{Q}_{in} = \bar{Q}_{out} \quad (14)$$

Because the design parameter of each restrictor which can be calculated first is used to express the flow rate through a restrictor, the dimensionless pressure of the recess $\bar{P}_r$ can be solved out by the flow continuity equation (14) directly without iteration, so that much computer time may be saved.

Using the super relaxation iteration technique, the pressures at all nodal points of the bearing mesh can be obtained by solving both the Reynolds' equation (2) and the flow continuity equation (14) simultaneously. Therefore, the bearing load capacity and the attitude angle can be calculated easily as follows: (see Fig.3)

$$\overline{DLOAD} = \sum_{j=1}^{NMJ}\bar{P}_{ij}\left(\frac{\Delta X_1+\Delta X_2}{2}\right)\left(\frac{\Delta Z_1+\Delta Z_2}{2}\right) \quad (15)$$

$$\overline{VLOAD} = -2\sum_{i=1}^{NMI}\overline{DLOAD}\,.\cos\theta_i \quad (16)$$

$$\overline{ULOAD} = -2\sum_{i=1}^{NMI}\overline{DLOAD}\,.\sin\theta_i \quad (17)$$

The dimensionless load in X direction is

$$\bar{W}x = \overline{VLOAD}\cdot\sin\varphi + \overline{ULOAD}\cdot\cos\varphi \quad (18)$$

and the total dimensionless bearing load capacity is

$$\bar{W} = \sqrt{\overline{ULOAD}^2 + \overline{VLOAD}^2} \quad (19)$$

The attitude angle is

$$\varphi = \text{arctg}\left(-\frac{\overline{ULOAD}}{\overline{VLOAD}}\right) \quad (20)$$

where, $\overline{ULOAD}$ and $\overline{VLOAD}$ are the dimensionless bearing load in U and V direction respectively.

The computer algorithm of finite difference method for analysis and optimum design of hydrostatic and hybrid bearings is given in Fig.4

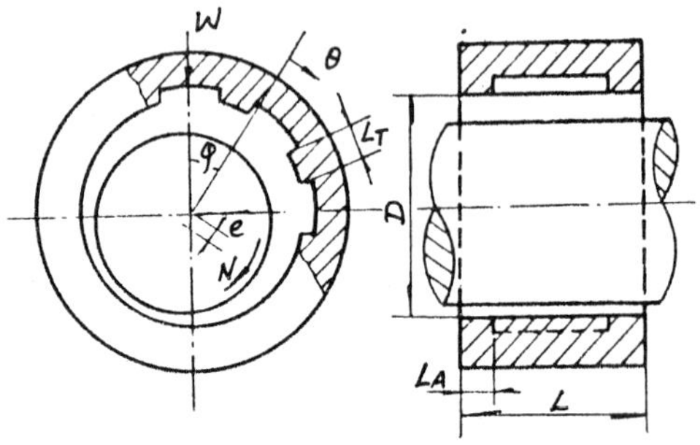

Fig.1

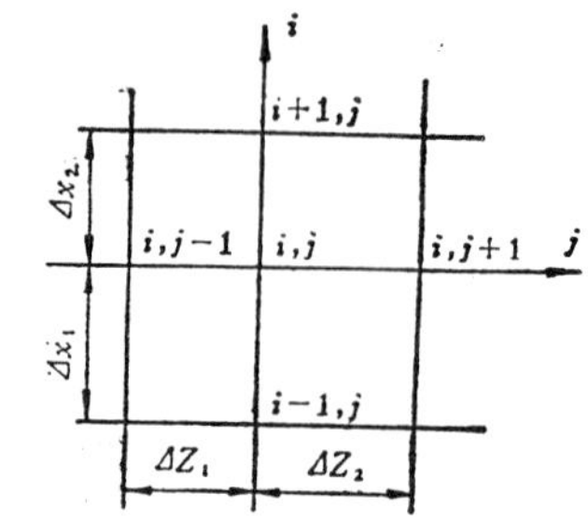

Fig.2

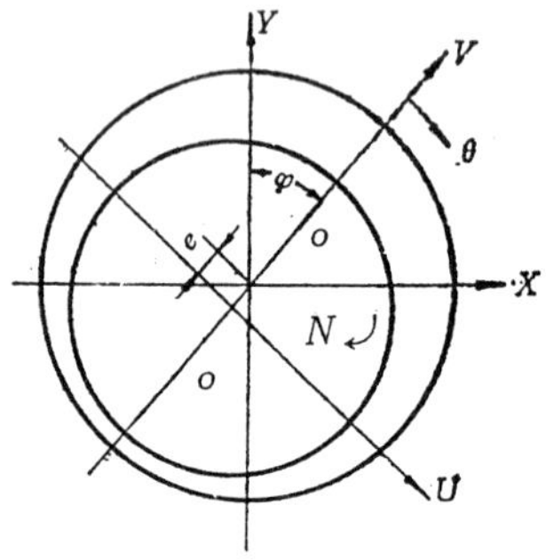

Fig.3

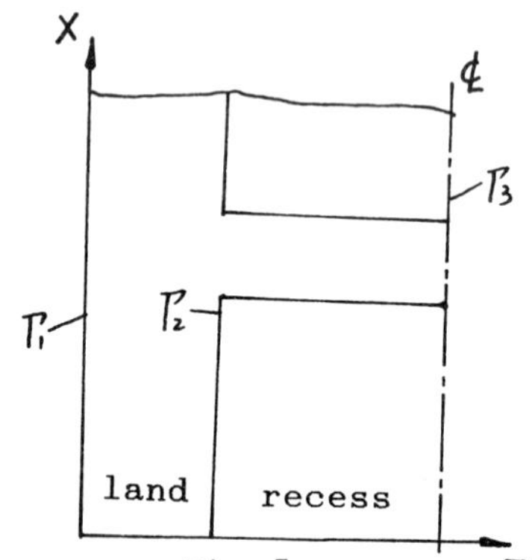

Fig.5

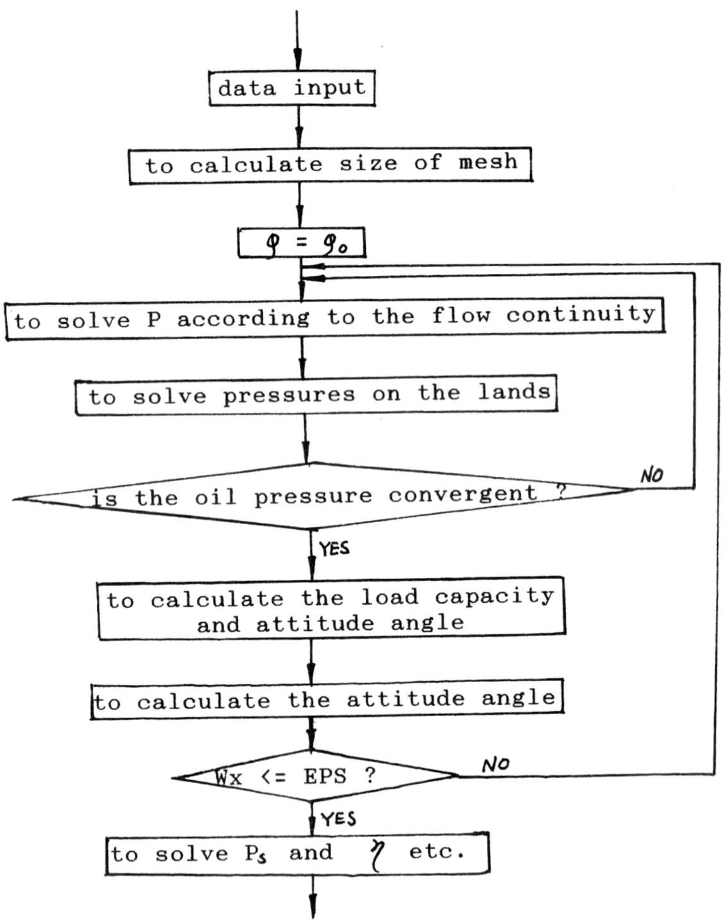

Fig.4 The Flow Chart For Finite Difference Solution

3.2 The finite element method

The finite difference method is easy and quite efficient, but it is difficult to coincide with the curve boundary of the solution domain. However, the finite element method has got a widespread application in lubrication field since the equivalent functional of Reynolds' equation was found out in the middle of 60's.

Because the configuration of bearing are various and the relationship between the pressure and the flow rate through the restrictor may be nonliear,the following functional of dimensionless Reynolds' equation is employed.

$$J(\bar{P})=\iint_{\Omega}\left[H^3\cdot\left(\frac{\partial\bar{P}}{\partial X}\right)^2+\left(\frac{D}{L}\right)^2\cdot H^3\cdot\left(\frac{\partial\bar{P}}{\partial Z}\right)^2-48\pi S_H\cdot\frac{\partial\bar{P}}{\partial X}\right]d\Omega \qquad (21)$$

its boundary conditions are:(see Fig.5)

(1). $\bar{P}\big|_{\Gamma_1}=0$,it means the pressure at the edge of circumferential lands is zero.

(2). $\bar{P}\big|_{\Gamma_2}=\bar{P}_r$,it means the pressure at the edge of a recess is equal to the recess pressure .

Where, the $\bar{P}_r$ can be made out by the equilibrium of flow rate into and out from a recess.

(3). $\frac{\partial\bar{P}}{\partial Z}\Big|_{\Gamma_3}=0$,it means the domain of bearing is symmetrical at the centre of bearing in axial direction.

According to the kind of element ,such as a triangular, quadratic or isoparametric element,we can define the suitable interpolation function and got the representative pattern of functional for each element. Then,minimizing the functional and assembling the element stiffness matrix and viscosity-flow rate matrix to the overall stiffness matrix and viscosity-flow rate matrix respectivelly, we have generated a linear equation group as follows:

$$[K]\{P\} = \{F\} \qquad (22)$$

Where, [K] is the overall stiffness or fluidity matrix, {P} is a column matrix of nodal pressure and {F} is the velosity-flow rate metrix.

It is practicable to solve this linear equation group when all the boundary conditions are taken into account.

The whole precedure to analyse and calculate the bearings by finite element method is shown in Fig.6. It is well known that a large amount of work in finite element analysis is the preparation of all the input data. This job is tedious,time-consuming and easy to make mistakes, espacially for the mesh containing hundreds or even thousands of nodes. To reduce the preparation time and improve the quality of input data, a versatile two-dimensional mesh generation with automatic bandwidth reduction is employed in the finite element analysis. It is only necessary to put in a little information which describes the geometric characteristics of domain , the kind of element and the density. The nodal numbering, nodal coordinates and some other information of element can be given out automatically by computer.

4. THE OPTIMIZATION METHOD

For a bearing design , there are a number of design parameters and performance . Besides , bearing performances (such as load capacity and power losses) is difficult to describe with a functional expression and more difficult to provide analytical function for its derivatives, therefore, the direct optimization method seems to be more suitable for this kind of problem.

Some of them are employed , such as Flexible Polyhedron Method and Complex Method and the SUMT which transforms a constrained optimization into a sequential unconstrained minimization. After some caculation and test, it is found that the Complex Method is the most efficient of these three methods owing to the characteristics of the mathematical model of the bearing . Its computer algorithm can be found easily in the most of books on optimization. Users can also connect this program with the library of optimization methods to select a much more effective algorithm.

The objective function and constraints have to be found by solving Reynolds' equation and it is well known that the analysis of the bearing must proceed with two iterations : one is for the attitude angle and the other is for the pressure P, which usually costs long time. Obviously, it takes much more computer time for optimization of a bearing.

In order to make the optimum design available for use , a series of technical approaches are applied to save computer time, such as the selection of the initial values and the modification of the super-relaxation factors automatically . The most effective one is the higher and/or lower accuracy iteration technique , that is, at the beginning of the optimization, the design variables are far from the optimum values ,so that the lower accuracy could be used in the analysis program to save computer time. Then, after certain iterations , they might be around the optimum values, the higher accuracy should be immediately used. In fact, more than 50% computer time is saved by this approach.

5. THE AUTOMATIC DRAWING

The standard graphic software package Auto-CAD is employed in this program. However, it is an effective program to edit a drawing . To apply

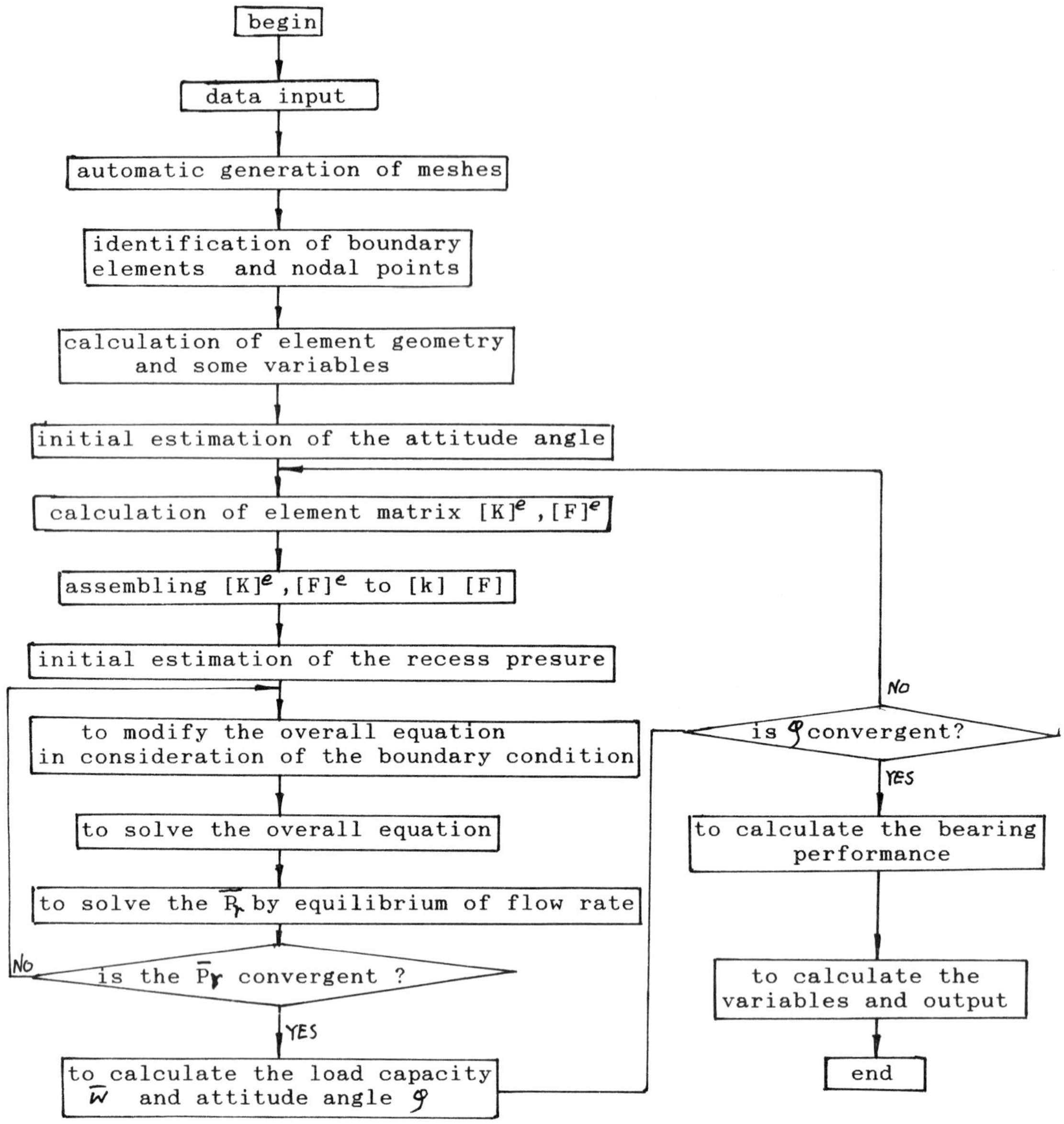

Fig.6 The flow chart for finite element solution

Table 1. A Comparision of The Parameters

	Unit	Original Design Parameters	Optimal Result
Supply pressure P_s	pa	0.20×10^7	0.167×10^7
Flow rate Q	l/min	0.552	0.676
Total power H_t	Kw	0.056	0.041
Bearing diameter D	mm	80.00	80.00
Bearing length L	mm	80.00	95.84
Axial flow land width L_T	mm	20.00	30.26
Circumferential flow land width L_A	mm	8.00	11.68
Attitude angle φ		22.16°	23.47°
Pressure ratio β		0.546	0.503
Power ratio K		2.364	1.154

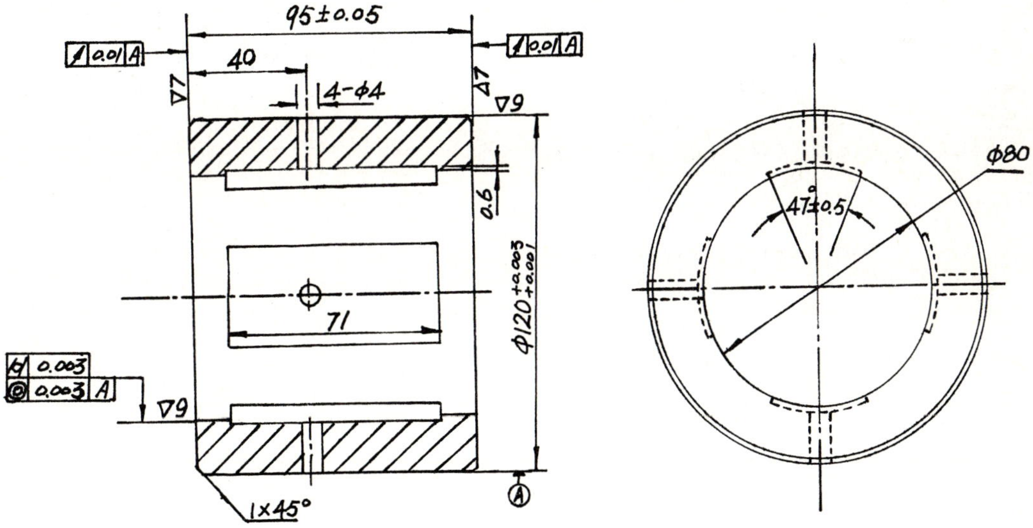

Fig. 7

to the computer-aided design ,it is very important to endow it with the capacity of "parametric drawing". That means what the user want to do is giving out the parameters, not the commands such as "line" , "text", etc. Therefore, the interface between the graphic software and the programming language should be carried out. In another situation, we can use this interface to develop and improve the drawing drafted by Auto-CAD using the programming languages.

Because the graphic data base of Auto-CAD is stored in a compressed pattern, the user can not get information directly by the program written in programming language. Therefore, to realize the exchange of Auto-CAD and drawing generated by programming language, it is necessary to use a concise and precise ASCII text to describe the drawing in Auto-CAD, that is "draft exchange file"(DXF). When we connect the Auto-CAD with programming languages, the internal data base of drawing should be changed into text in DXF first, then it is read out and developed by programming languages. After that, the DXF is sent back to Auto-CAD to generate the desired drawing.

According to the above principle, the interface between Auto-CAD and programming languages has been carried out. Now, we can generate the drawing of bearings by the software written in FORTRAN and modify the parameters of drawing freely according to the design parameters. Finally, all the files are transmitted to Auto-CAD to draft the desired drawing.

6. DISCUSSION AND CONCLUSIONS

Using the optimum design program developed by the authors, the analysis and the optimum design of a concrete bearing which is used in a high precision lathe have been carried out.

The specifications of the bearing are as follows:

Bearing type: Recessed bearing without axial grooves
Type of restrictor: Orifice restrictor
Number of recesses: No = 4
Bearing Diameter: D = 50.0 mm
Bearing Length : L = 50.0 mm
Eccentricity: Ep = 0.30

All the data above are stored in a data file, then to run the program, some results are given in Tab.1.

It has been shown that the total power loss has been decreased by 27% from 0.056 kW to 0.041 kW.

It is very convenient to design a hybrid bearing by using the dimensionless speed variable S_H . The iteration procedure of the flow rate can be omitted by means of the dimensionless design parameters of the restrictors.

Because a series of technical approaches are employed , much more computer time is saved. Therefore, it has become practical and available to do the optimum design of a bearing by solving the basic equation of the lubricated system -- Reynolds' equation directly.

The drawing of the bearing has been generated simultanuously.(see Fig.7)

The program has been tested by many concrete examples and it has been found that this program is correct , reliable and efficient . This program package can be applied to the calculation , analysis , optimum design and automatic drawing of a hydrostatic and hybrid bearing easily on a microcomputer. The program is versatile and can be provided for the designer directly.

REFERENCES

(1) W.B.Rowe, D.Koshal,'A New Basis for The Optimization of Hybrid Bearings' Wear 1980,vol.64

(2) M.K.Ghosh, B.C.Majumdar,'Design of Multirecess Hydrostatic oil Journal Bearings' Tribology Int. April, 1980

(3) M.El-Sherbiny, et al, ' Optimum Design of Hydrostatic Journal Bearings' Tribology Int. June 1984, Vol.17

(4) Xu Shangxian, 'Analysis and Optimum Design of Hydrostatic Bearings' (in Chinese) Journal of Nanjing Institute of Technology,No.1 1980

(5) Xu Shangxian, 'Hole-entry Hybrid Bearing and Its Optimum Design' Journal of Nanjing Institute of Technology, No.4 1983

(6) Xu Shangxian, ' Optimum Design of The Slot-entry Hybrid Bearing' Lubrication Engineering No.4 1984 (in Chinese)

(7) Xu Shangxian, Chen Baosheng, ' optimum design of hydrostatic and hybrid bearings' (in Chinese) Machine Tool & Hydraulics, No.5 1986

(8) Xu Shangxian, Chen Baosheng,' Finite Element Analysis of hydrostatic and hybrid bearings' (in Chinese) Machine Design, No.4 1986

(9) W.B.Rowe ,Xu Shangxian,F.S.Chong and W.Weston , 'Hybrid Bearings ---- with particular reference to hole-entry configuration' Tribology Int. Dec. 1982

(10) W.M.Newman, et al,'The principles of interactive computer graphics' McGraw-Hill , 1981

Paper XIV(ii)

Computer aided design of externally pressurized bearings

G. J. J. van Heijningen and C. M. Kalker-Kalkman

A method is presented for the selection and specifation of externally pressurized bearings, and calculation of the design-variables involved. An expert system is proposed to process the information about the bearings and to handle the different kinds of design problems. Two ways of implementing such a system are discussed, one for a personal computer and one for a CAD-system.

1 INTRODUCTION

The selection and calculation of suitable bearings for a given mechanical application is a significant problem for a mechanical engineer. He meets the difficulty of comparing different types of bearings and making the best choice from the most promising types.
The designer should be an expert on all competing types to make a meaningful comparison. There is,however,scope for making the necessary information available through a computer, which in effect acts as the expert and operates in an interactive consultation mode.
A research program has been set up at the Delft University of Technology which aims at introducing such a computer-aided technique into the bearing selection process. As a first example the bearing type: "externally pressurized bearing" has been selected.
Two programs have been developed to implement the system. In section 2 information is given about those implementations. The selection criteria can be reduced to a small number of basic items, in our case 8.
An account is being kept of existing bearings on the base of all those factors ,including the selection criteria and the names or the numbers of the drawings and necessary shape-parameters. This is presented in section 3.1.
When a bearing type has been chosen, it is presented on the screen by a simple drawing.
If the designer wants to go on with the selected bearing,he has to specify certain shape-parameters. This is treated in section 3.2.
Several variables have to be calculated in order to get an impression of the properties of the chosen bearing. The designer has a certain freedom in choosing the variables he wants to specify, and the variables to be calculated and plotted by the computer. This gives rise to a number of design problems, which have to be solved for all possible bearings. A method to do this as generally as possible is treated in section 3.3.
In section 3.4. a method is presented to select the problem.
In section 4 a conclusion and recommendation for further research is presented.

1.1 Notation

Symbol	Name
W	Load
R	Maximum outer dimension(radius)
S	Stiffness
Ps	Supply pressure
h	film-thickness
Q	flow rate
N	power
ΔT	maximum temperature rise
Re1	Reynolds number
Re2	Re1(h/1)
1	film-length
β	pressure ratio,Precess/Psupply
ρ	density lubricant
η	dynamic viscosity lubricant
c	specific heat lubricant
n	maximum number of choices for selection criteria
N_{appl}	number of selection criteria for application
N_{phys}	number of selection criteria for physical principle
a_i	line number selected from block i of application
f_i	line number selected from block i of physical principle
λ1 to λ5	shape-parameters
D1	outer diameter
D2	inner diameter

2 IMPLEMENTATION

Two programs have been written :

a. A program LAGERKEUZE written for a personal computer in Data Manipulation Language using the database system DBase III and using the spreadsheet-program SuperCalq for the calculations. A number of graphic functions has been used with the aid of a color-graphics card. Drawings and listings of results are produced by a plotter and printer.
b. A program BEARINGCHOICE in FORTRAN has been written and implemented on a TECHNOVISION CAD-system. With this system it is possible to file prepared drawings and call them by a Fortran program.

It is possible to set up a database (SIBAS-codasyl-database) with this system and to approach this database either interactively or with a Fortran-program.
The drawings and plots produced by the program can be filed and sent to a plotter for making a hard-copy.
The drawings prepared for this program can be written into IGES(Initial Graphics Exchange Specification) files. This feature makes it possible to implement the program in another system. It is not possible to implement the database calls. The program would have to be adapted to another database- system. However, it is a relatively simple matter to construct a file or data structure which can handle the same task.

3 GENERAL THEORY

In this section we present the methods employed in as general as possible a way and illustrate those methods with examples used in both programs.

3.1 Selecting procedure

We have to structure the expertise on bearings and put this into a Knowledge Base. In this Knowledge Base all factors and parameters by which a bearing is specified will have to be grouped in a convenient manner.
The selection criteria for choosing a bearing can be divided into two categories:

a. Selection with respect to the application of the bearing.
b. Selection with respect to the physical principle of its operation.

For the application a) we distinguish the following factors:

a1) the kind of load to be supported,
a2) the bearing geometry and the configuration of the machine in which the bearing is projected.

As to a1), it is important to know the direction of the load, which can be axial or radial or both. Moreover, the load can be constant or variable, so that opposed pads may be necessary.
As to a2), the shape of the bearing and the presence of slots have to be considered. The physical principle b) has to do with the way a load is supported and the type of pressure-control system to be used. We confine ourselves to externally pressurized bearings, having one of three different types of flow control devices, viz. constant flow, orifice or capillary restrictors.As to the shape of the film, we consider bearings with stepped, tapered and slotted films. The stiffness is an important factor, especially when tilting stiffness is required. The presence of recesses is also a factor of consideration. Summarizing all those factors we arrive at the selecting criteria given in fig.1, giving rise to 4 series of choices for both application and physical principle. We note that this number of selection criteria and the number of choices for each criterion are not relevant to the method presented here, and that it can be applied to any way of grouping bearing-properties. Looking at fig.1, we see that the designer can specify his wishes by making a choice from a menu in each block, a choice is given by a number in the block. Let us denote the maximum number of choices in all blocks by n. In figure 1 we have n=16, being the number of possible choices in the menu of the block bearing film stiffness. The total number of blocks belonging to Application we call N_{appl}, and the number of blocks belonging to Physical Principle is called N_{phys}. In the example of figure 1 both have the value 4. We now create two numbers in the number-system with base n:

$$K_{appl} = \sum_{i=1}^{N_{appl}} (a_i - 1)\, n^{i-1} \qquad 3.1.(1)$$

$$K_{phys} = \sum_{i=1}^{N_{phys}} (f_i - 1)\, n^{i-1} \qquad 3.1.(2)$$

where a_i and f_i denote the integers chosen from block number in the blocks for "Applications" and "Physical Principle" respectively.
K_{appl} and K_{phys} depend on the choices made.
For instance, if n=10 we have to do with the decimal system and if the choices 1,2,3,4,5,6,7,8 from the menus from left to right were made, then the decimal number

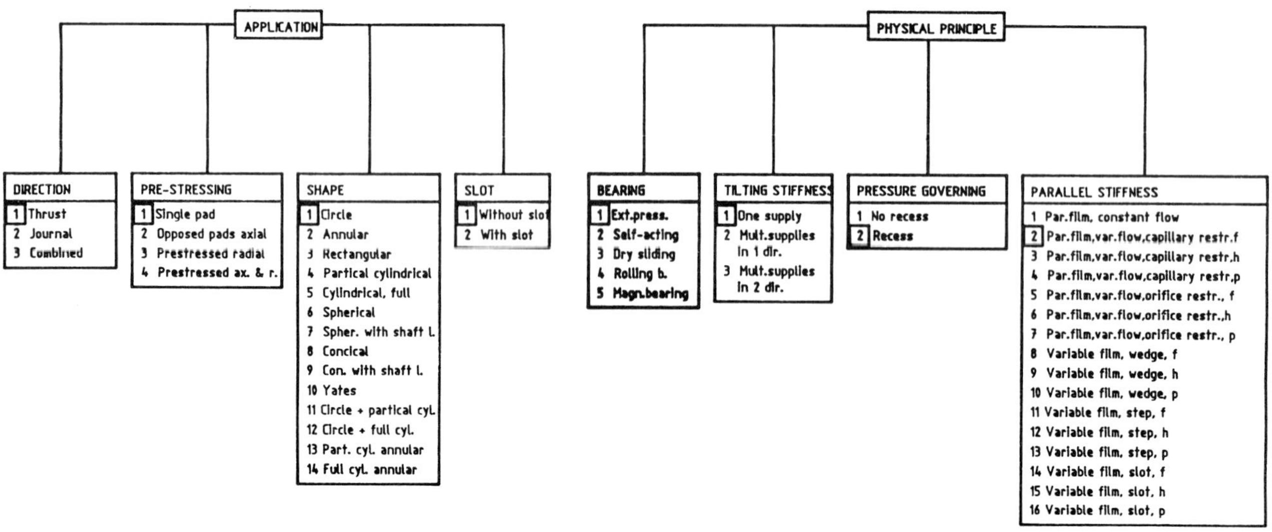

f: fixed valve
h: regulated by film-thickness h
p: pressure-regulated restriction

Fig.1 Selection criteria for externally pressurized bearings

01234567 would be constructed, which represents the selected bearing uniquely. When we number the blocks i in such a way that the number of choices in a block is highest for smallest i, the values for K_{appl} and K_{phys} are kept as small as possible. This is important for representing the numbers in the computer by single-length integers. In figure 1 we have n=16 and the maximum values for K_{appl} and K_{phys} resulting from the highest values of the chosen numbers are:

$$K_{appl} = 13 + 3.16 + 2.16^2 + 16^3 \qquad 3.1.(3)$$

$$K_{phys} = 15 + 4.16 + 2.16^2 + 16^3 \qquad 3.1.(4)$$

Both numbers are sufficiently small to be represented by an integer. We note that the numbers K_{appl} and K_{phys} are unique numbers for every n and represent a certain combination of choices. The combination of K_{appl} and K_{phys} can be used as a search-key for finding the corresponding bearing and the drawing of that bearing. This means that actually only one table is needed for finding a specific bearing and its drawing, that is the table containing all possible values and combinations of K_{appl}, K_{phys}, and drawing number or name. Conversely, the kind of bearing represented by the number in the table can be found by decomposing the numbers K_{appl} and K_{phys} in the sums 3.1.(1) and 3.1.(2). In the Fortran-program figure 1 was prepared as a drawing and called by the program. The designer is prompted to select a bearing by pointing to the numbers with a stylus on a tablet. From the selected numbers K_{appl} and K_{phys} were calculated and the corresponding bearing was found with this 'compound key'. When a combination of choices is made which has no physical meaning or that has not been implemented, it is simply not found in the database and the search-call results in an indicator being set. This indicator is detected by the program and the designer is prompted to modify his choice.

3.2 Drawings and shape-parameters

Once a bearing has been chosen with the aid of the selection procedure described in the previous section, a drawing is presented. In this drawing with text the choice of the designer is evident. With both implemented systems, these drawings have been prepared for all possible bearings and filed. For the Technovision system of Norsk Data, they can be stored in the archive of the system under an unique name, referred to as Drawing Name. The Fortran-interface of the ND-system makes it possible to call the drawing by name and present it on the screen. Hence, only one table is needed, containing N_{appl}, N_{phys} and the name of the drawing. In this table the combination of N_{appl} and N_{phys} is the search-key. For an example of a drawing, see fig.2. As many drawings are similar, one can utilize the possibility of copying frequently occurring parts into a drawing when preparing them. The drawings are qualitative, that is the shape and size of the bearing is not set yet. To specify these, we define so-called dimensionless shape-parameters. As an example, we see from figure 2 that in this case we have one shape-parameter, $\lambda_1 = D2/D1$ (ratio of diameters) This parameter can be chosen by the user. By preparing the drawing in a 'parametric way' one can redraw the bearing after λ_1 has been chosen, see fig.3. As we have more shape-parameters for other bearings, one feels the need to store information about those parameters for all the bearings in the same table as K_{appl}, K_{phys} and Drawing Name. In the evolved programs a maximum of 5 different shape-parameters λ_1 to λ_5 can be specified, their number and lower- and upper-boundaries were stored in the table for each tuple representing a bearing. The designer is prompted to specify the values of the shape-parameters which are relevant for the selected bearing. The bearing is then redrawn accordingly.

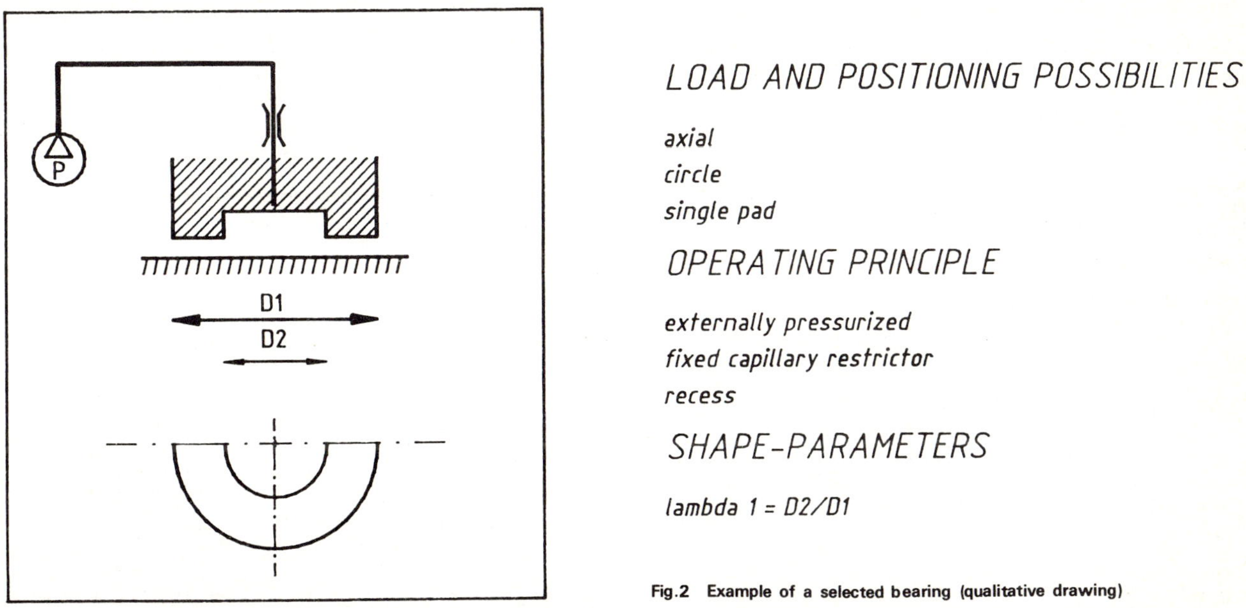

Fig.2 Example of a selected bearing (qualitative drawing)

Fig.3 Selected bearing , drawn with $\lambda 1=0.75$

Data machine design and operating conditions:

1. Load W (stationary) in N
2. Maximum outer dimension R in m
3. Wanted stiffness S in N/m
4. Pressure Ps in N/m2

Data Manufacturing:

5. Allowed film thickness in m

Optional:

6. Number of revolutions in rev./min.

Variables are calculated according to:

3W(1-ß)=hs
W=const.Ps.R2
const depends on bearing
ß=pressure ratio = P(recess)/P(source)

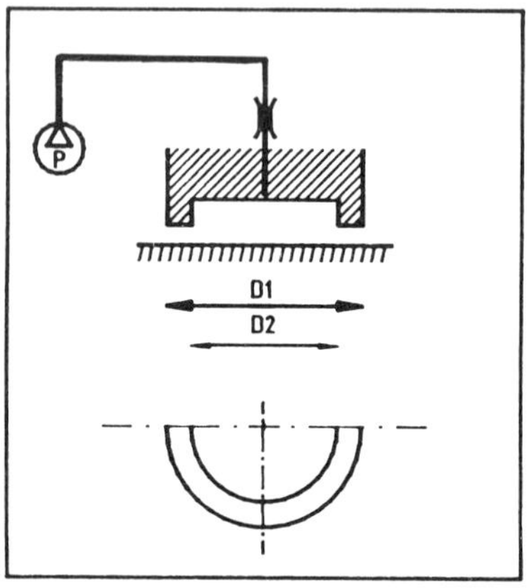

LOAD AND POSITIONING POSSIBILITIES

axial
circle
single pad

lambda 1 = 0.75

OPERATING PRINCIPLE

externally pressurized
fixed capillary restrictor
recess

SHAPE-PARAMETERS

lambda 1 = D2/D1

Lubricant: *oil*

3.3 Calculation of variables.

When the type of bearing has been selected and the shape has been specified with the aid of the shape-parameters, a number of variables has to be calculated in order to find out the properties of the bearing. In designing externally pressurized bearings, the following variables play a role:

Symbol	Name
W	Load (carrying capacity)
R	Maximum outer dimension (radius)
S	Stiffness
Ps	supply pressure
h	film-thickness
Q	flow rate
N	pumping power
ΔT	maximum temperature rise
Re1	Reynoldsnumber Re
Re2	Re(h/l)
β	pressure ratio, $0<\beta<1$, =Precess/Psupply
ρ	density of lubricant
η	dynamic viscosity of lubricant
C	specific heat of lubricant

The relations between those variables are kept as general as possible for the different bearings. For a large number of bearings, they can for instance be expressed in the following dimensionless factors (see [1]) :

$$\frac{W}{hS} = \frac{1}{3(1-\beta)}$$ Stiffness factor 3.3.(1)

$$\frac{W}{hS} = \frac{1}{3(1-\beta)}$$ Stiffness factor 3.3.(1)

$$\frac{W}{Ps\ R^2} = \text{const}\ \beta$$ Load factor, determines which part of supplied pressure is effective 3.3.(2)

$$\frac{Q}{h^3 Ps} = \text{const}\ \beta$$ Flow factor 3.3.(3)

$$\frac{N}{Q\ Ps} = 1$$ Power factor 3.3.(4)

$$\frac{\rho c\ \Delta T}{Ps} = 1$$ Maximum temperature rise 3.3.(5)

$$\frac{Re\ \eta R}{\rho Q} = \text{const}$$ Reynoldsnumbers, determine whether flow is laminar or turbulent 3.3.(6)

$$\frac{Re(h/l)\ \eta R^2}{\rho Q h} = \text{const}$$ (l=length film) 3.3.(7)

where the different constants indicated by "const" depend on the selected bearing and chosen shape-parameters. As ρ, η and c are properties of the lubricant used, various lubricants are presented to the designer to choose from. Once a lubricant has been chosen, the values of ρ, η and c are set. We are left with 10 variables which can be seen as functions of β, and 7 relations between them. A designer may want to know the behaviour of the variables under extreme situations, sometimes he wants to optimize one of the variables or he may want to know the sensitivity of a particular variable to

variations of another one. As these problems give rise to complex mathematical calculations, the developed programs were designed in such a way that the designer is prompted to specify 3 of the variables in order to calculate the other 7 variables as function of the pressure ratio β. When 3 of the variables are given, expressions can be derived to calculate the other 7 variables explicitly for any combination of chosen variables. In both programs, the user is given the choice of 3 of the first 5 variables W,R,S,Ps and h, see fig.3.
Choosing 3 variables out of 5 gives rise to $\binom{5}{3}$ = 10 different possibilities, called 'problem numbers'. The calculation of the problem number is discussed in the next section.
As the variables are considered as functions of β, it is easy to see from 3.3.(1) and 3.3.(2) that the choices of W,h,S and that of W,Ps,R both give rise to a fixed value of β. In those cases the problem is underdefined, and one has to specify one more variable in order to calculate the others. We see from 3.3.(1) to 3.3.(7) that is is easy to express all variables explicitly in β when 3 of them have been specified. Moreover it is easy to calculate the extremes of all those functions analytically. In the two programs that have been developed the extremes were determined numerically while calculating all values for plotting. The maxima and minima were used in order to scale the plotted functions. To keep the program as general as possible, a subroutine has been written, which has as input the selected bearing (by bearing number), the problem number, the specified variables, and has as output all calculated variables as functions of the pressure-ratio β. All variables are plotted on the screen, see fig.4. Both with the Norsk Data-system and with the P.C. it is possible to 'store' those plots, the drawing of the bearing and all other information written by the program on the screen in the files and make a hard-copy of it with the aid of the plotter. For the Norsk-Data computer the calculations were performed by the Fortran-program, whereas for the PC a spreadsheet program was used.

3.4 Selection of input-and output variables.

As mentioned in the previous section,it is possible to choose a 'problem-number' by specifying 3 variables out of 5.
To construct a problem number one can use the same method as in section 3.1

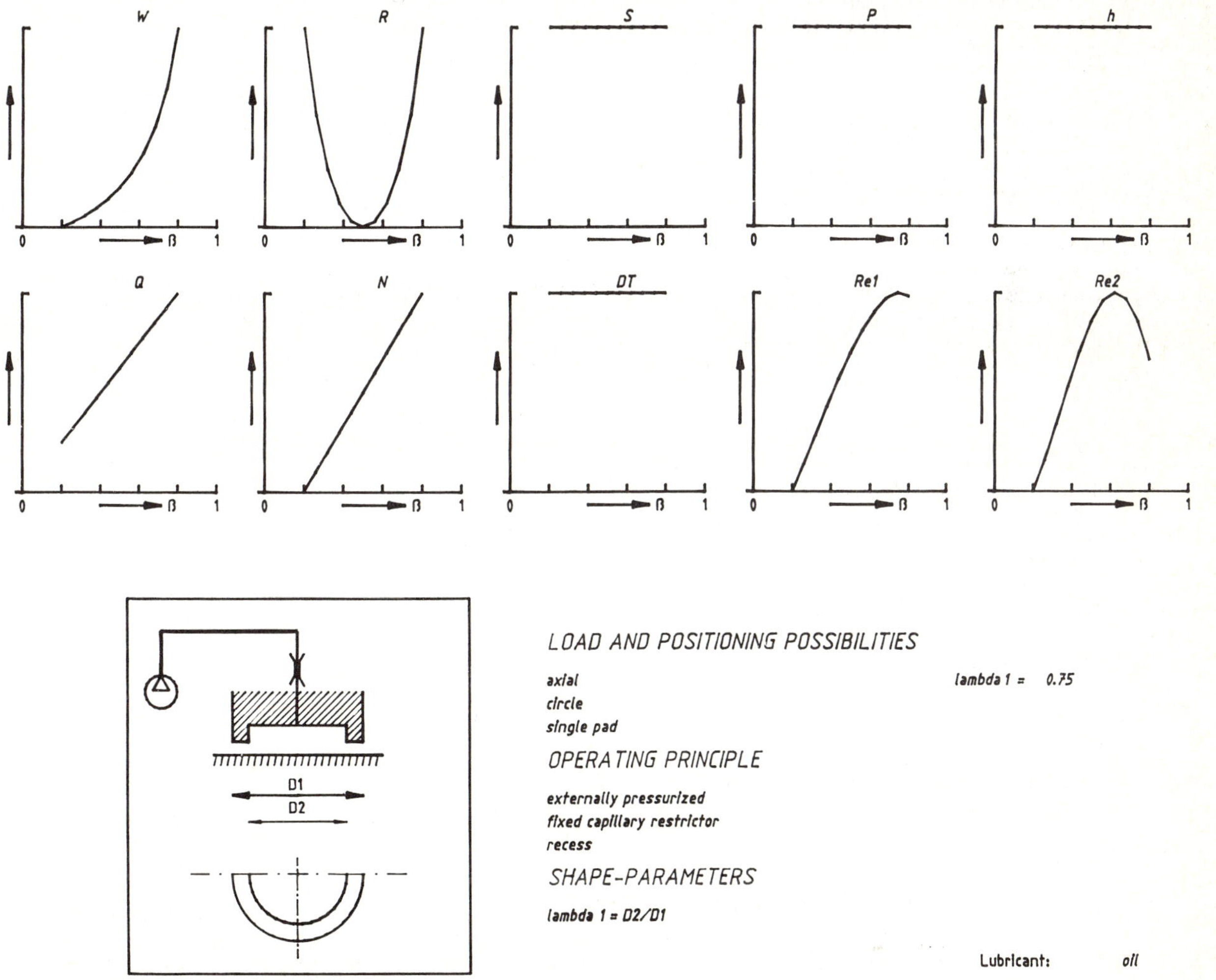

Fig.4 Plotting of the variables. The values of S,Ps and h are choosen, the others are calculated as a function of the pressure ratio, β

When out of 5 variables v1,v2,v3,v4,v5 an arbitrary number is chosen, an integer $C=\sum_{i=1}^{5} c_i 2^{i-1}$ is formed, where $c_i=1$ when variable i is not chosen $c_i=0$ when variable i is chosen.
C is an unique number.
When 3 variables are chosen out of 5, this gives rise to 10 different values of C, the highest being 24.
Putting those numbers in an array, a problem number 1 to 10 can be defined.
This method can be used for any number of variables and any number of choices.

4 CONCLUSION

Selecting a bearing and getting insight in the behaviour of the variables involved has become a relatively easy task when using an expert system as described. Further research is needed for more selection criteria and relations between the relevant variables. The authors are continuing their research into the matter of the optimalization of the variables. A method will be developed for preventing an unrealistic combination of choices. It is possible to implement such a system in various ways:
.with the aid of a database system
.using a data manipulation language
.using a spreadsheet
.using one table in a file or a data-structure
.using any programming language
For making the various drawings one can make use of standard graphic functions and a color-graphics card.
For working interactively with a light-pen, stylus and tablet or mouse it is helpful to have a CAD-system with an interface for a programming language and the possibility to manipulate drawings and graphic functions.

5 ACKNOWLEDGEMENT

The authors wish to thank A.Th.G.Maissan for his help in preparing the drawings of the bearings for the Norsk Data CAD-system.
Ing.A.J. Hoevenaar and H.Daane were helpful in programming the drawings for the personal computer.

References

[1] Rowe,W.B. 'Hydrostatic and Hybrid Bearing Design', First edition 1983 (Butterworths, London).

Paper XIV(iii)

A theoretical investigation of hybrid journal bearings applied to high speed heavily loaded conditions requiring jacking capabilities

D. Ives, W. Weston, P. G. Morton and W. B. Rowe

Previous work conducted on hybrid journal bearings has shown that a combination of hydrostatic and hydrodynamic lubrication principles leads to good load support for a range of speeds (including zero speed). Conventional generator bearings were identified as a typical application of high speed heavily loaded hybrid bearings, which require the jacking capability to reduce start-up wear. This work presents a theoretical investigation of a conventional jacking type generator bearing and an alternative slot entry configuration. The predictions show that optimisation of bearing geometry carried out in either the laminar or superlaminar flow regime produces the same basic bearing geometry. This indicates that simple laminar theory may be used for optimisation of bearing geometries, no matter which regime the bearing is to operate in. The work further shows that an increase in load support can be obtained by optimising the conventional generator bearing. The predictions show that further increases in load support could possibly be achieved by adopting an optimised slot entry configuration.

1 INTRODUCTION

The constant uprating of machinery, especially in the power generation industry, has led to increasing demands being placed upon journal bearing performance. Journal bearings are required to operate under conditions of increased load, whilst maintaining adequate minimum film thickness and low power consumption.

For conventional jacking type generator bearings in the start-up and shut-down operating modes, the jacking pockets are supplied with high pressure oil, to lift the journal and reduce wear. At full operating speed the jacking pressure is switched off for economy. The problem with jacking pockets is that they cause a reduction in load capacity at full speed, by reducing the area available for hydrodynamic load support and by allowing high pressure oil to escape to regions of low pressure.

An early conclusion from this work, was that the performance of this configuration could be improved by increasing the land separation between the two jacking pockets. This modification has already been introduced into some generator bearings.

The use of slot entry ports in hydrostatic bearings was originally proposed to combat dispersion problems associated with orifice controlled hydrostatic bearings [1]. However, it became apparent that the slot entry bearing could function equally well at speed. Rowe *et al* [2,3] conducted studies of the slot entry bearing operating in the hybrid mode. From these studies it was concluded that the slot entry journal bearing was superior to the recessed hydrostatic bearing, and comparable to the axial groove hydrodynamic bearing at high eccentricity ratio and high speed. In order that slot entry bearings could operate at optimum conditions, the following guidelines were given [3]:

length-diameter ratio	L/D	$= 1.0$
land-width ratio	a/L	$= 0.1$
pressure ratio	β	$= 0.5$
power ratio		$3 \leq K \leq 12$

The flow rate in plain hybrid journal bearings is usually fairly large, especially for small land width ratios, fig.2. This is due to a high proportion of the high pressure oil passing directly out of the bearing. For a generator bearing one of the operating requirements identified was that of a low flow rate. Thus if the slot entry bearing was to be considered for generator applications, a method of reducing loss of pressurised oil was required.

One method of reducing the loss of pressurised oil in a hybrid bearing is to reduce the number of pressurised inlets. Rowe *et al* [4] conducted a study of hole entry hybrid journal bearings, considering asymmetric inlet configurations. It was found that the flow rate could be greatly reduced by using asymmetric inlet configurations. However, it was suggested that it would be advantageous to include a low pressure axial groove at the maximum film thickness position, in order to prevent starvation in the central portion of the bearing. Ives and Rowe [5] studied asymmetric slot inlet configurations which included a low pressure axial groove. They concluded that the load support of these bearings was good at all speeds (including zero speed), and that the axial groove was a necessity in order to prevent severe starvation problems.

The asymmetric slot entry bearing with axial grooves is, in many ways, similar to a conventional jacking type generator bearing. Both types of bearing have axial grooves to ensure adequate lubrication and cool operation. Both types of bearing have pressurised sources in the loaded arc, to provide hydrostatic lift. However, the slot entry generator bearing would have external pressurisation applied continuously to the slot entry ports, to enhance the load support. The slot entry bearing obtains this advantage partly through maximising the bearing area available for hydrodynamic load support.

The aims of this work are as follows:

(1) To compare the results of optimisation studies carried out in laminar and superlaminar regimes.

(2) To compare the performance of a conventional jacking type generator bearing, fig.1(a), with that of an asymmetric slot entry bearing with a similar, but optimised geometry, fig.1(b).

2 NOMENCLATURE

h	Film thickness
h_o	Radial clearance
H	Dimensionless film thickness h/h_o
p	Pressure
P_s	Supply pressure
C_d	Diametral clearance
D	Bearing diameter
L	Bearing length
N	Rotational speed
V	Surface speed $2\pi N$
S_h	Speed parameter $((\eta N)/P_s)(D/C_d)^2$
W	Bearing load
Q	Bearing flow rate
H_t	Total power $H_f + H_p$
H_f	Friction power
H_p	Pumping power
K	Power ratio H_f/H_p
k_x, k_z	Viscosity factors for turbulence
Re	Reynolds number $(\rho Vh)/\eta$
Re_m	Mean Reynolds number $(2\rho U_m h)/\eta$
Re_e	Effective Reynolds number
Re_p	Poiseuille Reynolds number $(2\rho Uh)/\eta$
U_m	Mean fluid velocity (Couette and Poiseuille)
U	Mean fluid velocity (Poiseuille only)
x,z	Bearing coordinates
X,Z	Dimensionless bearing coordinates $x/D, z/L$
β	Concentric design pressure ratio
η	Fluid dynamic viscosity
ρ	Fluid density
τ_c	Couette shear stress
$\bar{\tau}_c$	Dimensionless Couette shear stress $(\tau_c h)/(\eta V)$

3 CONVENTIONAL GENERATOR BEARINGS

At the outset, it was necessary to identify the geometry and operating requirements of a conventional jacking type generator bearing. The geometry identified, fig.1(a), consists of a 120 degree load support arc in which two jacking pockets are located at a position corresponding to a land width ratio of $a/L=0.25$. At the start and finish of the bearing arc there are large low pressure inlet pockets, to minimise hot oil carry over and to cool the journal. The top portion of the bearing is relieved to reduce friction. The axial lands are necessary for containment of the journal in the event of reversed load.

The generator bearing operates under the following conditions:

(1) Start-up
High pressure oil is introduced via the jacking pockets, to separate the bearing and journal surfaces. The journal is run up to full speed, the high pressure jacking being switched off when the bearing is capable of supporting the load by pure hydrodynamic action.

(2) Running
The bearing operates at the required speed, as a purely hydrodynamic bearing. The bearing spends the major part of its life in this mode of operation.

(3) Shut-down
The high pressure jacking is switched on as the journal speed is reduced and the bearing can no longer support the load by pure hydrodynamic action. The journal is brought to rest and the jacking switched off.

For the particular case of a 457mm (18 inch) generator bearing as described above, the following operating requirements were identified:

(1) Ability to support a unidirectional load of 290 kN (65000 lbf).

(2) Ability to constrain the journal in overload conditions.

(3) Low power dissipation.

(4) Peak white metal temperatures not to exceed 100 degrees Centigrade.

(5) Peak pressure not to exceed 13.8 MPa (2000 psi).

(6) High stiffness.

(7) Low start-up torque.

(8) Moderate flow rates, around 1.26 l/s (20 gpm).

(9) Easy installation and replacement.

The dynamic characteristics during starting, stopping and running are as important as the static characteristics. Hybrid bearings provide variable characteristics which may be favourable from the dynamic view point [6,7]. This work is restricted to the static operating features.

4 THEORETICAL ANALYSIS

For the conventional generator bearing under steady conditions the magnitude of the supply pressure was taken as ambient. However, for a slot entry hybrid bearing, the performance is directly related to the magnitude of the supply pressure. For this reason, it was required that the conventional generator bearings and the slot entry bearings be analysed using different dimensionless notations for pressure, load and flow rate.

For the conventional generator bearing, the

usual hydrodynamic notation was adopted:

$$\text{Pressure} \quad P_{hd} = \frac{p}{\eta N}\left(\frac{C_d}{D}\right)^2$$

$$\text{Load} \quad W_{hd} = \frac{w}{\eta NLD}\left(\frac{C_d}{D}\right)^2$$

$$\text{Flow} \quad Q_{hd} = \frac{q}{RNh_oL} \qquad (1)$$

For the hybrid slot entry bearings, the usual hydrostatic/hybrid notation was used:

$$\text{Pressure} \quad P_{hb} = \frac{p}{p_s}$$

$$\text{Load} \quad W_{hb} = \frac{w}{p_sLD}$$

$$\text{Flow} \quad Q = \frac{q}{p_sh_o^3} \qquad (2)$$

Conveniently, both sets of dimensionless groups lead to almost identical forms of the Reynolds equation, equation (3).

$$\frac{\partial}{\partial X}\left(\frac{H^3}{k_x}\frac{\partial P}{\partial X}\right) + \left(\frac{D}{L}\right)^2 \frac{\partial}{\partial Z}\left(\frac{H^3}{k_z}\frac{\partial P}{\partial Z}\right) = 2\pi S_h \frac{\partial H}{\partial X} + S \qquad (3)$$

The terms S_h (hybrid speed parameter) and S (source term) were only required for hybrid slot entry bearing solutions, otherwise being set to one and zero respectively. The source term S for the hybrid bearings was derived by consideration of mass flow continuity for a control volume containing a slot entry port.

The parameters k_x and k_z are functions of the Reynolds number and may be obtained from eddy viscosity relations [8,9], mixing length concepts [10] or bulk flow theory [11]. The analysis presented here is based upon the work of Elrod and Ng [9], which can be applied to the hydrodynamic, hydrostatic or hybrid lubrication regimes.

For Couette dominant flow (pure hydrodynamic lubrication) k_x and k_z are functions of the Couette Reynolds number Re and were approximated by the following functions applied over the corresponding ranges of Reynolds number [12]:

$$k_x = 12 + k_x \, Re^{n_x} \qquad (4)$$

$$k_z = 12 + k_z \, Re^{n_z} \qquad (5)$$

where k_x, k_z, n_x and n_z were given by:

	k_x	n_x	k_z	n_z
Re < 5000	0.0039	1.06	0.0021	1.06
5000 < Re < 10000	0.0250	0.84	0.0089	0.88
10000 < Re < 50000	0.0250	0.84	0.0136	0.84
50000 < Re	0.0388	0.80	0.0213	0.80

For Poiseuille dominant flow the viscosity factors are functions of the Poiseuille Reynolds number R_{ep} and were given by the following:

$$Re_p < 3000 \qquad k = 12 + 0.0075\, Re_p$$

$$Re_p > 3000 \qquad k = \frac{Re_p^{0.681}}{6.8} \qquad (6)$$

In hybrid bearings there is a complex interaction between Couette and Poiseuille flows. Constantinescu [13] noted that this interaction of flows could be approximated by assuming k_x and k_z to be their respective values from equations (4) and (5), providing that k_x,k_z were greater than k evaluated from expression (6). However if this was not the case, k_x=k_z=k.

The transition region between laminar and turbulent flow was taken into account using the mean local Reynolds number concept, developed by Constantinescu *et al* [14]. The flow regime at any point in the fluid film, as follows:

$$Re_m \leqq T_c \qquad \text{for laminar flow}$$

$$T_c \leqq Re_m \leqq 2T_c \qquad \text{for transition flow}$$

$$Re_m \geqq 2T_c \qquad \text{for turbulent flow} \qquad (7)$$

where T_c is the critical Taylor number $(41.1/\sqrt{(C_d/D)})$.

This allowed thee formulation of an effective Reynolds number Re_e, which could be used in expressions (4) and (5).

$$Re_e = 0 \qquad \text{for laminar flow}$$

$$Re_e = \left(\frac{Re_m}{Re_c} - 1\right)\frac{\rho Vh}{\eta} \qquad \text{for transition flow}$$

$$Re_e = \frac{\rho Vh}{\eta} \qquad \text{for turbulent flow} \qquad (8)$$

relations similar to (8) being used to evaluate the Poiseuille Reynolds number Re_p, for the determination of k, equation (6).

The Reynolds equation was solved using finite difference techniques applied to a rectangular grid. Gauss-Seidel iteration with successive over relaxation was used. On convergence of the pressure distribution and viscosity factor distribution to a specified accuracy, the bearing parameters of load, flow rate, attitude angle and friction torque were calculated. Friction torque for the super-laminar case being calculated from a modified Couette shear stress, equation (9).

$$\bar{\tau}_c = \frac{\tau_c h}{\eta V} = 1 + 0.0012\, Re_e^{0.94} \qquad (9)$$

The following boundary conditions were adopted for the analysis:

(1) At the bearing edge the pressure was set to ambient (P = 0).
(2) At the bearing centreline the pressure gradient was set to zero (dP/dX = 0).
(3) The low pressure inlet pockets were set to ambient (P = 0).
(4) The Reynolds boundary condition for cavitation (dP/dX = 0 at breakdown).
(5) A full fluid film was assumed to exist over the top portion of the bearing.

The solution was restricted to the isothermal case, temperature effects being accounted for by assuming an effective viscosity. Inertia effects and recirculation in the supply pockets were not considered.

5 BASIS FOR COMPARISON OF RESULTS

The results for the conventional generator bearings and the slot entry bearings were based on different dimensionless groups. Thus it was necessary to formulate a basis on which to compare the performances of the bearings. As the efficiency of the bearings was an important aspect of the study, the ratio of load/total power was identified as the basis for comparing the performances of the different bearings. The usefulness of a load/total power ratio and a coefficient of merit was illustrated by Rowe and Koshal [15]. In the work presented here a similar load/total power ratio and coefficient of merit were used.

For the conventional generator bearings, the expression for the load is given in equations (1), and the total power is given by equation (10).

$$H_t = T\omega = \frac{\bar{T}\eta AUD}{C_d}\omega \qquad (10)$$

These expressions for load and power can be combined to give the expression for load/power ratio, in equation (11).

$$\left(\frac{W}{H_t}\right) = \frac{1}{2\pi^3 ND}\frac{\bar{W}}{\bar{T}}\left(\frac{D}{C_d}\right) \qquad (11)$$

Similarly using the load equation given in equations (2) and the total power given by equation (12), an expression for load/total power ratio can be found for the hybrid slot entry bearings, equation (13).

$$H_t = H_f + H_p = \frac{\bar{T}\eta AUD}{C_d}\omega + \frac{\bar{Q}p_s^2 h_o^3}{\eta} \qquad (12)$$

$$\left(\frac{W}{H_t}\right) = \frac{8\bar{W}S_h\left(\frac{L}{D}\right)}{ND\left[16\pi^3 S_h^2\left(\frac{L}{D}\right)\bar{T} + \bar{Q}\right]\left(\frac{C_d}{D}\right)} \qquad (13)$$

The coefficient of merit (M_t) is a comparison of the load/total power ratio for any bearing with the load/total power ratio for a reference bearing or reference condition, equation (14).

$$M_t = \frac{\left(W/H_t\right)}{\left(W/H_t\right)^*} \qquad (14)$$

where * refers to the reference bearing. For the purpose of this work, the reference bearing was defined as a conventional generator bearing, of length-diameter ratio L/D = 0.805, and land-width ratio a/L = 0.25.

6 OPTIMISATION OF THE CONVENTIONAL GENERATOR BEARING

The effect of the axial position of the jacking pockets on the load capacity of a conventional generator bearing using laminar and superlaminar analysis methods is shown in fig.3. As can be seen, the load capacities predicted by the two analysis methods are similar and substantial increases in the load capacity can be achieved by moving the pockets out to the edge of the bearing, especially at high eccentricity ratio. In order to assess whether this increase in load capacity is accompanied by an increase or decrease in efficiency, the variation of load/total power with land width ratio is presented in fig.4. This figure confirms that moving the jacking pockets out to the bearing edge increases the efficiency of the bearing on a load/total power basis for both laminar and superlaminar predictions. However the superlaminar predictions show a lower value of load/total power than the laminar predictions. This is due entirely to the increase in predicted friction torque when superlaminar flow is considered.

Further optimisation of the conventional generator bearing on a laminar basis is somewhat limited. Equation (11), for load/total power is independent of viscosity, indicating that variations in viscosity will not directly affect the efficiency on a load/total power basis. As shown in equation (11), varying the clearance directly affects the value of load/total power. Decreasing the clearance will increase the load/total power, however any decreases in clearance have to take into account minimum film thickness requirements.

When superlaminar effects are considered, it is possible to consider variations in Reynolds number, which can arise from either variations in viscosity or variations in clearance. As with the purely laminar case, variations in clearance have to take account of minimum film thickness requirements. From equation (11), it is clear that reducing the clearance will increase the value of load/total power. Reductions in clearance will also reduce the Reynolds number at which the bearing operates. Thus there is a reduction in friction power, and a further increase in the value of load/total power.

The datum bearing for all the optimisation studies was the conventional generator bearing with jacking pockets positioned at a land width ratio of a/L=0.25. When superlaminar effects were considered, the datum bearing was assumed to operate at a Reynolds number of Re=2000. Any value of coefficient of merit greater than M_t=1.0 indicates that the bearing under consideration is more efficient than the datum bearing on a load/total power basis. Fig.5 shows the variation of coefficient of merit based on varying viscosity, for a conventional generator bearing land width ratio a/L=0.1 at an eccentricity ratio of ε=0.8. If the Reynolds number of the bearing is reduced, by increasing the viscosity, then there is an increase in the coefficient of merit, indicating that the bearing is more efficient on a load/total power basis.

7 OPTIMISATION OF THE SLOT ENTRY GENERATOR BEARING

Optimisation of the slot entry generator bearing is a complex process due to the number of inter-related parameters which can be varied. Fig.6 shows the variation of coefficient of merit against land width ratio for various pressure ratios, for laminar and superlaminar analysis methods. For both analysis methods the slot entry generator bearing is more efficient if the land width ratio is a/L=0.1. It may be possible to further increase the efficiency of the bearing by further reductions in land width ratio, however, there is a limit to this imposed by machining requirements.

The effect of pressure ratio β on the efficiency of the bearing is dependent on the eccentricity ratio. At low eccentricity ratio,

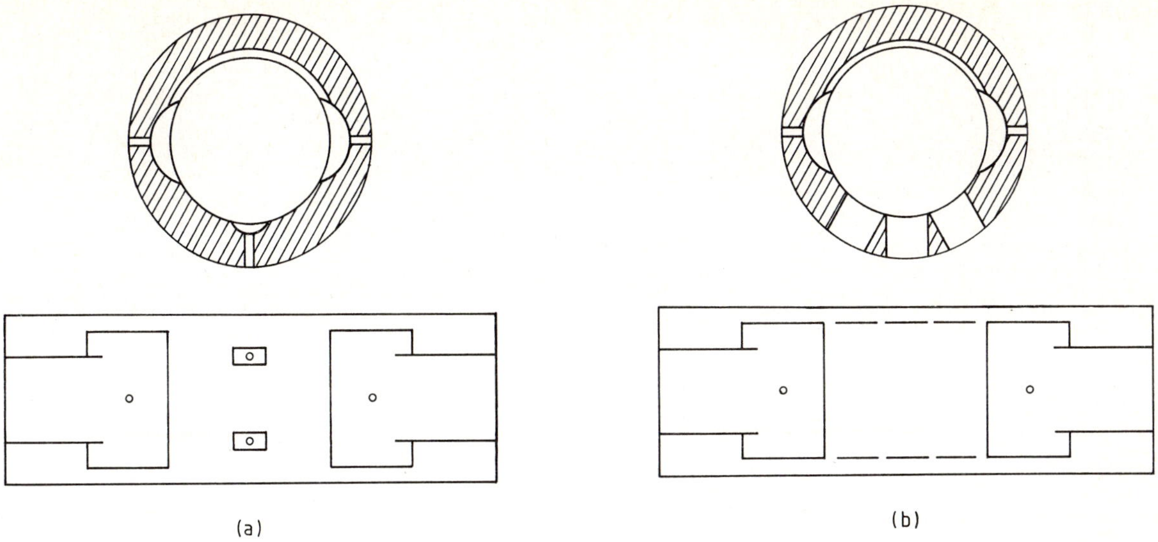

Fig.1 Bearing geometries:
(a) Conventional generator bearing.
(b) Slot entry generator bearing.

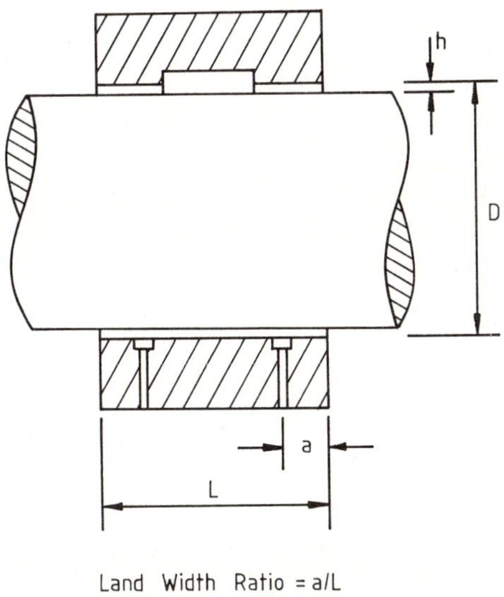

Fig.2 Bearing nomenclature.

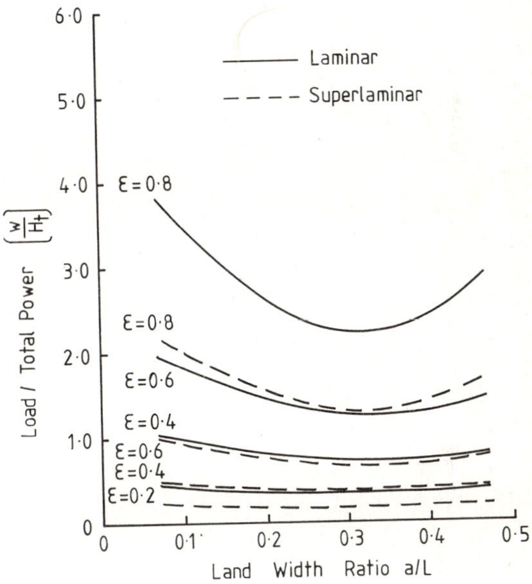

Fig.4 Load/total power against land width ratio for conventional generator bearing.

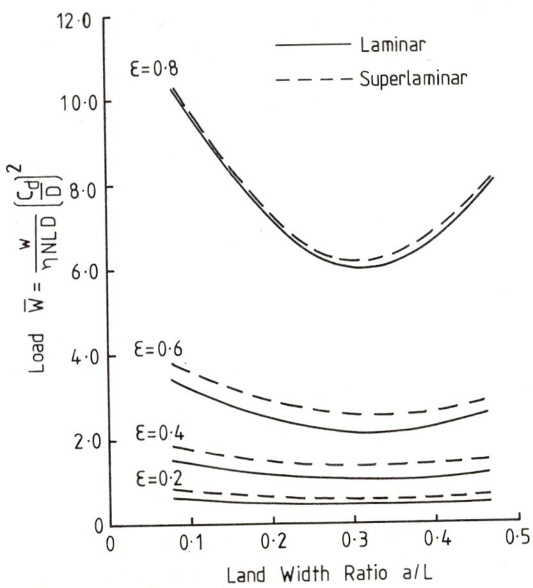

Fig.3 Load against land width ratio for conventional generator bearing.

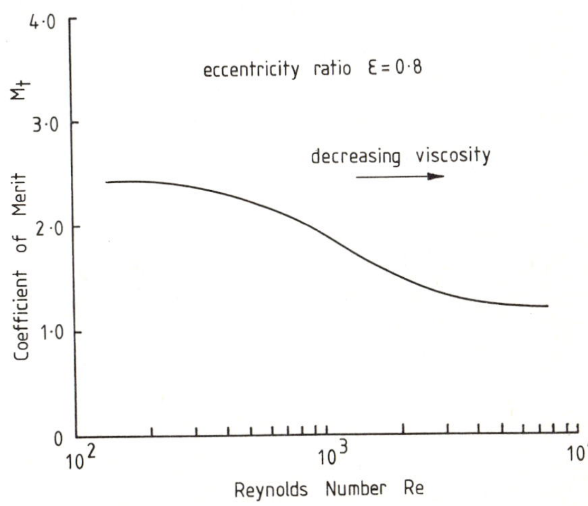

Fig.5
Coefficient of merit against Reynolds number for variations in viscosity, conventional generator bearing, superlaminar.

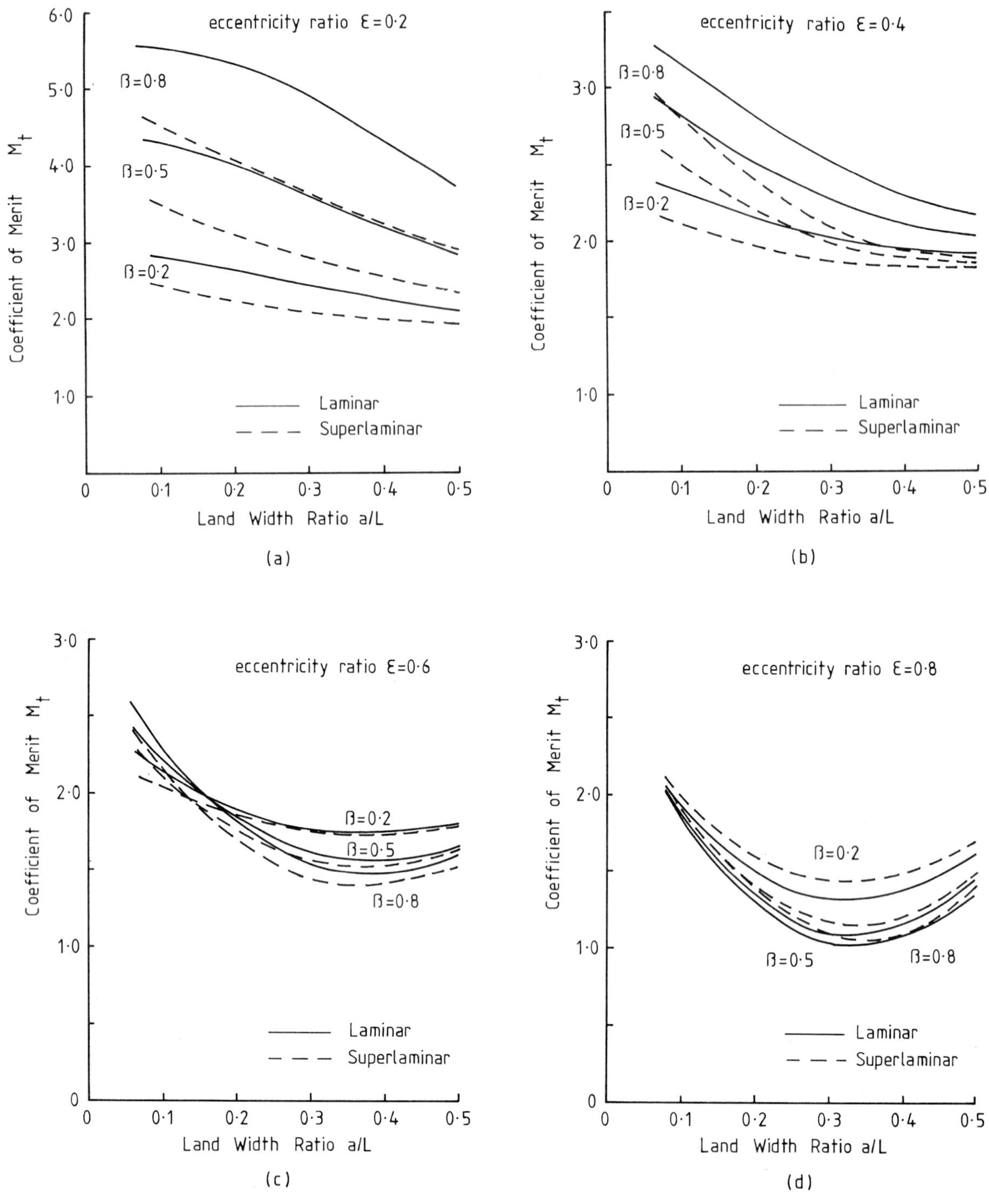

Fig.6 Coefficient of merit against land width ratio for slot entry generator bearing.

hydrostatic effects are apparent and it is advantageous to use a high pressure ratio. At high eccentricity ratio the effect of pressure ratio is greatly reduced, especially at a/L=0.1, due to the dominance of hydrodynamic effects.

Fig.6 indicates that the slot entry generator bearing is more efficient than the conventional generator bearing, and fig.7 shows that the slot entry generator bearing is capable of supporting the required load. As previously stated, the geometries of the bearings are identical, apart from the jacking sources. This results in negligible differences in friction power when operating under the same conditions. This, together with the fact that the total power for the slot entry generator bearing includes a term for the pumping power, indicates that the increase in load/total power for the slot entry generator bearing is due entirely to an increased load support compared to that of the conventional generator bearing.

The variation of coefficient of merit against speed parameter S_h for variations in clearance and variations in viscosity is shown in fig.8, for laminar theory. For variations in clearance, reducing the clearance increases the coefficient of merit. The clearance selected should be the minimum compatible with minimum film thickness requirements. Variations in viscosity indicate that significant increases in coefficient of merit can be achieved at the power ratios less than K=12, and that efficiency is increased by reductions in viscosity.

When superlaminar effects are considered, optimisation based on varying clearance, fig.9, shows the advantage of operating at a low Reynolds number. The Reynolds number is reduced by reducing the value of clearance, however, minimum film thickness requirements will still have to be taken into account.

Variations in viscosity, fig.10, again show the need to operate at low Reynolds number. Reducing the Reynolds number by varying the viscosity means increasing the viscosity, and this is contrary to the findings of the optimisation carried out using laminar flow theory.

8 DISCUSSION

Optimisation studies for the conventional generator bearing and the slot entry generator bearing have shown that the optimim geometry derived from laminar theory is identical to that derived from superlaminar theory. This is significant in that a bearing geometry can be optimised using simple laminar techniques. The results of that optimisation can then be applied to the superlaminar regime. It must be understood that this does not mean that the actual values of load/total power are identical for the two analysis methods. Indeed due to the increases in friction torque predicted by superlaminar theory the values of load/total power are significantly different, as illustrated in fig.4.

The optimisation studies for the conventional generator bearing showed that improvements could be made to the performance of the bearing by moving the jacking pockets away from the centre to a position of land width ratio a/L=0.1. This was in fact adopted and generator bearings with an improved performance have been manufactured.

The slot entry generator bearing was shown to be capable of supporting the required load and to be more efficient than the conventional generator bearing, for a wide range of operating parameters. The optimisation studies showed that reducing the clearance increases the efficiency of the bearing on a load/total power basis. The same conclusion was drawn by Rowe and Koshal [15] for the plain hybrid slot entry bearing. There was however, a discrepancy between laminar and superlaminar results when variations in viscosity were considered. From equation 12, it is seen that a major effect of reducing viscosity is to reduce the friction power. Thus for the laminar case, increases in efficiency are predicted. With superlaminar theory, turbulent shear stresses become a dominant feature and large increases in friction torque are predicted with increasing Reynolds number. For bearings operating in or near the transition and turbulent regimes, reductions in the magnitude of Reynolds number are of prime importance. It must be remembered that variations in clearance and viscosity may adversely affect the temperature rise through the bearing.

In order to indicate the level of power saving available by optimising the conventional generator bearing or using the slot entry generator bearing, a comparison was conducted for a specific case, and the results are presented in table 1. The bearings considered approximate to generator bearings of 457nm (18 inch) diameter, operating at a speed of 3000 rpm. For all cases the viscosity and clearance were kept identical in order to minimise any differences in temperature effects between the bearings. The table indicates that the slot entry generator is capable of meeting the general requirements whilst operating with a lower power loss. The actual values presented are different for laminar and superlaminar theories indicating the need to use superlaminar theory to obtain performance predictions.

Further improvements in the conventional generator bearing performance may be possible, by correct design of the jacking pocket restrictors and constant application of external pressurisation. It is considered that an optimised slot entry configuration will always have some level of superiority over conventional configurations, due to the maximisation of bearing area for hydrodynamic load support. The slot entry bearing has the further advantage of the ability to tailor the static and dynamic performance to suit the operating conditions. This is achieved by adjusting parameters such as supply pressure and pressure ratio. However, there are other commercial considerations which may offset these advantages. These include the need to provide an additional constant supply of pressurised oil and possible increases in complexity of manufacture.

9 CONCLUSIONS

The optimisation studies conducted on the conventional and slot entry generator bearings have indicated that the following conclusions may be drawn:

(1) The optimum geometry is consistent for both the laminar and superlaminar analysis methods, allowing bearing geometry to be optimised using simple

	LAMINAR			SUPERLAMINAR		
	Conventional		Slot Entry	Conventional		Slot Entry
Land Width Ratio a/L	0.25	0.1	0.1	0.25	0.1	0.1
Eccentricity Ratio ε	0.815	0.755	0.665	0.810	0.745	0.665
Flow Rate l/s	1.08	1.07	1.66	1.10	1.06	1.55
Total Power MW	0.115	0.098	0.090	0.198	0.180	0.172
Percentage Power Saving	—	14.8	21.7	—	9.1	13.1

TABLE 1

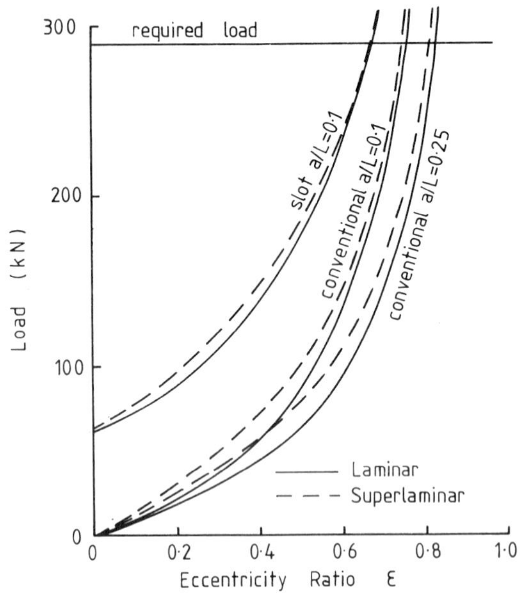

Fig.7 Load against eccentricity ratio for slot entry and conventional generator bearings.

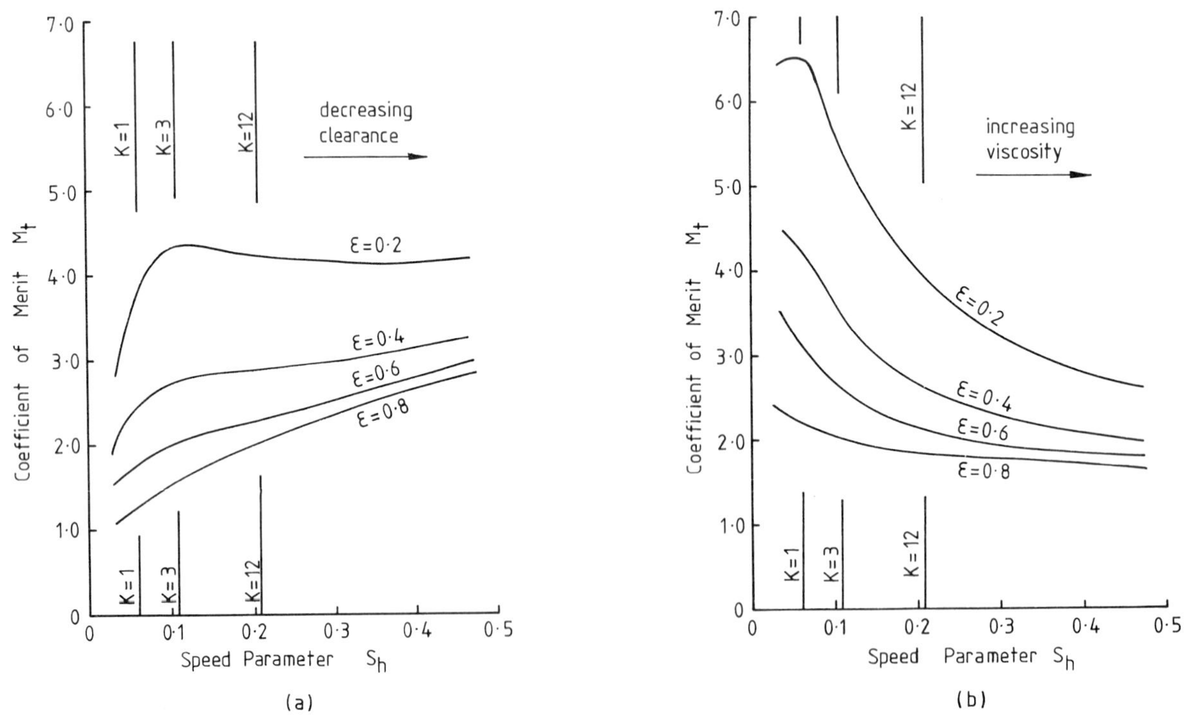

Fig.8 Coefficient of merit against speed parameter for
(a) variations in clearance and
(b) variations in viscosity.
Slot entry generator bearing, laminar.

laminar theory. However superlaminar theory is required for actual performance predictions.

(2) The performance of the conventional generator bearing can be improved by placing the jacking pockets at a land width ratio of a/L=0.1.

(3) Further savings in power consumption are possible by adopting an optimised slot entry configuration. The slot entry bearing also has the benefit of being able to tailor performance.

(4) The bearing should be designed with the minimum clearance, with due consideration for minimum film requirements.

(5) For bearings operating in or near the transition or turbulent regimes, reducing the magnitude of the Reynolds number is of prime importance for efficiency on a load/total power basis.

REFERENCES

(1) SHIRES, G.L. and DEE, C.W. 'Pressurised fluid bearings with inlet slots', Gas Bearing Symposium, 1967, Paper No.7, University of Southampton.

(2) ROWE, W.B., KOSHAL, D. and STOUT, K.J. 'Slot entry bearings for hybrid hydrodynamic and hydrostatic operation', J. Mech. Eng. Sci., 1976, 18, 73-78.

(3) ROWE,W.B. and KOSHAL, D. 'Fluid film bearings operating in a hybrid mode: Part 1 - Theoretical analysis and design', Trans. ASME, 1981, 103 (4), 558-565.

(4) ROWE, W.B., XU, S.X., CHONG, F.S. and WESTON, W. 'Hybrid journal bearings with particular reference to hole entry configurations'. Tribology International, 1982, 339-348.

(5) IVES, D. and ROWE, W.B. 'The effect of multiple supply sources on the performance of heavily loaded pressurised high speed journal bearings'. Int. Conf. on Tribology, Inst. Mech. Engrs., 1987, Paper No. C199/87.

(6) ROWE, W.B. and CHONG, F.S. 'Computation of the dynamic force coefficients for hybrid (hydrostatic/hydrodynamic) journal bearings by the finite disturbance technique and the perturbation technique', Tribology International, 1986, 19 (5), 260-271.

(7) ROWE, W.B., CHONG, F.S. and WESTON, W. 'A linearised stability analysis of rigid and flexible rotors in plain hybrid (hydrostatic/hydrodynamic) journal bearings', Proc. Conf. Vibrations in Rotating Machinery, York, Instn. Mech. Engrs., 1984, Paper C262/84.

(8) NG, C.W. and PAN, C.H.T. 'A linearised turbulent lubrication theory', Trans. ASME, J. Bas. Eng., 1965, 87 (3), 675-688.

(9) ELROD, H.G. and NG, C.W. 'A theory for turbulent films and its application to bearings', Trans. ASME, J. Lub. Tech., 1967, 346-362.

(10) CONSTANTINESCU, V.N. 'On turbulent lubrication', Proc. Instn. Mech. Engrs., 1959, 173 (38), 881-896.

(11) HIRS, G.G. 'A bulk flow theory for turbulence in fluid films', Trans. ASME, J. Lub. Tech., 1973, 95 (2), 137-146.

(12) TAYLOR, C.M. 'Turbulent lubrication theory applied to fluid film bearing design', Tribology Convention, Brighton, Proc. Instn. Mech. Engrs., 1970, 18 (A), 3L.

(13) CONSTANTINESCU, V.N. 'On the influences of inertia forces in hydrostatic turbulent lubrication', Rev. Roum. Sci. Tech-Mec. Appl., 1973, 18 (2), 282-310.

(14) CONSTANTINESCU, V.N., PAN, C.H.T. and HSING, F.C. 'A procedure for the analysis of bearings operating in the transition range between laminar and fully developed turbulent flow', Rev. Roum. Sci. Tech-Mec. Appl., 1971, 16 (5), 945-982.

(15) ROWE, W.B. and KOSHAL, D. 'A new basis for the optimisation of hybrid journal bearings', Wear, 1980, 64, 115-131.

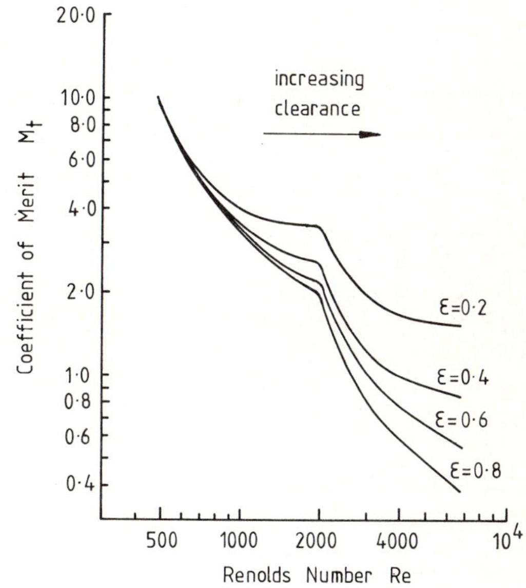

Fig.9 Coefficient of merit against Reynolds number for variations in clearance, slot entry generator bearing, superlaminar.

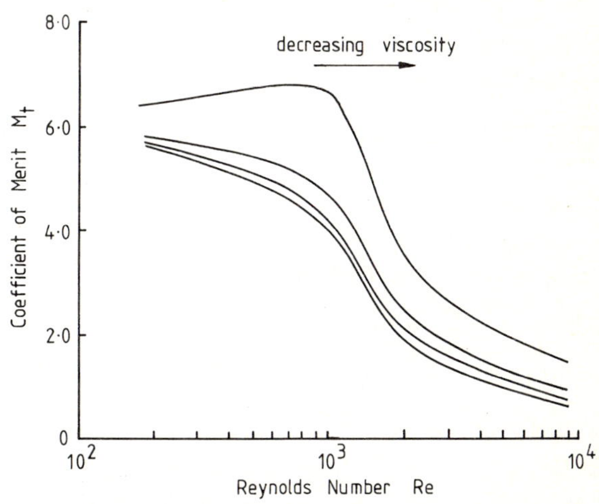

Fig.10 Coefficient of merit against Reynolds number for variations in viscosity, slot entry generator bearing, superlaminar.

Paper XIV(iv)

Behaviour of a high-speed hydrostatic thrust bearing with recess inserts and grooved lands

D. Ashman, E. W. Parker and A. Cowley

The programme of research reported in this paper was undertaken with the aim of improving the high-speed performance of a multi-recessed hydrostatic thrust bearing. The effects of high peripheral speeds are discussed and how recently proposed bearing modifications, in the form of grooved lands and changes in recess geometry, are used to reduce the frictional power consumption, lower the operating temperatures, and reduce unwanted hydrodynamic pressure variations. The results confirmed the merits of the new bearing design under high-speed conditions.

1. INTRODUCTION

In manufacturing industries, the speeds of many machining operations have increased steadily due to the introduction of cutting tools made from coated carbides, ceramics, polycrystalline diamonds and boron nitrides. In recent years, such new materials have made possible cutting speeds up to 20 m/s for steel and 60 m/s for aluminium (1).

During this period complementary developments in Computer Numerical Controlled (referred to as CNC) machining centres have demanded ever increasing speed and power requirements. Normally such machines have large diameter spindles that operate at rotational speeds in excess of 5,000 rpm for light machining of components and, in addition, carry high loads during heavy roughing operations at low speeds. These operations require bearing spindle systems with high stiffness and damping capacity, reliability and high life expectancy. The stage has been reached in CNC machining operations where the machine tool, and not the cutting tip, is the limiting factor. The load-carrying bearing elements in particular have become the focus of attention.

A typical hydrostatic bearing used in machine tool applications consists of deep pockets and thin lands. Lubricating oil is supplied to the bearing recesses by a hydraulic pump, via compensating elements, and thence over the lands. A major contribution to the subject was made by O'Donoghue and Rowe (2) who laid down design procedures and described the quasi-static operating characteristics for a variety of hydrostatic bearings.

The hydrostatic bearing has been exploited by the machine tool industry because of certain design characteristics, namely, high load-carrying capacity virtually independent of speed, no stick-slip characteristics and very low friction at low or zero speed, high stiffness and damping giving high machining accuracy and eliminating machining vibrations, and zero wear of bearing surfaces. However, under high-speed conditions, undesirable hydrodynamic, fluid inertia and temperature effects impair seriously the quasi-static design characteristics. Consequently, much research has concentrated on the thermal and inertial aspects of hydrostatic bearings. Such work takes the form of theoretical analyses of the effect of temperature and inertia (3, 4), together with several experimental investigations (5, 6). It was found that at high speed the elevated working temperatures reduce the absolute viscosity and, as a consequence, reduce torque at the expense of an increase in flow rate or a reduction of film thickness which could lead to failure. Also, fluid inertia effects in an annular recess alter the pressure distribution in the deep pockets and under the lands which could produce pressures below ambient causing aeration and again possible bearing failure.

Shinkle and Hornung (7) published a paper which described a simple theoretical analysis of the frictional torque resisting the motion of a conventional multi-recessed hydrostatic journal bearing, and backed this up with some experimental results. The effect of rotation on the bearing is shown in Fig. 1a. Their main conclusions were that:

(1) The linear velocity distribution in the pockets was modified by a pressure gradient along the length of the recesses producing a corresponding increase in drag.

(2) The pressure gradient under high-speed conditions may be large enough to produce pressures below ambient causing aeration of the fluid in the pockets.

(3) The pocket drag, which can be significant when fluid flow in the deep recesses is laminar, becomes dominant at high Reynolds numbers.

Other work, undertaken by Mohsin and Sharratt (1, 8, 9), involved the re-design of the land and pocket geometry of conventional hydrostatic bearings in order to correct the undesirable thermal, hydrodynamic and fluid inertia effects both under the lands and in the deep recesses. In the design shown in Fig. 1c the leading and trailing edges were connected by external conduits (referred to as an external recess flow system or ERFS) in order to reduce the pressure difference for a given speed, by allowing the flow to circulate around the system. Mohsin and Sharratt also produced flow and frictional drag coefficients for various orthogonal grooved lands (8). They concluded that:

(1) The harmful hydrodynamic effects in the recesses were reduced considerably by an ERFS using external conduits,

(2) A bearing with grooved lands and an ERFS reduced the frictional power consumption,

(3) The grooved lands produced a negligible level of hydrodyamic action thus eliminating cavitation, aeration and whirl.

(4) The dynamic stiffness was slightly reduced due to the larger pocket volume.

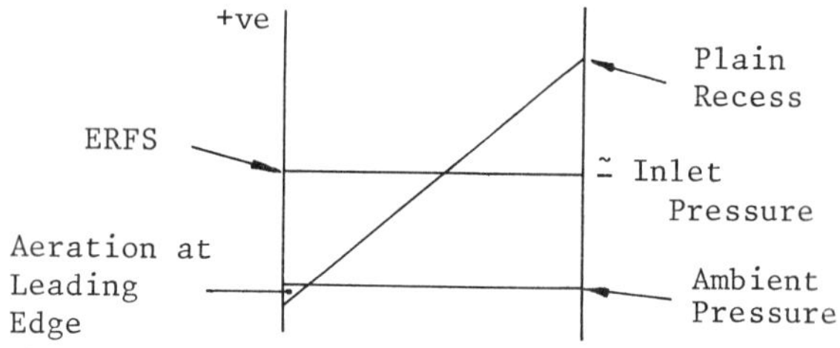

a) Pressure distributions

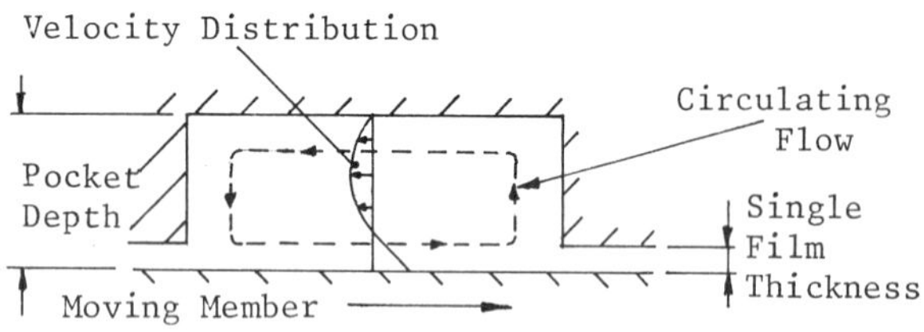

b) Geometry of a plain recess

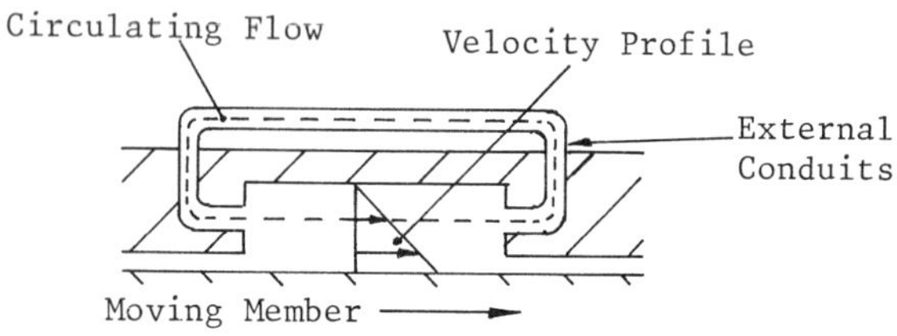

c) Geometry of an ERFS using external conduits

Fig.1 Operation of a plain recess and an ERFS with external conduits under speed conditions.

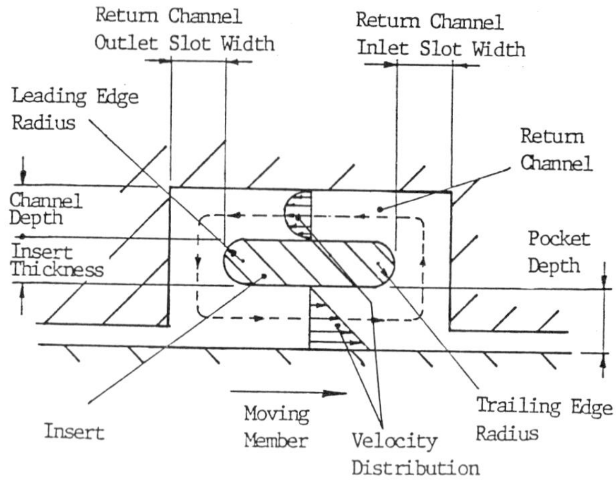

Fig.2 Geometry of an ERFS using recess insert.

An improved design was put forward to achieve an ERFS which involved the use of pocket inserts to obtain the same effects (1). Fig. 2 shows an ERFS containing a recess insert, which again was designed to allow fluid to circulate around the system. Preliminary analytical work was undertaken by Mohsin and Sharratt (1). It was shown that the pocket exhibited similar characteristics to the system using external conduits.

The work presented in this paper describes an experimental investigation of the loading and flow characteristics, and high-speed performance of an annular multi-recessed hydrostatic thrust bearing with various land and pocket geometries in which the former characteristics are compared with data from existing theoretical analyses. The project was carried out in order to assess the advantages of a bearing with an external return flow system using recess inserts and grooved lands over a hydrostatic bearing with plain recesses and flat lands, around the criteria of frictional power consumption, minimum temperature rise and reduction of undesirable hydrodynamic and inertia effects.

2 THE TEST RIG

A detailed description of the bearing rig is given in reference (10). Figs. 3 - 5 illustrate the components of the test bearing. The bearing blocks [items 10 and 15], each had three recesses. Spacer [item 26] and return flow inserts [item 27], which were placed in each pocket to change the recess geometry, were held in position by screws [item 25]. The pocket depth could be varied between 1.5 - 4.0 mm in steps of 1.25 mm. The return flow inlet and outlet slots could be varied between 1.5 - 4.0 mm in steps of 1.25 mm. However, for a given insert the return flow inlet and outlet slots were equal to the depth of the return flow channel. For a bearing configuration which contains an ERFS, the insert thickness, and leading and trailing edge radii, were 3.0 mm, 1.5 mm and 1.5 mm respectively. The oil inlet position to the bearing recess could also be varied i.e. in the pocket or return flow sections.

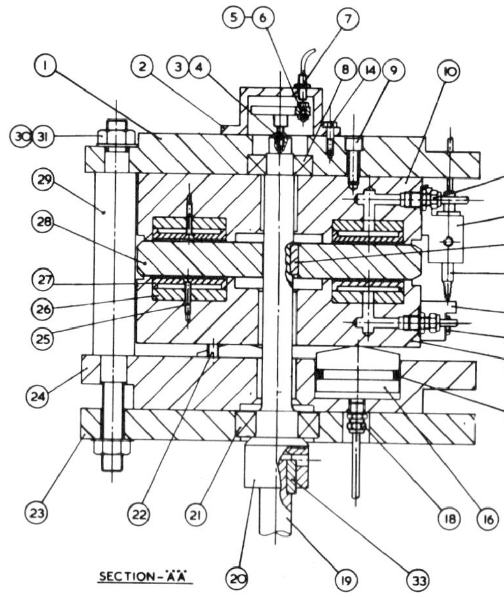

b) Axial cross-section

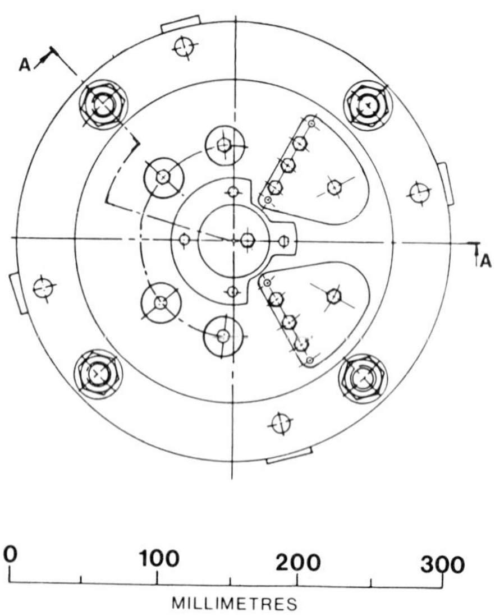

a) Plan view

Fig.3 Test Bearing Assembly

For convenience the numerous bearing pocket and land geometry configurations were given an identification format as follows:

s/v/w/z

where s = depth of recess (mm),
v = depth of return channel, and width of the return flow inlet and outlet slots (mm),
w = land geometry. (F for flat lands or G for grooved lands),
z = oil inlet position. (C for in the return flow channel or R for in the pocket).

The top bearing block was fixed to the top retaining plate [item 1] : the bottom bearing block was constrained against rotation only by rods [item 22]. The moving member [item 28] was keyed on the test bearing shaft [item 20] which was supported by two self-aligning ball bearings [items 8 and 21]. A diagram of the grooved land geometry incorporated in the design is given in Fig. 4.

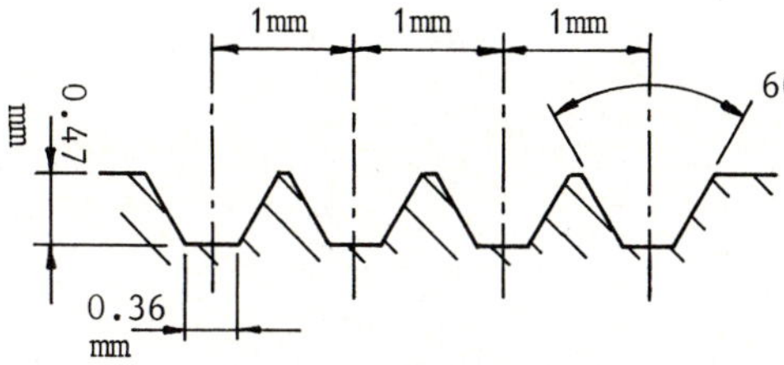

Fig.4 Definition of grooved-land geometry

The test bearings and loading pistons [item 16] were supplied by variable pressure supplies, the test bearing via six variable restrictors. The lubricant used throughout the experimental quasi-static loading and speed tests was Shell Tellus Oil 27.

Provision was made for the measurement of:

(1) The film thickness.

(2) The pressure and temperature distribution in the bearing recess and lands (see Fig. 5) and at the inlet and outlet of the compensating restrictors.

(3) The oil flow rate through the test bearing.

(4) The rotational speed.

(5) The power input to the electric motor.

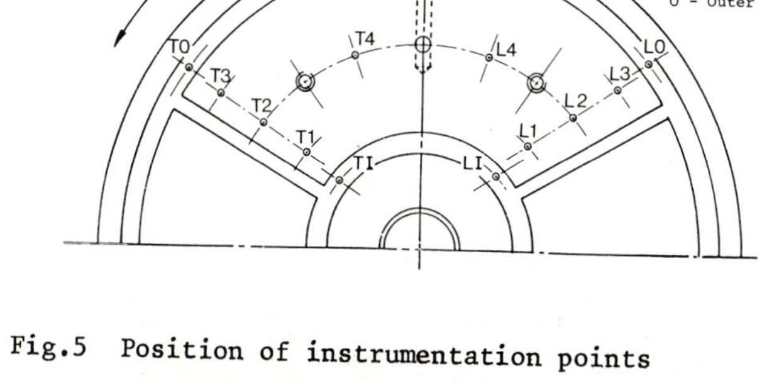

Fig.5 Position of instrumentation points

3 TEST PROGRAMME AND RESULTS

The test programme reported in this paper is made up of four sections, i.e. the study of:

(1) Quasi-static loading and flow characteristics,

(2) Frictional power consumption,

(3) Steady-state temperature distribution under speed conditions,

(4) Variation of pressure distribution under speed conditions.

Before each test, it was necessary to balance the restrictors to obtain the required pocket/supply pressure ratio for a design operating clearance, h_d. This was achieved by applying a load of several tonnes to the bearing via the static loading circuit and load rams, bringing the bearing faces together; the supply pressure, P_s, and pocket pressure, P_r, were then set to their design values, and the variable restrictors adjusted to obtain the required working clearance, h_d, on both bearings. Finally, the oil cooler water flow rate was regulated to maintain a constant bearing oil inlet temperature.

During the tests the correct values of pocket pressure and film thickness were obtained with both flat and grooved lands. The quasi-static characteristic quantities such as flow rate, supply and pocket pressures, double film clearance and oil temperatures were recorded. These characteristics were then compared with theoretical predictions, as calculated using the governing equations developed by O'Donoghue and Rowe (2) and Mohsin and Sharratt (8). Typical results are shown in Figs. 6 and 7.

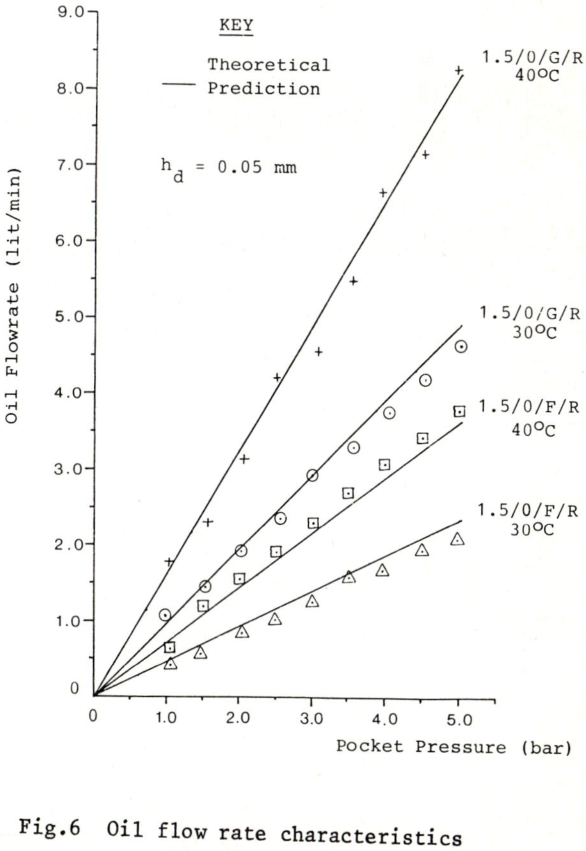

Fig.6 Oil flow rate characteristics

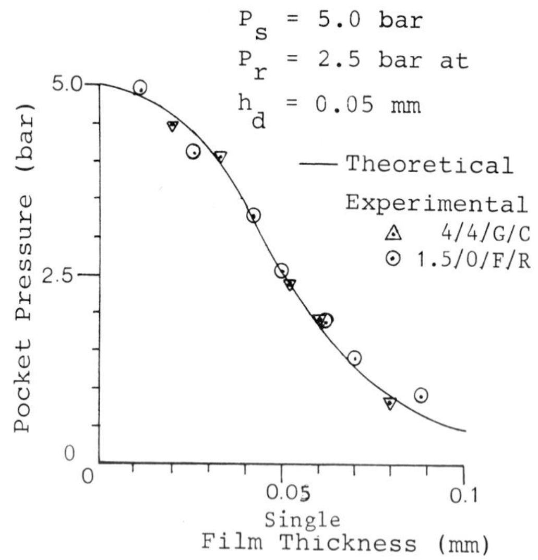

Fig.7 Variation of pocket pressure with film thickness.

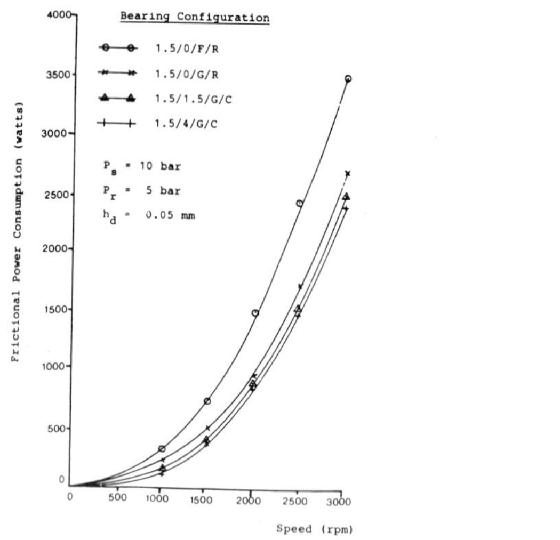

Fig.8 Frictional power consumption versus speed - pocket depth = 1.5 mm.

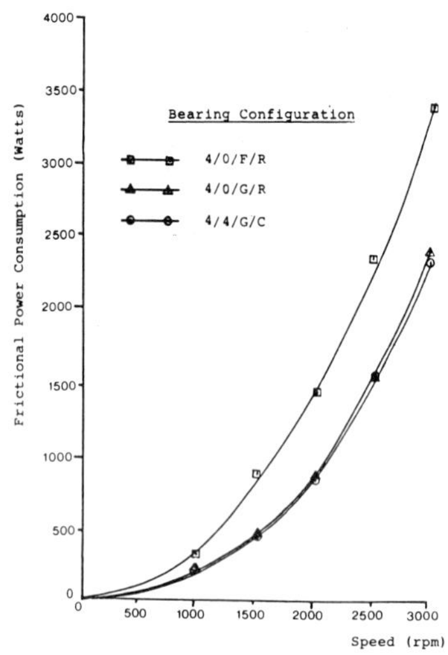

Fig.9 Frictional power consumption versus speed - pocket depth = 4.0 mm.

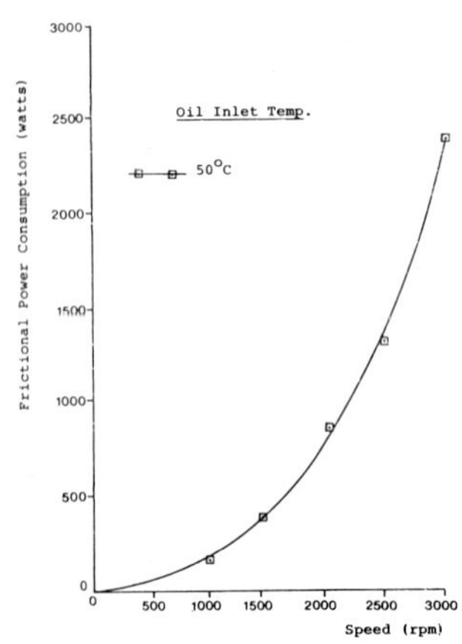

Fig.10 Frictional power consumption of an ERFS with oil feed into pocket section.

The frictional power consumption was obtained under isothermal conditions. This was achieved by setting the rotational speed of the bearing prior to testing, and measuring the power consumption of the bearing immediately after starting, to eliminate temperature variations in the deep recesses and under the lands. Throughout the programme an operating single film clearance of hd = 0.05 mm and bearing temperature of 50°C were maintained. The frictional power consumption of the bearing was obtained for a speed range of 0 - 3000 rpm. This speed range was chosen because at higher speeds, viscous and turbulent drag torque produced large bearing temperature rises, which reduced considerably the accuracy of the experimental results.

The final operating condition was obtained by rotating the bearing at a given speed until the bearing temperature distribution remained constant for 3 minutes. During this time the bearing was supplied with oil at a temperature of 40°c. The initial single film thickness for all the tests was 0.05 mm, and the experimental programme was based on a range of speeds between 0 - 4000 rpm. Tables 1 to 6 show temperature distributions for various bearing configurations.

The pressure distribution was also recorded when the average temperature in the pocket (as measured by the thermocouples embedded in the recess insert) had reached the required test value of 50°C for speeds within the range 0 - 5000 rpm. The initial single film clearance was adjusted to 0.065 mm as a safeguard against possible contact during operation, and to reduce the operating temperatures.

The test conditions and key to the speed test results are as follows:

(1) Test conditions:

P_s = 10 bar Load = 12.8 kN
P_r = 5 bar

(2) Key for Pressure Profile Results

◈ Stationary	△ 3000 rpm
+ 1000 rpm	⊡ 4000 rpm
⊙ 2000 rpm	× 5000 rpm

ROTATIONAL SPEED (rpm)	TEMPERATURE ALONG THE TRAILING EDGE AND AT THE ADJACENT POSITIONS IN THE INNER AND OUTER LANDS (^oC)						TEMPERATURE ALONG THE LEADING EDGE AND AT THE ADJACENT POSITIONS IN THE INNER AND OUTER LANDS (^oC)						AVERAGE DOUBLE FILM THICKNESS (mm)	FRICTIONAL POWER CONSUMPTION (Watts)
	T1	T2	T3	T4	TI	TO	L1	L2	L3	L4	LI	LO		
0	38	38	38	38	40	40	38	38	38	38	40	40	0.1	0
1000	40.5	41	42	38.5	42	43.5	38.5	38.5	38	38	41	43	0.098	280
2000	47	47.5	48	45	48	50	44	43	42	42	46.5	52	0.091	1790
3000	53.5	53.5	53	51	56	63	51	49	49	49	57	66	0.085	4860
4000	69.5	70	70	65	72	80	67	64	62	63	73	81	0.0787	6400

TABLE 1 Temperature distributions - bearing configuration 1.5/0/F/R

0	38	38	38	38	40	40	38	38	38	38	40	40	0.1	0
1000	39.5	40	40	38	40.5	41	39	38	38	38	41	42	0.095	160
2000	42.5	43	43.5	41	43.5	44	40	40	41	42	43	44	0.092	1120
3000	50	50	50	48	51	53	48	47	46	45	51	53	0.081	3080
4000	56	57	56	53	58	59	55	54	53	52	58	60	0.077	6240

TABLE 2 Temperature distributions - bearing configuration 1.5/0/G/R

0	38	38	38	38	39	40	38	38	38	38	39	40	0.1	0
1000	40	40	40	40	40.5	41	40	40	40	40	40.5	41	0.098	170
2000	41	41	41	41	41	42	41	42	41.5	41.5	41	42.5	0.097	1070
3000	45.5	46	46	46	46.5	48.5	45.5	46	45	46	46	48.0	0.086	2770
4000	52	52	53	53.5	53	57	49	52	52	52	53	58	0.076	5640

TABLE 3 Temperature distributions - bearing configuration 1.5/4/G/C

0	36	36	36	36	39	39	36	36	36	36	39	39	0.1	0
1000	36	37	37	36	40	40	36	36	35	35	39	40	0.096	120
2000	39	39	41	37	43	44	39	39	39	37	43	44	0.094	1110
3000	46.5	46.5	46.5	44.5	50	52	46	46	45	46	49	52	0.085	3000
4000	50	53	53	50	56	58	51	51	51	50	56	58	0.079	5710

TABLE 4 Temperature distributions - bearing configuration 4/0/G/R

0	38	38	38	38	39	39	38	38	38	38	39	39	0.1	0
1000	39	39	39	38	39	40	39	39	39	39	40	41	0.097	130
2000	41	41	41	41	41.5	42.5	41.5	41.5	41.5	41.5	41.5	42.5	0.093	980
3000	44	44	44	44	43.5	44	44	44	44	43.5	44.5	46.5	0.086	3190
4000	52	52	52	51	55	58	52	52	52	51	55	58	0.078	5750

TABLE 5 Temperature distributions - bearing configuration 4/4/G/C

BEARING CONFIG.	TEMPERATURE ALONG THE TRAILING EDGE AND AT THE ADJACENT POSITIONS IN THE INNER AND OUTER LANDS (C^o)						TEMPERATURE ALONG THE LEADING EDGE AND AT THE ADJACENT POSITIONS IN THE INNER AND OUTER LANDS (C^o)						AVERAGE DOUBLE FILM THICKNESS (mm)	FRICTIONAL POWER CONSUMPTION (Watts)
	T1	T2	T3	T4	TI	TO	L1	L2	L3	L4	LI	LO		
1.5/4/G/R	45.5	46	46	45	46	50	46	46	46	46	46	51	0.08	3300
2.75/4/G/R	44	44	44	43	44	47	44	44	44	44	45	48	0.086	3100
4/4/G/R	44.5	44.5	44.5	43.5	45	47	45	45	45	45	46	48	0.086	3000

Speed = 3000 rpm

TABLE 6 Temperature distributions - ERFS bearing configurations which had the oil inlet port in the recess section of the pocket

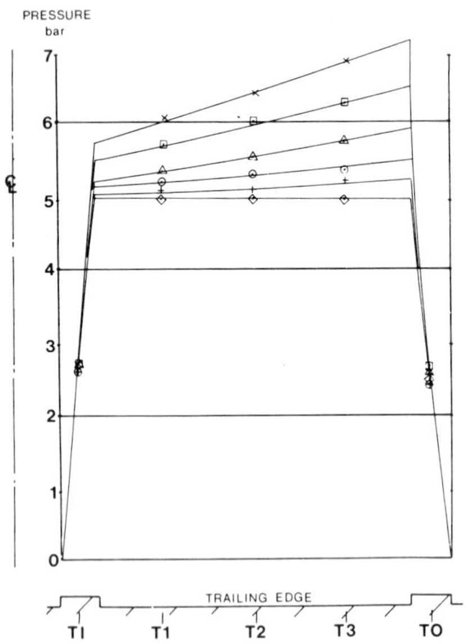

a) Trailing edge

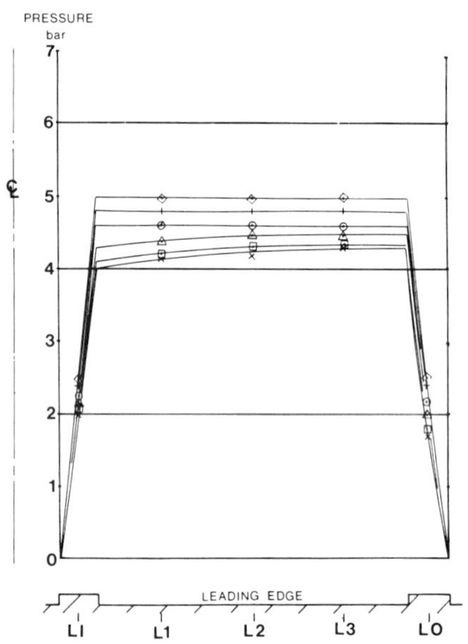

b) Leading edge

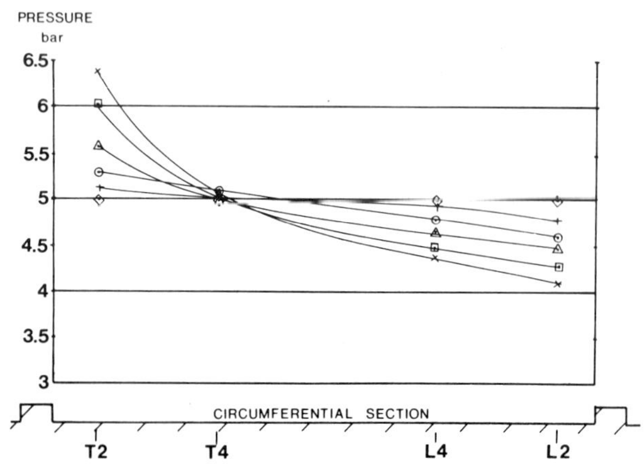

c) Circumferential section

Fig.11 Experimental pressure distributions - bearing configuration 1.5/0/F/R (50°C)

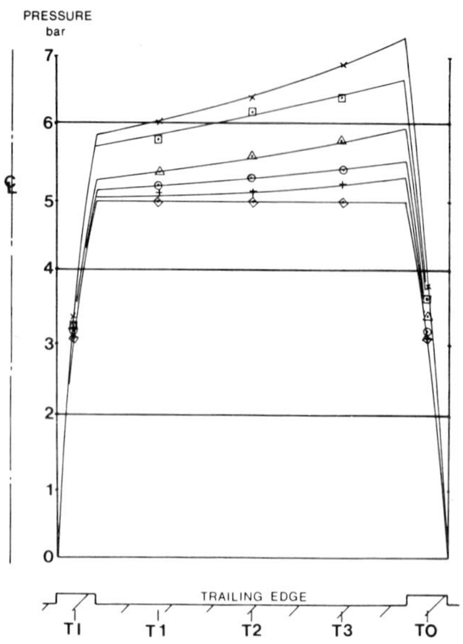

a) Trailing edge

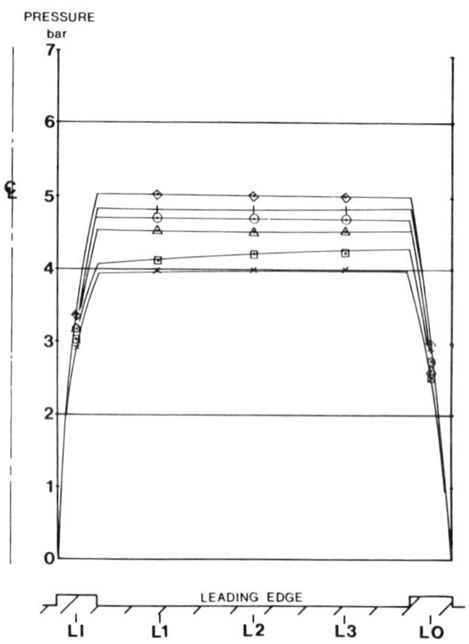

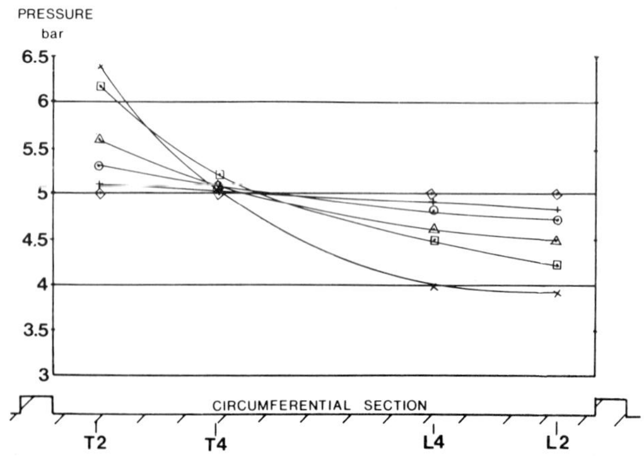

c) Circumferential section

Fig.12 Experimental pressure distributions - bearing configuration 1.5/0/G/R (50°C)

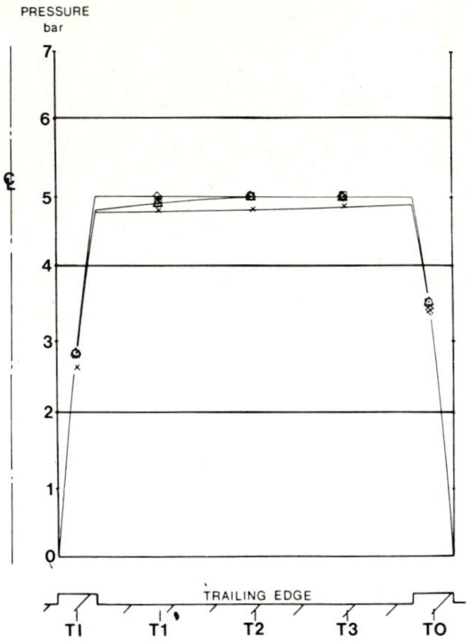

a) Trailing edge

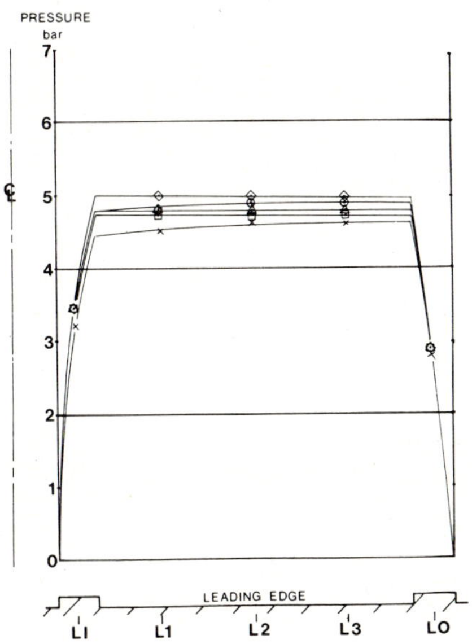

b) Leading edge

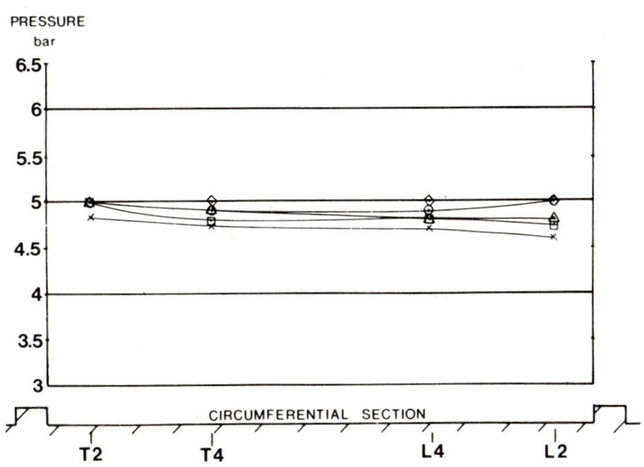

c) Circumferential section

Fig.13 Experimental pressure distributions - bearing configuration 1.5/4/G/C ($50^{o}C$)

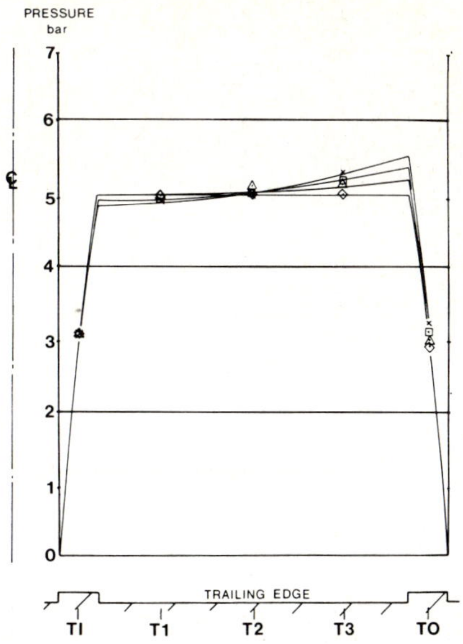

a) Trailing edge

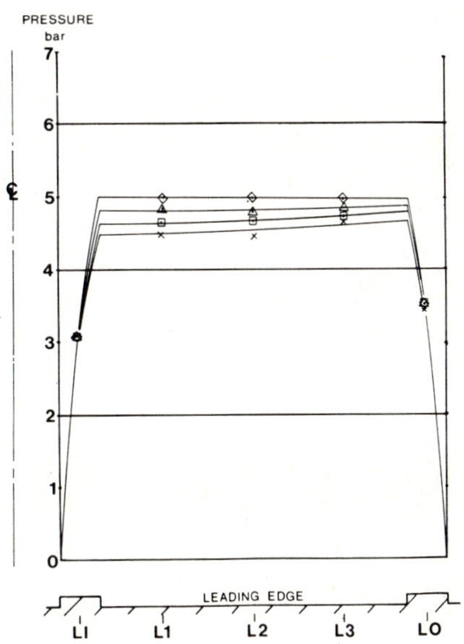

b) Leading edge

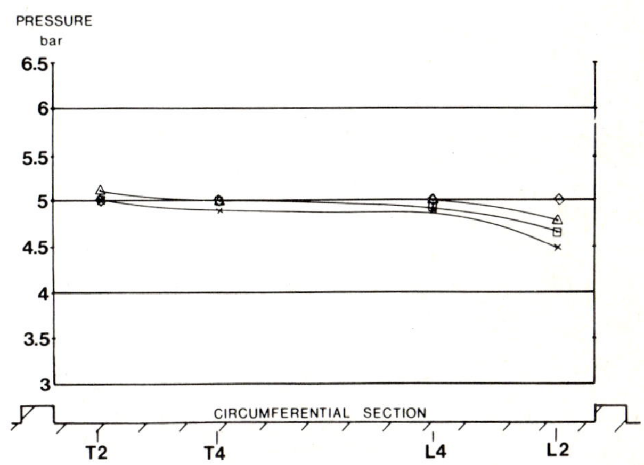

c) Circumferential section

Fig.14 Experimental pressure distributions - bearing configuration 1.5/4/G/R ($50^{o}C$)

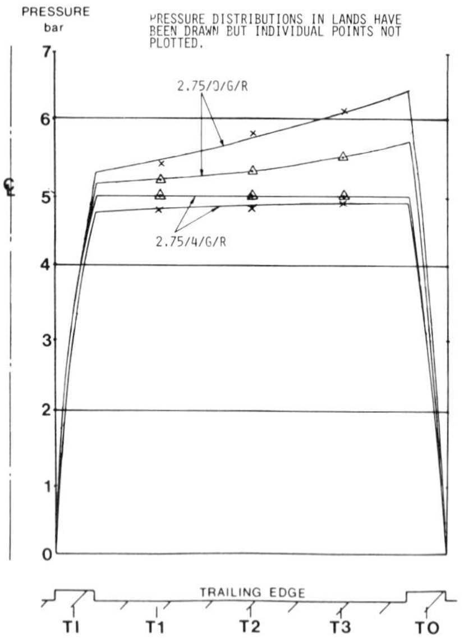

a) Trailing edge

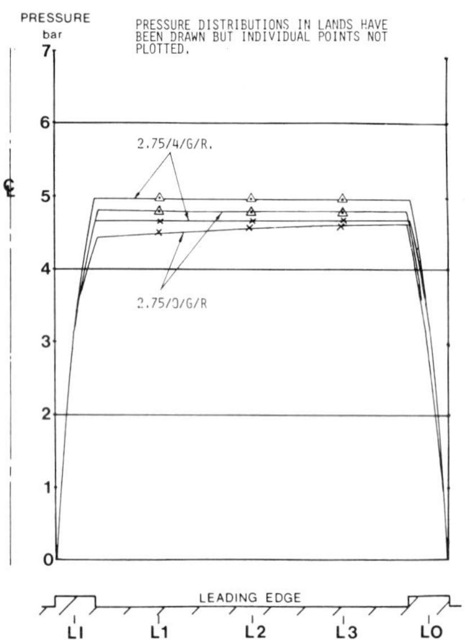

b) Leading edge

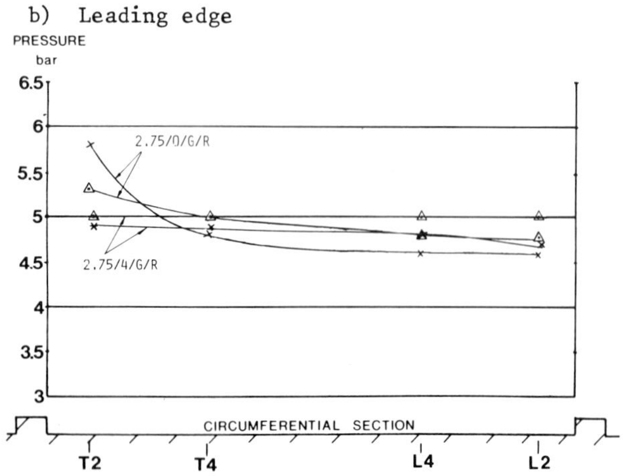

c) Circumferential section

Fig.15 Comparison of pressure distributions for bearing configurations 2.75/0/G/R and 2.75/4/G/R (50°C).

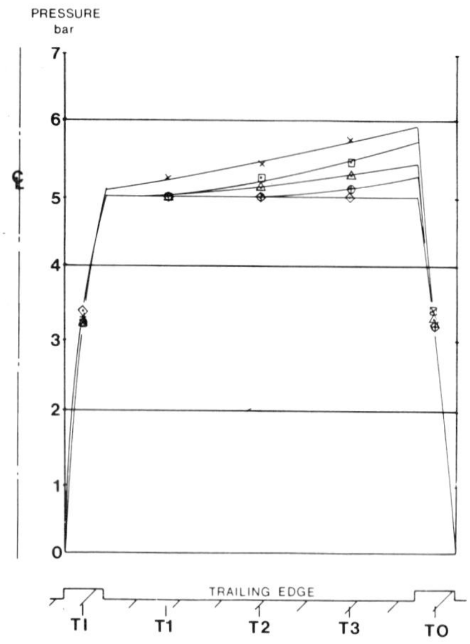

a) Trailing edge

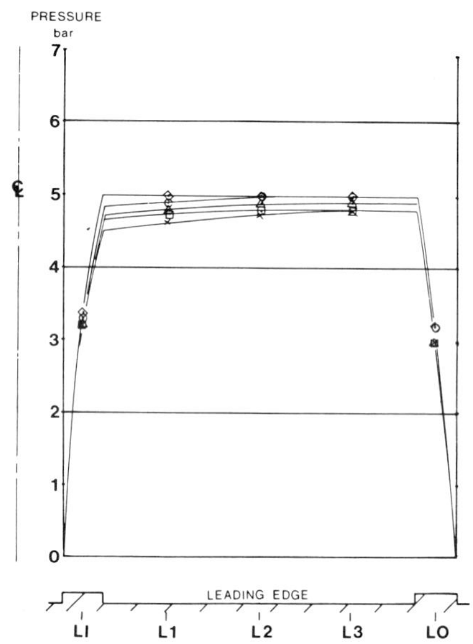

b) Leading edge

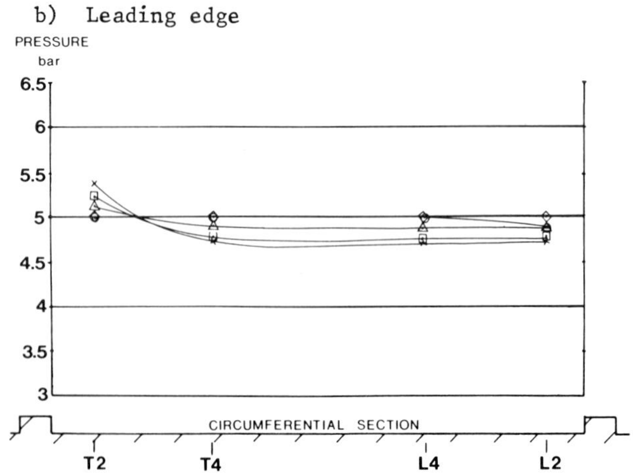

c) Circumferential section

Fig.16 Experimental pressure distributions - bearing configuration 4/0/G/R (50°C)

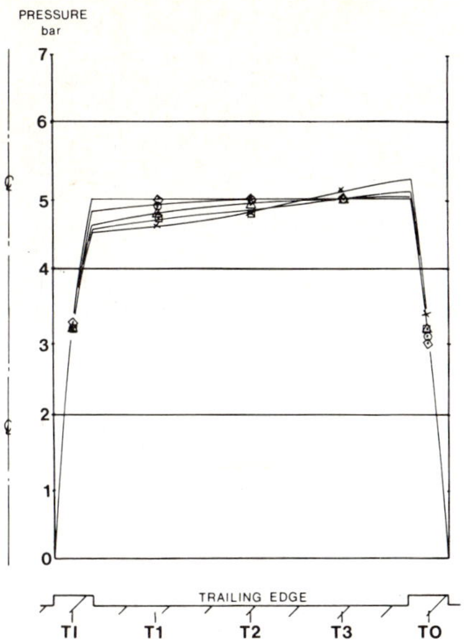

a) Trailing edge

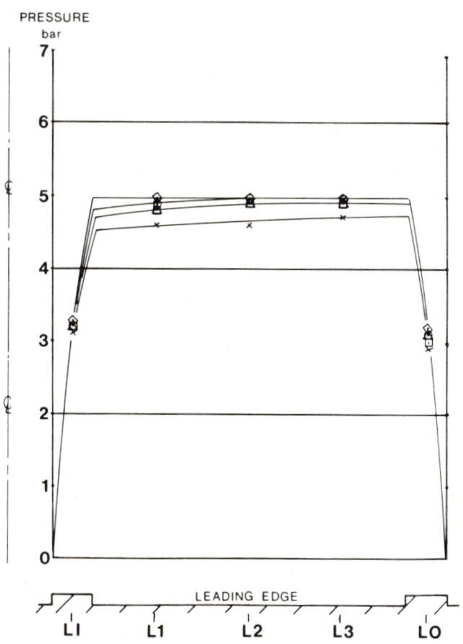

b) Leading edge

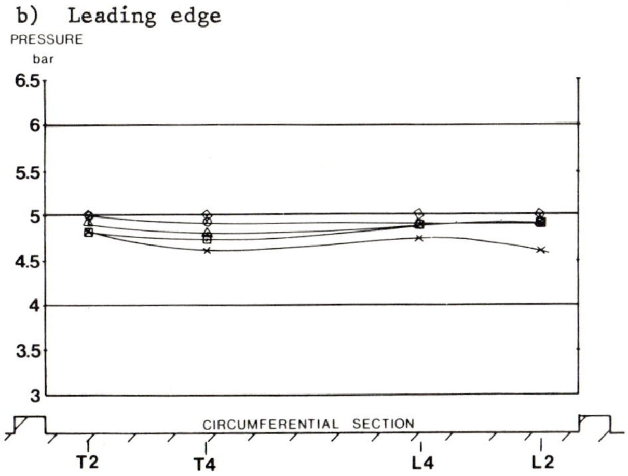

c) Circumferential section

Fig.17 Experimental pressure distributions - bearing configuration 4/4/G/C (50°C)

4 DISCUSSION OF RESULTS

The quasi-static results show close agreement with the theoretical predictions using the governing equations developed by O'Donoghue and Rowe (2) and Mohsin and Sharratt (8). The results include studies of the pocket pressure against operating clearance and flow rate characteristics. The experimental flow rate was sensitive to small variations of the operating clearance and oil inlet temperature. In the case of grooved lands, careful machining of the land profile was required in order to obtain reasonable agreement between the theoretical and experimental flow rate results. Results in Fig. 7 show no variation in static stiffness - as indicated by slopes of the graphs.

It can be seen from Fig. 8 that a reduction of the frictional power consumption was achieved, at a pocket depth of 1.5 mm, when the return flow depth was used. The return flow channel had the effect of decreasing the velocity gradient of the fluid at the moving surface, and therefore reducing the drag torque on the moving member as a result of a corresponding reduction of the induced pressure difference between the leading and trailing edges. The experimental results shown in Figs. 12 and 13 confirmed this hypothesis.

From examination of the experimental results in Figs. 16 and 17, the maximum pressure differences for the bearing configurations 4/0/G/R and 4/4/G/C at an operating temperature of 50°C and speed of 3000 rpm were 0.35 bar and 0.1 bar respectively. Consequently, the two bearing configurations had similar frictional power consumption characteristics.

During the experimental investigation it was observed that the oil inlet position had little effect upon the frictional power consumed by an ERFS. With plain pocket depths of 1.5 mm and 4 mm a reduction of the power consumption, observed in the case of grooved lands, substantiated the results of Mohsin and Sharratt (8, 9). When grooved lands were tested an increase of the oil flow rate was also recorded. This increased the pumping power required to operate the system. However, the frictional power consumption was predominant under high-speed conditions. For example, when the bearing configuration 1.5/0/G/R was tested with P_s = 10 bar, P_r = 5 bar, h_d = 0.05 mm and an oil inlet temperature of 50°C the experimental pumping power consumed by the bearing system (i.e. pumping power consumed = total flow x pressure drop across the system) was approximately 200 watts. Experimental data presented in Fig. 8 shows that under identical operating conditions the frictional power consumed by the bearing arrangement, when the moving member was rotated at a speed of 3000 rpm, was 2700 watts i.e. in excess of 13 times greater than the pumping power.

The operating temperature increased over the entire bearing surface when the rotational speed was increased. When flat lands were tested the temperatures in the inner and outer circumferential lands were considerably higher than that in the recess. When grooved lands were used, the temperature in the pocket and lands showed a marked reduction. This could be attributed to a decrease of the frictional power consumption, together with an increase in flow rate.

When plain recesses were tested, the final operating temperatures along the leading edge were lower than the temperatures along the trailing edge. This would suggest that the velocity distribution of the fluid in the recess was similar to the velocity distribution predicted by Shinkle and Hornung (7). The colder supply fluid flowed to the leading edge first, the lowest temperature in the main being at postion L4, and then to the trailing edge after viscous shearing of the fluid had caused an increase in temperature.

From examination of the results in Tables 2 and 3 for the bearing configurations 1.5/0/G/R and 1.5/4/G/C it can be observed that the use of an ERFS had little effect on the pocket temperature under low speed conditions.

However, during the isothermal frictional power consumption test programme it was found that under high-speed conditions, the ERFS reduced considerably the drag torque on the moving member. It can also be observed that the ERFS reduced the temperature difference between the leading and trailing edges. A reason for this effect could be that the ERFS produced thorough mixing of the oil as it circulated around the system.

Comparison of the results in Tables 4 and 5 with the bearing configurations 4/0/G/R and 4/4/G/C indicates that similar effects to the above were recorded, i.e. the temperature difference between the leading and trailing edges was reduced when the ERFS was incorporated into the design. Analysis of data using the above bearing configurations, produced identical frictional power consumption characteristics, and therefore, the configurations had similar experimental final operating average recess temperatures. Comparison of the experimental data presented in Table 6 with that given in Tables 3 and 5 suggests that the oil inlet position in the case of an ERFS has little effect on the final operating conditions.

The pressure difference between the leading and trailing edges increased radially outward and at higher rotational speeds under all operating conditions. However, the pressure difference decreased when an ERFS was used. For example, with the bearing configuration 1.5/0/G/R, the maximum induced pressure difference for an operating pocket temperature of 50°C and speed of 5000 rpm was 2.9 bar. However, when the 4.0 mm return flow channel was used, the maximum pressure difference was reduced to 0.7 bar under the same operating conditions, as shown in Fig. 14.

A reduction of the pressure difference was also obtained when deeper plain pockets were used. When the bearing configurations 1.5/0/G/R, and 4/0/G/R were tested, operating with a pocket temperature of 50°C and speed of 5000 rpm, the maximum pressure differences were 2.9 bar and 0.9 bar respectively. In the case of a pocket depth of 1.5 mm the pressure along the trailing edge increased radially outward, as a result of the increased velocity and pocket length, whilst the pressure along the leading edge remained constant. Fluid inertia effects, predicted by Coombs and Dowson (5) could be responsible for stabilising the pressure along the leading edge.

The tests suggest that the effects of turbulence and fluid inertia were both present in the bearing pockets. For the bearing configuration 1.5/0/G/R there was no substantial increase in the pocket pressure at position T4, over the range of speeds between 0 - 5000 rpm. In the case of the configuration 4/0/G/R, under the same operating conditions, a decrease in pressure was recorded. Moreover, using the bearing configuration 1.5/4/G/C, which incorporated in the design an ERFS supplied with oil in the return flow channel, the pressure was reduced at positions T1 - T4 as shown in Fig. 13. The decrease of pressure at the trailing edge and in the mid-pocket region was reduced by using the bearing configurations which had the oil inlet port in the recess section of the pocket.

The pressure in the lands was also influenced by high-speed effects. With bearing configurations which incorporated grooved lands in their design, the change of pressure in the inner and outer circumferential lands could be attributed to the pressure variations in the adjacent recess. As an example, in the case of the bearing configuration 1.5/0/G/R, the increase of pressure at the trailing edge increased the pressure at positions TI and TO. Similarly, a reduction in pressure at the leading edge produced a corresponding reduction of pressure at positions LI and LO. The pressure distribution when flat lands were tested shows similar characteristics to those observed in the case of grooved lands. However, in the case of flat lands, the pressure distribution could be further influenced by hydrodynamic effects. The effect of hydrodynamic action was uncertain as it was dependent on land geometry, film thickness and operating temperature.

5 CONCLUSIONS

The quasi-static operating characteristics of a multi-recessed hydrostatic thrust bearing with both flat and grooved lands, can be predicted by the available theory of O'Donoghue and Rowe (2) and Mohsin and Sharratt (8). The static stiffness of the test bearing is not affected by using grooved lands and an ERFS containing channelled inserts.

The high-speed performance of a multi-recessed hydrostatic thrust bearing was found to be improved considerably by the introduction of an ERFS and grooved lands. The modified bearing geometry lowered the frictional power consumption and operating temperatures and, in addition, reduced the hydrodynamic pressure variations in the deep pockets and under the lands. The experimental results indicated that turbulence and fluid inertia effects were present in the deep recesses.

6 ACKNOWLEDGEMENT

The authors gratefully acknowledge The Polytechnic, Wolverhampton for the funding which made the investigation in this paper possible.

References

(1) MOHSIN, M.E. and SHARRATT, A.H. 'The Behaviour of a Total Cross Flow Hydrostatic Journal Bearing', Proc. 23rd Int. M.T.D.R. Conf., Manchester, 14-15 September, Pages 13-24 (1982).

(2) O'DONOGHUE, J.P. and ROWE W.B. 'Hydrostatic Bearing Design', Tribology, V2, N1, February, Pages 25-71 (1969).

(3) TING, L.L. and MAYER, J.E., (Jr.) 'The Effects of Temperature and Inertia on Hydrostatic Thrust Bearing Performance', J. Lubr. Tech., ASME Trans., V93, April, Pages 307-312 (1971).

(4) DOWSON, D. 'Inertia Effects in Hydrostatic Thrust Bearings', ASME Trans., J. Basic Eng., Series D, V83, N2, June, Pages 227-234 (1961).

(5) COOMBS, J.A. and DOWSON, D. 'An Experimental Investigation of the Effects of Lubricant Inertia in a Hydrostatic Thrust Bearing', Inst. Mech. Engrs., London, Proc. Lubr. and Wear-Third Convention, Paper 12, V179, Pt.3J, Pages 96-108 (1964).

(6) KHER, A.K. and COWLEY, A. 'An Experimental Investigation into the Temperature Effects in Hydrostatic Journal Bearings', Tribology, V3, N1, August, Pages 165-167 (1970).

(7) SHINKLE, J.N. and HORNUNG, K.G. 'Frictional Characteristics of Liquid Hydrostatic Journal Bearings', ASME Trans., J. Basic Eng., Series D, V87, N1, March, Pages 163-169 (1965).

(8) MOHSIN, M.E. and SHARRATT, A.H. 'The Behaviour of Hydrostatic Pads with Grooved Lands', Tribology, V14, N1, February, Pages 33-45 (1981).

(9) MOHSIN, M.E. and SHARRATT, A.H. 'The Behaviour of a Total Cross Flow Hydrostatic Thrust Bearing', Proc. 21st Int. M.T.D.R. Conf., Swansea, Pages 449-459 (1980).

(10) ASHMAN, D. 'High-speed Performance of a Hydrostatic Thrust Bearing', Ph.D. Thesis, Wolverhampton Poly. (1987).

Paper XIV(v)

An experimental comparison between the performance of a 'total cross flow' and an equivalent conventional design hydrostatic journal bearing

M. Abdolmaleki, A. Skorin, F. P. Wardle and R. A. E. Wood

This paper presents the results of an experimental investigation into the advantages of a "Total Cross Flow" bearing design over an equivalent conventional hydrostatic bearing.

Practical designs of journal and thrust hydrostatic bearings embodying the features of "Total Cross Flow" (TCF), "External Return Flow System" (ERFS) and "Grooved Lands" are described. The bearings set-up in a journal and journal plus double thrust arrangement are evaluated. For a 75 mm diameter journal hydrostatic TCF bearing steady speeds up to 1 million DN have been reached so far. With the properties of good damping, stiffness and low temperature rise the TCF hydrostatic bearings are considered suitable for wide speed range machine tool spindle applications.

Experimental comparisons between the performance of TCF design and equivalent conventional pocket and flat land hydrostatic journal bearings clearly demonstrate the advantage of the former particularly in terms of much higher running speed capability with no penalty in temperature rise and shear power losses. Furthermore pocket pressure stability with respect to increased shaft speed for TCF journal bearings tested up to 1 million DN is over 4 times better than that for the equivalent conventional design tested up to 0.6 million DN with a similar temperature rise.

1 INTRODUCTION [1]

With advances in cutting tool technology, high speed machining to increase productivity has received a great deal of attention particularly for applications where the machine tool and not the cutting tip is the limiting factor. In manufacturing industries, the speeds of many machining operations have increased steadily due to new cutting tool materials such as coated carbides, ceramics, polycrystalline diamonds and boron nitrides. For steel and aluminium, cutting speeds up to 20 and 60 m/s [2] have been reached in recent years.

The ever increasing demands for heavy cuts at low speed and light cuts at high speed operations have led relevant industries to consider alternative spindle supports to conventional rolling element bearing systems. This is due to the fact that at such critical operations, complicated additional provisions are needed for rolling element bearing spindles and even then the bearing life is inevitably limited. Furthermore continuous developments in computer numerical controlled (CNC) machining also dictate the need for higher speed operations with good overall performance in terms of optimum power, stiffness and damping characteristics together with longer life requirements.

One of the most important components in a machine are the load carrying members, i.e. the bearings. Primarily due to their long life and good stiffness characteristics, hydrostatic bearings are used in machine tool applications. A typical conventional hydrostatic bearing (Figure 1) consists of a number of

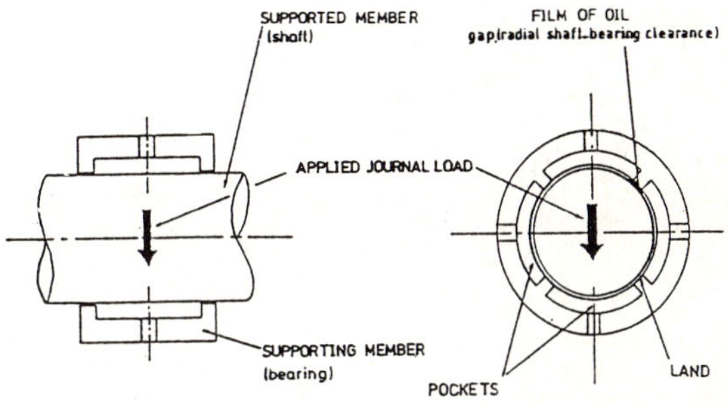

FIG.1 - DIAGRAM OF A TYPICAL 4 CIRCUMFERENTIAL POCKET CONVENTIONAL HYDROSTATIC BEARING

circumferentially recessed plain pockets and narrow lands. A hydraulic pump is required to supply pressurised lubricating oil into the bearing pockets and over the lands via compensating devices (usually termed as restrictors).

The hydrostatic bearing has certain meritorious designed characteristics, namely high load-carrying capacity virtually regardless of speed, no stick-slip and very low friction at low or zero speed, high stiffness and damping giving high machining accuracy and eliminating machining vibrations, and zero wear of bearing surfaces hence virtually unlimited life. At high-speed conditions, however, due to turbulence and cavitation in the pockets and hydrodynamic effects, the oil temperature rises critically which could lead to bearing surface touchdown. The critical rise in oil temperature reduces oil viscosity, consequently followed by a reduction in torque at the expense of an increase in flow rate or a reduction of oil film thickness which eventually leads to failure. Also, fluid inertia effects in an annular recess alter the pressure distribution in the pockets and under the narrow lands which could produce pressures below ambient causing aeration and again possible metal to metal contact failure.

The design of hydrostatically supported spindle systems for machine tools has received more attention in recent years and various optimisation procedures have been presented to generally improve the static stiffness and total power requirements as the main parameters of interest.

Considerable research and development work has been done at the Machine Tool Industry Research Association (MTIRA) presently known as the Advanced Manufacturing Technology Research Institute (AMTRI). Notes for designers and computer programs have been published to enable the design of hydrostatic bearings for specific applications. The procedures described by O'Donoghue and Rowe [3] and Stansfield [4] were aimed to optimise the hydrostatic bearing design with static stiffness and total power consumption as the main parameters of interest. Two books were written by Rowe [5] and Stansfield [4] to formulate and simplify design rules for hydrostatic bearings as a result of their invaluable experience. Cowley and Kher [6] investigated the dynamic performance of a hydrostatic journal bearing system. They also presented a computer-aided design procedure which takes into consideration both the static and dynamic aspects of performance. The work by Mohsin and Sharrat [7-9] involved the re-design of the pocket and land geometry of conventional hydrostatic bearings to improve the undesirable hydrodynamic effects at high speeds.

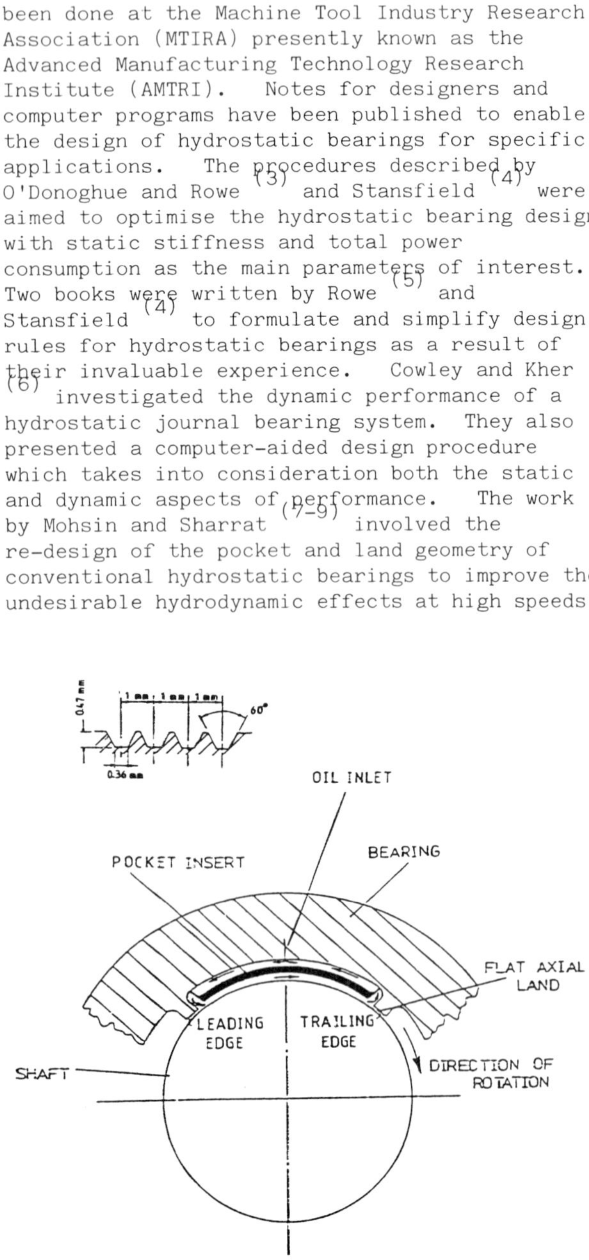

FIG.2-PRINCIPLE OF A TOTAL CROSS FLOW HYDROSTATIC BEARING WITH FEATURES OF ERFS AND ORTHOTROPIC LANDS.(A TYPICAL GROOVED-LAND GEOMETRICAL PROFILE IS SHOWN ABOVE)

The main object of this paper is to report the results of an independent practical investigation to confirm the advantages of a "Total Cross Flow" (TCF) hydrostatic bearing concept (Figure 2) with additional design features of pocket inserts and grooved lands (invented by Dr M E Mohsin, British patent number 2021705A, filed in May 1979). Two hydrostatic journal bearing spindle arrangements, one with TCF journal bearings and the other with conventional hydrostatic journal bearings having an equivalent flat land and plain pockets were designed, made and tested. The TCF bearing set-up comprising a journal and journal plus double thrust arrangement incorporated in a test rig was also evaluated experimentally.

1.1 The Hydrostatic Bearing

A conventional hydrostatic bearing unlike a hydrodynamic bearing has its sliding surfaces continuously and completely separated by a pressurised oil film under static and dynamic conditions within the design range. This type of bearing usually comprises a number of circumferential pockets and flat lands (Figure 1) and has the advantage of complete freedom from wear, hence making for reliability and durability. Apart from the inertia of the spindle the only resistance to sliding motion, even when starting from rest, is that due to the viscosity of the fluid. However, such an arrangement requires pumps and a cooler (an ancillary hydraulic system) to pump pressurised oil into the bearing pockets and adequate discharge and cooling facilities to return the oil back into a supply reservoir.

The main problem with conventional designs of hydrostatic bearings has been the undersirable hydrodynamic effects associated with the oil turbulence in the pockets under critical dynamic conditions. This is because friction losses due to the shearing of oil both in the pressurised pocket and under the lands can become excessive. At high speeds, oil temperature rises within the bearing can eventually result in deterioration of the oil film between the sliding surfaces and failure due to metal to metal contact. Research and design engineers have optimised and developed different designs and the most effective solution to enhance the performance of hydrostatic bearings so far is the total Cross Flow concept.

1.2 The Total Cross Flow Hydrostatic Bearing

Figure 2 illustrates the principle of operation for a TCF design bearing, the subject of European patent number 0,00500,624 B1. Figure 3 (a & b) shows the photographs of a journal bearing and a combined journal plus thrust bearing. The additional design features of pocket inserts and grooved lands are clearly visible on the journal and thrust members of the combined bearing. Figure 4(a) shows the end cover thrust bearing and these three bearing components together with the spindle form the horizontal head for the experimental rig. In addition to pocket inserts, each of the two bearings shown in figure 3 consists of three main parts. The centre parts are identical and

Fig.3(a) - Drive End TCF Journal Bearing

Fig.4(a) - End Cover Thrust Bearing

Fig.3(b) - Non-Drive-End Journal & Thrust Flange

Fig. 4(b) - Another TCF Journal Bearing Design for A Vertical Milling Machine

include the 4 journal pockets and inserts. The other two members embody the oil feed ports into the pockets, the circumferential grooved lands and the oil scavenge holes. The thrust flange member also comprises the axial pockets and inserts together with grooved axial lands. Figure 4(b) shows another TCF bearing design developed for a vertical milling machine at UMIST.

The novel features of external return flow system and grooved lands were added to the conventional hydrostatic bearing design to reduce the adverse effects of turbulence and cavitation in the bearing pockets and undesirable hydrodynamic action beneath the lands. The external return flow system makes it possible for the oil to circulate around the pocket insert from its trailing edge to the leading edge to reduce the pressure gradient of the couette type flow which would otherwise dominate inside a conventional type hydrostatic bearing pocket.

The other TCF bearing feature of grooved land reduces hydrodynamic effects on the shearing oil film beneath the lands. It is proved that a grooved wide land gives much less drag resistance than the equivalent narrow flat land and behaves almost entirely hydrostatic. Therefore employing the pocket inserts and grooved lands in the design of TCF hydrostatic bearings enhances the properties in terms of reduced temperature rise and lower shear power loss which are essential for high and ultra-high speed application.

2 THE 75MM DIAMETER SPINDLE TEST RIG

The general set-up of the rig is shown in figure 5(a). It comprises the main elements :-

(a) Test head, including hydrostatic bearings.
(b) Bank of external restrictors.
(c) Auxiliary equipment including high pressure oil supply, cooler and scavenge pump.
(d) Drive motor with inline pulley drive mechanism.
(e) Torque transducer and instrumentation.

The rig was designed to run at speeds up to 16,200 rpm, test bearings permitting, and provide continuous measurements of test head drive torque; bearing, oil in and oil out temperatures; oil flow rate; oil supply and bearing pad pressures. The test head (Figure 5(b) and Figure 6) contains a journal and journal plus opposed thrust bearing arrangement representative of a practical machine tool spindle bearing arrangement. Spacing of the journal bearings was optimised to give maximum

Fig.5(a) - General Set-up of the Hydrostatic Rig.

Fig.5(b) - Hydrostatic Test Head.

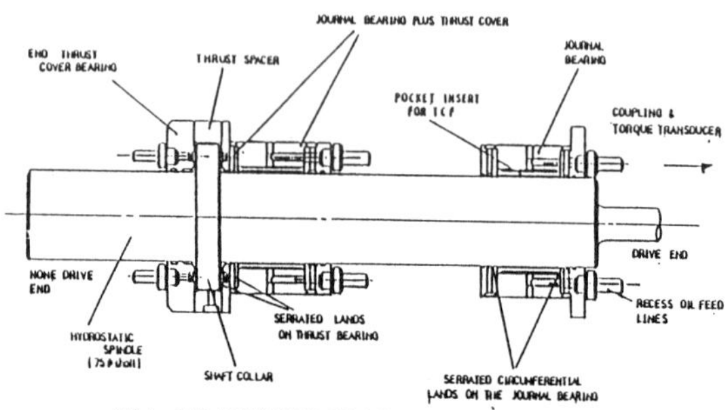

FIG.6 - RHP HYDROSTATIC SPINDLE INCORPORATING A JOURNAL AND JOURNAL PLUS OPPOSED PAD THRUST BEARINGS ARRANGEMENT.

spindle rigidity at an overhang of 138 mm. Diametral internal clearance of the journal bearings was set at 100 μm whilst the thrust bearings were set to have 100 μm total axial clearance. Static stiffness of each journal bearing was around 4×10^{8} N/m at an oil supply pressure of 70 bar (Appendix A). Although the bearing arrangement was optimised from a rigidity standpoint it was the lubrication and thermal aspects of performance at speed which were of primary interest.

Figure 7 shows a simple block diagram of the components for the closed loop hydraulic circuit. Shell Tellus oil R22 is pumped through a swash plate pump from a reservoir (capacity around 200 litres). The maximum oil delivery rate of this pump is 45 litres/minute at a maximum supply pressure of 70 bar.

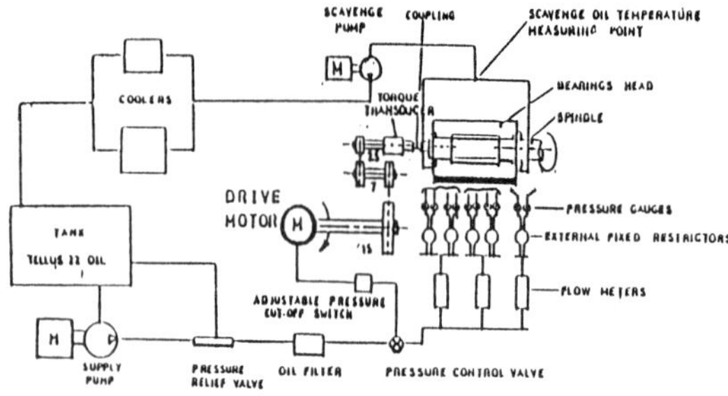

FIG.7 - SCHEMATIC DIAGRAM SHOWING THE CLOSED LOOP HYDRAULIC CIRCUIT BEARINGS HEAD & DRIVE

After a pressure relief valve the supply oil passes through a 25 micrometer absolute filter and a manually adjustable pressure reducing valve (pressure achieved is around 70 bar). The oil then flows to the external fixed annular gap restrictors which in turn deliver oil into the bearings. There are 10 feed lines to the bearings from the 5 back-back restrictor arrangements, eight of the feed lines supply the journal bearings; each line feeding one journal pocket. The other two feed lines deliver oil to the thrust bearings; each line feeding the 4 pockets of one thrust bearing. The oil pressure in each bearing pocket is measured downstream of the restrictors with the flow being measured prior to the oil entering the restrictors.

Figure 8 shows two fixed annular gap restrictors arranged back-back. They are used external to the bearings. The restriction is provided by a circular land and a designed gap set by a shim thickness. The parameters, gap size, inner and outer land radii are shown which were designed to give laminar oil flow and reduce oil supply pressure by about 50% for feeding to the bearings (Appendix B). This design of restrictor exhibits an oil flow - pressure relationship close to that expected from a simple analysis [5] over a wide temperature range of 0 to 70°C. The restrictors are also mechanically reliable.

From the bearing pockets the pressurised oil film between the surfaces of the spindle and the bearing lands is maintained and the oil is then discharged over the grooved lands into the bearing scavenge annulus at atmospheric pressure. The galleries in the housing block direct the discharge volume to the scavenge lines.

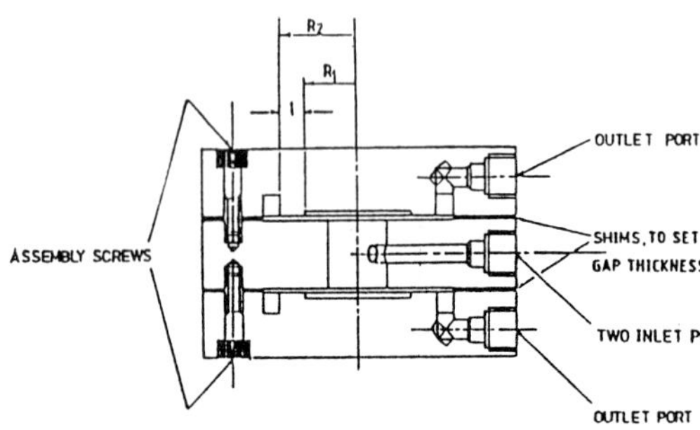

FIG.8 - DIAGRAM OF TWO EXTERNAL FIXED LAMINAR FLOW RESTRICTORS (BACK-BACK ASSEMBLY ARRANGEMENT)

The helical rotor scavenge pump, driven by a 0.55 kW motor at 920 rpm with adequate delivery removal rate, provides the back pressure necessary to discharge the oil out of the bearings through the galleries in the housing block to the oil cooler. The cooled oil is then discharged back into the tank; the oil start and termination point for the closed loop hydraulic circuit.

The spindle drive unit consists of a 37 kW continuously variable speed motor, belt driven pulleys and a flexible torque transmitted pin-drive coupling. The diameter ratios of the pulleys for the primary and secondary drives are 15:7 and 7:2.5 respectively. Therefore with the drive motor running at 2700 rpm, a 6 times speed of 16200 rpm can be achieved for the spindle.

The torque transducer is mounted axially in line with the spindle. It is a Vibrometer TG2 inductive transducer capable of transmitting 40 Nm at speed up to a maximum of 20,000 rpm. Bearing and oil temperatures are measured using J type thermocouples connected to a micron digital meter. Oil supply and bearing pad pressures are measured using analogue static pressure gauges whilst oil flow rate to the journal bearings is determined with a variable area flow meter.

3 TEST PROCEDURE AND MEASUREMENT

Journal and thrust bearing speed - torque measurements could be separated by first taking measurements for the complete bearing arrangement and then removing the thrust bearing housing to enable journal bearing speed - torque measurements to be obtained. Subtracting the latter from the former gives the thrust bearing speed - torque relationship.

The comparison between conventional and TCF bearings was made for journal bearings only. The equivalence between the TCF and conventional designs is based upon the same :-

(i) hydrodynamic impedance across the circumferential lands, i.e. 5.5 mm wide grooved land equivalent to 3.4 mm flat land;
(ii) width for axial lands of 3 mm between the journal pockets;
(iii) pocket depth of 1 mm (between shaft and pad insert for TCF);
(iv) diametral internal clearance, but in practice that of TCF journals was 100 μm and that for equivalent conventional journals was 78 μm due to production tolerance errors.

Experimental procedure consisted of setting spindle speed, allowing at least one hour for steady state conditions and then noting measurements. Speed was incremented to a maximum determined by failure of the oil temperature to stabilise. A maximum stable oil temperature of about 70°C could be achieved.

4 RESULT AND DISCUSSION

Comparison of scavenge oil temperature rise plotted against operating speed between TCF and conventional design journal bearings clearly revealed the much higher speed capability of the TCF bearings (Figure 9). For conventional journals, temperature rose steadily up to a speed of 6000 rpm followed by a 4.3 times sharper rate of increase up to a safe limiting speed of around 8000 rpm whereas for high speed TCF tests it rose at a relatively lower rate up to 8000 rpm followed by only 2.7 times sharper rate of increase to an optimum speed of 13,300 rpm. The significance of the change in gradients in both cases indicated onset of turbulence. The trend also revealed a much greater degree of turbulence for the flow in the conventional than the TCF design bearing pockets. The calculated Reynolds Numbers of the flow at the corresponding temperature and speed (Appendix C), shown in the brackets on Figure 9, confirmed similar trends. The critical values were 1071 and 1396 for conventional and TCF designs respectively. It therefore follows that the TCF added design features effectively delayed the onset of turbulence, moreover, usage of pocket inserts and grooved lands allowed a much higher operating speed due to maintaining low oil temperatures. However, it should be emphasised that the TCF additional features reduce but do not entirely eliminate the undesirable hydrodynamic effects under the bearing lands and turbulent flow and its associated whirl and cavitational effects in the bearing pockets and lands.

FIG.9-SCAVENGE OIL TEMPERATURE VS SPEED COMPARISON BETWEEN TCF & CONVENTIONAL BEARINGS

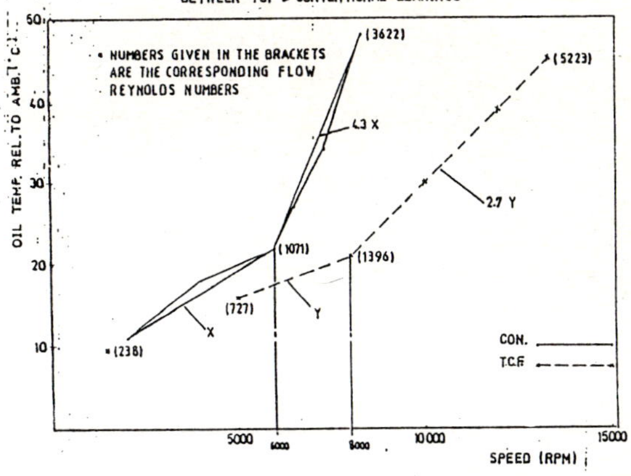

FIG10-PAD PRESSURE VARIATION WITH SPEED FOR CONVENTIONAL JOURNAL BEARING TESTS

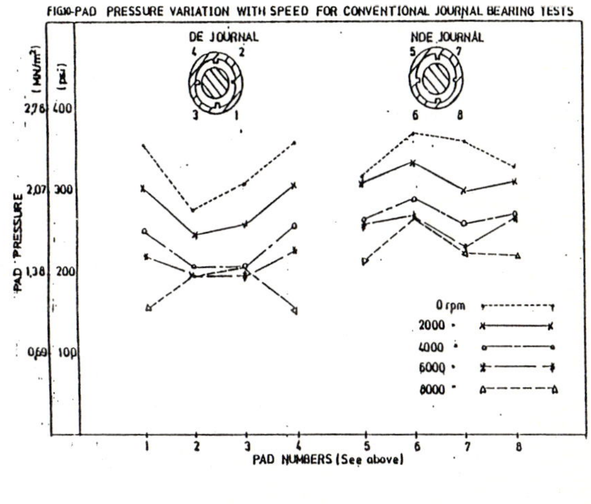

FIG.11-PAD PRESSURE VARIATION WITH SPEED FOR TCF JOURNAL BEARING TESTS

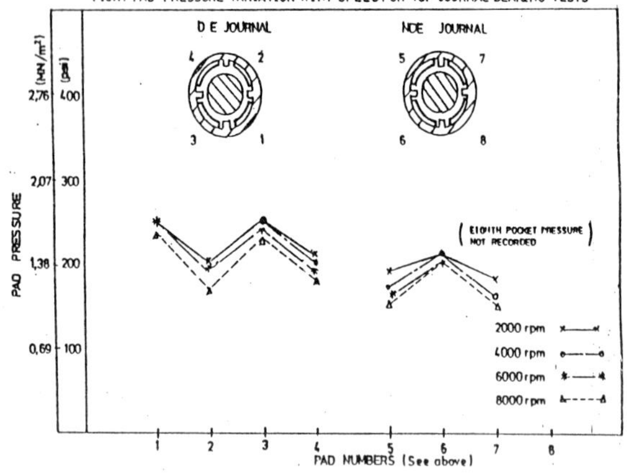

Perhaps the key difference between the performance of the two different journal bearing arrangements is the pocket pressure stability with respect to increasing speed. The results of the pocket (pad) measurements for conventional and TCF design journal bearings for tests up to 8000 rpm at speed intervals of 2000 rpm are summarised in Figures 10 and 11 where pad pressure variation with speed is plotted for the pockets of the drive and non-drive end journal bearings. The conventional bearing tests show significant drop in pocket pressures with increase in speed. The maximum drop was over 200 psi (1.38 Mpa) whereas that for the TCF design was only about 50 psi (0.34 Mpa). Therefore the pad pressure stability is far better for the TCF design. A much higher supply pressure at high speeds was also found to be required to maintain the conventional pocket pressures. This is better demonstrated in Figure 12 where pocket pressure variation with increasing speed (up to 13,300 rpm) is plotted.

FIG.12-HIGH SPEED RANGE PAD PRESSURE VARIATION FOR TCF JOURNAL BEARING TESTS

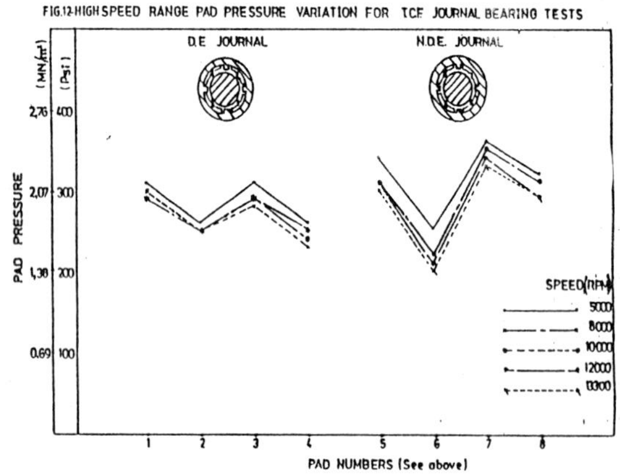

The significant delay in turbulence and much better bearing pocket pressure stability for the TCF design can be explained as follows : it is found that under high speed conditions, the oil temperature rose due to greater degree of viscous shearing which effectively reduced the oil viscosity. This caused the drop in the pocket pressure which was far worse for the plain pockets of the equivalent conventional bearings than TCF design with additional features.

With the reduction in viscosity at high operating speeds for conventional bearings the pocket pressure is reduced which results in an increased flow rate and reduction of film thickness which could ultimately lead to failure. This is most likely for conventional design hydrostatic bearings with lower speed capability and worse pocket pressure stability. The test head torque variation with speed is plotted against measuring time (Figure 13) for similar TCF and conventional bearing pocket pressures. The trend is more or less identical and no claim is made for better TCF torque characteristics.

FIG.13-TORQUE VS.TIME FOR TCF AND CONVENTIONAL JOURNAL BEARINGS AT 4 DIFFERENT SPEEDS FOR COMMON PAD PRESSURES

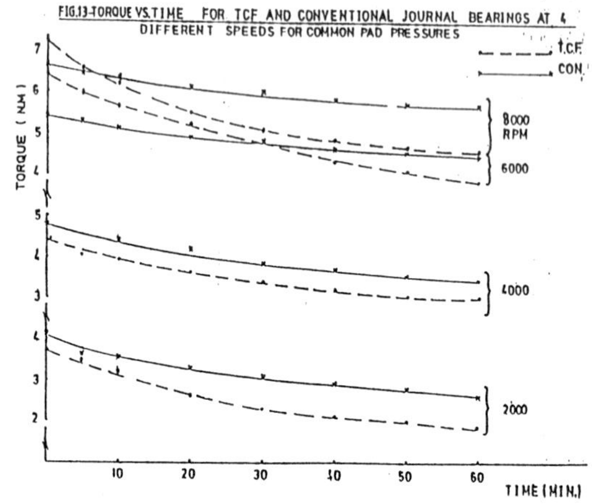

5 CONCLUSION

i. Temperature rise variation with speed is considerably lower for TCF design than equivalent conventional journal bearings. Moreover the onset of turbulence occurs at a relatively higher operating speed with a much reduced degree of turbulence compared to conventional bearings. The approximate critical Reynolds Numbers for transition from laminar to turbulent flow in the pockets of conventional and TCF bearing designs are 1071 and 1396 respectively.

ii. Pocket pressure stability with increasing speed is significantly better for TCF than equivalent conventional journal bearing tests.

iii. The much improved performance characteristics of TCF bearings are due to the pocket inserts and grooved lands which effectively reduce the undesirable hydrodynamic effects.

iv. With the superior performance of TCF bearings in terms of running at higher speed with acceptable temperature rise and stable pocket pressures under reduced power supply - there could be many industrial applications particularly in the machine tool industry where good spindle-bearing characteristics over a wide operating speed range are required.

6 ACKNOWLEDGMENT

Thanks are due to Dr C M Taylor for his expert advice and guidance.

REFERENCES

(1) Ashman, D.
High-speed Performance of a Hydrostatic Thrust Bearing,
Ph.D. Thesis, Wolverhampton Polytechnic, (1987).

(2) Mohsin, M. E. and Sharratt, A.
The behaviour of a Total Cross Flow Hydrostatic Journal Bearing,
Proc. 23rd Int. M.T.D.R. Conf., Manchester, (1982).

(3) O'Donoghue, J. P. and Rowe, W. B.
Hydrostatic Bearing Design,
Tribology, V2, N1, (1961).

(4) Stansfield, F. M.
The Design of Hydrostatic Bearings for Machine Tools and Similar Applications,
Machine Publ. Co. (1970).

(5) Rowe, W. B.
Hydrostatic and Hybrid Bearing Design,
Butterworth (1983).

(6) Cowley, A. and Kher, A. K.
The Dynamic Characteristics of a Hydrostatic Supported Spindle Bearing System, Proc. 10th Int. MTDR Conf.,
UMIST Manchester (1969).

(7) Mohsin, M. E.
A Hydrostatic Bearing for High Speed Applications (The Total Cross Flow Hydrostatic Bearing),
Tribology, V14, N1, (1981).

(8) Mohsin, M. E. and Sharratt, A.
The Behaviour of Hydrostatic Pads with Grooved Lands.
Tribology, V14, N1, (1981).

(9) Mohsin, M. E. and Sharratt, A.
The Behaviour of a Total Cross Flow Hydrostatic Thrust Bearing
Proc. 21st Int. M.T.D.R. Conf. (1980).

(10) Kher, A. K. and Cowley, A.
An Experimental Investigation into the Temperature Effects in Hydrostatic Journal Bearings,
Tribology, V3, N1, (1970).

(11) Baloglu, B.
A Theoretical Investigation of a High Speed Hydrostatic Spindle-Bearing System for a Milling Machine,
Ph.D. Thesis, UMIST (1984).

(12) Hall, G. S.
Investigation of RHP's Laminar Radial Flow Restrictors, Project Report, Department of Mechanical Engineering,
The University of Leeds (1988)

(13) Tsanis, K. and Leutheusser, J.
The Structure of Turbulent Shear-Induced Countercurrent Flow,
J Fluid Mech. V189 (1988).

(14) Taylor, C. M. and Dowson, D.
Turbulent Lubrication Theory Application to Design,
ASME. Paper No. 73-Lub 5-10 (1973)

(15) Wilcock, D. F.
Turbulence in High Speed Journal Bearings,
ASME Trans., V72 (1950).

APPENDIX A

TCF BEARING PERFORMANCE CALCULATIONS

The most important parameters in the study of a hydrostatic bearing are :-

(a) Load carrying capacity
(b) Stiffness
(c) Friction characteristics and power losses
(d) Flow rate
(e) Temperature rise
(f) Dynamic behaviour

Over the years, many studies have been presented by various authors, discussing one or several of the above aspects [3-6]. However, all such work has been pursued without taking the special feature of TCF bearings into account. Based mainly on the work by Kher and Cowley [6 & 10] on conventional hydrostatic bearings and taking into consideration the two major design changes in conventional bearing geometry made for TCF bearings, a theoretical analysis [11] was established to determine the TCF important design parameters in view of ERFS and grooved lands. This analysis includes the derivation of expressions to evaluate the above parameters for both TCF design journal and thrust bearings. Also, using these expressions, two computer programs were developed by B Baloglu [11] to evaluate the performance of each bearing for given dimensions.

It should be noted that the use of grooved lands would alter the damping capacity of a hydrostatic bearing. Therefore two shear and flow factors determined empirically were used for the analysis of a particular grooved land profile. Furthermore, due to the external return flow system provided by using inserts in the pockets of TCF bearings, provision for different shearing characteristics were made to that of a conventional hydrostatic bearing pocket. Moreover, using the finite-element packages available at UMIST, the dynamic and static performance of the TCF bearing spindle system was examined under expected practical cutting conditions.

The computer program for a TCF journal or thrust bearing receives input values such as bearing dimensions, supply pressure and viscosity. For eccentricity (deflection) ratios from 0 to 1 with increments of 0.1 and 0.05, it determines the load carrying capacity and stiffness for a given loading case. It also calculates the viscosity for the spindle speed 0 to 4500 rpm (low speed range) at increments of 500 and for 5000 to 15000 rpm with increments of 1000 rpm (i.e. high speed range). Hence for a computed viscosity, the following parameters are calculated for the zero-eccentricity case:-

Total pumping power
Total shear losses
Temperature rise
Damping constant

and the results listed for each spindle speed in a tabular form. A typical computer print for the 75 mm diameter TCF journal bearing is shown in Figure 14. Many important design parameters of TCF hydrostatic bearings are interrelated, making an exact choice almost impossible. Therefore, it is necessary to compromise between all design parameters concerned such as:
supply pressure, bearing dimensions and the grade of oil used (viscosity).

```
        DATE:12-1-89            JO
 1 RPM    (*1000)        =    0          | 13 DIAMETER            =   75
 2 SUPPLY.PRESS          =   70          | 14 DENSITY             =  875
 3 POCKET.DEPTH          =    1          | 15 VISC.CONST          =    0.8
 4 GAP    (MICRO.M)      =   50          | 16 VISC.T1             =  -20
 5 BRG.AXIAL.LENGTH      =   75          | 17 VISC.V1             = 1000
 6 CIRC.LAND.WIDTH       =    5.5        | 18 VISC.T2             =  120
 7 AXIAL.LAND.WIDTH      =    3          | 19 VISC.V2             =    3.1
 8 SERR.PORTION.WID      =    3          |[20]POCKET.CIRC.LEN     =   55.90485
 9 TEMP.INLET.OIL        =   25          |[21]EQV.CIRC.LAND.W     =    3.862
10 COEF.HEAT.CAP         = 2000          |[22]EQV.BRG.AX.LENG     =   71.724
11 COEF.SERR.FLOW        =    0.454      |
12 COEF.SERR.SHEAR       =    0.311      |
Eccentricity         Load capacity (N)   Stiffness (N/m)
 0.1                  1994.55410556       394509037.741
 0.2                  3945.09037741       381789758.442
 0.3                  5812.45168998       362133767.939
 0.4                  7566.4280568        337605002.455
 0.5                  9188.50171453       310660264.117
 0.6                  10673.030698        283852660.83
 0.7                  12027.0283228       259596721.224
 0.8                  13268.9979102       240042039.982
 0.9                  14427.4487227       227076205.462
 1                    15539.7599648

  RPM       Flow        Pump Power     Shear Power     Temp. rise     Damping
          [ m3/s ]       [ kW ]         [ kW ]           [ K ]        [ Ns/m ]
    0   1.3695E-04    9.58633E-01    0.000000E 00      4.0000      1.557129E 06
```

FIG 14 - A TYPICAL COMPUTER PRINT FOR THE 75 mm DIAMETER T.C.F. HYDROSTATIC JOURNAL BEARING AT 0 RPM.

APPENDIX B

DESIGN PROCEDURE FOR THE LAMINAR RADIAL FLOW RESTRICTOR

The three variables required to define the essential geometrical parameters of a fixed restrictor as shown in Figure 8 are gap (h), land radii $(R_1)(R_2)$ and land width (l) equal to R_2-R_1. Together with the viscosity of the oil flowing through the restrictor, the impedance (resistance to flow) is given by:-

$$I_r = \frac{6\mu \ln\left(\frac{R_2}{R_1}\right)}{h^3} = \frac{P_r}{q_r} \qquad \text{..........(1)}$$

where:

I_r = hydraulic impedance of the restrictor in $N.S.m^{-5}$
μ = dynamic viscosity of the oil in $N.S.m^{-2}$
P_r = supply inlet oil pressure in $N.m^{-2}$
q_r = flow through the restrictor in $m^3.S^{-1}$

The gap h is always a fraction of a millimetre and is expressed in metres. Consequently h^3 is an extremely small value and it is more convenient to express the distance travelled by the oil across the land as a ratio (λ) of land width to gap i.e. $= \frac{l}{h}$.
This is similar to the ratio used for oil flow in pipes; pipe length to pipe diameter $\frac{l}{d}$.
Experimental work on adjustable and fixed restrictors showed that for laminar flow to predominate λ should be between 100 and 200.

Experimental results showed that the relationship between pressure, flow and temperature deviated significantly from theoretical predictions (12) based on the formula (1) for simple lubrication theory as applied to an annular restrictor. The current theory does not adequately describe the practical physical situation in that it ignores, for example, the inertial pressure changes and elastic deformation.

To proceed with the design of a suitable restrictor it was necessary to make factorised adjustments, based on experimental work, to compensate for the deviations between theory and practice.

The design procedure adopted to determine the values of h, R_1 and R_2 to give a practical size to the restrictor is as follows :-

1. Determine the impedance of the bearing by dividing the required average pocket pressure by the oil flow rate through the bearing as :-

$$I_b = \frac{P_b}{q_b} \qquad \text{..................(2)}$$

2. Decide on the pressure ratio to be used, i.e. the ratio of the average bearing pocket pressures to the inlet oil supply pressure to the restrictor. For the case described a ratio of 0.5 was aimed at. However, experimental results showed that due to increasing temperature the pressure ratio was reduced. To compensate for an unexpected loss of pressure on the outlet side of a fixed restrictor of the type used, the presure ratio in the calculations was increased to 0.6.

3. Since there is one restrictor per bearing pad the required impedence of each restrictor was set at a ¼ of the bearing impedance :

$$I_r = 0.25\, I_b, \quad \ldots\ldots\ldots\ldots\ldots\ldots (3)$$

However, experimental results showed that oil flow rate, almost regardless of temperature, was approximately 75% of that calculated from theory. This was compensated for the practical model by reducing the impedance by a factor of 0.75.

4. Use a trial and error method for different values of h, R_1 and R_2 in equation (1) to give a practical size to the unit.

APPENDIX C

THEORY (COUNTERCURRENT OIL FLOW IN THE HYDROSTATIC BEARING POCKETS)

The oil flow beneath the pocket inserts of the TCF or under the pocket of the conventional design hydrostatic bearing consists of pressure and shearing shaft surface velocity induced components. This type of flow is termed as "countercurrent flow" which is a generalised plane couette flow. Under these circumstances a shear-induced drift current is opposed by a pressure- driven return flow, as illustrated in figure 2, such that the resulting mass flux is zero. This type of flow is encountered in both environmental fluid mechanics and tribology (elastohydrodymamic lubrication) such as the case discussed here.

The Reynolds number, expressed in terms of surface velocity and depth of flow, varies from 200 to 20 000 [13]; the critical Reynolds number of laminar to turbulent transition as determined herein are approximately 1071 and 1396 for the oil flow in a conventional and the TCF design pockets respectively. The area of transition from laminar to turbulent flow expressed in Reynolds Number (Re) values is $1000 < Re < 2000$ [14]. For the countercurrent oil flow at hand the appropriate Reynolds number is defined as:

$$Re = \frac{V_s . h}{\nu}$$

where V_s is the velocity of the shearing surface, h is the depth of the shearing thin layer of oil beneath the bearing pocket insert and ν is the kinematic viscosity of the oil determined graphically from the typical viscosity-temperature characteristics of Shell Tellus oils R chart.

Furthermore the corresponding temperature values for different operating speeds are given in figure 9 where the experimental scavenge oil temperature is plotted against the rotational speed of the shaft. The interesting phenomenon of increasing gradient was first reported by Wilcock [15] in 1950. His finding was reported similarly in terms of power loss plotted against rotational speed.

As a typical example, the critical Reynolds numbers for the conventional design and TCF pockets of two hydrostatic journal bearings were determined as follows:-

Case 1 - A conventional design pocket

Nominal pocket depth $= 1 \times 10^{-3}$ m
Critical shaft speed $= 6000$ rpm
Shaft nominal diameter $= 75 \times 10^{-3}$ m

The critical shaft speed is the speed at which laminar to turbulent transition occurs, therefore :-

$$V_s = \frac{\pi D N}{60}$$

where D is the nominal diameter of the shaft, N is the rotational shaft speed.

So $$V_s = \frac{\pi \times (75 \times 10^{-3}) \times 6000}{60}$$

$$V_s = 23.562 \text{ m/s}$$

The measured scavenge oil temperature at 6000 rpm is 41.5°C from figure 9. The kinematic viscosity of Tellus 22 oil from the Shell chart at this temperature is 22×10^{-6} m²/s.

thus : $$Re_{Crit,\,6000} = \frac{23.562 \times 1 \times 10^{-3}}{22 \times 10^{-6}}$$

$$Re_{Crit,6000} = 1071$$

Case 2 - A TCF design pocket

Nominal pocket depth (under the insert) $= 1 \times 10^{-3}$ m
Critical shaft speed $= 8000$ rpm
Shaft nominal diameter $= 75 \times 10^{-3}$ m

Then : $$V_s = \frac{\pi \times (75 \times 10^{-3}) \times 8000}{60}$$

$$V_s = 31.416 \text{ m/s}$$

The measured scavenge oil temperature at 8000 rpm for the TCF bearing test is 40.5°C and the corresponding kinematic viscosity at this temperature for the Tellus 22 oil is 22.5×10^{-6} m²/s, thus :-

$$Re_{Crit,8000} = \frac{31.416 \times 1 \times 10^{-3}}{22.5 \times 10^{-6}}$$

$$Re_{Crit,8000} = 1396$$

The other Reynolds numbers indicated on the temperature/speed graph (figure 9) were also similarly calculated.

The following considerations should be noted:-

1. Scavenge oil temperatures were measured in the scavenge pipe line about one metre away from the bearing pockets.

2. Transition from laminar to turbulent countercurrent flow of Tellus 22 oil was approximately determined from the experimental temperature-speed graph for conventional and TCF design bearing tests.

3. The nominal values for the pocket depth and shaft diameter were taken rather than the exact values.

4. The kinematic viscosity values were extracted from the Shell Tellus oils R chart.

SESSION XV
INFORMATION STORAGE AND RETRIEVAL/MAGNETIC BEARINGS

Chairman: Dr C M Taylor

Paper XV(i)

Review Paper: Tribological design - Information storage and retrieval

B. Bhushan

This paper presents the status of our current understanding of the tribology of head-medium interfaces in magnetic storage devices. To start out, designs and materials used in the construction of heads and media for tape, floppy, and rigid disk drives are presented. Theories of conventional friction, stiction, interface temperatures, wear, liquid/solid lubrication, and compressible hydrodynamic lubrication relevant to magnetic media are presented. Whenever necessary, experimental data are presented.

1.0 INTRODUCTION

Magnetic recording process is accomplished by relative motion between magnetic media against a stationary (audio and data processing) or rotating (video) read/write magnetic head. Under steady operating conditions, a load carrying air film is formed. There is a physical contact between the media and the head during starting and stopping. In modern high-end computer tape and rigid disk drives, the head-to-media separation ranges from about 0.1 to 0.4 μm. In floppy disk drives, head-to-media separation is even less (<0.2 μm) and physical contact generally occurs (Bhushan, 1989).

Need for higher and higher recording densities requires that surfaces be as smooth as possible and flying heights be as low as possible. Smoother surfaces lead to an increase in friction and interface temperatures and closer flying heights lead to occasional rubbing of high asperities and increased wear. Since interface failure even on a microscopic scale can lead to loss of valuable magnetic data recorded on a media, high interface reliability through the life of a magnetic storage becomes extremely important. A fundamental understanding of the tribology of magnetic storage systems is, therefore, crucial for the continued growth of the magnetic storage industry.

This paper presents the status of our understanding of tribology of head-medium interfaces.

2.0 MAGNETIC STORAGE DEVICES

Magnetic storage systems used for information storage and retrieval are: tape, floppy disk, and rigid disk drives. Magnetic media fall into two categories: particulate media, where magnetic particles are dispersed in a polymeric matrix and coated onto the polymeric substrate for flexible media (tape and floppy disks) or onto the rigid substrate (typically aluminum) for rigid disks; thin-film media, where continuous films of magnetic material are deposited onto the substrate by vacuum techniques. Typical materials used for various magnetic media and operating conditions in computer applications are shown in Table 1 (Klaus and Bhushan, 1985).

Magnetic heads used to date are either conventional inductive or thin-film inductive and magnetoresistive (MR) heads. Conventional heads are combination of a body forming the air bearing surface and a magnetic ring core carrying the wound coil with a read-write gap. In the film heads, the core and coils or MR stripes are deposited by thin-film technology. The air-bearing surfaces of most heads are made of hard ceramic materials – Ni-Zn ferrite, Mn-Zn ferrite, calcium titanate, and Al_2O_3-TiC (Klaus and Bhushan, 1985).

Schematic of head-medium interfaces for tape, floppy, and rigid disk drives are shown in Fig. 1.

3.0 ADHESION AND FRICTION

When two surfaces come in contact under load, the contact takes place at the tips of the asperities and the load is supported by the deformation of the contacting asperities. The proximity of the asperities results in adhesive contacts caused by either physical or chemical interactions. When these two surfaces (in contact) move relative to each other, there are two processes, deformation and adhesion, which contribute to "intrinsic" (or conventional) frictional resistance (Bhushan et al., 1984a, 1984b).

In addition, "stiction" can occur due to meniscus/viscous effects, microcapillary evacuation, and changes in surface chemistry (Bhushan et al., 1984a; Bradshaw and Bhushan, 1984; Bradshaw et al., 1986). We define the difference between stiction and conventional static and kinetic friction being that stiction requires a measurable normal force (several grams) to pull the two surfaces apart in the static conditions.

3.1 Conventional Friction

From Tabor's classical theory of adhesion, frictional force due to adhesion (F_A) is defined as follows (Bowden and Tabor, 1950):

$$\text{for dry contact, } F_A = A_r\tau_a \tag{1a}$$

$$\text{for lubricated contact, } F_A = A_r(\alpha\tau_a + (1-\alpha)\tau_\ell) \tag{1b}$$

and

$$\tau_\ell = \eta_\ell U/h \tag{1c}$$

where A_r is the real area of contact, α is the fraction of

Table 1

Typical Operating Conditions and Materials
Used in Different Magnetic Media for Computer Use
(Klaus and Bhushan, 1985)

Magnetic Media	Normal Pressure, kPa	Sliding Speed, m/s	Flying Height, μm	Substrate	Coating Binder System/ Thickness	Lubricant	Lubrication Application Method/Quantity
Tape	7-28*	2-4	0.1-0.3	PET 23 μm thick	Polyester-polyurethane 2-5 μm	Fatty-acid ester	Internal, 0.2 to 3% by weight
Floppy	14-70 (10-20 g)	1-10 (600-1800 rpm)	Partial Contact (<0.2)	PET 75 μm thick	Polyester-polyurethane and epoxy** 2-5 μm	Fatty-acid ester	Internal 0.2 to 3% by weight
Rigid	7-14 (9.5-15 g)	10-60 (3600 rpm)	0.15-0.4	Al-Mg 1.3 to 1.9 mm thick	Phenolic & epoxy** 1-2 μm	Perfluoroalkyl polyether	Topical 10-30 nm thick
				Al-Mg 1.3-1.9 mm thick with 10-20 μm electroless Ni-P	Thin-film (metal)*** 25-100 nm	Amorphous Carbon, SiO_2 and perfluoro-alkyl polyether	Sputtered/spin coated 20-40 nm Topical 0.5-4 nm thick
				Al-Mg 1.3 to 1.9 mm thick with 2-20 μm anodized layer	Thin-film (oxide)**** 100-150 nm	"	"

PET – Poly(ethylene terephthalate)

Magnetic particles used – γ-Fe_2O_3, Co-γ-Fe_2O_3, CrO_2, BaO.6Fe_2O_3, and Fe

*Interlayer pressure on a tape surface near the hub of a wound reel (end-of-tape) can be as high as 1.38 MPa

**Load-bearing Al_2O_3 particles are added to increase the wear resistance of the media.

***Magnetic materials – sputtered CoCr, Co-Pt-Ni, and CoCr-X alloys; electroless Co-P, etc.

****Magnetic materials – sputtered γ-Fe_2O_3

unlubricated area, τ_a and τ_ℓ are the shear strengths of the dry contact and of the lubricant film, respectively, η_ℓ is the absolute viscosity of the lubricant, U is the relative sliding velocity, and h is the lubricant film thickness.

The contacts can be either elastic or plastic which primarily depend on the surface topography and the mechanical properties of the mating surfaces. The expression for real area of contact for elastic (e) and plastic (p) contacts are (Greenwood and Williamson, 1966; Bhushan, 1984)

$$A_{re}/A_a p_a \sim 3.2/E_c(\sigma_p/R_p)^{1/2} \quad \text{for } \psi_p < 1.8 \text{ or } \psi < 0.6, \text{ elastic contact} \tag{2a}$$

$$A_{rp}/p_a A_a = 1/H \quad \text{for } \psi_p > 2.6 \text{ or } \psi > 1, \text{ plastic contact} \tag{2b}$$

and

$$\psi_p = (E_c/Y)(\sigma_p/R_p)^{1/2}, \text{ for polymers} \tag{2c}$$

$$\psi = (E_c/H)(\psi_p/R_p)^{1/2}, \text{ for metals/ceramics} \tag{2d}$$

where A_a is the apparent area of contact; p_a is the apparent pressure; E_c is the composite modulus of elasticity, H and Y are the hardness and yield strength of the softer material, and σ_p and R_p are the composite standard deviation and radius of curvature of the surface roughness.

Bhushan (1984) and Bhushan and Doerner (1988) have measured the mechanical properties and surface roughnesses of various particulate tapes and particulate and thin-film (metal and oxide) disks. Mechanical properties of magnetic coatings of tapes were measured by dynamic mechanical analysis (DMA) system and of rigid disks were measured by a nanoindentation apparatus. Surface roughness parameters of tapes and disks were measured by a noncontact optical profiler (Bhushan et al., 1988). Measured values were used in the contact model and they found that most contacts in typical head-medium interfaces are elastic. Experimental measurements of the real area of contact of magnetic tapes by optical-interference technique verified that most contacts are elastic (Fig. 2). Therefore, the real area of contact and friction is governed by the E_c and σ_p/R_p of the magnetic media surface. Figure 3(a) shows an example that the friction of various magnetic tapes depend significantly upon the complex modulus. Stable frictional behavior was exhibited only by those tapes which displayed a complex modulus of greater than 1.2 to 1.5 GPa. Figures 3b and 4 show the examples that the friction also strongly depends on the surface roughness. Typical contact diameters for tapes and rigid disks were found to be about 6 μm and 2 μm, respectively.

In the case of magnetic tapes, creep compliance and hydrolytic degradation characteristics of the binder also need to be optimized for sustained low friction after storage at high pressure (e.g., near end-of-tape on a reel) and high temperature/humidity (Bhushan, 1989).

3.2 Stiction

Stiction can arise from meniscus/viscous effects, microcapillary evacuation, or changes in surface chemistry (Bhushan et al., 1984a). In this section, we will only concentrate on the meniscus/viscous effects.

Generally, any liquid that wets or has a small contact angle on surfaces will condense from vapor in the form of an annular-shaped capillary condensate in the contact zone. The pressure of the liquid films of the capillary condensates or preexisting film of lubricant can significantly increase the adhesion between solid bodies. Liquid-mediated adhesive forces can be divided into two components: meniscus force (F_M) due to surface tension and a rate-dependent viscous force (F_V). The total tangential force F required to separate the surfaces by sliding is equal to an intrinsic force (F_A) and stiction force F_S (combination of friction force due to meniscus effect and the peak viscous force)

$$F = F_A + F_S = f_s(p_a A_a + F_M) + F_V \tag{3}$$

where f_s is true static coefficient of friction.

Figure 5 shows a model of contact region between smooth surfaces with different level of "fills" of the interface and it depends on the mean interplanar separation and the liquid levels (Matthewson and Mamin, 1988). Two are the extreme regimes in which either a small quantity of liquid bridges the surfaces around the tip of a contacting asperity (the "toe-dipping" regime) or the liquid bridges the entire surface (the "flooded" regime); and in the third regime, the liquid bridges around from few asperities to large fraction of the apparent area. The different regimes can be modelled and the expressions for F_M and F_V can be obtained. We note that in the toe-dipping regime, the adhesion force is independent of the apparent area and proportional to the normal load. However, the flooded regime shows the opposite tendencies. The pill box regime is intermediate and can exhibit either behavior at the extremes.

An increase in the lubricant thickness and its viscosity, relative humidity of the environment, rest period, and head-slider area generally increases the stiction of a head-medium interface (Bradshaw and Bhushan, 1984; Liu and Mee, 1983; Yanagisawa, 1985a). Figure 6 shows lubricant thickness dependence on a thin-film (metal) disk surface for two surface roughnesses. The coefficient of static friction increased significantly above critical thickness of liquid lubricants. The critical thickness depends on the surface roughness, namely, critical thickness values is about half of the rms roughness. Yanagisawa (1985a) has also shown that if the interface was in flooded regime, the static coefficient of friction increased linearly with slider area.

Miyoshi et al. (1988) have measured the effect of water vapor on adhesion of a Ni-Zn ferrite pin in contact with a flat of Ni-Zn ferrite or of magnetic tape A, Fig. 7. They found that the adhesive force (normal pull-off force) of ferrite-ferrite or ferrite-tape A contact remained low below 40%, the adhesion increased greatly with increasing relative humidity above 40%. Changes in the adhesion of contacts were reversible on humidifying and dehumidifying. They concluded that ferrites adhere to ferrites or tapes in a saturated atmosphere primarily from the surface tension (or meniscus effects of a thin-film of water adsorbed on the interface).

Matthewson and Mamin (1988) measured the static friction force as a function of start-up rate (proportional to sliding velocity) between smooth and lubricated surfaces, Fig. 8. They found that the static friction force increases linearly with an increase of the square root of the loading rate. This is attributed to viscous effects.

4.0 INTERFACE TEMPERATURES

In a sliding operation, almost all of the frictional energy input is directly converted to heat in the material close to the interface. During a sliding situation, asperity interactions results into numerous high temperature flashes. Bhushan (1987a) recently presented a detailed thermal analysis to predict the interface temperatures and applied it to predict temperatures in a head-medium interface (Bhushan, 1987b). He showed that the total flash temperature consists of temperature of an individual asperity contact and effect of other asperity contacts on an individual asperity temperature (interaction). The head-medium interface can be modelled as the case of sliding two equally rough surfaces and the relevant equations for the average and maximum asperity temperature rise of the interface are given as (Bhushan, 1987a):

$$\bar{\theta} = r_1\big[0.65\, f p_a (A_a/A_r)(U d_{max}/\kappa_1)^{1/2} / \rho_1 C_{p1} + f p_a (U\ell/\kappa_1)^{1/2} / \rho_1 C p_1\big]$$

$$\theta_{max} = r_1\big[0.95\, f p_a (A_a/A_r)(U d_{max}/\kappa_1)^{1/2} / \rho_1 C p_1 + 1.5\, f p_a (U\ell/\kappa_1)^{1/2} / \rho_1 C p_1\big]$$

and

$$r_1 \sim 1/[1 + (k_2 \rho_2 C p_2 / k_1 \rho_1 C p_1)^{1/2}] \tag{4}$$

where, ρCp is the volumetric heat capacity, κ is the thermal diffusivity, k is the thermal conductivity, d_{max} is the maximum contact diameter, and ℓ is the half length of the slider.

Average and maximum interface temperatures predicted for the assumed particulate media were 7° and 10°C, respectively (Bhushan, 1987b). These predictions compared fairly with the infrared measurements conducted at head-tape interface (Gulino et al., 1986). The asperity-contact temperatures at head-tape interface are relatively low because of its high real area of contact, as compared to that of metal-metal or ceramic-ceramic contacts. The interface temperature of 7-10°C rise can lead to high friction in some tapes because the transition temperature of some tapes' mechanical properties is within 5°C above the ambient temperature. In isolated cases, if the magnetic particles are exposed (or get exposed in a high-speed rub) and contact the head surface, the average and maximum instantaneous temperature rise could be about 600°C and 900°C, respectively. These temperatures potentially will cause a breakdown of the medium lubricant and a degradation of the medium binder leading to excessive friction and seizure of medium motion.

5.0 WEAR AND LUBRICATION

5.1 Heads

The wear of oxide magnetic particles and ceramic head body materials is different from metallic wear because of the inherent brittleness and the relatively low surface energy of ceramics. The first sign of ferrite head wear with a magnetic tape is the appearance of very small scratches on the head surface (Fig. 9a). The physical scale of scratches is usually very fine and scratches as small as 25 nm have been reported (Bhushan, 1985). Ferrite surface is microscopically removed in a brittle manner as stripes or islands, depending on the smoothness of the tape surface. Wear generally occurs by microfragmentation of the oxide crystals in the ceramic surface. Fragmentation is the result of cleavage and transgranular fractures, one dominated by intergranular fracture. We note that worn head surface is work hardened which reflects a shift from a mechanism dominated by transgranular fracture to one dominated by intergranular fracture.

Figure 9b shows a region where fracture and rupture have occurred. Such regions are commonly called "pullouts." Debris originating from these regions causes additional small scale plastic deformation and grooves (three-body abrasion) as shown in Fig. 9c. At the recording gap, chipping of ferrite may occur which depends on the width of the gap and type of epoxy used in the gap. An example of ferrite chipping adjacent to the recording gap is shown in Fig. 9d. We also note that ferrite head surface is work hardened with a large compressive stress field after wear which is detrimental to magnetic signal amplitude (Chandrasekar et al., 1987a, 1987b). Wear mechanisms of floppy and rigid disk head sliders against particulate disks are similar to those just discussed for tape heads. Head slider wear occurs by abrasive and fracture mechanisms.

Head sliders after usage are sometimes become coated with thin layers of a *new organic material* of high molecular weight called "tribopolymers." Friction is essential for the formation of these materials. Another requirement for the formation of friction polymers in a rubbing contact is that one of the surfaces, lubricant, or even a material nearby should be organic (Lauer and Jones, 1986). It seems clear that all friction polymers are products of chemical reaction, whether they derive initially from solid polymers or from organic liquids or vapors. Friction polymers are found on head surfaces. These result in discoloration of the head surface and give an appearance of brown or blue color, and, therefore are sometimes called brown or blue stains, respectively.

Head wear depends on the physical properties of head and medium materials, drive operating parameters, and environmental conditions. Wear data of common head materials against a γ-Fe_2O_3 tape is shown in Fig. 10. We observe a linear relationship between wear rate and material hardness for abrasive wear mode. Head wear as a function of surface roughness of tapes and isolated asperities on the tape surface are shown in Figs. 11 and 12. We note that head wear increases with an increase in the surface roughness or number of isolated asperities of the tape surface. Wear rate also increases humidity above about 40-60% relative humidity, Fig. 13. An increase in abrasive wear at high humidities is believed to be due to moisture-assisted fracture of the grains to yield finer particles (Bhushan, 1985; Bhushan, 1989).

5.2 Media

During contact of particulate tape with the head in contact start/stops or during partial contact in streaming, binder and magnetic particle debris is generated primarily by adhesive wear mode. Tape debris, loose magnetic particles, worn head material or foreign contaminants are introduced between the sliding surfaces and abrade, material off each. The debris that adheres to drive components lead to polymer-polymer contact, whose friction is higher than that of rigid material polymer contact and can lead to magnetic errors and sometimes to catastrophic failures.

Microscopic examinations of worn particulate disk show circumferential wear grooves and support either of the two-body or three-body abrasive wear (Fig. 14). In the case of thin-film disk, Gatzen et al. (1987) have shown that wear of the overcoat is primarily adhesive in nature. Once the overcoat is depleted, the wear of metallic film is adhesive in nature and is generally catastrophic. An example of wear track on dc sputtered amorphous carbon on plated Co-Ni-P magnetic film after head crash in contact start/stop (CSS) test is shown in Fig. 15.

An increase in overcoat hardness improves the wear resistance of the thin-film disks (Yanagisawa, 1985b). Figure 16a shows Vickers hardness dependence on baking temperature for SiO_2 films. Increase in hardness results in improvement in wear resistance as shown in Fig. 16b, where the normal load is 185 mN and the sliding velocity was 1.12 m/s. Ohta et al. (1987) studied the wear rate and effective hardness of the sputtered γ-Fe_2O_3 disks as a function of alumite underlayer thickness, Fig. 17. They found that microhardness increases and wear rate decreases, significantly if the alumite thickness is increased up to about 5μm.

Figure 18 shows the coefficient of static friction as a function of number of contact start-stops (CSS) passes for a carbon overcoated thin-film (metal) disk against Mn-Zn ferrite, $CaTiO_3$, and Al_2O_3-TiC sliders (Ishikawa et al., 1986; Doan and Mackintosh, 1988). We note that static friction increases with CSS and the rate of increase is highest in the case of Al_2O_3-TiC, next higher in case of $CaTiO_3$ and the least in case of Mn-Zn ferrite slider. Since Al_2O_3-TiC, is hardest (2300 kg/mm^2), it burnishes the disk surface more than $CaTiO_3$ (950 kg/mm^2), or Mn-Zn ferrite (600 kg/mm^2) and $CaTiO_3$ burnishes more than Mn-Zn ferrite. Examination of Mn-Zn ferrite shows that the scratches along the air-bearing surface suggesting that the Mn-Zn ferrite is slightly softer than the disk structure, therefore, ferrite is gentle to the disk surface. We believe that matching of slider and disk hardnesses is essential for low wear.

Liquid lubricants are used to reduce the wear rate of disks. Figure 19 shows the wear life of a particulate disk lubricated with two grades of perfluoropolyether-Fomblin Z-25 (nonpolar) and Fomblin AM2001 (polar with reactive end groups) as a function of lubricant thickness. We note that relative wear life increases with an increase in the lubricant thickness and polar lubricants have longer wear life than nonpolar lubricants. We have seen earlier that an increase in lubricant thickness increases the static friction, therefore, a compromise in lubricant thickness is necessary for optimum friction and wear performance.

Figure 20 shows the static friction as a function of storage time for disks with a nonpolar lubricant (perfluoropolyether or PFPE) and disks with dual lubricant consisting of polar (aminosilane) and nonpolar (PFPE) fraction. Increase in friction from aging the disks with dual lubricant film was found to be less than that for a disk with only nonpolar lubricant (Hoshino et al., 1988). Lubricant is also spun off with a disk rotation during use. Yanagisawa (1985a) and others have shown that polar lubricants spin off less than nonpolar lubricants.

Mechanism of interface failure was studied by Kawakubo et al. (1984) for particulate disks in CSS test. Friction force, acoustic emission (AE) signal (to monitor head-disk contact) and read back signal were measured during the test. The changes in read back signal, friction force, and AE signal are shown in Fig. 21. At the point of interface failure, the read back signal decreased to almost zero and friction force and AE signal rose significantly. This implies that head was virtually in contact even at full speed. Kawakubo et al. (1984) also videotaped the wear process through a transparent sapphire slider.

6.0 HYDRODYNAMIC AIR FILMS

Head-medium interface is designed so that the magnetic head is separated form the media by a thin air film during calculate the air film thickness. The Reynolds equation is based on the continuum theory of fluid mechanics. In today's magnetic storage devices the mean free path of the molecules is comparable to the film thickness, therefore, the gas does not behave entirely as a continuum fluid but exhibits rarefaction effects. In order to account for the rarefaction effects, Reynolds equation is modified by a Knudsen number (= local mean free path/film thickness). In some magnetic storage devices, film thickness is very small and is only 3 to 6 times the surface roughness. In that case, roughness effects must be included. These effects are accounted for by using the pressure flow factor approach (Bhushan and Tønder, 1989a, 1989b).

Figure 22 shows the results of head-tape spacing and pressure profiles over a cylindrical head. It takes about 8 ms for the film thickness profile to reach steady state. The film thickness reaches a minimum value near the trailing edge. Figure 23 shows the three-dimensional pressure contours of the air bearing in a typical 3370-type head-rigid disk interface. In the longitudinal direction, we note the two peaks, one at the end of the taper and another near the trailing edge. From leading edge towards the trailing edge, the pressure continues to drop because of side leakage. However, near the trailing edge, because of the pitching of the slider, the gap decreases resulting in smaller side leakage and recompression of the gas, consequently leading to an increase in the air pressure. In the transverse direction, the pressure profile is symmetric about the center line and is parabolic in shape. Mitsuya (1986) has shown that the load capacity of a head-disk interface is reduced, roughly by a factor of 2 if Knudsen number (M) is 1 for a smooth or rough surface (Fig. 24). Thus, rarefaction effects can be significant.

In some head-tape interfaces, the head-to-tape spacing (h) is significantly correlated with rms tape roughness, σ (Fig. 25). As h/σ becomes smaller (<6) for future high-density disk drives, the effect of roughness on h would become significant. Bhushan and Tønder (1989b) have calculated roughness-induced effect in shear- and squeeze-flow for magnetic storage devices. They found that roughness-induced changes in the shear flow cannot account for the measured changes in h. However, roughness-induced changes in the squeeze flow can account for the measured changes in h. Vertical motion of one of the surface necessary for squeeze effects to become dominant can be generated by the isolated asperities present of heights larger than the minimum film thickness, bearing load variations, and/or modulations of the interface from other instabilities.

Submicron film thicknesses in the steady and dynamic conditions have been measured by using capacitance and optical interference techniques (Mizoshita et al., 1985; Millman et al., 1986). Dynamic vibrations of less than 1 nm peak-to-peak have been measured from 0 to 100 kHz. Millman et al. (1986) measured slider dynamics during an actuator stroke from a 3380 head-disk assembly. They used a four-corner capacitance probe on a particulate disk, Fig. 26. They found that actuating has very little effect ($<10\%$ of the mean value) on the slider attitude (trailing-edge film thickness, pitch, and roll). The change in film thickness was about 40 nm p-p primarily at a frequency of 2.2 kHz.

7.0 SUMMARY

Analytical models and the experimental data available for the tribology of magnetic media are significant. A majority of the publications have appeared since 1984 after the author initiated the symposium on "Tribology and Mechanics of Magnetic Storage Systems" now held annually. As the film thicknesses and surface roughnesses of the head-medium interface reduce in the future, our understanding will have to be fine tuned to design a reliable product. We will require measurement techniques with better resolutions to characterize the interface. New materials and processes will be required to meet the new challenges.

References

(1) Bhushan, B., 1984, "Analysis of the Real Area of Contact Between a Polymeric Magnetic Medium and a Rigid Surface," J. Lub. Tech., Trans. ASME, Vol. 106, pp. 26-34.

(2) Bhushan, B., 1985, "Assessment of Accelerated Head-Wear Test Methods and Wear Mechanisms," Tribology and Mechanics of Magnetic Storage Systems, Vol. 2, Ed. by B. Bhushan and N. S. Eiss, SP-19, ASLE, Park Ridge, pp. 101-111.

(3) Bhushan, B., 1987a, "Magnetic Head-Media Interface Temperatures Part I – Analysis," J. Trib., Trans. ASME, Vol. 109, pp. 243-251.

(4) Bhushan, B., 1987b, "Magnetic Head-Media Interface Temperatures, Part II – Application to Magnetic Tapes," J. Trib., Trans. ASME, Vol. 109, pp. 252-256.

(5) Bhushan, B., 1989, Tribology and Mechanics of Magnetic Storage Devices, Springer Verlag (in press).

(6) Bhushan, B., Sharma, B. S., and Bradshaw, R. L., 1984a, "Friction in Magnetic Tapes I: Assessment of Relevant Theory," ASLE Trans., Vol. 27, pp. 33-44.

(7) Bhushan, B., Bradshaw, R. L., and Sharma, B. S., 1984b, "Friction in Magnetic Tapes II: Role of Physical Properties," ASLE Trans., Vol. 27, pp. 89-100.

(8) Bhushan, B., Wyant, J. C., and Meiling, J., 1988, "A New Three-Dimensional Digital Optical Profiler," Wear, Vol. 122, pp. 301-312.

(9) Bhushan, B. and Doerner, M. F., 1988, "Role of Mechanical Properties and Surface Texture in the Real Area of Contact of Magnetic Rigid Disks," J. Trib., Trans-ASME (in press).

(10) Bhushan, B. and Tønder, K., 1989a, "Roughness-Induced Shear and Squeeze-Film Effects in Magnetic Recording Part I: Analysis," J. Trib., Trans.-ASME (in press).

(11) Bhushan, B. and Tønder, K., 1989b, "Roughness-Induced Shear and Squeeze-Film Effects in Magnetic Recording Part II: Applications," J. Trib., Trans. ASME (in press).

(12) Bowden, F. P. and Tabor, D., 1950, Friction and Lubrication of Solids, Claraden Press, Oxford.

(13) Bradshaw, R. L. and Bhushan, B., 1984, "Friction in Magnetic Tapes Part III: Role of Chemical Properties," ASLE Trans., Vol. 27, pp. 207-219.

(14) Bradshaw, R. L., Bhushan, B., Kalthoff, C., and Warne, M., 1986, "Chemical and Mechanical Performance of Flexible Magnetic Media Containing Chromium Dioxide," IBM J. Res. Develop., Vol. 30, pp. 203-216.

(15) Calabrese, S. J., 1986, Rensselaer Polytechnic Institute, Troy.

(16) Chandrasekar, S., Shaw, M. C., and Bhushan, B., 1987a, "Comparison of Grinding and Lapping of Ferrites and Metals," J. Eng. for Indus., Trans. ASME, Vol. 109, pp. 76-82.

(17) Chandrasekar, S., Shaw, M. C., and Bhushan, B., 1987b, "Morphology of Ground and Lapped Surfaces of Ferrite and Metal" J. Eng. for Indus., Trans. ASME, Vol. 109, pp. 83-86.

(18) Doan, T. Q. and Mackintosh, N. D., 1988, "The Frictional Behavior of Rigid-Disk Carbon

Overcoats," Tribology and Mechanics of Magnetic Storage Systems, Vol. 5, Ed. by B. Bhushan and N. S. Eiss, SP-25, STLE, Park Ridge (in press).

(19) Gatzen, H. H., Smallen, M. J., and Tedrow, P. T., 1987, "Head-Media Wear in 5 1/4 in. Rigid Disk Drives," Tribology and Mechanics of Magnetic Storage Systems, Vol. 4, Ed. by B. Bhushan and N. S. Eiss, SP-22, STLE, Park Ridge, pp. 116-122.

(20) Greenwood, J. A. and Williamson, J. B. P., 1966, "Contact of Nominally Flat Surfaces," Proc. Roy. Soc. (Lond.), Vol. A295, pp. 300-319.

(21) Gulino, R., Bair, S., Winer, W. O., and Bhushan, B., 1986, "Temperature Measurement of Microscopic Areas Within a Simulated Head/Tape Interface Using Infrared Radiometric Technique," J. Trib., Trans. ASME, Vol. 108, pp. 29-34.

(22) Hahn, F. W., 1984, "Head Wear As a Function of Isolated Asperities on the Surface of Magnetic Tape," IEEE Trans. on Magn., Vol. Mag-20, pp. 918-920.

(23) Hendriks, F., 1988, "A Design Tool for Steady Gas Bearings Using Finite Elements," Tribology and Mechanics of Magnetic Storage Systems, Vol. 5., Ed. by B. Bhushan and N. S. Eiss, SP-25, STLE, Park Ridge (in press).

(24) Hoshino, M., Kimachi, Y., Yoshimura, F., and Terada, A., 1988, "Lubrication Layer Using Perfluoropolyether and Aminosilane for Magnetic Recording Media," Tribology and Mechanics of Magnetic Storage Systems, Vol. 5, Ed. by B. Bhushan and N. S. Eiss, SP-25, STLE, Park Ridge (in press).

(25) Ishikawa, M., Tani, N., Yamada, T., Ota, Y., Nakamiura, K., and Itoh, A., 1986, "Dual Carbon, A New Surface Protective Film for Thin-Film Hard Disks," IEEE Trans. Magn., Vol. Mag-22, pp. 999-1001.

(26) Kawakubo, Y., Ishihara, K., Seo, H., and Hirano, Y., 1984, "Head Crash Process of Magnetic Coated Disk During Contact Start/Stop Operations," IEEE Trans. on Magn., Vol. Mag-20, pp. 933-935.

(27) Kelly, J., 1982, "Tape and Head Wear," Magnetic Tape Recording for the Eighties, NASA Ref. Publ., 1075, pp. 7-22.

(28) Klaus, E. E. and Bhushan, B., 1985, "Lubricants in Magnetic Media – A Review," Tribology and Mechanics of Magnetic Storage Systems, Vol. 2, Ed. by B. Bhushan and N. S. Eiss, SP-19, ASLE, Park Ridge, pp. 7-15.

(29) Lauer, J. L. and Jones, W. R., 1986, "Friction Polymers," Tribology and Mechanics of Magnetic Storage Systems, Vol. 3, Ed. by B. Bhushan and N. S. Eiss, SP-21, ASLE, Park Ridge, pp. 14-23.

(30) Liu, C. C. and Mee, P. B., 1983, "Stiction at the Winchester Head-Disk Interface," IEEE Trans. Magn., Vol. Mag-19, pp. 1659-1661.

(31) Matthewson, M. J. and Mamin, H. J., 1988, "Liquid-Mediated Adhesion of Ultra-Flat Solid Surfaces," MRS Society (in press).

(32) Mizoshita, Y., Aruga, K., and Yamada, T., 1985, "Dynamic Characteristics of a Magnetic Head Slider," IEEE Trans. Magn., Vol. Mag-21, pp. 1509-1511.

(33) Millman, S. E., Hoyt, R. F., Horne, D. E., and Beye, B., 1986, "Motion Pictures of In-Situ Air Bearing Dynamics," IEEE Trans. Magn., Vol. Mag-22, pp. 1031-1033.

(34) Mitsuya, Y., 1986, "Stokes Roughness Effects on Hydrodynamic Lubrication, Part II – Effects Under Slip Flow Boundary Conditions," J. Trib., Trans. ASME, Vol. 108, pp. 159-166.

(35) Miyoshi, K., Buckley, D. H., Kusaka, T., Maeda, C., and Bhushan, B., 1988, "Effect of Water Vapor on Adhesion of Ceramic Oxide in Contact With Polymeric Magnetic Medium and Itself," Tribology and Mechanics of Magnetic Storage Systems, Vol. 5, Ed. by B. Bhushan and N. S. Eiss, SP-25, STLE, Park Ridge (in press).

(36) Ohta, S., Yoshimura, F., Kimachi, Y. and Terada, A., 1987, "Wear Properties of Sputtered $\gamma - Fe_2O_3$ Thin Film Disks," Tribology and Mechanics of Magnetic Storage Systems, Vol. 4, Ed. by B. Bhushan and N. S. Eiss, SP-22, STLE, Park Ridge, pp. 110-115.

(37) Stahl., K. J., White, J. W., and Deckert, K. L., 1974, "Dynamic Response of Self-Acting Foil Bearings," IBM J. Res. Develop., Vol. 18, pp. 513-520.

(38) Tanaka, K. and Miyazaki, O., 1982, "Wear of Magnetic Materials and Audio Heads Sliding Against Magnetic Tapes," Wear, Vol. 66, pp. 289-306.

(39) Yanagisawa, M., 1985a, "Lubricants on Plated Magnetic Recording Disks," Tribology and Mechanics of Magnetic Storage Systems, Vol. 2, Ed. by B. Bhushan and N. S. Eiss, SP-19, ASLE, Park Ridge, pp. 16-20.

(40) Yanagisawa, M., 1985b, "Tribological Properties of Spin-Coated SiO_2 Protective Film on Plated Magnetic Recording Disks," Tribology and Mechanics of Magnetic Storage Systems, Vol. 2, Ed. by B. Bhushan and N. S. Eiss, SP-19, ASLE, Park Ridge, pp. 21-26.

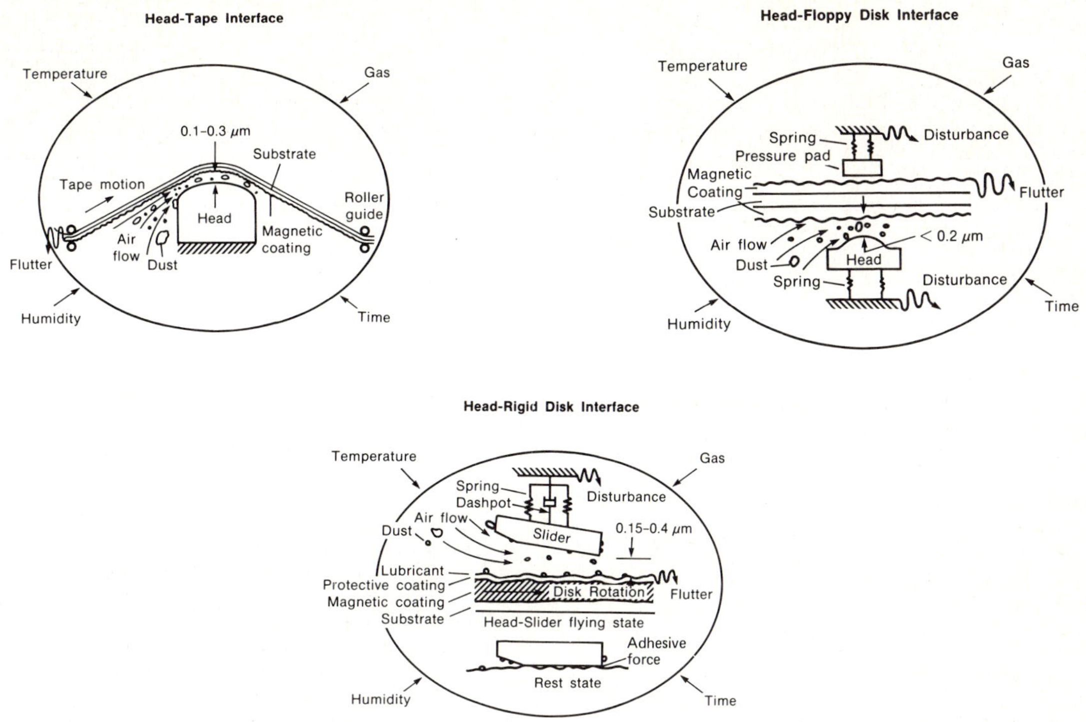

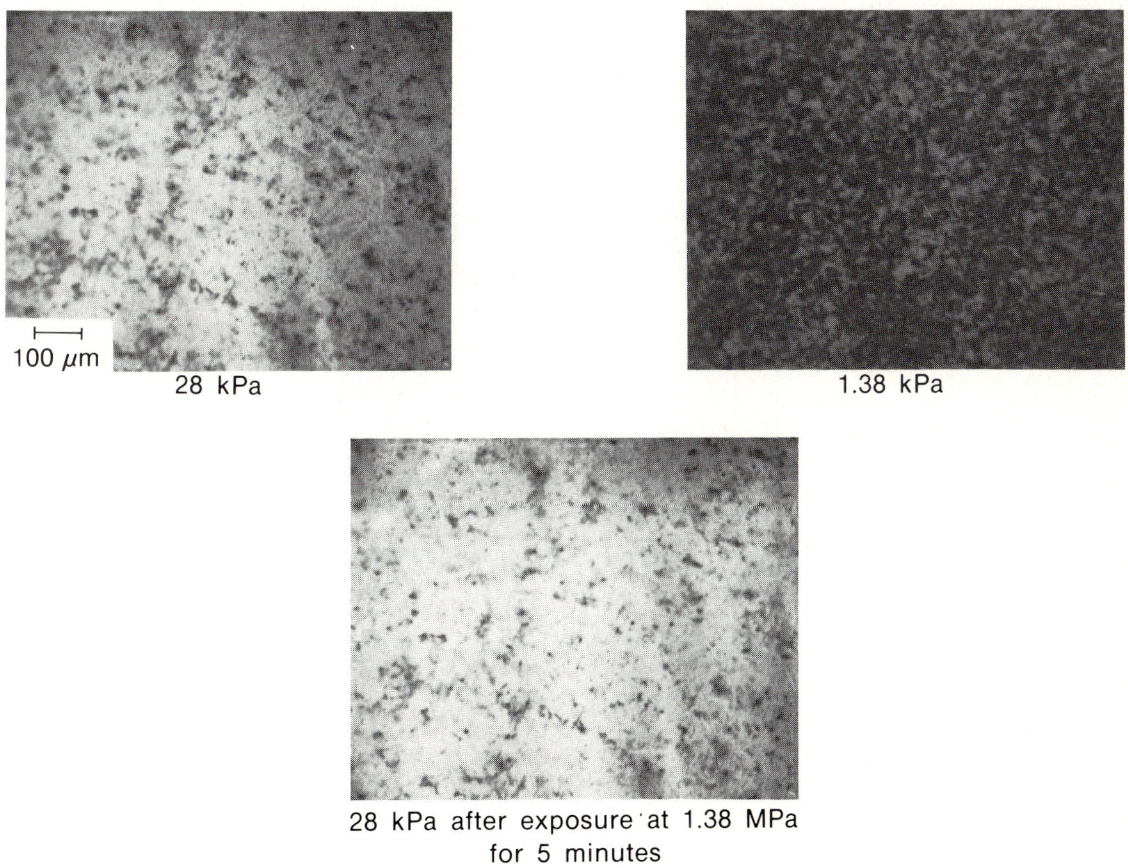

Figure 2. Optical interference photographs of tapes taken at 28 kPa; then subjected to higher pressures for short duration and brought back to 28 kPa and rephotographed. We see no change in the real area of contact implying an elastic contact (Bhushan, 1984).

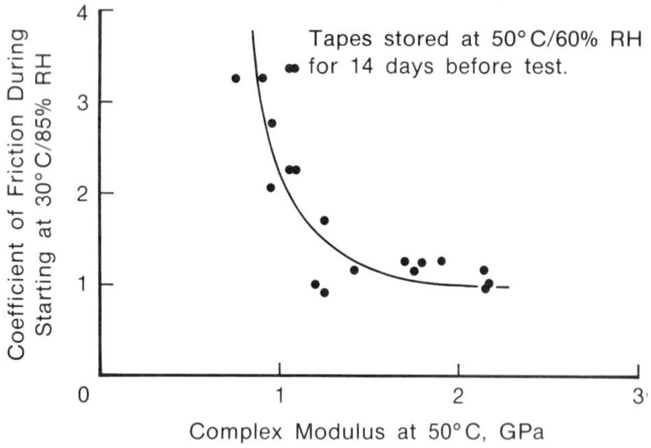

Figure 3a. Coefficient of friction during start at 30°C/85% RH measured on a commercial tape drive versus complex modulus at 50°C. The CrO_2 tapes were stored at 50°C/60% RH for 14 days before the tests (Bhushan et al., 1984b).

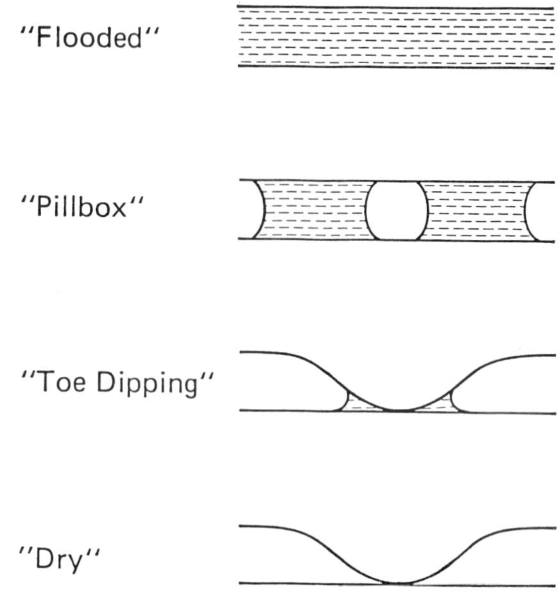

Figure 5. Regimes of different liquid levels in the interface between smooth surfaces (Matthewson and Mamin, 1988).

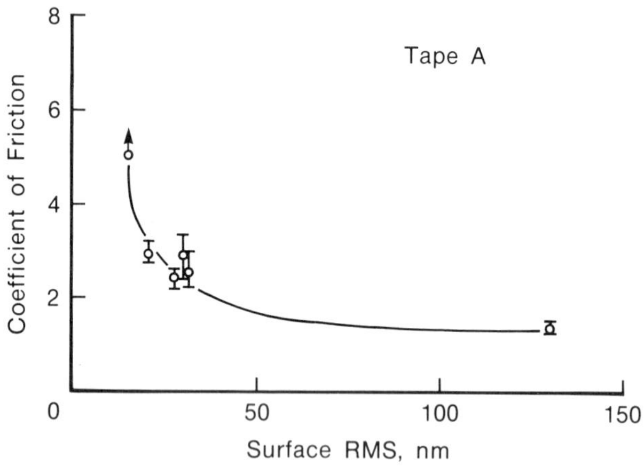

Figure 3b. Effect of surface roughness on coefficient of friction for CrO_2 particulate tapes (Bhushan et al., 1984b).

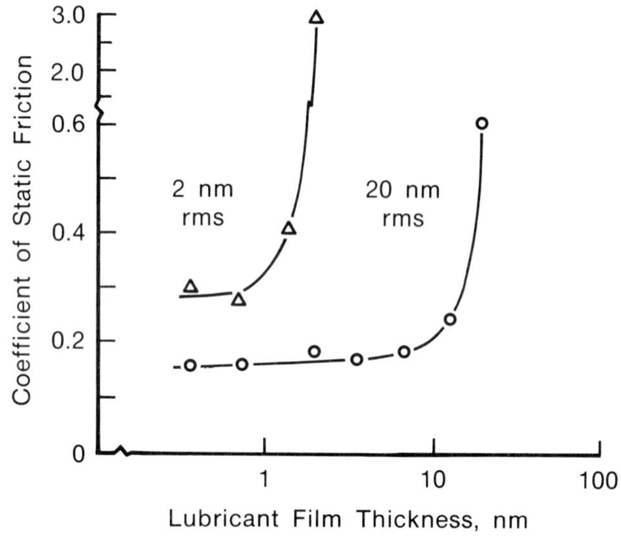

Figure 6. Lubricant thickness dependence on coefficient of static friction for liquid perfluoropolyether on a thin-film rigid disk at 2 and 20 nm rms surface roughness (Yanagisawa, 1985a).

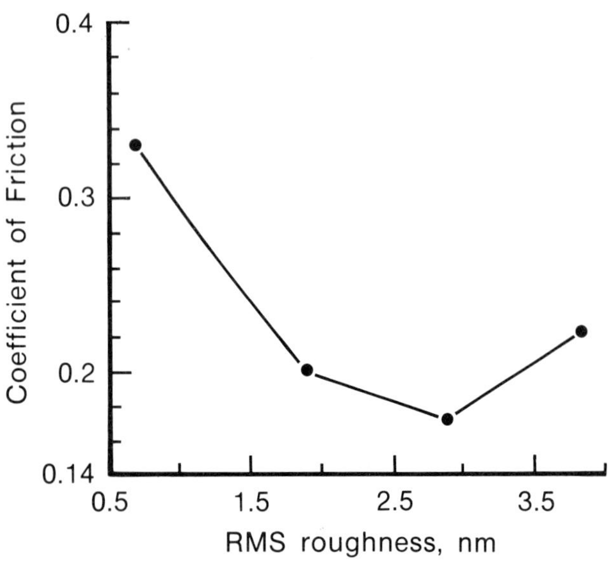

Figure 4. Coefficient of friction as a function of degree of disk texturing for a thin-film (metal) rigid disk with sputtered carbon overcoat against ferrite slider (Doan and Mackintosh, 1988).

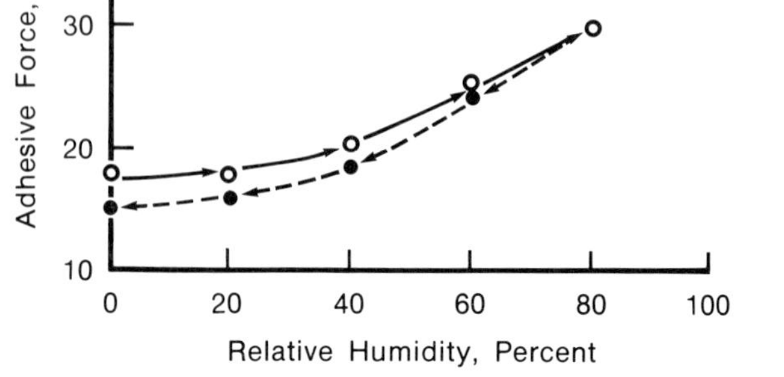

Figure 7. Effect of humidity on adhesion of CrO_2 tape A in contact with a Ni-Zn ferrite pin (Miyoshi et al., 1988).

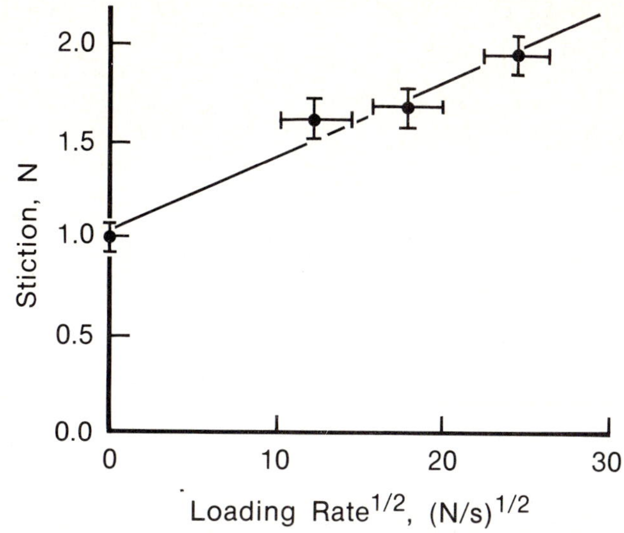

Figure 8. Static friction force as a function of loading rate (or velocity) during start up between smooth and lubricated surfaces (Matthewson and Mamin, 1988).

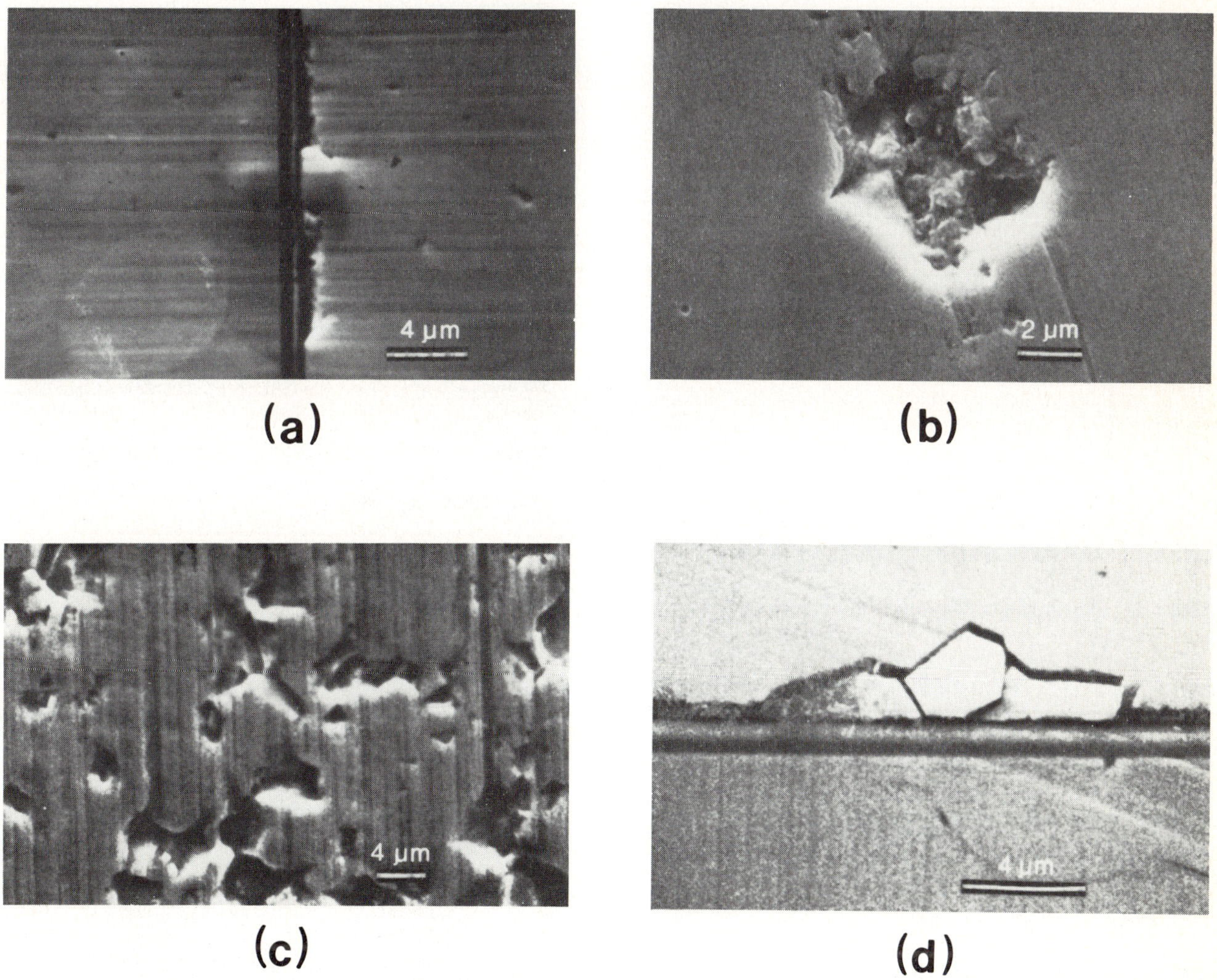

Figure 9. SEM micrograph of worn Ni-Zn ferrite head with a CrO_2 tape A (a) abrasion marks (direction of tape motion-left to right), (b) surface pull out, (c) extensive surface pull out and associated plastic deformation (direction of tape motion-top to bottom), (d) ferrite chipping adjacent to the recording gap (Bhushan, 1985).

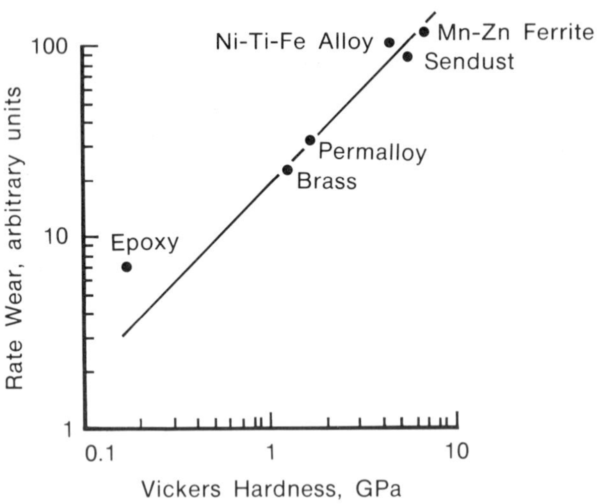

Figure 10. Wear rate of magnetic materials slid against a diamond cone as a function of Vickers hardness (Tanaka and Miyazaki, 1981).

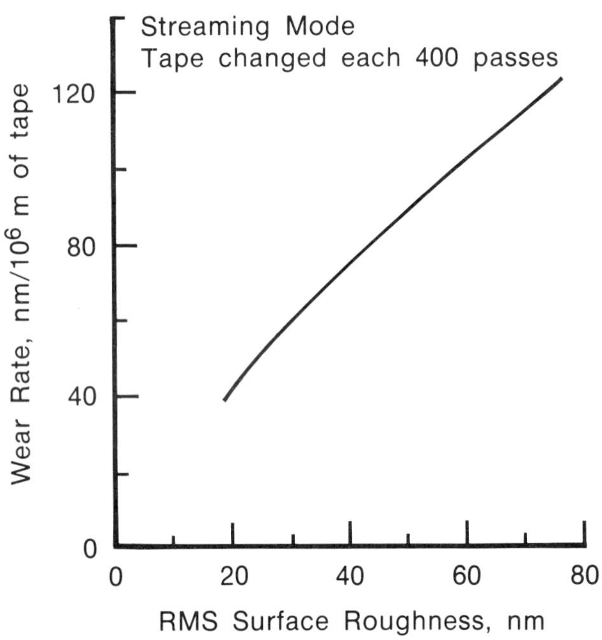

Figure 11. Ni-Zn ferrite as a function of rms surface roughness of a CrO_2 tape A (Bhushan, 1985).

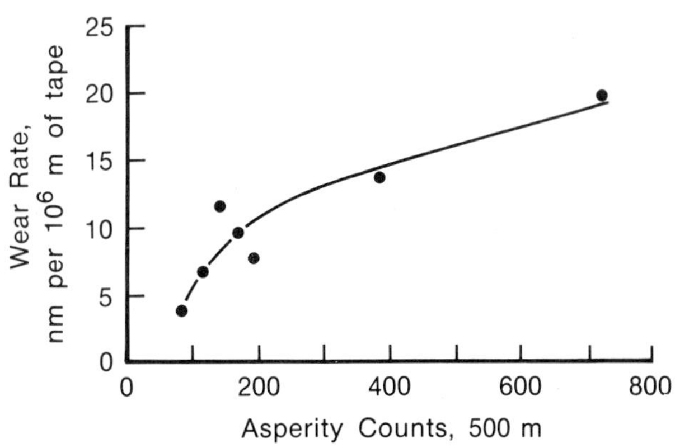

Figure 12. Ni-Zn ferrite head wear rate as a function of asperity counts on CrO_2 tape A (Hahn, 1984).

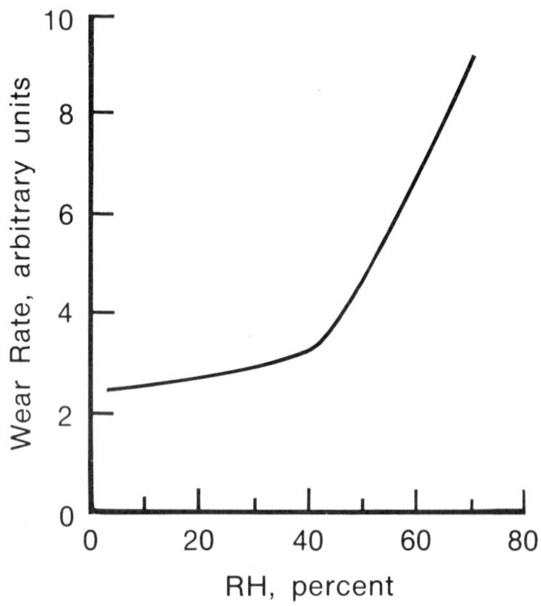

Figure 13. Head wear rate as a function of relative humidity (Kelly, 1982).

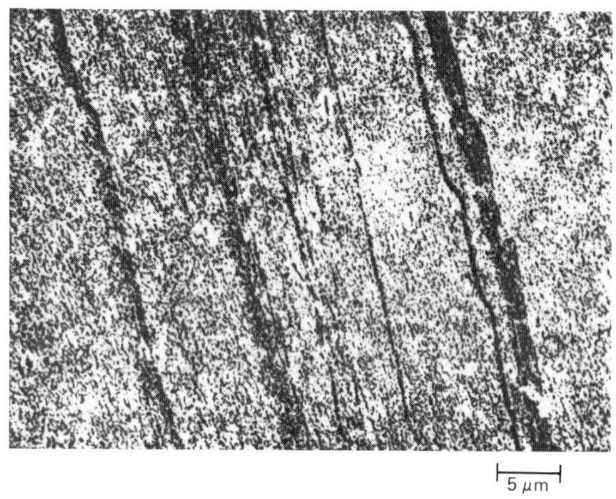

Figure 14. SEM micrograph of a worn particulate disk surface against $Al_2O_3 - TiC$ slider after a head crash in CSS (Calabrese, 1986).

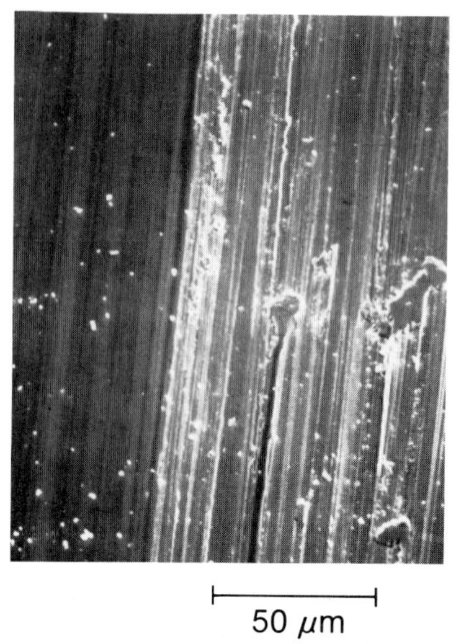

Figure 15. SEM micrograph of wear track on sputtered carbon film against Mn-Zn ferrite slider after 61000 CSS (Gatzen et al., 1987).

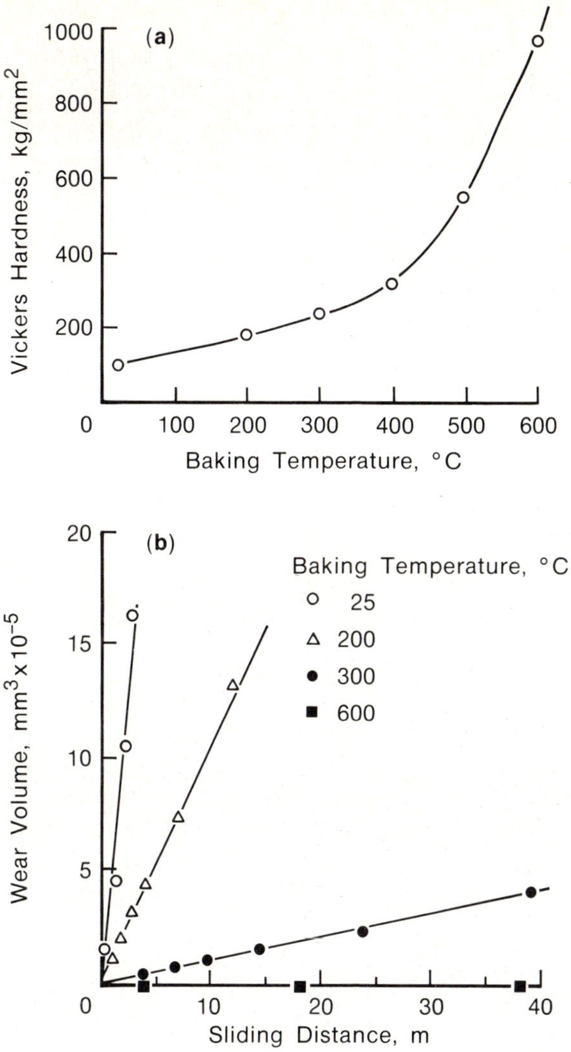

Figure 16. (a) Dependence of Vickers hardness on baking temperature for SiO_2 films (b) Relation between sliding distance and wear volume for SiO_2 films exposed to air, baked at various temperatures, slid against Al_2O_3-TiC slider (Yanagisawa, 1985b).

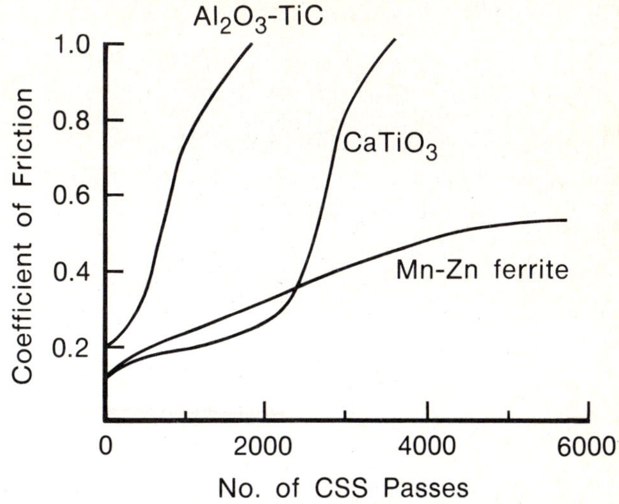

Figure 18. Change of the coefficient of static friction as a function of number of passes of Mn-Zn ferrite, $CaTiO_3$, Al_2O_3-TiC head sliders on a thin-film disk with carbon overcoat (Ishikawa et al., 1986; Doan and Mackintosh, 1988).

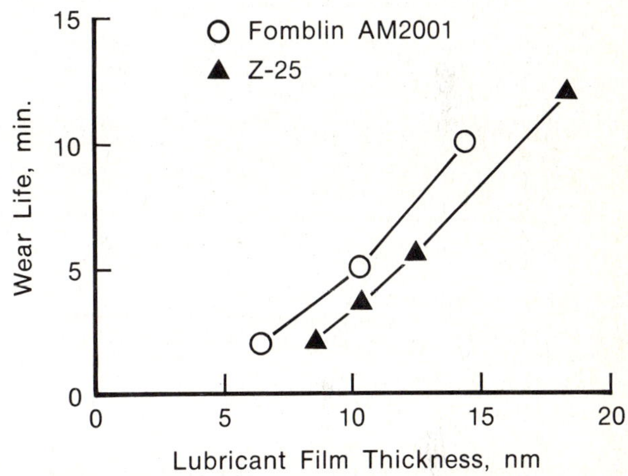

Figure 19. Wear life as a function of lubricant film thickness on a particulate disk slid against Mn-Zn ferrite slider (Scarati and Caporiccio, 1987).

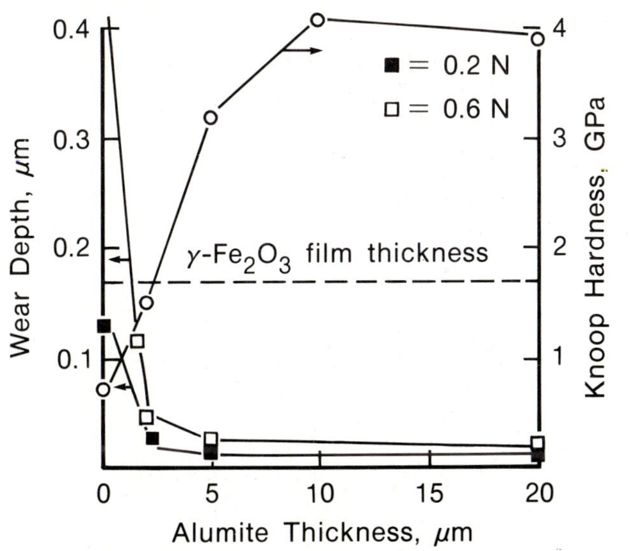

Figure 17. Wear depth and Knoop hardness of sputtered γ-Fe_2O_3 as a function of alumite (underlayer) thickness at two loads in a pin-on-disk wear experiments (Ohta et al., 1987).

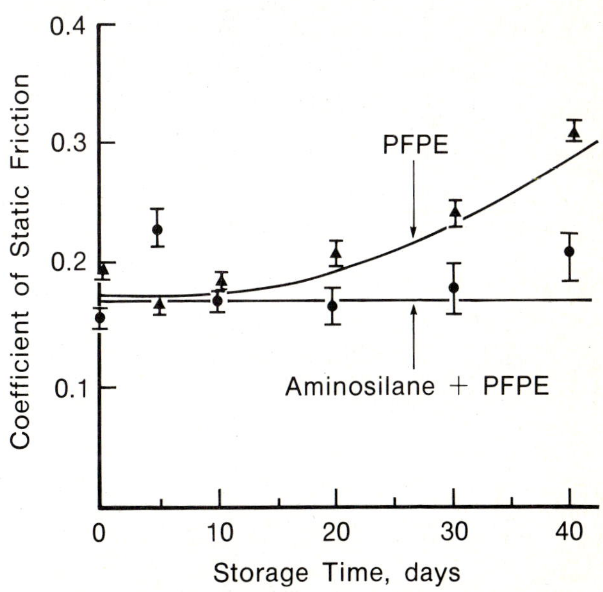

Figure 20. Coefficient of static friction as a function of exposure at 50°C and 10^{-3} torr for spin coated SiO_2 film on a sputtered oxide magnetic film (Hoshino et al., 1988).

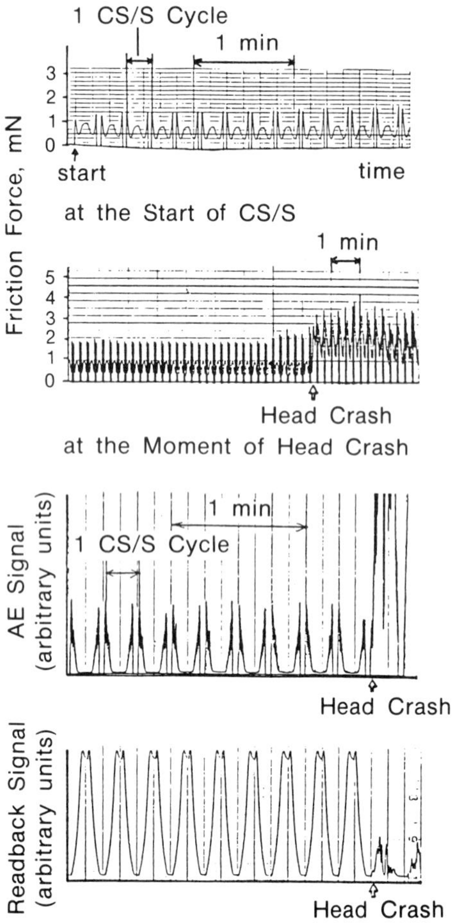

Figure 21. Friction force, AE signal and read back signal near the head crash in CSS test of a particulate disk (Kawakubo et al., 1984).

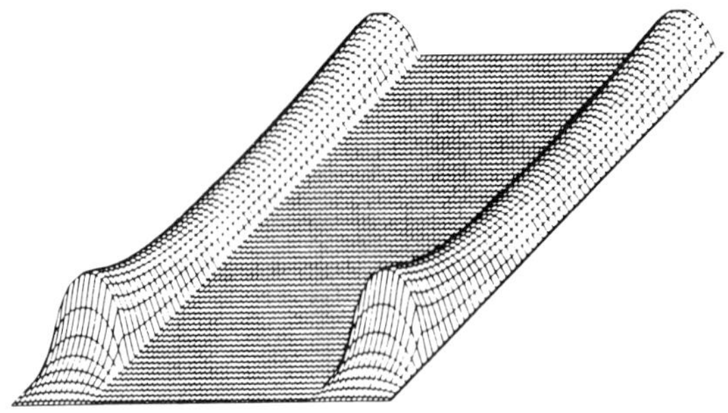

Figure 23. Three-dimensional pressure profiles for a 3370-type slider with zero skew (Hendriks, 1988).

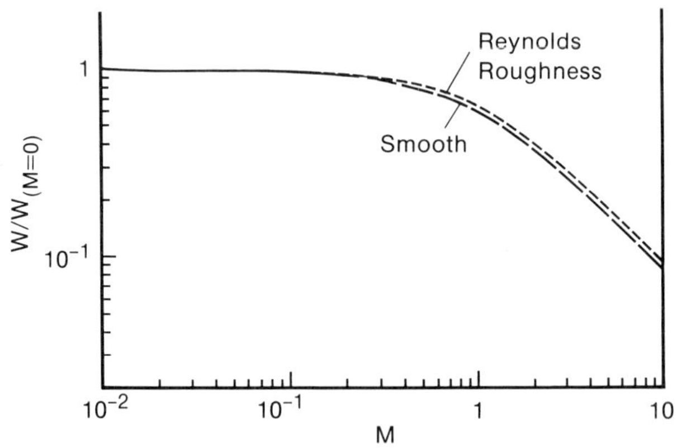

Figure 24. Effect of rarefaction on load capacity (W) for a head-disk interface (Mitsuya, 1986).

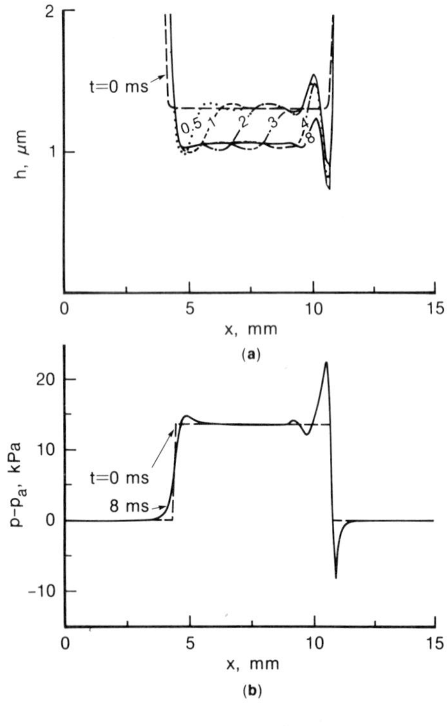

Figure 22. Film thickness and air-bearing pressure rise profiles of a tape over cylindrical head (Stahl et al., 1974).

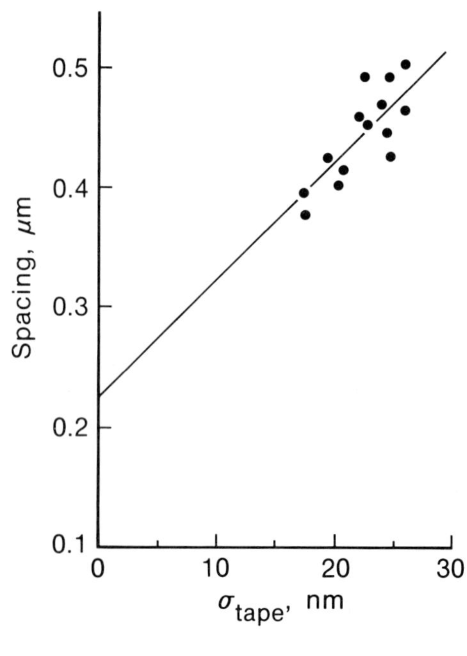

Figure 25. Effective head-to-(particulate)-tape spacing as a function of rms surface roughness of a tape for a data processing application (Bhushan and Tønder , 1989a).

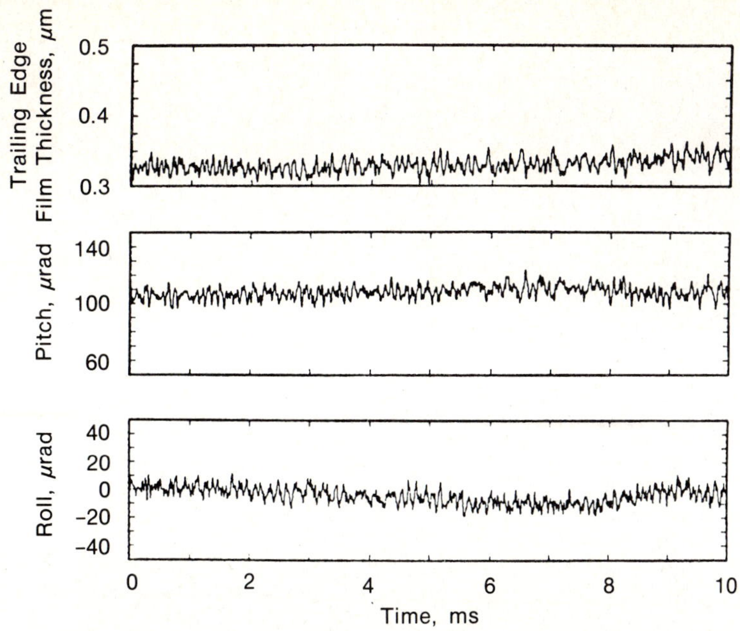

Figure 26. Film thickness, pitch angle and roll angle of a slider during the first quarter of an actuator stroke in a 3380 HDA (Millman et al., 1986).

Paper XV(ii)

Active magnetic bearing design methodology - A conventional rotordynamics approach

H. M. Chen

The essential components of an active magnetic bearing (AMB) i.e., the electromagnets, amplifier, controller, sensor, and their associated parameters are discussed and quantified using a technical language that is more acceptable to mechanical engineers. AMB dynamics are represented by a set of first-order differential equations in formulating a rotor-AMB system model. The system dynamics design guidelines are presented through use of a flexible rotor example and are consistent with conventional rotor-bearing design methodology.

1 INTRODUCTION

The recent advances of active magnetic bearing (AMB) technology [1,2] are best studied in the context of the well-developed rotordynamic experience base. A long-standing and currently routine analytical procedure used in the early design stage of a rotor-bearing system involves [3].

- reviewing an undamped critical speed map and modeshapes for potential unbalance and instability problems.

- performing stability and unbalance response analyses.

The AMB presents a new challenge to the rotor dynamists. This new type of bearing has many advantages over the conventional type, such as, no contact, low power loss, and most important, the feature of adjustable stiffness and damping. However, many rotor dynamists lack a thorough understanding of these new bearings because mathematical modeling of AMBs is traditionally performed in electrical engineering and control languages to which the mechanical engineer is not accustomed. To fill the communication gap, the author promotes the features of the AMB from a rotor dynamist's viewpoint [4]:

- AMBs should be treated as locally controlled devices similar to other types of bearings, not actuators.
- Like other bearings, the main functions of AMBs are to provide load capacity, stiffness and damping for rotor support.

A similar approach was adopted by Hustak, et al.[5]; however, a different emphasis is presented herein. First, the essential AMB components and parameters will be discussed using terminology more familiar to the mechanical engineer. Second, a rotor-AMB system modeling technique will be formulated for rigorous dynamics analysis.

1.1 Notation

AMB	active magnetic bearings
A_p	surface area per pole
a	phase-lead parameter
B	AMB damping coefficient
B_s	saturation flux density
b	phase-lead parameter
C	radial air gap
C_d	proportional feedback gain
C_v	derivative feedback gain
C_e	integral feedback gain
E	AMB controller output
f	magnetic pole constant for a given number of windings
F	AMB perturbed magnetic force
F_{max}	maximum load
G_a	amplifier gain
G_p	displacement probe sensitivity
L	inductance of electromagnets
L_p	AMB axial length
I	bias current
i	control current
j	$\sqrt{-1}$
K	AMB stiffness coefficient
K_i	current stiffness
K_m	magnetic (negative) stiffness
M	mass supported by AMB
N_t	number of winding turns per pole
Q	integrator output
S	Laplace variable
t	Time
V_s	a DC source
V_p	Y_p'
W	AMB static load
Y	AMB journal displacement in Y-direction
Y_p	displacement sensor output in Y-direction
Z	phase-lead circuit output
ω	excitation frequency
ω_c	displacement probe cut-off frequency
ω_n	amplifier cut-off frequency
ω_o	integrator cut-off frequency
ω_v	differentiator cut-off frequency
'	d/dt

2 AMB PRINCIPLE

The practical AMBs are mostly the attractive-force type, the principle of which was well explained by Habermann and Liard [1]. This presentation concentrates on radial AMBs. The radial AMBs generally adopt an 8-pole stator configuration as shown in Figure 1. Both the stator and journal are stacks of laminations of ferromagnetic material. The journal is shrunk on a shaft without windings. The use of laminations reduces eddy current, which not only causes a power loss, but degrades the dynamic performance of the bearing. The stator poles are separated into four quadrants [2]. In each quadrant, the electromagnetic windings are wound in such a way that the magnetic flux will mainly circulate inside the quadrants [6]. Therefore, each quadrant of poles can be controlled independently.

It is well known that the magnetic force is proportional to the current to air-gap ratio squared (Figure 2). To support a static load, W, in a controlled axis (Figure 3), unequal steady state or bias currents are induced in opposite pairs of poles, such that

$$W = f(I_1^2 - I_3^2)/C^2 \qquad (1)$$

where $I_1 > I_3$
C = radial air gap
f = a magnetic pole constant for a given number of windings.

The bias current produces an I^2R loss, which is a major power loss in an AMB. However, the total resistance including the windings and other parts in the current path is not large; the AMB power loss is in general insignificant compared to conventional oil-film bearings.

The journal floating in the magnetic field due to the bias currents alone is not stable. This is analogous to supporting a vertical rod at its base using an uncontrolled hinge. Linearized feedback control of the AMB is achievable by making the air gap large, relative to the journal normal vibration amplitude. To create stability at the AMB center, the journal motion must be sensed and corrected instantaneously and continuously by superimposing a small control current to each bias current. For example, as shown in Figure 3, when the journal moves upward off the center by a small displacement Y, the current in the top quadrant will be reduced by a small amount, i, and the bottom quadrant increased by i. The control currents produce a net downward force, F, which pulls the journal back to the center. From sensing Y to producing F, a series of AMB components are involved, namely, the sensor, controller, power amplifier and electromagnets.

3 AMB COMPONENTS

3.1 Electromagnets

For a given maximum load including static and dynamic loads, the AMB physical size is determined by the saturation flux density of the lamination material. For the 8-pole configuration, it is easily shown that

$$F_{max} \simeq 5.75 \times 10^5 A_p B_s^2 \qquad (2)$$

where, F_{max} = maximum load, N
A_p = surface area per pole, m^2
B_s = saturation flux density, Weber/m^2

The maximum value of the flux density in linear range which is about 90% of the actual B_s value should be used in applying Equation 2.

Choosing the axial length, L_p, the circumferential pole width is A_p/L_p. The radial dimensions can be determined outward from a given shaft diameter at the AMB. The sizing guidelines are:

- The cross-sectional area at any point of the flux path is not less than A_p.
- Adequate wiring space is provided.
- The axial length is no greater than the journal OD.
- As a rule, the air gap should be ten times the expected journal vibration.

Note that the pole surfaces are the most effective areas for heat dissipation by convection. The ampere-turns per pole is fixed for a given ferromagnetic material; the optimal choice of winding turns, N_t is a trade-off of total current and inductance load, L, to the power amplifiers. The latter is proportional to $N_t^2 A_p C$ which is a crucial parameter causing control delay and bearing instability.

More than 8 poles can be designed for the stator, such as, 16 or 24 poles evenly spaced. A large number of poles saves radial space because it better localizes flux circulation. The coil pairs can be in series or parallel to a power amplifier with the same trade-off.

3.2 Power Amplifiers

Converting a low power control voltage signal to a high power control current and actuating the electromagnets requires power amplifiers. Two types of power amplifiers are commercially available, the linear type and the pulse-width-modulation (PWM) type. The linear amplifier applies the control signal to a power transistor in an "active mode". The transistor continuously regulates the current through the windings from a DC source, V_s, with the current directly proportional to the control signal. The PWM type applies the control signal to generate high voltage pulses at a fixed frequency above audible range. The on-time period of each pulse is proportional to the input signal. The voltage pulse train produces current to the windings. The PWM type is electrically noisy and needs its own filters. The power transistors operate in a "saturation mode" with much less power loss. Schweitzer and Traxler [7] had indicated that the borderline in favoring one type over the other is about 0.5 kVA.

There are three limiting parameters in the power amplifiers. First, the control current, i, can not be larger than the bias current. Second, the inductance of the electromagnets causes the control current to diminish and delay above a certain frequency (cut-off frequency). The PWM amplifiers usually apply their own current feedback to increase this frequency. Third, the value of V_s/L, called the current slew rate limit, is the maximum amperes per second that the amplifier can provide. The amplifier frequency response as constrained by these interrelated parameters was well explained by Bradfield, et al [8]. To account for this in AMB dynamics, an approximate transfer function is shown in Equation 3.

$$i/E = G_a\omega_n/(S + \omega_n) \quad (3)$$

where ω_n = amplifier cut-off frequency, rad/sec
S = Laplace variable
E = controller output
G_a = amplifier gain, A/V

3.3 Controller

The function of the controller is to generate a low power voltage signal, E, to drive the power amplifier and achieve the control current, i, in the windings. If it is assumed that no delay occurs in the other components ($\omega_n = \infty$ and G_a = 1), the requirement of the controller can be described as follows:

At any rotor vibration mode, the AMB will see a dynamic mass, M. The AMB equation of motion is

$$M\ Y'' = K_i\ i + K_m\ Y \quad (4)$$

where Y is the displacement of the mass.

K_i and $-K_m$ are, respectively called the current stiffness and the negative spring of the magnetic field due to the bias current [6]. Without the small control current i, Equation 4 represents an unstable dynamic system. To make it stable, it is simple and sufficient to have

$$i = E = -C_dY - C_vY' \quad (5)$$

The small control current should be made proportional to the journal displacement and velocity. The minus sign is indicative of "negative feedback" - a common regulating mechanism. C_d and C_v are constants, called proportional and derivative control gains.

Substituting Equation 5 into Equation 4, we have

$$M\ Y'' + K_iC_v\ Y' + (K_iC_d - Km)\ Y = 0 \quad (6)$$

If C_d is large enough such that $K_iC_d > K_m$, we achieve a stable, one degree of freedom system in Y-axis, which has a damping coefficient K_iC_v and a stiffness coefficient $(K_iC_d - K_m)$.

If the static load W, is miscalculated, or there is an occasional slow varying load, as may occur from the gyroscopic effect in spacecraft manuvering, the journal will sag or drift off the bearing center. To avoid this potential problem, a third corrective mechanism called the integral control can be added to Equation 5:

$$E = -C_dY - C_vY' - C_e\int Ydt \quad (7)$$

As shown in Figure 4, the accumulated journal position error over a period of time will produce a part of the control voltage. The integral control does not respond to high frequency vibrations. Equation 7 represents a proportional integral derivative (PID) controller.

3.4 Sensors

Three displacement sensors prove to be practical, the capacitance probe, inductance probe, and eddy current probe. Each differs with its advantages and disadvantages, but all relate the small distance between the stationary sensor and the rotating shaft to an output electrical signal in volts. A low-pass filter is usually included in the sensor conditioning device to eliminate high frequency noise, including its own FM carrier. This filter, similar to the power amplifier cut-off charateristics, may cause a significant time delay in the frequency range of interest. A phase-lead circuit implanted in series in the feedback loop can reduce the delay.

An inexpensive and reliable sensor is not yet available for measuring the journal velocity, Y'. Different analog circuits, such as a differentiator with a low-pass filter, and phase-lead circuits have been used to produce a psuedo velocity from the displacement measurement.

An analog surrogate called a Velocity Observer is an alternative recently reported by Chen and Darlow [6]. Instead of differentiating displacement, it integrates journal force, (equivalent to acceleration) to obtain velocity. The output of any pseudo velocity circuit is a combination of displacement and velocity signals. Thus, its feedback not only produces damping, but also contributes to the stiffness.

4 AMB STIFFNESS AND DAMPING

From the previous discussion, a practical single axis control [9,10] can be represented by the block diagram of Figure 5. A radial AMB needs two independently controlled axes like this, while a thrust AMB needs only one.

A second-order (Butterworth) low-pass filter was assumed to be part of the sensor. It as easily could have been a fourth-order or other type of filter. G_p is the sensor sensitivity, e.g., 1000 V/in. (40 V/mm). Note that a phase-lead circuit is applied in series here for compensating the time delay mainly caused by the inductance loads to the power amplifiers. One may set the phase-lead parameter "a" to be equal to the amplifier cut-off frequency, ω_n. Thus a system "zero" cancels a system "pole" in the S-plane. This does not improve the current slew-rate of the amplifier, but does increase the damping-to-stiffness ratio around ω_n. The other phase-lead parameter "b" is set in the range $5a \le b \le 10a$.

The AMB stiffness, K, and damping, B, of this controlled axis can be calculated by using Equations 8 and 9 with S equal to $j\omega$.

$$-F/Y = K + j\omega B$$
$$= -K_i(i/Y) - K_m \quad (8)$$

$$i/Y = (Ts)(Tc)(Tp)(Ta) \quad (9)$$

where

ω = excitation frequency, rad/sec

$$Ts = G_a\omega_n^2/(S^2 + \sqrt{2}\,\omega_nS + \omega_n^2)$$

$$Tc = -C_d - C_e\omega_o/(S+\omega_o) - C_vS/(S+\omega_v)$$

$$Tp = (b/a)(S+a)/(S+b)$$

$$Ta = G_a\omega_n/(S + \omega_n)$$

Equations 8 and 9 indicate that both K and B are functions of excitation frequency ω, not rotational speed. Numerical example results are plotted in Figure 6 using the AMB data in Table 1. The frequency axis in this plot is normalized with respect to 50 Hz which is the average of two rigid-body critical speeds of a rotor. The amplitude is normalized with respect to $(G_pG_aK_iC_d - K_m)$.

At the low frequency range where the integral control dominates, the plot shows negative damping value. This should not cause alarm, however, since mechanical system resonances seldom exist in that low range. Yet, at the high frequency range, especially where the first two bending criticals exist, the negative damping can cause resonances. This will be discussed in an example later.

5 ROTOR-AMB SYSTEM DYNAMICS

5.1 System Design Guidelines

AMBs are generally less stiff than rolling element, or hydrodynamic oil-film bearings. Therefore, the first two system criticals have relatively rigid mode shapes, and their vibrations are easily controlled. The third and the fourth criticals with bending mode shapes must be given careful design considerations for high speed turbomachines.

Taking the rotor model in Figure 7 as an example, its critical speed map (Figure 8) shows that the rotor operates between the third and the fourth criticals. Two identical 8-pole AMBs are chosen to support the rotor with dimensions in Table 1. The first design issue is finding the best method for determining the stiffnesses. In this case, the stiffness per bearing can be made 1000 lb/in. or 10,000 lb/in. (1.75×10^5 N/m or 1.75×10^6 N/m). The answer depends on rotor shock load. To take 1 g shock, this rotor of approximate 100 lb (45 kg) moves radially 50 mil and 5 mil (1.25 mm and .125 mm) respectively, for the lower and higher stiffnesses. The catcher bearing is set at 10 mil (.25 mm) away from the rotor for a designed air gap of 20 mil (.5 mm). To avoid pounding the catcher bearing when shocked, the higher stiffness is chosen.

Reviewing the mode shapes at the chosen stiffness (Figure 9) reveals that there are sufficient relative displacements at the bearings for control of the first and second modes. The third and the fourth modes are lacking the displacement at one bearing. To help control the third mode, the displacements sensor is mounted at the outboard side of each AMB where the sensor sees more than only the AMB center motion.

The second design issue is to determine how many bending modes should be controlled. To keep the control electronics relatively simple, the frequency range with acceptable control response is limited by two factors, namely, the inductance load and the filtering delay. It is imperative to have an adequately damped bending mode below the operating speed (the third mode in this case), because of the unbalance excitation during traversing the critical. The bending mode immediately above the operating speed (the fourth mode in this case) should be 15 to 20% away in frequency. However, it still can be excited by harmonics as the rotor is going up in speed, or by a shock load. But, less damping is required for controlling this mode.

The higher bending modes normally are less likely to be excited. The rotor material damping is a source to resist the minor, or occasional excitation. Note that oil-film bearings always provide positive damping but there is no guarantee of this for AMBs. The control current at the high critical frequency may lag behind the displacement measurement, or the probe may be at the wrong side of the AMB. The AMB may become a small exciter for that mode. When it happens, a band-reject filter for the excitable mode can be implemented in series in the feedback loop to block the control at that modal frequency.

For the example herein, the power amplifier and the sensor low-pass filter are assumed to have the cut-off frequencies at 500 and 5000 Hz, respectively. Applying the normalized stiffness and damping of Figure 6, the normalized frequency of 1.0 is 50 Hz, which is between the first and the second criticals as read from the map. The values of $\omega B/K$ for the lower four modes range from 0.2 to 0.6, which is adequate for properly designed vibration modes. More damping can be achieved by increasing the value of C_v.

5.2 Rigorous Dynamic Analysis

After component sizing and cursory analysis, rigorous system analysis is needed to prove the expected rotor vibration behavior. Considering the fact that the AMB stiffness and damping are functions of excitation frequency, the state vector of the conventional rotor model is extended to include the state variables of the AMBs [11]. The control dynamics of each AMB axis are represented by a set of first-order linear differential equations as shown in Figure 5, in terms of the extended states (i, E, Q, Z, Y_p, V_p). The coupling terms between the rotor model and the AMB model exist in the AMB force equation, which is also shown in Figure 5. The mathematical rotor model is the same as the conventional model including sections of shaft with specified ID, OD, length, and concentrated masses and inertias. The model for each bearing would be the bearing station number, the measurement station number, and the key parameters of Table 1 (presented earlier).

Using this electromechanical model, the lower four damped frequencies of the example rotor running at 15,000 rpm were computed and presented in Table 2. The first three modes are adequately damped because the associated log decrement values are all significantly above 0.4. The latter is a damping value generally accepted for a rotor system supported in oil-film bearings. The fourth mode has a log decrement value of 0.06 without considering the rotor material damping. It should be acceptable since it is much higher in frequency than the third mode and the operating speed.

Note that in Table 2, the cross-coupling stiffness produces a destablizing seal effect at the wheel, but only affects the first mode damping. The reason for this is that only the first mode shape has significant lateral displacment at the wheel. The rotor/AMB system can sustain a value of 4000 lb/in. (7×10^5 N/m) before becoming unstable.

Figure 10 presents an unbalance response at the third critical speed using the same electromechanical model. It indicates that the response peak is well damped and far away from the operating speed of 15,000 rpm. The peak dynamic current of AMB #2 was calculated to be 0.45 A (0-peak) at 11,500 rpm. It specifies that a current slew rate no less than 550 A/sec must be provided by the power amplifier design.

6 CONCLUSIONS

The properties of active magnetic bearings and the essence of rotor-AMB system control have been described and quantified in a language that is more familiar to mechanical engineers. In doing so, the following viewpoints on AMB development have been emphasized:

1. The AMB should be treated as a locally controlled device similar to other types of bearings. The two axes of a radial AMB should be controlled independently.

2. Pitfalls exist in controlling the rotor bending critical modes; the reasons and design guidelines are explained by conventional rotordynamics terminology.

3. A modified rotordynamics analysis method with an extended state vector including the AMB state variables is needed for rigorous system performance prediction.

In presentation of this paper, the author aims to broaden the understanding of AMBs and thus, accelerate the development of AMB technology.

References

1. Habermann, H. & Liard, G., "An Active Magnetic Bearing System", Tribology International, April 1980, 85-89.

2. Weise, D. A., "Active Magnetic Bearings and Their Industrial Applications", 5th Annual Rotating Machinery and Controls Industrial Research Conference, San Antonio, Texas, 1985.

3. Malanoski, S. M., "Rotor-Bearing System Design Audit", Proc. 4th Turbomachinery Symposium, Texas A&M Univ., Oct. 1975, 65-70.

4. Chen, H. M. & Dill, J., "A Conventional Point of View on Active Magnetic Bearings", NASA Langley Workshop on Magnetic Suspension Technology, Feb. 2-4, 1988.

5. Hustak, J. F., Kirk R. G., & Schoeneck, K. A., "Analysis and Test Results of Turbocompressors Using Active Magnetic Bearings", Journal of ASLE, Lubrication Engineering, May 1987, 43, 356-362.

6. Chen, H. M. & Darlow, M. S., "Design of Active Magnetic Bearings with Velocity Observer", ASME 11th Biennial Conference on Mechanical Vibrations and Noise, Boston, September 1987.

7. Schweitzer, G. & Traxler, A., "Design of Magnetic Bearings", Proc. of Intl. Symp. on Design and Synthesis, Tokyo, Japan, July 1984, 423-428.

8. Bradfield, C. D., Roberts, J. B., & Karunendirkan, R., "Performance of an Electromagnetic Bearing for the Vibration Control of a Supercritical Shaft", Proc. Instn. Mech. Engrs., 1987, 201, No. C3, 201-211.

9. Fukata, S., et al., "Dynamics of Active Magnetic Bearings Composed of Solid Cores and Rotor" Memoirs of the Faculty of Engineering, Kyushu University, 1986, 46, No.3, 279-295.

10. Humphris R. R., et al., "Effect of Control Algorithms on Magnetic Journal Bearing Properties", 1986, ASME 86-GT-54.

11. Chen, H. M., 1988, "Magnetic Bearings and Flexible Rotor Dynamics", ASLE Annual Meeting at Cleveland, Ohio, May 1988.

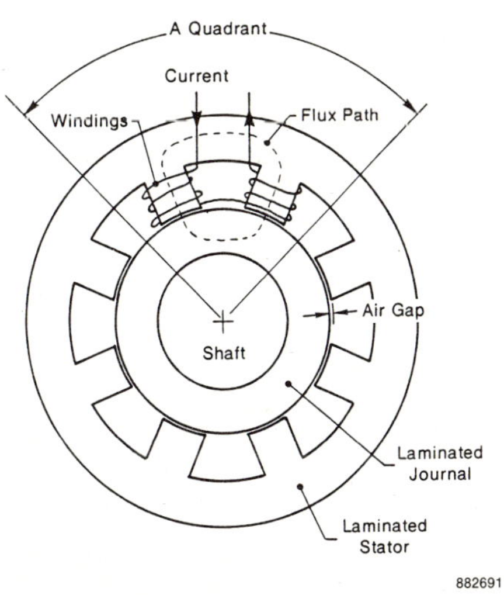

Fig. 1 An 8-pole configuration of an active magnetic bearing

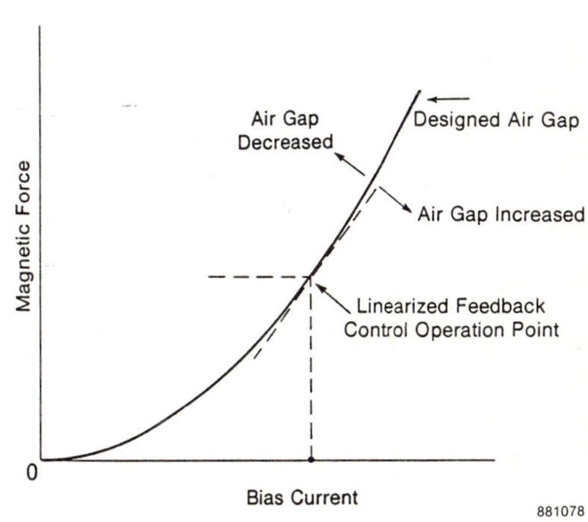

Fig. 2 Nonlinearity of magnetic force

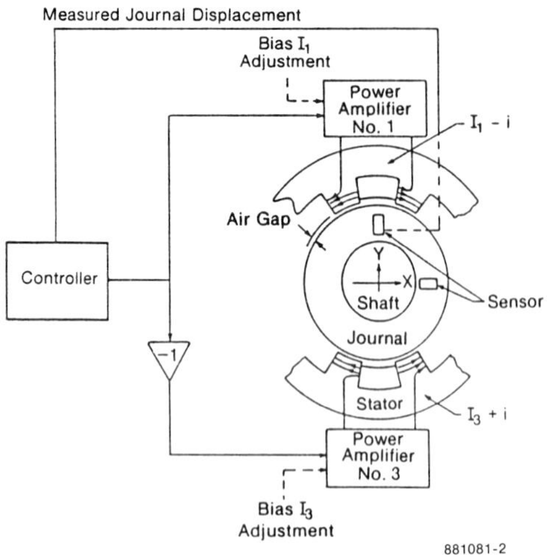

Fig. 3 An independently controlled axis

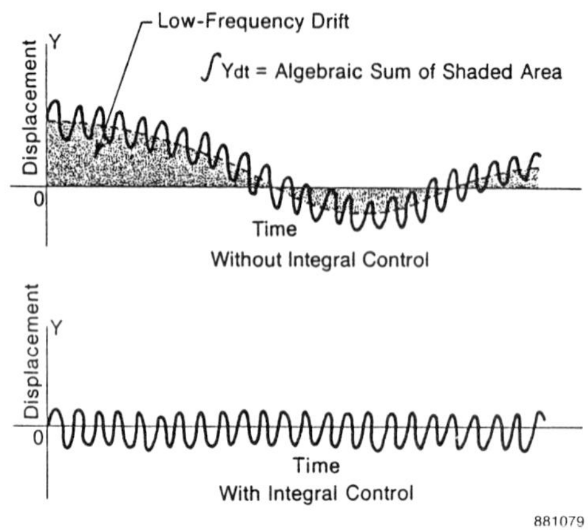

Fig. 4 Elimination of low-frequency drift through integral control

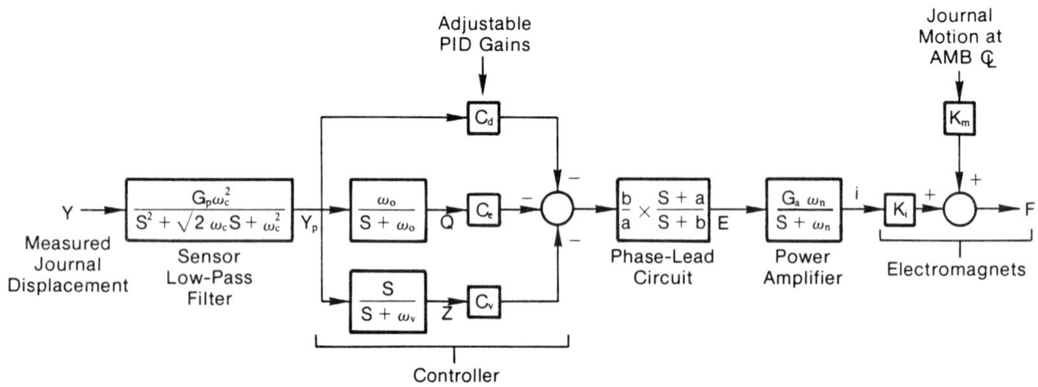

Rotor/AMB Force Coupling Equation

$$F = K_i\, i + K_m\, Y$$

AMB State Equations

$$i' + \omega_a i = G_a \omega_a E$$

$$E' + b E = -b/a\,[C_d Y_p' + C_c Q' + C_v Z' + a\,(C_d Y_p + C_c Q + C_v Z)]$$

$$Q' + \omega_o Q = \omega_o Y_p$$

$$Z' + \omega_v Z = Vp$$

$$V_p' = V_p$$

$$V_p' + \sqrt{2}\,\omega_c V_p + \omega_c^2 Y_p = G_p \omega_c^2 Y$$

882066

Fig. 5 A single-axis control diagram

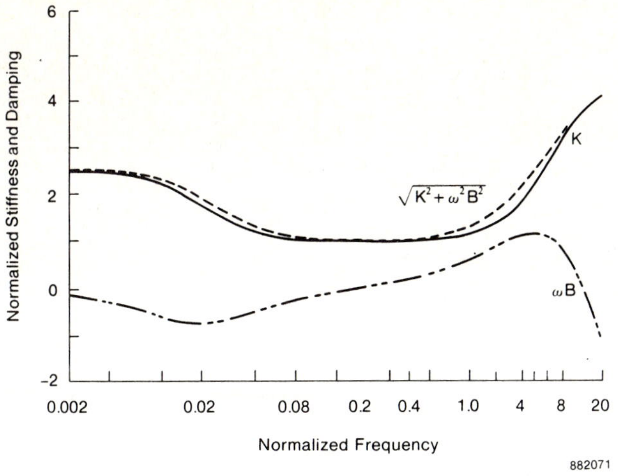

Fig. 6 AMB stiffness and damping - a numerical example

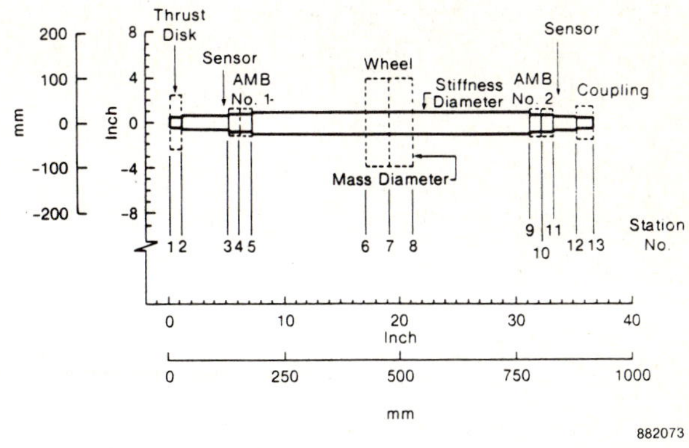

Fig. 7 A rotor model

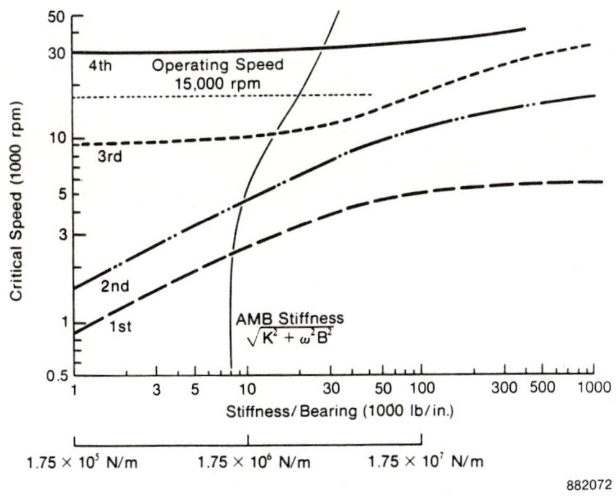

Fig. 8 Rotor critical speed map

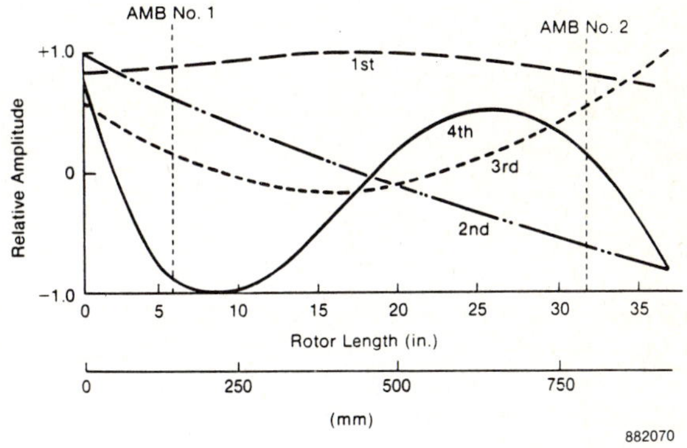

Fig. 9 Rotor critical mode shapes

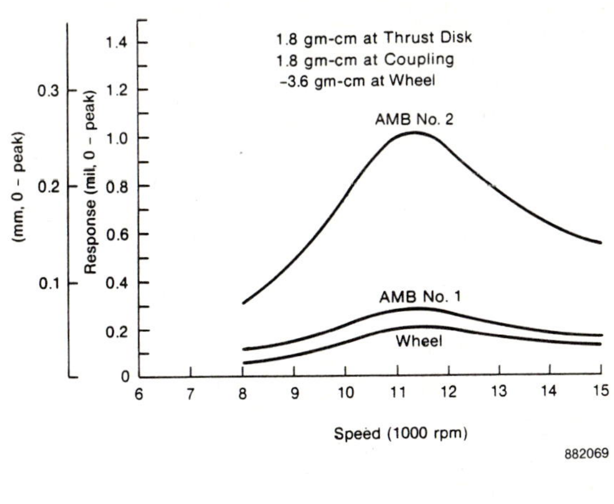

Fig. 10 Unbalance response at 3rd critical speed

Table 1 - AMB Dimensions and Parameters

L_p = 2.0 in. (50.8 mm)
D = 2.5 in. (63.5 mm)
C = 0.020 in. (0.5 mm)
A_p = 1.25 in.^{2} (8.06 x 10^{-4} m^2)
N_t = 100 turns
F_{max} = 200 lb (890 N)
I_1 = 3.5 A
I_3 = 2.0 A
K_i = 80 lb/A (356 N/A)
K_m = 12,500 lb/in. (2.19 x 10^6 N/m)
C_d = 0.26
C_v = 0.60
C_e = 1.00
G_p = 1000 V/in. (40 V/mm)
G_a = 1 V/A
ω_c = 5000 Hz
ω_n = 500 Hz
ω_o = 1 Hz
ω_v = 500 Hz

Table 2 - Damped Natural Frequencies of Forward Modes
rotor speed =15,000 rpm
cross-coupling stiffness (K_{xy}) at Wheel

	K_{xy} = 0		K_{xy} = -4000 lb/in. (-7 x 10^5 N/m)	
Mode	Frequency (cpm)	Log Decrement	Frequency (cpm)	Log Decrement
1st	2318	1.05	2374	-0.00
2nd	5263	2.20	5623	2.19
3rd	11,154	0.89	11,144	0.88
4th	34,156	0.06	34,156	0.06

SESSION XVI
KNOWLEDGE BASED SYSTEMS

Chairman: Dr C M Taylor

Paper XVI(i)

The incorporation of artificial intelligence in the design of herringbone journal bearings

K. Ishii, B. J. Hamrock and J. Klinger

Research was conducted to develop an intelligent computer program to aid in tribological design. Our focus is to integrate artificial intelligence (AI) and conventional numerical design techniques. As a vehicle for our effort, we focus on an example problem: the optimal design of herringbone journal bearings. We utilize an established numerical optimization method which, given the bearing parameters (requirements), finds the groove parameters (design variables) such that the radial load capacity of the bearing is maximized. AI techniques add another level of computer aid that incorporates qualitative issues as well as numerical. We use the method of design compatibility analysis (DCA) to form an outer, intelligent design loop that applies to the numerical optimization. DCA identifies any incompatibility between the design requirements and the design solution, explains the flaw, and suggests modifications to improve the design. Design issues considered include. (1) suitability of a hydrodynamic bearing, (2) safe minimum film thickness, (3) adequacy of load capacity and stiffness, (4) self-excited whirl instability ratio, and (5) flow rate, power loss, and attitude angle. We implemented both the numerical and AI programs on an IBM PC. The program uses Turbo-Prolog for DCA and Fortran for numerical optimization.

1. INTRODUCTION

Design is a creative process arrived at finding a solution to a particular problem. In all forms of design a problem may have many different solutions mainly because design requirements can be interpreted in many ways. For example, it may be desirable to produce

(1) The cheapest design
(2) Or the easiest to build with available materials
(3) Or the most reliable
(4) Or the one that is lightest in weight
(5) Or the one that takes the smallest space
(6) Or the best from any of a whole variety of possible standpoints

The task of the designer is therefore not clear cut, because he or she has to choose a reasonable compromise between these various requirements and then has to decide to adopt one of the possible designs that could meet this compromise.

The design of bearings is one of the critical ingredients of machine design. Designers need to understand the problem specification, to select the appropriate type of bearing, and then to assign values to the design variables such that all the functional requirements are satisfied. The difficulty arises from the varying nature of the specification. The problem specification involves not only functional requirements such as load capacity but also maximum bearing envelope, operational speed, bearing life, type of lubricant, etc. Note that some of these items could be considered as design variables over which designers have control (e.g., bearing envelope, speed, and lubricant type). In addition, the designer should pay attention to the manufacturing method. For example, the machining method dictates the surface finish of a bearing and a journal, which in turn, influences the allowable minimum film thickness.

The process of bearing design usually involves the following steps:

(1) Selecting a suitable type of bearing
(2) Estimating a bearing size that is likely to be satisfactory
(3) Analyzing the performance of the bearing to see whether it does in fact meet the requirements
(4) Modifying the design and operating parameters until the performance is near to whichever optimum is considered most important

The last two steps in the process can be handled fairly easily by someone who is trained in analytical methods and understands the fundamental principles of the subject. The first two steps, however, require some creative decisions to be made and for many people represents the most difficult part of the design process. Hence one can easily see that the design process is complex, requiring a thorough understanding of the field. It takes years of training to acquire the experience and intuition needed to be a good bearing designer.

Computers are a powerful tool for bearing designers. The recent development of personal computers has accelerated the trend of using computer programs to aid tribological designs. Most of the programs developed and used to date have focused on numerical requirements, either for analyzing or simulating the performance of a proposed design, or for optimizing the design parameters given an objective function and design constraints. These programs typically use procedural languages such as Fortran and Pascal and do not consider qualitative issues. The emerging field of AI offers techniques that add another level of computer aid to the already established numerical design programs. The symbolic computation ability allows AI programs to accommodate qualitative design issues as well as numerical issues. Among the many possibilities are:

(1) Help designers understand and organize problem specifications.
(2) Determine the most suitable type of bearing.
(3) Help designers identify limiting factors of design and use numerical tools.
(4) Evaluate the compatibility of a proposed solution at various stages of design.
(5) Identify incompatibilities in the proposed solution and provide suggestions.

Many researchers have applied AI to the design of machine elements. Dixon (1983, 1985) uses AI to guide designers of mechanical devices through relatively simple but iterative design problems such as the design of V-belts and heat fins. Ishii and Barkan (1987a) take an example of a hydraulic cylinder and developed the concept of Rule-based Sensitivity Analysis (RSA) that aids in optimization. Knowledge based approaches to bearing selection have been reported by Lin (1988), and Waldron (1987). These studies fall under step (2) above. Little work has been reported on the application of AI to the more detailed stages of bearing design that involve numerical programs. (steps 3, 4, and 5).

This research was conducted to develop an intelligent computer program for use in tribological design. Our focus was to integrate AI and conventional numerical design techniques. As a vehicle for our effort we selected an example problem: the optimal design of herringbone grooved journal bearings. This paper describes our development effort for this particular example. Specifically, we focus on the already established numerical optimization program, and the *evaluate-suggest* capability of AI techniques (steps 4 and 5 above).

1.1 Notation

b_1	width of groove, m
b_2	width of ridge, m
CKB	compatibility knowledge base
CKB(s)	the compatibility knowledge base concerning the element s
ckb-data	a data in CKB
H_o	film thickness ratio, h_{10}/h_{20}
h_{10}	film thickness in groove region when journal is concentric, m
h_{20}	film thickness in ridge region when journal is concentric, m
L	length of journal, m
L1	total axial length of groove, m
MCD	a set of matching compatibility data
MI	match index
MR	match range
M(s)	match coefficient
N	number of grooves
p_a	ambient pressure, N/m^2
R	bearing radius, m
U	bearing speed, m/s
W	bearing load, N
w_r	radial load capacity of herringbone journal bearing, N
α	groove width ratio, $b_1/(b_1+b_2)$
β	groove angle, degrees
γ	groove length ration, L1/L
Λ	dimensionless bearing number, $6\mu UR/p_a h_{20}^2$
μ	absolute viscosity, N s/m^2
σ	grooved or smooth member rotating

2. HERRINGBONE JOURNAL BEARING DESCRIPTION

A herringbone journal bearing is a fixed geometry journal bearing that has demonstrated good load capacity and stability characteristics for both incompressible and compressible lubrication. The bearing consists of a circular journal and bearing sleeve with shallow, herringbone-shaped grooves cut into either member. Figure 1 illustrates a partially grooved herringbone concentric journal bearing. Note the angled, shallow grooves in the journal surface. The grooves can be partial, as shown, or extend the complete length of the bearing. Also, the grooves can be placed in either the rotating or nonrotating surface. The purpose of these grooves is to pump the lubricant toward the center of the bearing, thereby raising the lubricant pressure in the bearing. This self-pressurization can increase the load capacity over that of a smooth bearing; it is also responsible for the herringbone bearing's good stability. The bearing shown in figure 1 is unidirectional; that is, it pumps inwardly for one direction of rotation.

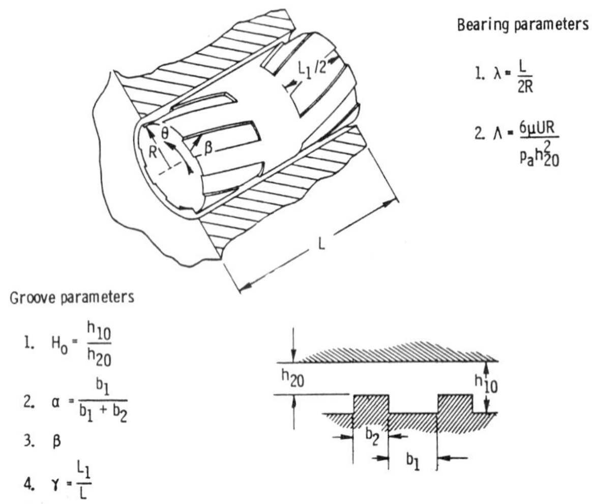

Fig. 1 Bearing and Groove Parameters

In figure 1 the number of grooves is six. However, the Vohr and Chow (1965) analysis used in this paper assumes essentially an infinite number of grooves. Reiger (1966) develops a criterion for the minimum number of grooves such that the infinite groove analysis yields valid results. This criterion indicates that the minimum number of grooves placed around the journal can be represented conservatively by

$$N \geq \frac{\Lambda}{5} \qquad (1)$$

where N = number of grooves

$$\Lambda = \frac{6\mu UR}{p_a h_{20}^2} = \text{bearing compressibilitity number}$$

In this paper we will adhere to the inequality given in (1). Figure 2 indicates the groove and bearing parameters. Hamrock and Fleming (1971) report a study on the optimization of self-acting herringbone journal bearings. Hamrock and Fleming developed a program that, given the bearing parameters λ and Λ (design requirements), finds the groove parameters (design variables) such that the radial load capacity of the bearing is maximized.

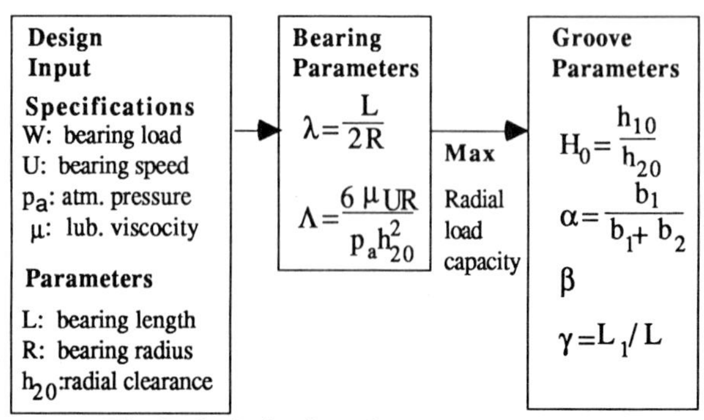

Fig. 2 Optimization of groove parameters

3. THE ROLE OF AI IN THE DESIGN PROCESS

As indicated previously, knowledge based techniques add another level of computer aid to a design synthesis program. Such techniques could identify incompatibilities between the design solution and factors not considered in the optimization program: bearing load requirement, method of surface finish, constraints on speed, etc. A knowledge based program also catches obvious overdesign, such as an unnecessarily large bearing envelope. Another important task for our program is to make intelligent suggestions to remedy the incompatibilities: use of a different lubricant or modify speed, radial clearance, or bearing envelope. The suggestions should also take into account the effectiveness (sensitivity) of each remedy and its implications for other design constraints.

Ishii and Barkan (1987b, 1988a) develops a framework for knowledge based design called design compatibility analysis (DCA) that accommodates the features described above. DCA focuses on the compatibility, both qualitative and quantitative, between design requirements and design solutions. DCA matches a given requirement and its solution with compatibility knowledge: good and bad templates of design. DCA combines the matching compatibility knowledge and gives a total evaluation. In addition, DCA provides justifications for the evaluation and suggestions for improvement that apply not only to the design solution, but also to the design specification.

This paper reports the use of DCA as a knowledge based design loop that applies to the optimization program developed by Hamrock and Fleming. The typical flow of a design session is as follows. The user of the program first enters the design requirements: bearing envelope (L and R), lubricant viscosity μ, bearing speed U, radial clearance h_{20}, and the bearing load W. DCA will perform a preliminary evaluation of the consistency of the input. Then the optimization program calculates the value of the groove parameters H_0, α, β, γ (i.e., the design solution). Then DCA asks additional questions (e.g., surface finish) and evaluates the compatibility of the design solution. If the design solution is unacceptable or is obviously an overdesign, the user can seek suggestions from DCA. DCA will suggest changes in bearing envelope, lubricant, bearing speed, or radial clearance. The overall program helps the designers to consider manufacturing concerns in their design and thus helps them make tradeoffs between performance and cost. (Section 4 gives a more detailed discussion and a flowchart.)

We implemented both the numerical and AI programs on an IBM PC. The program uses Turbo-Prolog for DCA and Fortran for numerical optimization. The overall program generates an executable file that the user can run without compilers or loaders.

4. OPTIMIZATION PROCEDURE FOR BEARINGS

The task completed by Hamrock and Fleming (1971) was to find the optimal herringbone journal bearing for maximum radial load capacity for various bearing parameters [Λ, λ, and whether the grooved or smooth member is rotating (σ)]. The radial load component is the component of the total load capacity in the direction of journal displacement. Therefore the problem reduces to

Given: Λ, λ and σ

Find: H_o, α, β, and γ, which statisfy

$$\frac{\partial w_r}{\partial H_0} = \frac{\partial w_r}{\partial \alpha} = \frac{\partial w_r}{\partial \beta} = \frac{\partial w_r}{\partial \gamma} = 0 \qquad (2)$$

the method used is a Newton-Raphson method of solving simultaneous equations.

5. DESIGN COMPATIBILITY ANALYSIS

5.1 Concept of Design Compatibility Analysis

One of the important tasks in engineering design is to ensure the compatibility of the elements of the proposed design with each other and with the design specifications (requirements, constraints, and production method). Major design decisions such as selection of components, determination of system type, and sizing of components must be made with the compatibility issue in mind. Some design features make a good match, but others may be totally incompatible for the required design specifications. Experienced designers usually know of a good combination of design choices for a particular situation. This knowledge can be viewed as good or bad templates of design. The knowledge can also be interpreted as an understanding of compatibility. For example, a rolling-element bearing is not suitable for an automobile connecting rod bearing. AI techniques provide us with a framework for storing this compatibility information in a knowledge base and for evaluating the overall degree of compatibility of a given design. Examples of such compatibility knowledge are expressed below in a rule format.

Example 1

(**Machining** of bearing is to be done by grind, lap, and superfinish)
(**Minimum allowable film thickness** for this machining process is between 0.0001 and 0.00025 in.)
(**Design film thickness** is 0.0002 in.)
=>
(Design of bearing using grind, lap, and superfinish is **EXCELLENT**)
Reason: (Grind, lap, and superfinish will meet the required surface finish.)
Suggestion: (none) (3)

Example 2

(The **load** specification for the bearing is 1 lb.)
(The **maximum load** from optimization is 10 lb.)
=>
(Design for load requirement is **POOR**)
Reason: (You have overdesigned your bearing for the required load)
Suggestions:
(Reduce the bearing envelope (length/diameter))
or (Increase the radial clearance)
or (Use a higher viscosity lubricant) (4)

Example 3

(The Reynolds number for this design is greater than 2000)
=>
(You are operating with Taylor vortex flow. Your design is **VERY BAD**)
Reason: (The optimization program is valid only for laminar flow)
Suggestions:
(Decrease the radial clearance)
or (Decrease the required speed) (5)

In addition to the task of expressing the compatibility itself, we must deal with the degree of confidence of such compatibility data. Some data may indicate total incompatibility; others may only suggest avoiding a certain type of design. We cannot altogether rule out undesirable choices or costly designs, since they may, in some cases, provide the only alternative given all the other constraints. Also, we need to weight the evaluation depending on the importance of each individual element. An undesirable combination of major components should be weighted higher in the evaluation than an undesirable combination of minor parts.

The theory of fuzzy measure (Ishii and Sugeno, 1985) provides a sound mathematical basis for the treatment of the degree of confidence. We can consider the evaluation of compatibility, or the match index (MI), as the utility of design. The utility of individual components can be defined as the normalized weight of importance. The match index essentially depends on the compatibility templates that apply to a given situation.

Design compatibility analysis (DCA) utilizes the compatibility knowledge in helping designers generate a good design. The fundamental strategy of DCA is as follows (figure 3). The user first defines the specifications and describes the proposed design . The proposed design may be a partial design of the entire system with some parts of the design left undecided. The expert system then infers some physical characteristics of the proposed design and some physical requirements from the user-provided data. Once all these data are complete, the expert system evaluates the compatibility of the proposed design and indicates it in terms of the utility of the entire design and the match index, and also provides a justification for the evaluation and suggestions for improvement.

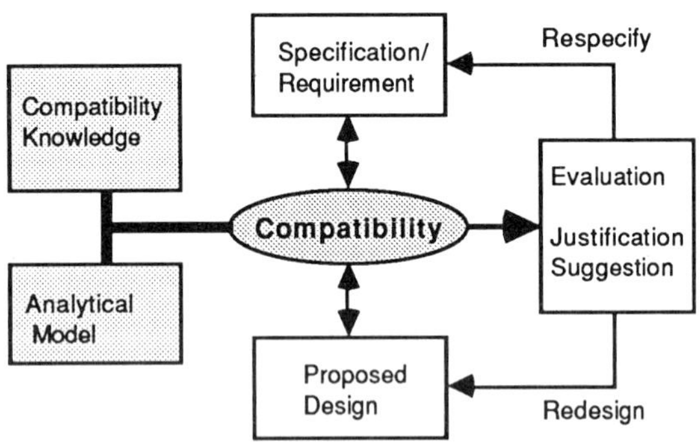

Fig. 3 Concept of DCA

The match index is a normalized scale between 0 and 1: An MI of 0 indicates an absolutely incompatible design, an MI of 1 is a perfectly balanced design, and an MI of 0.5 indicates that no compatibility information is available. Ideally, if compatibility KB is reasonably rich, the user would rarely obtain an MI of 0.5. The MI is a normalized compatibility evaluation of the proposed design; the physical interpretation of the match index depends on the user's interpretation of *compatibility*. If the knowledge base represents qualitative compatibility information between *components* and their specifications, then the MI is best interpreted as the *fractional part of the entire design that is compatible* and well matched. If the knowledge base represents the cost of the design elements, the MI is interpreted as *the normalized cost index with respect to a certain* ***standard*** *cost*. Overall, the designer can then utilize this information to update and improve the proposed design, or alternatively, to compromise on the specifications. (Note that DCA flags not only bad designs, but also inappropriate specifications.) The designer can start by eliminating obvious incompatibilities and then further seeking an improved design.

The advantages of taking this approach in DCA over the conventional decision-making methods, where each decision is made in a sequential fashion, are as follows:

(1) By evaluating the entire design, and detecting every incompatible aspect of design, DCA allows the designer to make flexible decisions that will remedy the incompatibilities, thereby making the decision process rapidly converge to a sound solution.

(2) The theory of fuzzy measure provides DCA with the match index as an indicator for the soundness of design. The computation of the MI is based on the utility of the design. The designer can control the weight of each element in DCA evaluations of the MI by assigning an appropriate utility value for each design element.

5.2 Measure of Compatibility (MI)

By focusing on the theory of fuzzy measure (Ishii and Sugeno, 1985), we define a quantitative measure for the compatibility of design, which we call the match index (Ishii and Barkan, 1987b). Figure 4 illustrates how DCA computes the match index and compiles justifications and suggestions.

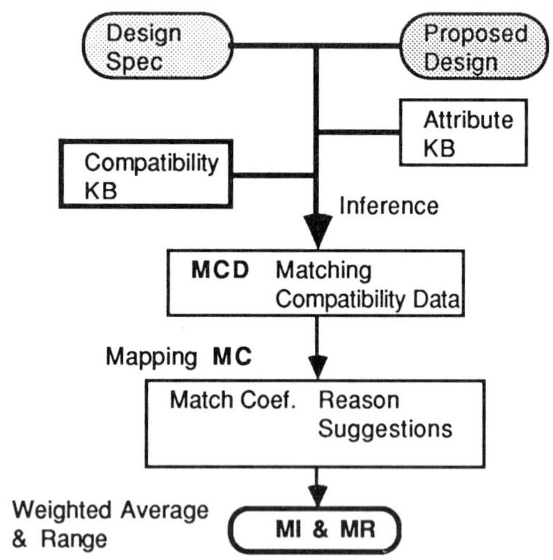

Fig. 4 The deduction process in DCA.

Given a description of the proposed design and the design specification, DCA uses the attribute knowledge base (KB) to establish more detailed requirements and physical characteristics of the proposed design. DCA next matches the situation with the compatibility KB and infers a set of matching compatibility data (MCD) for each design element. DCA then maps MCD into [0,1] to obtain a match coefficient (M(s)) compatibility evaluation for each element, according to a function MC. Finally, the *weighted sum* of the match coefficients yields the match index (MI). The range of match coefficients represents the match range (MR).

The following sections give a mathematical description of the deduction process of the ***match index*** and the associated ***justifications*** and ***suggestions***.

5.2.1 The Match Index

Def. 5.1: The match index (MI). Consider a design comprising a set of elements **K**.

$$MI = \Sigma_K \text{utility}(s) \cdot M(s), \quad s \in K \qquad (6)$$

where: utility (s) is the weight of the evaluation of a component *s*. (Σutility(s) = 1.0)
M(s) is the compatibility of an element **s** with the rest of the elements and the design specification.

Hence, the match index is essentially the weighted sum of the match coefficient (M(s)). The match coefficient (M(s)) is the compatibility evaluation of a design element with the rest of the design and with the specification (c.f.

section 5.2.2). Def. 5.1 gives the MI a meaning of overall utility. If all the elements are deduced to be totally compatible (i.e., M(s) = 1.0 for all s), then our MI is 1.0, indicating an excellent design. If there are incompatibilities or inadequacies associated with a certain element, the design loses some of its utility.

Def. 5.2: The Match Range (MR). Consider a design comprising a set of elements **K**.
$$MR = [\ \min\{M(s) \mid s \in K\}, \max\{M(s) \mid s \in K\}\ ] \quad (7)$$

The match index is only an averaged measure of compatibility and does not reflect situations where most design elements are compatible but some minor elements are not (low M(s)). This element may seriously jeopardize the entire design. The match range will indicate such unbalance in the design. Hence a good design gives a high MI and a narrow MR. Note that the designer should use these measures in conjunction with the deduced information: justifications for the evaluation and suggestions for improvement.

5.2.2 The Match Coefficient M(s)

In order to compute the match index, equation (6) requires the match coefficients (M(s)) of design elements *s*. The match coefficient is a [0,1] evaluation of the compatibility of the design element with respect to the rest of the design and the design specifications. For a given proposed design under a certain design specification, the match coefficients are derived from the compatibility knowledge base. The derivation comprises the following steps:

(1) Deducing the set of ***compatibility data*** that matches the situation.
(2) Mapping the deduced set of compatibility data into the range [0,1]. The nature of mapping depends on the field of application.

The compatibility knowledge base **CKB** contains the compatibility data in the following form:

ckb-data (<ID>,
<Elements>,
< Descriptor>,
< Reason>,
<Condition>) (8)

where
<ID> is the identification number of this compatibility data.
<Elements> is the set of design elements concerned.
<Descriptor> indicates the evaluation of compatibility. Descriptor could be an adjective describing the match (e.g., poor, good) or numerical values indicating relative cost, etc.
<Reason> is the justification for the match.
<Condition> is a list of preconditions for this clause.

First, for any given proposed design under certain design specifications and constraints, we can derive a set of matching compatibility data from the compatibility knowledge base. We label this deduction **CD**. For a given design element s,

$$\mathbf{CD: C \cup D \cup CKB(s) \Rightarrow MCD} \quad (9)$$

where
C is a conjunction of predicates describing the specifications.
D is a conjunction of predicates describing the design elements.
CKB(s) is the compatibility knowledge base concerning the element s, that is,
s ∈ (list of elements in **CKB**)
MCD a set of matching compatibility data.
A => B indicates that A implies B.

Next, we further map the set of matching compatibility data to a [0,1] assessment according to an application-specific mapping **MC**.

$$\mathbf{MC: MCD} \rightarrow [0,1]. \quad (10)$$

Def. 5.3: Match coefficient (M(s)) is a [0,1] scale derived by equations (9) and (10).
$$\mathbf{M(s)} = \mathrm{CD} \bullet \mathrm{MC}\,(\,\mathrm{C} \cup \mathrm{D} \cup \mathrm{CKB(s)}\,) \quad (11)$$

As mentioned before, the mapping **MC** depends on the application. Specifically, users should define **MC** according to how they describe compatibility in the compatibility data. The main reason behind applying **MC** is to normalize the evaluation of each design element and thus allow us to compare uniformly the compatibility of each element. In this paper, we use compatibility data with qualitative descriptors. Here, an adjective **w** describes the level of compatibility. For example, **w** ∈ **W**={excellent, very good, good, poor, bad, very bad}. Hence, the set of matching compatibility data (**MCD**) contains *n* numbers of adjective descriptors (i.e., **MCD** = W^n). In order to map W^n to [0,1], we use a function **Translate: W**→ [0,1] to each descriptor. For example,
{excellent, very good, good, poor, bad, very bad}
→ {1.0, 0.8, 0.6, 0.4, 0.2, 0.0}
Hence, we obtain n numbers (i.e., $[0,1]^n$). We further introduce a mapping **Bestinfo:** $[0,1]^n \rightarrow [0,1]$ to derive the match coefficient.

Def. 5.4: Bestinfo(V) is defined as follows:
(1) The maximum in the set **V** , if the set consists only of numbers greater than or equal to 0.5 (i.e., there is no "negative" comment about the design).
(2) The minimum in the set **V**, if the set contains at least one number less than 0.5 (i.e., if there is at least one "negative" comment about the design).
(3) 0.5 if the set **V** is empty, indicating neutral compatibility.

Def. 5.5: MC = bestinfo(translate **(MCD)**) (12)

Hence, we first take the matching compatibility data (**MCD**), which contains n numbers of adjective descriptors, map each of them to the range [0,1], and then apply **bestinfo** to obtain a [0,1] number.

6. INTEGRATED PROGRAM: HERRINGBONE-EXP

The integrated program called Herringbone-exp combines the concepts of DCA and the use of a herringbone groove journal bearing optimization program. Herringbone-exp uses the power of Prolog's object-oriented programing and recursion for DCA and the power of Fortran's procedural computation ability for the optimization. This combination provides a strong useful design tool, because in the design of any mechanical system both qualitative and quantitative issues are considered.

The DCA portion was implemented by using Turbo-Prolog; and the optimization section, by using standard Fortran 77. The integrated program runs on an IBM PC as an executable file. The user interaction with Herringbone-exp is shown in figure 5.

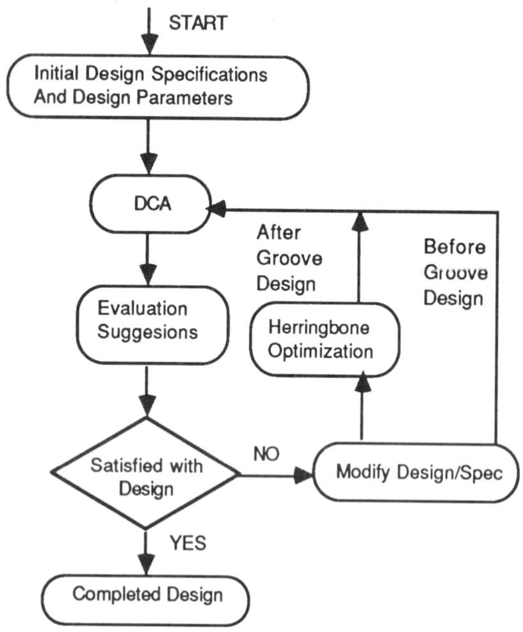

Fig. 5 Flow of design using Herringbone-exp

Users start by entering the design specification for the bearing, the required load, the bearing speed, and the lubrication used. Then the user enters a few design parameters, the bearing length and radius, the radial clearance, the machining process that will be used to manufacture the bearing, and whether the smooth or the grooved member will be rotating.

After the initial design specifications and design parameters are entered, the user does an initial DCA. If the evaluation is less than satisfactory, modifications can be made to the design parameters, the specifications, or both. If the initial design is satisfactory, the herringbone groove optimization is carried out on the initial design. The results of the optimization program are then used to do another DCA on the bearing. Suggestions are made for improvements to the design. Again the user can change the design parameters, the specifications, or both and continue the design cycle. The process is carried out until the user is satisfied with the results.

For this example of herringbone-grooved journal bearings, specific design and performance rules are contained in the knowledge base of the program. Examples of some of the compatibility knowledge were given in section 5.1.

The information contained in the examples 1 to 3 is very powerful to the user. The knowledge base when used with DCA will evaluate the design as excellent, very good, bad, etc., list the reasons for the evaluation, and list suggestions for improvements. This type of knowledge base will help the designer obtain the best possible solution for the design cycle of the bearing in a short time.

7. IMPLEMENTATION

The Herringbone-exp program comprises of three main modules that share a common data base. The top executive module's main purpose is to control the flow of data between the Prolog user interface and DCA module, and the Fortran herringbone optimization load module. (figure 6) When the program is run, the top executive module is initiated first and proceeds to give control to the Prolog user interface and DCA module.

The Prolog user interface and DCA module has two principal functions. The first is the user interface function. The user interface plays an important role in allowing the user to enter the design specifications and design parameters in a user-friendly way. The second function of the Prolog user interface and DCA module is to run the compatibility analysis. As previously discussed, DCA will evaluate the present design and give suggestions for improvement.

After DCA is completed, control will then go to the top executive module and the design parameters will be passed to the Fortran herringbone load optimization module. In the Fortran herringbone load optimization module the design parameters are used to find the groove configuration that will maximize the load capacity.

Once the optimum values are obtained, the top executive module will transfer the data to the Prolog user interface and DCA module. This interaction will continue between the three modules until a satisfactory design is achieved.

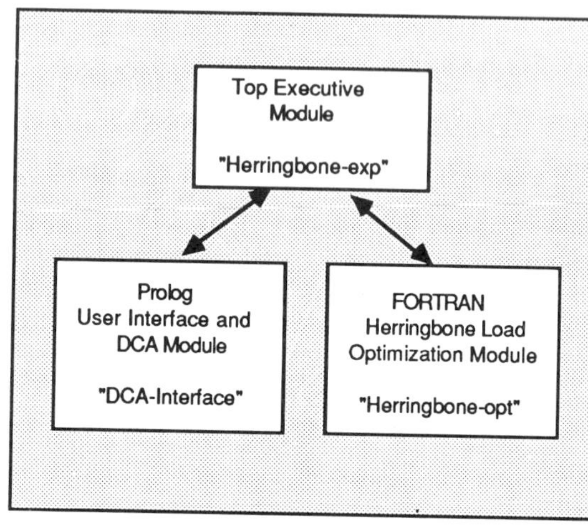

Fig. 6 Program structure of Herringbone-exp

8. ILLUSTRATIVE EXAMPLE

The power of the Herringbone-exp program is best observed through an example. Suppose we need a herringbone journal bearing with the following specifications:

load: 55.0 lb.
Speed: 38 000 rpm
Viscosity: Air

The initial design parameters are taken to be

Bearing length: 1.0 in.
Bearing radius: 0.5 in.
Radial clearance: 0.0007 in.
Grooved member rotating
Machining process: grind, lap, and superfinish

Figures 7 and 8 show the menus that the user will interact with when entering the information.

After the first DCA the match index yields 0.4 which is poor. The machining process is over specified for the required minimum film thickness. In order to improve the design, the designer changes the machining process to grind and lap. Again the DCA is done and our design seems satisfactory to this point, without taking into account the groove design from the optimization routine. Now the optimization routine is run. After the results from the optimization routine are obtained, the user will check DCA to see if the design is satisfactory, figure 9. The match index now indicates that our design is 0.01, which is very bad, figure 10. We have not met our load requirements. One suggestion is to decrease the radial clearance. The user changes the clearance to 0.0006 inches and runs DCA to make sure new incompatibilities do not arise. The optimization routine is then conducted on the new design, figure 11. Finally, the user checks the compatibility. This time The match index is 1.0, which is an excellent design. The user, being satisfied with the design ends the program.

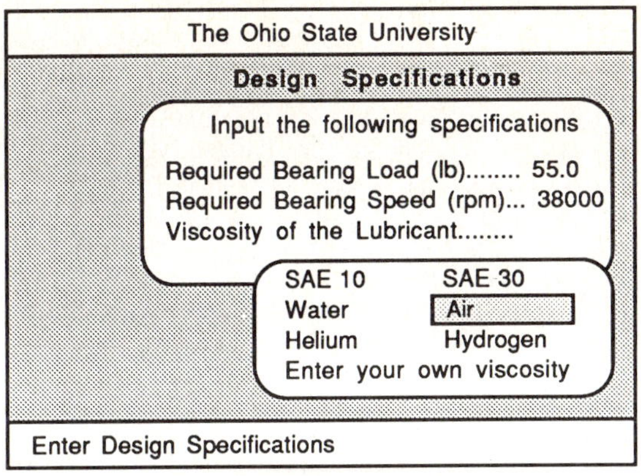

Fig. 7 Specification menu screen

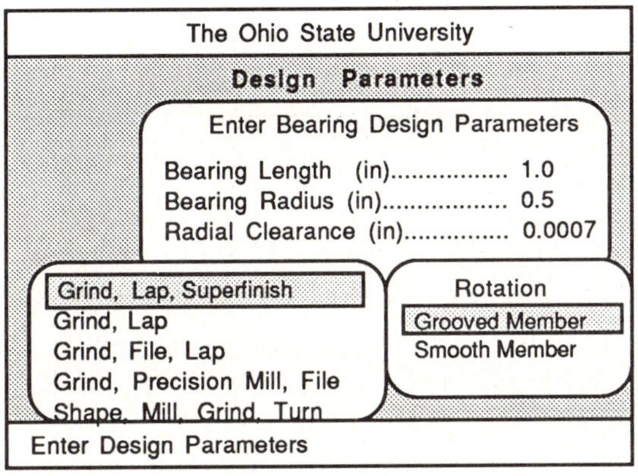

Fig. 8 Parameter input screen

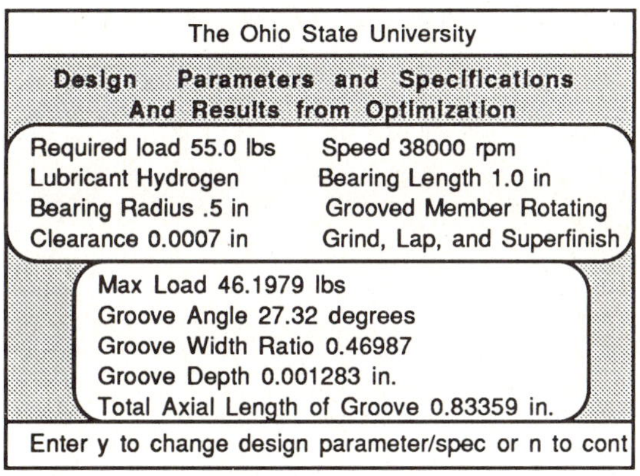

Fig. 9 Optimization output screen

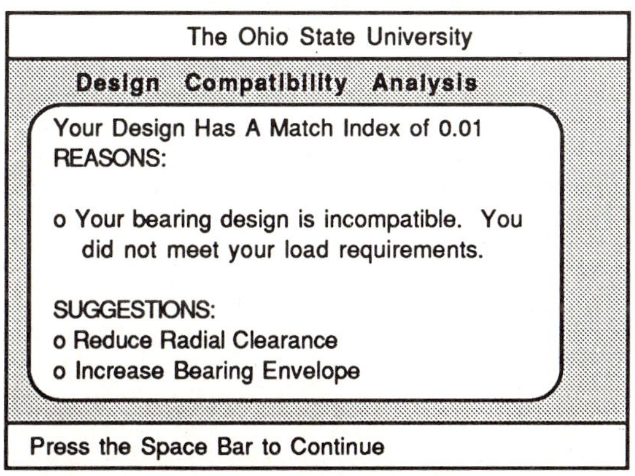

Fig. 10 DCA output screen

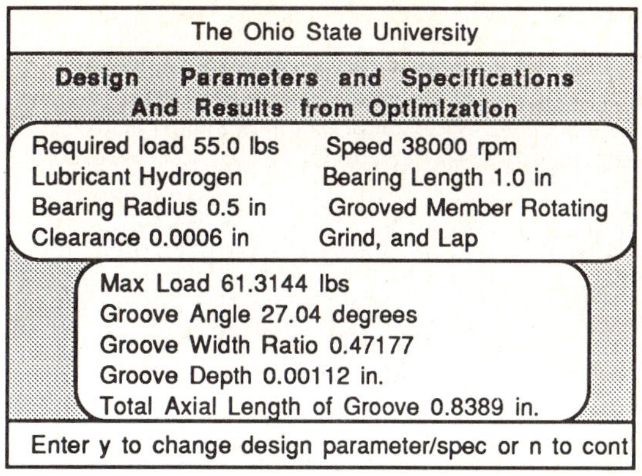

Fig. 11 Final result

9. CONCLUSION

A unique computer aid was developed to assist the engineer in designing a herringbone grooved journal bearing. Generally, software packages deal with only optimization or design rules. However, this program is unique because, it integrates both the numerical optimization methods required for the design of herringbone journal bearings, and AI techniques needed for life cycle design. We feel that by incorporating both optimization and AI that the design for herringbone journal bearings will be easier and a less expensive bearing will be produced.

Conventionally, the design decisions are made sequentially with an iterative process until a satisfactory design has been reached. Combining DCA with the optimization has the advantages of evaluating the overall design at once from various viewpoints. DCA handles not only totally acceptable and unacceptable compatibilities, but assesses the relative desirability of a possible choice.

In the future, we plan to expand this program to include additional design rules. (eg. material selection, type of bearing mounting, etc.) Also, we plan on incorporating the numerical optimization analysis for stability of externally pressurized herringbone grooved journal bearings. Finally, we would like to utilize the concept of Rule-based Sensitivity Analysis (RSA). RSA would review the suggested design changes from DCA and evaluate their effectiveness on the entire design. For example, DCA may suggestion changing the radial clearance or the bearing length. RSA would take this information and analyze the effects of the entire design when one of the suggested changes is made. Then, RSA would inform the designer of the best fix solution to the design.

We believe that by combining the power of both numerical optimization and DCA, we have created a useful tool for engineers. Our system will prevent subtle incompatibilities that may seriously undermine the design. This feature will help not only novice designers, but also experienced designers.

10. ACKNOWLEDGEMENTS

This research is funded partly by the National Science Foundation DMC-8810824, and the Ohio State University seed grant.

References

Dixon, J.R. and Simmons, M.K. (1983) Computers that design: expert systems for mechanical engineers. ASME Computers in Mechanical Engineering, Nov., pp.10-18.

Dixon, J.R. and Simmons, M.K. (1985) Expert system for mechanical design: a program of research. ASME paper No. 85-DET-70.

Grassam, N. and Powell, J. (1964) Gas Lubricated Bearings. Butterworths, London. pp. 27-84.

Hamrock, B.J. and Fleming, D.P. (1971) Optimization of self-acting Herringbone journal bearings for maximum radial load capacity. NASA Technical Note, NASA TN D-6351.

Ishii, K. and Sugeno, S. (1985) A model of human evaluation process using fuzzy measure. Int. J. of Man-Machine Studies, 22, pp.19-28.

Ishii, K. and Barkan, P. (1987a) Rule-based sensitivity analysis--a framework for expert systems in mechanical design, in Gero, J. (ed.), Expert systems in Computer -aided Design, North-Holland, Amsterdam, pp.1979-198.

Ishii, K and Barkan, P. (1987b) Design Compatibility Analysis--a framework for expert systems in mechanical design. ASME Computers in Engineering 1987. Vol. one, pp.95-102.

Ishii, K., Adler, R. and Barkan, P. (1988a) Knowledge-based Simultaneous Engineering using Design Compatibility Analysis. Artificial Intelligence in Engineering: Design, Computational Mechanics Publications, Southampton. pp.361-378.

Juvinall R. (1983) Fundamentals of Machine Component Design. John Wiley and Sons. pp 387-425.

Lin, G. (1988) A Knowledge-based Expert System for the Design of Bearings, MS Thesis, Department of Mechanical Engineering, The Ohio State University.

Rieger, N.F. (1966) Design of Gas Bearings. Vol. I: Design Notes. Mechanical Technology, Inc., p. 6.1.35.

Vohr, J.H. and Chow, C.Y. (1965) Characteristics of Herringbone-Grooved, Gass Lubricated Journal Bearings. J. Basic Engineering., vol. 87, no. 3, pp. 568-578.

Waldron, K.J. and Waldron, M.B. (1987) An Expert System for Initial Bearing Selection. ASME paper #86-DET-125.

Paper XVI(ii)

Bearing selection using a knowledge based system

R. T. Griffin, M. J. Winfield and S. S. Douglas

Industry is beginning to realize the need for the utilization of existing tribological knowledge via efficient, accessible and intelligible technology transfer techniques. One approach which may be suitable for providing such a transfer path is artificial intelligence (AI). To demonstrate its suitability, two prototype knowledge based systems (KBS) were developed for the application domain of rotary rolling-element bearing selection, one with an expert system (ES) shell, the other an artificial intelligence programming language.

The practicalities of acquiring knowledge about the domain from documentation and relevant authorities is outlined and the representation of the knowledge is described. The system design, operation and results are described and a comparison of the two systems developed is provided.

1 INTRODUCTION

It is recognized the transfer of tribological knowledge from academia to companies competing in the market place requires improvement (1). Recent developments in technology transfer utilize database techniques. Another approach which potentially exhibits favourable characteristics for effecting technology transfer is knowledge based systems (2). Knowledge based systems (3) have the ability to explicitly record knowledge in a form readily useful to users together with a computer program to manipulate and 'reason' with that knowledge.

The aim of this paper is to illustrate the suitability of knowledge based systems as a means to effect technology transfer. The application domain chosen to evaluate a knowledge based approach was the selection of rotary rolling-element bearings. Bearing selection was chosen for its importance to mechanical engineering, and has been limited to two supporting bearing locations which is the most common application.

Knowledge based systems are being used in many manufacturing and engineering applications (4,5), especially in the design domain (6,7,8). Bearing selection is frequently encountered within the design activity and techniques are available to assist in this process (9,10), giving functional and qualitative guidance. In order to fully transfer a technology it is our recommendation that functional, qualitative and quantitative aspects of the selection process need to be considered.

This paper reports the development of knowledge based systems designed to assist the selection of rotary rolling-element bearings.

2 KNOWLEDGE ENGINEERING

There are two phases associated with knowledge engineering; the first phase is the acquisition of knowledge and the second is the representation of knowledge within a formalism:

2.1 Knowledge acquisition

This is a demanding task and the following general difficulties were encountered:

(1) As a process it is time consuming.
(2) Access to relevant expert(s) was usually limited.
(3) Expert(s) tend to meander from specific points under discussion.
(4) The knowledge obtained from an expert is not usually in a form easily encoded within a knowledge based system.

Many knowledge acquisition techniques are currently established, the most commonly used being described by Hart (11). Three knowledge acquisition techniques were used for the bearing selection project. Firstly, documentation associated with bearings and bearing selection were studied (12). Secondly, the repertory grid technique (13) was used during discussions and interviews with a bearing selection expert. In essence, the grid produces a relief map of the important considerations of the expert during the selection process. Figure 1 illustrates a grid, here a number of bearing types are rated against a number of bearing attributes. An analysis performed on the grid is used to extract a hierarchy of bearing types and inference rules for selection. The information obtained using the repertory grid can be viewed as a set of examples of selection strategies. This enabled a third technique called rule induction (14) to be used which automatically generates inference rules from examples.

2.2 Knowledge representation

Many representation formalisms are available, the most common and popular being production rules (15), frames (16), semantic nets (17), and logic (18). Arising from the knowledge acquisition a hierarchy of considerations were identified which provided a procedural view of how the expert utilized his knowledge in the form of a tree structure similar to that of figure 2. This type of knowledge structure readily lends itself to translation into rules for a rule-based system, which uses knowledge constructs of the form shown in figure 3. If a set of conditions exist then a conclusion can be derived. Using such a representation formalism allows either an expert system shell such as Apes (19), which is a front-end to Micro-Prolog, or an artificial intelligence programming language such as Micro-Prolog (20) to be considered for system development. A shell is a complete system without a knowledge base. Shells are used for rapid development of prototypes by insertion of a knowledge base.

ELEMENT(E):- BEARING TYPE													
CONSTRUCT(C):-BEARING TYPE ATTRIBUTE													
PROJECT:-BEARING SELECTION				TOPIC:-BEARING TYPE SELECTION						DATE:-24/5/88			
KNOWLEDGE ENG.(S):-R.T. GRIFFIN				EXPERT(S):-J. SCHOFIELD						TIME:-40 MIN			
C	C-NAME	C-VALUE=1	E1	E2	E3	E4	E5	E6	E7	E8	E9	E10	C-VALUE=5	C
C1	RAD-LOAD	NONE	5	4	3	4	1	1	2	2	3		V-HIGH	C1
C2	AXL-LOAD	NONE	2	1	3	3	4	5	3	5	2		V-HIGH	C2
C3	2-WAY-AXL-LD	NO	5	1	1	5	1	5	5	5	5		YES	C3
C4	ROTARY-SPD	V-LOW	3	1	2	1	2	2	5	3	3		V-HIGH	C4
C5	SELF-ALIGN	NO	5	1	1	1	1	1	1	1	5		YES	C5
C6	IN-RING-SEP	NO	1	1	5	1	1	1	1	1	1		YES	C6
C7	OUT-RING-SEP	NO	1	1	5	1	1	1	1	1	1		YES	C7
C8	AXL-BND-RES	NONE	1	5	4	4	1	1	3	1	1		V-HIGH	C8
C9	DYN-LOAD-CAP	V-LOW	3	4	3	3	1	1	1	3	2		V-HIGH	C9
C10														C10

RADIAL_SPHERICAL_ROLLER
RADIAL_NEEDLE_ROLLER
SINGLE_ROW_RADIAL_TAPERED_ROLLER
DOUBLE_ROW_RADIAL_TAPERED_ROLLER
SINGLE_DIRECTION_THRUST_BALL
DOUBLE_DIRECTION_THRUST_BALL
SINGLE_ROW_DEEP_GROOVE_BALL
THRUST_SPHERICAL_ROLLER
SELF_ALIGNING_BALL

FIGURE 1-EXAMPLE REPERTORY GRID FOR BEARING SELECTION

Micro-Prolog is a software tool with which bespoke systems can be developed. It is a logic programming language and is available upon personal computers. Figure 4 shows examples of Micro-Prolog codification for the knowledge illustrated in figure 2.

3 SYSTEM SPECIFICATION

Following the knowledge engineering activity the system requirements were identified as:

(1) Where possible, select a previously identified bearing arrangement from a set of standard arrangements for

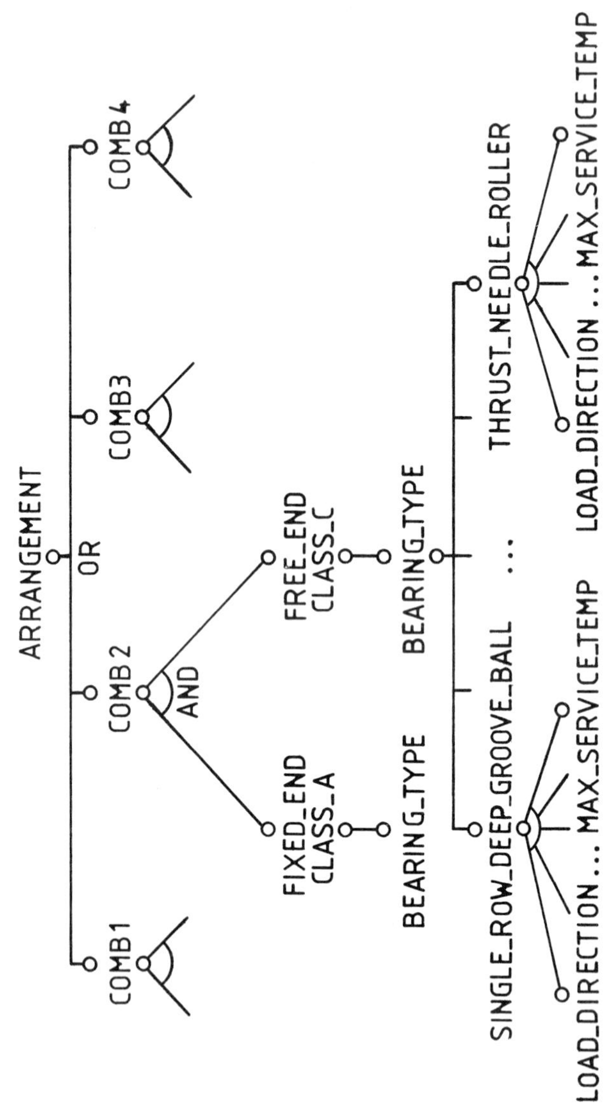

FIGURE 2-TREE STRUCTURE FOR BEARING SELECTION

```
IF   < CONDITION-1 >
AND  < CONDITION-2 >
 .          .
 .          .
 .          .
AND  < CONDITION-N >
THEN < CONCLUSION >
```

FIGURE 3-GENERAL FORM OF RULES

```
((ARRANGEMENT _COMB)
  (COMB1 _COMB) )
((ARRANGEMENT _COMB)
  (COMB2 _COMB) )
((ARRANGEMENT _COMB)
  (COMB3 _COMB) )
((ARRANGEMENT _COMB)
  (COMB4 _COMB) )

((COMB1 _COMB)
  (FIXED_END_CLASS_A _TYPE1)
  (FREE_END_CLASS_B _TYPE2)
  (EQ _COMB (_TYPE1 _TYPE2)) )
```

a) Micro-Prolog code for rules

```
  if the COMB1 is _COMB
then the ARRANGEMENT is _COMB
  if the COMB2 is _COMB
then the ARRANGEMENT is _COMB
  if the COMB3 is _COMB
then the ARRANGEMENT is _COMB
  if the COMB4 is _COMB
then the ARRANGEMENT is _COMB

  if the FIXED_END_CLASS_A is _TYPE1
 and the FREE_END_CLASS_B is _TYPE2
then the COMB1 is (_TYPE1 _TYPE2)
```

b) English translation for rules

FIGURE 4-RULES CODED IN MICRO-PROLOG

specific applications and select the individual bearing from the database, or

(2) Undertake a fundamental bearing selection study.

In essence the fundamental study must be capable of seeking the best possible arrangement of bearing types for an application, and ultimately convert the bearing type into a physical component extracted from a database of bearings.

To satisfy this, the system has three levels of information which it utilizes to infer a solution. On the first level the functional requirements associated with the application are established, an example being the requirement for a bearing to take both radial and axial loads. The second level identifies the qualitative aspects of the application. For example at this level cost and reliability could be considered. The third level identifies the quantitative aspects facilitating the database search. For example, bearing life, bearing bore diameter.

The domain has initially been constrained to a single generic arrangement of bearings, that of one or more bearing types at two supporting locations. The bearing types addressed are shown in figure 5.

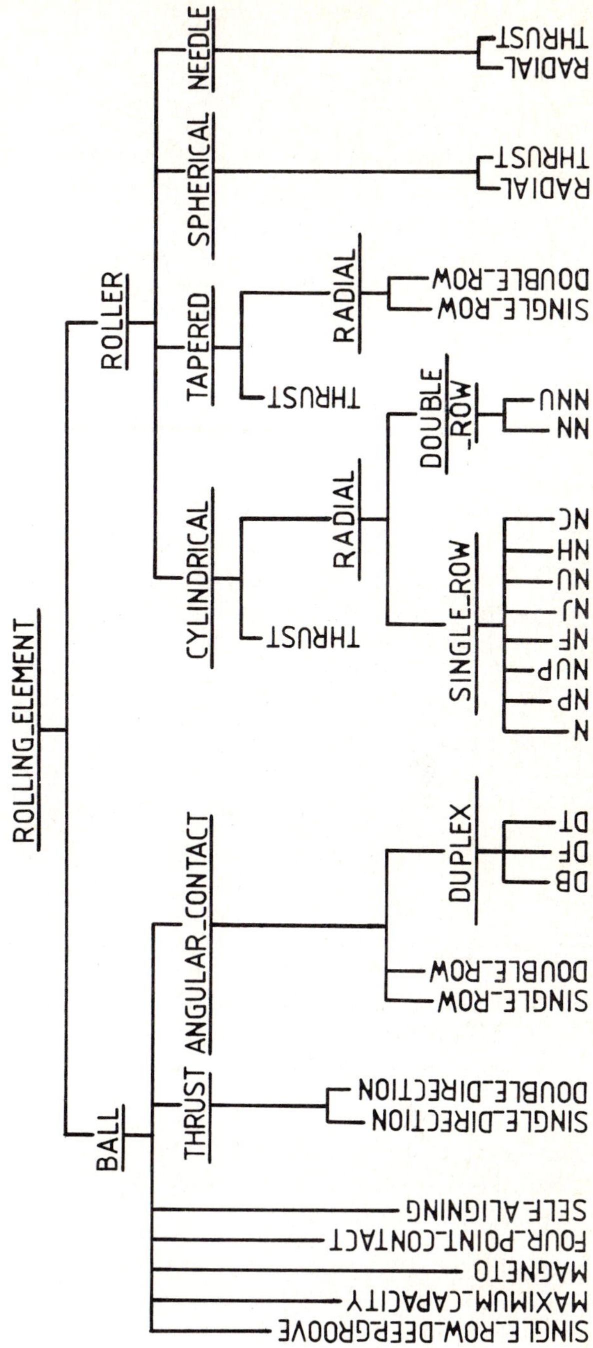

FIGURE 5-HIERARCHY OF BEARING TYPES WITHIN BEARING SYSTEM

4 SYSTEM DESIGN

The system has four major components, their relationships are shown in figure 6 and described as follows:

(1) The KNOWLEDGE BASE is a store for the domain knowledge. This includes: a dictionary of knowledge items, menus and forms; a set of facts about the domain; a set of constraints on the allowable combinations of user input data; a set of selection rules; and a set of procedures for numerical calculations.

(2) The PROBLEM SOLVER controls the search for a solution and the method by which inferences are made. The depth-first search and backward chaining inference strategy of Micro-Prolog has been selected. In this

strategy the problem solver is given an abstract goal to prove or otherwise, in a top-down approach. A goal is a configuration of bearings. The knowledge base and database is searched in a depth-first manner to identify and select a bearing type and component to match the requirements of the application.

(3) The INTERFACE provides a facility for communication between user and system using forms, windows and menus.

(4) The DATABASE is a store for constant domain data. This includes a set of components from which a solution can be selected, together with interpolation tables to support numerical calculations.

5 SYSTEM OPERATION AND RESULTS

The system operates at three levels as shown in figure 7. On the first level the user is interactively requested via a set of forms to enter data concerning the application under investigation. At the second level the problem solving activity which is the selection process is encountered. Here the problem solver interrogates the knowledge base to identify the bearing arrangement and bearing type for the application, and also selects the individual component from the database. During this activity, interactive questions are prompted to the user to supply additional information if and when the system requires it. The third level reports the solution to the user. Figure 8 illustrates a typical set of application requirements and a solution derived to satisfy the requirements.

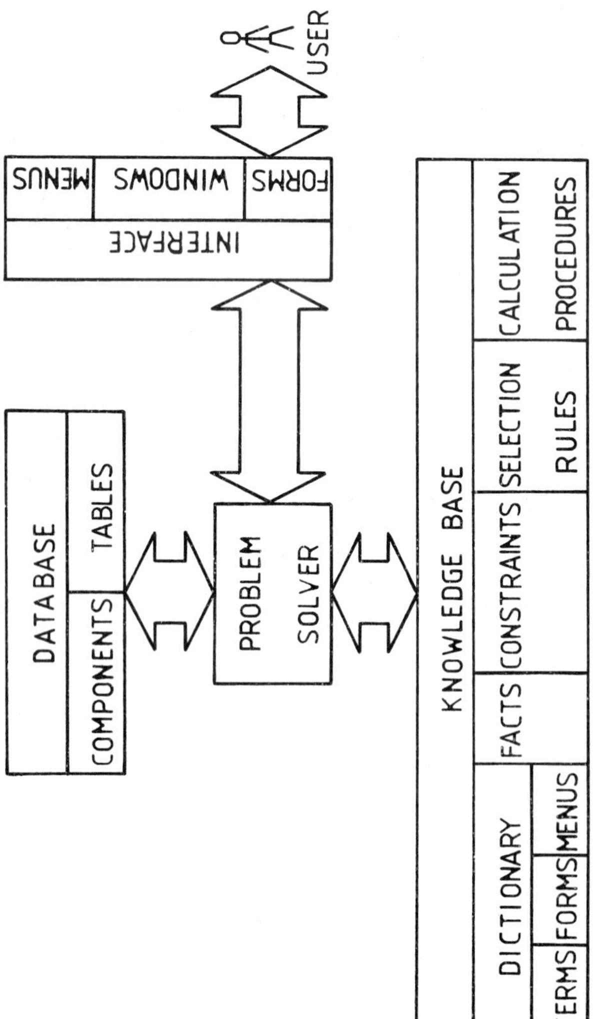

FIGURE 6 -COMPONENTS OF BEARING SYSTEM

6 CONCLUSIONS

Two systems have been developed, initially an expert system shell was used to test and validate some of the simpler knowledge structures. Unfortunately, as the knowledge associated with the selection process was built up the shell performance deteriorated to an unacceptable level. The second system was developed as a bespoke system using the Micro-Prolog programming language and encouraging results have been obtained. A disadvantage of the present system is the lack of an explanation facility to support the user during interactions with the system.

This work demonstrates that a form of technology transfer can be achieved through a knowledge based systems approach. An expert's knowledge is encapsulated in the form of rules within a computer system. The crux of such systems is the knowledge which they encompass. Therefore the quality of the knowledge base is a reflection of the knowledge engineering activity. A cavalier attitude to knowledge engineering will result in a system in which the user will have dubious confidence.

Future developments include the design of an explanation facility and the integration of graphics within the system. In addition, the

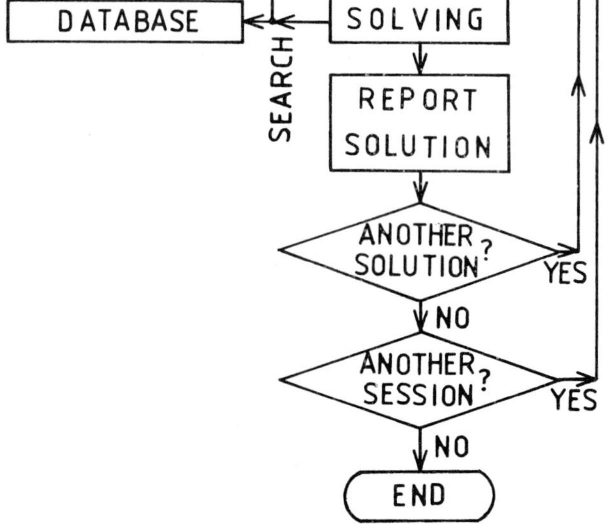

FIGURE 7 -OPERATION OF BEARING SYSTEM

```
Data:
application_domain = OTHER
env_temp_changes_negligible
dist_fixed_free_negligible
angular_misalignment = very_small
degree_of_rigidity = very_low
applied_radial_load_fixed_end = 3 kN
applied_radial_load_free_end = 2 kN
applied_axial_load = 2 kN
shaft_diameter_max = 500 mm
shaft_diameter_min = 258 mm
housing_bore_diameter_max = 600 mm
housing_bore_diameter_min = 510 mm
snap_ring_type = without
protection = none
rotational_speed = 750 rpm
environmental_temperature_max = 50 C
time_or_revs_life = revs
revs_life = 6 million revs

SOLUTION:
ARRANGEMENT:
  fixed_end = free_end
BEARING TYPE:
  single_row_deep_groove_ball
FEASIBLE COMPONENT NUMBERS:
  6068, 6072, 6976, 6076, 6980, 6080
```

FIGURE 8-SOLUTION INFERRED USING DATA

operation of the system will be enhanced to include the recording and analysis of selection failures (9) given that the user's requirements are too stringent and no solution is found.

7 ACKNOWLEDGEMENT

The authors gratefully acknowledge Jack Schofield of Liverpool Polytechnic and Mike Goodwin of North Staffordshire Polytechnic for their participation during knowledge acquisition sessions. We also thank Dr. Conway-Jones of The Glacier Metal Co. Ltd. together with Tony Cuff and Gerry Madden of NSK Bearings Europe Ltd. for their useful comments concerning the system operation.

References

(1) SIBLEY, L.B., PETERSON, M.B. and LEVINSON, T. 'An Assist for Tribological Design', Mechanical Engineering, September 1986, 68-74.

(2) TALLIAN, T.E. 'Tribological Design Decisions Using Computerized Databases', Journal of Tribology - Transactions of the ASME, July 1987, Volume 109, 381-387.

(3) WATERMAN, D.A. 'A Guide to Expert Systems', 1986 (Addison-Wesley).

(4) BERNOLD, T. (Ed) 'Artificial Intelligence in Manufacturing', 1987 (North-Holland).

(5) SRIRAM, D. and ADEY, R. (Eds) 'Applications of AI in Engineering Problems', 1986 (Springer-Verlag).

(6) GERO, J.S. (Ed) 'Knowledge Engineering in CAD', 1985 (North-Holland).

(7) SMITH, A. (Ed) 'Knowledge Engineering and Computer Modelling in CAD', 1986 (Butterworths).

(8) GRIFFIN, R.T., DOUGLAS, S.S. and WINFIELD, M.J. 'SELECTOR - A Knowledge Based System for Mechanical Engineering Design', 12th International Congress for Statistics, Scientific Computations, Social and Demographic Research, Ain Shams University, Cairo, Egypt, March 28 - April 2, 1987.

(9) DEWAN, D. and LARSEN, K.A. 'DABKON - a knowledge based system for the selection of bearings', 6th International Workshop on Expert Systems and their Applications, Avignon, France, 28-30 April, 1986, Volume 2, 1237-1250.

(10) FAGAN, M.J. 'Expert systems applied to mechanical engineering design - experience with bearing selection and application program', Computer-Aided Design, Sept. 1987, Volume 19, Number 7, 361-367.

(11) HART, A. 'Knowledge Acquisition for Expert Systems', 1986 (Kogan Page).

(12) NSK 'NSK Ball and Roller Bearings Catalogue', 1987 (Nippon Seiko K.K.).

(13) BEAIL, N. (Ed) 'Repertory Grid Technique and Personal Constructs: Applications in Clinical and Educational Settings', 1985 (Croom Helm Ltd.).

(14) BLOOMFIELD, B.P. 'Capturing expertise by rule induction', The Knowledge Engineering Review', March 1987, Volume 2, Number 1, 55-63.

(15) DAVIS, R. and KING, J. 'An Overview of Production Systems', Machine Intelligence 8, 1977 (Ellis Horwood - Elcock and Michie Eds.), 300-332.

(16) FIKES, R.E. and HEHLER, T.P. 'The Role of Frame-Based Representation in Reasoning', 1985 (Intellicorp Technical Article).

(17) BRACHMAN, R.J. 'What's in a Concept: Structural Foundations for Semantic Networks', International Journal of Man-Machine Studies, 1977, Volume 9, 127-152.

(18) HAMMOND, P. and SERGOT, M. 'Logic Programming for Expert Systems', 1984 (Chapman and Hall).

(19) HAMMOND, P. and SERGOT, M. 'Apes: Augmented Prolog for Expert Systems', 1987 (Logic Based Systems Ltd.).

(20) CONLON, T. 'Learning Micro-Prolog: A Problem Solving Approach', 1985 (Addison-Wesley).

Paper XVI(iii)

Tribology aids for designers

C. J. Thijsse

This paper deals with a systematic approach of the tribological knowledge with the intention to render it easily accessible for designers. Delimitation of the knowledge that is suited for this purpose is required. Tribological evaluation of a construction takes place via a checklist. It has been indicated where it is necessary to implement the knowledge in readily applicable computer programs. In those cases where insufficient data is available or where use has to be made of empirical facts, the tribological knowledge has been implemented in the TRIBEXSY expert system. The set-up of this system is given.

1. INTRODUCTION

Within the Philips Concern, R&D activities are carried out at a central level by approximately 5000 people. Roughly 4000 people, distributed between eight laboratories in six countries, are occupied in research. Of these, 2400 work at the Research Laboratories in Eindhoven.

Apart from the Research Laboratories, Eindhoven also has the Centre For manufacturing Technology (CFT), which operates as a centre of knowledge within Philips.

The CFT has about 1000 employees and provides technical support in the form of turnkey projects, research projects and consultation to the various Product Divisions, where the products are manufactured.

Tribological research is carried out at both the Research Labs and the CFT. In view of its set-up, the CFT's research must be aimed at finding solutions for tribological problems. Within the CFT, in the Mechanics and Mechanisms Department, there is a tribology group of eight people. The objective of this group is the reduction of tribological problems in both products and production means. This can only be achieved if the tribological knowledge available at the CFT and the Research Labs is transferred in a user-friendly form to the countless designers at Philips.

A designer designs a mechanical construction from a list of specifications. The designer therefore thinks more in functions than in physical principles. In this train of thought, it is not important to know, for example, whether the construction lifetime is limited by adhesive or abrasive wear. It is more important to determine beforehand whether a certain design will attain the desired lifetime.

This means that clear ways must be indicated for the evaluation of the tribological aspects of a design. Seeing that tribology is merely one aspect of the design, the evaluation may not take up too much time and the knowledge required for the evaluation must be easily accessed.

If we look at the above, we see that a subdivision of tribology into full film lubrication, mixed and boundary lubrication and dry running systems, although desirable from a research point of view, is of little use to the designer. The subdivision is too strongly based on physical principles. A better approach is obtained by subdivision into construction, lubrication, (surface of) material and environment.

The most important mechanical specifications for Philips' constructions are static and dynamic behaviour (in time) and, where cleanrooms are concerned, the dust production (the quantity and type of wear particles).

In view of the fact that dustfree construction is a still-developing discipline, we have not yet started developing tribological evaluation aids. The activities are at the moment concentrated in the research phase. We therefore limit ourselves to the static and dynamic behaviour of the construction during its lifetime. For the parts that move with respect to each other, we translate the design specifications directly into constructive aspects, such as loading, type of movement, shape and permissible wear height. The specifications with respect to static and dynamic behaviour are therefore enclosed in the construction.

For an actual realisation of the design, one requires building blocks which are formed by:

a) the lubrication film, whereby the type of pressure build-up and the lubricant are important parameters.
b) the (surface of) materials, with elasticity, hardness and roughness as parameters.

The environment forms the operating boundary conditions in which the system has to work.

This allows us to give the following problem definition (which is useful for designers): "Which sort of building blocks has to be selected so that the design specifications can be satisfied under the environmental boundary conditions?"

It is necessary to indicate which failure criteria have to be included in the tribological evaluation and which (simple to handle) aids have been developed. This is explained in more detail in the following sections.

2. FAILURE CRITERIA

The tribological failure criteria can be subdivided into two main groups: wear and friction.

2.1 Failure criteria with respect to wear

The most frequently occurring wear mechanisms in Philips' constructions are:

- plastic deformation:
- adhesive wear;
- abrasive 2-body wear;
- abrasive 3-body wear;
- surface fatigue;
- fretting corrosion.

The evaluation means available to the designer are now described in succession.

2.1.1 Plastic deformation

Most Philips products are small and therefore have small contact surfaces. This means high contact pressures, with resulting high wear.

In production machines with cam to cam-follower roller drives, high contact pressures occur. It may be assumed that, for reliable operation, the contact materials (seen at macro scale) may not undergo any plastic deformation in most cases.

We use the Hertzian contact stress as a measure of the contact pressure and have developed a computer (PC) program for a quick calculation of this contact stress. For doubly curved surfaces that are pressed together with a certain force, this program can calculate the Hertzian contact tension, the size of the contact ellipse and the contact stiffness. Literature information from Horowitz [6] have been included in the program.

Should a traction force be present along the surfaces, a proposed estimated correction from Bayer and Ku [1] is added.

By correctly comparing the actual contact stress with the permissible stress for various materials, an insight is obtained into the applicability of certain materials for a selected contact geometry. If the correct materials cannot be found, it will be necessary to modify the contact geometry.

2.1.2 **Adhesive and abrasive 2- and 3-body wear**

As already mentioned, a designer is more interested in the lifetime of a construction (i.e. speed of wear) than in the physical background of the wear. For this reason, a construction is evaluated with the aid of the wear factor (k-value) proposed by Archard.

If we involve the influence of the wear on the static and dynamic behaviour, then the only important parameter is the change in shape of the wearing surfaces. This change is closely related to the wear height.

For general usability, however, wear values are mostly presented in the form of wear factors. To be able to quickly assess the performance of the construction on the basis of the permissible wear height with the aid of actual wear factors, it is necessary to make a simple translation from wear height to wear factor.

For this reason, a computer (PC) program has been written to calculate the **maximum permissible wear factor** of:

- the permissible wear height
- the desired lifetime
- the movement pattern of the surfaces with respect to each other;
- the contact point with respect to the surfaces;
- the contact force;
- the shapes of the contacting parts.

This value must be compared with the **probable actual wear factor** that is found under various tribological circumstances. In this way, we can determine from which materials the wearing surfaces have to be made and what the desired lubrication regime must be.

To quickly obtain a first impression of the applicable material lubrications and lubrication regimes, use is made of the graph (see fig. 1) from Verbeek [10]. This graph is composed from the many experiments carried out in our laboratory and shows the relationship between the probable wear factor and the various tribological systems.

If one enters the permissible wear factor (determined by the previously-mentioned computer program) for a selected construction on the y-axis, and the selected tribological system on the x-axis, then a quick insight is obtained into the size of the wear problems due to a low permissible wear factor to be expected.

In the area above the indicated band, no wear problems are expected; in the area below the band, the chance of wear problems is virtually 100%. In the area inside the band, wear problems can be expected.

If one comes out inside or below the band, then the design can be improved either by increasing the permissible wear factor (modifying the construction) and/or by selecting another "more to the left" tribology system (different materials, different lubrication conditions).

Should it appear that a certain measure of lubrication is required to fulfil the required lifetime (e.g. full film lubrication), then it must be possible to easily assess whether this can be achieved with the aid of self acting bearing systems.

For contra-shaped contacts, we have therefore developed a PC program that calculates the lubrication conditions for various lubricants, on the basis of the work by Hamrock, Dowson and Higginson [2], [5] and by Johnson [7].

For conformal situations, although more than sufficient knowledge is available, we have not yet developed a user-friendly program for the designer. This must be one of our activities for the near future.

Should the required lubrication regime not appear feasible, then externally pressurised bearing systems must be considered. If these bearing systems are not used, then the construction will have to be modified in such a way that the maximum permissible wear factor is increased. In most cases, this will mean a reduction of the contact pressure.

If the construction has an acceptable maximum wear factor, then a correct material choice can be made for the contact, with the aid of a database.

This database, which operates on a PC, is filled with the data of roughly 600 test series. The tests (mainly pin-disc and pin-ring) have been carried out in our laboratory under varying conditions for:

- the constructional aspects (load, contact shape, speed);
- the lubrication (dry running, boundary lubrication, type of lubrication);
- the surfaces (metal, plastic, surface treatment, coating, roughness) and
- the environment (temperature, humidity, type of gas).

An identical database with literature data is being built.

Plastic-metal systems can be tested for the permissible pv- and pt-value with a computer program based on Erhard's and Strickle's formulae [4].

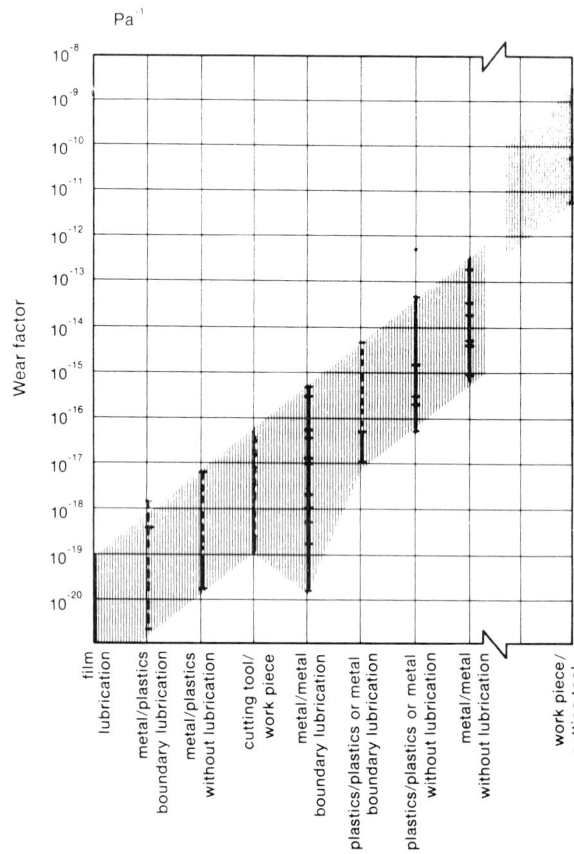

fig.1 Tribo systems and their specific wear factors

2.1.3 **Surface fatigue**

If certain parts of a design are subject to a varying contact load, than a check will have to be made to see whether the lifetime of the parts is limited by surface fatigue.

Empirical formulae have been compiled for example of data from the literature [3], [8] and [9]. They express the relationship between the magnitude of the contact stress, the number of changes in loading, lubrication conditions and hardness. These formulae have proved their use in many places, both for our products and for our production means.

2.1.4 **Fretting corrosion**

Fretting corrosion does not only occur in joints where the components unintentionally vibrate with respect to each other. In precision engineering constructions, we often see a situation where two components have a low amplitude oscillating movement with respect to each other.

If the double amplitude is smaller than the length of the contact and if the contacting components are made of metal, this almost always leads to fretting corrosion problems.

Take note that the contact area can increase in contra-shaped situations.

We maintain, therefore, the simple rule that constructions must be so designed that the contact length is (much) less than the double amplitude of the contacting components. This has the dangers of plastic deformation and of a too small permissible wear factor.

After redesign, the contact must therefore always be assessed according to the procedures described in sub-paras. 2.1.2 and 2.1.3. If problems ensure, then the amplitude with respect to the rest of the construction is mostly very small and the solution can be found in the application of elastic elements, such as leaf springs or intermediate layers of plastic.

2.2 Failure criteria with respect to friction

The problems involved with friction arise because the friction is:

- too high;
- too low;
- unstable.

To obtain an insight into the size of the friction coefficient that occurs under various tribological circumstances, we use the database mentioned in sub-para. 2.1.2.

In practice, however, it appears that a deviation of 50% with respect to the values found in the database can be expected under not too unfavourable circumstances. The lowest and highest values of the coefficient of friction can therefore differ by a factor 3.

With fluctuating environmental conditions, the value of this value can increase to 6 or more.

The only advice we can give our designers is the rule that the friction requirements that is placed on a guide is always such that the coefficient of friction is either higher or lower than a certain value.

Material combinations and lubrication conditions can then be selected such that the nominal value of the coefficient of friction is sufficiently far away from the danger zone. The integration of functions in a design which creates the requirement that the magnitude of the friction must be between two values must be avoided because it gives too little room for play.

A stable dynamic behaviour is essential, in connection with a high positional accuracy requirement on both our products and production means. This means that stick-slip phenomena may not occur.For this reason, we have included Stribeck curves for many types of bearings. These types of curves can also be included for other guides than radial bearings.

If the working area of a guide is too far to the left of the Stribeck curve (see fig. 2), the friction force along the surfaces will decrease as the speed of the surfaces with respect to each other increases.

The guide displays the behaviour of a damper with a negative damping coefficient. If the rest of the construction contains insufficient damping, unstable behaviour (stick-slip) occurs.

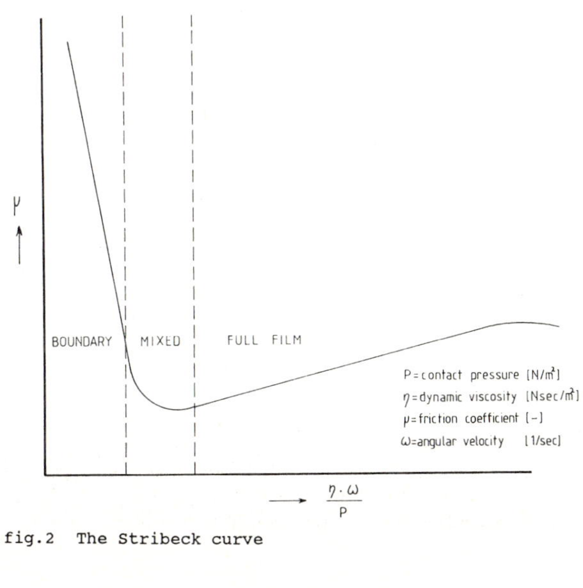

fig.2 The Stribeck curve

To the right of the Stribeck curve we find the desirable situation with positive damping coefficient. This is the reason why the following rule applies: "The lower the relative speed in a guide, the lower the contact pressure and the higher the viscosity of the lubricant to be applied." If these attempts to end up sufficiently to the right of the Stribeck curve are without success, use will have to be made of externally pressurised bearings because full film lubrication always shows a positive damping behaviour.

3. CHECKLIST

For the designer, the above-mentioned facts have been summarized in a checklist as an aid in the design of guides.

1) Because of all kinds of constructional boundary conditions, among them those from the viewpoint of friction, a first selection of the principle of guidance of force is determined by the wearing surfaces (e.g. solid, liquid, elastical element, magnetic).
2) Next, every mechanical contact has to be post-calculated and possibly adapted so as not to exceed the permissible contact stress (by means of the Hertz PC program).
3) In the case of oscillating movements, check whether the risk of fretting corrosion is present. If so, adapt the design according to the rules given in sub-para 2.1.4 and proceed to point 2.
4) Determine (by means of PC program "k-value") the permissible wear factor of the selected construction. By means of the graph in fig. 1 and the empirical formulae for surface fatigue, estimations of the lifetime can be made. In case the required lifetime is greater than the estimated lifetime, the selected tribological system (contact material, lubrication) and/or the construction (load, movement, shape) require(s) adaptation.
5) If the conclusion from 4) is that lubrication is inevitable, it will be necessary to examine how far it is possible to build a lubrication film (PC program "EHL" and probably also externally pressurised bearings). If it turns out to be impossible to build a satisfactory lubrication film, the design will require renewed study and it will be necessary to return to point 1.
6) Because of points 2, 3, 4 and 5, the main dimensions of the wearing surfaces and the selection of the tribological system are fixed.
7) Make a choice from the two wearing surface materials (bulk material and possibly surface layer) and the lubricant (by means of the PC databases).
8) If the environment is too much of a limiting factor in finding the correct solution, the environment must be adapted; if necessary through application of seals each having its own tribological problems.
9) In case full film is applied, optimization calculations can be performed.

We will not deal any further with the optimization calculations mentioned sub 9 since this topic is discussed in a separate contribution to this Leeds-Lyon conference.

4. EXPERT SYSTEM

Although the checklist with the mentioned evaluation aids appears to be very useful in practice, it was felt that a more extensive system was needed for the following reasons.

- The way of evaluation mentioned above is rather complex so that automation will increase its user-friendliness.
- Upon the start of the design process, the designer does not always have all the requested data at his disposal. This asks for a system that can work with these uncertainties; an expert system.
- The evaluation tools given above are based on knowledge that can be comprised fairly well in models. However, more tribological knowledge is available. But this knowledge is based more on separate facts known from practice. It is impossible to process all this

knowledge by hand during an evaluation cycle. Only the knowledge relevant to a certain design must be included in the evaluation and must thus be so stored that only the relevant data will be presented in a simple way.

- In those cases in which the knowledge is applied for an existing design that presents tribological problems in practice, it will usually be more effective both for making the diagnosis and for finding the correct solutions to discover the physical principles that play a role. Here it is possible to think of a solution model in which physics are represented but do not at all come to the fore, or in view of a learning effect, only to a limited extent.

4.1 System type

The reasons cited above have led to the development of a prototype expert system. To achieve this we have filled a commercially available shell (ENVISAGE of System Designers) with tribological knowledge. The first question to be answered is what the set-up of the system should be like; a tool during the design proces or a diagnosis/repair system for problems occurring in practice. One set-up does not automatically exclude the other.

General experience with expert systems has taught us that it is many times more difficult to build a design system than to build a diagnosis/repair system (fig. 3). Consequently, we started with a diagnosis/repair system in which we have tried to implement the knowledge in such a way that in future it can be used in an expert system for the design stage.

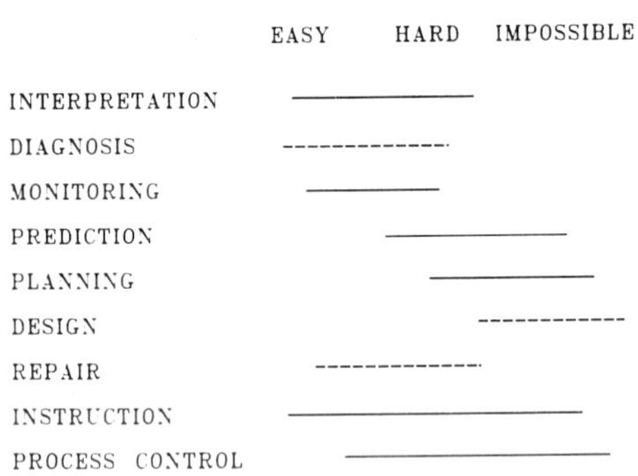

fig.3 Feasibility of application classes

4.2 Failure criteria

Expert systems are only useful if the implemented knowledge is considered to be sufficiently established by experts. There must also be sufficient knowledge available on the subject. This is the reason for implementing the wear problems occurring most frequently in our environment, i.e. the failure criteria mentioned subpara 2.1. As mentioned above, the knowledge on friction is more limited. That is why it has not been implemented in the actual system but it does play a role as a user constraint of possible remedies.

4.3 TRIBEXSY a tribological expert system

The structure of the expert system is shown in fig. 4. Via a number of questions on the construction, the environment, the materials applied and the lubrication conditions, the chance of one or more failure mechanisms is determined. If one wants to use the system in the design stage, this is possible on the condition that a concrete elaboration of the system exists so that the answer need not often be too "Unknown", otherwise the probability of a correct outcome becomes very low. If the system is applied to find solutions for problems that have already occurred in practice, supplementary questions on the ambient conditions will be asked, while information must be given on the nature of the damage through comparison of the appearance of the actual damaged surfaces with photographs of surfaces that are representative of the various types of wear. The knowledge has been so implemented that symptom solving is avoided as much as possible. If the diagnosis is, for instance, 3-body abrasive wear, the system will look for other wear mechanisms as a possible source of third bodies, both in the contact to be considered itself and, in case of a central lubrication system, in other contacts. It is important that the system knows its limitations. In case it is not possible to make the correct diagnosis, the telephone number of the CFT tribological group appears on the screen.

If a diagnosis is made, which may also imply that the problem comprises more than one failure mechanism, a subset relevant to the diagnosis is taken from the total set of remedies present. The total set of remedies consists of approximately 100 constructive, materials and lubrication solutions. Next, this subset is passed through two filters.

The first filter consists of process constraints. If a possible solution should consist of the application of a CVD-TiN layer, it will first of all be necessary to check whether the dimensions of the part correspond with the dimensions of common CVD reactors. Moreover, other parts of the construction element that the wearing surface must be able to withstand the process temperature, whereas it must also be taken into account that the entire part will be provided with a TiN layer. If necessary, the possibility of building the part from smaller subparts will have to be examined.

The second filter is formed by user constraints. If, for example, the mechanical point of contact also acts as an electric contact, full film lubrication will not be directly applicable. As previously mentioned, this filter also includes the requirements with respect to friction.

The complete system comprises the above-mentioned checklist with evaluation tools, and further, about 200 empirical rules in the form of "IF A AND IF B THEN C". Here C represents an algorithm which performs an operation on the chance that something occurs or that something is possible.

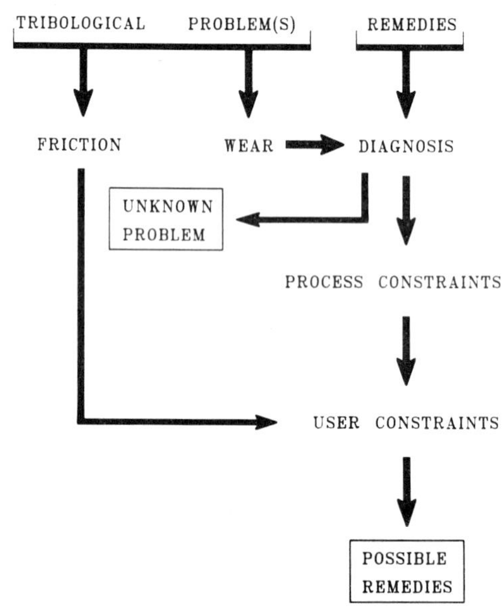

fig.4 TRIBEXSY: general setup

5. CONCLUSIONS

To render tribological knowledge usable for designers it will not only require systemization but also simplification.

Since tribology contains a multitude of factual knowledge, it will be necessary to implement this knowledge in an automated system in view of the accessibility.

The tribological behaviour of a construction is subject to many external influences, thus rendering it virtually impossible to make forecasts with a 100% certainty. It is more a weighing of chances. An expert system is perfectly suited for laying down this type of knowledge.

During the development of the expert system, it proved to be an advantage to be able to start from a certain definition of the problems. The system has been built for use within the Philips concern. As a result, a fairly good picture can be obtained of the conditions under which tribological problems may occur and it is thus not necessary to model "the whole world".

With the approach described in this paper, the most trivial wear problems can be avoided; in the case of very specific tribological problems the human tribological experts will have to be on stand-by to lend a hand.

REFERENCES

[1] BAYER, R.G. and KU, T.C. 'Handbook of analytical design for wear', 1964 (mcGregor, Plenum Press, New York).

[2] DOWSON, D. and HIGGINSON, G.R. 'Elastohydrodynamic Lubrication', 1966 (Pergamon Press, Oxford).

[3] DUDLEY, D.W. 'Wear Control Handbook', Chapter 'Gear Wear', 1980, A.S.M.E. 755-830.

[4] ERHARD, G. and STRICKLE, F. 'Gleitelemente aus Thermoplasten', Kunstoffe, 1972, 62.

[5] HAMROCK, B.J. and DOWSON, D. 'Minimum Film Thickness in Elliptical Contacts for Different Regimes of Fluid Film Lubrication', N.A.S.A. Techn. Paper 1342, Oct. 1978.

[6] HOROWITZ, A. 'A contribution to the engineering design of machine elements involving contrashaped contacts', Israel Journal of Technology, 1971, Vol.9, No.41.

[7] JOHNSON, K.L. 'Regimes of Elastohydrodynamic Lubrication', Journ. Mech. Eng. Science, 1970, Vol.12, No.1.

[8] NIEMANN, G. 'Maschinenelemente', Zweiter Band 'Getriebe', 1960 (Springer Verlag, Berlin).

[9] ROWE, C.N. and ARMSTRONG, E.L. 'Lubricant Effects in Rolling - Contact Fatigue', Lub. Eng., January 1982, Vol.38, No.1.

[10] VERBEEK, H.J. 'Tribological systems and wear factors', Wear 56, 1979, 81-92.

WRITTEN DISCUSSIONS AND CONTRIBUTIONS

WRITTEN DISCUSSIONS AND CONTRIBUTIONS

DISCUSSION

SESSION IV - CAMS

'Lubrication and Fatigue Analysis of a Cam and Roller Follower' B A GECIM

Mr H J van Leeuwen (Eindhoven University of Technology, The Netherlands)

This paper covers many interesting aspects of the tribological behaviour of a cam and roller follower contact, and I am looking forward to the release of the full paper. I would like to ask the following questions:

1. What is the origin of the very steep gradient in the film thickness at the transition between cam base circle and cam flanks? Can the author give some more information on the influence of the squeeze effect (how has it been modelled, what influence does it have)?

2. The Johnson diagram for film thckness representations applies to entrainment effects only. When is it applicable for cam/follower EHD lubrication?

3. At higher speeds the traction shows a sign change at 0° cam angle. Is it due to deceleration of the roller?

Reply by Dr B A Gecim (General Motors Research Labs, Warren, U S A)

The author would like to thank Dr van Leeuwen for his interest in the paper.

The steep change in film thickness at the transition points is caused by the change in load between that exerted by the hydraulic lash adjuster on the base circle and that exerted by the valve spring on the cam flanks. The squeeze film effects were predicted by solving the transient Reynolds equation in the inlet region of the Hertzian contact, at one degree intervals of the cam-shaft rotation. The transient solution and the quasi-stationary solution yielded almost identical results, except at the transition points. This indicated that the entrainment effects were dominant. Hence, the lubrication regimes chart was used to determine the elastic and viscous characteristics of the contact.

At high speeds, roller inertia becomes dominant. The roller accelerates during valve opening and decelerates during valve closing, hence the traction forces act in opposite directions during these events. The position of sign change for the traction force is slightly after the top of the lift where the needle bearing torque is just enough to balance the inertia of the decelerating roller.

'Prediction of Cam Wear Profiles' R H FRIES and C A ROGERS

Dr J C Bell (Shell Research Ltd., Chester, UK)

I have a comment on the presentation of Dr Fries relating to the usefulness of the Archard wear model. At Thornton Research Centre we have been applying this model to automotive pivoted-follower valve train systems, and including a boundary/EHD lubrication transition model. Some results for finger follower wear profiles were presented at a previous Symposium (Ref 1) showing good agreement of positions of wear maxima with those of experimental wear profiles. Recently Dr T A Colgan and I have applied this model to cam wear profiles with similarly good agreement with experimental wear maxima, and exhibiting similar features to those of Dr Fries. We conclude from this that:

1. The simple Archard wear model is proving to have validity for the prediction of cam and follower wear profiles.

2. The most critical parameters appear to be the velocities of the contact over the cam (V_C) and follower (V_F) surfaces.

3. Hydrodynamic factors can influence wear but are less important than V_C and V_F for most operating conditions.

These observations help to explain why maximum wear does not occur at the same points in the operating cycle for the cam and the follower, except under catastrophic conditions when damage is transferred from one surface to the other.

Referring to Dr Willermet's paper in Session II, it is suggested that V_C and V_F could be of use as simple valve train wear criteria, though clearly not to be used in isolation. The selection of design geometries that minimise the periods over which V_C and V_F have low values would be expected to lead to better wear performance on this basis.

Reference

1. J C Bell, P T Davies and W B Fu, "Prediction of automotive valve train wear patterns with a simple mathematical model", Proc. 12th Leeds-Lyon Symposium on Tribology, Lyon, September 1985.

Mr B A Shotter (Westland Helicopters, Yeovil, U K).

From your description it sounds as though your technique can be applied to conditions of abrasive wear, but I seem to remember seeing cam and follower degradation by contact fatigue. Could you please comment on the implications of such failure mechanisms.

Reply by Professor R H Fries and C A Rogers (Virginia Polytechnic Institute and State University, Blacksburg, USA).

The authors wish to thank Mr Shotter and Dr Bell for their comments and discussions. Our method is general in nature, and it can accommodate virtually any wear model that can be described mathematically. We illustrated the method using an Archard wear model that is appropriate for abrasive wear. Had we elected to illustrate the method with a cam on a roller follower, then a fatigue wear model may have been more appropriate. Our wear prediction method is in no way restricted to the use of an Archard wear model.

We have applied this method to the problem of wear profile prediction in rail vehicle wheels. In the wheel wear work, we have used both abrasive wear models and fatigue wear models (1). Interestingly, the predicted profiles are very insensitive to the selection of wear model. In addition, the predicted wear profiles closely resemble actual profiles of service worn wheels. In addition, the predicted wear profiles closely resemble actual profiles of service worn wheels. The cam wear prediction work is in its infancy, and we cannot yet draw similar conclusions about influence of the wear model on the predicted profiles.

Citation of the experimental work by Dr Bell and his colleague, Dr Colgan, is gratifying to the authors. We made our predictions entirely without benefit of experimental work, and we are pleased to know that experimental work exists that exhibits features similar to our predictions. We expect to continue this work, and we hope to make comparisons between our wear profile predictions and the experimental work of others and ourselves.

Reference

1. Fries, R H and Davila, C G., 'Wheel Wear Predictions for Tangent Track Running', J Dynamic Systems, Measurement and Control, Vol 109, December 1987, pp 397-404.

SESSION V - BELTS

'Power Rotating of Flat Belt Drives - A Wear Approach' B G GERBERT

Dr D A Boffey (University of Edinburgh, UK)

Figure 1 in the Synopsis is perhaps an oversimplifcation of the competing modes of failure. It is (I think) normal in the catalogue selection of belts to specify a maximum centre distance for a given linear speed, presumably to avoid undesirable cyclical lateral belt vibrations excited by longitudinal fluctuations in belt tension, leading to fatigue. The problem is, however, an extremely complex one as the belt is moving laterally. Provided the belt satisfies the above criterion then Figure 1 would be valid for steady-state induced fatigue. Would the author like to comment?

Reply by Professor B G Gerbert (Chalmers University of Technology), Gothenberg, Sweden)

The simple natural frequency of lateral vibration of a moving belt is

$$f_N = \frac{1}{2 L_B} \frac{F - m V^2}{\sqrt{F m}}$$

where

F = belt tension
m = mass of belt per unit length
V = belt velocity
L_B = length of free span of belt

Equation (A3) in the paper gives the bending frequency to be,

$$f_B = z V/L$$

The sum of the two is the resulting bending frequency f of a vibrating belt. Thus by substituting f by f_B in equation (21) for the quantity B we are able to consider the influence of vibration on fatigue. In the diagrams in Figure 4 this implies that the "Fatigue" curve moves to the right.

To my knowledge influence of vibration is not considered in the design procedure in the catalogues.

SESSION VI - GEARS

'The Relationship between Uneven Tooth Contact Loading and Surface Durability in Flexible Gear Designs' J F HARROP and A TAM.

Mr B A Shotter (Westland Helicopters, Yeovil, Somerset, UK).

Concerning high contact ratio gears, whilst I accept that their noise can be less than conventional designs, I have always been uncertain about their load distribution. In aircraft engineering one has to be able to define loads as accurately as possible if one wants to achieve the very high reliabilities needed. Thus the uncertainties induced by trying to maintain a minimum of two teeth in contact seemed to be a retrograde step. You have demonstrated that the complexity of the problem is even greater with the rim flexural distortion. Thus I feel that you have justified my concern; would you agree with this point of view?

A further comment concerns an aspect of the asymmetry induced by helical gears which could be relevant in this case. Because the instantaneous contact lines on the teeth sweep across the facewidth the oil tends to be swept towards one face of the gearbox. With high speed, highly loaded gears, this can result in thermal distortion due to the combined energy losses from churning and contact sliding being passed to one end of the gear case. Did you considered this effect in your studies?

Reply by Messrs J F Harrop and A Tam (Pratt and Whitney, Mississauga, Canada).

High contact ratio (HCR) spur teeth are being used successfully in both engine and helicopter gearboxes. They provide increased tooth bending fatigue capacity, reduce the dynamic loads in the gear teeth and decrease noise and vibration. However, suitable tip modification to compensate for tooth flexing, edge relief and high quality surface finish (< .4 μm) are essential. Provided that the gear rim is designed to normal aircraft standards, rim flexing does not present a problem on HCR spur gears.

Rim flexing/distortion of helical gears is more serious due to the effect of induced axial forces and moments at the gear teeth. It is the tangential deflection of the gear teeth that causes the problem. HCR teeth are probably more sensitive to these deflections than LCR teeth, i.e., the smooth transmission of loading, where more than two teeth are in contact, is perhaps lost.

HCR teeth would probably be acceptable in a helical gear mesh if rim stiffness was sufficient to minimise distortion.

The effect of possible thermal distortion due to energy losses was not included in our analysis. There has never been any sign of overheating in the gear teeth. Baffles, to prevent "oil hiding" in the gearbox, help retain oil mist in the gear mesh cavity and contribute towards uniform lubrication and cooling of the gear teeth.

'Temperature and Pressure Measurements in Gear Contacts with Thin-Film - Transducers' H PEEKEN and P AYANOGHU

Mr H J van Leeuwen (Eindhoven University of Technology, The Netherlands).

To the discusser's knowledge, this is the first successful application of thin film micro-transducers in gear contacts. The authors are to be commended for their ingeneous testing device, which allows for sputtering processes on tooth flanks. I would like to address the following questions:

(1) Why did the authors return to manganine pressure transducers? It was concluded in [1] that NiCr has to be preferred.

(2) It appears that the pressure and temperature distribution curves are presented in real time. As the radius of curvature and the load of the tooth flank is changing along the line of action, the maximum Hertzian stress will change accordingly. Hence, to be able to compare the measured pressure to the Hertzian stress distribution, the pressure data have to be transformed to the same point in time. This necessitates many transducers, or a variable transducer position on the tooth flank. The time dependency may also partly explain the difference between nominal and measured load.

(3) Another explanation could be misalignment of the flanks. If I am correctly informed, it amounts to 15 μm over a length of 20 mm, which is quite a bit. This can easily be checked by locating the transducers in different axial positions. Misalignment affects the temperature, film thickness and pressure distribution considerably [2]. The best rig would improve from a self aligning device.

(4) Is the pressure zero or negative in the cavitating area (outlet)? and is the pressure gradient finite or zero at the onset of cavitation? This in interesting, because most theoretical treatments in EHD lubrication assume that the pressure gradient is zero (Reynolds' boundary condition).

[1] Peeken H and Köhler A. "Determination of pressure and load distribution in a roller bearing with evaporated transducers", in Friction and Traction, Proceedings 7th Leeds-Lyon Symposium on Tribology, 1981, pp 186-191.

[2] Van Leeuwen H., Meyer H and Schouten M. 'Elastohydrodynamic Film Thickness and Temperature Measurements in Dynamically Loaded Concentrated Contacts: Eccentric Cam - Flat Follower", in: Fluid Film Lubrication - Osborne Reynolds Centenary, Proceedings 13th Leeds-Lyon Symposium on Tribology, 1987, pp 611-625.

Reply by Professor Dr Ing H Peeken and Dr P Ayanoglu (Institute fur Maschinenelemente und Maschinengestaltung, Aachen, West Germany).

(1) At the time the sputtering apparatus was delivered, the manganine target was the only one available. We have now acquired a NiCr target and are experimenting with it. Manganine transducers have nevertheless proven to be quite satisfactory, with pressure and temperature coefficients of $1.6 \cdot 10^{-6}$/bar and $\pm 5.0 \cdot 10^{-6}$/K respectively.

(2) For valid comparisons, the calculated and measured values for the same point on the line of action are to be considered. In fact, the given nominal loads are those for the point on the line of action in which the transducer is positioned. The Hertzian distribution seen on Fig. 10 is calculated from the actual force on the transducer, this being the integral of the corresponding measured pressure curve, while the curvature is again that on the point of the transducer. The intention is in this case to compare the measured and Hertzian pressure distributions and not the magnitude of the curves.

We intend to use differently positioned transducers to get the distributions for the whole line of action, but these have to be placed on the same tooth flank to avoid errors resulting from no two tooth flanks being exactly the same.

(3) We in fact intent to produce laterally

offset transducers, or two of them alongside, to get an idea of the lateral load distribution. We are also working on a new test rig for much higher loads and with the provision for alignment. But the whole problem is not solved by aligning the axles, because there is still the misalignments of single teeth and not of the whole gear to be considered.

(4) Recently taken pressure curves show a negative value in the cavitating area with a finite gradient, but it is too early to make definite statements about it. For one thing, it could simply be an overshooting of the measuring system - we will first have to investigate its high frequency behaviour. What's more, the relative positions of the pressure and temperature curves are not known, so that this could only be the temperature influence on the transducer. This is on the other hand not so probable, as the temperature should have fallen in the outlet region. The curves were taken quite recently and such an effect was not noticeable in previous measurements, but rather a hysteresis of the transducer. We do not want to come to any conclusions about it yet.

SESSION VII - ROLLING ELEMENT BEARINGS (1)

'The Palmgren - Miner Rule Derived' J J KAUZLARICH.

Professor H Blok (Rijswijk, The Netherlands)

In radially loaded rolling-element bearings one should distinguish between two kinds of duty cycles, here to be called the 'short' and the 'long' ones and which are imposed internally and externally, respectively. In such bearings the 'short' cycles relate to the variation of the load on each rolling element individually during it planetary motion. On the other hand, the 'long' cycles are associated with the variations of load and speed of the bearing in total during non-stationary operation, such as occurs in automobile practice.

Now, the Palmgren-Miner role has long proved itself for stationary operation, such as that on test benches for routine fatigue testing where only the 'short' cycles will occur. But it would appear questionable whether this rule may still be considered reliable when it come to non-stationary operation.

In fact, extensions of the present rule to non-stationary operation are usually made more or less tacitly subject to the assumption that the order or succession of severe and mild duty would be immaterial. However, from practice with aircraft structures this assumption has long been known to be invalid. The question may therefore be raised whether this assumption may nevertheless be justified for rolling-element bearings. If not, the application of the original Palmgren-Miner role, as well as that of the author's refinement thereof, would have to be limited to stationary operation only.

Reply by Professor J J Kauzlarich (University of Virginia, Charlottesville, USA).

I certainly agree with Professor Blok that the order of succession of severe and mild duty is important, and that the Palmgren-Miner Rule set to a constant equal to 1 is incorrect. The simplicity of the Rule has led to continued use, and one way to correct the rule for a particular duty cycle is to determine a proper constant by experiment.

As shown by Atkins and Mai (1, p 539), as well as others, fatigue crack growth can be divided into 3 stages, with Stage 1 concerned mainly with the initiation of crack growth, Stage II the continuum region of 'long crack' extension where fracture mechanics is rather successful, and Stage III a short time event where fracture takes place. Atkins and Mai (1, p 581) show that in Stage II the Palmgren-Miner rule may be 'proved', and that the Rule is independent of 'order of succession of severe and mild duty'. The difficulty lies with Stage I associated with crack initiation to the point of 'long crack' extension.

By modifying the P-M Rule based on L_{10} life, as in the paper, this has the effect of setting the Rule to a constant that has been determined by experiment. Collins (2, p 242) points out that "If the various cyclic stress amplitudes are mixed in the sequence in a quasi-random way, the experimental Miner's sum more nearly approaches unity at the time of failure". Some experimental work on this subject is needed.

[1] Atkins, A G and Mai Y W, 'Elastic and Plastic Fracture', Wiley, 1985.

[2] Collins, J A, 'Failure of Materials in Mechanical Design', Wiley, 1981.

Dr J D Summers-Smith (Guisborough U K).

The designer uses the concept of L_{10} life as the basis for the application of rolling bearings. The definition of "life" in L_{10} calculation is based on the somewhat subjective idea of the "first detectable fatigue pit", though this is merely a point in a process that ultimately ends in failure - collapse or seizure - that is the concern of the user rather than L_{10} life.

"Life" in L_{10} sense can be detected in service by an increase in noise or vibration, but clearly this is a warning rather than an indication of immediately impending failure. Can any of the authors give guidance on how much of the ultimate life has been expended by the time the first fatigue pit has been detected? It is appreciated that this is not a simple question but depends on many factors; nevertheless, the user must take such a decision and any guidance that can improve the quality of the decision-making process must be of benefit.

(This question was asked to all the authors in Session VII).

Reply by Dr E Ioannides, Professor B Jacobson and Dr J H Tripp (SKF Engineering and Research Centre, Nieuwegein, The Netherlands).

If this is a question, it certainly is "not a simple question", although one which is bound to arise in practice. Perhaps, with due deference to its asker, a re-statement will produce a question to which an answer can be supplied. The important issue is: how relevant is the manufacturer's definition of "failure", i.e. the first small spall, to the particular installation? This question has the following practical aspect. If the first spall is not taken as the criterion of failure but a given user can tolerate this spall, an appropriate extended L_{10} life can be considered for which failure suitable to the user's definition may occur. This will vary from application to application according to the precision and requirements of the application. The resulting extension of life should then be used to gauge the interval for planned maintenance, nothing more. Thus, the question turns on the relative costs of downtime and machine repair. If failure is catastrophic but maintenance is cheap, then the "appropriate" damage criterion should be a very slight one (first spall) and vice versa. The result clearly depends on the particular application. It is the kind of "rule of thumb" which the experienced engineer, perhaps unwittingly, already applies.

SESSION VIII - PLAIN BEARINGS (1)

'Axially Profiled Circular Bearings and Their Potential Application in High Speed Lubrication' S BASRI and D T GETHIN.

Mr F A Martin (London, UK).

Dr Gethin has produced some interesting and useful information on his paper on axial profiled bearings. It is noticeable, however, that the effect of different profiles have generally been considered on the basis of the same eccentricity ratio. An alternative, and probably a more practical way of comparing results, is to consider performance on the basis of using the same load. This can either be accomplished on the computer with further reiteration or simply produced graphically as shown in the accompanying figure (developed from the authors paper). This figure (opposite) highlights the possible difference in trends which can occur, depending on the basis used. For the particular case considered, profile 3 has a higher maximum film temperature than profile 1 when using the same eccentricity ratio. However, the trends are reversed when using the same load (the latter being common practice in the design process).

Reply by Drs S Basri and D T Gethin (University College of Swansea, UK).

The authors thank Mr Martin for his kind remarks and are fully aware of the advantages of presenting the information in the way suggested. However, the presentation method in the paper readily enables the evaluation of bearing performance for a prescribed load which is the design practice commonly encountered. As pointed out, under this circumstance, profile 1 gives the highest operating temperature since the thinnest film will occur with this profile (due to its inferior load carrying ability) which will be reflected in locally higher shear stress and reduced lubricant flow.

'Elapsed Time for the Decay of Thermal Transients in Fluid Film Bearings' C M M ETTLES, H HESHMAT and K R BROCKWELL

Professor J B Medley (University of Waterloo, Canada)

Do the authors anticipate extending the analysis of thermal transients to consider specific bearings quantitatively such as large sector pad thrust bearings and finite width journal bearings or do they consider their exponential correlation sufficient for treating thermal transients in design and practice?

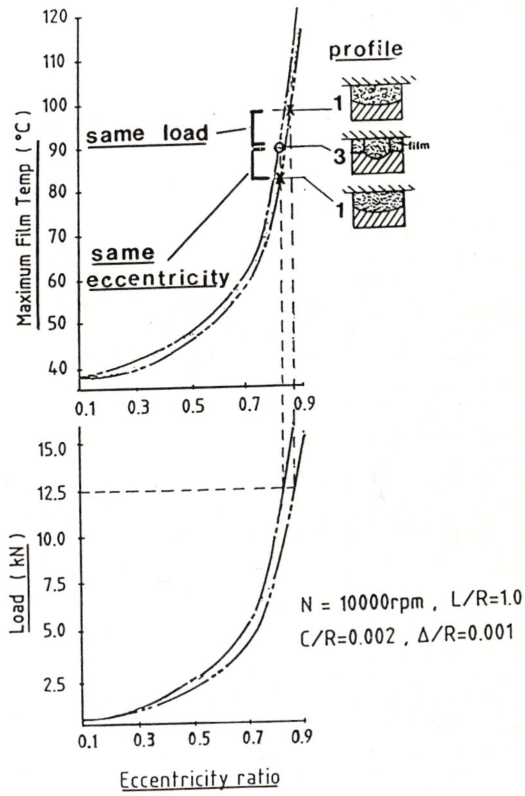

Higher temperatures with profile 1 on a constant load basis

Can the authors give examples of cases where considerations for thermal transients are critically important?

Reply by Dr C M M Ettles (Rensselaer Polytechnic Institute, Troy, USA), Dr H Heshmat (Mechanical Technology Inc., Albany, USA) and Mr K R Brockwell (National Research Council, Vancouver, Canada).

We do not currently have plans to extend the analysis to finite width thrust bearings or journal bearings. For the boundary conditions to be properly applied, we believe that the spatial coordinates in such an analysis should be three-dimensional, which would require extensive organization and, probably, long execution times. The two-dimensional analysis (Ref. 5 in the paper) gives the principal trends.

In journal bearings thermal transients can have important effects on the clearance. This is discussed briefly in the paper. Suppose that a journal bearing runs satisfactorily under steady running conditions A and also satisfactorily under steady running conditions B. When changing from A to B, the clearance may temporily reduce to a value which is unstable. This could occur if the thermal inertia of one of the components is much larger than the other, so that the thermal strain of one of the components lags significantly behind the other.

Thermal transients in thrust bearings may be important in very large bearings where the support system is designed such that over a large portion of the shoe, deformation is in the opposite sense to elastic deformation. Disc supported shoes when the disc radius is large are an example. If the face temperature is suddenly reduced (following, for example, a restart with cold lubricant) both modes of deformation can act in the same sense, giving a concave film which will not balance about the pivot. This can lead to surface damage, since in severe cases the bearing has no load capacity.

SESSION XI - PLAIN BEARINGS (2)

'Dynamically Loaded Journal Bearings: A Modal Approach to EHL Design Analysis' A KUMAR, J F BOOKER and P K GOENKA

Mr F A Martin (London, UK)

The authors are to be commended on their new modal approach. One of the problems with this type of work is how to present the results. It is difficult to present distorted bearing shapes without exaggerating or falsely distorting the real shape.

An alternative method is to look at limiting clearance shapes. The discusser has shown this previously, relating to Dr Goenka's work (using the same connecting rod and operating conditions as in the present paper). These clearance shapes are shown in the accompanying figure (D1) for crank angle positions 18° and 386°.

It would be of interest to see some of this new work plotted on the same basis, for it gives an added visual dimension, showing for instance the stretching effect, the journal position, and minimum film thickness. What are the authors' views on this?

Professor H Blok (Rijswijk, The Netherlands)

In one of their slides the authors showed a set of three curves, all depicting the cyclic variation of minimum film thickness, h_{min}. As a datum line one of these curves has been made to relate to the hypothetical case of perfect rigidity of the surface of both the bearing and journal. Another curve results from the authors' "previous" numerical elastohydrodynamic evaluation, i.e. that to be considered their most realistic one. In any case, this evaluation is more accurate than their "present" one for the remaining, third curve which is primarily meant for quick estimates.

Now, the second curve, marked "previous", shows a smallest h_{min} which amounts to only about one half that of the aforementioned curve for perfect rigidity. So, at least under the operating conditions concerned, the cyclic variation of the elastic distortion of the bearing surface, or say the "flapping action" (Ref. 1), would appear to be rather detrimental as regards the risk of metallic contact.

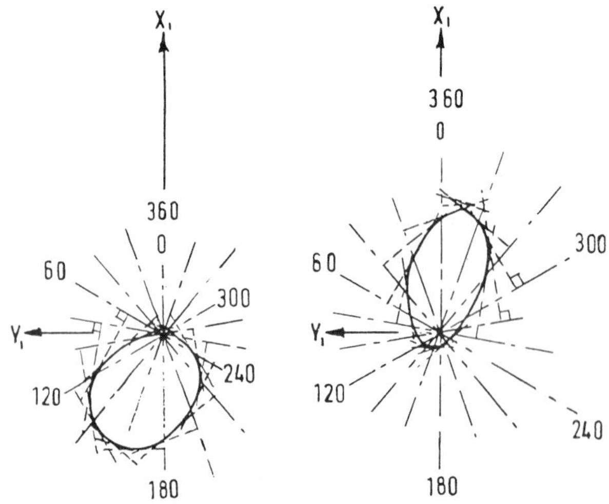

a) 18 deg. crank angle b) 386 deg. crank angle

Fig. D1 Limiting clearance shapes

In contrast, the third curve, marked "present", shows the quick estimate of the smallest h_{min} to exceed that for perfect rigidity. Even though the gain thus estimated amounts to a mere 10 or 15 percent, for design calculations it would appear unrealistic in that it is overoptimistic by over a factor of two. In other words, contrarily to the more accurate result of the second, "previous" curve the quick estimate is falsely indicative of a beneficial effect of the "flapping action".

Finally, in view of the above-mentioned detrimental effect to be expected of the "flapping action" in the bearing of the present design and under the operating conditions concerned, it would appear worthwhile to redesign this bearing by optimization of its elastic compliance. In this connection it should be observed that, judging from the basic "flapping" model in the above-mentioned paper (Ref. 1), potentialities inherent in such optimizations even include the rather paradoxical case where during an increase of the load the minimum film thickness will increase too.

Reference

(1) Blok H. 1975, 'Full Journal Bearings under Dynamic Duty: Impulse Method of Solution and Flapping Action'; Trans. ASME, J. Lubric. Tech., Vol. 97, Series F, No 2, pp 168-179. For Errata, see Vol 99 (1977) of the same journal, page 223.

Reply by Dr A Kumar, Professor J F Booker (Cornell University, Ithaca, USA) and Dr P K Goenka (General Motors Research Laboratories, Warren, USA).

We certainly agree with Mr Martin that presentation is the unsolved problem for EHL of dynamically loaded journal bearings (in which both rigid body and elastic displacements can vary circumferentially, axially and temporally).

The "dynamics clearance space" concept proposed by Mr Martin seems both intuitively helpful and computationally feasible. (For example, determination of tool paths and cam contours are classic problems in computer-aided design). As Mr Martin notes, the clearance space shape is one of the few graphic displays which is not distorted by an arbitrary change of scale(s).

Unfortunately, the 3-D character of the clearance spaces in our problems requires a level of complexity somewhat beyond such 2-D presentations as those suggested in this Discussion.

The same limitation applies to the excellent animated and narrated computer-generated video study of an elastic connecting rod bearing recently produced by B Fantino, J Frêne and J du Parquet. Effective presentation of the full dimensionality of transient EHL results will probably require future efforts much like this.

It is unfortunate that Professor Blok did not have access to the full text of our paper at the time of preparing his Discussion, particularly since our Introduction opens with citation of his classic paper (Blok [1975]) introducing the concept of "flapping action".

In his present Discussion Professor Blok notes that the effects of "flapping action" can be detrimental or beneficial, alluding to "the rather paradoxical case where during an increase of the load the minimum film thickness will increase too". We have previously observed a related phenomenon in the study of transient "squeeze films" (Kumar [1988]).

The table Below shows a steady external load without journal rotation for the idealistic bearing of Section 4.1.

Axis	X	Y	Z
load (kN)	0	-10	0
rotation (rev/s)	0	0	0

Idealistic Bearing - Steady State Without Rotation

Figure 1 shows transient variation of film thickness and pressure extrema computed using modal subsets m = 2,6,8,10 to form the transformation matrix. In every case modal displacement 2 (loading axis rigid body eccentricity ratio) is initially 0.95, while all other (elastic) modal displacements are initially null.

Figure 1 shows that (for rigid/elastic models m = 6,8,10) the resulting minimum film thickness actually increases for a time (though, presumably, the average film thickness always decreases), a qualitative result entirely consistent with elastic "wraparound".

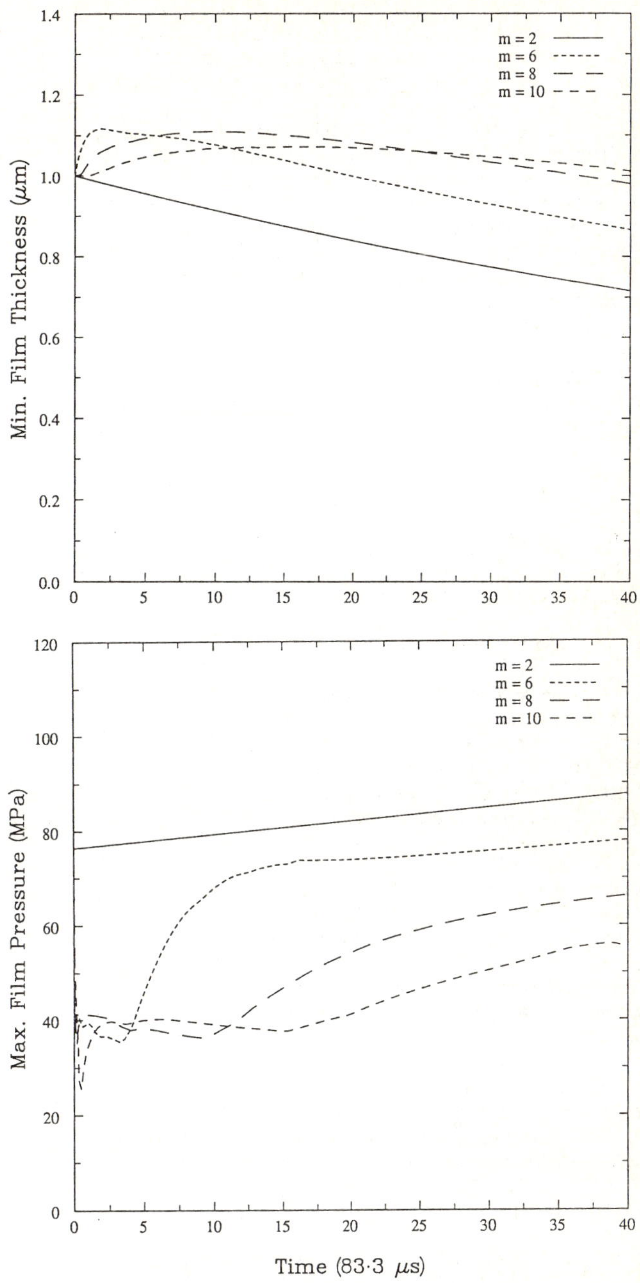

Fig. 12 Idealistic bearing - (steady load without rotation) film thickness and pressure transient extrema.

We believe that modal representation (not necessarily coupled with modal computation as here) can provide further insight into this and other evidences of the "flapping action" mechanism(s). We hope to elaborate on this approach in a later publication.

Unfortunately, our present paper is already somewhat dated; more recent results suggest that it is probably unduly conservative as to the accuracy possible with the present modal approach. For example, in the realistic bearing example of Section 4.2, the accuracy at the critical extreme points for m = 8 is much better than that shown in Figure 1 for m = 6. With this minor enhancement, results from the present program FEMEHL are very nearly identical to previous results by Goenka and Oh [1986b], as we will report in a forthcoming publication examining the computing cost/ numerical accuracy trade-offs in some detail.

Professor Blok calls attention to the potential for optimization of elastic compliance in such cases as this. General interest in such an approach is also evidenced by a very recent trade journal article by Murray [1988] reviewing the previous structural optimization work of Goenka and Oh [1986b]. We believe that the increased computational speed and limited design space of the modal approach will assist in studies of this sort.

Future Work.

Other possibilities for future work are inclusion of structural inertia and damping, temperature- and pressure-viscosity effects, and dynamic (mass-conserving) cavitation, as well as generalization of the mobility/ impedance method(s) for higher order systems via problem-independent (rather than problem-dependent) modal transformation matrices. The list of such possible enhancements and extensions is nearly endless.

Reference.

(1) Murray, C J. (1988), 'Compliant Connecting Rod Reduces Bearing Stresses', Design News, 1988, 44, n 19, 278-279.

SESSION XIV - HYDROSTATIC BEARINGS

'A Theoretical Investigation of Hybrid Journal Bearings Applied to High Speed Heavily Loaded Conditions Requiring Jacking Capabilities', D IVES, W WESTON, P G MORTON and W B ROWE.

Professor J B Medley (University of Waterloo, Canada).

How did you deal with thermal effects in your study?

Reply by Mr D Ives (Liverpool Polytechnic, UK), Mr W Weston (Huddersfield Polytechnic, UK), Mr P G Morton (General Electric Company, Stafford, UK) and Professor W B Rowe (Liverpool Polytechnic, UK).

It is realised that thermal effects in large high speed bearings are of importance. For the work presented it was considered that an isothermal model would be adequate at this stage, the temperature rise through the bearing being accounted for by assuming an effective viscosity. An investigation into thermal aspects of the performance of slot entry hybrid journal bearings is presently in progress.

GENERAL DISCUSSION

Dr M Stanojevic (Vandervell Ltd., Maidenhead, UK). (A contribution to SESSION XI - PLAIN BEARINGS (2).

Oil flow and power loss models in dynamically loaded bearings

1. Notation

(F - force, L - length, T - time)

Symbol	Name	Dim
Q	Bearing oil flow	[L3/T]
Q_p	Pressure flow	[L3/T]
Q_h	Hydrodynamic or velocity flow	[L3/T]
0	Index for plain bearings	-
f	Index for fully grooved bearings	-
d	Shaft diameter	[L]
b	Bearing length	[L]
ε	Eccentricity ratio	[-]
u	Surface velocity of the journal	[L/T]
c_d	diametral clearance	[l]
η	lubricant dynamic viscosity	[ft/l2]
h	average film thickness	[l]
p	lubricant feed pressure into bearing	[f/l2]
pl	bearing power loss	[fl/t]

2. Introduction

it is necessary to make estimates of oil flow through engine bearings when specifying engine pump capacity. prediction of bearing power loss is important in order to estimate engine power lost in bearing friction. furthermore, bearing temperature rise and effective operating oil viscosity can be estimated when oil flow and power loss are known.

journal bearings used in internal combustion engines are normally:

- plain i.e. without grooves or
- with a full central circumferential groove or
- with a partial central circumferential groove

journal bearing oil flow models have been described in reference 1, section 7.9 for oil flow with central circumferential groove and section 7.10 for oil flow with single hole. these models based upon an average eccentricity ratio over the engine cycle have been used for dynamically loaded bearings and are described in reference 4.

a simple journal bearing power loss model has been described in reference 2, section 7. reference 3, sections 12.16 to 12.18 considers similar models. these models based upon an average eccentricity ratio over an engine cycle have been adapted and used for engine bearings in references 4 and 5.

the above simple models are summarised below:

(a) oil flow through a plain bearing

$$Q = Q_{p0} + Q_{h0}$$

$$= \left[\frac{p\, c_d^3}{24\eta} \tan^{-1}\left(\frac{\pi d}{b}\right) \right] (1 + 1.5\,\varepsilon^2) + \frac{u\, c_d\, b\varepsilon}{2} \qquad (1)$$

(b) Oil flow through a fully grooved or a partially grooved bearing

$$Q = Q_{pf} + Q_{hf}$$

$$= \frac{p\,c_d^3}{24\eta}\,\frac{\pi d}{b}\,(1 + 1.5\,\varepsilon^2) + \frac{u\,c_d\,b\varepsilon}{4} \qquad (2)$$

it is seen that the total flow in the above formulae is a simple addition of pressure and hydrodynamic components. pressure flow component has been analysed in reference 6, and different total flow models have been discussed in reference 7 and 8.

c) power loss for plain and grooved bearings

$$PL = \frac{\eta\,d^3\,b\,\omega^2}{4\,c_d}\,\frac{\pi}{\sqrt{1-\varepsilon^2}}\left[2 + \frac{3\varepsilon^2}{2+\varepsilon^2}\right] \qquad (3)$$

Power loss defined above has been compared to several other models from different countries in reference 9.

In equations (1), (2) and (3) average eccentricity ratio is calculated from average film thickness h by

$$\varepsilon = \frac{c_d - 2h}{c_d}$$

Bearing length b is an effective length and it excludes groove width both for fully grooved and partially grooved bearings. The above formulae are characterised by the eccentricity ratio being averaged over the engine cycle.

3. Improvements to Simple Models from References 4 and 5

The following improvements to the simple models described in references 4 and 5 have been carried out:

- Oil flow and power loss are to be calculated at every crank angle increment and then averaged over the engine cycle instead of using the average eccentricity ratio in the equations (1), (2) and (3).

- It is possible to differentiate between fully grooved and partially grooved bearings by considering shaft locus during the engine cycle.

3.1 Improvements to the Bearing Oil Flow

Plain bearings without grooves have been considered in reference 10 where the improved method above has been compared to methods described in references 3 and 6. The feed hole diameter considered in reference 10 was small and the resulting ratios of velocity flows to pressure flows were large. The resulting differences in total flows were small for the models considered.

Further work is required in this area, but this section concerns an improved method for oil flow calculations of partially grooved bearings.

In the improved method it is assumed that the pressure flow is dominated by large clearances within the bearing circumference and that the hydrodynamic flow is dominated by small clearances. Suppose that the partial groove is in the upper half of the bearing circumference. The following relationships are then deduced, refer to the Figure below.

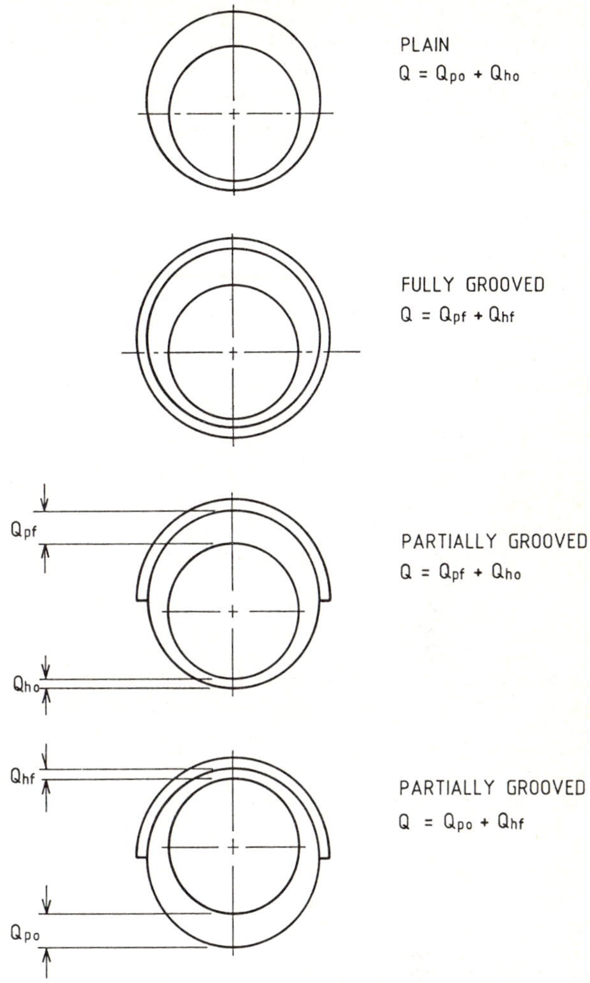

If the shaft centre locus is in the lower half of the clearance circle, the bearing oil flow consists of the following components:

(a) fully grooved bearing pressure flow
(b) plain bearing hydrodynamic flow, i.e.

$$Q = Q_{pf} + Q_{ho} \qquad (5)$$

- If the shaft centre locus is in the upper half of the clearance circle, the flow consists of the following components:

(a) plain bearing pressure flow
(b) fully grooved bearing hydrodynamic flow, i.e

$$Q = Q_{po} + Q_{hf} \qquad (6)$$

3.2 Improvements to the Bearing Power Loss

Improvements to the bearing power loss model as described in references 4 and 5 refer to partially grooved bearings. Again suppose that the partial groove is in the upper half of the bearing circumference.

- If the shaft centre locus is the lower half of the clearance circle, then the groove width is not subtracted from the bearing length b.

- If the shaft centre locus is in the upper

half of the clearance circle, then the groove width is excluded from the bearing length b.

4. Some Results

As an example a geometry of a main bearing of a four cylinder engine was analysed and the journal orbit was run for a plain, fully grooved and partially grooved bearing at two speeds. Results of oil flow and power loss using the improved models described above are shown in Table 1. Diametral clearances assumed in these calculations are 0.074 mm both for oil flow and power loss.

	Plain	Fully Grooved	Partially Grooved
ℓ/min @ 2500 r/min	0.37	1.10	0.91
kW @ 2500 r/min	0.07	0.12	0.08
ℓ/min @ 4500 r/min	0.58	1.20	1.03
kW @ 4500 r/min	0.24	0.39	0.26

Table 1 Bearing Oil Flow and Power Loss Using the Improved Method

From Table 1 it is seen that partially grooved bearings have smaller flow rates and power losses than fully grooved bearings.

Plain bearings have the smallest flow rates and power losses out of the three types of bearings considered.

The improved methods are compared to methods of references 4 and 5 in Table 2.

	Method of Refs. 4 and 5		Improved Method	
	Plain	Fully Grooved	Plain	Fully Grooved
ℓ/min @ 2500 r/min	0.40	1.13	0.37	1.10
kW @ 2500 r/min	0.08	0.11	0.07	0.12
ℓ/min @ 4500 r/min	0.62	1.24	0.58	1.20
kW @ 4500 r/min	0.28	0.38	0.24	0.39

Table 2 Comparison of the oil flow and power loss calculation methods of references 4 and 5 to the improved methods.

From Table 2 it follows that oil flow in the improved method is marginally lower for both fully grooved and plain bearings. Power loss of the improved method is marginally higher for fully grooved bearings whilst it is marginally lower for plain bearings. It is apparent that the differences between the two methods are small.

5. Discussion

The new models of bearing oil flow and power loss are improved in relation to the models described in references 4 and 5 because oil flow and power loss are calculated at every crank angle increment and these models consider shaft locus in the engine cycle.

In the new model for the oil flow of partially grooved bearings, pressure flow is governed by large clearances within bearing circumference and velocity flow is governed by small clearances.

In the new power loss model of partially grooved bearings, power loss is governed by small clearances within the bearing circumference.

Improved oil flow and power loss programs are run after the journal orbit programs and they use eccentricity-attitude data files as input. They may be incorporated into the journal orbit programs, so that oil flow and power loss are evaluated every time the orbit program is run.

References

(1) M C Shaw; E F Macks, 'Analysis and Lubrication of Bearings', McGraw-Hill, New York 1949.

(2) D D Fuller, 'Theory and Practice of Lubrication for Engineers', John Wiley, New York 1956.

(3) A Cameron, 'The Principles of Lubrication', Longmans, London 1966.

(4) P E Vickery, 'Friction Losses in Automotive Plain Bearings - A Practial and Thoeretical Study', SAE Paper 750052.

(5) R H Spikes; S M Robinson, 'Engine Bearing Design Up-to-Date', Instn. Mech Engrs Conference 'Tribology - Key to the Efficient Engine', Paper C1/82, Jan 1982.

(6) F A Martin; C S Lee, 'Feed-Pressure Flow in Plain Journal Bearings', ASLE Transactions, Vol. 26, No 3, pp 381-92, 1983.

(7) F A Martin, 'Developments in Engine Bearing Design', Tribology International, Volume 16, No 3, pp 147-64, 1983.

(8) G J Jones; C S Lee; F A Martin, 'Crankshaft Bearings: Advances in Predictive Techniques Incorporation the Effects of Oil Holes and Grooving', Paper No 1, AE Symposium 1982.

(9) F A Martin, 'Friction in Internal Combustion Engine Bearings', IMechE Conference "Combustion Engines - Reduction of Friction and Wear', Paper C67/85, 1985.

(10) M Stanojevic, 'Oil Flow Through a Plain Bearing', Vandervell Internal Engineering Note, May 1987.

15th LEEDS-LYON SYMPOSIUM ON TRIBOLOGY

THE TRIBOLOGICAL DESIGN OF MACHINE ELEMENTS

6th - 9th SEPTEMBER 1988

LIST OF AUTHORS

TITLE	NAME	AFFILIATION/ADDRESS
Mr	M Abdolmaleki	RHP Industrial Bearings Ltd P O Box 18 Northern Road Newark Notts NG24 2JF U K
Dr	D Ashman	Lucas Aerospace Ltd Engine Systems Division Hall Green Birmingham West Midlands B28 8SW U K
Dr	P Ayanoglu	Institute fur Maschinenelemente und Maschinengestaltung Technische Hochscule SchinkelstraBe 8 5100 Aachen West Germany
Dr	S Bair	Georgia Institute of Technology The George W Woodruff School of Mechanical Engineering Atlanta Georgia 30332-0405 U S A
Mr	A D Ball	The University of Leeds Institute of Tribology Department of Mechanical Engineering Leeds LS2 9JT U K
Dr	S Basri	University College of Swansea Department of Mechanical Engineering Singleton Park Swansea SA2 8PP U K
Prof	D Berthe	Institut National des Sciences Appliquees de Lyon, Laboratoire de Mecanique des Contacts Bâtiment 113 20 Avenue Albert Einstein 69621 Villeurbanne Lyon, France
Dr	B Bhushan	IBM, Almaden Research Centre K64/802, 650 Harry Road San Jose CA 95120-6099 U S A
Dr	P Boch	Ecole Nationale Superieure de Ceramique, Industrielle 47 & 73 Avenue Albert-Thomas 87065 Limoges Cedex France
Mr	S Boedo	Borg-Warner Automotive Inc Ithaca New York U S A
Prof	J F Booker	Cornell University Mechanical & Aerospace Engineering Upson Hall Ithaca New York 14853 U S A
Dr	F Bremer	Philips Research Laboratories 5600 JA Eindhoven The Netherlands
Dr	B J Briscoe	Department of Chemical Engineering & Technology, Imperial College of Science and Technology London SW7 2BX U K
Mr	K R Brockwell	National Research Council 3650 Wesbrook Mall Vancouver B C V6S 2L2 Canada
Dr	H M Chen	Mechanical Technology Inc 968 Albany-Shaker Road Latham New York 12110 U S A
Mr	B Chen	Nanjing Institute of Technology Nanjing, Jiangsu The People's Republic of China
Prof	H S Cheng	Northwestern University Technological Institute Evanston Illinois 60201 U S A
Dr	T H C Childs	University of Bradford Postgraduate School of Mechanical & Manufacturing Systems Engineering Bradford BD7 1DP U K
Dr	R J Chittenden	The University of Leeds Industrial Unit of Tribology Department of Mechanical Eng. Leeds LS2 9JT U K
Dr	T C Chivers	C E G B Berkeley Nuclear Laboratories Berkeley Glos. GL13 9PB U K
Mr	G A Clayton	The University of Leeds Department of Mechanical Engineering Institute of Tribology Leeds LS2 9JT U K
Mr	E W Cowking	Friction, Lubrication & Wear Group GEC Engineering Research Centre Cambridge Road Whetstone Leicester LE8 3LH U K
Dr	A Cowley	Department of Mechanical Engineering Manufacturing & Machine Tools Division, U M I S T, P O Box 88 Manchester M60 1Q U K
Dr	P E Dale	Mechanical Research Department Ontario Hydro 800 Kipling Avenue, Toronto Ontario M8Z 5S4 Canada

TITLE	NAME	AFFILIATION/ADDRESS
Dr	J A Dominy	Rolls-Royce plc Manager - Transmissions Research P O Box 31 Derby DE2 8BJ U K
Dr	S S Douglas	Liverpool Polytechnic Faculty of Engineering Byrom Street Liverpool L3 3AF U K
Prof	D Dowson	The University of Leeds Institute of Tribology Department of Mechanical Eng. Leeds LS2 9JT U K
Dr	C M M Ettles	Rensselaer Polytechnic Institute Department of Mechanical Eng. Troy N Y 12181 U S A
Dr	T S Eyre	Brunel University of Materials Technology, Department of Materials Technology Uxbridge Middlesex UB8 3PH U K
Dr	E Felder	Ecole Nationale Superieure des Mines de Paris, Centre de Mise en Forme des Materiaux, Sophia Antipolis 06560 Valbonne France
Prof	J Frene	Universite de Poitiers Laboratoire de Mecanique des Solides, 40 Avenue du Recteur Pineau, 86022 Poitiers Cedex France
Prof	H Fries	Virginial Polytechnic Institute and State University Department of Mechanical Eng. Blacksburg Virginia 24061 U S A
Mr	A Gabelli	SKF Engineering & Research Centre BV, Postbus 2350, 3430 DT Nieuwegein The Netherlands
Dr	B A Gecim	Senior Research Engineer Fluid Mechanics Department General Motors Research Labs. 30500 Mount Road, Warren MI 48090-9055 U S A
Prof	B G Gerbert	Chalmers University of Tech. Division of Machine Elements S-41296 Göteborg Sweden
Dr	D T Gethin	University College of Swansea Department of Mechanical Engineering Singleton Park Swansea SA2 8PP U K
Mr	O Giles	BW Mechanical Seals Long Beach, CA 90801 U S A
Dr	P K Goenka	General Motors Research Laboratories 30500 Mount Road Warren MI 48090-9055 U S A
Dr	R J Gozdawa	Advanced Bearing Technology Ltd Brunel University Science Park Uxbridge Middlesex UB8 3PH U K
Mr	R T Griffin	Birmingham Polytechnic Faculty of Computing and Information Studies Perry Barr Birmingham B42 2SU U K
Mr	J C Hamer	Imperial College of Science and Technology, Department of Mechanical Engineering Exhibition Road London SW7 2BX U K
Prof	B J Hamrock	The Ohio State University Department of Mechanical Eng 206 West 18th Avenue Columbus Ohio 43210 U S A
Dr	J F Harrop	United Technologies Pratt and Witney Canada Inc Box 4080, Mississauga Ontario L5A 3Z4 Canada
Dr	G J J van Heijningen	Delft University of Technology Faculty of Mechanical Eng. and Marine Engineering Section Tribology Delft The Netherlands
Dr	H Heshmat	Project Manager Sr Project Engineer Mechanical Technology Inc 968 Albany-Shaker Road Latham, New York 12110 U S A
Dr	P L Holster	Philips Research Laboratories 5600 JA Eindhoven The Netherlands
Dr	E Ioannides	S K F Engineering & Research Centre B V, Postbus 2350, 3430 DT Nieuwegein The Netherlands
Mr	K Ishii	The Ohio State University Department of Mechanical Eng. 206 West 18th Street Columbus Ohio 43210 U S A
Mr	D Ives	Liverpool Polytechnic Faculty of Engineering Byrom Street Liverpool L3 3AF U K
Prof	B Jacobson	SKF Engineering & Research Centre BV, Postbus 2350, 3430 DT, Nieuwegein The Netherlands
Ir	C M Kalker-Kalkman	Delft University of Technology Faculty of Mechanical Engrg. and Marine Engineering Section Tribology Delft The Netherlands

TITLE	NAME	AFFILIATION/ADDRESS
Ir	A F C Kanters	Eindhoven University of Tech. Institute for Power Transmission and Tribology, W-HOOG 3.110 P O Box 513, 5600 MB Eindhoven The Netherlands
Mr	G Kapelski	Ecole National Superieure de Ceramique, Industrielle 47 & 73 Avenue Albert Thomas 87065 Limoges Cedex France
Prof	J J Kauzlarich	University of Virginia Mechanical Engineering Department Thornton Hall Charlottesville VA 22903 U S A
Mr	W E Key	BW Mechanical Seals Temecula CA 92390 U S A
Dr	J Klinger	The Ohio State University Department of Mechanical Eng 206 West 18th Avenue Columbus Ohio 43210 U S A
Mr	C N Ko	S K F Engineering & Research Centre B V, Postbus 2350, 3430 DT Nieuwegein The Netherlands
Prof Dr Eng	H Krzeminski-Freda	Politechnika Lodzka Instytyt Konstrukcji Maszyn 90-924 Lodz ul. B Stefanowskiego 1/15, Lodz Poland
Mr	A Kumar	Cornell University Mechanical & Aerospace Engineering Upson Hall Ithaca New York 14853 U S A
Mr	M Kuno	Materials Research Laboratory NTN Toyo Bearings Co Ltd Kuwana 511 Japan
Mr	M Li	Tribology Research Institute Tsinghua University Beijing Peking The People's Republic of China
Mr	P Maspeyrot	Universite de Poitiers Laboratoire de Mecanique des Solides, 40 Avenue du Recteur Pineau, 86022 Poitiers Cedex France
Dr	J O Medwell	University College of Swansea Department of Mechanical Eng. Singleton Park Swansea SA2 8PP U K
Dr	A v Montfoortland	Philips Research Laboratories 5600 JA Eindhoven The Netherlands
Mr	P G Morton	GEC Research Principal Research Associate Engineering Research Centre Mechanical Laboratory P O Box 36 Lichfield Road Stafford ST17 4LN

TITLE	NAME	AFFILIATION/ADDRESS
Prof Dr	E A Muijderman	Philips Research Laboratories 5600 JA Eindhoven The Netherlands
Mr	R A J van Ostayen	Technische Universiteit Eindhoven Instituut voor Aandrfit-en Tribolotechniek, W-HHOG 4,103 Postbus 513, 5600 MB Eindhoven The Netherlands
Dr	E W Parker	Wolverhampton Polytechnic School of Engineering Wolverhampton West Midlands WV1 1LY
Mr	I K Parker	University of Bradford Postgraduate School of Mechanical & Manufacturing Systems Engineering Bradford BD7 1DP UK
Dr	T G Pearce	British Railways Board Research Training Officer Research Division Railway Technical Centre London Road Derby DE2 8UP U K
Prof Dr Ing	H Peeken	Institut fur Maschinenelemente und Maschinengestaltung Technische Hochscule SchinkelstraBe 8, 5100 Aachen West Germany
Mr	F Platon	Ecole Nationale Superieure de Ceramique, Industrielle 47 & 73 Avenue Albert-Thomas, 87065 Limoges Cedex France
Dr	C Pritchard	British Railways Board Research Training Officer Research Division Railway Technical Centre London Road Derby DE2 8UP U K
Prof	C A Rogers	Virginia Polytechnic Institute and State University Department of Mechanical Eng. Blacksburg, Virginia 24061 U S A
Dr	E W Roberts	National Centre of Tribology U K A E A, Risley, Warrington WA3 6AT U K
Prof	W B Rowe	Liverpool Polytechnic Department of Mechanical Marine and Production Engineering Byrom Street Liverpool L3 3AF U K
Dr	R A Rowntree	National Centre of Tribology U K A E A, Risley Warrington WA3 6AT U K
Dr	Ph Sainsot	Institut National des Sciences Appliquees de Lyon, Laboratoire de Mecanique des Contacts Bâtiment 113 20 Avenue Albert Einstein 69621 Villeurbanne LYON France

TITLE	NAME	AFFILIATION/ADDRESS
Prof	**R F Salant**	Georgia Institute of Technology The George W Woodruff School of Mechanical Engineering Atlanta Georgia 30332-0405 U S A
Prof	**S Sasaki**	Mechanical Engineering Lab. Namiki 1-2, Sakura-Mura, Isukuba Sciencen City 305 Ibaraki 305 Japan
Dr	**R S Sayles**	Imperial College of Science and Technology Department of Mechanical Eng. Exhibition Road London SW7 2BX U K
Mr	**A Skorin**	RHP Industrial Bearings Ltd Northern Road Newark Notts NG24 2JF U K
Ir	**M J L Stakenborg**	Technische Universiteit Eindhoven Instuut voor Aandrift-en Tribolotechniek, W-HOOG 4.103 Postbus 513, 5600 M B Eindhoven The Netherlands
Dr	**T A Stolarski**	Brunel University of West London Department of Mechanical Eng. Uxbridge, Middlesex UB8 3PH U K
Dr	**J D Summers-Smith**	Tribology Consultant Merlewood, The Avenue Guisborough, Cleveland TS14 8EE U K
Mr	**A Tam**	Pratt and Witney Canada Inc. Box 4080, Mississauga Ontario L5A 3Z4 Canada
Dr	**A G Tangena**	Philips Research Laboratories 5600 JA Eindhoven The Netherlands
Dr	**C M Taylor**	The University of Leeds Institute of Tribology Department of Mechanical Eng. Leeds LS2 9JT U K
Mr	**M J Todd**	National Centre of Tribology U K A E A, Risley Warrington WA3 6AT U K
Dr	**J H Tripp**	SKF Engineering & Research Centre BV, Postbus 2350 3430 DT, Nieuwegein The Netherlands
Ms	**R Tristani**	McMaster University Hamilton Ontario Canada
Dr	**P J Tweedale**	Imperial College of Science and Technology Department of Chemical Eng. and Technology London SW7 2BX U K
Dr	**C J Thijsse**	Nederlandse Philips Bedrijven BV Postbus 218 5600 MD Eindhoven The Netherlands
Dr	**Ph Velex**	Institut National des Sciences Appliquees de Lyon Laboratoire de Mecanique des Contacts, Bâtiment 113 20 Avenue Albert Einstein 69621 Villeurbanne Cedex France
Ir	**M Visscher**	Eindhoven University of Tech. Institute for Power Transmission and Tribology W-HOOG 3.110, P O Box 513 5600 MB Eindhoven The Netherlands
Dr	**B Warda**	Politechnika Lodzka Instytyt Konstrukcji Maszyn 90-924 Lodz ul. B Stefanowskiego 1/15 Lodz Poland
Dr	**F P Wardle**	RHP Industrial Bearings Ltd Northern Road Newark Notts NG24 2JF U K
Dr	**R B Waterhouse**	The University of Nottingham Department of Metallurgy and Materials Science University Park Nottingham NG7 2RD U K
Prof	**S Wen**	Tribology Research Institute Tsinghua University Beijing Peking The People's Republic of China
Mr	**W Weston**	Huddersfield Polytechnic Department of Mechanical Eng. Queens Gate Huddersfield HD1 3DH
Dr	**P A Willermet**	The Ford Motor Company Research Laboratory P O Box 2053, Dearborn MI 48121 U S A
Prof	**W O Winer**	The Georgia Institute of Technology The George W Woodruff School of Mechanical Engineering Atlanta Georgia 30332-0405 U S A
Dr	**M Winfield**	Birmingham Polytechnic Faculty of Computing and Information Studies Perry Barr Birmingham B42 2SU U K
Mr	**R A E Wood**	RHP Industrial Bearings Ltd Northern Road Newark Notts NG24 2JF U K
Prof	**S Xu**	Department of Mechanical Eng. Southwest University Nanjing The People's Republic of China

15th LEEDS-LYON SYMPOSIUM ON TRIBOLOGY

THE TRIBOLOGICAL DESIGN OF MACHINE ELEMENTS

6th - 9th SEPTEMBER 1988

LIST OF DELEGATES

TITLE	NAME	AFFILIATION/ADDRESS
Mr	M Abdolmaleki	RHP Bearings P O Box 18 Newark Notts U K
Mr	A Alberdi	Tekniker Asociacion de Investigacion c/Isasi s/n Eibar Guipùzcoa Spain
Mr	A F Alliston-Greiner	Plint & Partners Ltd Fishponds Road Wokingham RG11 2QG U K
Dr	B S Andersson	Volvo Car Corporation Department 96320 PV3C S-40508 Göteborg Sweden
Dr	D Ashman	Lucas Aerospace Ltd Engine Systems Division Shaftmoor Lane, Hall Green Birmingham B28 8SW U K
Mr	D Auger	The University of Waterloo Department of Mechanical Eng. Waterloo Ontario Canada N2L 3G1
Mr	P Ayanoglu	Institut für Maschinenelemente und Maschinengestaltung RWTH Aachen Schinkelstr. 8, 5100 Aachen, W Germany
Mr	A D Ball	The University of Leeds Department of Mechanical Eng. Institute of Tribology Leeds LS2 9JT U K
Mr	T J Banks	Ricardo Consulting Engineers plc Bridge Works Shoreham-by-Sea West Sussex BN4 6JQ U K
Mr	O Beaurepaire	SIC Ste Industrielle des Coussinets, B P 73-4 rue de la Liberte 74009 Annecy France
Dr	J C Bell	Shell Research Ltd Thornton Research Centre P O Box 1 Chester CH1 3SH U K
Mr	G Berger	SEP Vernon France
Dr	W J Beswarick	Hatfield Polytechnic Hatfield Herts AL10 9AB U K
Dr	B Bhushan	K64/802 IBM Almaden Research Center, 650 Harry Road San Jose, CA 95120-6099 U S A
Mr	G Binnington	16 Clarence Street Bowburn Co Durham DH6 5BB U K
Prof	H Blok	Consultant Dr H Colijnlaan 4, Fl 19, 2283 XM Rijswijk The Netherlands
Mr	S Boedo	Borg-Warner Automotive Inc 770 Warren Road Ithaca, New York 14850 U S A
Dr	D A Boffey	The University of Edinburgh Department of Mechanical Eng. Kings Buildings Mayfield Road Edinburgh EH9 3JL U K
Prof	J F Booker	Cornell University School of Mechanical & Aerospace Engineering, Ithaca, New York 14853 U S A
Dr	B Bou-Said	Institut National des Sciences Appliquees de Lyon Laboratoire dés Mécanique des Contacts, Bâtiment 113 20 Avenue Albert Einstein 69621 Villeurbanne, Lyon, France
Mr	M Bouvier	Institut National des Sciences Appliquees de Lyon Laboratoire dés Lyon Laboratoire dés Mécanique des Contacts, Bâtiment 113 10 Avenue Albert Einstein 69621 Villeurbanne Lyon, France
Ir	F Bremer	Philips Research Laboratories CFT/SAQ 2100 5600 MD Eindhoven The Netherlands
Mr	K R Brockwell	National Research Council 3650 Wesbrook Mall Vancouver, BC V6S 2L2 Canada
Mr	A Burrows	Butterworth Scientific Ltd P O Box 63 Guildford Surrey GU2 5BH U K
Dr	P Cann	Imperial College of Science and Technology, Tribology Section London SW7 2BX U K
Prof	H S Cheng	Northwestern University Technological Institute Evanston, Illinois 60201 U S A
Dr	T H C Childs	The University of Bradford Department of Mechanical and Manufacturing Engineering Bradford West Yorkshire BS7 1DP U K

TITLE	NAME	AFFILIATION/ADDRESS
Dr	R J Chittenden	Industrial Unit of Tribology The University of Leeds Leeds LS2 9JT U K
Dr	T C Chivers	Central Electricity Generating Board Berkeley Nuclear Laboratories Berkeley Glos GL13 9PB U K
Mr	G Clayton	The University of Leeds Department of Mechanical Eng. Institute of Tribology Leeds LS2 9JT U K
Mr	S N Collins	The University of Leeds Institute of Tribology Department of Mechanical Eng. Leeds LS2 9JT U K
Mr	E W Cowking	GEC Engineering Research Centre, Cambridge Road Whetstone Leicester LE8 3LH U K
Dr	D A Crolla	The University of Leeds Institute of Tribology Department of Mechanical Eng. Leeds LS2 9JT U K
Dr	P Dale	Ontario Hydro 800 Kipling Ave. Toronto Ontario M82 5S4 Canada
Mr	K Dalgano	The University of Bradford Department of Mechanical and Manufacturing Engineering Bradford W Yorks BD7 1DP U K
Dr	A J Day	The University of Bradford Department of Mechanical and Manufacturing Engineering Bradford W Yorks BD7 1DP U K
Mr	P De Laine	Clextral, B P 10 2I De Chazeau, 3270 Firminy Cedex France
Mrs	FdM Barragan De Ling	University College Department of Mechanical Eng. Newport Road Cardiff CF2 1TA U K
Dr	J Dominy	Rolls-Royce plc P O Box 31 Derby DE2 8BJ U K
Dr	S Dougas	Liverpool Polytechnic Department of Engineering & Technology Managerment Byrom Street Liverpool L3 3AF U K
Prof	D Dowson	The University of Leeds Institute of Tribology Department of Mechanical Eng. Leeds LS2 9JT U K
Ass. Prof	D Duan	Gear Research Institute Taiyuan University of Tech Taiyuan Shanxi People's Republic of China
Ms	M-C Dubourg	Institut National Des Sciences Appliquees de Lyon, Laboratoire des Mecanique des Contacts Batiment 113 20 Avenue Albert Einstein 69621 Villeurbanne Lyon, France

TITLE	NAME	AFFILIATION/ADDRESS
Prof	C M M Ettles	Rensselaer Polytechnic Institute JEC 2034 Troy N Y 12180-3590 U S A
Mr	D C Evans	The Glacier Metal Co Ltd Premier House 1 Canning Road Harrow Middlesex HA3 7TS U K
Dr	H P Evans	University College Department of Mechanical Eng. Newport Road Cardiff CF2 1TA U K
Dr	T S Eyre	Brunel University Kingston Lane Uxbridge Middlesex UB8 3PH U K
Mr	F M H Ezzat	The University of Leeds Institute of Tribology Department of Mechanical Eng. Leeds LS2 9JT U K
Dr	E Felder	Ecole des Mines CEMEF Sophia Antipolis F 06560 Valbonne France
Dr	D N Fenner	Kings College Department of Mechanical Eng. The Strand London WC2R 2LS U K
Mr	J Fieldhouse	Head of Department The Polytechnic Department of Mechanical and Production Engineering Queensgate Huddersfield GD1 3DH U K
Dr	J Fisher	The University of Leeds Department of Mechanical Eng Institute of Tribology Leeds LS2 9JT U K
Prof	J Fréne	University of Poitiers Laboratoire de Meanique des Solides 40 Ave du Recteur Pineau 86022 Poitiers Cedex France
Prof	R H Fries	Virginia Polytechnic Institute and State University Department of Mechanical Eng Blacksburg Virginia 24061 U S A
Dr. Ing.	A Gabelli	SKF Engineering & Research Centre BV, Postbus 2350 3430 DT Nieuwegein The Netherlands
Dr	B A Gecim	General Motors Research Labs. Fluid Mechanics Department Warren, MI 48090-9055 U S A
Prof	B G Gerbert	Chalmers University of Technology Division of Machine Elements 41296 Göteborg Sweden
Dr	D Gethin	University College Department of Mechanical Eng. University Park Swansea SA2 8PP U K

TITLE	NAME	AFFILIATION/ADDRESS
Prof	M Godet	Institut National Des Sciences Appliquees de Lyon, Laboratoire des Mecanique des Contacts Batiment 113 20 Avenue Albert Einstein 69621 Villeurbanne Lyon, France
Mr	R T Griffin	City of Birmingham Polytechnic Department of Computing Perry Bar Birmingham B42 2SU U K
Mr	D Hallgren	Mobil Oil Box 502 S-18215 Danderyd Sweden
Mr	C Hamer	Imperial College of Science and Technology, Tribology Section London SW7 2BX U K
Dr	G M Hamilton	Reading University Department of Engineering Whiteknights Reading U K
Mr	R T Harding	The University of Leeds Department of Mechanical Eng. Institute of Tribology Leeds LS2 9JT U K
Mr	J F Harrop	Pratt & Whitney Canada Box 4080, Mississauga Ontario L5A 3Z4 Canada
Mr	S S Hassan	The University of Leeds Institute of Tribology Department of Mechanical Eng. Leeds LS2 9JT U K
Dr Ir	G J J van Heijningen	Technical University Delft Fac det Werktuigbouwkunde en Maritieme Techniek Mekelweg 2, 2628 CD Delft Holland
Dr	W A Hilder	Esso Petroleum Company Ltd Esso Research Centre Milton Hill Abingdon Oxon OX13 6AE U K
Ing.	P L Holster	Philips Research Laboratories POB 80.000 5600 JA Eindhoven The Netherlands
Mr	M Homayounieh	The University of Leeds Institute of Tribology Department of Mechanical Eng. Leeds LS2 9JT U K
Dr	C J Hooke	The University of Birmingham Department of Mechanical Engineering, P O Box 363, Birmingham B15 2TT U K
Dr	D Horner	Michell Bearings Scotswood Road Newcastle upon Tyne NE15 6LL U K
Dr	E O Ioannides	SKF Engineering & Research Centre BV, POB 2350, 3430 DT Nieuwegein The Netherlands
Dr	K Ishii	The Ohio State University Department of Mechanical Eng. 206 West 18th Avenue Columbus Ohio 43210 U S A
Mr	D Ives	Liverpool Polytechnic School of Engineering and Technology Management Byrom Street Liverpool L3 3AF, U K
Prof	Bo Jacobson	SKF Engineering & Research Centre POB 2350 3430 DT Nieuwegein The Netherlands
Mr	Z M Jin	The University of Leeds Institute of Tribology Department of Mechanical Engrg Leeds LS2 9JT U K
Mr	B Jobbins	The University of Leeds Institute of Tribology Department of Mechanical Engrg. Leeds LS2 9JT U K
Mr	D A Jones	The University of Leeds Institute of Tribology Department of Mechanical Eng. Leeds LS2 9JT U K
Dr	H Peter Jost CBE	K S Paul Product Ltd Angle Lodge Laboratories Nobel Road London N18 3DB U K
Mrs	C M Kalker	Technical University Delft Department of Mechanical Eng. Fac. der Werktuigbouwkunde en Maritieme Techniek Mekelweg 2 2628 CD Delft Holland
Mr	A F C Kanters	Eindhoven University of Technology, Department of Mechanical Engineering, WH 03 110, P O Box 513 5600 MB, Eindhoven The Netherlands
Mr	G Kapelski	E.N.S.C.I., 47a 73 Avenue Albert Thomas, 87065 Limoges Cedex France
Prof	J J Kauzlarich	University of Virginia Department of Mechanical Engineering, Thornton Hall Charlottesville, VA 22903 U S A
Mr	C N Ko	SKF Engineering & Research Centre BV POB 2350 3430 DT Niewegein The Netherlands
Dr	E M Kopalinsky	The University of New South Wales Department of Mechanical Eng. Kensington N S W 2033 Australia
Prof Dr	H Krzeminski-Freda	Instytut Konstrukcji Maszyn Politechnika Lódska, 90-924 Lódz, ul. Stefanowskiego 1/15 Poland
Dr	M Kuno	NTN Toyo Bearing Co Ltd Material Research Laboratory Kuwana 511, Japan
Dr	C C Kweh	University College Department of Mechanical Eng. Newport Road Cardiff CF2 1TA U K

TITLE	NAME	AFFILIATION/ADDRESS
Prof	O K Kwon	Director, Division of Mechanical Engineering and Electro-Physics, Kaist, P O Box 131, Cheongryang, Seoul, Korea
Mr	J Laucirica	Tekniker Asociacion de Investigacion c/Isasi s/n Eibar Guipùzcoa Spain
Mr	R Lee	Jaguar Cars Limited Advanced Engineering Group Jaguar Engineering Centre Abbey Road Whitley Coventry CV3 4LF U K
Mr	H J van Leeuwen	Eindhoven University of Technology, Department of Mechanical Engineering, WH 04-102 P O Box 513, 5600 MB Eindhoven The Netherlands
Prof	H J Leutheusser	University of Toronto Department of Mechanical Eng. Toronto, Ontario M5S 1A4 Canada
Mr	J-M Leroy	Institut National Des Sciences Appliquees de Lyon, Laboratoire des Mecanique des Contacts Batiment 113 20 Avenue Albert Einstein 69621 Villeurbanne Lyon, France
Mr	H Lindgren	Division of Machine Elements Chalmers University of Tech. 41296 Göteborg Sweden
Mr	J McDonald	Superintendent, Quality Engineering Test Establishment Department of Naval Defence Ottawa Ontario KIA OK2 Canada
Mr	J D C McIvor	Kings College Department of Mechanical Eng. The Strand London WC2R 2LS U K
Dipl.Ing	N Madzia	Steyr-Bearings, P O Box 120, 4400-Steyr, Austria
Mr	P Marchand	Institut Francais du Petrole BP 311, 92506 Rueil, Malmaison France
Mr	H Marginson	Imperial College of Science and Technology, Tribology Section London SW7 2BX U K
Mr	F A Martin	Consultant 98 Grove Avenue Hanwell London W7 3ES U K
Dr	J B Medley	The University of Waterloo Department of Mechanical Eng. Waterloo Ontario Canada N2L 3G1
Mr	D Mello	U S Department of Energy Office of Conservation Washington D C 20585 U S A
Dr	A A S Miranda	Universidade do Minho Av D. Afonso Henriques 4800 GuimarÃes Portugal
Dr	N K Mishkin	Institute of Mechanics and Polymer Metals Minsk U S S R
Dr	A J Moore	BP Research Centre Chertsey Road Sunbury-on-Thames Middlesex TW16 7LN U K
Dr	S L Moore	B P Research Centre Chertsey Road Sunbury on Thames Middlesex U K
Mr	P G Morton	G E C Engineering Research Centre P O Box 30 Lichfield Road Stafford ST17 4LN U K
Dr Ir	E A Muijderman	Philips Research Laboratories POB 80.000 5600 JA Eindhoven The Netherlands
Mr	A J Munday	The University of Southampton Department of Mechancal Eng. Southampton SO9 5NH U K
Dr	J P Naylor	CETIM, 52 Avenue Felix Louat 60304 Senlis France
Mr	D Nelias	Turbomeca Seirice Nisc au Point Bondes, 64320 Bizanos France
Prof Dr.Sc.	Tuan Anh Nguyen	Hanoi Polytechnic Institute Truòng DaiHoc Bách Khoa Hã Noi Vietnam
Mr	D Nicholson	Imperial College of Science and Technology, Tribology Section London SW7 2BX U K
Dr	G Nielsen	Technical University Denmark Department of Machine Elements 2800 Lyngby Denmark
Mr	J Olsson	Chalmers University of Tech. Division of Machine Elements S-41296, Göteborg Sweden
Dr	A V Olver	Westland Helicopters Ltd Yeovil Somerset BA20 2YB U K
Mr	B Osborne	Industrial Lubrication and Tribology Peterson House Northbank Berryhill Industrial Estate Droitwich Worc WR9 9BL U K
Mr	I K Parker	The University of Bradford Department of Mechanical and Manufacturing Engineering Bradford W Yorks BD7 1DP U K
Dr	A Parmar	John Crane U K Ltd Crossbow House Liverpool Road Slough SL1 4QX U K

TITLE	NAME	AFFILIATION/ADDRESS
Dr	C Pritchard	B R Research Division Room 305 K, Railway Technical Centre, London Road, Derby DE2 8UP U K
Mr	S J Radcliffe	Central Electricity Generating Board, Tribology Section Berkeley Nuclear Labs. Berkeley Glos GL13 9PB U K
Mr	B Rigaut	CETIM 52 Av. Felix Louat 60304 Senlis France
Ing	G Righes	Fiat Research Centre Strada Torino 50 10043 Orbassano Torino Italy
Dr	W H Roberts	National Centre of Tribology U K A E A Risley Warrington WA3 6AT U K
Dr	R A Rowntree	UK Atomic Energy Authority Risley Technical Centre Risley Warrington Cheshire WA3 6AT U K
Mr	Ph Sainsot	Laboratoire de Mécanique des Contacts - Bâtiment 113, 20 Avenue Albert Einstein 69621 Villeurbanne, Lyon France
Prof	R F Salant	Georgia Institute of Technology School of Mechanical Engineering Atlanta, Georgia 30332-0405 U S A
Mr	S Sasaki	Mechanical Engineering Lab. Namiki 1-2, Tsuluba-shi Ibaraki 305 Japan
Dr	R S Sayles	Imperial College of Science and Technology, Tribology Section London SW7 2BX U K
Dr	P A J Scott	Pilgrim Engineering Developments Balmoral House Longmore Avenue Bentley Mill Walsall WS2 ODA Staffs U K
Mrs	A Seyed-Harraf	The University of Leeds Institute of Tribology Department of Mechanical Eng. Leeds LS2 9JT U K
Mr	N P Sheldrake	The University of Leeds Institute of Tribology Department of Mechanical Eng. Leeds LS2 9JT U K
Prof	Wen Shizhu	Tsinghua University Tribology Research Institute Beijing People's Republic of China
Mr	B A Shotter	Westland Helicopters (Box 245) Engineering Division Yeovil Somerset BA20 2YD U K
Dr	M Stanojevic	Vandervell Ltd Vanwall Business Park Maidenhead Berks SL6 4BG U K
Dr	T A Stolarski	Brunel University Department of Mechanical Eng. Uxbridge, Middlesex UB8 3PH U K
Dr	J D Summers-Smith	Merlewood, The Avenue Guisborough Cleveland TS14 8EE U K
Mr	M Takeda	Technical Centre Europe Karkorstr 15, 4030 Ratingen West Germany
Dr	A G Tangena	Philips Research Laboratories POB 80.000 5600 JA Eindhoven The Netherlands
Dr	C M Taylor	The University of Leeds Institute of Tribology Department of Mechanical Engrg. Leeds LS2 9JT U K
Ir	C J Thijsse	Nederlandse Philips Bedrigven B V Centre for Manufacturing Technology - Tribology Building SAQ-2113 5600 MD Eindhoven The Netherlands
Dr	J Tripp	SKF Engineering & Research Centre BV POB 2350 3430 DT Nieuwegein The Netherlands
Dr	P J Tweedale	Imperial College Department of Chemical Eng. and Technology, Prince Consort Road, South Kensington London SW7 2AZ, U K
Mr	J M Valli	Or WP Ceramics Ltd Hämeentie 135 D 00560 Helsinki Finland
Mr	L C A Vandertak	GEC Avionics Ltd FARL, Airport Works, Rochester, Kent ME1 2XX U K
Dr	Ph Velex	Institut National des Sciences Appliquees de Lyon, Laboratoire de Mecanique des Contacts Bâtiment 113, 20 Avenue Albert Einstein, 69621 Villeurbanne Lyon, France
Mr	M Visscher	Eindhoven University of Tech. Department of Mechanical Eng. WH 03.110, P O Box 513 5600 MB, Eindhoven The Netherlands
Mr	D J Wall	C.E.G.B., Bedminster Down Bridgwater Road Bristol BS13 8AN, U K
Dr	F P Wardle	RHP Industrial Bearings Ltd P O Box 18 Northern Road Newark Notts NG24 2JF U K
Dr	W Weston	Head of Department The Polytechnic Queensgate Huddersfield HD1 3DH U K
Dr	P A Willermet	Ford Motor Company Research Laboratory Mail Drop 3179, P O Box 2053 Dearborn, MI 48121, U S A

TITLE	NAME	AFFILIATION/ADDRESS
Mr	W Wilson	SEPCO Unit 9 Surburton Street Attercliffe Common Sheffield S9 2DN U K
Prof	W O Winer	Georgia Institute of Technology George W Woodruss School of Mechanical Engineering Atlanta, GA 30332-0405, U S A
Mr	A J Winn	The University of Leeds Institute of Tribology Department of Mechanical Eng. Leeds LS2 9JT U K
Dr	H Xu	Lancashire Polytechnic School of Mechanical and Production Engineering Preston PR1 2TQ U K
Prof	S Xu	South East University Department of Mechanical Eng Nanjing People's Republic of China
Mr	J Q Yao	The University of Leeds Institute of Tribology Department of Mechanical Eng. Leeds LS2 9JT U K
Mr	M Zarrebini-Esfahani	The University of Leeds Institute of Tribology Department of Mechanical Eng. Leeds LS2 9JT U K
Mr	R Zechel	Technische Akademie Esslingen Postfach 12 69, D-7302 Osfildern